Ecologically Based Weed Management

Ecologically Based Weed Management

Concepts, Challenges, and Limitations

Edited by

Nicholas E. Korres
University of Ioannina
Kostakii, Greece

Ilias S. Travlos
Agricultural University of Athens
Athens, Greece

Thomas K. Gitsopoulos
Hellenic Agricultural Organization
Demeter, Greece

WILEY

This edition first published 2024

Registered Office
John Wiley & Sons, Inc., 111 River Street, Hoboken, NJ 07030, USA

For details of our global editorial offices, customer services, and more information about Wiley products visit us at www.wiley.com.

Wiley also publishes its books in a variety of electronic formats and by print-on-demand. Some content that appears in standard print versions of this book may not be available in other formats.

Library of Congress Cataloging-in-Publication Data
Names: Korres, Nicholas E., editor. | Travlos, Ilias S., editor. | Gitsopoulos, Thomas K., editor.
Title: Ecologically based weed management : concepts, challenges, and limitations / edited by Nicholas E. Korres, Ilias S. Travlos, Thomas K. Gitsopoulos
Description: First edition | Hoboken, NJ : John Wiley & Sons, 2024 | Includes bibliographical references and index.
Identifiers: LCCN 2023045924 (print) | LCCN 2023045925 (ebook) | ISBN 9781119709664 (hardback) | ISBN 9781119709725 (pdf) | ISBN 9781119709756 (epub) | ISBN 9781119709763 (ebook)
Subjects: LCSH: Weeds--Biological control. | Weeds--Control. | Weeds--Integrated control.
Classification: LCC SB611.5 .E36 2024 (print) | LCC SB611.5 (ebook) | DDC 632/.5--dc23/eng/20231031
LC record available at https://lccn.loc.gov/2023045924
LC ebook record available at https://lccn.loc.gov/2023045925

Cover Images: Courtesy of Nicholas E. Korres
Cover Design: Wiley

Set in 9.5/12.5pt STIXTwoText by Integra Software Services Pvt. Ltd, Pondicherry, India

Contents

Preface

The necessity of human and environmental protection, along with the evolution of herbicide-resistant weeds, intensifies the need for weed management approaches based on ecological principles. Ecologically based weed management emphasizes the use of ecological principles and practices to minimize weed infestations and crop loses or damages while maintaining and/or enhancing ecosystem health. There are several key components to ecologically based weed management. First, it involves understanding the ecology of the weeds and the ecosystem in which they are growing. This includes factors such as soil type, moisture levels, plant community structure, and disturbance history. By understanding these factors, managers can identify the conditions that favor weed growth and develop strategies to disturb those conditions.

Second, ecologically based weed management seeks to prevent weed infestations from occurring in the first place. This can involve a range of practices, including promoting healthy plant communities, minimizing soil disturbance, using cover crops and other plant-based strategies to suppress weed growth, and implementing early detection and rapid response programs to quickly address new weed outbreaks.

Third, ecologically based weed management emphasizes the use of nonchemical and low-impact control methods whenever possible. This can include manual weed removal, cultural practices such as crop rotation and intercropping, and biological control methods such as the introduction of natural enemies of weeds.

Finally, it recognizes the significance of herbicides or other chemical control methods, although these methods should be used only as an integral part of weed management program and should be applied in a judicious manner to minimize their impact on nontarget species and the surrounding environment. Overall, ecologically based weed management represents a holistic and integrated approach to weed management that balances the needs of crop production with the health and resilience of natural ecosystems.

Ecologically based weed management offers provisional, regulatory, cultural, and supportive services for human wellbeing but also protects the environment. Weed management practices have become closely linked to social and economic, rather than biological, factors, particularly in conventional agriculture, where economic pressures have led to simplification of cropping systems and the replacement of alternate methods of weed management with synthetic chemical options. As a result, the evolution of agroecosystems and weed management strategies, an important part of the agricultural activities, is not progressing in parallel.

This is where this book becomes invaluable. It discusses weed-management practices under the frame of ecological and agro-ecological principles and highlights the benefits and future challenges but also the limitations that the ecologically based weed management approach must overcome. The wide diversity of the topics, along with the important issues in weed science, which are thoroughly discussed in this book, makes each chapter a unique case study.

Chapter 1 discusses the principles of ecologically based weed management, supported by successful case studies. **Chapter 2** discusses the fundamental of ecological services and focusses on the services provided when ecological principles are considered in weed control programs, while it refers to specific examples. Another interesting chapter relating ecologically based weed management with climate change is **Chapter 3**, whereas an extensive reference to nonchemical weed control is made in **Chapter 4**. **Chapter 5**, in line with previous chapters, discusses the important role of cultural practices and successfully concludes their positive contribution to ecologically based weed management.

Ecologically based weed management recognizes the significance of herbicides or other chemical control methods in our battle against weeds, although these methods should be used only as a last option and should be applied in a targeted and judicious manner to minimize their impact on nontarget species and the environment. For this reason **Chapter 6**, which discusses the role of world leading agri-chemical companies such as Syngenta, Bayer, and FMC, on weed management and what they are doing to protect the environment is most interesting. **Chapter 7** examines how ecologically based weed management supports pollination, an important constituent for biodiversity conservation and food production within agro-ecosystems. **Chapters 8** and **9** address an important, although less investigated, method of weed control – the use of biological agents and biological-based products. **Chapter 8** focuses on weed control using arthropods as biological agents and **Chapter 9** focuses on plant pathogens and critically assesses their role for weed suppression. It also refers to nanoformulations as a mean to deliver active ingredients for weed suppression. Cover crops, intercrops, and mulches have started to regain interest as effective weed control methods. **Chapter 10** extensively analyses the effects and practicalities of these techniques offer in terms of ecologically based weed management.

"*One year's seeding, seven years weeding*," says an old proverb. The focus should be on how to prevent weeds to build a large soil seedbank up, to minimize future problems. **Chapter 11** discusses important issues of weed soil seedbank from an ecological perspective with emphasis on soil seedbank management. **Chapter 12** further expands our options to reduce soil seedbank and discusses the implications of tillage systems on crop competitiveness, herbicide efficacy, and organic farming. Finally, **Chapters 13, 14, 15, 16,** and **17** investigate the application of ecologically based weed management on vegetable and specialty crops, row crops, orchards, and pastures. The information provided in these chapters is exceptional and can be used for a wide range of cropping and farming systems.

This book will be an invaluable source of information for scholars, growers, consultants, researchers, and other stakeholders dealing with agronomic, horticultural, and grassland-based production systems. The uniqueness of this book comes from the coverage of the most suitable ecologically based weed management practices that secure ecosystem services to humans and the environment. It reviews the available information critically and suggests solutions that are not merely feasible but also optimal. Readers will gain an in-depth knowledge on ecosystems services and weed practices. They will also be able to learn the principles of ecologically based weed control management, which are needed now more than ever.

Despite the great effort that authors, editors, and reviewers have invested in this work, mistakes may have been made. We would like to ask readers to inform us of any mistakes or omissions they find, as well as suggestions for future improvements by mailing us at the following email addresses, with "Ecologically Based Weed Management. Concepts, Challenges, and Limitations" in the subject line.

Nicholas E. Korres, PhD
Ilias S. Travlos, PhD
Thomas K. Gitsopoulos, PhD

List of Contributors

Ignacio Aspiazú
Department of Agricultural Sciences
State University of Montes Claros
Janaúba, Brazil

Susan M. Boyetchko
Deceased. Formerly Research Scientist
Saskatoon Research and Development Centre
Agriculture and Agri-Food Canada
University of Saskatchewan
Saskatoon, Canada

Parminder Chahal
Field Development Representative
FMC Corporation
University of Nebraska
Lincoln, Nebraska, USA

Raghavan Charudattan
President and CEO, BioProdex, Inc.
Emeritus Professor, University of Florida,
Gainesville, Florida, USA

Bhagirath S. Chauhan
Queensland Alliance for Agriculture and Food Innovation (QAAFI) and School of Agriculture and Food Sciences (SAFS)
The University of Queensland
Gatton, Queensland, Australia

Germani Concenço
Brazilian Agricultural Research Corporation
Capão do Leão, Brazil

Michael D. Day
Department of Agriculture and Fisheries
Brisbane, Queensland, Australia

Steve Fennimore
Department of Plant Sciences
University of California–Davis
St. Salinas, California, USA

Evander Alves Ferreira
Institute of Agricultural Sciences
Federal University of Minas Gerais
Montes Claros, Brazil

Karla L. Gage
School of Agricultural Sciences/School of Biological Sciences
Southern Illinois University–Carbondale
Carbondale, Illinois, USA

Leandro Galon
Universidade Federal da Fronteira Sul
Chapecó, SC, Brazil

Ioannis Gazoulis
Agricultural University of Athens
Athens, Greece

Roland Gerhards
Department of Weed Sciences, Institute of Phytomedicine
University of Hohenheim
Stuttgart, Germany

Clevison Luiz Giacobbo
Universidade Federal da Fronteira Sul
Chapecó, SC, Brazil

Thomas Gitsopoulos
HAO-Demeter, Institute of Plant Breeding and Genetic Resources
Thermi – Thessaloniki, Greece

Katherine M. Jennings
Department of Horticultural Science, North Carolina State University
Raleigh, USA

Panagiotis Kanatas
University of Patras (Teaching Staff P.D. 407/80)
Mesolonghi, Greece

Filitsa Karamaouna
Scientific Directorate of Pesticides Control and Phytopharmacy
Benaki Phytopathological Institute
Kifissia, Greece

Vaya Kati
Laboratory of Agronomy, School of Agriculture
Faculty of Agriculture, Forestry and Natural Environment
Aristotle University of Thessaloniki,
Thessaloniki, Greece
and
Scientific Directorate of Pesticides Control and Phytopharmacy
Benaki Phytopathological Institute
Kifissia, Greece

Nicholas E. Korres
Department of Agriculture
University of Ioannina
Kostakii, Arta, Greece

Stevan Z. Knezevic
Professor of Integrated Weed Management
Department of Agronomy and Horticulture,
University of Nebraska-Lincoln, Lincoln,
Nebraska, USA

Eduardo S. Leguizamon E.S
Rosario National University
Rosario, Republic of Argentina

Natalie P. Lounsbury
Department of Agriculture, Nutrition, and Food Systems
College of Life Sciences and Agriculture
University of New Hampshire, Durham,
USA

Jonathan W. McLachlan
School of Environmental and Rural Science
University of New England
Armidale, Australia

Victor Martins Maia
Department of Agricultural Sciences
State University of Montes Claros
Janaúba, Brazil

Miriam Messelhäuser
Department of Weed Sciences, Institute of Phytomedicine
University of Hohenheim
Stuttgart, Germany

Andrea Monroy-Borrego
Department of NanoEngineering
University of California–San Diego
San Diego, California, USA

Adusumilli Narayana Rao
Consultant Scientist
Jubilee Hills, Hyderabad, India

Georg Naruhn
Department of Weed Sciences, Institute of Phytomedicine
University of Hohenheim
Stuttgart, Germany

Iraj Nosratti
Department of Plant Production and Genetics
Faculty of Agriculture, Razi University
Kermanshah, Iran

Samuel A. Palmer
Department of Natural Resources and the Environment
University of New Hampshire, Durham, USA

Gerassimos Peteinatos
Centre for Automation and Robotics (CSIC)
Madrid, Spain
and
Hellenic Agricultural Organization – DIMITRA
Department of Agricultural Engineering, Soil and Water Research Institute
Athens, Greece

David Reiser
Department of Technology in Crop Production
Institute of Agricultural Engineering
University of Hohenheim,
Stuttgart, Germany

Erin N. Rosskopf
Research Microbiologist, USDA-ARS
U.S. Horticultural Research Laboratory
Fort Pierce, Florida, USA

Aritz Royo-Esnal
Department of Agricultural and Forest Science and Engineering
ETSEAFIV-Agrotecnio Centre, Universitat de Lleida
Lleida, Spain

Lauren M. Schwartz-Lazaro
Blue River Technology
Sunnyvale, California, USA

Matthias Schumacher
Department of Weed Sciences, Institute of Phytomedicine
University of Hohenheim
Stuttgart, Germany

Brian M. Sindel
School of Environmental and Rural Science
University of New England
Armidale, Australia

Alexandre Ferreira da Silva
Brazilian Agricultural Research Corporation
Capão do Leão, Brazil

Milena Simić
Maize Research Institute "Zemun Polje"
Belgrade–Zemun, Serbia

Richard G. Smith
Department of Natural Resources and the Environment
University of New Hampshire, Durham, USA

Michael Spaeth
Department of Weed Sciences, Institute of Phytomedicine
University of Hohenheim
Stuttgart, Germany

George Andrade Sodré
State University of Santa Cruz,
Neighborhood Salobrinho, BA, Brazil

Nicole F. Steinmetz
Institute of Engineering in Medicine
University of California–San Diego
San Diego, California, USA
Department of Radiology,
Moores Cancer Center, University of California-San Diego,
La Jolla, USA

Joel Torra
Department of Agricultural and Forest Science and Engineering
ETSEAFIV-Agrotecnio Centre, Universitat de Lleida
Lleida, Spain

Ilias Travlos
Agricultural University of Athens
Athens, Greece

Vijay K. Varanasi
Bayer Crop Science
St. Louis, Missouri, USA

Ioannis Vasilakoglou
Department of Agronomy – Agrotechnology
University of Thessaly
Thessaly, Greece

Vasileios P. Vasileiadis
Head of Regenerative Agriculture, Europe, Africa, and Middle East
Syngenta Crop Protection
Athens, Greece

Kaydene T. Williams
U.S. Horticultural Research Laboratory
Fort Pierce, FL
and University of Florida-Gulf Coast Research and Education Center
Wimauma, Florida, USA

Rachel L. Winston
MIA Consulting
Shelley, Idaho, USA

Arne B. R. Witt
CABI
Nairobi, Kenya

Rosa Witty
Department of Weed Sciences, Institute of Phytomedicine
University of Hohenheim
Stuttgart, Germany

List of Reviewers

Albert T. Adjesiwor Assistant Professor and Extension Specialist, Department of Plant Sciences, University of Idaho, Kimberly Research & Extension Center, Kimberly, USA.

Warwick Badgery Research Leader Rangelands and Tropical Pastures, NSW Department of Primary Industries, Australia.

Barbara Baraibar Researcher, University of Lleida, Spain.

Lammert Bastiaans Professor, Wageningen University, Centre for Crop Systems Analysis, The Netherlands.

Milan Brankov Researchers, Maize Research Institute "Zemun Polje", Serbia, Ioannis Gazoulis, Research Assistant, Agricultural University of Athens, Greece.

Mehmet Nedim Dogan Professor, Adnan Menderes University, Turkey

Stephen Duke Adjunct Research Professor, Thad Cochran Research Center, Mississippi State University, USA.

Eric Gallandt Professor, Weed Ecology and Management, University of Maine, USA.

Thomas Gitsopoulos Senior Researcher, Institute of Plant Breeding and Genetic Resources, ELGO-DIMITRA, Thessaloniki, Greece.

Kerry Harrington Associate Professor, School of Agriculture and Environment, Massey University, New Zealand.

Panagiotis Kanatas Teaching Staff, University of Patras, Greece.

Vaya Kati Assistant Professor, School of Agriculture, Faculty of Agriculture, Forestry and Natural Environment, Aristotle University of Thessaloniki, Thessaloniki, Greece.

Ioannis Kazoulis Associate Professor, Agricultural University of Athens, Greece

Khawar Jabran Associate Professor, Plant Production & Technologies Department, Nigde Omer Halisdemir University, Nigde, Turkey.

Nicholas E. Korres Associate Professor, Dept. of Agriculture, School of Agriculture, University of Ioannina, Kostakii, Arta, Greece.

Spyridon Mantzoukas Research Fellow, Department of Agriculture, University of Ioannina, Ioannina, Arta, Greece.

Maor Matzrafi Senior Researcher, Volcani Institute, Newe-Ya'ar Research Center, Israel

Fabian Menalled Professor, Land Resources and Environmental Sciences, Montana State University, USA.

Husrev Mennan Professor, Ondokuz Mayıs University, Agriculture Faculty, Department of Plant Protection, Samsun, Turkey.

Mario Luiz Ribeiro Mesquita Professor, Departamento De Ciencias Agrarias – Bacabal, Brazil.

Sheeja K. Raj Assistant Professor (Agronomy), Department of Organic Agriculture, Kerala Agricultural University, College of Agriculture, Vellayani, Thiruvananthapuram, Kerala, India.

Ilias Travlos Associate Professor, Agricultural University of Athens, Greece.

N. T. Yaduraju Formerly: Director, ICAR - Directorate of Weed Research, Jabalpur, National Coordinator, National Agriculture Innovation Project (ICAR) and Principal Scientist ICT4D, ICRISAT, Hyderabad, India.

Ioannis Vasilakoglou Professor, Department of Agriculture-Agrotechnology, University of Thessaly, Larissa, Greece.

Costas Zachariades Senior Researcher, Agricultural Research Council, South Africa, ARC, Plant Protection Research Institute, South Africa.

1

Ecologically Based Weed Management (EbWM)

Enabling and Reinforcing the Approach

Leguizamon Eduardo S.[1,*]*, Royo-Esnal Aritz*[2]*, and Torra Joel*[2]

[1] *Cajaraville, Rosario, República Argentina*
[2] *Department of Agricultural and Forest Science and Engineering, ETSEAFIV-Agrotecnio-CERCA Centre, Universitat de Lleida. Alcalde-Rovira Roure, Lleida, Spain*
[*] *Corresponding author*

1.0 Introduction

Managing food production systems on a sustainable basis is one of the most critical challenges for the future of humanity. Being fundamentally dependent on the world's atmosphere, soils, water, and genetic resources, these systems provide the most essential ecosystem services on the planet. They are also the largest global consumers of land and water, threats to biodiversity through habitat change, and significant sources of air and water pollution in several regions on Earth (Naylor 2008).

The increase in the world population is necessarily associated with a greater demand for food produced by crops, among other approaches (e.g. reducing food wastes or synthetic food). Currently, there are limited possibilities of achieving crops with superior yields, and incorporating new territories into agriculture is not a realistic option. Under these grounds, it is clear that one factor that favors the increase in crop productivity is the management of species considered pests. In this context, weeds are one of the most important biotic constraints.

During the last 70 years, intensive measures have been taken for crop protection against pests through the widespread use of chemical pesticides in order to reduce the loss of agricultural yield. Although mainly chemical-based, crop protection practices have reduced the overall potential losses of 50% to actual losses of about 30%, with crop losses due to pests still varying from 14% to 35% depending on the considered crop and country (Oerke 2006). Consequences of this massive-intensive chemical use in the agroecosystems are increasingly studied as concerns rocketed all over the world.

Integrated pest management (IPM) was proposed 70 years ago by Stern et al. (1959), who outlined a simple but sophisticated idea of pest control in order to manage insect pests while reducing reliance on synthetic pesticides. Briefly, IPM is based on four elements:

1) Knowledge of the thresholds to determine the need for control
2) Necessary population sampling to determine critical densities (economic damage)
3) The biological control capacity in the system
4) Use of insecticides or selective methods compatible to biological control enhancement (Thill et al. 1991)

Later on, Swanton and Weise (1991) after other precursors, proposed the use of integrated weed management (IWM) as a similar approach for weed management in agroecosystems. IWM was then inspired by IPM, as a long-term management strategy that uses a combination of strategies to reduce the population size of weeds to a tolerable level, being economically affordable and also as a tool to reduce undesired environmental effects of herbicides. However, in most crop production systems, generalized recommendations include just a combination of management tactics (cultural + chemical). After more than 30 years, IWM remains in its infancy, since the implementation of IWM has been poor, with little evidence of its sustainability (e.g. reductions in herbicide use). Moreover, nonchemical methods (mechanical) are often adopted as a means of compensating for reduced herbicide efficacy, due to increasing resistance, rather than as alternatives to herbicides. Reluctance to adopt nonchemical methods may be due not only to a lack of knowledge, but also to a lack of

Ecologically Based Weed Management: Concepts, Challenges, and Limitations, First Edition. Edited by Nicholas E. Korres, Ilias S. Travlos, and Thomas K. Gitsopoulos.

farmer motivation and action and/or risk aversion (Moss 2018). Justifiably, herbicides are often seen as the easiest and still most effective option since their convenience outweighs the increased complexity, costs and management time associated with nonchemical alternatives.

To bring numbers to the statements already said, surveys recently made by crop advisors in Argentina (Satorre 2015) within high technology agricultural entreprises concerning the weed problems they faced in the last decade, revealed the following:

- There is great concern about weeds, considered the main adversity, especially in summer crops.
- Weed management technologies used in the last decades have only been based on herbicides and, surprisingly, they are still led by glyphosate.
- Increasing number of species exhibit a range of herbicide tolerant / resistant responses.
- It is only recently that some farmers (less than 10% acreage) began to include cultural practices (such as cover crops).
- There is a lack of incentives for application of IWM practices and/or pesticides reduction programs.
- 70% of summer crops are planted on short-term leased land, where actions are mostly oriented toward annual productivity, neglecting possible future problems such as contamination or the appearance of herbicide resistant weeds.

Thus, although IWM is frequently advertised/proclaimed as a dominant concept associated to sustainability, the difficulty of evaluating the benefits derived from alternative approaches, ignorance, and/or a low use of available knowledge of weed biology present a severe barrier to changes: up to the present time, a majority of the farmers have failed in a massive implementation of IWM.

Similar results emerged from a recent survey made in the cornbelt from the USA. To elucidate the causes or barriers that prevent huge adoption of IWM, Al-Mamun (2018) identified "failures of institutional context declining government policies, counteracted by multinational private companies as main actors." A further contribution (Wilson et al. 2009a) states that "agrochemical supplier companies impose the massive use of their products through business marketing strategies, preventing producers from being oriented in the use of more sustainable practices," and that a key aspect is that "crop advisors should manage to transmit sufficient trust toward the farmers to be able to address the real problem in a clear and concrete way." For this, it is necessary to create a strong link between the different actors through intensive two-way communication and a greater understanding of the way in which producers perceive this problem (Wilson et al. 2009b). A further issue that should be revised is stewardship programs systematically launched and advertised by the agrochemical industry and also by plant protection sellers and distributors in general, with a rather shallow view in relation to the principles of IWM and advanced available knowledge.

An example on a low-profile marketing campaign was that made about 30 years ago when Roundup-Ready soybeans were launched in Argentina. Adverts in media and in rural roads, boosted the use of *the simplest solution* to tackle weed problems: just glyphosate. Think of glyphosate in the crops – first in soybeans, then in corn, and later on in cotton and others – plus glyphosate in the fallow. In fact, the tremendous success of direct drilling in Argentina was only possible when two hard technological bottlenecks were overcome: how to place seeds in an undisturbed soil by mouldboard-plough + harrowing (using newly designed planters) and how to get rid of weeds without mechanical tactics (using a novel and very effective herbicide: glyphosate).

Not only companies but also advisors and educators may fail to promote IWM within the frame of farmers' experience and belief structure. Targeted communication efforts that address key misperceptions, and highlight the cost-effective nature of integrated approaches may increase adoption of IWM and ultimately increase sustainability of the agroecosystem. Unfortunately, IWM systems have been perceived as unreliable, resulting in increased risk of weed control failure. The acceptance of IWM by growers will depend on their risk perception of management, individual management capability and environmental interactions that will influence the economic viability of the crop system. The adoption of IWM is usually hindered by the fact that chemical means are often growers' first and only choice, as synthetic herbicides are perceived as an effective, rapid and cost-effective solution for weed management. However, the consequences of intensive pesticide usage in agriculture are now quite well known and fortunately widely and progressively studied, being the increasingly widespread herbicide-resistant weed biotypes as the most striking example of the unsustainability of current plant protection strategies.

1.1 Basis for a Sucessful Ecologically Based Weed Management Approach

In 2008, Bastiaans et al. reflected on the possibilities and limitations of ecological approaches in weed control practices, highlighting the need for research in order to provide clear insight in effectiveness and applicability of the utilization of ecological knowledge translated into practical strategies of weed management. If we do agree that the maintenance of resilience and diversity are key issues for the agriculture to be sustainable (and even more under the intensification process already started), then we should address for a reinforced and enlarged ecologically based weed management (EbWM) definition, backed up by complementary major related disciplines. It must be pointed out that ecology provides the theoretical basis for weed science, much as physics provides a theoretical basis for engineering and biology the theoretical basis for medicine (Liebman et al. 2001). Although much research has been focused on the ecological relationships of weeds within agroecosystems in recent years, substantial gaps in knowledge relevant to weed management still exist, since weed management strategies must include multiple points of intervention in their life cycles (Liebman and Gallandt 1997).

Production models successfully developed in the last decades in various regions of Europe, America, and elsewhere, clearly demonstrate that the coexistence of production systems conducted under different and conceptual views is possible. What is clear is that goverment policies should ensure the success of each of them, whether implemented by individual producers, by organizations, by cooperatives, or by companies, designing provisions and control and promotion mechanisms with policies that allow their free implementation within a framework of respect and sustainability in the broadest sense. Such is the case of the European standards for the creation and use of transgenic varieties, of organic productions and even more, of crop productions rescueing ancestral varieties.

Simultaneously, weed scientists should shift their focus from trying to create prescriptive ways to manage weeds to developing ways for farmers to gain site-specific knowledge that will allow them to decrease the uncertainty of greater reliance on natural weed population regulating mechanisms. Most "weed problems" are really "people problems" evolved from poor land management and a deep lack of ecological insight.

1.2 Enabling and Reinforcing EbWM Principles in All Crop Production Systems

Kleijn et al. (2019) warn that large-scale adoption of ecological intensification requires stronger evidence than is currently available. Future research should therefore not only address ecological, agronomic, and economic aspects of ecological intensification but also the sociological aspects. To contribute to this goal, knowledge may contribute to reinforce and spread the application of ecological principles in a variety of crop systems, including those who depend on herbicide technology.

It is envisaged that the following six pillars can contribute to reach the objectives for an extended and deeper usage of EbWM principles in any kind of crop system.

1.2.1 Systems Approach

> *"Rotation of crops...is the most effective means yet devised for keeping land free of weeds. No other method of weed control, mechanical, chemical, or biological, is so economical or so easily practiced as a well-arranged sequence of tillage and cropping."*
>
> C.E.R. Leighty 1938. Yearbook of Agriculture. USA.

Box 1 Population Biology of Plants

A fundamental issue and the core of EbWM is the consideration that not only the crop but also the weeds are organized at the level of the population. In plants, growth is modular: the "sections" or "modules" (the phytomer) are repeated over and over again, allowing the increase in size and biomass: plants (crops and weeds) are modular organisms. Growth based on repeated modular structures (the phytomer) provides great plasticity and has a profound significance in the competitive capacity, fertility, propagation, and persistence of a given genotype. Plastic responses should be taken into account when studying and modeling weed-crop interactions and also when evaluating the effects of management factors affecting density (e.g. herbicides, competition with the crop, interactions with cover crops). From the population dynamics point of view, each state that defines a weed life history begins with the germination of the seed and ends with the production of new seeds by adult plants, their subsequent dispersal and posterior seedbank incorporation. The number of variations of any stage in the life cycle of a plant (e.g. seedlings, plants, seeds) can be traced using demographic tools. The responses of vital rates in relation to the environment, determine the dynamics of populations in an ecological time and the evolution of life histories, in an evolutionary time. When calculating vital rates during the life cycle, demographics take into account both the dynamics and the structure of populations. The goal of a successful weed management program is to reduce the rate of population change, that is, lambda (λ) calculated as the ratio of the next to the current population size.

from Harper, J.L. 1979

EbWM should be the core of weed management in any cropping system, whatever it is intensive, extensive, organic or industrial. Gage et al. (2019) define the concept of Systems Approach as managing weeds by combining practice and knowledge with the goals of increasing yield and minimizing economic loss, minimizing risks to human health and the environment, and reducing energy requirements and off-target impacts. The reliance on herbicides in modern cropping systems should shift the management focus from requiring intimate knowledge of biology, ecology, and ecological systems to herbicide chemistry, mixtures, and rotations, application technology, and herbicide-tolerant crop traits. Prevention of spread, seedbank management, crop rotations, tillage, cover crops, competitive cultivars, or biological weed control all require to fill identification of knowledge gaps where research advancements may be possible. Then, an ecological systems approach may provide improved stewardship of new herbicide technologies and reduce herbicide resistance evolution through diversification of selection pressures. Interestingly, several authors include the need of a careful planning and setting of experiments lay-out according to the objectives, focusing on scale considerations.

A total system approach may contribute to the necessary reduction of the heavy use of pesticides by using the knowledge provided by ecology and related disciplines. Lewis et al. (1997) proposed a diagram to illustrate the necessary shift to a total system approach to pest management through a greater use of inherent strengths based on a good understanding of interactions within an ecosystem while using therapeutics as backups: an upside-down pyramid reflected the unstable conditions under heavy reliance on pesticides, and an upright pyramid reflected sustainable qualities of a total system strategy.

1.2.2 Increased Biodiversity in the System

Several experiments have been made with this perception in the last 20 years. For example, Bastiaans et al. (2007) focused their research on enhancing the diversity to manage weeds in very different cropping systems in the Netherlands (e.g horticultural) by intercropping slow growing vegetables such as onion, carrot and leek, and sequential use of cover crops when the main crop is absent (stubble).

The complementary exploitation of resources by combined extensive crops has also been studied during three consecutive years in soybean-corn strips by Verdelli et al. (2012). These authors demonstrated that corn yield in the strips significantly increased as compared to that in the monocultures due to increased yields in corn plants of the border rows of the strips, which was highly correlated to an increased radiation interception, allowing higher crop growth rates at critical crop stages. Conversely, soybean yields in the strips were lower than that in the monocultures; however, the strip-crops system overyielded monocultures. Authors emphasize that the use of more appropriate genotypes may contribute to increase the differences and then ease the spread of this technique in actual massive monocultured agricultural systems of Argentina.

Getting back to experimental design considerations, Petit et al. (2018) appoint that further advances in the understanding of biodiversity-based options and their performance for weed biocontrol require farm-scale experimental trials. In this sense, the evaluation of the influence of herbicide-resistant crops on biodiversity (e.g invertebrates and vegetation of field margins) made by Roy et al. (2003) across English countryside fields, may be a good experimental and theoretical vision as they, early on, highlighted the importance of butterflies evaluation as key indicator species in the study of agroecosystems. In the same line, Alignier et al. (2020) have found that crop heterogeneity increases within-field plant diversity.

Petit et al. (2018) have studied several biodiversity-based options for arable weed management since they questioned that IWM currently recommends agronomic practices for weed control, but it does not integrate the use of biodiversity-based options that enhance the biological regulation of weeds. In their contribution, they alert and describe existing knowledge related to three potentially beneficial interactions, crop–weed competition, weed seed predation, and weed interactions with pathogenic fungi. They found that promoting cropped plant–weed competition by manipulating cropped cover could greatly contribute to weed reduction; that weed seed predation by invertebrates may significantly reduce weed emergence; and that a wide range of fungi may be pathogenic to various stages of weed development. Again, they warn about the necessary requirement of farm-scale experimental trials. Hails (2002) also points out the careful design of key elements of long monitoring biodiversity.

In a very interesting approach, Petit et al. (2015) use weeds as a model for exploring management options relying on the principle of ecological intensification in 55 experimental farm fields. The authors use weeds because they can cause severe crop losses, contribute to farmland functional biodiversity and are strongly associated with the generic issue of pesticide use. They monitor the impacts of herbicide reduction following a causal framework starting with less herbicide inputs triggering changes in (i) the management options required to control weeds, (ii) the weed communities and functions they provide, and (iii) the overall performance and sustainability of the implemented land management options. Interestingly, the reduction of herbicide use was not antagonistic with crop production, provided that alternative practices are put into place. Outcomes suggest that sustainable management could possibly be achieved through changes in weed management, along a pathway starting with herbicide reduction. Humans should increase biodiversity in human-dominated landscapes. Science provides robust foundations for predictions on human land-use trends and species-area relationships.

1.2.3 Inclusion of the Spatial Scale in the System: From the Field to the Landscape

The intensification of agricultural practices and the increase of area under agricultural production, which was accompanied by a destruction of perennial habitats, made agriculture one of the main causes of biodiversity losses. Though annual arable weed populations outlast with their seedbank, they can also benefit from the seed rain from the surrounding landscape. Thus, it is not only important to support the survival conditions within the fields (e.g. by extensive management), but also consider the structure of the landscape (Solé-Senan et al. 2014). The role of arable weeds in cereal aphid-natural enemies' interactions was analyzed in Roschewitz's PhD thesis (2005).

The decision-making regarding weed management in agricultural systems is influenced by a wide range of factors that operate at variable spatio-temporal scales. In 1997, Rabbinge proposed a simple scheme combining the two leading factors in the ecological consideration of living organisms of the agroecosystem (Figure 1.1): space (x-axis) and time (y-axis). It helps farmers figure out where to focus their activities and how to envisage the issues and dynamics concerning the time scale (e.g. a weed seed in the soil) or the space scale (the movement of a seed by the wind a long way from the mother plant). Interestingly, the author superimposed succesive demographic, geographic, and ecological levels where the experiments and research may be located.

EbWM grounded on ecological systems approach may maximize yield and minimize risks to farmers' health and to the environment, while reducing energy requirements and additional effects ("externalities") such as the maintenance of biodiversity by managing borders and margins. Thus, tactics used should modulate the processes strongly influencing weed population size (e.g. seedbank management, fertility management, postdispersal), crop rotation, tillage, service crops, and the competitive ability of crop cultivars. Herbicide evaluation should consider not only the efficacy but also the fecundity of the uncontrolled population, consequently affecting the seedbank. The scale should be strongly replaced, shifting from "weed management of the field" to "management of weeds in the production systems of the region." Under this broad conception, we may find *healthy* agroecosystems, where traditional low-input activities are performed, with diverse weed communities that contribute to resilience. For example, several studies identify a wide range of taxa, including birds and mammals, invertebrates, and arable flora that benefit from organic management compared to conventional agriculture (Hole et al. 2005). In other cases, plant species richness can change with altitude because less intensified agriculture is associated with higher elevations, as was found in Central Europe (Pysek et al. 2005) and Spain (Cirujeda et al. 2011). First, these agroecosystems should be preserved, of course, and second, there are very important principles that can be learned from them to fuel the application of EbWM tools.

There are other agroecosystems where weed diversity has been reduced to very invasive species (*fast weeds*) often resistant to herbicides, and where arable less-competitive plants (*slow weeds*) have disappeared. In these agroecosystems, applying EbWM is more difficult. The high densities and high competitive capacity of these weeds require a first step that might take some years of effective IWM strategy applications to be able to reduce their populations trying

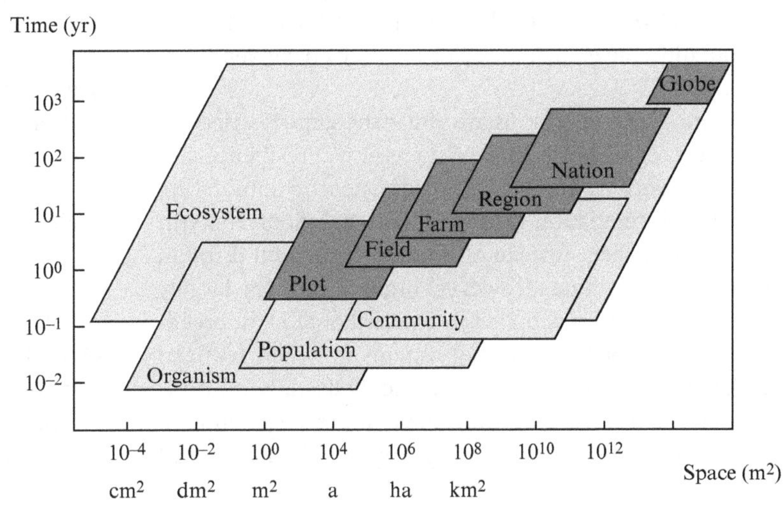

Figure 1.1 Examples of the spatial and temporal scale for investigations of hierarchical levels within natural (light colored) and agricultural systems (dark colored) (Dalgaard et al. 2003).

to deplete their soil seedbank. Once these populations have been reduced (not eradicated), the ecological niche is receptive to establishing other species populations that could be managed with EbWM principles.

There is then a range of agroecological approaches with variable performances, but win-win scenarios are demonstrated, where both environment and profitability can be reinforced. Among them, sustainable intensification or agroecological intensification (AEI) stands out. Here, intensification involves improvement of farm and system performance through the implementation of agroecological principles, rather than intervention (Elliot et al. 2013).

Finally, it must be pointed out, as MacLaren et al. (2020) did, that the design and implementation of EbWM strategies at agroecosystem level is complex, because an understanding of the ecological interactions is required, as well as the theoretically relevant practices that could match the different environments and farming systems to achieve sustainable, healthy, and environmentally friendly food production systems. Moreover, the competition capacity of a weed community against crops is affected by its composition and diversity, as well as its capacity to support biodiversity and provide ecosystem services (MacLaren et al. 2020). Thus, in each situation (field-crop-landscape) knowledge of the type of weeds present and their relative abundance, in addition to total weed density or biomass, is needed in order to apply effective EbWM strategies.

1.2.4 Significant Improvement in the Objectives of the Crop Breeding Programs

Lamichhane et al. (2017) advocate a need for suitable breeding approaches to boost a more sustainable management since European farmers do not have access to a sufficient number and diversity of crop species/varieties. This prevents them from designing more resilient cropping systems to abiotic and biotic stresses. These authors propose a new breeding paradigm called *breeding for integrated pest management (IPM)*, which could easily be extended to EbWM with an ultimate goal of reducing reliance on conventional pesticides.

Organic farming systems are under the same restrictions (Lammerts et al. 2002): these systems aim at resilience and buffering capacity in the farm ecosystem by stimulating internal self-regulation through functional biodiversity in and above the soil, instead of external regulation through chemical protectants. However, organic farmers largely depend on varieties supplied by conventional plant breeders and developed for farming systems in which artificial fertilizers and agrochemicals are widely used. Until now, many of the desired crop traits have not received enough priority in conventional breeding programs. The proposed organic crop ideotypes may benefit not only from organic farming systems but also from conventional systems that move away from high inputs of nutrients and chemical pesticides.

The conditions briefly described above illustrate the issue: a major contribution for the enhancement of crop competitive ability should be made within crop breeding programs. Interestingly, early steps should be part of a whole-broader scope of breeding policy, which should consider yield maximization and/or the incorporation of resistance to pests while contributing to the purpose of more sustainable crop systems. Westwood et al. (2018) reviewed that enhanced weed-competitive crops based on morphological traits has not resulted in the knowledge required by plant breeders to reliably enhance the competitive ability of crops against weeds. Thus, these authors propose to focus research on the molecular, physiological, and morphological mechanisms of both interspecific and intraspecific competition. In the same line, Pester et al. (1999) pointed out that plant breeders need basic and applied information to identify favorable crop-weed competitive traits in order to enhance or incorporate those traits into crop cultivars. Accelerated research on competitive crops against weeds should lead to more economic, effective, and feasible IWM programs for all crops. Similarly, Gibson et al. (2003) investigated the weed suppression ability of rice cultivars, since resistance to herbicides and the lack of efficient control options have led to an interest in increasing the role of crop competition as a weed management tool in water-seeded rice production. These authors saw that an indirect selection program, based on traits that can be identified early in the season under weed-free conditions, has a great potential for developing more competitive rice crop cultivars.

Competitive crop cultivars offer a potentially cheap option to be included in EbWM strategies. Although cultivars with high competitive potential have been identified among cereal crops, competitiveness has not traditionally been considered a priority for breeding or farmer cultivar choice. However, the challenge of managing herbicide-resistant weed populations has renewed interest in cultural weed control options, including competitive cultivars (Andrew et al. 2015). Breeding programs usually seek to optimize key agronomic traits, such as seed quality and quantity, biomass production, and pest and disease resistance. Clearly, this objective should persist when a crop mixture is the breeding target, but other traits should also be incorporated and optimized, critically including an ability to live and perform with others. Interestingly, community ecology has recently made important progress in this issue, helped by theoretical advances in trait-based ecology. Ecology theory provides the key for the maintenance and optimized application of heterogeneous covers. Thus, some authors call for ecological assembly rules as the foundations for a novel paradigm in plant-breeding programs (Litrico and Violle 2015).

A further advance in enhancing crop competition against weeds by increasing the shading has been explored by Colbach et al. (2018), using a novel experimental technique (virtual experiments) with which several runs were made in diverse regions, cropping systems, and weed floras. They found that plant-morphology and shade-response parameters were related to crop production, but there was a trade-off between yield promotion and weed suppression traits. With similar purposes, Worthington and Reberg-Horton (2013), stated that the combined effects of allelopathy and competition determine the weed suppressive potential of a given cultivar. Both allelopathy and competitive ability are complex, quantitatively inherited traits that are heavily influenced by environmental factors. Again, good experimental designs and sound breeding procedures are essential to achieve genetic gains. Weed suppressive rice cultivars are now commercially available in the United States and China, as a result of three decades of research. Furthermore, a strong foundation has been laid during the past 10 years for the breeding of weed-suppressive wheat and barley cultivars.

A broader breeding scope has been recently used in Italy: it is focused on the unravelling of competitive ability traits (Lazzaro et al. 2018) in order to identify the most suitable combinations of competitiveness and production traits, which often show trade-offs that led to the identification of accessions with reduced grain yield to plant height trade-off. These authors characterized 160 common wheat (*Triticum aestivum* L.) accessions cultivated since the nineteenth century for four traits linked to competitive resistance to weeds (above-ground biomass before stem elongation, tillering index, plant height, and flag leaf morphology), and for two production-related traits (grain yield and thousand-kernel weight).

1.2.5 Use of Herbicides Only Based on Dose-Response Technology

Applying herbicides at lower rates than the label recommendation has been the rule rather than the exception in Denmark since the late 1980s. This tendency emerged after a mandatory law to reduce agrochemicals by 50%. Dose response curves were built for multiple weed stages and a wide range of herbicides. The susceptibility of dominant weeds to herbicides has been argued to justify reduced herbicide rates. For example, even reduced rates would result in maximum effects, when the growth stage of weeds, crop vigor and climatic conditions were optimum and, thus, promote the activity of the herbicide, which would allow this management (Kudsk 2014).

Weed control is generally considered to be essential for crop production. Herbicides have become the main method for weed control in developed countries, but concerns about harmful environmental consequences have led to strong pressure for farmers to reduce their use. Gaba et al. (2016) analyzed the relationship between weeds, herbicides, and winter wheat yields using data from 150 winter wheat fields in western France. A Bayesian hierarchical model was built considering farmers' behavior, implicitly including their perception of weeds and weed control practices on the effectiveness of treatment. No relationship was detected between crop yields and herbicide use. Herbicides were found to be more effective at controlling rare plant species than dominant and harmful weed species. These results suggest that reducing the use of herbicides by up to 50% could maintain crop production, a result confirmed by previous studies, while encouraging weed biodiversity. Food security and biodiversity conservation may, therefore, be achieved simultaneously in intensive agriculture simply by reducing the use of herbicides.

1.2.6 Calculation of Pesticide Load in Each Field

The design of specific policies for the rational use of pesticides and care for the environment, which have been successfully applied in Denmark and other European countries, is a clear example of rationality, comprehensiveness and common sense that deserves imitation. Several pesticide risk indicators have been developed over the years. Recently, a new pesticide risk indicator, the pesticide load (PL), was introduced in Denmark. The PL consists of three subindicators for human health, ecotoxicology, and environmental fate. For each of the three subindicators, PL is calculated and expressed as the PL per unit commercial product (kg, L). PL for human health is based on the risk phrases on the product label, while PL for ecotoxicology is calculated based on the dose values of the active ingredients for acute toxicity to mammals, birds, fish, daphnia, algae, aquatic plants, earthworms and bees, and NOEC (No Observed Effect Concentration) values for chronic toxicity to fish, daphnia and earthworms. PL for environmental fate is calculated on the basis of the half-life in soil and bioaccumulation and groundwater concentration indexes. PL reflects the relative risks associated with the use of pesticides. Besides using PL for monitoring the yearly trend in pesticide use and load, PL was also used for setting up a new pesticide tax scheme and for setting quantitative reduction targets. In Denmark, it is now compulsory for farmers to upload their pesticide use data (e.g. the annual pesticide statistics) and PL can be calculated based on pesticide use data rather than sales data that may not reflect actual use by farmers (Kudsk et al. 2018).

1.3 Projects / Experiments Where EwBM Principles Are Being Tested

1.3.1 Example 1

Replacing heavy reliance on herbicides with integrated strategies employing diverse sets of complementary tactics involved weed management in a large-scale, long-term cropping system experiment in the US cornbelt (Davis et al. 2012). This experiment included a conventionally managed corn–soybean rotation and a more diverse corn–soybean–cereal/alfalfa–alfalfa system receiving lower amounts of herbicides. The reductions in herbicide use in the more diverse system resulted from applying herbicides only in bands over corn and soybean rows, rather than broadcast spraying; using an interrow cultivator in the unsprayed areas between corn and soybean rows; and using mowing and hay removal rather than herbicides to control weeds in cereal stubble and alfalfa.

Empirical measurements of weed seed population densities in the soil of the experimental plots over a nine-year period indicated that they declined for both the simple corn–soybean system under conventional, full-herbicide-rate management, and the more diverse four-crop system treated with less herbicides. The more diverse system also matched or exceeded the crop yields and profitability of the simpler conventional system. These results are consistent with those from a set of on-farm experiments conducted in Italy, Germany, and Slovenia in which blending mechanical and chemical weed control tactics was found to be effective for suppressing weeds in corn with greatly reduced reliance on herbicides, while maintaining yields and economic returns. Importantly, modeling studies predict that more diverse management systems integrating chemical and nonchemical tactics can not only keep weed population densities lower but also slow down evolution of herbicide resistance.

1.3.2 Example 2

The possibility of farming without glyphosate is becoming an important research and development issue for the agri-food sector. Contingency plans need to be formulated in the event that glyphosate is banned. Lamichhane et al. (2017) summarized international events that have led to this possible situation, described current glyphosate usage in major agronomic field crops worldwide, outlined possible alternatives to glyphosate in two agroregions and performed bioeconomic model scenarios of southern Australian cropping systems without the herbicide. Model predictions suggest that farming may be done profitably without glyphosate by consistently using key nonherbicidal weed management practices combined with robust pre-emergence soil residual herbicide treatments. However, maintaining low weed seedbanks may be challenging.

1.3.3 Example 3

In order to support farmers in defining integrated management strategies, IWMPRAISE (2020) project designed a framework consisting of five pillars for IWM. Each pillar contains a list of tactics. The tactics affect one or more parts of the weed life cycle or the weed-crop interaction. Experiments under IWMPRAISE-EEC Project have been deployed in several European countries.

1.3.4 Example 4

Lolium multiflorum (annual Italian ryegrass) and other grass weeds are an increasing problem in cereal cropping systems in Denmark. Grass weeds are highly competitive and an increasing number of species develop resistance against the most commonly used herbicide modes of action. A diverse management strategy provides a better overall control of grass weeds and decreases herbicide reliance. The bio-economic decision support system, DK-RIM (Denmark-Ryegrass Integrated Management), was developed to assist integrated management of *L. multiflorum* in Danish cropping systems, based on the Australian RIM model. DK-RIM provides long-term estimations (10-year period) and visual outputs of *L. multiflorum* population development, depending on management strategies. The dynamics *of L. multiflorum* plants within the season and of the soil seedbank across seasons are simulated. The user can combine cultural weed control practices with chemical control options. Cultural practices include crop rotation, seeding density, sowing time, soil tillage system, and cover crops. Those scenarios with increasing crop rotation diversity or different tillage strategies were evaluated. DK-RIM aims at being an actual support system, aiding farmers taking decisions and encouraging discussions among stakeholders on alternative management strategies (Sonderskov et al. 2020).

1.3.5 Example 5

A final but not less important issue to be considered is economic: when planning a crop (or for the best, a crop sequence) agrochemicals (plus spraying or harvest) are considered as a cost (in the list are also other supplies, such as crop seeds, fertilizer, etc). At least in the case of pests, the only ones that have persistent offspring are the weeds (the seedbank). Thus, for example, if a graminicide causes a great impact (e.g. 90% control) in *Sorghum halepense* rhizome populations under present crop, the size of the rhizome population in the following cycle will be clearly less as compared to a nontreated population. In other words, having plant populations regulatory mechanisms for controlling its size, assigning the herbicide costs to just the year of spraying would not be proper.

Provided that enough knowledge on the population dynamics of the weed is available, net present value (NPV) is the tool to estimate "induced benefits" or benefits in futures years. NPV is a measure of calculating returns over the long term. In this instance, future gross margins are summed and discounted back to a present day value. The discounted average annual return, obtained by dividing the NPV by the time period, can also be used. The term *discounting* means converting future gross margins to a present day monetary value so as to account for factors such as inflation and the opportunity cost of capital. This approach is able to account for important economic factors such as changes to the weed seedbank from one year to the next due to weed management actions and herbicide resistance. The benefits of agronomy targeting weed control (e.g. a change in crop sequence) and IWM tactics (e.g. green manuring where there is a loss of income in the year of activity) could then be included (GRDC-IWM 2018).

Finally, if using an EbWM approach, and there would be yield losses, should *less yield* be proportional to *less economic income*? Considering that *less yield* in EbWM is a consequence of saving inputs (e.g. fertilization and herbicide costs), the answer will probably be yes but just for a single season. However, if crop rotation is included, coupled with site-specific inputs (e.g. fertilizers or chemicals), the approach can lead to economic gains in the mid- to long term.

1.4 Concluding Remarks

To produce the food we need yet ensure the landscape we want, we should minimize environmental impacts: one future vision of the agriculture involves the farmer as a steward of the countryside, putting aside the ideological boundaries that are often set up between different agricultural systems. We then should straightforward focus in the "corpus" of principles and theoretical knowledge provided by ecology, particularly focused at the population level, since agronomists, farmers, and crop advisers deal with this organization level in a range of organisms (crops, pests, weeds, cattle, insects, bees). To achieve this purpose, we should be able to set aside definitions and euphemisms such as, among others, agroecology, IPM, IWM and transdisciplinary research. In the end, the craddle that gave sense and founded all of them is *ecology*.

Also, a proper focus should be applied to what is researched and how knowledge is disseminated, as was questioned by Moss (2008). He argued that the overall balance and current direction of much weed research was wrong, with too much emphasis on scientific impact at the expense of practical application, and that could be one of the reasons why, despite considerable research effort, IWM has not been widely adopted by farmers. A lack of appreciation of the difficulty and costs involved in scaling up experimental results to be applicable at a realistic field scale in real farming systems and a lack of awareness of the complexities and resources needed to translate research results into actions to farmers were mentioned as the major reasons explaining farmers' reluctance to adopt IWM strategies. He wisely noted that whatever great impact a publication has, it achieves nothing in terms of improving the ability to manage weeds until the results are put into practice.

One of the most relevant changes to address is the philosophy of weed science itself. We need to accept that simple answers for *weed control* will inevitably fail if used too often and for too long (MacLaren et al. 2020). However, the requirement for short and secure responses are grounded in the risk aversion: interviews with 839 focus groups in 28 US states suggest that farmers largely attribute the introduction and movements of weeds to factors beyond their control, which are therefore unavoidable (e.g environment, plant characteristics). And although they frequently cite IWM as important, and have information related to the biology of weeds and the attributes that make them successful in agroecosystems, their approach is much more directed to control than to prevent them.

The status of land tenure would not be significant in decisions' adoption, and – much more important – is a sort of *technological optimism*, the tendency to have great confidence that science will provide innovative, effective, and inmediate solutions. Interestingly, most farmers claimed that any new herbicide would be overused, similar to what has happened in recent decades, leading to further resistance. In other words, there is a dominant ontology in American agriculture that emphasizes simplicity, ease, independence, and decision-making from year to year. This ideological component reflects the structural difficulties farmers face in adopting a variety of sustainable agricultural practices. A high percentage of farmers also felt that seed and chemical companies should do more to respond to these issues.

A further severe constraint to the adoption of IWM practices is the ownership of arable land: in Canada, over 40% of growers rent or lease land, and usually manage it only for short-term duration, which can negatively affect long-term sustainability. The same applies to growers in the United States. In Argentina, about 60% of soybean fields are sown and managed under less than one-year contracts among owners and tenants, usually big corporations, seeding pools, and/or agricultural service providers. In many European countries such as Denmark and Romania, arable land is rented to growers for only a few years. In Denmark, it has been estimated that up to 25% of the land is handed over to another tenant every year. Consequently, growers make decisions based on short-term profits, and therefore rarely consider long-term benefits.

In summary, EbWM may be a powerful tool for helping the achievement of a great challenge: the reconciliation of agricultural productivity with the environmental integrity (Robertson and Swinton 2005). Agriculture's main challenge for the coming decades will be to produce sufficient food and fiber for a growing global population at an acceptable environmental cost. This challenge requires an ecological approach to agriculture that is largely missing from current management and research portfolios. To create agricultural landscapes that are managed for multiple services, in addition to food and fiber, will require integrative research, both ecological and socioeconomic, as well as policy innovation and public education.

References

Alignier, A., Solé-Senan, X.O., Robleño, I. et al. (2020). Configurational crop heterogeneity increases within-field plant diversity. *Journal of Applied Ecology* 57: 654–663. https://doi.org/10.1111/1365-2664.13585.

Al-Mamun, M.M. (2018). Land tenure, farmers' perception and adoption of weed resistance management practices. MSc thesis. University of Arizona.

Andrew, I.K.S., Storkey, J., and Sparkes, D.L. (2015). A review of the potential for competitive cereal cultivars as a tool in integrated weed management. *Weed Research* 55: 239–248.

Bastiaans, L., Paolini, R., and Baumann, D. (2008). Focus on ecological weed management: what is hindering adoption? *Weed Research* 48: 481–491.

Bastiaans, L., Zhao, D.L., Den Hollander, N.G. et al. (2007). Chapter 21. Exploiting diversity to manage weeds in Agro-ecosystems. In: *Scale and Complexity in Plant Systems Research: Gene-Plant-Crop Relations* (ed. J.H.J. Spiertz, P.C. Struik, and H.H. van Laar), 267–284.

Cirujeda, A., Aibar, J., and Zaragoza, C. (2011). Remarkable changes of weed species in Spanish cereal fields from 1976 to 2007. *Agronomy for Sustainable Development* 31: 675–688.

Colbach, N., Gardarin, A., and Moreau, D. (2018). The response of weed and crop species to shading: which parameters explain weed impacts on crop production? Colbach, N. Agroécologie, AgroSup Dijon, INRA, Univ. Bourgogne, Univ. Bourgogne Franche-Comté, F-21000 6 Dijon.

Dalgaard, T., Hutchings, N.J., and Porter, J.R. (2003). Agroecology, scaling and interdisciplinarity. *Agriculture, Ecosystems and Environment* 100: 39–51. https://www.researchgate.net/publication/222574951.

Davis, A.S., Hill, J.D., Chase, C.A. et al. (2012). Increasing cropping system diversity balances productivity, profitability and environmental health. *PLoS One* 7: 1–8.

Elliot, E.J., Firbank, L.G., Drake, B. et al. (2013). Exploring the concept of sustainable intensification. ADAS/Firbank, LUPG Commissioned Report.

Gaba, S., Gabriel, E., Chadœuf, J. et al. (2016). Herbicides do not ensure for higher wheat yield, but eliminate rare plant species. *Scientific Reports* 6: 30112. https://doi.org/10.1038/srep30112.

Gage, K.L. and Schwarz-Lazaro, L.M. (2019). Shifting the paradigm: an ecological systems approach to weed management. Review. *Agriculture* 9: 179.

Gibson, K., Fischer, A.J., Foin, T.C., and Hill, J.E. (2003). Crop traits related to weed suppression in water-seeded rice (*Oryza sativa* L.). *Weed Science* 51 (1): 87–89.

GRDC-IWM. (2018). Economic benefits of adoption of Integrated Weed Management. Integrated Weed Management Manual. Section I, 16–26.

Hails, R.S. (2002). Assessing the risks associated with new agricultural practices. *Nature* 418: 685–688.

Harper, J.L. (1979). *Population Biology of Plants*. Academic Press, 890 p.

Hole, D.G., Perkins, A.J., Wilson, J.D. et al. (2005). Does organic farming benefit biodiversity? *Biological Conservation* 122: 113–130.

IWMPRAISE. (2020). Summary report. H2020-SFS-2016-2017/H2020-SFS-2016-2 European Union's Horizon 2020 research and innovation programme. Grant agreement No. 727321.

Kleijn, D., Bommarco, R., Fijen, T.P.M. et al. (2019). Ecological intensification: bridging the gap between science and practice. *Trends in Ecology and Evolution* 34: 154–166.

Kudsk, P. (2014). Reduced herbicide rates: present and future. *26th German Conference on Weed Biology and Weed Control* (11–13 March 2014). Braunschweig, Germany: Julius-Kühn-Archiv, 443.

Kudsk, P., Jorgensen, L.N., and Orum, J.E. (2018). Pesticide Load-A new Danish pesticide risk indicators with multiple applications. *Land Use Policy* 70: 384–393.

Lamichhane, J.R., Yann Devos, H.J., Beckie, M.D.K. et al. (2017). Integrated weed management systems with herbicide-tolerant crops in the European Union: lessons learnt from home and abroad. *Critical Reviews in Biotechnology* 37: 459–475. https://doi.org/10.1080/07388551.2016.1180588.

Lammerts Van Bueren, E.T., Struik, P.E., and Jacobsen, E. (2002). Ecological concepts in organic farming and their consequences for an organic crop ideotype. *Netherlands Journal of Agricultural Science* 50.

Lazzaro, M., Bàrberi, P., Dell'Acqua, M. et al. (2018). Unraveling diversity in wheat competitive ability traits can improve integrated weed management. *Agronomy for Sustainable Development* 39: 6.

Leighty, C.E. (1938). Crop rotations. In: *Soils and Men*, 406–430. United States Department of Agricutlture. A Yearbook of Agriculture.

Lewis, W.J., van Lenteren, J.C., Phatak, S.C., and Tumlinson, J.H., III. (1997). Perspective. A total system approach to sustainable pest management. *Proceedings of the National Academy of Sciences of the United States of America* 94: 12243–12248.

Liebman, M. and Gallandt, E.R. (1997). Chapter 9: many little hammers: ecological management of crop-weed interactions. *Physiological Ecology* series. In: *Ecology in Agriculture* (ed. L.E. Jackson), 472 p. Academic Press.

Liebman, M., Nguyen, H.T.X., Woods, M.M. et al. (2001). Weed seedbank diversity and sustainability indicators for simple and more diverse cropping systems. *Weed Research* 61: 164–177.

Litrico, I. and Violle, C. (2015). Diversity in plant breeding: a new conceptual framework. *Trends in Plant Science* 20: 604–612. https://doi.org/10.1016/j.tplants.2015.07.007.

MacLaren, C., Storkey, J., Menegat, A. et al. (2020). An ecological future for weed science to sustain crop production and the environment. A review. *Agronomy for Sustainable Development* 40: 24. https://doi.org/10.1007/s13593-020-00631-6.

Moss, S. (2018). Integrated Weed Management (IWM): why are farmers reluctant to adopt non-chemical alternatives to herbicides? *Pest Management Science* 75 (5). https://doi.org/10.1002/ps.5267.

Moss, S.R. (2008). Weed research: is it delivering what it should? *Weed Research* 48: 389–393.

Naylor, R.L. (2008). Managing food production systems for resilience. In: Chapter 12 in *Principles of Natural Resource Stewardship: Resilience-Based Management in a Changing World* (ed. F.S. Chapin, G.P. Kofinas, and C. Folke). New York: Springer.

Oerke, E. (2006). Crop losses to pests. *The Journal of Agricultural Science* 144 (1): 31–43.

Pester, T.A., Burnside, O.C., and Orf, J.H. (1999). Increasing crop competitiveness to weeds through crop breeding. *Journal of Crop Production* 2: 31–58.

Petit, S., Cordeau, S., Chauvel, B. et al. (2018). Biodiversity-based options for arable weed management. A review. *Agronomy for Sustainable Development* 38: 48.

Petit, S., Munier-Jolain, N., Bretagnolle, V. et al. (2015). Ecological intensification through pesticide reduction: weed control, weed biodiversity and sustainability in arable farming. *Environmental Management* 56: 1078–1090.

Pysek, P., Jarosik, V., Kropacc, Z. et al. (2005). Effects of abiotic factors on species richness and cover in Central European weed communities. *Agriculture, Ecosystems and Environment* 109: 1–8.

Rabbinge, R. (1997). Integrating policy and technical issues for international research on agriculture and the environment using systems approaches. In: *Applications of Systems Approaches at the Farm and Regional Levels*, vol. 1 (ed. P.S. Teng and M.J. Kropff), 249–262. Dordrecht: Kluwer Academic Publishers.

Robertson, G.P. and Swinton, S.M. (2005). A grand challenge for agriculture. *Frontiers in Ecological Environment* 3: 38–46.

Roschewitz, I. (2005). Farming systems and landscape context: effects on biodiversity and biocontrol. Dissertation zur Erlangung des Doktorgrades an der Fakultät für Agrarwissenschaften der Georg-August-Universität Göttingen, 152 p.

Roy, D.B., Bohan, D.A., Haughton, A.J. et al. (2003). Invertebrates and vegetation of field margins adjacent to crops subject to contrasting herbicide regimes in the Farm Scale Evaluations of genetically modified herbicide-tolerant crops. *Phylosophical Transactions of the Royal Society B: Biological Sciences* 358: 1879–1898.

Satorre, E.H. (2015). El rol de las tecnologías sobre la producción de soja, su variabilidad y la sostenibilidad de los sistemas de los sistemas de producción: Análisis y reflexiones. Mesa Tecnológica de Oleaginosos (MTO), Montevideo, Uruguay (Conferencia ppt).

Solé-Senan, X.O., Juárez-Escario, A., Conesa, J.A. et al. (2014). Plant diversity in Mediterranean cereal fields: unravelling the effect of landscape complexity on rare arable plants. *Agriculture, Ecosystem and Environment* 185: 221–230.

Sonderskov, M., Somerville, G.J., Lacoste, M. et al. (2020). DK-RIM: assisting integrated management of *Lolium multiflorum* Italian Ryegrass. *Agronomy* 10: 856. https://doi.org/10.3390/agronomy10060856.

Stern, V.M., Smith, R.F., van den Bosch, R., and Hagen, K.S. (1959). The integrated control concept. *Hilgardia* 29: 81–101.

Swanton, C.J. and Weise, S.F. (1991). Integrated weed management: the rationale and approach. *Weed Technology* 5: 657–663. https://doi.org/10.2307/3987055.

Thill, D.C., Lish, J.M., Callihan, R.H., and Bechinsk, E.J. (1991). Integrated weed management: a component of integrated pest management. A critical review. *Weed Technology* 3: 648–656.

Verdelli, D., Acciaresi, H.A., and Leguizamón, E.S. (2012). Corn and soybeans in a strip intercropping system: crop growth rates, radiation interception, and grain yield components. *International Journal of Agronomy* 2012: 980284.

Westwood, J.H., Charudattan, R., Duke, S.O. et al. (2018). Weed management in 2050: perspectives on the future of weed science. *Weed Science* 66: 275–285.

Wilson, R., Hooker, N., Tucker, M. et al. (2009b). Targeting the farmer decision making process: a pathway to increased adoption of integrated weed management. *Crop Protection* 28: 756–764. https://doi.org/10.1016/j.cropro.2009.05.013.

Wilson, R.S., Tucker, M.A., Hooker, N.H. et al. (2009a). Perceptions and beliefs about weed management: perspectives of Ohio grain and produce farms. *Weed Technology* 22: 339–350.

Worthington, M. and Reberg-Horton, C. (2013). Breeding cereal crops for enhanced weed suppression: optimizing allelopathy and competitive ability. *Journal of Chemical Ecology* 39: 213–231. https://doi.org/10.1007/s10886-013-0247-6.

2

Ecologically Based Weed Management

Implications and Agroecosystem Services

Nicholas E. Korres

Department of Agriculture, School of Agriculture, University of Ioannina, Kostakii, Arta, Greece

2.0 Introduction

Agricultural intensification along with consumer's preferences for eco-friendly food and fiber production systems necessitate changes in the traditional model of productivism agricultural era (Wilson 2007) in favor of a post-productivism agricultural production systems that focuses on environmental management and "production of nature" (Marsden 1999). In Europe, the new Common Agricultural Policy (CAP) 2023–2027 promotes the compliance with certain environmental standards such as the reduction of pesticides use, the protection of biodiversity, and the adoption of new technologies that aim to a sustainable agricultural production. Holmes (2006) reported that the transition of agricultural sector from productivism to post-productivism model can be achieved through its multifunctional role. Marsden and Sonnino (2008) considered multifunctional agriculture as a part of a sustainable development paradigm within an agro – friendly production model. In relation to weed management, Vidaller and Dutoit (2022), Deguine et al. (2023), Bastiaans et al. (2008), and Chauhan et al. (2012) stated that the most successful weed management programs will be based on the foundation of ecological principles. Weed management practices that focused exclusively on weed control, given the interrelationship between all life forms within an agro-ecosystem, are inaccurately formulated and they lack a long-term strategic vision. Consequently, understanding the weed-crop relationships will inevitably increase the efficacy of weed control (Chauhan et al. 2012). Ecologically based weed management emphasizes the use of ecological principles and practices to minimize weed infestations and crop loses or damages while maintaining and/or enhancing ecosystem health.

There are several key components to ecologically based weed management. First, it involves understanding the ecology of the weeds and the ecosystem in which they are growing. This includes factors such as soil type, moisture levels, plant community structure, and disturbance history. By understanding these factors, managers can identify the conditions that favor weed growth and develop strategies to disturb those conditions. Second, ecologically based weed management seeks to prevent weed infestations from occurring in the first place. This can involve a range of practices, including promoting healthy plant communities, minimizing soil disturbance, using cover crops and other plant-based strategies to suppress weed growth, and implementing early detection and rapid response programs to quickly address new weed outbreaks. Third, ecologically based weed management emphasizes the use of nonchemical and low-impact control methods whenever possible. This can include manual weed removal, cultural practices such as crop rotation and intercropping, and biological control methods such as the introduction of natural enemies of weeds. Finally, it recognizes the significance of herbicides or other chemical control methods on weed control, although these methods should be used only as an integral part of weed management program and should be applied in a judicious manner to minimize their impact on nontarget species and the surrounding environment. Overall, ecologically based weed management represents a holistic and integrated approach to weed management that balances the needs of crop production with the health and resilience of natural ecosystems. In summary, ecological weed management implicates the practice of various information types and weed control approaches, thus the exposure of weeds to multiple stresses. It is important to mention here that the use of herbicides is not excluded

Ecologically Based Weed Management: Concepts, Challenges, and Limitations, First Edition. Edited by Nicholas E. Korres, Ilias S. Travlos, and Thomas K. Gitsopoulos.

from ecologically based weed management systems but rather is treated as an option and not as an absolute prerequisite. Therefore, the aim of this chapter is to discuss how weed management practices affect ecosystem services and which of them could potentially enhance them.

2.1 Agro- and Natural Ecosystem Services

An ecosystem is characterized by the interaction of plants and animals with their physico-chemical environment; thus, it can be considered as a functional unit where biogeochemical processes occur simultaneously. Consequently, is a dynamic open system that constantly changes through the "activities" of its components (Schulze et al. 2019). Therefore, ecosystem services, due to the dynamic nature of ecosystems *per se*, are also changing. Ecosystem services, a term that emerged from the interface between economics and ecology, defined by Deguine et al. (2020) as a direct or indirect provision of goods and services that humans can acquire from ecosystems for their welfare through the reorganization of cropping practices and the improvement of agroecosystem sustainability by connecting its ecological functions. Hence, the optimization of interactions between plant, animal, and microbial communities within and around agroecosystems is essential.

In agroecosystems, ecological services co-provided by both nature and human activities. Food production, a provisioning service for example, is a result of activities manipulated by humans but also is promoted by ecological processes such as pollination. Agroecology mobilizes ecosystem services for both agricultural production and for societal needs (e.g. landscapes, water quality). As abovementioned ecosystems provide many services that facilitate human needs such as food, fiber, landscape, income, and leisure through certain eco-system functions. Agricultural soil ecosystems, for example, serve as structural mediator (a medium for plant growth), filter and reservoir facilitator (clean and store water), fertility regulator (nutrient cycling), biodiversity conservator (reservoir for biological diversity), and climate regulator (carbon sequestration and greenhouse gas regulation) (Dominati et al. 2010). Likewise, freshwater ecosystems, also important to humans' well-being, provide drinking water, food resources, nutrient flow, improve water quality, modulate climate, serve as aesthetic landscapes and cultural activities (Wetzel 2001; Korner 2002; Van Donk and Van de Bund 2002) but they also support aquatic and nonaquatic organisms (Soininen et al. 2015). Several studies have reviewed services provided by various production systems such as organic (Sandhu et al. 2010) or conservation (Palm et al. 2014) agriculture. However, few studies have focused on ecosystem services provided by conventional agricultural systems (Vidaller and Dutoit 2022) and even less with the weed management approaches within these systems. A generic framework for classifying and quantifying ecosystem services (i.e. supporting, cultural, regulating, and provisioning) has been proposed by Daily (1997), Dominati et al. (2010), Vidaller and Dutoit (2022) based on which ecosystem service can be linked with inherent properties as well as properties that are influenced by human actions such as management and farming practices (Figure 2.1).

In summary, agricultural ecosystems and ecosystem services provide (i) supporting, e.g., as a medium for plant growth, fertility and nutrient cycling, biodiversity conservation and as reservoir for biological diversity etc., (ii) regulation of climate, carbon sequestration and greenhouse

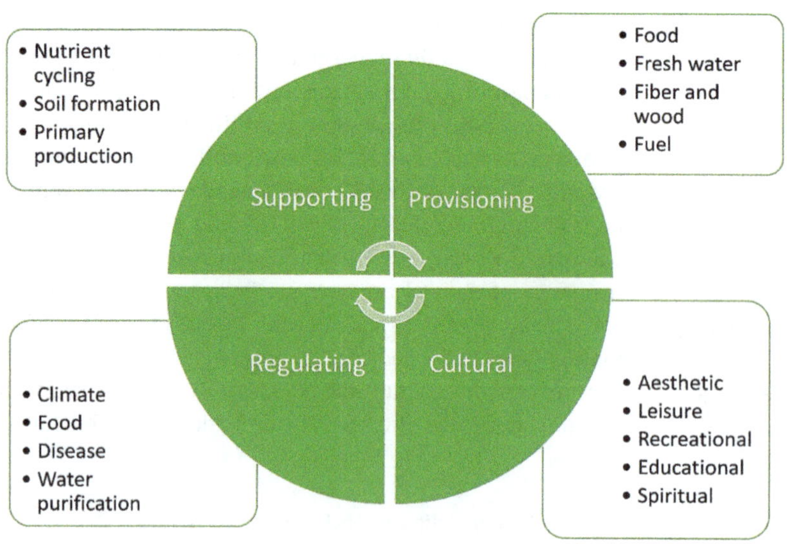

Figure 2.1 Ecosystem services as a basis for human welfare (based on Schulze et al. 2019).

gas, water purification etc., (iii) provisioning of food, fiber, wood, fuel, (iv) cultural such as aesthetic, leisure, educational (Figure 2.1). As mentioned previously, ecosystem services emerged from the interface between economics and ecology. According to Costanza et al. (1997) an estimated minimum value of these services accounts for $33 trillion per year whereas the annual global national product at that time was $18 trillion.

2.2 Do Weed Management Practices Negatively Affect Ecosystem Services?

Weed management practices have become closely linked to social and economic rather than biological factors particularly in conventional agriculture where economic pressures have led to simplification of cropping systems and the replacement of cultural methods of weed management with synthetic chemical alternatives. More specifically, in early stages of agriculture, when agroecosystems were evolving according to their biological constituents, new plant communities appeared in response to human disturbances and new types of management were applied in response to those new plant communities (Korres 2005). Additionally, the domestication of some species within these new communities, if they proved useful to human needs for food, was encouraged. In contrast, in modern agriculture, agroecosystems have diversified in scale of influence since management practices (including weed management systems) are increasingly unrelated to biological feedback from the ecosystem and are strongly linked to social and economic criteria. While biological systems are regulated mainly by biophysical processes, human activities related to modern agriculture are influenced strongly by socioeconomic activities, which are developed and maintained by human input of fossil energy and regulated by cultural, scientific, and technological information (Korres 2005). As a result, the evolution of agroecosystems and weed management strategies, as a part of the agricultural activities, is not progressing in parallel (Korres 2005). Continuous agricultural practices and recurrent selective pressures on weeds have enhanced their weeding characteristics such as discontinuous germination, plasticity, and ability to compete. Additionally, introgressive hybridization among species and hybrid speciation, which interrupts normal allopatric mechanisms of speciation have been developed. The enhancement of species weediness created the need for more effective and powerful methods of weed management such as improved mechanical implements (e.g. various types of harrows, the mouldbroad plough), development of herbicides and herbicide resistant crops, which led to modern agroecosystems that represent a closed system in which inputs are greater than outputs (Korres 2005). Co-evolution between the agroecosystem and management processes almost ceased. Agriculture is not limited by the characteristics of the biome anymore since new technology has made possible an enormous expansion of agriculture resulting in the creation of a vast, almost homogeneous, habitat suitable for a relatively small number of crop species. Under these circumstances, the deterioration of such a system is inevitable unless if it is subsidized by means of external inputs (e.g. fertilizers, manure, irrigation, and pesticides) with all the consequent implications (e.g. soil deterioration and erosion, agricultural pollution, reduction in biodiversity, alterations in weed-crop associations, development of herbicide resistance) (Korres 2005). Weed management practices, as an integral part of agricultural production systems, when mishandled, contribute, although to a lesser extent compared with other crop husbandry practices such as for example harvesting or fertilizing, negatively on the environment and could negatively affect the ecosystem services (Korres et al. 2010). There are various tactics to control weeds (Figure 2.2), some of which have a potential effect on agroecosystem services via soil and water resources, biodiversity, and other important ecosystem constituents (Arts and Hanson 2019; Capinera 2019; Tyler and Locke 2019).

Agricultural practices including weed management techniques, such as conventional tillage can negatively influence soil ecosystem services through, for example, increased soil erosion, hence reduced soil fertility, and soil carbon storage (Benayas et al. 2009; Landis 2017; Tyler and Locke 2019).

Direct application of herbicides can also affect soil services negatively, although without herbicide incorporation into the soil, impacts may be small and limited to surface residues (Tyler and Locke 2019). In addition, herbicides can have a major impact on soil ecosystems through their effects on soil microorganisms (i.e. microbial biomass, microbially mediated processes, microbial community composition), a major contributor to soil functions and nutrient cycling (Tyler and Locke 2019). However, the impact of herbicide application on soil health varies greatly depending on the chemical used, and a variety of other factors, including soil texture, pH, organic matter, and existing dominant plant species. Similarly, freshwater ecosystems are important to society, but this has not prevented freshwater systems from being threatened by a wide range of stressors resulting in eutrophication, invasiveness, habitat loss and overuse of the resource. In agricultural settings, plant protection products can enter freshwater ecosystems not only by spray drift (i.e. the physical movement of pesticide out of the target area to any nontarget site at the time of the spraying operation or soon thereafter), drainage, and

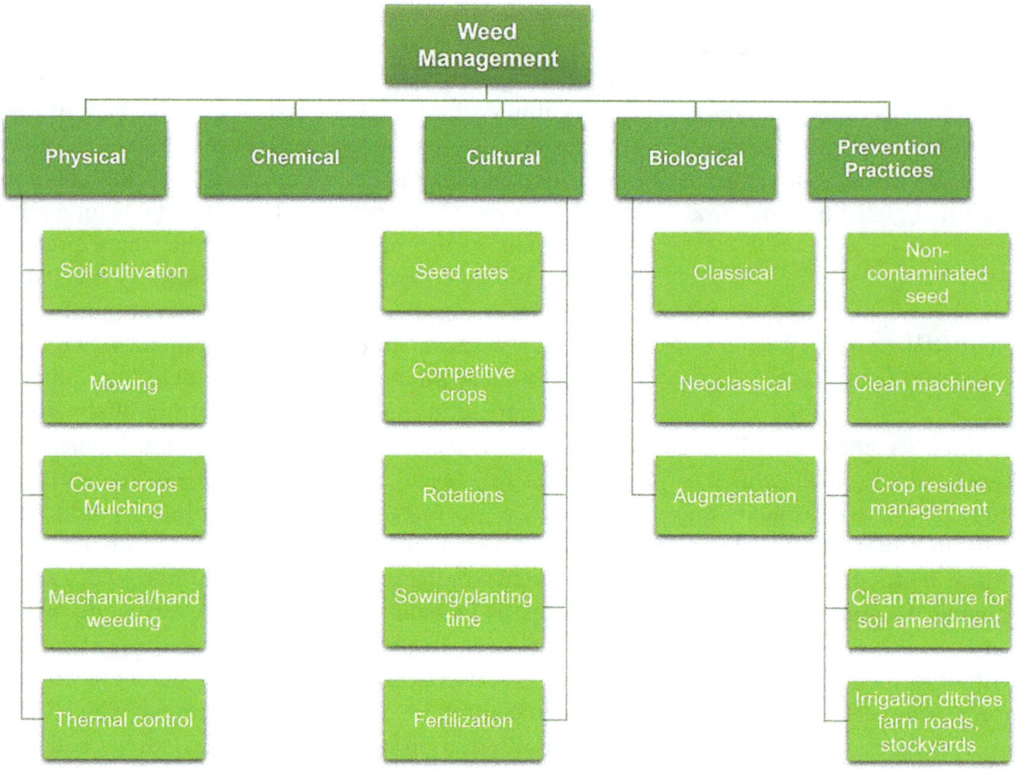

Figure 2.2 Weed management approaches (adapted from Korres 2018).

runoff from intensively cropped areas (Figure 2.3), but also through direct application in special cases, such as spraying for the control of invasive species (Schafer et al. 2012).

Herbicide drift causes several hazards:

- Adjacent susceptible crops can be damaged (Figure 2.4). For example, drift of an herbicide for broad-leaved weeds will also damage vegetable, legume, and other cash crops. A few years back, dicamba off-target movement in the United States caused damages to adjacent susceptible to dicamba crops of many million US dollars (Korres et al. 2018). Additionally, drifting can contaminate crops ready for harvest, leaving unacceptably high residues on the harvested product.
- Water contamination, when herbicide drift onto ponds, streams, and other water bodies occurs may cause alterations on physico-chemical characteristic of the water. Furthermore, the contamination of water bodies could be a source of future crop damage if the water is used for irrigation purposes but also could cause damages to aquatic organisms.
- Improper pesticide application may endanger humans, pollinators and other nontargeted animals.
- The amount of herbicide applied to the target area, is reduced below the recommended quantity due to herbicide drift resulting in reduced herbicide efficacy (Korres 2005).

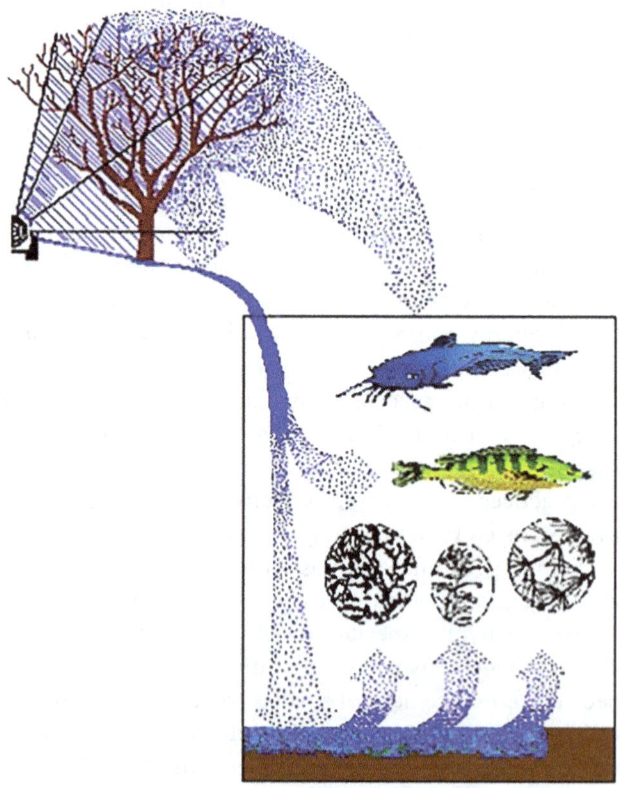

Figure 2.3 Potential drift and subsequent contamination of water bodies when foliage application of herbicide is used without justification (adapted from Korres 2005).

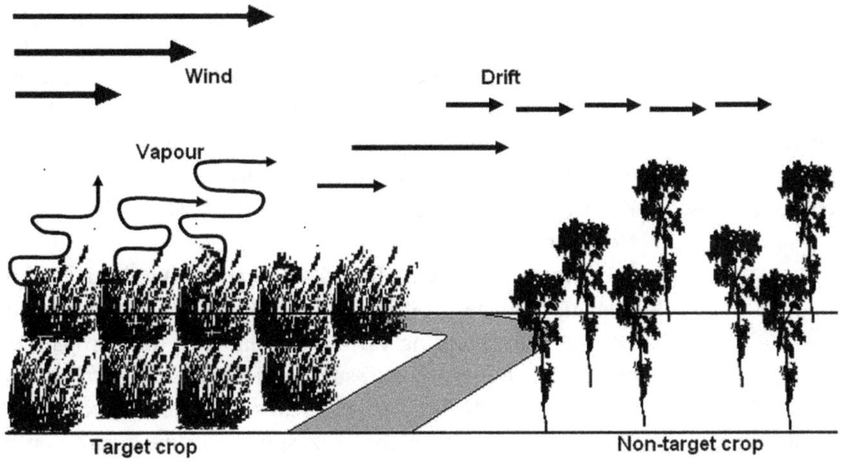

Figure 2.4 Herbicide vapors via herbicide drift can cause damages to adjacent crops and ecosystems (adapted from Korres 2005).

It is known that weed management using mostly herbicides has increased over the last several decades due to their efficiency and lower cost compared to other weed control methods as well as the recent introduction of genetically modified herbicide-resistant crops (Green 2014). Herbicide sales are worth of US$25 billion worldwide (Swanton et al. 2015; Willis 2017). Increase use of few single herbicide mechanisms of action resulted in the development, and the rapid spread, of herbicide-resistant weeds (Figure 2.5). This was intensified using monocultures and the negligence of other weed control (Mortensen et al. 2012).

Nowadays, despite all the benefits gained through herbicide usage, we are facing a higher threat from weeds compared to a few decades ago, since modern agriculture has selected for few competitive weed species, which are rapidly becoming resistant to new herbicide sites of action herbicides (Storkey et al. 2012; Tehranchian et al. 2017; Korres 2018; Fogliatto et al. 2020; Vidotto et al. 2021). In addition, the overreliance on herbicides and pesticides caused a decline of many farmland species, including those that contributing to the control of the parasitic populations feeding on weed seeds and crop pests (Virili 2023). The beneficial species in most cropping systems depend on natural areas within the landscape for food and shelter (Marshall et al. 2003; Navntoft et al. 2009). Absence of these habitats or "biodiversity hot-spots" that host these species results in a continuous declining population of the beneficial species in arable and horticulture cropping systems (Virili 2023). However, through integrated crop management, of which integrated weed management is

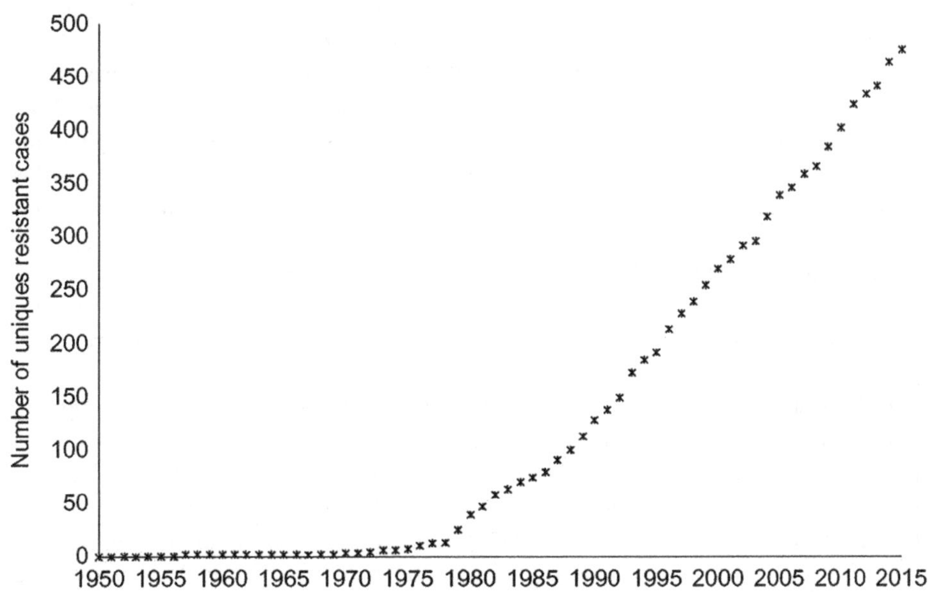

Figure 2.5 Global increase in unique cases of herbicide-resistant weeds. Note: a unique case is a species a single site of action case (Heap 2023).

a vital part, efforts to reinstate the equilibrium between the natural ecosystem and agroecosystem have initiated. The integration of techniques such as cultivar selection, manipulation of crop density, reduced herbicide applications, cultivation at night, flaming, biological control, crop rotations, organic farming have stimulated a new approach for the development of economically and environmentally friendly agroecosystems (Korres 2005).

2.3 Weed Management Practices that Enhance Ecological Services

The necessity for a modified ecology-based weed management approach can offer solutions to lessen problems of herbicide resistance, environmental pollution, loss of biodiversity, weed invasion, and consequently yield losses (Korres and Froud-Williams 2002; Bajwa 2014; Korres et al. 2019a, 2019b; Kumari et al. 2023a, 2023b) but also enhance ecosystem services. Despite the potential hazardous effects some of the weed management practices may impose to ecosystem services there are many others that enhance them such as improved tillage, crop nutrient management, cover crops, crop rotations, precision weed management.

Improved tillage, for example, by optimizing the role of tillage in weed control serves as mean to soil erosion prevention. For this reason, no-tillage or reduced-tillage systems, known as conservation systems, came in action (Phillips et al. 1980) to maintain soil quality and moisture availability (Kaspar et al. 2001). The use of herbicide-resistant crops favored cropping systems that were based on conservation tillage which in combination with various herbicide sites of action used effectively to suppress weeds (Vencill et al. 2012). However, weeds with small seeds or these that had developed herbicide-resistance became a major challenge in adoption of conservation tillage systems (Price et al. 2011; Bajwa 2014). Consequently, integrated weed management strategies need to be implemented to suppress the evolution of herbicide resistance and disturb the establishment of weed populations without reducing potential crop yield and agro-ecosystems services. In this case, the use of high-residue cover crops such as cereal rye (*Secale cereale*) proved effective in suppressing weeds such as *Amaranthus palmeri* (Palmer amaranth), *Senna obtusifolia* (sicklepod), *Ipomoea* spp. (morningglory), *Eleusine indica* (goosegrass), and *Cyperus* spp. (nutsedge) in soybean (*Glysine max*) (Kumari et al. 2023b). Similarly, cereal rye suppressed weeds such as *A. palmeri*, *Portulaca oleraceae* (common purslane), *Chamaesyce maculata* (spotted spurge), *Ipomoea lacunose* (pitted morningglory), *Digitaria sanguinalis* (large crabgrass), *Mollugo verticillate* (carpetweed), and *Eleusine indica* (goosegrass) in cotton (Korres and Norsworthy 2015). In addition, the inclusion of a cover crop in a production system has been shown to improve soil chemical, biological, and physical properties in various cropping systems, thus enhance agro-ecosystem services. The ecological services offered by cover crops regarding soil moisture conservation, soil erosion control, improvements of soil chemical, biological, and physical properties and soil fertility are well documented (Munawar et al. 1990; Alberts and Neibling 1994; Dabney et al. 2001; Korres 2005; Moyer and Blackshaw 2009; Price and Norsworthy 2013).

However, Korres et al. (2020) reported a negative relationship between soil moisture and soil NO_3-N content at 28 days after planting various vegetable legume crops (i.e. edamame (*Glycine max*), snap and lima beans (*Phaseolus vulgaris*) in early terminated cereal rye used as a cover crop). Nitrogen losses from the rhizosphere of the legume crop due to higher moisture content, especially during the early growth stages of the crop after the termination of the cover crop, can negatively affect the growth of poor N fixers such as common bean (Hardarson and Zapata 1984; Lindemann and Glover 2003). Therefore, the use of a cover crop must be chosen carefully. Crops such as *Vicia villosa* (winter vetch) or rye can provide uniform and dense ground cover when properly managed, while crops like *Coronilla varia* (crown vetch) can provide long-term soil management. Other crops that could be used as cover crops are clovers such as *Trifolium incarnatum* (crimson clover), *Trifolium pratense* (red clover), and *Trifolium repens* (white clover), *Pisum* spp. (peas), *Lotus corniculatus* (bird's-foot trefoil), *Avena sativa* (common oat), *Lolium* spp. (ryegrasses), *Festuca* spp. (fescue), *Poa* spp. (bluegrasses), *Bromus inermis* (smooth brome), *Phleum pratense* (timothy grass), and *Dactylis glomerata* (orchard grass) (Korres 2005).

As mentioned previously, the inclusion of cover crops in crop rotations except for weed suppression improves soil fertility. Nitrogen fertilization in cotton and maize, for example, varies between 120–170 and 75–250 kg of N fertilizer ha^{-1}, respectively (Ali 2015; Mueller and Vyn 2016). The use of clovers could add extra N into the soil (Shaner and Beckie 2014). It has been reported that N fixation by clover species fluctuates between 10 and 250 kg N ha^{-1} $year^{-1}$ if clover establishment is between 60% and 80% (Woodmansee 1978; Moller-Hansen et al. 2002; Brockman and Wilkins 2003). Clover in crop rotation could reduce chemical N fertilizer application by approximately 75 kg N ha^{-1} (Korres et al. 2010, 2023) or 400 kg CO_2 eq. ha^{-1} in terms of greenhouse gas emissions (Korres et al. 2010, 2023).

Due to high economic benefits, continuous cropping is the dominant production system even though causes higher land-use intensity and aggravates the problem of climate change and influences ecological environment (Yu et al. 2022). Crop rotation is an important tool for improving the climate resilience of agriculture and solves the inadequacies of the current continuous crop practices. It also ensures food security, ecological environment development, and rural revitalization (Yu et al. 2022). Also, crop rotation facilitates the control of problematic weeds by exploiting the morphophysiological differences and production practices of the crop grown and by allowing rotation of the herbicides used in the cropping sequence, hence reducing the selection pressure for herbicide resistance through stress modifications and mortality factors that affect weed population dynamics (Harper 1956; Schwartz-Lazaro et al. 2022). Schwartz-Lazaro et al. (2022) showed that a three-year crop rotation including corn, soybean, and cotton in combination each crop's herbicides was an effective tactic in reducing the emergence and soil seedbank of the naturally occurring weed flora dominated by *A. Palmeri*, *Sida spinosa* (prickly sida), *Ipomoea* spp. (morningglories), *Digitaria sanguinalis* (large crabgrass), *Echinochloa crus-galli* (barnyardgrass), and *Sorghum halepense* (johnsongrass). They concluded that reductions of the weed seedbank cannot be achieved based on one management practice but requires the use of a multitactic approach with various control methods. In addition, crop diversification positively affects biodiversity, ecological function, and crop resilience to climate change (Marques et al. 2020; Wang et al. 2020; Yang et al. 2020; Zhang et al. 2020; Beillouin et al. 2021; Garland et al. 2021).

The use of organic amendments in agriculture such as compost is a common practice and they have shown a considerable potential to improve soil quality and increase soil health (Souza Krupek et al. 2022) and to suppress weeds (Korres et al. 2023). Soil physicochemical properties such as water content, organic matter, exchangeable cations, and cation exchange capacity can also be improved by adding organic amendments (Mohd Din et al. 2017; Xu and Geelen 2018; Waqas et al. 2019). Composting organic material prior to land application reduces GHG emissions in relation to the application of uncomposed organic material as they increase plant productivity while the use of chemical fertilizer can be reduced by 25% (Agegnehu et al. 2016; Mondal et al. 2017). Optimal application rate and cheapest incorporation of low-cost organic amendments can achieve a sustainable solution for enhancing crop productivity, soil organic carbon sequestration, and ensuring farmer income in infertile farming regions (Sciubba et al. 2014; Xu et al. 2020). This practice can help to control surface water and groundwater diffusion or point contamination. However, the application of organic amendments to agricultural soil involves a variety of risks for environmental and human health, for example, organic amendments often contain a range of pollutants, including heavy metals, persistent organic pollutants, potential human pathogens, and emerging pollutants (Goss et al. 2013). Further, a strong adsorption of herbicides by organic amendments may lead to their inactivation (Graber et al. 2012), compromising weed control, or may potentially increase herbicide damage in rotational crops due to its accumulation in the soil. It is therefore important to understand the potential impact of amendments on herbicides to achieve a balance between herbicide efficacy and soil amendments application (Kookana 2010), before advocating their widespread use in agricultural field soils.

Crop residues that could be integrated into weed management tactics and possibly ease herbicide usage include maize gluten meal, a by-product of the maize milling process with relatively high nitrogen content, Brassicaceae seed meal, by-products of the oil seed extraction process from Brassicaceae plants, and abrasive grit, and abrasive grit (i.e. grits derived from agricultural residues e.g. maize cob, walnet, almond shell, grape seed, olive seede, poultry manure, sand, soybean meal) (Korres et al. 2019a). When maize gluten was used as a feed substitute for cattle, the corresponding GHG emissions were reduced by 43.9–62.4% for various ration to maize gluten compared to the energy that could have been consumed to produce an equivalent amount of feed to that of maize gluten (Korres et al. 2023). Brassicaceae seed or oilseed meals have exhibited herbicidal activities against several weed species (Korres et al. 2019a). Additionally, grits can contribute to crop fertilization by adding 35–105 kg N ha^{-1} whereas direct N_2O-N emissions savings from this amount of substituted N equals to 0.6–2 kg N_2O-N ha^{-1}, an equivalent GHG emission reduction of approximately 150–600 kg CO_2 eq. ha^{-1} (Korres et al. 2023).

2.4 Conclusions

Ecologically based weed management offers provisional, regulatory, cultural, and supportive services to agroecosystems. Cover crops, crop rotations, and conservation tillage, soil amendments in combination with the justified use of herbicides as constituents of a weed management program are the most important tactics that could possibly offer these services. In addition, crops that have exposed and/or evolved under various environmental conditions can form the basis of the post-productivism agriculture era considering climate change and food production. These crops will require less inputs (e.g. herbicides, fertilizer) and will allow for more biodiversity in production systems, thus more ecosystem services.

References

Agegnehu, G., Bass, A.M., Nelson, P.N., and Bird, M.I. (2016). Benefits of biochar, compost and biochar-compost for soil quality, maize yield and greenhouse gas emissions in a tropical agricultural soil. *Science Total Environment* 543: 295–306.

Alberts, E.E. and Neibling, W.H. (1994). Influence of crop residues on water erosion. In: *Managing Agricultural Residues* (ed. P.W. Unger), 19–39. Ann Arbor, MI: Lewis Publishing.

Ali, N. (2015). Nitrogen utilization features in cotton crop. *American Journal of Plant Sciences* 6: 987–1002.

Arts, G. and Hanson, M. (2019). Effects of herbicides on fresh water ecosystems. In: *Weed Control. Sustainability, Hazards, and Risks in Cropping Systems Worldwide* (ed. N.E. Korres, N.R. Burgos, and S.O. Duke), 62–75. Boca Ratoon, FL: CRC Press.

Bajwa, A.A. (2014). Sustainable weed management in conservation agriculture. *Crop Protection* 65: 105–113.

Bastiaans, L., Paolini, R., and Baumann, D. (2008). Focus on ecological weed management: what is hindering adoption? *Weed Research* 48: 481–491.

Beillouin, D., Ben-Ari, T., Malezieux, E. et al. (2021). Positive but variable effects of crop diversification on biodiversity and ecosystem services. *Global Change Biology* 27: 4697–4710.

Benayas, J.M.R., Newton, A.C., Diaz, A., and Bullock, J.M. (2009). Enhancement of biodiversity and ecosystem services by ecological restoration: a meta-analysis. *Science* 325: 1121–1124.

Brockman, J.S. and Wilkins, R.J. (2003). Grassland. In (ed. R.J. Halley and R.J. Soffe): *Primrose McConnell's the Agricultural Notebook*, 131–176. London, UK: Blackwell Science Ltd.

Capinera, J.L. (2019). Direct and indirect effects of herbicides on insects. In: *Weed Control. Sustainability, Hazards, and Risks in Cropping Systems Worldwide* (ed. N.E. Korres, N.R. Burgos, and S.O. Duke), 76–91. Boca Ratoon, FL: CRC Press.

Chauhan, B.S., Singh, R.G., and Mahajan, G. (2012). Ecology and management of weeds under conservation agriculture: a review. *Crop Protection* 38: 57–65.

Costanza, R., D'Arge, R., De Groot, R. et al. (1997). The value of the world's ecosystem services and natural capital. *Nature* 387: 253–260.

Dabney, S.M., Delgado, J.A., and Reeves, D.W. (2001). Using winter cover crops to improve soil and water quality. *Communications in Soil Science Plant Annals* 32: 1221–1250.

Daily, G. (1997). *Nature's Services*. Washington, DC: Island Press.

Deguine, J.P., Aubertot, J.N., Bellon, S. et al. (2023). Agroecological crop protection for sustainable agriculture. *Advances in Agronomy* 178: 1–59.

Deguine, J.-P., Ratnadass, A., Robin, M.-H. et al. (2020). Agroecological crop protection. Definition. *Dictionary of Agroecology*. https://dicoagroecologie.fr/en/encyclopedia/agroecological-crop-protection.

Dominati, E., Patterson, M., and Mackay, A. (2010). A framework for classifying and quantifying the natural capital and ecosystem services of soils. *Ecological Economics* 69: 1858–1868.

Fogliatto, S., Ferrero, A., and Vidotto, F. (2020). Current and future scenarios of glyphosate use in Europe: are there alternatives? *Advances in Agronomy* 163: 219–278.

Garland, G., Edlinger, A., Banerjee, S. et al. (2021). Crop cover is more important than rotational diversity for soil multifunctionality and cereal yields in European cropping systems. *Nature Food* 2: 28–37.

Goss, M.J., Tubeileh, A., and Goorahoo, D. (2013). A review of the use of organic amendments and the risk to human health. *Advances in Agronomy* 120: 275–379.

Graber, E.R., Tsechansky, L., Gerstl, Z., and Lew, B. (2012). High surface area biochar negatively impacts herbicide efficacy. *Plant and Soil* 353: 95–106.

Green, J.M. (2014). Current state of herbicides in herbicide-resistant crops. *Pesticide Management Science* 70: 1351–1357.

Hardarson, G. and Zapata, F. (1984). Effect of plant genotype and nitrogen fertilizer on symbiotic nitrogen fixation by soybean cultivars. *Plant and Soil* 82: 397–404.

Harper, J.L. (1956). The evaluation of weeds in relation to resistance to herbicides. *Proceedings of the 3rd Brighton Weed Control Conference*, Brighton, UK, 179–188.

Heap, I. (2023). *International survey of herbicide resistant weeds*. http://www.weedscience.org.

Holmes, J. (2006). Impulses towards a multifunctional transition in rural Australia: gaps in the research agenda. *Journal of Rural Studies* 22: 142–160.

Kaspar, T.C., Radke, J.K., and Laflen, J.M. (2001). Small grain cover crops and wheel traffic effects infiltration, runoff, and erosion. *Journal of Soil Water Conservation* 56: 160–164.

Kookana, R.S. (2010). The role of biochar in modifying the environmental fate, bioavailability, and efficacy of pesticides in soils: a review. *Australian Journal of Soil Research* 48: 627–637.

Korner, S. (2002). Submerged macrophytes-an important tool for lake restoration in Germany? *Wasser und Boden* 54: 38–41.

Korres, N.E. (2005). *Encyclopaedic Dictionary of Weed Science. Theory and Digest*. Andover, UK and Lavoisier SA, Paris, France: Intercept Ltd., 675.

Korres, N.E. (2018). Agronomic weed control: a trustworthy approach for sustainable weed management. In: *Non-chemical Weed Control* (ed. K. Jabran and B.S. Chauhan), 97–114. Elsevier: Academic Press.

Korres, N.E., Burgos, N.R., Vurro, M. et al. (2019a). New directions for integrated weed management: modern technologies, tools and knowledge discovery. *Advances in Agronomy* 155: 243–319.

Korres, N.E. and Froud-Williams, R.J. (2002). Effects of winter wheat cultivars and seed rate on the biological characteristics of naturally occurring weed flora. *Weed Research* 42: 417–428.

Korres, N.E. and Norsworthy, J.K. (2015). Influence of a rye cover crop on the critical period for weed control in cotton. *Weed Science* 63: 346–352.

Korres, N.E., Norsworthy, J.K., and Lazaro, M.L. (2018). Effects of simulated dicamba and 2,4-D tank contamination on Roundup Ready 2 and Xtend soybean injury and yield. *Southern Weed Science Society Meeting*, Atlanta, GA (21–24 January 2018).

Korres, N.E., Norsworthy, J.K., and Mauromoustakos, A. (2019b). Effects of Palmer amaranth (*Amaranthus palmeri*) establishment time and distance from the crop row on biological and phenological characteristics of the weed: implications on soybean yield. *Weed Science* 67: 126–135.

Korres, N.E., Norsworthy, J.K., Mauromoustakos, A., and Williams, M.M. (2020). Soybean density and Palmer amaranth (*Amaranthus palmeri*) establishment time. Effects on weed biology, crop yield and economic returns. *Weed Science* 68: 467–475.

Korres, N.E., Singh, A., Nizami, A.S., and Murphy, J.D. (2010). Is grass biomethane a sustainable transport biofuel? Biofuels. *Bioproducts and Biorefining* 4: 310–325.

Korres, N.E., Singh, A., and Prasad, S. (2023). Agricultural residues management: life cycle assessment implications for sustainable agricultural practices and reduction of greenhouse gases emissions. *Advances in Agronomy* 180. ISBN: 9780443192647.

Kumari, A., Price, A.J., Korres, N.E. et al. (2023a). Effect of crimson clover on the critical period of weed control in conservation tillage corn. *Frontiers in Agronomy* 4: 1068365.

Kumari, A., Price, A.J., Korres, N.E. et al. (2023b). Influence of a cereal rye cover crop on the critical period for weed control in soybean. *Weed Technology*. https://doi.org/10.1017/wet.2022.100.

Landis, D.A. (2017). Designing agricultural landscapes for biodiversity-based ecosystem services. *Basic and Applied Ecology* 18: 1–12.

Lindemann, W. and Glover, C. (2003). Nitrogen fixation by legumes. New Mexico State University. *Cooperative Extension Services*. http://www.soilcropandmore.info/crops/alfalfa/Nitrogen-Fixation-by-Legumes-a-129.pdf.

Marques, E., Kur, A., Bueno, E., and von Wettberg, E. (2020). Defining and improving the rotational and intercropping value of a crop using plant soil feedbacks approach. *Crop Science* 60: 2195–2203.

Marsden, T. (1999). Rural futures: the consumption countryside and its regulation. *Sociologia Ruralis* 39: 501–520.

Marsden, T. and Sonnino, R. (2008). Rural development and the regional state: denying multifunctional agriculture in the UK. *Journal of Rural Studies* 24: 422–431.

Marshall, E.J.P., Brown, V.K., Boatman, N.D. et al. (2003). The role of weeds in supporting biological diversity within crop fields. *Weed Research* 43: 77–89.

Mohd Din, A.R.J., Cheng, K.K., and Sarmidi, M.R. (2017). Assessment of compost extract on yield and phytochemical contents of Pak Choi (Brassica Rapa cv. Chinensis) grown under different fertilizer strategies. *Communication in Soil Science Plant and Analysis* 48: 274–284.

Moller-Hansen, E., Hogh-Jensen, H., and Djurhuus, J. (2002). Biological nitrogen fixation in a grazed perennial grass/clover ley and correlation with herbage and soil variables. *European Journal of Agronomy* 16: 309–320.

Mondal, T., Datta, J.K., and Mondal, N.K. (2017). Chemical fertilizer in conjunction with biofertilizer and vermicompost induced changes in morpho-physiological and biochemical traits of mustard crop. *Journal of Saudi Society Agricultural Sciences* 16: 135–144.

Mortensen, D.A., Egan, J.F., Maxwell, B.D. et al. (2012). Navigating a critical juncture for sustainable weed management. *Bioscience* 62: 75–84.

Moyer, J.R. and Blackshaw, R.E. (2009). Fall-seeded cover crops after dry bean and potato in southern Alberta. *Canadian Journal of Plant Science* 89: 133–139.

Mueller, S.M. and Vyn, T.J. (2016). Maize plant resilience to N stress and post-silking N capacity changes over time: a review. *Frontiers in Plant Science* 7: 53.

Munawar, A., Blevins, R.L., Frye, W.W., and Saul, M.R. (1990). Tillage and cover crop management for soil water conservation. *Agronomy Journal* 82: 773–777.

Navntoft, S., Wratten, S.D., Kristensen, K., and Esbjerg, P. (2009). Weed seed predation in organic and conventional fields. *Biological Control* 49: 11–16.

Palm, C., Blanco-Canqui, H., DeClerck, F., et al. (2014). Conservation agriculture and ecosystem services: An overview. *Agriculture, Ecosystems and Environment* 187: 87–105.

Phillips, S.H., Blevins, R.L., Thomas, G.W., et al. (1980). No-Tillage Agriculture. *Science* 208: 1108-1112–1113.

Price, A.J., Balkcom, K.S., Culpepper, S.A. et al. (2011). Glyphosate-resistant Palmer amaranth: a threat to conservation tillage. *Journal of Soil Water Conservation* 66: 265–275.

Price, A.J. and Norsworthy, J.K. (2013). Cover crops for weed management in southern reduced tillage vegetable cropping systems. *Weed Technology* 27: 212–217.

Sandhu, H.S., Wratten, S.D., and Cullen, R. (2010). Organic agriculture and ecosystem services. *Environmental Science & Policy* 13: 1–7.

Schafer, R.B., Bundschuh, M., Rouch, D.A. et al. (2012). Effects of pesticide toxicity, salinity and other environmental variables on selected ecosystem functions in streams and the relevance for ecosystem services. *Science Total Environment* 415: 426–435.

Schulze, E.D., Beck, E., Buchmann, N. et al. (2019). Ecosystem characteristics. In: *Plant Ecology* (ed. E.D. Schulze, E. Beck, N. Buchmann et al.), 461–479. Germany: Springer-Verlag GmbH.

Schwartz-Lazaro, L.M., Korres, N.E., Bararpour, M.T. et al. (2022). Population dynamics of naturally occurring weed flora in response to crop rotation and HPPD-based herbicide programs. *Weed Technology* 36: 426–435.

Sciubba, L., Cavani, L., Negroni, A. et al. (2014). Changes in the functional properties of a sandy loam soil amended with biosolids at different application rates. *Geoderma* 221–222: 40–49.

Shaner, D.L. and Beckie, H.J. (2014). The future for weed control and technology. *Pest Management Science* 70: 1329–1339.

Soininen, J.P., Bartels, J., Luoto, H.M., and Hillebrand, H. (2015). Toward more integrated ecosystem research in aquatic and terrestrial environments. *Bioscience* 65: 174–182.

Souza Krupek, F., Redfearn, D., Eskridge, K.M., and Basche, A. (2022). Ecological intensification with soil health practices demonstrates positive impacts on multiple soil properties: A large-scale farmer-led experiment. *Geoderma* 409: 115–594. https://www.sciencedirect.com/journal/geoderma

Storkey, J., Doring, T., Baddeley, J. et al. (2015). Engineering a plant community to deliver multiple ecosystem services. *Ecological Applications* 25: 1034–1043.

Storkey, J., Meyer, S., Still, K.S., and Leuschner, C. (2012). The impact of agricultural intensification and land-use change on the European arable flora. *Proceedings of the Royal Society B: Biological Sciences* 279: 1421–1429.

Swanton, C.J., Nkoa, R., and Blackshaw, R.E. (2015). Experimental methods for crop–weed competition studies. *Weed Science* 63: 2–11.

Tehranchian, P., Norsworthy, J.K., Powles, S. et al. (2017). Recurrent sublethal-dose selection for reduced susceptibility of Palmer amaranth (*Amaranthus palmeri*) to dicamba. *Weed Science* 65: 206–212.

Tyler, H.L. and Locke, M.A. (2019). Effects of weed management on soil ecosystems. In: *Weed Control. Sustainability, Hazards, and Risks in Cropping Systems Worldwide* (ed. N.E. Korres, N.R. Burgos, and S.O. Duke), 32–61. Boca Ratoon, FL: CRC Press.

Van Donk, E. and Van de Bund, W.J. (2002). Impact of submerged macrophytes including charophytes on phyto- and zooplankton communities: allelopathy versus other mechanisms. *Aquatic Botany* 72: 261–274.

Vencill, W.K., Nichols, R.L., Webster, T.M. et al. (2012). Herbicide resistance: toward an understanding of resistance development and the impact of herbicide resistant crops. *Weed Science* 60: 2–30.

Vidaller, C. and Dutoit, T. (2022). Ecosystem services in conventional farming systems. A review. *Agronomy for Sustainable Development* 42: 22.

Vidotto, F., Dalla Valle, N., Fogliatto, S. et al. (2021). Rapid increase of herbicide resistance in *Echinochloa* spp. consequent to repeated applications of the same herbicides over time. *Archives of Agronomy and Soil Science* 67: 620–632.

Virili, A. (2023). *Investigating weed-crop interactions and potential weed-mediated agro-ecosystem services in horticulture*. PhD Thesis, Sant' Anna, School of Advance Studies, Pisa, Italy, 146.

Wang, C.H., Wu, L., Wang, Z. et al. (2020). Characterizing changes in soil microbiome abundance and diversity due to different cover crop techniques. *PLoS One* 15: e0232453.

Waqas, M., Korres, N.E., Khan, M.D. et al. (2019). Advances in the concept and methods of seed priming. In (ed. M., HazanuzzamanV. Fotopoulos): *Priming and Pretreatment of Seeds and Seedlings: Implication in Plant Stress Tolerance and Enhancing Productivity in Crop Plants*, 11–41. Singapore: Springer.

Wetzel, R.G. (2001). *Limnology: Lake and River Ecosystems*. Monograph. Academic Press.

Willis, J. (2017). *State of the World's Plants*. Kew: Royal Botanics Gardens.

Wilson, G.A. (2007). *Multifunctional Agriculture. A Transition Theory Perspective*. Wallingford: CAB International.

Woodmansee, R.G. (1978). Additions and losses of nitrogen in grassland ecosystems. *Bioscience* 28: 448–453.

Xu, C., Wang, J., Wu, D. et al. (2020). Optimizing organic amendment applications to enhance carbon sequestration and economic benefits in an infertile sandy soil. *Journal of Environmental Management* 303: 114–129.

Xu, L. and Geelen, D. (2018). Developing biostimulants from agro-food and industrial by-products. *Frontiers in Plant Science* 9: 1567.

Yang, T., Siddique, K.H., and Liu, K. (2020). Cropping systems in agriculture and their impact on soil health-a review. *Global Ecology Conservation* 23: e01118.

Yu, T., Mahe, L., Li, Y. et al. (2022). Benefits of crop rotation on climate resilience and its prospects in China. *Agronomy* 12: 436.

Zhang, J., Heijden, M.G., Zhang, F., and Bender, S.F. (2020). Soil biodiversity and crop diversification are vital components of healthy soils and agricultural sustainability. *Frontiers in Agricultural Science and Engineering* 7: 236–242.

3

Climate Change and Ecologically Based Weed Management

Adusumilli Narayana Rao (A.N. Rao)[1] *and Nicholas E. Korres*[2]

[1] *Consultant Scientist, Jubilee Hills; Hyderabad, India*
[2] *Department of Agriculture, University of Ioannina, Kostakii, Arta, Greece*

3.0 Introduction

Agriculture is facing major challenges of meeting the food and nutritional security of a rapidly growing global population (Hickey et al. 2019) while improving the livelihoods of farmers owning the world's 570 million farms (Lowder et al. 2016) and facing challenges such as climate change, soil degradation, water pollution, biodiversity loss (Rial-Lovera et al. 2017; Barrett and Rose 2020), physical constraints, and biological constraints such as weeds (Rao et al. 2017). It was estimated that about 50% more food must be produced by 2050 in order to feed the increasing world population (FAO 2018). The sustainable intensification of agriculture (Firbank et al. 2018) was suggested to meet these challenges as it aims at increasing productivity while minimizing degradation to the environment and providing social benefits (Rose et al. 2020). In global agriculture, farmers spend more time and labor in removing weeds in their crop fields than in any other farm operation (Chikoye et al. 2001). Worldwide, weeds constitute a major constraint to the crops production as many crops are poor competitor to weeds because of slow growth rate at early stages of crop growth (Rao and Ladha 2011). Hence, losses could be substantial when optimum weed control during critical period is not achieved (Oerke 2006; Gharde et al. 2018; Rodenburg et al. 2019). The losses caused by weeds in agro-ecosystems could vary from site to site or year to year depending on the crop species, the type of weeds, the prevailing weed species, level of weed infestation, soil type, soil moisture content, climate and management practices adopted by farmers. Weeds have several characteristics with wide ecological amplitudes, which give them advantages over associated crops to survive more successfully in the disturbed habitat and changed climatic conditions (Chandrasena 2009; Ziska and Dukes 2011). Thus, the climate change is predicted to have profound impacts on weed invasion, losses caused, management and interactions with other biological organisms.

The major climate change processes influencing agriculture are global warming, elevated CO_2 and increased risks of extreme events (droughts, floods, heat waves, cyclones, etc.). Human activities have increased atmospheric CO_2 concentrations from 280 ppm in the pre-industrial era to >400 ppm (IPCC 2008), and it may surpass 550 ppm by 2050 and 700 ppm by 2100 (Rodrigues et al. 2016). Since 1900, the global average surface temperature has increased by about 1°C (1.8°F), which is mainly a result of the increased concentrations of CO_2 and other greenhouse gases (NAS and RS 2020). Under a "no action" greenhouse gas [GHG] emission scenario, the projected globally averaged warming is 2.4–6.4 °C by the end of the century (McDonald et al. 2009). This temperature rise will result in continued increases in sea level and overall rainfall, changes in rainfall patterns and timing, and decline in snow cover, land ice, and sea ice extent. It is very likely that the Earth will experience a faster rate of climate change in the twenty-first century than is seen in the last 10,000 years (Backlund et al. 2008). Climate change will impact many activities, but its effects on agricultural production could be acute (Korres et al. 2016). Estimates of annual damages in agriculture due to temperature increase or extended periods of drought will be more costly than damages in other activities. Yield losses are caused both by direct effects of climate change on crops and by indirect effects such as increased inputs in crop production, including those for weed management. In an analysis, Vilà et al. (2021) found that the effect on crops was additive when impacts of weeds and environmental change were combined.

Ecologically Based Weed Management: Concepts, Challenges, and Limitations, First Edition. Edited by Nicholas E. Korres, Ilias S. Travlos, and Thomas K. Gitsopoulos.

The change in atmospheric CO_2, rainfall and temperature will affect weed communities, weed species distribution, prevalence in agroecosystems (McDonald et al. 2009), increase the yield losses caused by weeds (Ziska 2000), and impacts weed management (Ziska 2016). Several reviews have addressed the crops and weed/crop interactions and climate change (Patterson and Flint 1980; Patterson 1993, 1995a, 1995b; Bunce and Ziska 2000; Ziska and Dukes 2011; Kristian et al. 2014). The climate change was expected to cause the expansion in geographical range of weeds and thus may cause greater losses to productivity of crops (Patterson 1995a, 1995b; Parmesan and Yohe 2003). The crop production losses due to weeds may enhance in future due to the climate change caused changes in herbicide susceptibility (Varanasi et al. 2016). The effect will be more critical when the worst weeds in a given crop are similar in growth habit or photosynthetic pathway and when weed species are "wild" species of cultivated species, such as red rice or weedy rice in rice (both *Oryza sativa*) (Ziska and McClung 2008); wild oat *(Avena fatua)* in oat (*Avena sativa*) (Ziska et al. 2019); shatter cane (*Sorghum bicolor*) in sorghum (*Sorghum vulgaris*) (Sahoo et al. 2010); and crop mimics (*Echinochloa* species) in rice (Rao and Moody 1992).

Climate change may also necessitate adaptation of agronomic practices, which in turn influence weed growth and proliferation of certain weed species. Weed management operations (e.g. chemical and mechanical) could be influenced by climate change and the current weed management options are known to enhance greenhouse gas (Das et al. 2011; Shi et al. 2020), cause water pollution (Lewis et al. 2009), reduce biodiversity (Schütte et al. 2017), and intensify weed menace with evolution of herbicide resistant weeds (Délye et al. 2013) and emergence of difficult to control weeds due to weed shifts (Rao et al. 2007). The current challenge is to design alternative ecologically based sustainable weed management systems that use minimal external inputs, reduced losses due to weeds, maintain sustainable food production systems with minimal environmental footprint. The ecologically based weed management (EWM) systems aim at an understanding of weed ecological processes, identifying and incorporating in agroecosystems the ecological factors that minimize weeds negative impact while enhancing crop growth and productivity.

In this review, the information on impact of climate change on weeds and their management in agroecosystems and possible EWM approaches for enhancing crop productivity while mitigating the adverse impact of climate change in a sustainable manner will be synthesized based on the published research information.

3.1 Climate Change and Weeds

Global warming and other climatic changes will affect the growth, phenology, and geographical distribution of weeds (Patterson 1995a, 1995b). The weeds may have a greater growth and reproductive response capacity than crops, in the climate change scenario, due to their greater genetic diversity than most crops (Ziska and Dukes 2011). The weed species with broad ecological niches may adapt to climate change caused changed environmental conditions that differ from their endemic regions (Roger et al. 2015) and may cause greater negative effects on agriculture on a global scale (Neve et al. 2009; Ziska 2016). If climate change projections are realized, cropping systems are likely to experience significant geographic range transformations among damaging endemic weed species and new vulnerabilities to exotic weed invasions (McDonald et al. 2009). To anticipate these changes and to devise management strategies for proactively addressing them, it is necessary to characterize the environmental conditions that make specific weed species abundant, competitive, and therefore damaging to particular crops (McDonald et al. 2009). The response to climate change of different weeds, belonging to different photosynthetic groups, vary (Table 3.1). In this part of the review, a brief summary of the responses of weeds to climate change is given as it may help guide future enhancements to EWM.

3.1.1 CO_2 Enrichment

Atmospheric carbon dioxide (CO_2) has risen from 278 ppm in 1750 to 409.8 ± 0.1 ppm in 2019 (Dunn et al. 2020), as a result of continuous anthropogenic activities. The elevated CO_2 caused greater stimulation of the non-native species than native species in the region (Sasek and Strain 1991), which was attributed to plant architectural differences (Ziska 2003b) and altered reproductive allocation (Smith et al. 2000).

The response to higher CO_2 levels varies by weed species due to photosynthetic biochemistry as it stimulates photosynthesis and weed growth (Patterson 1995a, 1995b). The weeds with C_3 photosynthetic pathway (about 95% of all plant species) (Ziska and Dukes 2011) are likely to respond more strongly than weeds possessing the C_4 photosynthetic pathway as their photosynthetic rates are saturated at current, ambient CO_2. (Ziska 2003a) suggested that the recent increases in atmospheric CO_2 during the twentieth century may have been a factor in the selection of most invasive *Cirsium arvense* (L.) Scop., *Convolvulus arvensis* L., *Euphorbia esula* L., *Sonchus arvensis* L., *Centaurea maculosa* Lam., and *Centaurea solstitialis* L. in USA. The invasive weeds that reproduce by vegetative means may show

Table 3.1 The impact of climate change on a few arable land weeds.

Weed	Photosynthesis type	Climate change effect	References
Parthenium hysterophorus	$C_3 - C_4$ intermediate	Rising temperatures and atmospheric CO_2 and changing rainfall patterns are favorable for its growth, reproductive output, future spread and infestation. It may become more aggressive and grow more rapidly under elevated temperature and CO_2 conditions as well as reduced soil moisture in the future.	Bajwa et al. 2016
Parthenium hysterophorus	$C_3 - C_4$ intermediate	Life span reduced by up to 43% and growth enhanced by 20% at low soil moisture condition due to modifications in the plants vegetative and reproductive biology.	Nguyen 2011
Parthenium hysterophorus	$C_3 - C_4$ intermediate	Soil moisture percentage of >60% or <40% is deleterious for its seeds sprouting and monthly average temperatures of >34°C are detrimental.	Kathiresan et al. 2005.
Parthenium hysterophorus	$C_3 - C_4$ intermediate	Despite having C_4 rosette leaves, the main vegetative part of the parthenium weed is C_3 and shows significant improvement in growth and biomass production under a higher CO_2 concentration (480 ppm) than at ambient 360 ppm	Navie et al. 2005
Parthenium hysterophorus	$C_3 - C_4$ intermediate	Showed improved photosynthetic rate, water use efficiency, and better growth under a high CO_2 concentration (700 ppm) and a temperature of 25–35°C.	Pandey et al. 2003
Parthenium hysterophorus	$C_3 - C_4$ intermediate	A substantial increase of 52, 55, 62, 120, 94, and 400% in plant height, biomass, branching, leaf area, photosynthesis, and water use efficiency, respectively, at an elevated CO_2 concentration (550 ppm) than at ambient 380 ppm.	Shabbir et al. 2014
Parthenium hysterophorus	$C_3 - C_4$ intermediate	A remarkable increase in parthenium growth occurred under water-deficit conditions together with high temperature and elevated CO_2 levels.	Belgeri 2013; Nguyen 2011
Parthenium hysterophorus	$C_3 - C_4$ intermediate	Predictive CLIMEX modelling studies to estimate its potential spread in different parts of the world and under a changing climate indicated that it could invade further parts of Australia, Pakistan, and other South Asian countries especially under a changing climate of 3°C.	Shabbir 2012; Kriticos et al. 2015; Mainali et al. 2015
Parthenium hysterophorus	$C_3 - C_4$ intermediate	Predictive CLIMEX modelling studies suggested that sub-Saharan African countries were the most at risk of parthenium weed invasion in future. It also predicted that most parts of the Asian-Pacific region and some European countries were at risk including Portugal, Italy, Spain, and France.	McConnachie et al. 2011
Parthenium hysterophorus	$C_3 - C_4$ intermediate	With CO_2 enrichment: the wheat could gain biomass against *Phalaris minor* but under water stress conditions, *P. minor* had advantage over wheat. Hence, Parthenium problem could be aggravated with water scarcity.	Naidu 2011
Parthenium hysterophorus	$C_3 - C_4$ intermediate	The parthenin concentrations in invasive biotypes were 49% higher at ~400 ppm (recent) than at ~300 ppm (mid-twentieth-century) CO_2 concentrations and might be potentially contributing to Parthenium's invasive success.	Rice et al. 2021
Ambrosia artemisiifolia	C_3	Potential distribution projections under future climatic change scenarios suggested an averaged percentage of suitable area (2.21%) and habitat gain (1.49%) in *A. artemisiifolia* distribution, with further expansion to environmentally favorable locations in southeast coastal regions, northern Taiwan, and the Beijing–Tianjin–Tangshan area in northern China; northern parts of Hong Kong and perimeter locations of Taiwan. In contrast, suitable habitat was projected to decrease by 0.15% in southern China. Hence management priorities should be focused.	Qin et al. 2014

(Continued)

Table 3.1 (Continued)

Weed	Photosynthesis type	Climate change effect	References
Ambrosia trifida	C_3	Potential distribution projections under future climatic change scenarios suggested, central Beijing and the southern portions contiguous to Hebei regions would still provide suitable environmental conditions for this species and show an increase of 1.2% in area. Small portions of several northeastern provinces (0.87%) were also predicted to be favorable for its establishment by the year 2050. The effective control strategies may be optimized by concentrating efforts on those relatively fewer regions of China where the species is currently abundant.	Qin et al. 2014
Avena fatua	C_3	Enrichment of CO_2 has strong stimulatory effect on its growth, biomass, seed production, and reduced the level of dormancy in after-ripened caryopses, which increases competitiveness, causing greater crop loss.	O'Donnell and Adkins 2001
Avena fatua	C_3	It may become less serious in the regions where more precipitation is projected, but more serious in areas that will become drier.	Wang and Mohan 2008
Convolvulus arvensis	C_3	Total biomass production increased by 60% from a 284 mmol mol CO_2 level to 380 mmol mol level and by another 36% at 719 mmol mol CO_2 level, which is projected for the end of the twenty-first century.	Ziska 2003a
Weedy rice (*Oryza sativa* L.)	C_3	Responds more strongly than cultivated rice to rising CO_2 level with greater competitive ability, suggesting that it may become a more problematic weed in the future.	Ziska et al. 2010
Amaranthus retroflexus	C_4	The leaf photosynthesis increased by 30% when CO_2 increased from 380 to 690 mmol mol but biomass responsiveness to elevated CO_2 was smaller, yet significant.	Ziska and Bunce 1997
Amaranthus retroflexus	C_4	It allocated more biomass under elevated CO_2 to support structures, such as stems, and could potentially enhance its ability to shade competitors and hence reduce the competitive ability for light by associated crop.	Tremmel and Patterson 1993
Amaranthus retroflexus	C_4	Increase of CO_2 from 350 to 700 mmol mol delayed the onset of its reproduction.	Dippery et al. 1995
.*Amaranthus retroflexus*	C_4	Its total seed production decreased from 0.25 g per plant at ambient CO_2 to 0.19 g per plant at 700 mmol mol.	Bazzaz and Garbutt 1988
Amaranthus retroflexus	C_4	The soyabean yield loss it caused was higher at ambient CO_2 (745%) than at elevated CO_2 (730%), indicating pigweed's reduced competitiveness against soyabean at higher CO_2.	Ziska 2000
Cyperus rotundus	C_4	The allelopathic effects will increase as atmospheric CO_2 continues to rise because of the stimulating effects of elevated CO_2 on below-ground production.	Wang 2007; Wang et al. 2008
Cynodon dactylon	C_4	Ability of its rhizomes to grow as deep as 1 m in to the soil (Holm et al. 1991) ability to maintain higher rates of photosynthesis, stomatal conductance, and water-use-efficiency (WUE) at relatively low water content in the leaves (Carmo-Silva et al. 2007) could increases its adaptive ability to extreme climatic conditions, such as drought; endurance, competitiveness in the regions that will become drier in the future (Wang and Mohan 2008).	Holm et al. 1991; Carmo-Silva et al. 2007; Wang and Mohan 2008
Eleusine indica	C_4	Its competitive advantage against important cereal crops and seriousness in future agriculture increases, particularly in the area where temperature and soil N fertility are expected to increase.	Wang and Mohan 2008

a strong response to increases in atmospheric CO_2 (Ziska and George 2004).

The elevated CO_2 was observed to have positive effects on C_3 crop rice competitiveness against C_4 weeds (Patterson et al. 1999; Fuhrer 2003), in addition to tolerance to infection of *Striga* (Watling and Press 2000). The crop-weed competition studies have assessed the effects of elevated ambient CO_2 on comparative physiology and growth (Saebo and Mortensen 1998; Ziska 2000, 2001, 2002, 2003a, 2003b; O'Donnell and Adkins 2001), including interactions with factors such as soil nitrogen status (Zhu et al. 2008). In weed/crop competition studies where the photosynthetic pathway is the same, weed growth increases as CO_2 increases (Ziska and Bunce 1998; Ziska and Runion 2006). Several weeds have shown a greater stimulating response to elevated CO_2 concentrations relative to crops species, and they are likely to contribute to increased risk of greater crop loss from weed pressure (Walthall et al. 2013). The rising CO_2 could alter yield losses due to weed competition and weed management will be crucial in realizing any potential increase in the yield of C_3 crops such as soybean, as climate change occurs (Ziska 2000). Many weed species have the C_4 photosynthetic pathway but the crop competes with both C_3 or C_4 weeds that are associated. The growth, competition, and yield potential of C_4 crops could be reduced by associated C_3 weeds as atmospheric carbon dioxide increases (Ziska 2002a, 2002b). CO_2 enrichment enhanced the growth of C_3 weeds: *Xanthium strumarium* and *Abutilon theophras* and C_4 weed: *Amaranthus retroflexus*, leading to increased competition and losses in sorghum (*Sorghum bicolor*) (C_4) (Ziska 2001, 2003b).

Changes in leaf surface characteristics and excess starch accumulation in the leaves of C_3 weeds may interfere with herbicidal control (Patterson 1995a, 1995b). Ziska et al. (1999) observed that glyphosate loses its efficacy on weeds grown at CO_2 levels that likely will occur in the coming decades. Other studies also demonstrates that herbicide efficacy can be reduced at elevated CO_2 (Harris and Hossell 2001; Ziska and Teasdale 2001; Ziska et al. 2004).

The rising CO_2 may increase/double below-ground root growth relative to above-ground shoot growth (Ziska et al. 2004). Perennial weeds may become more difficult to control under elevated CO_2 as increased photosynthesis stimulates greater production of rhizomes and other storage organs (Patterson 1995a, 1995b). The asexual propagation from small root segments is commonly observed in agroecosystems. In the developing countries, mechanical control is still the predominant means to manage weeds in crops. Tillage, a common form of weed control, cuts and discs roots, and increasing tillage as a control measure would lead to additional plantpropagation for perennial weed species in a higher CO_2 environment.

3.1.2 Increased Temperature

The wider gene pool of weeds, compared with crops, enables weeds to adapt to diverse environment conditions and to show greater range of response to increased temperatures. The habitable zone of many weed species is largely determined by temperature (Patterson et al. 1999). The germination rates of *Amaranthus albus* L., *A. palmeri* L., *Digitaria sanguinalis* (L.) Scop., *Echinochloa crus-galli* (L.) P. Beauv., *Portulaca oleracea* L., and *Setaria glauca* (L.) Beauv., *Hirschfeldia incana* (L.) Lagr., *Sonchus oleraceus* L., and *Conyza canadensis* (L.) Cronquist increased linearly with increased temperatures between 15°C and 30°C. The germination peak was recorded at 25/20°C for *Amaranthus tuberculatus* L. and at 35/30°C for *Amaranthus retroflexus* L. and *Amaranthus palmeri* L. (Guo and Al-Khatib 2003). These weed species could potentially become greater competitors with crop plants in the increased temperature environment (Steinmaus et al. 2000). The increased temperatures could extend the distribution limits of parasitic weeds by several hundred miles (Mohamed et al. 2006) and some invasive weeds could increase in distribution and importance (Ahmed and Wardle 1991).

Most (14 of the 18) of the "world's worst weeds" are C_4 plants (Holm et al. 1991) while most of the crops feeding the world (12 out of 15) are C_3 (Ziska and Dukes 2011). The increased temperature will favor C_4 plants than C_3 plants especially under elevated CO_2 conditions that are expected during climate change (Alberto et al. 1996). Higher temperatures were reported to cause increase in photorespiration and decrease in net photosynthesis and growth of C_3 weeds (O'Donnell and Adkins 2001; Valerio et al. 2013) while it was noted to stimulate photosynthesis and growth of C_4 weeds at high CO_2 (McDonald et al. 2009; Valerio et al. 2011).

The significant CO_2 by temperature interactions revealed that CO_2 responses were smallest at higher temperatures (Tremmel and Patterson 1993). The beneficial effects of elevated CO_2 to C_3 crops including more successful competition with C_4 weeds may be lost, to some extent, in a warmer climate as warming accelerates plant development and reduces grain-fill, reduces nutrient-use efficiency, increases crop water consumption, and favors C_4 weeds over C_3 crops (Fuhrer 2003; Lee 2011). Tomato crop losses due to weed competition were higher at 26/18°C than at 21/12°C (Valerio et al. 2013). It was predicted that by the end of the century in the US Corn Belt, cold-tolerant species like *Abutilon theophrasti* Medicus may be of minor importance whereas *Sorghum halepense* (L.) Pers., a predominantly southern US weed species at present, may become common and damaging to maize production (McDonald et al. 2009).

3.1.3 Elevated CO_2 and Temperature

The understanding is inadequate about interactive impacts associated with temperature and CO_2 interaction on invasive weeds biology and their influence on agricultural productivity. The rising temperature and CO_2 levels impact were reported to make crops less competitive with weeds (Rosenzweig and Hillel 1998) or benefit weeds more than cash crops (Wolfe et al. 2008), depending on other agroecological conditions. Increased atmospheric CO_2 levels are likely to be accompanied by higher-temperature favoring C_4 weeds over C_3 crops (Fuhrer 2003). C_3 crops like rice might compete better against a C_4 weed at elevated CO_2 alone, simultaneous increases in CO_2 and temperature could favor a C_4 species (Alberto et al. 1996). With increases in temperature and CO_2 *Ambrosia artemisiifolia* is expected to grow faster, flower earlier, and produce more pollen, with implications for public health (Wayne et al. 2002; Ziska et al. 2003). A warming of 3 °C (day night temperature increase from 26/20 to 29/23°C) caused increased biomass and leaf area of *Rottboelliia cochinchinensis* by 88% and 68%, respectively (Patterson et al. 1979) indicating the probability of its spread to warmer areas.

3.1.4 Precipitation Extremes and Water (Drought and Flood)

Climate change also cause variability in moisture availability and precipitation extremes that bring constraints and risks for agriculture (Lemmen et al. 2008). The agricultural production is already getting affected by rising temperature and precipitation changes due to climate change (Long et al. 2006; Lobell et al. 2008) and are expected to affect by reducing global crop yield by 2050 (Long et al. 2006) and may affect weed growth, biology, geographical distribution, reproduction, and competitive abilities of weeds and weed management (Patterson 1995a, 1995b; Ziska et al. 1999; Ziska and George 2004; Ziska and Dukes 2011). Weeds adapted to conditions of precipitation and soil water extremes will flourish as they will have a higher competitive advantage in agroecosystems.

The soil moisture is a key factor required for weed seed germination and establishment. Weed competition for water is an important cause of crop failure or decreases in crop yields worldwide. Changes in precipitation and availability of water also affect seed production and dispersal (Rodenburg et al. 2011) in addition to overall growth of weeds. A recent analysis revealed the mitigation of negative effects of weeds on crops with elevated CO_2 and warming but added to the negative effect of drought (Vilà et al. 2021). Under prolonged drought conditions, composition of weed species will change, leading to a shift in the vegetation. The growth and dynamics of both C_3 and C_4 plants are regulated by precipitation. C_4 plants expansion is promoted by dry winters and wet summers, while abundance of C_3 plants may increase under wet winters and dry summers (Niu et al. 2005). The response to water stress differs among C_3 and C_4 weeds. The C_4 plants may have greater competitive advantage than C_3 plants under water-stressed situations because of their higher water use efficiency characteristic (Osmond et al. 1982) and their ability to produce more biomass, root growth, and seeds than C_3 plants (Long 1999). The incidence of sedges increases under reduced flooding regimes in lowland rice (Haden et al. 2007).

Available information shows that high temperature extremes, drought, and floods substantially affect ecosystems. Increasing gaps and overall contraction of the distribution range for species habitat could result from increases in the frequency of large-scale disturbances due to extreme weather and climate events (Opdam and Wascher 2004). Assessment of Fischlin et al. (2007) revealed that 20% to 30% of studied plant and animal species may be at an increased risk of extinction if global warming exceeds 2–3 °C above the preindustrial level.

The phenotypic plasticity and rapid genetic change can enhance the risk of invasion to new areas under climate change by weed species like *I. cylindrica* and *Solidago canadensis* (Clements and Ditommaso 2011; Wan et al. 2020; Wang and Wan 2020). *Sorghum halepense* was reported to spread in many agricultural habitats in Europe (Weber and Gut 2005), and a progressive change of *S. halepense* habitat affiliation and a dispersal to other areas of central Europe at higher temperatures was reported (Follak and Essl 2013). The effects of climate change on weed expansion on a global scale need to be monitored continuously.

Greater-intensity tropical cyclones and projected increases in wind speeds may increase the likelihood of wind-borne incursions by weeds that occur in countries nearby (Scott et al. 2014) and invaded weeds may become serious threats to agricultural productivity, export markets, or the environment. For example, *Digitaria insularis* (Poaceae) and *Chromolaena odorata* produce large volumes of wind-dispersed seed. It was predicted that *C. odorata* poses a very significant risk of introduction to Australia via wind to the islands of the Torres Strait, Cape York Peninsula and the Northern Territory (Kriticos et al. 2005).

Under climate change in many agricultural areas, crops selected for certain conditions are expected to experience increased heat stress and moisture stress, and in certain cases increased pressure from weeds more adopted to changed climate. Human responses to climate change will affect how well weeds are controlled. These include deliberate adaptation choices like changed farming practices and opportunistic cropping to take advantage of years with good rain, introduction of new crop and pasture species better able to tolerate extreme conditions, changes in choice of sown crops, and development of new cropping regions (Scott et al. 2014). The imposition of selection pressure on weeds due to climate change (Ziska et al. 2019) and associated responses will make weeds adapt rapidly to changes through alterations in their biology, physiology, and phenology (Korres et al. 2016; 2017). These changes will lead to newer weed communities that require appropriate weed management strategies development and adoption.

3.2 Climate Change and Weed Management

Higher temperatures could give invasive weed species (Dukes and Mooney 1999) a competitive advantage over crop production because of their greater genetic diversity (Chandrasena 2009). Hence the current weed management approaches need to be revisited and approaches for future need to be evolved to meet the aggressive weed competition challenges in the changing climatic conditions. Understanding the climate change effect on weed growth and efficacy of different current weed management methods is critical to optimizing those methods for effective weed management. Increased CO_2 simulated growth shortens the weed seedling stage, which is most sensitive to postemergence herbicides leading to reduced herbicide efficacy. Also, enhanced atmospheric CO_2 concentration could increase C_3 weeds' glyphosate tolerance (Ziska et al. 1999). Increased tolerance of glyphosate was observed with perennial C_3 weed, quackgrass [*Elytrigia repens* (L.) Nevski] due to continuous greater photosynthesis and growth in perennial weeds under increased atmospheric CO_2 concentration (Ziska and Teasedale 2001). The herbicide uptake, translocation and overall effectiveness was affected by anatomical (increased leaf thickness), morphological (reduced stomatal number) and physiological (decreased conductance) changes due to increased CO_2 making weed management difficult and costly (Patterson 1995a, 1995b). The stimulation of belowground root/rhizome and tuber growth, relative to above ground shoot growth, of perennial C_3 weeds (Oechel and Strain 1985) under high CO_2 environments (Ziska 2003b) make perennial weeds control more difficult (Patterson et al. 1999). Increased tolerance C_4 grass species to glyphosate under elevated CO_2 was observed (Ziska and Goins 2006; Manea et al. 2011). Decreasing stomatal conductance with increasing CO_2 could also increase assimilation of weeds and reduce the transpiration (Jennings et al. 1977), which may lead to reduced uptake of soil applied herbicides reducing herbicide efficacy. As water availability reduces due to climate change, the use of maintaining a flood-water layer to suppress weeds is likely to become increasingly difficult in irrigated rice ecosystems (Rao et al. 2007). Water-saving production methods such as dry direct-seeded rice, which is becoming popular among farmers in response to water-scarcity, address this, but weed competition tends to increase severely (Rao et al. 2007), so improved alternate weed management approaches are needed.

Drought due to climate change may also pose serious challenge to the use of herbicides, as many herbicides lose effectiveness during dry periods or drought conditions. The effectiveness of soil-incorporated herbicides is best when soils have reasonable moisture levels after their soil incorporation. Preemergence herbicides also are more effective when adequate soil moisture conditions prevail at herbicide application time. Increases in temperature can also pose a variety of direct and indirect effects on herbicides such as limiting herbicide uptake either from the soil or foliage due to reduction in moisture in both soil and plant; herbicides either causing crop damage or not being effective; increased herbicide volatility and carryover that may harm susceptible crops that come into rotation (Patterson et al. 1999; Bailey 2004). Under intensified frequent rainfall regimes, the herbicides may be diluted and cease to be effective (Kanampiu et al. 2003). The environmental conditions have appreciable influence on herbicide efficacy, particularly foliage-applied herbicides (Kudsk and Kristensen 1992). The efficacy of glyphosate on *Cirsium arvense*, *Elytrigia repens*, and *Chenopodium album* was reduced under elevated CO_2 (Ziska and Teasdale 2001). The CO_2 effect on the glyphosate efficacy on *Amaranthus retroflexus* is unclear. The efficacy of fenoxaprop-p-ethyl / fluazifop-p-butyl in *Avena fatua* is reduced by approximately 50% under a doubling of ambient CO_2 concentrations (Archambault et al. 2001). Increasing daytime temperatures from 23°C to 29°C caused decrease in the efficacy of fenoxaprop-p-ethyl / fluazifop-p-butyl and glufosinate ammonium, but not imazamethabenz on *Avena fatua*. Climate change effects will reduce efficacy of metabolism-based herbicides on weeds and increase the evolution of nontarget site herbicide resistance (Matzrafi et al. 2016). Depending on the amount, intensity, and timing of the rain after herbicide application, rainfall can also impact the need for multiple applications due to washing away of the herbicide.

The response of weed species to herbicides under elevated CO_2 were found to be species-specific, with C_3 grasses tended to be the most sensitive than C_4 grasses (Waryszak et al. 2018). The impact of climate change factors on herbicide efficacy varies within herbicide modes of action and also among herbicides belonging to same mode of action making any generalized assumptions on its impact difficult (Varanasi et al. 2016). Herbicide use is expected to increase in the near future and with it more resistant weed ecotypes (Ziska et al. 1999). Research should focus on understanding the impact of climate change factors and their interactions on herbicides belonging to different groups to understand the implications for weed management in future climate scenario.

3.3 Ecologically Based Weed Management

3.3.1 Need for Ecologically Based Weed Management

Weed populations are able to evolve rapidly in response to intense, human-derived selection pressures that weed species are selected by the very cultural practices, which were originally designed to suppress them. The local adaptation of evolved weed populations to agroecosystems depends on genetic variation in weed populations, selection pressure imposed by local climatic and environmental conditions, and crop and weed management practices. Climatic and/or environmental selection is likely to result in regional patterns of differentiation between weed populations (Neve et al. 2009). Thus, weeds respond to the prevailing environment and other factors operating in the ecosystem and continuously adapt relatively faster (Clements and Ditommaso 2011) by evolving needed adoptive characteristics to enable them to compete with the crops grown in agroecosystems. Changes in climate may influence the weed population and dynamics, depending on local environmental factors that may change and their ecological interaction with weed community. Thus, an understanding of ecological responses of weeds to climate change and diversifying weed management in both time and space across different agroecological zones will help minimizing the losses caused by weeds. It is important to gain information on factors conducive as well as detrimental to weeds growth to ecologically manage weeds in a sustainable manner. However, the research focus and technologies developed in developed countries and currently in developing countries is mostly herbicides based. Herbicides were found effective and cheaper by farmers because of increased cost of labor in all developed countries and most of developing countries also. Hence, farmers focused mainly on herbicides for managing weeds in agroecosystems. However, long-term sustainable weed management was not achieved by herbicide-based weed control (Mortensen et al. 2000) as weeds continue to be major loss causing constraints in croplands and farmers still lose considerable crop yield due to weeds (Rao et al. 2007). Furthermore, herbicide resistance is making weed management more critical with increasing farm production costs. In addition, the heavy use of herbicides in weed management has several environmental and health risks and potential negative impacts of herbicide compounds on food safety, nontarget organisms, beneficial species, quality of surface and groundwater (Barbash et al. 2001; Mendes et al. 2019), public health, and the environment (Kropff and Walter 2000; Liebman 2004). Thus, a broader perspective to weed management is being looked into rather than relying only on herbicides. Hence, the EWM strategies are needed for effective and economical weed management reduce environmental negative effects of inputs like herbicides, minimize reliance on costly inputs, increase the input use efficiency, prevent build of difficult to control weeds during weed shifts and herbicide resistant weeds, to combat the adverse impact of climate change and make cropping systems profitable in a sustainable manner (Liebman 2004). The EWM tactics are developed on a foundation of adequate ecological understanding. The insights into factors affecting weed occurrence, colonization, growth, competitive interactions, reproduction, dissemination, mortality, local extinction, and other processes knowledge forms the basis of ecologically based sustainable weed management strategies and methods. The profitability of cropping is the major concern of farmers across the globe due to increasing cost of all the inputs. The EWM strategies help to improve profitability through reduced cost of crop production as they make better use of ecological processes (Liebman 2004).

3.3.2 Progress

An understanding of weed diversity, distribution, and identifying weed species that have the greatest potential to impact (D'Antonio and Chambers 2006) is critical to develop weed management programs in any ecosystem. Information on the ecology and population dynamics of the weeds, important population parameters and life-stage transitions help understanding the factors that most likely suppress weeds. It provides a scientific basis for developing integrated EM strategies. However, such efforts in agroecosystems are relatively less. The research progress made on weed management approaches based on EWM principles are discussed briefly.

3.4 Managing Weed Soil Seedbank Using Preventive Measures

The soil seedbank builds up through weed seed production and their dispersal, while it depletes through weed seedling recruitment after weed seed germination, predation, and decay. Weed seedbank may be depleted by adopting preventive measures. Rao et al. (2017), while reviewing preventive measures for direct seeded rice (DSR), concluded that there are several potential preventive measures for managing weeds:

- Minimize weed seed production, both in the cultivated and fallow fields, on the bunds, irrigated channels to manage weed seedbanks.
- Minimize dispersal of weed seeds into fields as contaminants in crop seeds or through irrigation canals.
- Promote seed predation by management of factors that promote predator's activity.
- Promote seed decay wherever feasible.

- Adopt strategies that stimulate fatal germination of weed seeds.

The preventive weed control measures alone are unlikely to be sufficient for the effective and economical management of weeds in agroecosystems, but their integration with other components of EWM approaches should reduce weed management costs and increase farmer incomes.

3.4.1 Weed Seed Elimination/Destruction at Crop Harvest

The population growth rates of annual weed species are affected by seed survival (Davis 2006). Many weeds retain the majority of their seeds on the plant at maturity, and when those weeds' maturity coincides with that of crop, use of combine harvesters or unintentional manual harvest and threshing of weeds along with crop results in weed seed contamination of crop seed resulting in weed seed dissemination with crop seed (Rao and Moody 1990). The use of crop seed free of weed seed is recognized as the nonmonetary input in managing weeds (Norsworthy et al. 2012).

The weedy patch expansion in crop fields is influenced by seed dispersal by crop harvesting machines (Mohler 2001). Harvest weed seed control (HWSC) technology is being developed and popularized by targeting weed seed at crop harvest using special machinery that trails behind the combine harvester to trap and grind weed seeds picked up during crop harvesting and thus reducing the weed seed addition to soil seedbank (Harrington and Powles 2012; Walsh et al. 2012, 2013; Somerville et al. 2018), minimize the weed infestation area and decrease the dissemination of weed seeds from one location to another or from one cropping system to another. The chaff carts, bale-direct systems, narrow windrow burning, chaff lining, chaff tramlining, and integrated Harrington Seed Destructor are a few of HWSC currently implemented (Schwartz-Lazaro et al. 2017; Walsh et al. 2018).

3.4.2 Weed Seed Burial by Tillage

The tillage depth and inversion affect the weed seed movements in the soil profile (Mohler and Galford 1997). Deep tillage can reduce weed seedling emergence from weed seedbank by burying the bulk of the weed seed below maximum weed seed germination and weed seedling emergence depth. The burial of smaller wild lettuce seeds at a depth of at least 4 cm will significantly reduce their emergence as its germination significantly decreased as burial depth increased (Chadha et al. 2019). However, the larger seeds have greater food reserves, which allow plants to emerge from deeper burial depths than smaller seeds (Hoyle et al. 2013). Thus, this method is useful for specific weed species. The repeated tillage will bring deeply buried weed seeds back to the surface again. Tillage practices may also be used by increasing seed mortality to fatal germination by tilling to stimulate germination at the period that is not favorable for weeds and at conditions that favor fungal pathogens (Davis and Renner 2007).

3.4.3 Utilizing Stale Seedbed Technique

In stale seedbed weeds are encouraged to germinate prior to seeding the crop then eliminated with appropriate method (Rao et al. 2007). Stale seed bed technique (SSB) is one of the cultural management approaches that can be used before crop establishment to reduce the number of weed seedlings emerging with crop and weed seedbank. The weed seedlings emergence is promoted by tillage and irrigation or rain prior to final land preparation and crop seeding. The emerged weed seedlings are managed by herbicide or shallow tillage with minimal further soil disturbance and it results in reduced in weed density in the crop (Dogan et al. 2009; Kanatas et al. 2020) and reduced weed seedbank.

3.5 Application of Principles of Conservation Agriculture for EWM

The minimum soil disturbance, permanent soil cover, and crop rotation are the three major components included in conservation agriculture (CA) (Hobbs 2007). These practices, which result in the maintenanceof crop residues on the soil surface, thus avoiding water losses by evaporation, are now being used increasingly throughout the world. The CA is considered as effective approach for sustainable increase in crop productivity with efficient resource use, when all the three components of CA are integrated and CA was suggested as an important climate-change adaptation strategy for global ever-drier regions (Pittelkow et al. 2015). The weed control options available and farm machinery now allow many crops to be grown with minimal tillage (reduced tillage) or without tillage (Zero-till). The ways of using the three components of CA for managing weeds ecologically are discussed.

3.5.1 Minimal Soil Disturbance

The normally followed conventional tillage, entire soil surface is disturbed once or many times during the crop growing season, while no tillage is practiced in zero tillage and in minimum tillage, the soil disturbance is minimal. Since soil disturbance tends to stimulate soil carbon losses through enhanced decomposition and erosion, reduced- or no-till agriculture normally results in soil carbon gain (West and Post 2002). The conventional tillage redistributes seeds throughout the soil profile while in zero tilled, seeds get accumulated in the top 0 to 5 cm (Hoffman et al. 1998), even though a few may infiltrate the soil through cracks and other means. The seeds near the soil are exposed to greater mortality risks through weather variability and predation, variability in weather and germination due to untimely rains followed by dry weather as exists in tropics. In comparison to minimum or conventional tillage, in zero tilled plots, the seedling emergence of *Digitaria ciliaris, Echinochloa colona*, *Eleusine indica*, *Ageratum conyzoides*, *Eclipta prostrata*, and *Portulaca oleracea* was greater and these weeds can be managed by timing and adoption of appropriate weed management methods (Chauhan and Johnson 2009). Assuming no weed seed input into the system, viable weed seedbanks under no-till decrease more rapidly than under conventional tillage due to greater predation because of their availability on soil surface to predators (Westerman et al. 2006), whose population tend to increase with lesser soil disturbance, greater desiccation and harsh weather caused loss in viability (Anderson 2005), difficulty in seedling emergence and resulting mortality (Mohler 2001), reduced dispersal of weed seeds by machinery due to their minimal use of machinery which are known to disperse the weed seed (Heijting et al. 2009). In zero-tillage, the seedbank that was in deeper soil layers does not come to the surface soil and new seeds are not ploughed in to the soil deeper layer weed seedbank. Reducing tillage may shift weed communities from annual dicots to grassy annuals and perennials (Erenstein and Laxmi 2008) but practicing zero tillage in rotation with conventional in two or more crops rotations did not increase perennial weeds (Tuesca et al. 2001). The zero tillage systems were also proved to result in desired greater weed species diversity (Murphy et al. 2006). Thus, depending on location specificity, the zero or minimum tillage may be used for minimizing weed seedbank while maintaining greater weed biodiversity.

3.5.2 Soil Cover by Retaining Crop Residues in the Crop Field

Conventionally, the crop residues are incorporated in the soil by conventional tillage. In CA systems, the residues are retained on the soil surface in the crop field. The surface residue retention provides multiple benefits, including soil moisture conservation, improvement in soil organic matter and soil structure, and suppression of weeds (Singh et al. 2007; Ladha et al. 2016). Soil residues were proven to influence the weed seed germination and emergence by physical hinderance (Bosy and Reader 1995); modifies soil microclimate, which inhibits seed germination (Teasdale 1997); reduces soil temperature that slows down germination (Liebman and Mohler 2004); physically intercepts solar radiation and shades the seedlings making them less vigorous (Wilson et al. 1995); enhances weed seed decay due to increased micro-flora activity (Yang et al. 2013); decomposing residues emit allelopathic substances that inhibit germination (Prati and Bossdorf 2004) and weed growth (Sathishkumar et al. 2020), particularly of small seeded weeds (Liebman and Davis 2000); reduces weed seedling emergence (Buhler et al. 1996), weed infestation and helps crop to have lesser competition from weeds during initial growth stage (Queiroz et al. 2010). The growth and weed seed production also will be affected due to combined effect of all the above effects resulting reduced weed seed addition to soil weed seedbank (Franke et al. 2007). The ground cover with crop residues or litter could restrict weed seedling emergence by providing a habitat for seed predators (Reader 2011). However, the crop residues need to be spread uniformly to be effective in suppressing weeds and not to provide conducive environment for emergence of certain weeds (Nikolić et al. 2021).

3.5.3 Crop Rotations

Crop rotations were proven to reduce weeds (Weisberger et al. 2019) and increase crops yield (Zhao et al. 2020). The crop production practices (tillage, sowing date, herbicides usage, mechanical weeding used, competition period, crop duration, crop residues amount, etc.) in each of the crop in crop rotations will help disrupt different phases of the weed species' life cycle (Derksen et al. 2002). Thus, crop rotation helps in diversifying the selection pressures on weeds and favors diverse weed flora establishment due to varying management tactics, alternating patterns, and timings of tillage, light transmission, nutrients use, weed control methods practiced and reduced input use, including herbicides use and the costs (Liebman et al. 2014; Mishra et al. 2019; Liebman and Nichols 2020).

The cropping systems and crop production technologies will have to change, adjust, or be altered based on the understanding of the impact of climate change. The changing climatic conditions may provide growth benefits for certain weeds, at the same time as some crops are under additional stress related to variable heat and/or moisture. This may provide an enhanced competitive advantage for the weeds and hence there is a strong need for more resilient weed management strategies to be evolved and adopted.

3.5.4 Weed and Crop Diversity

The recent decline in weed diversity due to selection and predominance of fewer dominant weed species with similar resources requirement to crop may be attributed to continuous adoption of similar crop rotations in homogenized habitats, use of similar herbicides, inorganic fertilizers, and other inputs (Borgy et al. 2012). It was hypothesized that greater weed community diversity will lead to lesser predominance of competitive, adapted, herbicide resistant species and that the diversity of the weed seedbank indicates the overall sustainability of the cropping system (Storkey and Neve 2018). Diversified weed communities limited the negative effect of competitive and dominant species on crop productivity (Adeux et al. 2019).

The diversified cropping systems have the advantages of improved weed control (Goplen et al. 2018), reducing soil loss, nutrient discharge (Hunt et al. 2019), lesser greenhouse gas emissions (Hunt et al. 2020), and reduced herbicide-related aquatic toxicity (Hunt et al. 2017). The diversified cropping system may include multiple genotypes of the same crop or different crops grown in polyculture, legumes inclusion in cereal-dominated cropping systems and temporal and spatial crop rotations (Kremen and Miles 2012). The production systems diversification offers adaptation opportunities to climate change, which can be achieved through promotion of neglected and underutilized species (Mbow et al. 2019) such as buckwheat that grows fast and suppresses weeds (Adhikari et al. 2017). The species-rich forage mixtures suppress weeds (Sanderson et al. 2013). The increase in crop diversity exposes weeds to a number of stresses reducing the reliance on external chemicals (Smith and Gross 2007) for weed control in a cropping system. The crop diversification aimed at weed management should include crop sequences, species, and their management practices, which expose weeds to different stress and mortality factors while maximizing the availability of light, nutrients, water, and other resources by crop to reduce the losses caused by weeds (Liebman and Staver 2004).

The weed and crop diversity and their differences may form a basis for developing weed management strategies (Dekker 1997). The ecological approach focuses on the development of multiple-year diversified cropping systems that minimizes weeds. Thus, enhancing weed diversity, the characterization of weedy genetic and phenotypic diversity along with spatial distribution of weed populations; enhancement of location specific crop, cropping system, and agroecosystem diversity may be used for developing location specific EWM strategies.

3.5.5 Weed Suppressive Cover Crops Inclusion in the Cropping Systems

Cover crops are grown in the period between two main crops and are an important component of a system-oriented EWM approach (Kruidhof et al. 2008; Rao and Ladha 2011). Cover crops interfere with the development of weed populations through prevention of weed growth and development through niche preemption and occupying the space and utilizing the resources that would otherwise be available to weeds and by inhibiting or retarding germination and establishment of weeds with incorporated or soil surface-placed cover crop residues (mulches) due to both allelopathic and physical effects (Liebman and Davis 2000; Kruidhof et al. 2009). Among cover crops, grass cover species were found to provide greater weed suppression compared to broadleaved cover crops (Osipitan et al. 2019). The cover cropping in association with soil disturbance was found to cause the seedbank losses to germination (Mirsky et al. 2010). The consistent decrease in weed seedbank was observed with cover cropping system involving summer fallow and the yellow mustard/buckwheat/winter rape mixture as it stimulated weed seed germination and the germinated seedlings were either suppressed by cover crop or managed by tillage of three to four times throughout crop growth period (Mirsky et al. 2010).

The mixture of cover crop species with different characteristics performs better when composed of highly competitive species. The weed suppression by cover crops depends on choice of cover crop (Osipitan et al. 2019), seeding time (Baraibar et al. 2018), diversity (Finney et al. 2017; Sharma et al. 2021), termination timing (Wallace et al. 2017), and cover crops strategy. Cover crop-mediated EWM chiefly depends on cover crop species and management, the environment, cropping system, and the farmers' preferences. The major characteristics to consider before selecting cover crops for inclusion in a given cropping system include high initial growth rates for height and leaf area development to develop a close dense canopy and to cover the ground completely with dense vegetation for better competitive ability and weed suppression.

3.5.6 Intercropping

Traditionally in developing countries, two or more crop species or genotypes are cultivated together and coexist for a time. Intercrops yield similar or more crop yield than sole crops utilizing reduced inputs while reducing the weed menace by better using resources (space, water, light, nutrients) by crops. Sorghum in semi-arid and arid tropics intercropped with short-duration legume crops such as cowpea and greengram (*Vigna radiata*

L.) provides higher yield and reduced weed density and biomass due to quicker canopy development and greater light interception by intercrop (Rao and Shetty 1981). In rice, both rice and Sesbania crops are seeded together as intercrop and allowed to grow for 30 days. Subsequently, Sesbania intercrop is knocked down with 2, 4-D at 500 g ha^{-1} (Singh et al. 2007). The plant residues are left standing. This technology, called *brown manuring*, reduces weed density by nearly half without any adverse effect on rice yield. Sesbania surface mulch decomposes very fast to supply N and other recycled nutrients. The parasitic weed Striga was controlled when peanut (*Arachis hypogaea* L.), bean (*Phaseolus vulgaris* L.), chickpea (*Cicer arietinum* L.), and soybean were intercropped with corn due to lower light, lower temperature, and higher humidity (Oswald et al. 2002). In pearl millet-cowpea (1:1) intercropping, the lowest Striga density and shoot biomass was observed due to the suicidal germination of *Striga hermonthica* L. seeds resulting in Striga seedbank depletion of 46% and enhanced soil fertility. Hence, pearl millet intercropping with cowpea (1:1) was suggested for resource poor farmers to manage Striga and maximize grain yields in Striga endemic areas of Ghana (Asodewine 2020). An aqueous solution of silverleaf (*Desmodium uncinatum* L.) inhibited haustoria development of *S. hermonthica*, which was controlled when intercropped with *D. uncinatum* (Khan et al. 2002) indicating the role of allelopathy in intercrops efficacy in managing weeds. The infestation of *Orobanche crenata* L was reduced when berseem clover (*Trifolium alexandrinum* L.) was intercropped with pea and lentil (Fernández-Aparicio et al. 2010).

3.5.7 Enhancing Crop Competitiveness against Weeds

Improved crop competitiveness can be used in integration with other methods for EWM in agroecosystems. Enhanced crop competitiveness against weeds may be achieved by selecting the proper method of crop establishment, growing weed-competitive cultivars (possessing faster early growth and allelopathy potential), and using optimal crop density, seed rate, and proper direction of planting to make efficient use of resources in terms of space, light, moisture, and nutrients by the crop rather than weeds.

3.5.8 Method of Crop Establishment

Transplanted crop seedlings were found to be more competitive against weeds resulting in lower associated crop yield losses than direct-seeded crop seedlings as observed with rice (Rao et al. 2007). The puddling of soil, followed by stagnation of water, coupled with greater competitiveness of transplanted vigorous rice seedlings, provide effective weed control in the early vegetative rice growth stage. Thus, depending on resources availability and socioeconomic conditions prevailing, crop establishment method may help as method to enhance crop competitiveness. However, the transplanted rice is not environmentally benign as it uses more water, energy and leaves more carbon footprint (Ladha et al. 2016).

3.6 Crop Competitiveness

3.6.1 Competitive Crop Cultivars

Crop cultivars with greater seedling vigor, fast early growth, and rapid canopy closure have greater ability to suppress weeds through resource competition and allelopathy. Hence, they may play a key role as components of integrated EWM. Several of the crops have been reported to release allelochemicals and produce suppressive effects on weeds (Dilday et al. 1994; Weston 1996). Sorghum suppresses weeds under the field conditions due to its root hairs exudation of bioactive quantities of sorgoleone (Czarnota et al. 2001) and sorghum is most studied on its phytoallelopathic potential at molecular, physiological, and agroecological scales (Weston et al. 2013). The momilactone B secreted from rice roots into the neighboring environments during its entire life cycle at phytotoxic levels seems to account for the majority of the observed rice allelopathy (Noguchi and Peters 2013). The studies on genetic basis of allelopathy in rice revealed that allelopathy is quantitatively inherited (Dilday et al. 1998), polygenic, and a quantitative trait (Kong et al. 2011), and involves several loci and probably some levels of epistasis (Jensen et al. 2001). Traditional varieties of crops are viewed as more potent producers of allelochemicals (Masum et al. 2016) and hence are being used in breeding programs of wheat, rice, and other crops (Quader et al. 2001; Worthington and Reberg-Horton 2013) to explore their allelopathic potentiality for practical use (Seal et al. 2005). Allelopathic interactions in plants are mediated by chemical signaling mediated recognition mechanisms (Kong et al. 2018) and by recognition of neighboring, competing plant species, the allelochemical synthesis or enhanced exudation (allelobiosis) (Li et al. 2016), as observed with wheat in the presence of certain weeds (Kong et al. 2018). Detailed understanding of signaling compounds that mediate such inter species specific crop weed allelopathic interactions is essential (Hickman et al. 2021) to make the best practical use of allelopathy under field conditions.

The crop cultivars characterized by greater weed competitive ability include greater plant height, early canopy cover, high tillering nature, vertical leaf orientation, high biomass accumulation at the early crop stage, high leaf area index, and specific leaf area during vegetative growth, early vigor, and greater root biomass and volume. The crop competitiveness may be integrated with other methods, such as the use of herbicides and manual weeding for successfully managing weeds (Ramesh et al. 2017). Crop cultivars need to be bred for crop characteristics with increased weed-suppressive ability, high yielding potential, and an ability to thrive in changing climatic conditions (Korres et al. 2016) to integrate them in climate change adaptation technologies.

3.6.2 Quicker Canopy Closure

Quicker canopy closure controls light penetration to weed seeds at or near the soil surface, the emergence of many weed seedlings will be reduced (Bradley 2006). The narrow row spacing (Sardana et al. 2017), increasing seeding rate (Mhlanga et al. 2016), and doubling crop density (Lemerle et al. 2004) cause quick crop canopy cover with significant decrease in weed growth in the crops.

As the weeds compete with crop for same resource above and below ground, weed suppression could be achieved with balanced resource use depending on the prevailing location specific ecological conditions. The fertilizer application and time of planting (Fodor et al. 2008), water (Kaur et al. 2018) were reported to increase the competitiveness of the crops and appropriate fertilizer and water management (Kaur et al. 2018) was suggested for use as a component of integrated EWM in agroecosystems.

3.7 Soil Solarization

Soil solarization is a method of heating the soil's surface by using transparent low-density polyethylene (LDPE film) sheets placed on the soil's surface to trap solar radiation (Khan et al. 2003). The crop field is covered with transparent plastic sheet during summer months to increase the soil temperatures (Rubin and Gamliel 2018), which kills weed seed on soil surface and thereby reduces soil weed seedbank, germination of weeds, and overall weed density (Khan et al. 2003; Das and Yaduraju 2008), indicating that soil solarization may be used as component of EWM strategy. Some species of weeds are known to be quite tolerant to solarization and hence will affect the decision to use solarization as a component.

3.8 Mechanical Weed Management

The weed removal by hand tools or improved machinery proved to be an effective method of managing weeds. Environmental degradation and pollution caused by herbicides use is reduced by the use of mechanical weeder in addition to overcoming of the low effective operation, high work effort, and high time requirement for different types of hoes or cutlass (Reddy et al. 2018). The reduced time use will also reduce the environmental footprint of weeding operation. The innovative machines developed for physical weed control help to reduce costs and increase farmers income (Cloutier et al. 2007). The development of automatic weeding machinery may help in mechanical weed control through the selective use of weeders in crops. Advanced smart-machine technology for automatic weed control has been demonstrated in certain row crops (Fennimore et al. 2014). Robotic systems combined with computer vision and machine learning hold the potential in site-specific mechanical weed management. Currently commercial robotic site-specific weed management is being used in fallow land by using infra-red reflectance to perform "green-on-brown" detection. Efforts are being made to extend autonomous weeding capabilities to "green-on-green" detection – that is, detecting weeds amongst crop (Bender et al. 2020).

3.9 Biocontrol

The regulation of weeds by natural enemies is an important ecosystem service in arable lands (Shields et al. 2019). Classical biological control is the predominant method in weed biocontrol as it offers the safe, economic, and environmentally sustainable solution (McFadyen 1998). Thirty-six fungal pathogens have been authorized for introduction across 18 countries for the classical biological control of weeds (Morin 2020). The arthropods (insects, mites) and plant pathogens are widely used for targeted weed control (Charudattan 2010; Winston et al. 2014; Schwarzländer et al. 2018). Predators like carabids are shown to consume a wide range of weed species seeds in field studies (Kulkarni et al. 2015). The usage of ducks (Furuno 2001) and water birds (Kendig et al. 2003) was also found effective in managing weeds and hence may be used as a component of EWM.

The climate and associated variability in weather events in future may result in variability in weed control efficacy due to the changes in timing of bioagent activity and abundance (Harms and Cronin 2020). The climate change may have both positive and negative effects on biocontrol efficacy due to expected rapid evolution in both biocontrol agents and their target weeds under climate change (Sun et al. 2020). This necessitates ecological and evolutionary responses predicting framework that captures responses both bioagent and target weed by using approaches such as modeling and experimental evolution studies as it helps to predict and utilise the efficacy of biocontrol as a component of EWM.

3.10 Herbicide Use and EWM

EWM does not exclude the use of herbicides, but minimizes their use with the creation of weed-suppressive habitat (Liebman 2004). EWM utilizes all available information on biology and ecology of weeds, integrates all available weed management options including safe herbicide use based on sustainable ecological principles to enable farmer to make location specific decisions. The use of herbicides reduced GHG emissions by reducing the fuel consumption of farming machinery that would otherwise be used to manage weeds (Gianessi 2013). However, it was reported that the application of herbicides (quizalofop-p-ethyl and bentazone) on alfalfa increased soil GHG emissions (Shi et al. 2020). In contrast, Jiang et al. (2015) observed reduced N_2O emissions with the use of butachlor and bensulfuron-methyl in winter wheat. In rice fields, bensulfuron-methyl and pretilachlor reduced GHG emissions when applied alone while the combination of these two herbicides resulted in increase in the emissions of N_2O and CH_4 (Das et al. 2011). Research needs to be intensified on the impact of herbicides in GHG emissions and herbicides with lower environmental footprint need to be identified (Giudice et al. 2008) and recommended for use in different crops and cropping systems.

The rapid and widespread adoption of genetically modified (GM) crops tolerant to herbicides in the USA since 1996 to 2012 resulted in economic (equal to $21.7 billion additional farm income and a 225 million kg reduction in herbicide active ingredient use) and environmental benefits that farmers have derived from its use (a 225 million kg reduction in herbicide active ingredient use during 1996–2012) (Brookes and Barfoot 2017). The associated EIQ (indicator of environmental impact), is about 30% lower (Brimner et al. 2005) when compared to what can reasonably be expected if the area planted in GM HT crops reverted to conventional production methods (Brookes 2014). HT crops promote no-till or conservation tillage due to which even after the mortality of weeds and the no-till soil helps protects the soil from washoff as with less or no tilling, there would be less soil erosion with greater water retention and lesser greenhouse gas emissions (https://www.isaaa.org/resources /publications/pocketk/57/default.asp). Inclusion of GM crops tolerant to herbicides as a component of EWM should be based on the farmers need, socioeconomic aspects of the respective country in the world and after taking measures for strict adherence to recommended stewardship guidelines.

Weeds distribution within a crop field is not uniform. The blanket application of herbicides may lead to increased cost and associated adverse impacts of herbicide use. Hence, if the weeds presence is mapped and herbicides are applied only in the areas of weeds presence or where their density exceeded threshold density, large reductions in herbicide use would be possible (Mortensen et al. 1998). The site-specific precision weed management by efficient application of herbicide to weed patches only within the crop field (Jensen et al. 2012) has resulted in herbicide savings for winter oilseed rape (19–22%), sugar beet (46–57%), and winter wheat (60–77%) production (Gutjahr and Gerhards 2010). The use of on-farm sensors for real time management of crop performance (Aqeel-Ur-Rehman et al. 2014) may enhance EWM efficiency. In the fallow fields, using "green on brown" technology with WEED-It™ or WeedSeeker™ sprayers is becoming more widespread in Australia, and herbicide savings of up to 90% have been reported (Beckie et al. 2019). Real-time weed detection/recognition and control in crops ("green on green") requires seamless integration and high performance of sensors, data processing, and actuation systems (Beckie et al. 2019). The efforts are in progress on detection, mapping, guidance, and control technologies to pave the way for robotic weeds control by identifying weeds from both ground and aerial photographyand mechanically remove or to spray weeds with postemergence herbicides on the scale of individual plants (Olsen et al. 2019). These site-specific technological knowledge intensive weed management methods are will play key role in EWM in this era of climate change.

3.11 Conclusions

Ecological management of weeds stresses shifting the crop-weed balance in favor of crops by adapting an integrated strategy that modifies the weeds microenvironment placing stress at multiple growth stages in weeds life cycle to kill weed seeds, to increase the competitive ability of the crop for attaining optimum crop productivity, to minimize or eliminate weed seed production and dispersal, to accelerate weed seedbank depletion, and to judiciously use herbicides as the last resort rather than as the only resort (Rao and Ladha 2011). The cultural practices that are nonconducive to weeds growth based on weed biology, tillage and cultural practices, soil conditions manipulation, diversified crop rotations, competitive allelopathic crop cultivars, biocontrol agents, and other factors can be used to minimize weed germination, seedling emergence, growth, competition, reproduction, dispersal, and ultimately the adverse impact in agroecosystems.

In agroecosystems, shifts in weed communities would be the most important outcome of climate change (Peters et al. 2014). Thus, future EWM strategies must consider the management aspects of probable new invasive weeds that may replace currently important weed species. An understanding of the long-term ecological and evolutionary responses of weed populations to changing weed management, agricultural environments, and global climate will be beneficial to evolve future EWM technologies. It is essential for farming community to shift focus from heavy reliance on herbicides to reducing herbicide use by increasing their use efficiency by integrating their use with more eco-friendly inputs and EWM practices by cropping system-level redesign to enhance habitat characteristics and ecological interactions that favor crop growth rather than weed growth and reduce cost of cultivation while increasing the net profit. The ecological weed management tactics must be integrated with other agronomic farming practices that optimize health, crop productivity, and resources use efficiency; minimize the impacts of other pests and changing climatic factors influence; and reduce economic risks.

References

Adeux, G., Vieren, E., Carlesi, S. et al. (2019). Mitigating crop yield losses through weed diversity. *Nature Sustainability* 2: 1018–1026. https://doi.org/10.1038/s41893-019-0415-y.

Adhikari, L., Hussain, A., and Rasul, G. (2017). Tapping the potential of neglected and underutilized food crops for sustainable nutrition security in the mountains of Pakistan and Nepal. *Sustainability* 9 (2): 1–18. https://doi.org/10.3390/su9020291.

Ahmed, M. and Wardle, D.A. (1991). Increasing temperatures may enhance emergence and seedling growth of nodding thistle, summer grass and spiny emex. *Proceedings 44th New Zealand Weed and Pest Control Conference*, 288–291. https://doi.org/10.30843/nzpp.1991.44.10792.

Alberto, A., Ziska, L., Cervancia, C., and Manalo, P. (1996). The influence of increasing carbondioxide and temperature on competitive interactions between a C_3 crop, rice (*Oryza sativa*) and a C_4 weed (*Echinochloa glabrescens*). *Functional Plant Biology* 23 (6): 795–802. https://doi.org/10.1023/A:1005947802161.

Anderson, R.L. (2005). A multi-tactic approach to manage weed population dynamics in crop rotations. *Agronomy Journal* 97: 1579–1583. https://doi.org/10.2134/agronj2005.0194.

Aqeel-Ur-Rehman, Abbasi, A.Z., Islam, N., and Shaikh, Z.A. (2014). A review of wireless sensors and networks' applications in agriculture. *Computer Standards & Interfaces* 36: 263–270. https://doi.org/10.1016/j.csi.2011.03.004.

Archambault, D.J., Li, X., Robinson, D. et al. (2001). The effects of elevated CO_2 and temperature on herbicide efficacy and weed/crop competition. Prairie Adaptation Research Collaborative. https://calgarycmmc.com/E-books/E%20Books%20A-B-C/Co2%20Humidity.pdf. http://www.parc.ca (accessed 11 March 2021).

Asodewine, P.A. (2020). Effects of pearl millet (*Pennisetum glaucum* L.)–cowpea (Vigna unguiculata L.) intercrop on *Striga hermonthica* (DEL.) Benth, yield components and grain yield in the Sudan savannah zone of Ghana. M.Sc. Thesis submitted to the department of agronomy, faculty of agriculture, University for development studies, Ghana. accessed March 11, 2021, at https://afribary.com/works/effects-of-pearl-millet-pennisetum-glaucum-l-cowpea-vigna-unguiculata-l-intercrop-on-striga-hermonthica-del-benth-yield-components-and-grain-yield-in-the-sudan-savannah-zone-of-ghana.

Backlund, P., Schimel, D., Janetos, A. et al. (2008). Introduction. In: *The Effects of Climate Change on Agriculture, Land Resources, Water Resources, and Biodiversity*. A Report by the U.S. Climate Change Science Program and the Subcommittee on Global Change Research, 362. Washington, DC, USA. https://digital.library.unt.edu/ark:/67531/metadc12030/m2/1/high_res_d/sap4-3-final-all.pdf.

Bailey, S.W. (2004). Climate change and decreasing herbicide persistence. *Pest Management Science* 60 (2): 158–162. https://doi.org/10.1002/ps.785.

Bajwa, A.A., Chauhan, B.S., Farooq, M., and Adkins, S.W. (2016). What do we really know about alien plant invasion? A review of the invasion mechanism of one of the world's worst weeds. *Planta* 244: 39–57. https://doi.org/10.1007/s00425-016-2510-x.

Baraibar, B., Mortensen, D.A., Hunter, M.C. et al. (2018). Growing degree days and cover crop type explain weed biomass in winter cover crops. *Agronomy for Sustainable Development* 38: 65. https://doi.org/10.1007/s13593-018-0543-1.

Barbash, J.E., Thelin, G.P., Kolpin, D.W., and Gilliom, R.J. (2001). Major herbicides in ground water: results from the National Water-Quality Assessment. *Journal of Environmental Quality* 30 (3): 831–845. https://doi.org/10.2134/jeq2001.303831x.

Barrett, H. and Rose, D.C. (2020). Perceptions of the fourth agricultural revolution: what's in, what's out, and what consequences are anticipated? *SociologiaRuralis*. https://doi.org/10.1111/soru.12324.

Bazzaz, F.A. and Garbutt, K. (1988). The response of annuals in competitive neighbourhoods: effects of elevated CO_2. *Ecology* 69: 937–946. https://doi.org/10.2307/1941249.

Beckie, H.J., Ashworth, M.B., and Flower, K.C. (2019). Herbicide resistance management: recent developments and trends. *Plants* 8: 161. https://doi.org/10.3390/plants8060161.

Belgeri, A. (2013). Assessing the impact of parthenium weed (*Parthenium hysterophorus* L.) invasion and its management upon grassland community composition. Ph.D. thesis. The School of Agriculture and Food Sciences, The University of Queensland, Australia.

Bender, A., Whelan, B., and Sukkarieh, S. (2020). A high-resolution, multimodal data set for agricultural robotics: a Ladybird's-eye view of Brassica. *Journal of Field Robotics* 37 (1): 73–96. https://doi.org/10.1002/rob.21877.

Borgy, B., Gaba, S., Petit, S., and Reboud, X. (2012). Non-random distribution of weed species abundance in arable fields. *Weed Research* 52: 383–389. https://doi.org/10.1111/j.1365-3180.2012.00920.x.

Bosy, J.L. and Reader, R.J. (1995). Mechanisms underlying the suppression of forb seedling emergence by grass litter. *Functional Ecology* 9: 635–639. https://doi.org/10.2307/2390155.

Bradley, K.W. (2006). A review of the effects of row spacing on weed management in corn and soybean. *Plant Management Network*. Online. *Crop Management*. https://doi.org/10.1094/CM-2006-0227-02-RV.

Brimner, T., Gallivan, G., and Stephenson, G. (2005). Influence of herbicide-resistant canola on the environmental impact of weed management. *Pest Management Science* 61: 47–52. https://doi.org/10.1002/ps.967.

Brookes, G. (2014). Weed control changes and genetically modified herbicide tolerant crops in the USA 1996–2012. *GM Crops & Food* 5: 321–332. https://doi.org/10.4161/21645698.2014.958930.

Brookes, G. and Barfoot, P. (2017). GM crops: global socio-economic and environmental impacts 1996–2015. *GM Crops and Food* 8: 117–147. https://doi.org/10.1080/21645698.2017.1309490.

Buhler, D.D., Mester, T.C., and Kohler, K.A. (1996). The effect of maize residues and tillage on emergence of *Setaria faberi*, *Abutilon theophrasti*, *Amarathus retroflexus* and *Chenopodium album*. *Weed Research* 36: 153–165. https://doi.org/10.1111/j.1365-3180.1996.tb01811.x.

Bunce, J.A. and Ziska, L.H. (2000). Crop ecosystem responses to climatic change: crop/weed interactions. In: *Climate Change and Global Crop Productivity* (ed. K.R. Reddy and H.F. Hodges), 333–348. New York: CABI Publishing. https://doi.org/10.1079/9780851994390.0333.

Carmo-Silva, A.E., Soares, A.S., Da Silva, J.M. et al. (2007). Photosynthetic responses of three C_4 grasses of different metabolic subtypes to water deficit. *Functional Plant Biology* 34: 204–213. https://doi.org/10.1071/FP06278.

Chadha, A., Florentine, S., Chauhan, B.S. et al. (2019). Environmental factors affecting the germination and seedling emergence of two populations of an emerging agricultural weed: wild lettuce (*Lactuca serriola*). *Crop and Pasture Science* 70: 709–717. https://doi.org/10.1071/CP18594.

Chandrasena, N. (2009). How will weed management change under climate change? Some perspectives. *Journal of Crop and Weed* 5: 95–105. ISSN: 0974-6315.

Charudattan, R. (2010). A reflection on my research in weed biological control: using what we have learned for future applications. *Weed Technology* 24: 208–217. https://doi.org/10.1614/WT-D-09-00012.1.

Chauhan, B.S. and Johnson, D. (2009). Influence of tillage systems on weed seedling emergence pattern in rainfed rice. *Soil and Tillage Research* 106: 15–21. 10.1016/j.still.2009.10.004.

Chikoye, D., Ekeleme, F., and Udensi, E.U. (2001). *Imperata cylindrica* suppression by intercropping cover crops in *Zea mays*/*Manihot esculenta* systems. *Weed Science* 49: 658–667. https://doi.org/10.1614/0043-1745(2001)0490658:CSBICC2.0.CO;2.

Clements, D.R. and Ditommaso, A. (2011). Climate change and weed adaptation: can evolution of invasive plants lead to greater range expansion than forecasted? *Weed Research* 51: 227–240. https://doi.org/10.1111/j.1365-3180.2011.00850.x.

Cloutier, D.C., Van der Weide, R.Y., Peruzzi, A., and Leblanc, M.L. (2007). Mechanical weed management. In: *Nonchemical Weed Management. Principles, Concepts and Technology* (ed. M.K. Upadhyaya and R.E. Blackshaw), 111–134. Wallingford, UK: CAB International. ISBN-13: 978-1-84593-290-9.

Czarnota, M.A., Paul, R.N., Dayan, F.E. et al. (2001). Mode of action, localization of production, chemical nature, and activity of sorgoleone: a potent PSII inhibitor in *Sorghum* spp. root exudates. *Weed Technology* 15: 813–825. https://doi.org/10.1614/0890-037X(2001)0150813:MOALOP2.0.CO;2.

D'Antonio, C.M. and Chambers, J.C. (2006). Using ecological theory to manage or restore ecosystems affected by invasive plant species. In: *Foundations of Restoration Ecology* (ed. D.A. Falk, M.A. Palmer, and J.B. Zedler), 260–279. Washington: Society for Ecological Restoration International, Island Press.

Das, S., Ghosh, A., and Adhya, T.K. (2011). Nitrous oxide and methane emission from a flooded rice field as influenced by separate and combined application of herbicides bensulfuron methyl and pretilachlor. *Chemosphere* 84 (1): 54–62. https://doi.org/10.1016/j.chemosphere.2011.02.055.

Das, T.K. and Yaduraju, N.T. (2008). Effect of soil solarization and crophusbandry practices on weed species competition and dynamics in soybean wheat cropping system. *Indian Journal of Weed Science* 40: 1–5. ISSN: 0253-8050.

Davis, A.S. (2006). When does it make sense to target the weed seed bank? *Weed Science* 54: 558–565. https://doi.org/10.1614/WS-05-058R.1.

Davis, A.S. and Renner, K.A. (2007). Influence of seed depth and pathogens on fatal germination of velvetleaf (*Abutilon theophrasti*) and giant foxtail (*Setaria faberi*). *Weed Science* 55: 30–35. https://doi.org/10.1614/W-06-099.1.

Dekker, J. (1997). Weed diversity and weed management. *Weed Science* 45 (3): 357–363. https://doi.org/10.1017/S0043174500092985.

Délye, C., Jasieniuk, M., and Le Corre, V. (2013). Deciphering the evolution of herbicide resistance in weeds. *Trends in Genetics* 29: 649–658. https://doi.org/10.1016/j.tig.2013.06.001.

Derksen, D.A., Anderson, R.L., Blackshaw, R.E., and Maxwell, B. (2002). Weed dynamics and management strategies for cropping systems in the Northern Great Plains. *Agronomy Journal* 94 (2): 174–185. https://doi.org/10.2134/agronj2002.1740.

Dilday, R.H., Lin, J., and Yan, W.G. (1994). Identification of allelopathy in the USDA-ARS rice germplasm collection. *Australian Journal of Experimental Agriculture* 34 (7): 907–910. https://doi.org/10.1071/EA9940907.

Dilday, R.H., Yan, W.G., Moldenhauer, K.A.K., and Gravois, K.A. (1998). Allelopathic activity in rice for controlling major aquatic weeds. In: *Allelopathy in Rice* (ed. M. Olofsdotter), 7–26. Manila, Philippines: International Rice Research Institute. ISBN 971-22-0101-5.

Dippery, J.K., Tissue, D.T., Thomas, R.B., and Strain, B.R. (1995). Effects of lowand elevated CO_2 on C_3 and C_4 annuals. *Oecologia* 101: 13–20. https://doi.org/10.1007/BF00328895.

Dogan, M.N., Ünay, A., Boz, Ö., and Ögüt, D. (2009). Effect of pre-sowing and pre-emergence glyphosate applications on weeds in stale seedbed cotton. *Crop Protection* 28: 503–507. https://doi.org/10.1016/j.cropro.2009.01.013.

Dukes, J.S. and Mooney, H.A. (1999). Does global change increase the success of biological invaders? *Trends in Ecology and Evolution* 14: 135–139. https://doi.org/10.1016/S0169-5347(98)01554-7.

Dunn, R.J.H., Stanitski, D.M., Gobron, N., and Willett, K.M. (eds.) (2020). Global climate. In State of the climate in 2019. *Bulletin of the American Meteorological Society* 101 (8): S9–S127. https://doi.org/10.1175/BAMSD-20-0104.1.

Erenstein, O. and Laxmi, V. (2008). Zero tillage impacts in India's rice–wheat systems: areview. *Soil Tillage Research* 100: 1–14. https://doi.org/10.1016/j.still.2008.05.001.

FAO. (2018). The future of food and agriculture: alternative pathways to 2050. Food and Agriculture Organization of the United Nations (FAO), Rome, Italy, 228 p. ISBN 978-92-5-130989-6.

Fennimore, S.A., Smith, R.F., Tourte, L., LeStrange, M., and Rachuy, J.S. (2014). Evaluation and economics of a rotating cultivator in bok choy, celery, lettuce, and radicchio. *Weed Technology* 28: 176–188.

Fernández-Aparicio, M., Emeran, A.A., and Rubiales, D. (2010). Inter-cropping with berseem clover (*Trifolium alexandrinum*) reduces infection by *Orobanche crenata* in legumes. *Crop Protection* 29: 867–871. https://doi.org/10.1016/j.cropro.2010.03.004.

Finney, D.M., Murrell, E.G., White, C.M. et al. (2017). Ecosystem services and disservices are bundled in simple and diverse cover cropping systems. *Agriculture Environment Letters* 2: 170033. https://doi.org/10.2134/ael2017.09.0033.

Firbank, L.G., Elliott, J., Field, R.H. et al. (2018). Assessing the performance of commercial farms in England and Wales: lessons for supporting the sustainable intensification of agriculture. *Food and Energy Security* 7: e00150. https://doi.org/10.1002/fes3.150.

Fischlin, A., Midgley, G.F., Price, J. et al. (2007). Ecosystems, their properties, goods, and services, Chapter 4. In: *Climate Change 2007: Climate Change Impacts, Adaptation and Vulnerability* (ed. M. Perry, O. Canziani, J. Palutikof, P. van der Linden, and C. Hanson), The IPCC Fourth Assessment Report. Cambridge: Cambridge University Press. ISBN 978 0521 88010-7.

Fodor, L., Lehoczky, E., Fodorne, F.E. et al. (2008). Crop competitiveness influenced by seeding dates and top-dress nitrogen rates. *Communications in Agricultural and Applied Biological Sciences* 73 (4): 945–950. PMID:19226847.

Follak, S. and Essl, F. (2013). Spread dynamics and agricultural impact of *Sorghum halepense*, an emerging invasive species in Central Europe: spread dynamics and impact of *Sorghum halepense*. *Weed Research* 53: 53–60. https://doi.org/10.1111/j.1365-3180.2012.00952.x.

Franke, A., Singh, S., McRoberts, N. et al. (2007). *Phalaris minor* seedbank studies: longevity, seedling emergence and seed production as affected by tillage regime. *Weed Research* 47: 73–83. https://doi.org/10.1111/j.1365-3180.2007.00533.x.

Fuhrer, J. (2003). Agroecosystem responses to combinations of elevated CO_2, ozone, and global climate change. *Agriculture, Ecosystems & Environment* 97: 1–20. https://doi.org/10.1016/S0167-8809(03)00125-7.

Furuno, T. (2001). *The Power of Duck. Integrated Rice and Duck Farming*, 1e. Australia: Tagari Publications, 1–17. ISBN: 0908228120.

Gharde, Y., Singh, P.K., Dubey, R.P., and Gupta, P. (2018). Assessment of yield and economic losses in agriculture due to weeds in India. *Crop Protection* 107: 12–18. https://doi.org/10.1016/j.cropro.2018.01.007.

Gianessi, L.P. (2013). The increasing importance of herbicides in worldwide crop production. *Pest Management Science* 69 (10): 1099–1105. https://doi.org/10.1002/ps.3598.

Giudice, B.D., Massoudieh, A., Huang, X., and Young, T.M. (2008). A stochastic simulation procedure for selecting herbicides with minimum environmental impact. *Environmental Science and Technology* 42: 354–360. https://doi.org/10.1021/es070959n.

Goplen, J.J., Coulter, J.A., Sheaffer, C.C. et al. (2018). Economic performance of crop rotations in the presence of herbicide-resistant giant ragweed. *Agronomy Journal* 110: 260–268. https://doi.org/10.2134/agronj2016.09.0536.

Guo, P. and Al-Khatib, K. (2003). Temperature effects on germination and growth of redroot pigweed (*Amaranthus retroflexus*), Palmer amaranth (*Amaranthus palmeri*), and common waterhemp (*A. rudis*). *Weed Science* 51: 869–875. https://doi.org/10.1614/P2002-127.

Gutjahr, C. and Gerhards, R. (2010). Decision rules for site-specific weed management. In: *Precision Crop Protection – The Challenge and Use of Heterogeneity* (ed. E. Oerke, R. Gerhards, G. Menz, and R.A. Sikora), 223–240. Dordrecht: Springer. https://doi.org/10.1007/978-90-481-9277-9_14.

Haden, V.R., Duxbury, J.M., Ditommaso, A., and Losey, J.E. (2007). Weed community dynamics in the system of rice intensification (SRI) and the efficacy of mechanical cultivation and competitive rice cultivars for weed control in Indonesia. *Journal of Sustainable Agriculture* 30: 526. http://dx.doi.org/10.1300/J064v30n04_03.

Harms, N. and Cronin, J. (2020). Biological control agent attack timing and population variability, but not density, best explain target weed density across an environmental gradient. *Scientific Reports* 10: 11062.

Harrington, R.B. and Powles, S.B. (2012). Harrington seed destructor: a new nonchemical weed control tool for global grain crops. *Crop Science* 52: 1343–1347. https://doi.org/10.2135/cropsci2011.11.0608.

Harris, D. and Hossell, J.E. (2001). Weed management constraints under climate change. In: *Weed Management Constraints under Climate Change* (ed. C.R. Riches). Farnham, UK: British Crop Protection Council. ISBN: 1901396614.

Heijting, S., Van der Werf, W., and Kropff, M. (2009). Seed dispersal by forage harvesterand rigid-tine cultivator in maize. *Weed Research* 49: 153–163. https://doi.org/10.1111/j.1365-3180.2008.00681.x.

Hickey, L.T., Hafeez, A.N., Robinson, H. et al. (2019). Breeding crops to feed 10 billion. *Nature Biotechnology* 37: 744–754. https://doi.org/10.1038/s41587-019-0152-9.

Hickman, D.T., Rasmussen, A., Ritz, K. et al. (2021). Allelochemicals as multi-kingdom plant defence compounds: towards an integrated approach. *Pest Management Science* 77: 1121–1131. https://doi.org/10.1002/ps.6076.

Hobbs, P.R. (2007). Conservation agriculture: what is it and why is it important for future sustainable food production? *The Journal of Agricultural Science* 145: 127–137. https://doi.org/10.1017/S0021859607006892.

Hoffman, M.L., Owen, M.D., and Buhler, D.D. (1998). Effects of crop and weedmanagement on density and vertical distribution of weed seeds in soil. *Agronomy Journal* 90: 793–799. https://doi.org/10.2134/agronj1998.0002196200900060013xa.

Holm, L.G., Pancho, J.V., Herberger, J.P., Plucknett, D.L. (1991). *A Geographic Atlas of World Weeds*. Malabar, FL: Krieger Publishing Company.

Hoyle, J.A., McElroy, J.S., and Guertal, E.A. (2013). Soil texture and planting depth affect large crabgrass (*Digitaria sanguinalis*), Virginia buttonweed (*Diodia virginiana*), and cock's-comb kyllinga (*Kyllinga squamulata*) emergence. *Horticultural Science* 48: 33–636. https://doi.org/10.21273/hortsci.48.5.633.

Hunt, N., Hill, J., and Liebman, M. (2017). Reducing freshwater toxicity while maintaining weed control, profits, and productivity: effects of increased crop rotation diversity and reduced herbicide usage. *Environmental Science and Technology* 51: 1707–1717. https://doi.org/10.1021/acs.est.6b04086.

Hunt, N., Hill, J., and Liebman, M. (2019). Cropping system diversity effects on nutrient discharge, soil erosion, andagronomic performance. *Environmental Science and Technology* 53: 1344–1352. https://doi.org/10.1021/acs.est.8b02193.

Hunt, N.D., Liebman, M., Thakar, S.K., and Hill, J.D. (2020). Fossil energy use, climate change impacts, and air quality-related human health damages of conventional and diversified cropping systems in Iowa, USA. *Environmental Science and Technology* 54: 11002–11014. https://doi.org/10.1021/acs.est.9b06929.

IPCC (Intergovernmental Panel on Climate Change). (2008). Climate change 2007: synthesis report. Contribution of Working Groups I, II and III to the Fourth Assessment Report of the Intergovernmental Panel on Climate Change (ed. Core Writing Team, R.K. Pachauri, and A. Reisinger), 104 p. Geneva, Switzerland: IPCC. ISBN 92-9169-122-4.

Jennings, D.H., Cowan, I.R., and Farquhar, G.D. (1977). Stomatal function in relation to leaf metabolism and environment. In: *Integration of Activity in the Higher Plants* (ed. D.H. Jennings), 471–505. Cambridge, UK: Cambridge Univ Press. ISBN 0521216176.

Jensen, H.G., Jacobsen, L.B., Pedersen, S.M., and Tavella, E. (2012). Socioeconomic impact of widespread adoption of precision farming and controlled traffic systems in Denmark. *Precision Agriculture* 13: 661–677. https://doi.org/10.1007/s11119-012-9276-3.

Jensen, L.B., Courtois, B., Shen, L. et al. (2001). Locating genes controlling allelopathic effects against barnyardgrass in upland rice. *Agronomy Journal* 93 (1): 21–26. https://doi.org/10.2134/agronj2001.93121x.

Jiang, J.J., Chen, L.M., Sun, Q. et al. (2015). Application of herbicides is likely to reduce greenhouse gas (N_2O and CH_4) emissions from rice-wheat cropping systems. *Atmospheric Environment* 107: 62–69. https://doi.org/10.1016/j.atmosenv.2015.02.029.

Kanampiu, F.K., Kabambe, V., Massawe, C. et al. (2003). Multi-site, multi season field tests demonstrate that herbicide seed-coating herbicide-resistance maize controls *Striga* spp. and increases yields in several African countries. *Crop Protection* 22 (5): 697–706. https://doi.org/10.1016/S0261-2194(03)00007-3.

Kanatas, P., Travlos, I., Papastylianou, P. et al. (2020). Yield, quality and weed control in soybean crop as affected by several cultural and weed management practices. *Notulae Botanicae Horti Agrobotanici Cluj-Napoca* 48 (1): 329–341. https://doi.org/10.15835/nbha48111823.

Kathiresan, R.M., Gnanavel, I., Anbhazhagan, R. et al. (2005). Ecology and control of parthenium invasion in command area. *Proceedings of the Second International Conference on*

Parthenium Management, Bangalore, India (5–7 December 2005). 2: 77–79.

Kaur, S., Kaur, R., and Chauhan, B.S. (2018). Understanding crop-weed-fertilizer-water interactions and their implications for weed management in agricultural systems. *Crop Protection* 103: 65–72. https://doi.org/10.1016/j.cropro.2017.09.011.

Kendig, A., Williams, B., and Smith, C.W. (2003). Rice weed control. In: *Rice, Origin, History, Technology and Production* (ed. C.W. Smith and R.H. Dildey), 457–472. New Jersey: Wiley Pub. ISBN: 978-0-471-34516-9.

Khan, A.R., Srivastava, R.C., Ghorai, A.K., and Singh, S.R. (2003). Efficient soil solarization for weed control in the rain-fed upland rice ecosystem. *International Agrophysics* 17: 99–103. ISSN:0236-8722.

Khan, Z.R., Hassanali, A., Overholt, W. et al. (2002). Control of witch weed *Striga hermonthica* by intercropping with *Desmodium* Spp., and the mechanism defined as allelopathic. *Journal of Chemical Ecology* 28: 1871–1885. https://doi.org/10.1023/A:1020525521180.

Kong, C.H., Chen, X.H., Hu, F., and Zhang, S.Z. (2011). Breeding of commercially acceptable allelopathic rice cultivars in China. *Pest Management Science* 67 (9): 1100–1106. https://doi.org/10.1002/ps.2154.

Kong, C.H., Zhang, S.Z., Li, Y.H. et al (2018). Plant neighbour detection and allelochemical response are driven by root secreted signalling chemicals. *Nature Communications* 9: 3867. doi: 10.1038/s41467-018-06429-1.

Korres, N.E., Norsworthy, J.K., FitzSimons, T. et al. (2017). Differential response of Palmer amaranth (*Amaranthus palmeri*) gender to abiotic stress. *Weed Science* 65: 213–227. https://doi.org/10.1017/wsc.2016.34.

Korres, N.E., Norsworthy, J.K., Tehranchian, P. et al. (2016). Cultivars to face climate change effects on crops and weeds: a review. *Agronomy for Sustainable Development* 36: 12. https://doi.org/10.1007/s13593-016-0350-5.

Kremen, C. and Miles, A. (2012). Ecosystem services in biologically diversified versus conventional farming systems: benefits, externalities, and trade-offs. *Ecology and Society* 17 (4): 40. https://doi.org/10.5751/ES-05035-170440.

Kristian, P., Laura, B., and Bärbel, G. (2014). Impact of climate change on weeds in agriculture: a review. *Agronomy for Sustainable Development* 34: 707–721. https://doi.org/10.1007/s13593-014-0245-2.

Kriticos, D.J., Brunel, S., Ota, N. et al. (2015). Downscaling pest risk analyses: identifying current and future potentially suitable habitats for *Parthenium hysterophorus* with particular reference to Europe and North Africa. *PLoSOne* 10: e0132807. https://doi.org/10.1371/journal.pone.0132807.

Kriticos, D.J., Yonow, T., and McFadyen, R.E. (2005). The potential distribution of *Chromolaena odorata* (Siam weed) in relation to climate. *Weed Research* 45: 246–254. https://doi.org/10.1111/j.1365-3180.2005.00458.x.

Kropff, M.J. and Walter, H. (2000). EWRS and the challenges for weed research at the start of a new millennium. Weed Research 40: 7–10. https://doi.org/10.1046/j.1365-3180.2000.00166.x.

Kruidhof, H.M., Bastiaans, L., and Kropff, M.J. (2008). Ecological weed management by cover cropping: effects on weedgrowth in autumn and weed establishment in spring. *Weed Research* 48: 492–502. https://doi.org/10.1111/j.1365-3180.2008.00665.x.

Kruidhof, H.M., Bastiaans, L., and Kropff, M.J. (2009). Cover crop residue management for optimizing weed control. *Plant and Soil* 318: 169–184. https://doi.org/10.1007/s11104-008-9827-6.

Kudsk, P. and Kristensen, J.L. (1992). Effect of environmental factors on herbicide performance. *Proceedings of the First International Weed Control Congress*, Melbourne, 1: 173–186.

Kulkarni, S.S., Dosdall, L.M., and Willenborg, C.J. (2015). The role of ground beetles (Coleoptera: Carabidae) in weed seed consumption: a review. *Weed Science* 63: 355–376. https://doi.org/10.1614/WS-D-14-00067.1.

Ladha, J.K., Rao, A.N., Raman, A. et al. (2016). Agronomic improvements can make future cereal systems in South Asia far more productive and result in a lower environmental footprint. *Global Change Biology* 22 (3): 1054–1074. https://doi.org/10.1111/gcb.13143.

Lee, J.S. (2011). Combined effect of elevated CO_2 and temperature on the growth and phenology of two annual C_3 and C_4 weedy species. *Agriculture, Ecosystems & Environment* 140: 484–491. https://doi.org/10.1016/j.agee.2011.01.013.

Lemerle, D., Cousens, R., Gill, G. et al. (2004). Reliability of higher seeding rates of wheat for increased competitiveness with weeds in low rainfall environments. *The Journal of Agricultural Science* 142 (4): 395–409. https://doi.org/10.1017/S002185960400454X.

Lemmen, D.S., Warren, F.J., and Lacroix, J. (2008). Synthesis. In: *From Impacts to Adaptation: Canada in a Changing Climate 2007* (ed. D.S. Lemmen, F.J. Warren, J. Lacroix, and E. Bush), 1–20. Ottawa, ON: Government of Canada. https://www.nrcan.gc.ca/sites/www.nrcan.gc.ca/files/earthsciences/pdf/assess/2007/pdf/synth_e.pdf.

Lewis, S.E., Brodie, J.E., Bainbridge, Z.T. et al. (2009). Herbicides: a new threat to the Great Barrier Reef. *Environmental Pollution* 157: 2470–2484. https://doi.org/10.1016/j.envpol.2009.03.006.

Li, Y.H., Xia, Z.C., and Kong, C.H. (2016). Allelobiosis in the interference of allelopathic wheat with weeds. *Pest Management Science* 72: 2146–2121. https://doi.org/10.1002/ps.4246.

Liebman, M. (2004). Weed management: a need for ecological approaches. In: *Ecological Management of Agricultural Weeds* (ed. M. Liebman, C.L. Mohler, and C.P. Staver),

1–39. CambridgeCB2 2RU, UK: Cambridge University Press. ISBN 0-511-01764-2.

Liebman, M. and Davis, A. (2000). Integration of soil, crop and weed management in low-external-input farming systems. *Weed Research* 40: 27–48. https://doi.org/10.1046/j.1365-3180.2000.00164.x.

Liebman, M., Miller, Z.J., Williams, C.L. et al. (2014). Fates of *Setaria faberi* and *Abutilon theophrasti* seeds in three crop rotation systems. *Weed Research* 54: 293–306. https://doi.org/10.1111/wre.12069.

Liebman, M. and Mohler, C. (2004). Weeds and the soil environment. In: *Ecological Management of Agricultural Weeds* (ed. M. Liebman, C. Mohler, and C.P. Staver), 210–268. Cambridge, UK: Cambridge University Press. ISBN 0-511-01764-2.

Liebman, M. and Nichols, V.A. (2020). Cropping system redesign for improved weed management: a modeling approach illustrated with giant ragweed (*Ambrosia trifida*). *Agronomy* 10: 262. https://doi.org/10.3390/agronomy10020262.

Liebman, M. and Staver, C.P. (2004). Crop diversification for weed management. In: *Ecological Management of Agricultural Weeds* (ed. M. Liebman, C.L. Mohler, and C.P. Staver), 322–374. CambridgeCB2 2RU, UK: Cambridge University Press. ISBN 0-511-01764-2.

Lobell, D.B., Burke, M.B., Tebaldi, C. et al. (2008). Prioritizing climate change adaptation needs for food security in 2030. *Science* 319: 607–610. https://doi.org/10.1126/science.1152339.

Long, S.P. (1999). Environmental responses. In: *C4 Plant Biology* (ed. R.F. Sage and R.K. Monson), 215–249. San Diego: Academic Press. ISBN:978-0-12-614440-6.

Long, S.P., Ainsworth, E.A., Leakey, A.D.B. et al. (2006). Food for thought: lower-than-expected crop yield stimulation with rising CO_2 concentrations. *Science* 312 (5782): 1918–1921. https://doi.org/10.1126/science.1114722.

Lowder, S.K., Skoet, J., and Raney, T. (2016). The number, size, and distribution of farms, smallholder farms, and family farms worldwide. *World Development* 87: 16–29. https://doi.org/10.1016/j.worlddev.2015.10.041.

Mainali, K., Dhileepan, K., Warren, D. et al. (2015). Projecting future expansion of invasive species: comparing and improving methodologies. *Global Change Biology* 21: 4464–4480. https://doi.org/10.1111/gcb.13038.

Manea, A., Leishman, M.R., and Downey, P.O. (2011). Exotic C_4grasses have increased tolerance to glyphosate under elevated carbon dioxide. *Weed Science* 59: 28–36. https://doi.org/10.1614/WS-D-10-00080.1.

Masum, S.M., Hossain, M.A., Akamine, H. et al. (2016). Allelopathic potential of indigenous Bangladeshi rice varieties. *Weed Biology and Management* 16: 119–131. https://doi.org/10.1111/wbm.12103.

Matzrafi, M., Seiwert, B., Reemtsma, T. et al. (2016). Climate change increases the risk of herbicide-resistant weeds due to enhanced detoxification. *Planta* 244: 1217–1227. https://doi.org/10.1007/s00425-016-2577-4.

Mbow, C., Rosenzweig, C., Barioni, L.G. et al. (2019). Food security. In: *Climate Change and Land: An IPCC Special Report on Climate Change, Desertification, Land Degradation, Sustainable Land Management, Food Security, and Greenhouse Gas Fluxes in Terrestrial Ecosystems* (ed. P.R. Shukla, J. Skea, E. Calvo Buendia et al.). https://www.ipcc.ch/site/assets/uploads/sites/4/2020/02/SRCCL-Chapter-5.pdf.

McConnachie, A.J., Strathie, L.W., Mersie, W. et al. (2011). Current and potential geographical distribution of the invasive plant *Parthenium hysterophorus* (Asteraceae) in eastern and southern Africa. *Weed Research* 51: 71–84. https://doi.org/10.1111/j.1365-3180.2010.00820.x.

McDonald, A., Riha, S., DiTommaso, A., and DeGaetano, A. (2009). Climate change and the geography of weed damage: analysis of U.S. maize systems suggests the potential for significant rangetransformations. *Agriculture, Ecosystems, and Environment* 130: 131–140. https://doi.org/10.1016/j.agee.2008.12.007.

McFadyen, R.E.C. (1998). Biological control of weeds. *Annual Review of Entomology* 43: 369–393. https://doi.org/10.1146/annurev.ento.43.1.369.

Mendes, K.F., Régo, P.J.A., Takeshita, V., and Tornisielo, V.L. (2019). Water resource pollution by herbicide residues. In: *Organic Pollutants*. IntechOpen. https://doi.org/10.5772/intechopen.85159.

Mhlanga, B., Chauhan, B.S., and Thierfelder, C. (2016). Weed management in maize using crop competition: a review. *Crop Protection* 88: 28–36. https://doi.org/10.1016/j.cropro.2016.05.008.

Mirsky, S.B., Gallandt, E.R., Mortensen, D.A. et al. (2010). Reducing the germinable weed seedbank with soil disturbance and cover crops. *Weed Research* 50: 341–352. https://doi.org/10.1111/j.1365-3180.2010.00792.x.

Mishra, J.S., Kumar, R., Kumar, R. et al. (2019). Weed density and species composition in rice-based cropping systems as affected by tillage and crop rotation. *Indian Journal of Weed Science* 51: 116–122. https://doi.org/10.5958/0974-8164.2019.00027.3.

Mohamed, K.I., Papes, M., Williams, R. et al. (2006). Global invasive potential of 10 parasitic witchweeds and related Orobanchaceae. *Ambio* 35: 281–288. https://doi.org/10.1579/05-r-051r.1.

Mohler, C.L. (2001). Mechanical management of weeds. In: *Ecological Management of Agricultural Weeds* (ed. M. Liebman, C.L. Mohler, and C.P. Staver), 139–192. Cambridge, UK: Cambridge University Press. ISBN 0-511-01764-2.

Mohler, C.L. and Galford, A.E. (1997). Weed seedling emergence and seeds survival: separating the effects of seed position and soil modification by tillage. *Weed Research* 37 (3): 147–155. https://doi.org/10.1046/j.1365-3180.1997.d01-21.x.

Morin, L. (2020). Progress in biological control of weeds with plant pathogens. *Annual Review of Phytopathology* 58: 201–223. https://doi.org/10.1146/annurev-phyto-010820-012823.

Mortensen, D.A., Bastiaans, L., and Sattin, M. (2000). The role of ecology in the development of weed management systems: an outlook. *Weed Research* 40: 49–62. https://doi.org/10.1046/j.1365-3180.2000.00174.x.

Mortensen, D.A., Higley, L.G., Dieleman, J.A. et al. (1998). Ecological principles underlying integrated weed management systems. *Proceedings Weed Science Society of America* 38: 62.

Murphy, S.D., Clements, D.R., Belaoussoff, S. et al. (2006). Promotion of weed species diversity and reduction of weed seedbanks withconservation tillage and crop rotation. *Weed Science* 54: 69–77. https://doi.org/10.1614/WS-04-125R1.1.

Naidu, V.S.G.R. (2011). Impact of elevated CO_2 and temperature on crop-weed interaction. In: *Climate Change Impacts and Adaptations in Crop Plants* (ed. M. Singh, S. Khetrapal, R. Pandey, and P. Kumar), 157–168. New Delhi, India: Today and Tomorrow Printers and Publishers.

NAS and RS (The National academy of Sciences and The Royal Society). (2020). Climate change. Evidence & causes. Update 2020. An overview from the Royal Society and the US National Academy of Sciences. National academy of Sciences, U.S.A. and The Royal Society, U.K. https://royalsociety.org/~/media/royal_society_content/policy/projects/climate-evidence-causes/climate-change-evidence-causes.pdf.

Navie, S.C., McFadyen, R.E., Panetta, F.D., and Adkins, S.W. (2005). The effect of CO_2 enrichment on the growth of a C_3 weed (*Parthenium hysterophorus* L.) and its competitive interaction with a C_4 grass (*Cenchrus ciliaris* L.). *Plant Protection Quarterly* 20: 61–66.

Neve, P., Aiub, M.V., and Roux, F. (2009). Evolutionary thinking in agricultural weed management. *New Phytologist* 184: 783–793. https://doi.org/10.1111/j.1469-8137.2009.03034.x.

Nguyen, T.L.T. (2011). The invasive potential of parthenium weed (*Parthenium hysterophorus* L.) in Australia. Ph.D. Thesis, The School of Agriculture and Food Sciences, The University of Queensland, Australia. https://espace.library.uq.edu.au/view/UQ:259481.

Nikolić, N., Loddo, D., and Masin, R. (2021). Effect of crop residues on weed emergence. *Agronomy* 11: 163. https://doi.org/10.3390/agronomy11010163.

Niu, S., Yuan, Z., Zhang, Y. et al. (2005). Photosynthetic responses of C_3 and C_4 species to seasonal water variability and competition. *Journal of Experimental Botany* 56: 2867–2876. https://doi.org/10.1093/jxb/eri281.

Noguchi, H.K. and Peters, R.J. (2013). The role of momilactones in rice allelopathy. *Journal of Chemical Ecology* 39: 175–185. https://doi.org/10.1007/s10886-013-0236-9.

Norsworthy, J.K., Ward, S.M., Shaw, D.R. et al. (2012). Reducing the risks of herbicide resistance: best management practices and recommendations. *Weed Science* 60: 31–62. https://doi.org/10.1614/WS-D-11-00155.1.

O'Donnell, C.C. and Adkins, S.W. (2001). Wild oat and climate change: the effect of CO_2 concentration, temperature, and water deficit on the growth and developmentof wild oat in monoculture. *Weed Science* 49: 694–702. https://doi.org/10.1614/0043-1745(2001)0490694:WOACCT2.0.CO;2.

Oechel, W.C. and Strain, B.R. (1985). Native species responses to increased atmospheric carbon dioxide concentration. In: *Direct Effects of Increasing Carbon Dioxide on Vegetation*. (ed. B.R. Strain and J.D. Cure). Honolulu, HI: University Press of the Pacific. https://doi.org/10.2172/6134866.

Oerke, E. (2006). Crop losses to pests. *The Journal of Agricultural Science* 144: 31–43. https://doi.org/10.1017/S0021859605005708.

Olsen, A., Konovalov, D.A., Philippa, B. et al. (2019). Deep weeds: a multiclass weed species image dataset for deep learning. *Scientific Reports* 9: 2058. https://doi.org/10.1038/s41598-018-38343-3.

Opdam, P. and Wascher, D. (2004). Climate change meets habitat fragmentation: linking landscape and biogeographical scale levels in research and conservation. *Biological Conservation* 117: 285–297. https://doi.org/10.1016/j.biocon.2003.12.008.O.A.

Osipitan, J.A., Dille, Y., Assefa, E. et al. (2019). Impact of cover crop management on level of weed suppression: a meta-analysis. *Crop Science* 59: 833–842. https://doi.org/10.2135/cropsci2018.09.0589.

Osmond, C.B., Winter, K., and Ziegler, H. (1982). Functional significance of different pathways of CO_2 fixation in photosynthesis. In: *Physiological Plant Ecology II* (ed. O.L. Lange, P.S. Nobel, C.B. Osmond, and H. Ziegler), 479–547. Berlin Heidelberg: Springer. ISBN:978-3-642-68150-9.

Oswald, A., Ransom, J.K., Kroschel, J., and Sauerborn, J. (2002). Intercropping controls Striga in maize based farming systems. *Crop Protection* 21: 367–374. https://doi.org/10.1016/S0261-2194(01)00104-1.

Pandey, D.K., Palni, L.M.S., and Joshi, S.C. (2003). Growth, reproduction, and photosynthesis of ragweed parthenium (*Parthenium hysterophorus*). *Weed Science* 51: 191–201. https://doi.org/10.1614/0043-1745(2003)051[0191:GRAPOR]2.0.CO;2.

Parmesan, C. and Yohe, G. (2003). A globally coherent fingerprint of climate change impacts across natural systems. *Nature* 421: 37–42. https://doi.org/10.1038/nature01286.

Patterson, D.T. (1993). Implications of global climate change for impact of weeds, insects and plant diseases. In: *International Crop Science 1* (ed. D.R. Buxton, R. Shibles, R.A. Forsberg et al.), 273–280. 1 Crop Science Society of America, Inc. Book Series: ASA, CSSA, and SSSA Books. https://doi.org/10.2135/1993.internationalcropscience.c42.

Patterson, D.T. (1995a). Weeds in a changing climate. *Weed Science* 43: 685–701. https://doi.org/10.1017/S0043174500081832.

Patterson, D.T. (1995b). Effects of environmental stress on weed/crop interactions. *Weed Science* 28: 71–75. https://doi.org/10.1017/S0043174500081510.

Patterson, D.T. and Flint, E.P. (1980). Potential effects of global atmospheric CO_2 enrichment on the growth and competitiveness of C_3 and C_4 weed and crop plants. *Weed Science* 28: 71–75. https://doi.org/10.1017/S0043174500027806.

Patterson, D.T., Meyer, C.R., Flint, E.P., and Quimby, P.C., Jr. (1979). Temperature responses and potential distribution of itchgrass (*Rottboellia exaltata*) in the United States. *Weed Science* 27: 77–82. https://doi.org/10.1017/S0043174500043538.

Patterson, D.T., Westbrook, J.K., Joyce, R.J.V. et al. (1999). Weeds, insects, and diseases. *Climatic Change* 43 (4): 711–727. https://doi.org/10.1023/A:1005549400875.

Peters, K., Breitsameter, L., and Gerowitt, B. (2014). Impact of climate change on weeds in agriculture: a review. *Agronomy for Sustainable Development,* 34: 707–721. 10.1007/s13593-014-0245-2.

Pittelkow, C.M., Liang, X., Linquist, B.A. et al. (2015). Productivity limits and potentials of the principles of conservation agriculture. *Nature* 517: 365–368. https://doi.org/10.1038/nature13809.

Prati, D. and Bossdorf, O. (2004). Allelopathic inhibition of germination by *Alliaria petiolata* (Brassicaceae). American Journal of Botany 91: 285–288. https://doi.org/10.3732/ajb.91.2.285.

Qin, Z., Ditommaso, A., Wu, R.S., and Huang, H.Y. (2014). Potential distribution of two *Ambrosia* species in China underprojected climate change. *Weed Research* 54: 520–531. https://doi.org/10.1111/wre.12100.

Quader, M., Daggard, G., Barrow, R. et al. (2001). Allelopathy, DIMBOA production and genetic variability in accessions of Triticum speltoides. *Journal of Chemical Ecology* 27: 747–760. https://doi.org/10.1023/A:1010354019573.

Queiroz, L.R., Galvão, J.C.C., Cruz, J.C. et al. (2010). Weed suppression and organic green corn production in no tillage system. *Planta Daninha* 28 (2): 263–270. https://doi.org/10.1590/S0100-83582010000200005.

Ramesh, K., Rao, A.N., and Chauhan, B.S. (2017). Role of crop competition in managing weeds in rice, wheat, and maizein India: a review. *Crop Protection* 95: 4–21. https://doi.org/10.1016/j.cropro.2016.07.008.

Rao, A.N., Brainard, D.C., Kumar, V. et al. (2017). Preventive weed management in direct-seeded rice: targeting the weed seedbank. *Advances in Agronomy* 144: 45–142. https://doi.org/10.1016/bs.agron.2017.02.002.

Rao, A.N., Johnson, D.E., Sivaprasad, B. et al. (2007). Weed management in direct-seeded rice. *Advances in Agronomy* 93: 153–255. https://doi.org/10.1016/S0065-2113(06)93004-1.

Rao, A.N. and Ladha, J.K. (2011). Possible approaches for ecological weed management in direct-seeded rice in a changing world. *Proc. 23rd Asian-Pacific Weed Science Society Conference*, The Sebel Cairns, Australia, 444–453. ISBN Number: 978-0-9871961-0-1.

Rao, A.N. and Moody, K. (1990). Weed seed contaminationin rice seed. *Seed Science and Technology* 18: 139–146. ISSN 0251-0952.

Rao, A.N. and Moody, K. (1992). Competition between *Echinochloa glabrescens* and rice (*Oryza sativa*). *Tropical Pest Management* 38: 25–29. https://doi.org/10.1080/09670879209371640.

Rao, A.N. and Shetty, S.V.R. (1981). Investigation on weed suppressing ability of smother cropping systems in relation to canopy development and light interception. *Proc. 8th Asian-Pacific wee Science Society Conference*, Bangalore, India, 357–364. https://apwss.org/documents/8th%20Asian-pacific%20Weed%20Science%20Society%20Conference%20Proceedings%20Volume1.pdf.

Reader, R. (2011). Control of seedling emergence by ground cover: a potential mechanism involving seed predation. *Canadian Journal of Botany* 69: 2084–2087. https://doi.org/10.1139/b91-260.

Reddy, G.T., Harish, P., Nikil, A., and Asish, Ch. (2018). Power weeder. *Journal of Emerging Technologies and Innovative Research* 5: 573–582.

Rial-Lovera, K., Davies, P., and Cannon, N.D. (2017). Implications of climate change predictionsfor UK cropping and prospects for possible mitigation: a review of challenges and potential responses. *Journal of the Science and Food of Agriculture* 97: 17–32. https://doi.org/10.1002/jsfa.7767.

Rice, C., Wolf, J., Fleisher, D.H. et al. (2021). Recent CO_2 levels promote increased production of the toxin parthenin in an invasive *Parthenium hysterophorus* biotype. *Nature Plants* 7: 725–729. https://doi.org/10.1038/s41477-021-00938-6.

Rodenburg, J., Johnson, J.M., Dieng, I. et al. (2019). Status quo of chemical weed control in rice in sub-Saharan Africa. *Food Security* 11: 69–92. https://doi.org/10.1007/s12571-018-0878-0.

Rodenburg, J., Meinke, H., and Johnson, D.E. (2011). Challenges for weed management in African rice systems in a changing climate. *The Journal of Agricultural Science* 149: 427–435. https://doi.org/10.1017/S0021859611000207.

Rodrigues, W.P., Martins, M.Q., Fortunato, A.S. et al. (2016). Long-term elevated air CO_2 strengthens photosynthetic functioning and mitigates the impact ofsupra-optimal temperatures in tropical *Coffea arabica* and *C. canephora* species. *Global Change Biology* 22: 415–431. https://doi.org/10.1111/gcb.13088.

Roger, E., Duursma, D.E., Downey, P.O. et al. (2015). A tool to assess potential for alien plant establishment and expansion under climate change. *Journal of Environment Management* 159: 121–127. https://doi.org/10.1016/j.jenvman.2015.05.039.

Rose, D.C., Wheeler, R., Winter, M. et al. (2020). Agriculture 4.0: making it work for people, production, and the planet. *Land Use Policy* 100: 104933. https://doi.org/10.1016/j.landusepol.2020.104933.

Rosenzweig, C. and Hillel, D. (1998). *Climate Change and the Global Harvest*. Oxford: Oxford University Press, 1–324.

Rubin, B. and Gamliel, A. (2018). Soil solarization: a sustainable method for weed management. In: *Integrated Weed Management for Sustainable Agriculture* (ed. R.L. Zimdahl), 303–318. Cambridge, UK: Burleigh Dodds. ISBN: 978 1 78676 164 4.

Saebo, A. and Mortensen, L.M. (1998). Influence of elevated atmospheric CO_2 concentrationon common weeds in Scandinavian agriculture. *Acta Agriculturae Scandinavica* 48: 138–143. https://doi.org/10.1080/09064719809362491.

Sahoo, L., Schmidt, J.J., Pedersen, J.F. et al. (2010). Growth and fitness components of wild x cultivated *Sorghum bicolor* (Poaceae) hybrids in Nebraska. *American Journal of Botany* 97 (10): 1610–1617. https://doi.org/10.3732/ajb.0900170.

Sanderson, M.A., Brink, G., Stout, R., and Ruth, L. (2013). Grass–legume proportions in forage seed mixtures and effects on herbage yield and weed abundance. *Agronomy Journal* 105: 1289–1297. https://doi.org/10.2134/agronj2013.0131.

Sardana, S., Mahajan, G., Jabran, K., and Chauhan, B.S. (2017). Role of competition in managing weeds: an introduction to the special issue. *Crop Protection* 95: 1–7. https://doi.org/10.1016/j.cropro.2016.09.011.

Sasek, T.W. and Strain, B.R. (1991). Effects of CO_2 enrichment on the growth and morphology of a native and an introduced honeysuckle vine. *American Journal of Botany* 78: 69–75. https://doi.org/10.2307/2445229.

Sathishkumar, A., Srinivasan, G., Subramanian, E., and Rajesh, P. (2020). Role of allelopathy in G. Weed management: a review. *Agricultural Reviews* 41: 380–386. https://doi.org/10.18805/ag.R-2031.

Schütte, G., Eckerstorfer, M., Rastelli, V. et al. (2017). Herbicide resistance and biodiversity: agronomic and environmental aspects of genetically modified herbicide-resistant plants. *Environmental Sciences Europe* 29 (1): 5. https://doi.org/10.1186/s12302-016-0100-y.

Schwartz-Lazaro, L.M., Norsworthy, J.K., Walsh, M.J., and Bagavathiannan, M.V. (2017). Efficacy of the Integrated Harrington Seed Destructor on weeds of soybean and rice production systems in the Southern United States. *Crop Science* 57: 2812–2818. https://doi.org/10.2135/cropsci2017.03.0210.

Schwarzländer, M., Hinz, H.L., Winston, R.L., and Day, M.D. (2018). Biological control of weeds: an analysis of introductions, rates of establishment and estimates of success, worldwide. *Biocontrol* 63: 319–331. https://doi.org/10.1007/s10526-018-9890-8.

Scott, J.K., Murphy, H., Kriticos, D.J. et al. (2014). Weeds and climate change: supporting weed management adaptation. CSIRO, Australia. https://adaptnrm.csiro.au/wp-content/uploads/2014/08/Adapt-NRM_M2_WeedsTechGuide_5.1_LR.pdf.

Seal, A.N., Pratley, J.E., and Haig, T.J. (2005). Evaluation of rice varieties for allelopathic effects on Australian rice weeds – linking laboratory to field. In Establishing the Scientific Base. *Proceedings of the 4th World Congress on Allelopathy*, Wagga Wagga, New South Wales, Australia (21–26 August 2005), 164–167. ISBN (Print)1864671688.

Shabbir, A. (2012). Towards the improved management of parthenium weed: complementing biological control with plant suppression. PhD thesis. The School of Agriculture and Food Sciences, The University of Queensland, Australia.

Shabbir, A., Dhileepan, K., Khan, N., and Adkins, S.W. (2014). Weed-pathogen interactions and elevated CO_2: growth changes in favour of the biological control agent. *Weed Research* 54: 217–222. https://doi.org/10.1111/wre.12078.

Sharma, G., Shrestha, S., Kunwar, S., and Tseng, T.M. (2021). Crop diversification for improved weed management: a review. *Agriculture* 11: 461. https://doi.org/10.3390/agriculture11050461.

Shi, L., Guo, Y., Ning, J. et al. (2020). Herbicide applications increase greenhouse gas emissions of alfalfa pasture in the inland arid region of northwest China. *PeerJ* 25 (8): e9231. https://doi.org/10.7717/peerj.9231.

Shields, M.W., Johnson, A.C., Pandey, S. et al. (2019). History, current situation and challenges for conservation biological control. *Biological Control* 131: 25–35. https://doi.org/10.1016/j.biocontrol.2018.12.010.

Singh, S., Ladha, J.K., Gupta, R.K. et al. (2007). Evaluation of mulching, intercropping with *Sesbania* and herbicide use for weed management in dry-seeded rice (*Oryza sativa* L.). *Crop Protection* 26: 518–524. https://doi.org/10.1016/j.cropro.2006.04.024.

Smith, R.G. and Gross, K.L. (2007). Assembly of weed communities along a crop diversity gradient. *Journal of Applied Ecology* 44: 1046–1056. https://doi.org/10.1111/j.1365-2664.2007.01335.x.

Smith, S.D., Huxman, T.E., Zitzer, S.F. et al. (2000). Elevated CO_2 increases productivity and invasive species success in an arid ecosystem. *Nature* 408: 79–82. https://doi.org/10.1038/35040544.

Somerville, G.J., Powles, S.B., Walsh, M.J., and Renton, M. (2018). Modelling the impact of harvest weed seed control on herbicide-resistance evolution. *Weed Science* 66: 395–403. https://doi.org/10.1017/wsc.2018.9.

Steinmaus, S.C., Prather, T.S., and Holt, J.S. (2000). Estimation of base temperatures for nine weed species. *Journal of Experimental Botany* 51 (343): 275–286. https://doi.org/10.1093/jexbot/51.343.275.

Storkey, J. and Neve, P. (2018). What good is weed diversity? *Weed Research* 58 (4): 239–243. https://doi.org/10.1111/wre.12310.

Sun, Y., Ding, J., Siemann, E., and Keller, S.R. (2020). Biocontrol of invasive weeds under climate change: progress, challenges and management implications. *Current Opinion in Insect Science* 38: 72–78. https://doi.org/10.1016/j.cois.2020.02.003.

Teasdale, J.R. (1997). Cover crops, smother plants, and weed management. In: *Integrated Weed and Soil Management* (ed. J.L. Hatfield, D.D. Buhler, and B.A. Stewart), 247–270. Chelsea, MI: Ann Arbor. ISBN 9781575040691.

Tremmel, D.C. and Patterson, D.T. (1993). Responses of soybean and five weeds to CO_2 enrichment under two temperature regimes. *Canadian Journal of Plant Science* 73: 1249–1260. https://doi.org/10.4141/cjps93-164.

Tuesca, D., Puricelli, E., and Papa, J.C. (2001). A long-term study of weed flora shifts indifferent tillage systems. *Weed Research* 41: 369–382. https://doi.org/10.1046/j.1365-3180.2001.00245.x.

Valerio, M., Tomecek, M., Lovelli, S., and Ziska, L. (2011). Quantifying the effect of drought on carbon dioxide-induced changes in competition between a C_3 crop (tomato) and a C_4 weed (*Amaranthus retroflexus*). *Weed Research* 51: 591–600. https://doi.org/10.1111/j.1365-3180.2011.00874.x.

Valerio, M., Tomecek, M., Lovelli, S., and Ziska, L. (2013). Assessing the impact of increasing carbon dioxide and temperature on crop-weed interactions for tomato and a C_3 and C_4 weed species. *European Journal of Agronomy* 50: 60–65. https://doi.org/10.1016/j.eja.2013.05.006.

Varanasi, A., Vara Prasad, P.V., and Jugulam, M. (2016). Impact of climate change factors on weeds and herbicide efficacy. *Advances in Agronomy* 13: 107–146. https://doi.org/10.1016/bs.agron.2015.09.002.

Vilà, M., Beaury, E.M., Blumenthal, D.M. et al. (2021). Understanding the combined impacts of weeds and climate change on crops. *Environmental Research Letters* 16: 034043. https://doi.org/10.1088/1748-9326/abe14b.

Wallace, J., Williams, A., Liebert, J. et al. (2017). Cover Crop-based, organic rotational no-till corn and soybean production systems in the Mid-Atlantic United States. *Agriculture* 7: 34. https://doi.org/10.3390/agriculture7040034.

Walsh, M.J., Broster, J.C., Schwartz-Lazaro, L.M. et al. (2018). Opportunities and challenges for harvest weed seed control in global cropping systems. *Pest Management Science* 74: 2235–2245. https://doi.org/10.1002/ps.4802.

Walsh, M.J., Harrington, R.B., and Powles, S.B. (2012). Harrington seed destructor: a new nonchemical weed control tool for global grain crops. *Crop Science* 52: 1343–1347. https://doi.org/10.2135/cropsci2011.11.0608.

Walsh, M.J., Newman, P., and Powles, S. (2013). Targeting weed seeds in-crop: a new weed control paradigm for global agriculture. *Weed Technology* 27: 431–443. https://doi.org/10.1614/WT-D-12-00181.16.

Walthall, C.L., Anderson, C.J., Baumgard, L.H. et al. (2013). Climate change and agriculture in the United States: effects and adaptation. *Geological and Atmospheric Sciences Reports* 1. https://lib.dr.iastate.edu/ge_at_reports/1.

Wan, J., Oduor, A.M.O., Pouteau, R. et al. (2020). Can polyploidy confer invasive plants with a wider climatic tolerance? A test using *Solidago canadensis*. *Ecology and Evolution* 10: 5617–5630. https://doi.org/10.1002/ece3.6303.

Wang, X.Z. (2007). Effects of species richness and elevated carbondioxide on biomass accumulation: a synthesis usingmeta-analysis. *Oecologia* 152: 595–605. https://doi.org/10.1007/s00442-007-0691-5.

Wang, X.Z. and Mohan, J.E. (2008). Effects of global environmental changes on weeds. *CAB Reviews* 3: 067. https://doi.org/10.1079/PAVSNNR20083067.

Wang, X.Z., Ngigi, A.R., Smith, D.L., and McPeek, T.M. (2008). Biomass responses to intraspecific competition differ between wild species and crops grown at elevated CO_2. *Journal of Plant Ecology* 1: 25–32. https://doi.org/10.1093/jpe/rtm007.

Wang, X.Z. and Wan, J. (2020). Assessing the habitat suitability of 10 serious weed species in global crop lands. *Global Ecology and Conservation* 23: e01142. https://doi.org/10.1016/j.gecco.2020.e01142.

Waryszak, P., Lenz, T.I., Leishman, M.R., and Downey, P.O. (2018). Herbicide effectiveness in controlling invasive plants under elevated CO_2: sufficient evidence to rethink weeds management. *Journal of Environmental Management* 226: 400–407. https://doi.org/10.1016/j.jenvman.2018.08.050.

Watling, J.C. and Press, M.C. (2000). Infection with the parasitic angiosperm *Striga hermonthica* influences the response of the C_3 cereal *Oryza sativa* to elevated CO2. *Global Change Biology* 6: 919–930. 10.1046/j.1365-2486.2000.00366.x.

Wayne, P.S., Foster, S., Connolly, J. et al. (2002). Production of allergenic pollen by ragweed (*Ambrosia artemissiifolia* L.) is increased in CO_2-enriched atmospheres. *Annals of Allergy, Asthma andImmunology* 80: 669–679. https://doi.org/10.1016/S1081-1206(10)62009-1.

Weber, E. and Gut, D. (2005). A survey of weeds that are increasingly spreading in Europe. *Agronomy for Sustainable Development* 25: 109–121. https://doi.org/10.1051/agro:2004061.

Weisberger, V., Nichols, V., and Liebman, M. (2019). Does diversifying crop rotations suppress weeds? A meta-analysis. *PLoS ONE* 14: e0219847. https://doi.org/10.1371/journal.pone.0219847.

West, T.O. and Post, W.M. (2002). Soil organic carbon sequestration by tillage and crop rotation: a global data analysis. *Soil Science Society of America Journal* 66: 1930–1946. https://doi.org/10.2136/sssaj2002.1930.

Westerman, P.R., Liebman, M., Heggenstaller, A.H., and Forcella, F. (2006). Integrating measurements of seed availability and removal to estimate weed seed losses due to predation. *Weed Science* 54: 566–574. https://doi.org/10.1614/WS-05-067R.1.

Weston, L.A. (1996). Utilization of allelopathy for weed management in agroecosystems. *Agronomy Journal* 88 (6): 860–866. https://doi.org/10.2134/agronj1996.00021962003600060004x.

Weston, L.A., Alsaadawi, I.S., and Baerson, S.R. (2013). Sorghum allelopathy—from ecosystem to molecule. *Journal of Chemical Ecology* 39: 142–153. https://doi.org/10.1007/s10886-013-0245-8.

Wilson, B.J., Wright, K.J., Brain, P., Clements, M., and Stephens, E. (1995). Predicting the Competitive Effects of weed and crop density onweed biomass, weed seed production and crop yield in wheat. *Weed Research* 35: 265–278. 10.1111/j.1365-3180.1995.tb01789.x.

Winston, R.L., Schwarzländer, M., Hinz, H.L. et al. (eds.) (2014). *Biological Control of Weeds: A World Catalogue of Agents and Their Target Weeds*, 5e. Morgantown, WV: USDA For. Serv. For.Health Technol. Enterp. Team. https://www.ibiocontrol.org/catalog/index.cfm.

Wolfe, D.W., Ziska, L., Petzoldt, C. et al. (2008). Projected change in climate thresholds in the North eastern U.S.: implications for crops, pests, livestock, and farmers. *Mitigation and Adaptation Strategies for Global Change* 13: 555–575. https://doi.org/10.1007/s11027-007-9125-2.

Worthington, M. and Reberg-Horton, S.C. (2013). Breeding cereal crops for enhanced weed suppression: optimizing allelopathy and competitive ability. *Journal of Chemical Ecology* 39: 213–231. https://doi.org/10.1007/s10886-013-0247-6.

Yang, Q., Wang, X., Shen, Y., and Philp, J. (2013). Functional diversity of soil microbial communities in response to tillage and crop residue retention in an eroded loess soil. *Soil Science and Plant Nutrition* 59: 311–321. https://doi.org/10.1080/00380768.2013.775004.

Zhao, J., Yang, Y., Zhang, K. et al. (2020). Does crop rotation yield more in China? A meta-analysis. *Field Crops Research* 24: 107659. https://doi.org/10.1016/j.fcr.2019.107659.

Zhu, C., Zeng, Q., Ziska, L. et al. (2008). Effectof nitrogen supply on carbon dioxide–induced changes in competition between rice and barnyardgrass (*Echinochloa crus-galli*). *Weed Science* 56 (1): 66–71. https://doi.org/10.1614/WS-07-088.1.

Ziska, L.H. (2000). The impact of elevated CO_2 on yield loss from a C_3 and C_4 weed in field-grown soybean. *Global Change Biology* 6: 899–905. https://doi.org/10.1046/j.1365-2486.2000.00364.x.

Ziska, L.H. (2001). Changes in competitive ability between a C_4 crop and a C_3 weed with elevated carbon dioxide. *Weed Science* 49 (5): 622–627. https://doi.org/10.1614/0043-1745(2001)0490622:CICABA2.0.CO;2.

Ziska, L.H. (2002). Influence of rising atmospheric CO_2 since 1900 on early growth and photosynthetic response of a noxious invasive weed, Canada thistle (*Cirsium arvense*). *Functional Plant Biology* 29: 1387–1392. https://doi.org/10.1071/FP02052.

Ziska, L.H. (2003a). Evaluation of the growth response of six invasive species to past, present and future atmospheric carbon dioxide. *Journal of Experimental Botany* 54 (381): 395–404. 10.1093/jxb/erg027.

Ziska, L.H. (2003b). Evaluation of yield loss in field sorghum from a C_3 and C_4 weed with increasing CO_2. *Weed Science* 51 (6): 914–918. https://doi.org/10.1614/WS-03-002R.

Ziska, L.H. (2016). The role of climate change and increasing atmospheric carbon dioxide on weed management: herbicide efficacy. *Agriculture, Ecosystems & Environment* 231: 304–309. https://doi.org/10.1016/j.agee.2016.07.014.

Ziska, L.H., Blumenthal, D., and Franks, S. (2019). Understanding the nexus of rising CO_2, climate change, and evolution in weed biology. *Invasive Plant Science and Management* 12 (2): 79–88. https://doi.org/10.1017/inp.2019.12.

Ziska, L.H. and Bunce, J.A. (1997). Influence of increasing carbon dioxide concentration on the photosynthetic and growth stimulation of selected C_4 crops and weeds. *Photosynthesis Research* 54: 199–208. https://doi.org/10.1023/A:1005947802161.

Ziska, L.H. and Bunce, J.A. (1998). The influence of increasing growthtemperature and CO_2 concentration on the ratio of respiration to photosynthesis in soybean seedlings. *Global Change Biology* 4 (6): 637–643. https://doi.org/10.1046/j.1365-2486.1998.00179.x.

Ziska, L.H. and Dukes, J.S. (2011). *Weed Biology and Climate Change*. Hoboken, NJ: Wiley-Blackwell. ISBN: 978-0-470-95867-4.

Ziska, L.H., Faulkner, S., and Lydon, J. (2004). Changes in biomass and root: shootratio of field-grown Canada thistle (*Cirsium arvense*),a noxious, invasive weed, with elevated CO_2: implications for control with glyphosate. *Weed Science* 52 (4): 584–588. https://doi.org/10.1614/WS-03-161R.

Ziska, L.H., Gebhard, D.E., Frenz, D.A. et al. (2003). Cities as harbingers of climate change: common ragweed, urbanization, and public health. *The Journal of Allergy and Clinical Immunology* 111: 290–295. https://doi.org/10.1067/mai.2003.53.

Ziska, L.H. and George, K. (2004). Rising carbon dioxide and invasive, noxious plants: potential threats and consequences. *World Resource Review* 16: 427–447. https://www.ars.usda.gov/ARSUserFiles/80420520/FullTextPublicationspdf/Publications/ziska/potentialthreats.pdf.

Ziska, L.H. and Goins, E.W. (2006). Elevated atmospheric carbon dioxide and weed populations in glyphosate treated soybean. *Crop Science* 46: 1354–1359. https://doi.org/10.2135/cropsci2005.10-0378.

Ziska, L.H. and McClung, A. (2008). Differential response of cultivated and weedy (red) rice to recent and projected increases in atmospheric carbon dioxide. *Agronomy Journal* 100 (5): 1259–1263. https://doi.org/10.2134/agronj2007.0324.

Ziska, L.H. and Runion, G.B. (2006). Future weed, pest and disease problems for plants. In: *Agroecosystems in a Changing Climate* (ed. P.C.D. Newton, A. Carran, G.R. Edwards, and P.A. Niklaus), 262–279. Boston: CRC. ISBN: 9780849320.

Ziska, L.H. and Teasdale, J.R. (2001). Sustained growth and increased tolerance to glyphosate observed in a C_3, quackgrass (*Elytrigia repens*), grown at elevated carbon dioxide. *Functional Plant Biology* 27 (2): 159–166. https://doi.org/10.1071/PP99099.

Ziska, L.H., Teasdale, J.R., and Bunce, J.A. (1999). Future atmospheric carbon dioxide concentrations may increase tolerance to glyphosate. *Weed Science* 47: 608–615. https://doi.org/10.1017/S0043174500092341.

Ziska, L.H., Tomecek, M.B., and Gealy, D.R. (2010). Competitive interaction between cultivated and red rice as a function of recent and projected increases in atmospheric carbon dioxide. *Agronomy Journal* 102 (1): 118–123. https://doi.org/10.2134/agronj2009.0205.

4

The Ecological Base of Nonchemical Weed Control

Iraj Nosratti[1,*] *and Bhagirath S. Chauhan*[2]

[1] *Department of Plant Production and Genetics, Faculty of Agriculture, Razi University, Kermanshah, Iran*
[2] *Queensland Alliance for Agriculture and Food Innovation (QAAFI) and School of Agriculture and Food Sciences (SAFS), The University of Queensland, Gatton, Queensland, Australia*
[*] *Corresponding author*

4.0 Introduction

Historically, a variety of methods have been employed to control weeds. However, in the last decades, chemical control of weeds has become the most important and common method of weed control worldwide. Nevertheless, there are few areas where herbicides have not been used (Gianessi 2013).

Initially, it was thought that herbicides were the ultimate and definitive solution to weed control in agricultural and nonagricultural lands. However, only a few years after the introduction of herbicides, several important weed species evolved resistance to the herbicides. This unfavorable result was then intensified by the widespread use of herbicides (Beckie 2020).

What has made the chemical management of herbicide-resistant biotypes of weeds virtually impossible is the emergence of species with multiple and cross-resistance to herbicides. The International Herbicide-Resistant Weed Database has recorded more than 100 weed species globally that have become resistant to at least two different herbicide sites of action. For example, there are some species that are resistant to 12 herbicide sites of action (Heap 2020).

In addition, the growing public demand for organic products and environmental concerns have led to serious concerns about the continued widespread use of herbicides (Chhokar and Sharma 2008). However, the most important reason for the adoption of nonchemical methods is that some important weeds can no longer be controlled by any herbicides. Therefore, long-term and sustainable weed control solutions should include nonchemical control methods in the weed management programs. Nonchemical practices can be split into two groups: direct and indirect.

The direct weed control method is divided into two main categories namely physical and biological. Some major instances of physical control measures in practice include mulching, as well as thermal and mechanical (hand and machinery weeding) weed control. Using insects, microorganisms, and their products and allelopathy are some instances of biological weed control methods. Obviously, applying such methods is easy and can target weeds invading field crops. The main challenge in implementing direct weed control means is the safety of crops, in which the level of crop injury varies, depending on the type of crop and environmental conditions. Furthermore, the crop loss due to weed competition during early weed growth and utilizing weed control methods further increase the cost of crop production. Generally, the choice of direct control methods depend on many factors such as the type of crop and soil, weed species and its density, the labor required, and the purchase costs required for the control method.

Direct control method should be accompanied by long-term prevention methods so that weed densities can be maintained at an acceptable level. In many cases, weed control measures need to be repeated due to late emergence of weeds, as well as regrowth in perennial weeds, especially in low-competitive crops. Biological weed control also has its advantages and disadvantages, among which its disadvantages are the impossibility of its use in all agricultural systems, including in the one-year crop cultivation system. Collectively, the most important disadvantages of direct control methods are related to the high cost of these methods and the risk of failure to control weeds in all conditions (Bond et al. 2003).

The indirect weed control method is any operation that changes the environment against weeds while favoring

Ecologically Based Weed Management: Concepts, Challenges, and Limitations, First Edition. Edited by Nicholas E. Korres, Ilias S. Travlos, and Thomas K. Gitsopoulos.

crop growth, cultivation operations including tillage, crop agronomy (crop cultivar, density, planting date, arrangement, etc.), cover crops, and sanitation measures are all indirect tools that can be adopted for managing weeds. Therefore, indirect weed control is more economical and environmentally friendly. In this method, the balance between weeds and crops is changed so that cultural practices are in favor of the crop (Bond et al. 2003). Understanding how different nonchemical methods affect weed ecology can play a critical role in organizing a more effective weed management strategy. Research directed at weeds has revealed important insights into different dimensions of weed ecology, including population equilibria, density-dependent mortality and growth stages. These insights have been used to significantly improve, on occasion, the efficacy of nonchemical weed management practices. Therefore, the characteristics of the research programs that are required to enhance the understanding of the ecological principles underlying nonchemical methods, particularly cultural methods, have to be examined to determine how to best improve the design of integrated weed management plans.

The primary goal of cultural control method is to increase crop competitiveness against weeds (Blackshaw et al. 2007). It has been well established that there are large numbers of weed seeds (or asexual reproductive organs) in the soil that act as the most important source of weed contamination in agricultural fields (Christoffoleti and Caetano 1998; Haring and Flessner 2018). Cultivating operations, such as tillage and crop characteristics, may stimulate germination of weed seeds and lead to severe infestation of the field (Ball 1992).

This chapter will examine various research into weed ecology to elucidate new strategies to address weed management problems and will investigate the possibility of linking the effects of non-chemical methods on weed control using ecological principles.

4.1 Physical Weed Control

Mechanical, thermal, and mulching strategies are the three main methods of physical weed control, as they directly target weeds. In contrast to other pests, selective control of weeds within crops by direct physical control means is challenging.

4.1.1 Mechanical Weed Control

A wide range of implements, from hand-hoeing to progressive cultivators, are used for mechanical control of weeds. Despite recent advances in weed control techniques, hand weeding still is a common weed control approach for many organic agronomic and horticultural crops, especially those of high economic values (Yamada and Nemoto 2020). However, this traditional method is time-consuming, hard to practice, inefficient against perennial weeds, and uneconomic (Rose 2019). Hence, mechanization of weed control can lead to offsetting the drawbacks associated with hand weeding, while still taking advantage of nonchemical weed control methods (Alba et al. 2020). The main goal of implementing cultivators is to control seedlings in the earliest development stages, but they do not work against perennial weeds or stronger plants (Armengot et al. 2013).

Nowadays, rotary hoes or harrows are the most popular nonchemical control options for the early season removal of small-seeded weed species in organic farming (Leblanc et al. 2006; Armengot et al. 2013). The mechanism of harrowing is uprooting of weakly established and small weeds, while utilizing the inter-row cultivators is common for control of bigger weeds. The inter-row cultivator is able to bury weeds, dig them out, or break them apart (Alba et al. 2020).

Cultivating between rows in wide row crops, like maize, soybean, sunflower, and sugar beet, reduces herbicide use substantially across the globe, as these crops are cultivated on a large scale (Loddo et al. 2020). In addition to weed control purposes, mechanical methods can incorporate fertilizers and reduce evaporation of water, increase soil aeration, and induce mineralization to enhance crop productivity (Barbosa et al. 2019).

4.1.2 Harrows and Rotary Hoes

The challenge to weed harrowing and rotary hoeing is avoiding crop damage, which limits the selectivity of these implements. So, in many parts of the world, harrows and rotary hoes are commonly used pre-emergence, or in very early growth stages (Place et al. 2010). Differences between the size of the crop and weed plants (large crop plants and small weeds), as well as the soil or climate conditions can help with the selective removal of weeds within the fields (Kurstjens and Perdok 2000; Rasmussen et al. 2008). Relatively dry soil, fine soil structure, light soil texture, absence of stones, and planar topography are favorable for shallow tillage operations by harrows and rotary hoes (Hussain et al. 2018).

Harrowing controls weeds by burying and uprooting small seedlings. Hence, the less area covered by the crop, the higher the selectivity of the harrow (Rasmussen 1991; Kurstjens and Kropff 2001). Likewise, rotary hoes are usually applied to row crops like in maize and soybean. Rotary hoes are less destructive in comparison with harrows. Rotary hoes control weeds by penetrating the hoes into soil and digging out small weeds. Therefore, if the crop plants are rooted deeper than weed seedlings, this equipment can selectively remove weeds (Upadhyaya and Blackshaw 2007).

4.1.3 Inter-row Cultivation

The main removal mechanisms of inter-row comprise of burying and uprooting the weeds (Kolb et al. 2010; Loddo et al. 2020). However, the effectiveness of inter-row cultivation, as for other mechanical weed control methods, is dependent on soil and climate conditions (Melander et al. 2005). Furthermore, the inter-row cultivators work best between crop rows of at least 17 cm distance (Testa et al. 2016; Peruzzi et al. 2017). Various forms of inter-row cultivators, such as sweeps, horizontal rotating brushes, and shovels, have been used to control weeds in row crops like sugar beets and vegetables for many years (Upadhyaya and Blackshaw 2007). These mechanical tools remove weeds between rows of crops to a depth of 5 cm, so there is no time limit for using this equipment and it can be implemented at any growth stage of the crops.

4.1.4 Intra-row Cultivation

Mechanical methods for controlling weeds within crop rows is difficult, but new equipment have emerged, such as finger-weeders, torsion-weed finger weeders, torsion weeders, and cycloid hoe for intra-row weed control (Van der Weide et al. 2005, 2008; Rueda-Ayala et al. 2010; Peruzzi et al. 2017). The cycloid hoe is equipped with sensors for plant recognition, while the other three machines did not profit from the high technology and are simplified (Rueda-Ayala et al. 2010).

Generally, regrowth of weeds after treatment, or late germination of the weeds, makes repetition of mechanical weeding inevitable in row crops, particularly in those with open canopy structures and low competitive ability (Van der Weide et al. 2008). In confirmation of this content, Alba et al. (2020) evaluated the single and combined effects of rotary hoeing, harrowing, and inter-row cultivation on weed control in lentil. They found that across the four site-years and mechanical devices, increasing the number of mechanical weed control operations resulted in increased weed control (Figure 4.1). On average, applying one, two, and three mechanical weed control operations resulted in 48%, 64%, and 74% reductions in weed biomass, respectively (Figure 4.1).

4.1.5 Innovative Implements for In-row Crops

Removal of weeds between crop rows by mechanical cultivation is simple and efficient. However, selectively killing weed seedlings within the row (intra-row weeding) is difficult as the hoe must choose weeds among the weed and crop mix population (Pallottino et al. 2018). Modern technologies, such as digital imaging, sensing, robotics, as well as remote and proximal sensing, have been used to distinguish weeds from crops (Srinivasan 2006; Burgos-Artizzu et al. 2010). Real-time proximal sensing using optical sensors can be used for both detecting and targeting weeds (Lee et al. 1999). Remote sensing is reliant on maps for site-specific weed control and requires intensive post-acquisition image processing, which makes it uneconomic to small farms. In recent years, precision agriculture technology has worked in the proximal sensing mode, predominantly with optical sensors (Adamchuk et al. 2018). For site-specific control on a small scale, the use of digital cameras, either spectral or optical sensor systems, is more interesting for monitoring weeds (Downey et al. 2004; King 2017). These systems offer greater flexibility for a broad range of applications and can increase the accuracy of agricultural practices and enhance the performance of specific tasks, such as weeding (Pallottino et al. 2018).

4.1.6 Cutting and Mowing

Cutting involves the mechanical removal of the aerial parts of weed plants (Hatcher and Melander 2003). In wide-row crops like maize, soybean, and potato, weed mowing can be implemented easily (Van der Weide et al. 2008). It is well established that cutting or mowing weeds frequently is needed to efficiently control perennial weeds (Graglia et al. 2006; Rueda-Ayala et al. 2010). Cutting typically is not effective against mature plants as it cannot inhibit reproduction of seeds (Mirsky et al. 2009).

In general, mechanical weed control plays a unique role in nonchemical weed management strategies and is applied at very early weed growth stages. This makes its efficacy and selectivity challenging as crop plants are also

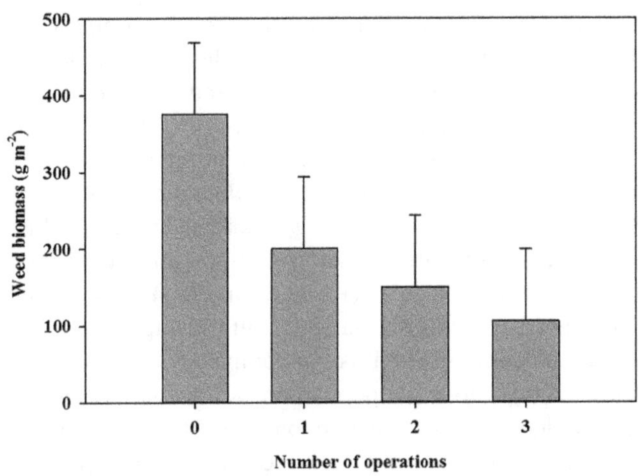

Figure 4.1 The relationship between the number of mechanical weed control operations and weed biomass in lentil averaged over all site-years at Kernen Crop Research Farm and Goodale Research Farm in 2016 and 2017. Adapted from Alba et al. (2020).

sensitive to weed control operations. Upper soil layer characteristics, weed densities, growth stages, and weather conditions influence mechanical weed control efficacy (Hatcher and Melander 2003). Mechanical tools are not able to effectively control weeds growing close to crop plants or within high weed densities (Nichols et al. 2015). The effectiveness of mechanical weed control, particularly by cultivators, on weeds is greatly dependent on the relative biomass of both the crop and the weed seeds. Crops with large seeds are usually competitive with weeds under this control measure because the initial rate of growth is fast enough that they escape from being buried. In addition, with large-seeded crops, it is possible to sow seeds deeper so that annual weeds germinating in the upper layer of the soil can be destroyed by mechanical means without harming the crop.

4.2 Soil Tillage

Soil tillage before planting (primary tillage) and during crop season (secondary tillage) is considered an early weed control measure (Blackshaw et al. 2007; Schutte et al. 2014; Armengot et al. 2016). However, the adoption of reduced tillage practices, e.g., noninversion tillage, has increased worldwide in recent years. The reason for this increase is owing to reducing soil erosion, improving soil fertility and quality, reducing water runoff, and decreasing cost of tillage. All of these benefits eventually make crop production more sustainable and environmentally safe in the long term (Armengot et al. 2016; Santín-Montanyá and Sombrero Sacristán 2019; Govindasamy et al. 2020).

Based on the results of previous research and surveys conducted in a wide range of crops, soils, and climate conditions, reducing or omitting soil inversion usually increases weed infestations and changes the composition of weed community (Børresen and Njøs 1994; Anderson 1998; Håkansson et al. 1998; Bàrberi and Lo Cascio 2001; Mas and Verdú 2003; Meurer et al. 2018; Woźniak and Rachoń 2019). Changes in weed communities (both weed richness and relative abundance) could not be attributable to tillage system alone, as it may be affected by factors like crop rotation, climate and soil conditions, and the moving of seeds and propagules (Pardo et al. 2019). Despite the abundance of such information about the impact of tillage on weed communities, it is not completely clear whether these changes in weed populations are a consequence of nonrandom filters on the functional attributes of weed species and may thus influence the potential competition between weeds and crops.

Furthermore, it was demonstrated that a higher weed density under no- or reduced-tillage compared to conventional tillage systems does not always result in increased crop yield losses (Sans et al. 2011). Hence, the understanding of the ecological aspects of tillage on weeds can be beneficial in weed control. Therefore, there must be an understanding in how ecological mechanisms behind tillage and cultivation affect weeds, as this knowledge is essential for improving and optimizing the effectiveness of soil tillage in weed management. In addition, under reduced or no-tillage circumstances, mechanical weeding is typically not allowed. Therefore, the only ecological control options left to deal with the problem of weed management would be agronomic practices.

Generally, soil tillage affects weeds in three main ways, including moving weed seeds and unsexual reproduction propagules vertically and horizontally in soil layers as well as changing the soil environment for germination and establishment of weeds. Each of these mechanisms, in turn, determine the stand of weeds present in the field. Being aware of such mechanism would beneficial in regulating the weed population as each weed species react to tillage operations in a different manner.

4.2.1 Effect of Tillage on Weed Seeds

In agricultural lands, after their logging off, weed seeds have different fates. The first scenario is remaining on the soil surface wherein they may be eaten by predators (Gallandt et al. 2005). In addition, if not being dormant, environmental conditions may stimulate their germination and subsequent emergence (Nosratti et al. 2019).

The second scenario for shed seeds is to be buried immediately under the soil by tillage or soil cracking, and depending on the weed type and the depth of seed placement in the soil profile, several fates will be waiting for seeds (Schutte et al. 2014). Therefore, both tillage timing and intensity, as well as their interactions with environmental factors, have a critical effect on prevalence of weed species that are reproduced by seeds buried in the soil (Mas and Verdú 2003).

Tillage moves weed seeds vertically and horizontally within the soil profile and consequently affects seed germination and survival as well as seedlings emergence. Furthermore, tillage modifies the soil environment in ways that can either promote or inhibit germination and the establishment of weeds. The effects of tillage on weed seed movement and their changing soil environments are described in below.

4.2.2 Vertical Weed Seed Distribution

Long-term experiments, such as seedbank studies, can provide foundation knowledge regarding the effects of tillage operations on future weed management plans (Forcella 1992). Long-term field experiments have shown that about

50% to 90% of the seedbank may be deposited between 0 and 5 cm soil depth and the weed seed density decrease logarithmically with an increase in soil depth (Figure 4.2) (Sosnoskie et al. 2006; Woźniak and Rachoń 2019; Govindasamy et al. 2020). Conventional tillage, like moldboard plowing, buries seeds deeper into the soil profile where germination conditions are limiting and seed dormancy is most likely induced (Figure 4.2). Under these circumstances, the buried seeds could remain viable in the soil for a long time and subsequent tillage result in unearthing the weed seedbank, may stimulate seed germination (Gruber and Claupein 2009). Therefore, the seedbank can become depleted by restricting tillage depth and decreasing crop infestation by weeds. However, seeds with a long resting period may germinate even after decades of resting in the soil (Woźniak 2018). Hence, interpretation of the increase and decrease in weed populations assumes great caution.

Changing from conventional tillage to conservation tillage by altering the vertical placement of weed seeds in the soil, and subsequently seedling emergence, can affect weed population (Young and Thorne 2004). In reduced tillage systems, most of the weed seeds are accumulated in the topsoil layer, compared to conventional plowing (Figure 4.2). This is favorable to small-seeded weeds, as they can emerge easily from a shallow soil profile, as a deeper depth would hamper their emergence (Swanton et al. 1999). This may explain why adoption of no-tillage operations typically leads to higher seedbank densities in the topsoil layer. In addition, the higher organic carbon content in no-tillage systems promotes the germination of small-seeded weeds compared to conventional tillage systems (Govindasamy et al. 2020). In addition, small-seeded weeds produce a greater amount of seeds (e.g. more than 190,000 seeds plant^{-1} for

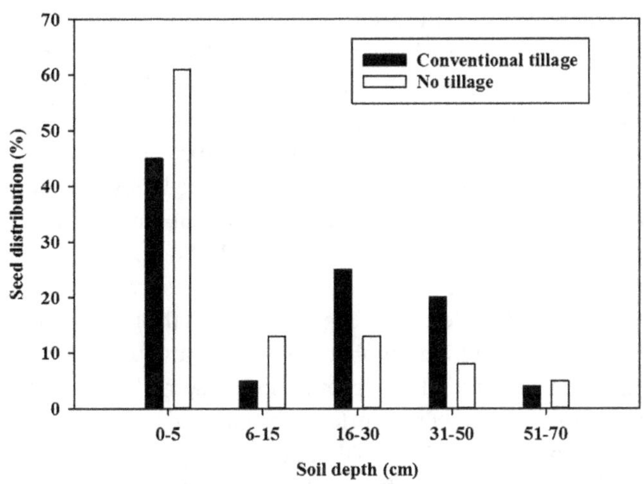

Figure 4.2 Vertical distribution of viable weed seeds as affected by 36 years of conventional tillage (a) or no-tillage (b) in a grain sorghum experiment, Texas, USA. Adopted from Govindasamy et al. (2020).

Amaranthus retroflexus L. and *Portulaca oleracea* L.), resulting in an enlarged population of small-seeded species under reduced tillage systems (Baker 1965).

In another example, Skuodienė et al. (2018) examined the effect of primary soil tillage, including conventional plowing (20–25 cm), shallow plowing (10–12 cm), and shallow plowless tillage (8–10 cm) and interactions with deep soil loosening (35–40 cm) at the location of weed seeds within the soil profile for 8 years. During the investigation period, seed numbers in the soil at the sites subjected to shallow plowing or shallow plowless tillage was higher by 52% and 67%, respectively, compared with conventional ploughing. Generally, the type of tillage was the main driver for the vertical distribution of weed seeds within the soil profile, so that decreasing soil tillage intensity led to weed seed accumulation in the upper soil layer. Increasing of the abundance of weed seeds in the upper topsoil would result in synchronized germination periods, which would enhance the seedling densities quickly (Ghersa and Martınez-Ghersa 2000). It should be kept in mind that the germination of seeds that remained on the soil surface typically would be reduced due to lack of sufficient moisture (Gardarin et al. 2010a; Cordeau et al. 2015).

In general, both weed species diversity and species richness in the noninverted soil trial are greater than in tilled soils. These differences have been attributed to some characteristics of no tillage, such as the relatively stable environment, longer persistence of nonincorporated weed seeds, and higher soil moisture content (Sosnoskie et al. 2006). These circumstances, combined with changes in the vertical distribution of weed seeds in the soil profile, may give a general overview of why weed pressure is drastically increased in fields that practice no-tillage systems.

4.2.3 Modifying Seed Germination Environment

The emergence period of each weed species is unique in indicating species-specific responses to soil temperature and moisture germination requirements (Guillemin et al. 2013). Ambient temperature is the most important factor influencing weed emergence, followed by soil moisture content (Grundy and Mead 2000). Therefore, timing of tillage in relation to rainfall/irrigation and environmental factors will determine overall weed occurrence (Teasdale et al. 2018a).

Cordeau et al. (2017) by assessing 61 weed species across the northeastern USA, found that the germination of spring-emerging species was positively correlated to temperatures in seeded grasses. In contrast, late summer-emerging species that responded to decreasing ambient temperatures were C_3 perennial broadleaf species, which had smaller seed sizes. Therefore, an effective weed control strategy could be to till during the fall using a moldboard plow that buries the small seeds of broadleaf perennials to a depth from where they cannot emerge efficiently (Schutte et al. 2014).

Tillage prior to crop planting as well as watering field and therefore formation of stale seedbeds are two mechanical approaches, which can be used to reduce the seed density in the soil-seedbank (Schutte et al. 2014). The top 10 cm of the soil layer is most frequently exposed to pre-plant seedbed preparation and other forms of tillage (Schutte et al. 2014). In the stale seedbed technique, weed seed germination is promoted in shallow soil layers and then the freshly germinated plants are killed by using shallow plowing (Kanatas et al. 2020). In addition, only tilling the superficial layer of the soil will not lead to an explosion of seed germination, as deeply sown seeds are not exposed to stimulant factors, such as exposure to light. However, there are serious constraints to the effectiveness of such weed control measures, including prolonged germination windows and particularly broad ecological niches with regard to germination and emergence (Chauhan and Johnson 2010).

Therefore, despite the importance of tillage on seed germination, the final emergence would be determined by the local conditions (Schutte et al. 2014). On the other hand, there are seed traits such as seed reservoir content, dormancy, and size that are known affect germination and emergence (Gardarin et al. 2010c). There is strong evidence that the tillage system affects each weed species in relation to its attributes, such as weed ecology. Indeed, the functional attributes of the species allow for the analysis of shifts in the community composition beyond the changes that may be imposed by the geographic context or the high variability in the local occurrence of weeds. Generally, changes in a weed population is the result of nonrandom filters acting on the local pool of species, based on their functional attributes (Gardarin et al. 2010c). It could be concluded, therefore, that genetic factors that regulate germination in response to soil inversion may be among the collection of traits detrimental to weed populations in agricultural environments.

Reduction in the intensity of soil tillage has been shown to increase the amount of organic matter in the topsoil (Meurer et al. 2018) and decrease organic matter mineralization (Peigné et al. 2007). By definition, any tillage operation that leaves more than 30% of crop residues on the soil surface after planting the next crop is called conservation tillage (Govindasamy et al. 2020). However, increasing crop residue on the top soil has several consequences affecting weed seed germination and seedling emergence behavior.

First, these changes in soil fertility would result in reduced nitrogen availability, which in turn diminishes the germination of nitrophilic weed seeds (Golmohammadzadeh et al. 2015). This may explain way nitrophilous weed species are not predominant under reduced tillage conditions. In addition, dominance of weeds with low nutrient requirements may be decreased under reduced tillage compared with conventional tillage (Armengot et al. 2016).

Second, high crop residues left on the soil surface in on-tillage systems, which reflect solar radiation and assist in maintaining soil moisture, result in a reduction in surface soil temperature, particularly during early season (Cox et al. 1990; Fabrizzi et al. 2005). Probably for these reasons, Govindasamy et al. (2020) observed a substantial delay in the emergence of C_4 weed species *Sorghum halepense* and *Amaranthus tuberculatus* in nontilled soils.

Third, crop residues can act as a physical barrier against emerging of weed seedlings and restrict light penetration required for their early growth (Teasdale and Mohler 2000). Reduction in light intensity has an inhibitory effect on small-seeded annual weeds compared with large-seeded crops (Mohler 1996). Indeed, the use of crop residues as an effective control method (mulch) to prevent weed emergence is dependent on the size difference between crop and weed seeds. Seedlings of large-seeded crops, as well as plants sprouting from the underground parts of perennial weeds, have an increased ability to push through the physical barrier created by the additional layer of crop residues.

Fourth, remaining residues on or near the soil surface could release allelopathic compounds into the surrounding soil, which may alter the chemical environment of the weed seed (Liebman and Davis 2000). Research has shown that in large-seeded crops, allelochemical selectively reduces weed seed germination and emergence (Narwal 2000). The reason behind this selectivity is germinating crops below the soil being infested by allelochemicals while relatively small seeds of weeds germinate near the surface, where surrounded by allelopathic compounds. Furthermore, small seed adsorb more allelochemical because of having a higher proportional surface area than large ones. Therefore, incorporating crop residue into the top soil layer is more effective against weeds than leaving them on the soil surface (Roth et al. 2000).

In addition to the previously mentioned physical and chemical changes in the seed environment caused by the tillage system, seed distribution through the vertical profile of the soil induced by tillage affects the exposure of seeds to light as well (Buhler 1997). Light stimulates the germination of the seeds of a high number of species, particularly small-size seeds. Nevertheless, germination of some large-seeded weed species has been reported to be induced by exposure to light (Nosratti et al. 2019). Tillage can aggregate photoblastic seeds on or near the surface of the soil where light can easily access and induce germination in the weed seeds (Benvenuti 1995). Furthermore, seeds receiving light during soil inversion by moldboard plowing can stimulate germination, as exposure to light for only a fraction of second is sufficient to induce germination in photoblastic seeds (Scopel et al. 1994; Robert and John 1998). Therefore, conducting tillage at night or covering tillage equipment to deprive photoblastic seeds from light in a conventional system is recommended for limiting the germination of many light-sensitive weed seeds *Amaranthus spp.*, *Chenopodium album* L., and *Ambrosia artemisiifolia* L. (Bastiaans et al. 2008). Being aware of how the present weed flora in the field responds to light gives producers the opportunity to adopt particular tillage systems based on the detection of certain weeds to avoid prevalence of a few weed species in the field (Santín-Montanyá and Sombrero Sacristán 2019).

4.2.4 Weed Seed Viability

Once in the soil, weed species tend to create stocks, better known as seedbanks. Seeds in the soil seedbanks are the primary source of weed infestation and thereby the most frequently investigated weed characteristics are those related to persistence and longevity of seeds in the soil (Dekker 1999). Longevity of dormant seeds in the soil seedbank is dependent on weed species characteristics, soil microorganisms, soil conditions, and management (Dekker 1999; Chee-Sanford et al. 2006). The most fundamental aspects of conservation agriculture is the permanent soil cover created with crop residues and live mulches, as well as crop rotation (Munkholm et al. 2013). All of these strategies increase soil organic matter, which can lead to soil microorganism proliferation. In addition, tillage always changes the soil structure and the distribution of nutrients, soil pH, and conductibility, which in turn can alter soil microbial communities (Kennedy 1999). Soil microorganisms alter the soil environment by decomposing organic matter and decrease or increase bioavailable nutrients, which can either directly or indirectly affect weed seeds (Nelson 2018).

The seed decay of six different weed species commonly found in the northern Italy was evaluated by Nikolić et al. (2020) in two adjacent sites, a field with minimum tillage and a buffer zone (completely undisturbed) (Figure 4.3).

In general, buried seeds (at a depth of 12 cm) experienced increased degradation than the seeds exposed to no-till conditions. Among them, *Digitaria sanguinalis* and *Alopecurus myosuroides* showed the highest susceptibility to degradation (Figure 4.3). In contrast, *Abutilon theophrasti* and *S. halepense* had the highest intact seed number under the no-till condition. The most and least degraded species in the buffer strip were *D. sanguinalis* (70%) and *S. halepense* (19%), respectively (Figure 4.3).

Furthermore, the results of Nikolić et al. (2020) indicated that soil microbial activity was higher in the no-tillage field relative to the buffer strip. It is reported that the physical properties of seeds determine the level of degradation. For example, *A. theophrasti* and *S. halepense* resisted degradation by having a thicker seed coat while *D. sanguinalis*, *A. myosuroides*, with thinner seed coats, were more susceptible to degradation and lost their viability sooner.

Such results have been reported by other researchers who have correlated seed persistence in soil seedbanks with seed coat characteristics (Davis et al. 2008; Gardarin et al. 2010b). Soil microorganisms feed on the lignin and cellulose of the seed coat (Bewley et al. 2012). It should be noted that long-term adoption of a reduced-tillage system would result in the soil containing higher levels of organic matter within the soil when compared to the conventional tillage system (Singh et al. 2020).

In contrast to results of the previously mentioned study (Figure 4.3), it is argued that soils containing higher levels of organic matter have a higher weed suppression ability, as they include greater numbers of microbes invading weed seeds (Kremer and Li 2003). This controversial result may contribute to the depth of the weed seed placement in this study (12 cm), which is much higher than that of a real, no-tillage system.

In the surface layer of soil, microbial damage of the seed coat may be beneficial for weeds as it could ease seed germination by weakening the seed coat. After weed seed coat degradation by microorganisms, the physiological seed dormancy mechanism determines the seed fate in the soil (Gómez-Maqueo and Gamboa-deBuen 2016). Therefore, it is vital to be aware of the fate of different weed seeds in the topsoil layer wherein seed germinates to understand weed seedbank dynamics under no-tillage conditions.

Under water deficit conditions, the presence of crop residue on the soil surface leads to an insulated soil atmosphere boundary that will decrease evaporative loss and maintain humidity, which in turn will protect seeds from desiccation. In contrast, under humid circumstances, residue by increasing micrograms activity can enhance seed decay (Yang et al. 2013). However, the presence of a hard seed coat, tillage strategy, and soil nutrient profiles are involved in the susceptibility of seeds to decomposition due to conditions caused by plant residues (Nichols et al. 2015).

4.2.5 Effect of Tillage Practice on the Growth and Establishment of Seedlings

Generally, conservation tillage that improve soil quality and water use efficiency may be better than conventional tillage for growth of both crops and weeds (Shahzad et al. 2017). Soil under a reduced tillage system has a high amount of crop residues on or near the soil surface, which resist against the weed root penetration and seedling emergence (Armengot et al. 2015). Dicotyledonous plants have taproots, which have restricted root penetration into non-tilled

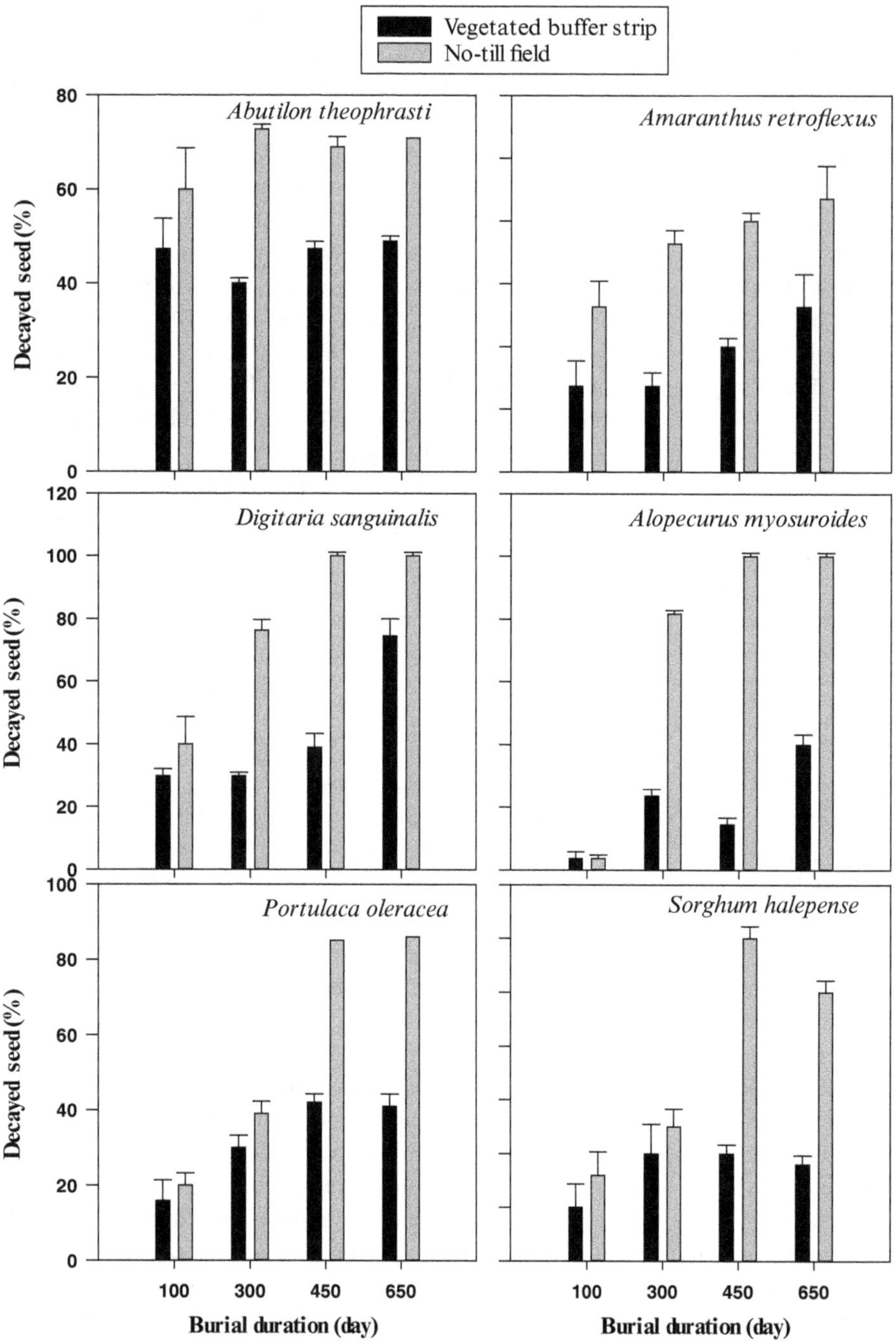

Figure 4.3 Decayed seeds (%) of different weed species after different days of burial in the soil of the no-till (NT) field and vegetated buffer strip (BS). Adopted from Nikolić et al. (2020).

soil, while roots of monocotyledons weeds are always fibrous (Baker et al. 2012). Reducing tillage can restrict the emergence of monocotyledons weeds, particularly those with an epigeal emergence pattern (Froud-Williams et al. 1981; Gardarin et al. 2010a). On the contrary, coleoptile of grasses helps them to emerge from non-tilled soil and may assist thick crop residue conditions (Rebetzke et al. 2005).

These weed characteristics may partially explain why the adoption of conservation tillage increases the population of grasses and decreases the population broadleaf weeds. Modification of weed functional traits of weed communities in different tillage systems has been documented. Armengot et al. (2016) noted that many weed species in conservation tillage had lower competitive ability as they were shorter and had less affinity for nutrients when compared with conventional tillage systems. This highlights the importance of implementing other weed suppressive methods for efficient weed control, particularly in low input fields under conservation tillage systems (Cooper et al. 2016; Hofmeijer et al. 2019).

4.2.6 Effect of Tillage on Asexual Reproduction Organs of Perennial Weeds

It has been demonstrated that conventional tillage, including high-depth moldboard plowing, can control perennial weeds substantially (Børresen and Njøs 1994; Håkansson et al. 1998). On the contrary, tillage systems with reduced cultivation depths, conducted by tine or discs, commonly result in increased populations of rhizomatous perennials as the rhizomes get broken up and dispersed partially (Peigné et al. 2007). Results of many long-term and large-scale studies indicate that the reduction in the intensity of the soil tillage commonly increases the abundance of perennial species, but the trends are less clear in relation to weed diversity (Fried et al. 2012; Armengot et al. 2015, 2016). Accordingly, a 14-year study recorded the highest density of *Convolvulus arvensis* within a no-tillage system followed by chisel, moldboard, and ridge plowing (Buhler et al. 1994).

Brandsæter et al. (2011) evaluated the effect of plowing depth on perennial weeds, including *Cirsium arvense* (L.) Scop., *Elymus repens* (L.) Gould and *Sonchus arvensis* L., found that applying deep tillage (25 cm) resulted in decline in shoot biomass by 50% compared to shallow tillage (15 cm). In their study, *S. arvensis* was more sensitive to deep plowing, and its control was higher than 90% compared with shallow plowing. They attributed this significant reduction in shoot biomass of *S. arvensis* to arising shoots from intact root systems, located just beneath the tillage depth. Therefore, an integration of plowing in spring and planting competitive crops after tillage operations would lead to sufficient control of *S. arvensis*. In opposition, Gruber and Claupein (2009) and Brandsæter et al. (2012) showed that implementing shallow/rotary tillage in autumn resulted in the highest control of perennial weeds such as *C. arvense*.

Environmental conditions in the spring enhance regrowth in perennial weeds, which in turn increases the depletion of nutrient stores in underground parts of these weeds (Håkansson 2003). Hence, conducting tillage during the spring is the preferable way to suppress perennial weeds by stimulating new shoots from underground reproductions organs. Nevertheless, Njøs and Ekeberg (1980) found that there were no difference between the effectiveness of tillage on *E. repens* in autumn versus spring.

In another example, it is a general belief in the Nordic countries that conducting tillage in the spring is better than in the fall to control *C. arvense* and *S. arvensis* (Pedersen and Gustavsson 2003). A possible explanation for the lower germination effect from tillage in autumn is the dormancy of the innate root bud, which develops during the autumn in Nordic countries (Brandsæter et al. 2010).

Brandsæter et al. (2017) divided the perennial weed species in their study in two main groups based on the weed response to tillage. The first group, including *C. arvense* and *S. arvensis*, was sensitive to tillage timing, while the second group, which had *E. repens*, was more sensitive to stubble cultivation. They reported that soil tillage in the spring had higher control of *C. arvense* compared to autumn, which was similar to results of previous works (Brandsæter et al. 2011).

Furthermore, Thomsen et al. (2013) demonstrated that a significant number of shoots of *C. arvense* emerged from the creeping root system below the tillage layer which was not fragmentized by tillage operations. Therefore, it could be concluded from this information that spring moldboard plowing would result in shoots produced by roots located in deep soil layers, which led to an increased depletion of food reserves in regenerative structures (creeping roots). In addition, shoots originating from deeper soil layers will give crops enough time to overshadow newly emerged weed plants. Nevertheless, this conclusion is not true where soil is sandy and emerging shoots do not struggle with high soil resistance (Melander et al. 2012).

Boström et al. (2013) reported that removal of the aerial parts of *E. repens*, *Equisetum arvense* L. and *Tussilago farfara* L. early in the autumn omitting the vascular system interferes with the transferring of food reservations to the rhizomes, which in turn would lead to weakening the rhizomatous weeds. Brandsæter et al. (2017) by evaluating the effect of tillage time and type on control of mixed populations of *C. arvense*, *S. arvensis* and *E. repens* concluded that stubble cultivation in autumn and moldboard plowing in the next spring would result in the best weed control and highest spring cereal yields.

Chovancova et al. (2019) by assessing the impact of soil tillage systems in maize crops on weed infestation intensity and weed species composition found that minimal tillage was dominated by late spring species. In opposition, perennial and overwintering species had the highest population in no-tillage systems and early spring weed species frequently observed in conventional tillage. In addition to disturbing perennial underground structures, tillage can also be an effective strategy for controlling perennial weeds by exposing underground organs to sunlight and desiccation (Govindasamy et al. 2020).

In conclusion, the weakness of poor perennial weed control hinders the widespread adoption of conservation tillage. Hence, integration of other control means, particularly cultural ones such as crop rotation, can overcome this challenge. Ecologically, frequently tilled soil is a simple environment with low weed species diversity and is consequently free of niches, which makes systems fragile to invasive species. Soil tillage is not the only driver determining weed population in the agricultural fields, as other weed control approaches and weather conditions are of great importance on weed occurrence (Plaza et al. 2011; Boscutti et al. 2015).

4.3 Thermal Weed Control

In the thermal weed control method, weeds are controlled by transferring heat to the target weed and subsequently increasing the temperature beyond that of the tolerance of plant cells (Astatkie et al. 2007). Several studies have shown that lethal effect on the weed is species-specific (Bauer et al. 2020). The most tolerant weed species to thermal control are those with growth points covered with soil such as grasses before reaching tillering growth stage (Abou Chehade et al. 2018). Generally, older weeds (Ulloa et al. 2012) and weeds with higher leaf thickness, such as *P. oleracea* L. and *Bellis perennis* L. (Leon and Ferreira 2008), are less sensitive to heat.

Spot-flaming (Raffaelli et al. 2013) and hot water (liquid or steam) (Rask et al. 2013) are the two main thermal weed control methods. Lack of sufficient selectivity (i.e. individual weed plants to be killed without harming nearby crops) is the weakness of the spot-flaming technique against weeds (Abou Chehade et al. 2018). Both thermal methods are not effective enough on perennial weeds, as they act only on aerial parts of weeds, which makes repeating treatment unavoidable to achieve satisfactory control (Astatkie et al. 2007). Soil solarization is another thermal weed control technique, in which well-watered soils are tilled and covered with transparent polyethylene plastic mulch during the sunny days of the year for at least one month. This results in higher temperatures under the plastic coverage due to solar radiation, which can kill weed seeds (Stapleton 2000; Samtani et al. 2017).

The fast development in new thermal methods, such as laser and microwave technologies provide new opportunities for weed control. Laser beams by warming up the plant tissue thorough high-density energy on selected spots can led to the death of the target plant (Sartorato et al. 2006; Brodie 2018).

In a greenhouse study, Heisel et al. (2002) by using a laser beam to cut stem of broadleaf weeds showed that stem should be cut under the meristem and near to the soil surface to prevent regrowth of weeds.

Mathiassen et al. (2006) demonstrated significant growth reduction in weed species, including *Stellaria media* (L.) Vill., *Tripleurospermum inodorum* Sch. Bip., and *Brassica napus* L. by applying laser treatment on the apical meristems at the cotyledon stage. They also attributed the efficacy of laser treatment on weeds to wavelength, exposure time, laser spot size, and laser power as well as weed species. In another study, Xiong et al. (2017) examined a robot equipped with machine vision and a laser pointer to recognize the crop and weeds on the cotyledons stage in indoor conditions. They indicated that the robot was able to identify weeds and controlled them by laser successfully. Similar to other technologies, cost is the main disadvantage of laser, microwave, and electric arc for weed control.

Microwave method uses electrical energy to kill weeds in which the whole plant and weed seeds can be destroyed (Sahin 2021). Electric field and thermal effect of microwave cause cell death in living organisms such as plant (Banik et al. 2003). The extent of heat by microwave imposed greatly depends on the dielectric properties of the material The water content of materials substantially influences their dielectric properties (Kuang and Nelson 1998). Some studies have reported that longer-term usage of microwave can reduce weed seed germination. However, energy cost of the microwave method would reduce its adoption by farmers as weed control tool (Sahin and Yalınkılıc 2017).

4.4 Mulching

After the weed seeds germinate, the initial growth of the seedlings depends on the energy stored in the endosperm. Therefore, successful emergence depends on factors such as the depth of burial, the soil structure, and the size of the seed (Vleeshouwers 1997). The use of mulch increases the physical layer against weed germination, resulting in reduced weed establishment or delaying its emergence (Jabran 2019). Plastic (usually black color) and straw mulches are the main types of mulches, which are using to control weeds in fruit, ornamental plant nurseries, and vegetable production systems (Steinmetz et al. 2016).

Black-colored plastic mulches work by obstructing sunlight for weed plant's photosynthetic activity, and have been found to be more efficient in suppressing weeds than other colored plastic mulches (Brault et al. 2002; Ngouajio and Ernest 2004). Black plastic mulch can control annual weeds completely, but its efficacy against perennial weeds has shown some inconsistence (Lament 1993; Gul et al. 2009). Using degradable or biodegradable plastic mulches has been suggested for overcoming negative impacts of plastic mulch on the environment (Cirujeda et al. 2012; Prill 2017).

When live plants in the field are ploughed into the soil and allow to decay to add to soil fertility, the process is green manuring (Cherr et al. 2006). Green manure by depriving weed seed from sun and air reduce weed population. Soil improvement and amendment are other benefits of using green manure is agricultural systems. Suppressing.

Furthermore, decay of plant materials in the soil would reduce the germination and early growth of weed seedlings (Lee et al. 2012).

Sorkheh et al. (2020) demonstrated that application of wheat residues and cowpea green manure integrated with minimum tillage could be an effective way to increase corn yield and reduce weed pressure. Results of five-year study showed he usage of green manure can diminish weed seed numbers exponentially with the increase in years after termination of the green manure crop (Melander et al. 2020). Field observations on rotating faba bean (*Vicia faba*) as green manure crop with maize demonstrated its ability in reducing both density and biomass of weeds, especially dicotyledon like *A. retroflexus, C. album*, and *Solanum nigrum* (Álvarez-Iglesias et al. 2018).

Keeping crop residues (straw mulch) over the soil surface can control weeds via both releasing allelopathic compounds and by providing a physical barrier (Farooq et al. 2011a; Jabran and Chauhan 2018). Residue from the previous crop can be used as mulch, or occasionally allow the crop to grow and then kill these to use as mulch for weed control (Tursun et al. 2018).

It has been reported that mulch of numerous plant species, including *Avena sativa*; *Fagopyrum esculentum*, *Phacelia tanacetifolia*, *Vicia villosa* (Tursun et al. 2018), *Vigna unguiculata* L. (Walp.), *Triticum aestivum* L (Monks et al. 1997), *Zea mays* L. (Anzalone et al. 2010), *Lysiloma latisiliquum* L. (Benth.), *Canavalia ensiformis* L. (DC.) (Caamal-Maldonado et al. 2001), *Oryza sativa* L. (Mahajan et al. 2007), and *Sorghum bicolor* L. (Cheema et al. 2000) can control weeds. These examples suggest that the residue of most crops in conservation agriculture can provide satisfactory weed control, while improving the soil properties.

4.5 Biological Weed Control

Biocontrol agents are any living organisms that can reduce the growth of weeds, which comprise classical weed biocontrol, bioherbicides, conservative measures, and allelopathy (Schwarzländer et al. 2018). Broad-spectrum biological control is when weeds are controlled by grazing animals, birds, phytophagous fish, etc. (Bond et al. 2003). Nevertheless, many countries are still reluctant to adopt weed biological weed control as they argued that it is too risky or doesn't suppress weeds effectively, particularly in intensified cropping systems.

4.5.1 Classical Weed Biocontrol

Classical weed biocontrol, which also referred as to inoculative or importation, has been used the most and includes controlling a specific weed in a region by intentionally releasing the natural enemies of the weed from the weed's origin (Schwarzländer et al. 2018). This method is economic and can be used for suppressing weeds across a wide range of environment conditions. After introduction of the biocontrol agent (predominantly insects) into the region, and its subsequent establishment and population increment, it can control weeds.

The potential harmful effects of introducing nonindigenous species for biological control purposes is a lack of host specificity, as they can pose a threat to nontargeted plant species. The prediction of behavior of biological agents in the environment is extremely complicated (Simberloff and Stiling 1996).

4.5.2 Bioherbicides

Bioherbicides (inundative or augmentative approach), which are made from living microorganisms, directly invade current weeds. Indeed, bioherbicides (containing microorganisms) are formulated and packaged as mass doses of pathogen inoculum that is applied in the region or field where the target weed needs to be controlled (Morin 2020). The mycoherbicides (using fungi as control agent) has shown promising results globally. Another form of bioherbicides is a natural herbicide, which is made by isolating and applying the phytotoxin produced by microorganisms, as opposed to the living organism itself, for weed control. Allelochemicals from allelopathic plants or synthetic derivatives could be used as natural herbicides (Morin 2020).

Fungal *Bipolaris bicolor* strain SYNJC-2-2 has showed potential to be used as a bioherbicide for controlling goosegrass (*Eleusine indica*), especially in tea gardens (Xiao et al. 2022). Hasan et al. (2021) by examining 1,300 bacterial strains, found that only *Streptomyces anulatus* strain-329 exhibited desirable bioherbicidal traits such as having contact and selective activity, especially target-specific toxicity on the selected weed plants. This bacteria produces herbicidal metabolites, which may provide a new molecule for a more efficient herbicidal effect to control weeds. When compared SC64-based bioherbicide isolated from fungus *Sclerotium rolfsii* with chemical herbicides (glyphosate), it was more effective in improving the weed community structure and dramatically increased the biodiversity in habitats invaded by *Solidago canadensis* (Zhang et al. 2019).

In a recent study, various fungi were isolated from plants with symptoms of fungal diseases for using as bioherbicide against *Urochloa plantaginea*, *Euphorbia heterophylla* and *Bidens pilosa* weeds in soybean and corn crops (Júnior et al. 2019). Results showed that *Fusarium oxysporum*, *Fusarium ploriferatum*, and *Trichoderma koningiopsis* have bioherbicide potential. Among these selected fungai, isolate of *T. koningiopsis* showed the highest performance on *Euphorbia heterophylla* without any phytotoxicity against corn crop.

In general, the inconsistency of bioherbicide performance is a primary factor limiting their extensive application, which is in turn influenced by factors such as bioactive compound content, weed control spectrum, formulation, and application method (Hasan et al. 2021).

4.5.3 Conservative Measures

In this indirect biological approach, the natural population of the pests feeding on particular native weeds are maintained at an optimum level. Hence, being aware of the ecology of the target weeds and the control agent is required to make a balance between weed population and biological agent and subsequently successful weed suppression during long time. Two main methods of conservation biological control includes greater predation, or an increase in natural decay, for reducing the soil seedbank and subsequently future weed populations (Sarabi 2019; Nikolić et al. 2020).

Hundreds of microorganisms in the microbiome are a potential source for biological control of many common weeds (Li et al. 2003). Using these bio-agents is still being researched, it does not require any operation and should be strengthen by conservation methods as in conservation agriculture. Reducing usage of agrochemicals such as fertilizers, and various pesticides, which are lethal for most microorganisms will improve the likelihood of their survival (Nichols et al. 2015).

Weed seed predators, including insects, rodents, and birds, feed on weed seeds either predispersal or postdispersal, which can be primary factors of weed seed mortality in agricultural ecosystems (Connolly et al. 2014). Under well-designed weed management programs, seed predators can significantly minimize the size of the soil weed seedbank (Sarabi 2019). The accumulation of weed seed on the soil surface will enhance accessibility of seeds for predators, explaining why under no-tillage systems, the likelihood of weed seed predation increases (Blubaugh and Kaplan 2015). Furthermore, mortality of granivores would be diminished under no-till soils (Shearin et al. 2014).

4.6 Allelopathy

Allelopathy, the direct or indirect effect of released allelochemicals from live or dead parts of a plant on the growth of the adjacent plants, can be considered as a component of biological control (Lovett 1991). Possible sources of allelochemicals are the mucilage around a germinating seed, the aerial parts of plants, exudates from plant roots, volatile emissions from a growing plant, and decomposing plant residues (Farooq et al. 2020). Practical usage of allelopathy in weed management strategies include either applying organic mulches, cover crops, or both to reduce weed density (Gerhards and Schappert 2020). The mechanisms of weed suppression involved in these two approaches are smothering and releasing phytotoxins from either living plants or decomposing plant materials. The secreted toxins, however, may have side effects on crop germination and early growth, particularly in small-seeded crops (Aslam et al. 2017; Macías et al. 2019).

Rye (*Secale cereale* L.) is among the most important allelopathic crops in which benzoxazinones [2,4-dihydroxy-1,4(2H)-benzoxazin-3-one (DIBOA) and 2(3H)-benzoxa zolinone (BOA)] are the most important allelochemicals responsible for its allelopathic potential. Rye is usually applied as cover crop in agricultural system for weed-control proposes, though it can be as a rotational crop or mulch (Schulz et al. 2013).

Sorghum (*Sorghum* spp.) is another important allelopathic crop from poaceae, which expresses its allelopathic activity through the production of phenolics and acyanogenic glycoside (dhurrin) allelochemicals. However, hydrophobic p-benzoquinone (sorgoleone) is its most allelochemical exuded by its roots. Sorghum can be used for weed control as mulch, cover crop and intercrop, and crop rotation (Weston et al. 2013).

It is well known that plants belonging to the *Brassicaceae* family can control weeds efficiently in different cropping systems by containing the allelochemical glucosinolate (Haramoto and Gallandt 2004). Phenolic acids have been distinguished as the main allelopathic compound in sunflower cultivars responsible to their weed control properties (Alsaadawi et al. 2012). Cover crops can be established quickly and have a high capability for biomass accumulation by increasing soil organic matter, enhancing soil structure, reducing soil erosion risk, and suppressing weeds (Teasdale et al. 2004; Sainju and Singh 2008). *Fabaceae* are popular cover crops because of their ability to convert atmospheric nitrogen into plant available forms.

Several species of the genus *Vicia* are cultivated for different proposes worldwide (Tate and Enneking 2006). *Vicia villosa* originating from Europe and Western Asia is the second-most important vetch in the world (Francis et al. 1999). Globally, *V. villosa* is the most important legume that has been widely adopted as a cover crop. Common vetch (*Vicia sativa* L.) is another important forage legume of genus *Vicia*, which can grow across a broad environmental and soil conditions (Mikić et al. 2013; Fırıncıoğlu 2014). *V. villosa* is considered as the only species from *Vicia* genus that can survive moderate to harsh winter conditions. In comparison with legume cover crops, in general, *V. villosa* is characterized by more winter hardy (Dabney et al. 2001), making this species a stronger legume competitor with fall weeds. Several species of the genus *Trifolium* are commonly adopted for cover cropping, due to their rapid growth and allelopathic activity (containing phenols and isoflavonoids) on weeds (Liu et al. 2013; Scavo et al. 2020). The genus *Trifolium* contains both perennial and annual species. Studies indicated that annual species are advantageous over perennials for utilization as cover crops as annual clover produce high aboveground biomass while in perennials root biomass is abundant (Ross et al. 2001). *Trifolium pratense* is the most commonly cultivated perennial forage legume in the world agricultural systems (Sato et al. 2005).

4.7 Cultural Weed Control

Cultural weed control relies on the designing and manipulation of cropping systems to reduce weed pressure. A wide range of cultural control can be used for weed control proposes which can be divided into groups including improving the competitive ability of crops (Gause's competitive exclusion principle) and diversifying the cultivated crop species.

4.7.1 Enhancing the Competitive Ability of Crops

Crop attributes, including crop rotation, cultivar, density, intercropping, and spatial arrangement, as well as fertilizer and water management, are the main cultural components influencing crop-weed competition (Jha et al. 2017). Hence, studies analyzing crop–weed competition can provide information on the merits of various components of a cropping system contributing more in weed suppression (Teasdale et al. 2018b).

4.7.2 Water and Fertilizer Management

Weeds have a higher ability to absorb water and nutrients than do crops (Allan et al. 2015). Therefore, any action that deprives weeds of water and nutrients will help control them, especially in soils with low fertility levels. Fertilizer timing, placement, and source can be manipulated so that weeds are denied access to nutrients while crops are fertilized (Khamare et al. 2020). Banding the fertilizers deep and close to the crop root can reduce weed growth compared to broadcasting (Di Tomaso 1995; Derksen et al. 2002).

Soil moisture can be managed in favor of weed control. For instance, in dryland wherein the top layer of the soil is dry, sowing large-seeded crops into deeper soil layers can provide crops with an initial advantage over weeds (Liebman et al. 2001). The type of irrigation (surface irrigation versus drip irrigation) can also be used to restrict water for weeds and thereby reduce weed growth.

It is well demonstrated that weed abundance, not diversity, is enhanced after fertilization (e.g., nitrogen, phosphorous, potassium, etc.) (Bilalis et al. 2010; Tang et al. 2014; Kakabouki et al. 2015). Likewise, organic farming by limiting control options usually increases species richness when compared to conventional practices (van Elsen 2000). In contrast to organic fertilization, the fast release of nitrogen by inorganic compounds induces seed germination in some weeds and subsequently increases their density (Little et al. 2015).

4.7.3 Crop Density and Arrangement

Studies have shown that increasing the uniformity of the crop arrangement can increase weed growth suppression (Colbach et al. 2014; Lu et al. 2020). Indeed, planting in uniform arrangement allows crop plants to occupy a large amount of space and reduces weeds, depending on water nutrient conditions, crop-to-weed-height ratio, and the timing of crop versus weed emergence (Marín and Weiner 2014). For mechanized production systems, it may not be feasible to adapt closer inter and intra row spacing. It is demonstrated that the more homogeneous the crop spatial arrangement is, the higher the weed control percentage is (Olsen et al. 2012).

Based on Olsen et al. (2012) a more spatially uniform distribution of plants, as opposed to a row-structured canopy, maximizes crop competition against weeds as the earlier canopy closure associated with uniform plant spacing.

Evers and Bastiaans (2016) demonstrated that a uniform planting pattern of the crop improved the weed-suppression ability of the crop canopy. Further, they showed differences in weed suppressiveness between planting patterns were largest with weed emergence before crop emergence while the effect of the crop was only marginal. Eventually, Evers and Bastiaans (2016) concluded with simultaneous emergence a uniform planting pattern was 8% and 15% more competitive than a row and a random planting pattern, respectively.

Olsen et al. (2005) found that planting wheat in a grid pattern with two perpendicular passes at half the normal seeding rate is a simple way to increase planting uniformity. A grid planting arrangement may increase the rate at which the ground is covered, and reduce intracrop competition. Increased crop density combined with a more uniform planting distribution should enable crops to compete more successfully with weeds (Weiner et al. 2001).

Boyd et al. (2009) by determining the effect of seeding rate and planting arrangement on rye growth and weed control also showed the positive effect of grid planting pattern on increasing crop growth and weed suppression.

In most crops, increasing the seed rate has a negative effect on weed density and growth in the field (Korres et al. 2020). The mechanism involves optimizing shoot and root competition for light, water, and nutrition, therefore enhancing crop yield and nutrition use efficiency (Ona et al. 2018). However, increasing the seed rate demonstrated more success than narrowing the row spacing, albeit, there are more opportunities to adjust crop spacing (Teasdale et al. 2018b). At relatively low crop densities, crop cover early in the growing season is low, resulting in a larger amount of resources available for the weeds, thus enabling them to establish and grow quickly (Lemerle et al. 2001).

Harker et al. (2003) demonstrated that increasing hybrid canola seeding rate from 100 to 150 seeds/m^2 reduced weed populations and increased canola yield by more than 40%. In another study, Lemerle et al. (2011) pointed out that a higher seeding rate improved the competitiveness of weakly competitive canola cultivars against volunteer wheat.

Nevertheless, for each crop there is an optimum plant density at which the highest crop yield would be produced, and increasing the crop density beyond that will negatively decrease either the crop quality or quantity (Kanatas et al. 2020). Exceptions in this case are forage plants and small-grain cereals, in which increasing seeding rate has no serious consequences and crop yield is more or less maintained (Olsen et al. 2005).

4.7.4 Date of Crop Establishment

The most important factor influencing the weed–crop competition is the time of weed emergence relative to the crop (Ona et al. 2018). Weeds that appear concurrently or before the crop have a more detrimental effect on the crop yield than those emerging later (Knezevic et al. 1997).

Asaduzzaman et al. (2014) showed that planting canola earlier by four weeks can significantly reduce the pressures of *Sisymbrium orientale*, *Capsella bursa-pastoris*, and *Hordeum leporinum*. In their study, weed pressure for each tested genotype was higher in the late planting and weed biomass was significantly varied between early and late sowing, depending on genotype. Hence, they concluded that earlier seeding of canola crops can improve their ability to compete with weeds, as well as grain quality and yield.

Under low weed infestation, earlier seeding of winter fall-seeded crops can be advantageous for yield potential while in field heavily infested with weeds delayed sowing can reduce early crop competition via management of early germinating weeds prior to onset of planting operations. For example, research in South Australia showed that a sowing delay of three weeks decreased in-crop annual ryegrass by an average of 52%, and the quantity of weed seed produced by 21% (Matthews et al. 1996).

Some later emerging weeds, although are less competitive, can interfere with crop harvesting or crop quality. It is demonstrated that weeds emerge seasonally, indicating their affinity to the timing of planting of specific crops (Vidotto et al. 2016). Avoiding the peak flush of weeds can be achieved by modifying the date of crop establishment (Kolb and Gallandt 2012). This avoidance could be done via two ways; regulating crop emergence before weeds, subsequently exposing newly emerged weeds to crop competition (allowing the crop to establish before emergence of the still dormant weeds), and allowing weeds to germinate and then removing weed seedlings before (e.g., stale seedbed) or during crop planting (Chauhan et al. 2012). Hence, it could be argued that the potential weed suppression offered by altering crop planting is derived from the truth that weeds germinate during specific periods. When compared with delayed planting, generally, the earlier planting dates have a more positive effect in controlling weeds (Chauhan and Mahajan 2012) as it is well shown that shortening the growth season will result in reduced crop yield regardless of weed infestation (Nichols et al. 2015). Similarly, successful application of the stale seedbed technique is dependent on cropping delay until after the main flush of weed emergence, which is dependent on moisture availability (Gallandt 2006). Under rainfed conditions, such a delay in sowing time may cause the loss of the first rain necessary for the crop to emerge in the fall and may cause the crop germination face cold conditions during the winter.

4.8 Crop Diversification for Weed Management

4.8.1 Intercropping

Intercropping, by definition, is the planting of two or more crops at the same time and space of land (Liebman and Dyck 1993). The most dominant intercropping system comprises legumes in cereal and other crops (Ghosh et al. 2006). Increasing and maintaining soil fertility, reducing soil erosion, and the suppression of weeds are some of the main benefits of intercropping different crops (Bedoussac et al. 2015). The reduction in the weed composition and frequency in intercropping systems could be due to the differences in the plant's architecture in the cropping system, resulting in a higher plant density per unit area. Furthermore, intercropping reduces niches available to weeds and reduces weed populations compared to a monoculture systems (Berdjour et al. 2020). Despite the many benefits of intercropping, the use of this cultural method to control weeds is only possible on small acreages because large-scale intercropping prevents the use of machinery for planting, harvesting and other cultural practices.

Advantages of intercropping include better usage of land, fertilizer, and water, which would result in improvement of crop yield and economic returns (Khan et al. 2012).

Furthermore, intercropping different crops, particularly those with allelopathic properties, can offer an alternative to applying herbicide in crops. Results of a field study showed that intercropping of fodder legumes in corn resulted in reduction of infestation of *Striga hermonthica* Del. In corn (Farooq et al. 2011b). The Population of *Cyperus rotundus* L. in cotton crop was significantly reduced with triple intercropping of sesame (*Sesamum indicum* L.), soybean, and sorghum on alternate rows (Abraham and Singh 1984).

The level of weed suppression by intercropping as a means to address weed control greatly depend on the genotypes of both cash and cover crop, plant density, and plant arrangement. There are three main types of intercropping: fully mixed (without a specific arrangement), relay (with a temporal separation between crops), and strip (two or more crops are separated in the space by cultivating them in strips). Strip intercropping is predominant, since it allows more interactions between crops and facilitates their cultivation (Brooker et al. 2015).

Cereal–legume intercropping is the most common example of allelopathic intercropping as many allelopathic crops are suitable for cover cropping, both in the *Poaceae* and *Fabaceae* families, such as *Sorghum bicolor* / *Vicia spp.*, *Trifolium spp.*, *S. cereale*, and *Lathyrus sativus* (L.) (Rad et al. 2020).

The selection of right intercropping ratios is crucial in increasing the weed control and improving quali-quantitative traits of soil as well as cover crop biomass (Blaise et al. 2020). Vrignon-Brenas et al. (2018) by intercropping 12 different crop species over 5 years in cotton indicated more than 90% reduction of weed biomass.

Cheriere et al. (2020) reported that a trade-off between soybean yield and weed control indicated lowest yield and highest weed control by intercropping sorghum and buckwheat with sunflower. They argued that the reduction in soybean yield can be alleviated by combining associated species choice and spatial arrangement. Such solution ways include planting alternate rows in sorghum and buckwheat intercrops.

Intercropping density influences the weed–crop allelopathic interactions and thus weed control. For instance, Pouryousef et al. (2015) found that increasing the density of *Trigonella foenum-graecum* L. resulted in higher weed control percentage, although the final level of weed suppression was not satisfactory. They related this insufficient weed control to lack of allelopathic properties in *Trigonella foenum-graecum*. Generally, cover crop mixtures performed better than single cover crops in reducing weed pressure (Ch et al. 2016; Florence et al. 2019). Some authors suggested synergisms and additive between different allelochemicals can increase the successful likelihood in using mixture of allelopathic crops in reducing weeds in intercropping (Florence et al. 2019). Meanwhile, cover crop mixtures by compensate for the temporal and spatial variations in growing conditions led to competitive traits and increased weed-suppressive capacity when compared with single crops (Gerhards and Schappert 2020). Furthermore, allelochemicals affect weed species differently (Sturm et al. 2018).

4.8.2 Crop Cultivar (Genotype)

There is strong evidence that crop cultivars differ in their ability to compete with weeds (Mwendwa et al. 2020). Through stronger competition for limited resources, competitive crop cultivars suppress weeds throughout the period that this grow. Furthermore, some crop cultivars can minimize the growth of neighboring weeds through allelopathy (Colbach et al. 2019). High competitive crop indices produce high biomass, plant height, and leaf area, and uptake higher amounts of nutrients and moisture (Sardana et al. 2017).

Since the onset of the Green Revolution (1968), plant breeders have not sought to develop crops that have either high competitiveness against weeds or allelopathic properties. Instead, their goal is to produce high-yielding, low height, and erectophile leaf-arranged cultivars (Mwendwa et al. 2020). Obviously, these modifications in crop morphology contribute to reducing crop competitiveness of developed varieties against weeds (Sardana et al. 2017). Therefore, instead of weed suppressive crop cultivars, crops with improved competitiveness from modifications at the population level should be a priority, such as crop sowing date and density. Arranging and planting time of cleaning or smothering plants should be such as to help sensitive plants compete effectively with weeds. One such method is to use cleaning or to smother plants before planting crops with low competitiveness.

Competitive cultivars often have better access to light, nutrients, and water resources in limited space due to crop architectural traits, thus suppressing the growth and reproduction of nearby weed species. Differences in weed control ability of some of the Australian commercial winter wheat cultivars have been attributed to crop architecture and phenology early in the growing season (Mwendwa 2019).

It has been shown that canola cultivars with early seedling emergence, seedling vigor, rapid root growth and rate of leaf expansion, early root and shoot biomass accumulation and canopy closure and plant height have higher ability against weeds and can outcompete them successfully (Mwendwa et al. 2019)

Therefore, planting cultivars with higher weed competitive ability in conjugation with other weed methods by maintaining acceptable yields and controlling weeds is highly recommended.

4.8.3 Crop Rotation

Crop rotations that comprise crops with variable patterns of resource competition, allelopathic characteristics, and weed-reducing related agronomic practices can suppress weeds and hinder the dominance of a group-specific weed (Li et al. 2019). Crop rotation that creates variation in plant growth resources (nutrients, water, and light) would determine where and when the soil is suitable for seed germination (Koocheki et al. 2009).

Preventing weed seed production is one of the most important mechanisms of weed control by crop rotation. Weeds are successful in producing seeds when they conform to the life cycle of the crop plant (Coulter et al. 2011). When weed growth stages do not match those of the crop, weed seed production is disrupted because the weed may be harvested at the vegetative growth stage and before seed production. The most important benefit of diversifying crop is, therefore, the prevention of repeated seed setting of a specific type of weed (Li et al. 2019). Dominance of a few weeds is the consequence of monoculture (Nichols et al. 2015). However, crop choice is complicated as other factors, like the market, determine the choice of crops for planting, rather than it being based on the weed management strategy. The type of crop in the sequence by mediating meteorological effects (creating micro-climate) like moisture, sunlight, and temperature on weeds indirectly influences weed control efforts (Teasdale et al. 2018b). This technique by altering agronomic activity, and choosing cultivars with different sowing dates (winter versus spring crops) would be beneficial to manage and control weeds.

Crop rotation prevents the creation of specialized and invasive weed species. However, it develops a weed community characterized by low densities for each species (Buhler et al. 2000) and it could not eradicate troublesome weeds (Scavo and Mauromicale 2021). In general, including an allelopathic crop (e.g. wheat, oat, corn, alfalfa, sunflower) within rotational sequences can improve weed control (Scavo et al. 2019). Significant weed control results from crop rotation containing allelopathic crops can achieved in zero-tillage systems where, in addition to weed control, it enhances soil conservation (Scavo and Mauromicale 2021).

4.8.4 Cover Crops

Cover crops, which is planting herbaceous plants either for part of or an entire year for improving crop yield, have proven potential to be used an effective ecologically based weed suppressive tool in various agricultural systems (Johnson et al. 1993; Alonso-Ayuso et al. 2018; Büchi et al. 2020). Using cover crops for weed control in organic production systems under no-till system condition play a unique role as conducting tillage and herbicide application (two main component of weed control in conventional farming) are no longer available options (DeLaune 2020; Mennan et al. 2020).

Living mulch, cover crop residue, catch crop, smother crop, and green manure are considered as main ways by which cover crops can be used in agricultural fields (Teasdale et al. 2007; Kumar et al. 2020). Hence, allelopathy, competition, physical properties (presence of cover crop residues on the soil surface) are main mechanism underlying the weed control property of cover crops (Teasdale et al. 1998; Tursun et al. 2018). In addition, cover crops by providing habitat for seed predators as well as changing environmental factors, such as light, temperature, and soil moisture indirectly influence weed population in the field (Moonen and Barberi 2004; Ward et al. 2011; Schipanski et al. 2014).

Accordingly, a cover crop species with high biomass accumulation, early-season emergence, is more efficient in managing weeds. Further, the cover crops producing higher concentration of allelochemicals are more toxic against weeds (Osipitan et al. 2018). Attributing weed control mechanism offered by cover crops to an exact mechanism is hard and mainly is due to either competition or allelopathy, or both of them (Lawley et al. 2012).

Many cover crops, from cereals, legumes, to *Brassicaceae*, are widely used in various cropping systems in different parts of the world and have shown good results for weed suppression (Haramoto and Gallandt 2005; Altieri et al. 2011; Ch et al. 2016). Cover crops (applied as living mulch, catch crop, smother crop, companion crop, or green manure) intensively compete for resources (i.e. light, water, space, and nutrients) and release allelochemicals to control weeds (see Mulching and Allelopathy sections) (Kunz et al. 2016).

In addition to competition and allelopathy, weed suppression through cover crops is achieved by its physical effect on weeds (Mirsky et al. 2013). Hence, a cover crop having high biomass production (at least 8 t ha^{-1}), particularly in early seasons, is likely to result in satisfactory and sustainable weed suppression (Mennan et al. 2009). Modifying the light quality is the other mechanism through which cover crops are likely to suppress weeds (e.g. reducing weed biomass and seed production by reflecting far-red light depletion of the weed seedbank) (Mirsky et al. 2010; Cressman et al. 2011). Additionally, live cover crops change the quality of light reaching the soil surface, which may restrict the germination of many weed species (Mennan et al. 2020).

The weed suppressive ability of cover crops has received much attention (Norsworthy et al. 2012; Wallace et al. 2019). Furthermore, cover crops are an excellent option in conservation agriculture, where use of tillage is restricted for seedbed preparation and weed control (Samarajeewa et al. 2006).

In contrast to single-species monocultures, implementing species-diverse mixtures of cover crops is more efficient in soil productivity, stability of the ecosystem, resource capture, weed suppression, and pest suppression (Tilman et al. 2014; MacLaren et al. 2019; Wallace et al. 2019).

Despite several beneficial ecosystem functions and weed control properties of cover crops, they have some negative aspects, such as promoting some insect pests, increased disease transmission, inability to outcompete perennial weeds, competing with the cash crop, and low persistence of its allelopathic substances in the soil (Kruidhof et al. 2009; Mennan et al. 2020).

Weed control relying only on cultural methods is likely to be much more challenging in small-seeded crops. Furthermore, cultural methods are usually less effective on perennial weeds, which have underground reserves and resist applied control measures.

Despite effective weed suppression and ecosystem services provided by cover crops in agroecosystems there are some constraints, which hamper cover crop implements in agricultural ecosystems. Information obtained on such barriers is useful for finding solution ways and thereby maximizing the weed suppression ability of cover crops and their other ecosystem benefits.

One of the main problems related to using cover crops in agriculture systems is the persistence of allelochemicals in the soil that is not consistent and commonly is less than the desired period for effective inhibition of weed seed germination (Grünwald et al. 2000; Mennan et al. 2020). Furthermore, reducing the growth and yield of the main crop exposed by cover crop competition is considered as another serious concern related to these weed management strategy (Masiunas 1998).

Cover crops do have some negative aspects. These include promoting pest invasion (Grünwald et al. 2000), increasing germination or seedling growth of weeds (Kruidhof et al. 2009), being ineffective against intra-row and perennial weeds (Reberg-Horton et al. 2012), disturbing the seed sowing operation, inhibiting crop establishment (Mirsky et al. 2012), and creating phytotoxicity of cover crop allelochemicals against the main crop (Tabaglio et al. 2013).

4.9 Conclusions

Nonchemical methods for in-crop weed control are not effective enough for the suppression of dense populations of weeds. Hence, herbicides are still needed for satisfactory weed control in fields. Nevertheless, there is increasing public pressure to use alternative methods to control weeds due to the environmental side effect of herbicides, the increasing evolution of herbicide resistance in main agricultural weeds across the globe, and health concerns regarding pesticides. Therefore, a deep understanding and finding ways of enhancing the efficacy of nonchemical weed control methods have a critical role in agricultural production systems.

References

Abou Chehade, L., Fontanelli, M., Martelloni, L. et al. (2018). Effects of flame weeding on organic garlic production. *HortTechnology* 28: 502–508.

Abraham, C. and Singh, S. (1984). Weed management in sorghum-legume intercropping systems. *The Journal of Agricultural Science* 103: 103–115.

Adamchuk, V., Ji, W., Rossel, R.V. (2018). Proximal soil and plant sensing. In: *Precision Agriculture Basics* (eds. D.K. Shannon, D.E. Clay, and N.R. Kitchen). https://doi.org/10.2134/precisionagbasics.2016.0093. ACSESS.

Alba, O.S., Syrovy, L.D., Duddu, H.S.N., and Shirtliffe, S.J. (2020). Increased seeding rate and multiple methods of mechanical weed control reduce weed biomass in a poorly competitive organic crop. *Field Crops Research* 245: 107648.

Allan, E., Manning, P., Alt, F. et al. (2015). Land use intensification alters ecosystem multifunctionality via loss of biodiversity and changes to functional composition. *Ecology Letters* 18: 834–843.

Alonso-Ayuso, M., Gabriel, J.L., García-González, I. et al. (2018). Weed density and diversity in a long-term cover crop experiment background. *Crop Protection* 112: 103–111.

Alsaadawi, I.S., Sarbout, A.K., and Al-Shamma, L.M. (2012). Differential allelopathic potential of sunflower (Helianthus annuus L.) genotypes on weeds and wheat (Triticum aestivum L.) crop. *Archives of Agronomy and Soil Science* 58: 1139–1148.

Altieri, M.A., Lana, M.A., Bittencourt, H.V. et al. (2011). Enhancing crop productivity via weed suppression in organic no-till cropping systems in Santa Catarina, Brazil. *Journal of Sustainable Agriculture* 35: 855–869.

Álvarez-Iglesias, L., Puig, C.G., Revilla, P. et al. (2018). Faba bean as green manure for field weed control in maize. *Weed Research* 58: 437–449.

Anderson, R. (1998). Seedling emergence of winter annual grasses as affected by limited tillage and crop canopy. *Weed Technology* 12: 262–267.

Anzalone, A., Cirujeda, A., Aibar, J. et al. (2010). Effect of biodegradable mulch materials on weed control in processing tomatoes. *Weed Technology* 24: 369–377.

Armengot, L., Berner, A., Blanco-Moreno, J.M. et al. (2015). Long-term feasibility of reduced tillage in organic farming. *Agronomy for Sustainable Development* 35: 339–346.

Armengot, L., Blanco-Moreno, J.M., Bàrberi, P. et al. (2016). Tillage as a driver of change in weed communities: a functional perspective. *Agriculture, Ecosystems & Environment* 222: 276–285.

Armengot, L., José-María, L., Chamorro, L., and Sans, F.X. (2013). Weed harrowing in organically grown cereal crops avoids yield losses without reducing weed diversity. *Agronomy for Sustainable Development* 33: 405–411.

Asaduzzaman, M., Luckett, D.J., Cowley, R.B. et al. (2014). Canola cultivar performance in weed-infested field plots confirms allelopathy ranking from in vitro testing. *Biocontrol Science and Technology* 24: 1394–1411.

Aslam, F., Khaliq, A., Matloob, A. et al. (2017). Allelopathy in agro-ecosystems: a critical review of wheat allelopathy-concepts and implications. *Chemoecology* 27: 1–24.

Astatkie, T., Rifai, M., Havard, P. et al. (2007). Effectiveness of hot water, infrared and open flame thermal units for controlling weeds. *Biological Agriculture & Horticulture* 25: 1–12.

Baker, H.G. (1965). Characteristics and modes of origin of weeds. In: *The Genetics of Colonizing Species* (ed. H.G. Baker and G.L. Stebbins), 147–172. New York, NY, USA: Academic Press.

Baker, J.M., Dalton, F.N., Foy, C.D. et al. (2012). *Limitations to Plant Root Growth*, vol. 19. Springer Science & Business Media.

Ball, D.A. (1992). Weed seedbank response to tillage, herbicides, and crop rotation sequence. *Weed Science* 40: 654–659.

Banik, S., Bandyopadhyay, S., and Ganguly, S. (2003). Bioeffects of microwave—a brief review. *Bioresource Technology* 87: 155–159.

Bàrberi, P. and Lo Cascio, B. (2001). Long-term tillage and crop rotation effects on weed seedbank size and composition. *Weed Research* 41: 325–340.

Barbosa, L.C., Magalhães, P.S.G., Bordonal, R.O. et al. (2019). Soil physical quality associated with tillage practices during sugarcane planting in south-central Brazil. *Soil and Tillage Research* 195: 104383.

Bastiaans, L., Paolini, R., and Baumann, D. (2008). Focus on ecological weed management: what is hindering adoption? *Weed Research* 48: 481–491.

Bauer, M.V., Marx, C., Bauer, F.V. et al. (2020). Thermal weed control technologies for conservation agriculture—a review. *Weed Research* 60.

Beckie, H.J. (2020). *Herbicide Resistance in Plants.* Multidisciplinary Digital Publishing Institute.

Bedoussac, L., Journet, E.-P., Hauggaard-Nielsen, H. et al. (2015). Ecological principles underlying the increase of productivity achieved by cereal-grain legume intercrops in organic farming. A review. *Agronomy for Sustainable Development* 35: 911–935.

Benvenuti, S. (1995). Soil light penetration and dormancy of Jimsonweed (Datura stramonium) seeds. *Weed Science* 43: 389–393.

Berdjour, A., Dugje, I.Y., Dzomeku, I.K., and Rahman, N.A. (2020). Maize–soybean intercropping effect on yield productivity, weed control and diversity in northern Ghana. *Weed Biology and Management* 20: 69–81.

Bewley, J.D., Bradford, K., and Hilhorst, H. (2012). *Seeds: Physiology of Development, Germination and Dormancy.* Springer Science & Business Media.

Bilalis, D., Papastylianou, P., Konstantas, A. et al. (2010). Weed-suppressive effects of maize–legume intercropping in organic farming. *International Journal of Pest Management* 56: 173–181.

Blackshaw, R.E., Anderson, R., and Lemerle, D. (2007). Chapter 3. Cultural weed management. In: *Non-chemical Weed Management: Principles, Concepts and Technology* (ed. M.K. Upadhyaya and R.E. Blackshaw), 35–48. Wallingford, UK: CAB International (www.cabi.org).

Blaise, D., Manikandan, A., Verma, P. et al. (2020). Allelopathic intercrops and its mulch as an integrated weed management strategy for rainfed Bt-transgenic cotton hybrids. *Crop Protection* 135: 105214.

Blubaugh, C.K. and Kaplan, I. (2015). Tillage compromises weed seed predator activity across developmental stages. *Biological Control* 81: 76–82.

Bond, W., Turner, R., and Grundy, A. (2003). A review of non-chemical weed management. HDRA, the Organic Organisation, Ryton Organic Gardens, Coventry, UK, 81.

Børresen, T. and Njøs, A. (1994). The effect of ploughing depth and seedbed preparation on crop yields, weed infestation and soil properties from 1940 to 1990 on a loam soil in south eastern Norway. *Soil and Tillage Research* 32: 21–39.

Boscutti, F., Sigura, M., Gambon, N. et al. (2015). Conservation tillage affects species composition but not species diversity: a comparative study in northern Italy. *Environmental Management* 55: 443–452.

Boström, U., Andersson, L., Forkman, J. et al. (2013). Seasonal variation in sprouting capacity from intact rhizome systems of three perennial weeds. *Weed Research* 53: 387–398.

Boyd, N.S., Brennan, E.B., Smith, R.F., and Yokota, R. (2009). Effect of seeding rate and planting arrangement on rye cover crop and weed growth. *Agronomy Journal* 101: 47–51.

Brandsæter, L.O., Bakken, A.K., Mangerud, K. et al. (2011). Effects of tractor weight, wheel placement and depth of ploughing on the infestation of perennial weeds in organically farmed cereals. *European Journal of Agronomy* 34: 239–246.

Brandsæter, L.O., Fogelfors, H., Fykse, H. et al. (2010). Seasonal restrictions of bud growth on rhizomes of Elymus repens and roots of Cirsium arvense and Sonchus arvensis. *Weed Research* 50: e109.

Brandsæter, L.O., Mangerud, K., Helgheim, M., and Berge, T.W. (2017). Control of perennial weeds in spring cereals through stubble cultivation and mouldboard ploughing during autumn or spring. *Crop Protection* 98: 16–23.

Brandsæter, L.O., Thomsen, M.G., Wærnhus, K., and Fykse, H. (2012). Effects of repeated clover undersowing in spring cereals and stubble treatments in autumn on Elymus repens, Sonchus arvensis and Cirsium arvense. *Crop Protection* 32: 104–110.

Brault, D., Stewart, K., and Jenni, S. (2002). Optical properties of paper and polyethylene mulches used for weed control in lettuce. *HortScience* 37: 87–91.

Brodie, G. (2018). The use of physics in weed control. In: *Non-chemical Weed Control.* 33–59. London: Elsevier.

Brooker, R.W., Bennett, A.E., Cong, W.F. et al. (2015). Improving intercropping: a synthesis of research in agronomy, plant physiology and ecology. *New Phytologist* 206: 107–117.

Büchi, L., Wendling, M., Amossé, C. et al. (2020). Cover crops to secure weed control strategies in a maize crop with reduced tillage. *Field Crops Research* 247: 107583.

Buhler, D.D. (1997). Effects of tillage and light environment on emergence of 13 annual weeds. *Weed Technology* 11: 496–501.

Buhler, D.D., Liebman, M., and Obrycki, J.J. (2000). Theoretical and practical challenges to an IPM approach to weed management. *Weed Science* 48: 274–280.

Buhler, D.D., Becker, R.L., and Gunsolus, J.L. (1994). Perennial weed populations after 14 years of variable tillage and cropping practices. *Weed Science* 42 (2): 205–209.

Burgos-Artizzu, X.P., Ribeiro, A., Tellaeche, A. et al. (2010). Analysis of natural images processing for the extraction of agricultural elements. *Image and Vision Computing* 28: 138–149.

Caamal-Maldonado, J.A., Jiménez-Osornio, J.J., Torres-Barragán, A., and Anaya, A.L. (2001). The use of allelopathic legume cover and mulch species for weed control in cropping systems. *Agronomy Journal* 93: 27–36.

Ch, K., Sturm, D.J., Varnholt, D. et al. (2016). Allelopathic effects and weed suppressive ability of cover crops. *Plant, Soil and Environment* 62: 60–66.

Chauhan, B.S. and Johnson, D.E. (2010). The role of seed ecology in improving weed management strategies in the tropics. *Advances in Agronomy* 105: 221–262. Elsevier.

Chauhan, B.S. and Mahajan, G. (2012). Role of integrated weed management strategies in sustaining conservation agriculture systems. *Current Science* 103: 135–136.

Chauhan, B.S., Singh, R.G., and Mahajan, G. (2012). Ecology and management of weeds under conservation agriculture: a review. *Crop Protection* 38: 57–65.

Cheema, Z.A., Asim, M., and Khaliq, A. (2000). Sorghum allelopathy for weed control in cotton (Gossypium arboreum L.). *International Journal of Agriculture and Biology* 2: 37–41.

Chee-Sanford, J.C., Williams, M.M., Davis, A.S., and Sims, G.K. (2006). Do microorganisms influence seed-bank dynamics? *Weed Science* 54: 575–587.

Cheriere, T., Lorin, M., and Corre-Hellou, G. (2020). Species choice and spatial arrangement in soybean-based intercropping: levers that drive yield and weed control. *Field Crops Research* 256: 107923.

Cherr, C., Scholberg, J., and McSorley, R. (2006). Green manure approaches to crop production: a synthesis. *Agronomy Journal* 98: 302–319.

Chhokar, R.S. and Sharma, R.K. (2008). Multiple herbicide resistance in littleseed canarygrass (Phalaris minor): a threat to wheat production in India. *Weed Biology and Management* 8: 112–123.

Chovancova, S., Neudert, L., and Winkler, J. (2019). The effect of three soil tillage treatments on weed infestation in forage maize. *Acta Agrobotanica* 72.

Christoffoleti, P. and Caetano, R. (1998). Soil seed banks. *Scientia Agricola* 55: 74–78.

Cirujeda, A., Aibar, J., Anzalone, Á. et al. (2012). Biodegradable mulch instead of polyethylene for weed control of processing tomato production. *Agronomy for Sustainable Development* 32: 889–897.

Colbach, N., Collard, A., Guyot, S.H. et al. (2014). Assessing innovative sowing patterns for integrated weed management with a 3D crop: weed competition model. *European Journal of Agronomy* 53: 74–89.

Colbach, N., Gardarin, A., and Moreau, D. (2019). The response of weed and crop species to shading: which parameters explain weed impacts on crop production? *Field Crops Research* 238: 45–55.

Connolly, B., Pearson, D., and Mack, R. (2014). Granivory of invasive, naturalized, and native plants in communities differentially susceptible to invasion. *Ecology* 95: 1759–1769.

Cooper, J., Baranski, M., Stewart, G. et al. (2016). Shallow non-inversion tillage in organic farming maintains crop yields and increases soil C stocks: a meta-analysis. *Agronomy for Sustainable Development* 36: 22.

Cordeau, S., Guillemin, J.P., Reibel, C., and Chauvel, B. (2015). Weed species differ in their ability to emerge in no-till systems that include cover crops. *Annals of Applied Biology* 166: 444–455.

Cordeau, S., Smith, R.G., Gallandt, E.R. et al. (2017). How do weeds differ in their response to the timing of tillage? A study of 61 species across the northeastern United States. *Annals of Applied Biology* 171: 340–352.

Coulter, J.A., Sheaffer, C.C., Wyse, D.L. et al. (2011). Agronomic performance of cropping systems with contrasting crop rotations and external inputs. *Agronomy Journal* 103: 182–192.

Cox, W.J., Zobel, R.W., van Es, H.M., and Otis, D.J. (1990). Tillage effects on some soil physical and corn physiological characteristics. *Agronomy Journal* 82: 806–812.

Cressman, S.T., Page, E.R., and Swanton, C.J. (2011). Weeds and the red to far-red ratio of reflected light: characterizing the influence of herbicide selection, dose, and weed species. *Weed Science* 59: 424–430.

Dabney, S., Delgado, J., and Reeves, D. (2001). Using winter cover crops to improve soil and water quality. *Communications in Soil Science and Plant Analysis* 32 (7–8): 1221–1250.

Davis, A.S., Schutte, B.J., Iannuzzi, J., and Renner, K.A. (2008). Chemical and physical defense of weed seeds in relation to soil seedbank persistence. *Weed Science* 56: 676–684.

Dekker, J. (1999). Soil weed seed banks and weed management. *Journal of Crop Production* 2: 139–166.

DeLaune, P. (2020). Managing cover crops in no-till farming systems. In: *No-till Farming Systems for Sustainable Agriculture* (ed. Y. Dang, R. Dalal, and N. Menzies), 47–63. Cham: Springer.

Derksen, D.A., Anderson, R.L., Blackshaw, R.E., and Maxwell, B. (2002). Weed dynamics and management strategies for cropping systems in the northern Great Plains. *Agronomy Journal* 94: 174–185.

Di Tomaso, J.M. (1995). Approaches for improving crop competitiveness through the manipulation of fertilization strategies. *Weed Science* 43: 491–497.

Downey, D., Giles, D., and Slaughter, D. (2004). Weeds accurately mapped using DGPS and ground-based vision identification. *California Agriculture* 58: 218–221.

Evers, J.B. and Bastiaans, L. (2016). Quantifying the effect of crop spatial arrangement on weed suppression using functional-structural plant modelling. *Journal of Plant Research* 129: 339–351.

Fabrizzi, K.P., García, F.O., Costa, J.L., and Picone, L.I. (2005). Soil water dynamics, physical properties and corn and wheat responses to minimum and no-tillage systems in the southern Pampas of Argentina. *Soil and Tillage Research* 81: 57–69.

Farooq, M., Flower, K., Jabran, K. et al. (2011a). Crop yield and weed management in rainfed conservation agriculture. *Soil and Tillage Research* 117: 172–183.

Farooq, M., Jabran, K., Cheema, Z.A. et al. (2011b). The role of allelopathy in agricultural pest management. *Pest Management Science* 67: 493–506.

Farooq, N., Abbas, T., Tanveer, A., and Jabran, K. (2020). Allelopathy for weed management. In: *Co-evolution of Secondary Metabolites* (ed. J.M. Mérillon and K. Ramawat), Reference Series in Phytochemistry. 505–519. Cham: Springer. https://doi.org/10.1007/978-3-319-96397-6_16.

Fırıncıoğlu, H.K. (2014). A comparison of six vetches (V icia spp.) for developmental rate, herbage yield and seed yield in semi-arid central Turkey. *Grass and Forage Science* 69: 303–314.

Florence, A., Higley, L., Drijber, R. et al. (2019). Cover crop mixture diversity, biomass productivity, weed suppression, and stability. *PloS One* 14: e0206195.

Forcella, F. (1992). Prediction of weed seedling densities from buried seed reserves. *Weed Research* 32: 29–38.

Francis, C.M., Enneking, D., and Abd El Moneim, A.M. (1999). When and where will vetches have an impact as grain legumes? In: *InLinking Research and Marketing Opportunities for Pulses in the 21st Century 2000* (ed. R. Knight), 375–384. Dordrecht: Springer. https://doi.org/10.1007/978-94-011-4385-1_34.

Fried, G., Kazakou, E., and Gaba, S. (2012). Trajectories of weed communities explained by traits associated with species' response to management practices. *Agriculture, Ecosystems & Environment* 158: 147–155.

Froud-Williams, R.J., Chancellor, R.J., and Drennan, D.S.H. (1981). Potential changes in weed floras associated with reduced-cultivation systems for cereal production in temperate regions. *Weed Research* 21: 99–109.

Gallandt, E.R. (2006). How can we target the weed seedbank? *Weed Science* 54: 588–596.

Gallandt, E.R., Molloy, T., Lynch, R.P., and Drummond, F.A. (2005). Effect of cover-cropping systems on invertebrate seed predation. *Weed Science* 53: 69–76.

Gardarin, A., Dürr, C., and Colbach, N. (2010a). Effects of seed depth and soil aggregates on the emergence of weeds with contrasting seed traits. *Weed Research* 50: 91–101.

Gardarin, A., Dürr, C., Mannino, M.R. et al. (2010b). Seed mortality in the soil is related to seed coat thickness. *Seed Science Research* 20: 243.

Gardarin, A., Guillemin, J.-P., Munier-Jolain, N., and Colbach, N. (2010c). Estimation of key parameters for weed population dynamics models: base temperature and base water potential for germination. *European Journal of Agronomy* 32: 162–168.

Gerhards, R. and Schappert, A. (2020). Advancing cover cropping in temperate integrated weed management. *Pest Management Science* 76: 42–46.

Ghersa, C. and Martınez-Ghersa, M. (2000). Ecological correlates of weed seed size and persistence in the soil under different tilling systems: implications for weed management. *Field Crops Research* 67: 141–148.

Ghosh, P., Manna, M., Bandyopadhyay, K. et al. (2006). Interspecific interaction and nutrient use in soybean/sorghum intercropping system. *Agronomy Journal* 98: 1097–1108.

Gianessi, L.P. (2013). The increasing importance of herbicides in worldwide crop production. *Pest Management Science* 69: 1099–1105.

Golmohammadzadeh, S., Zaefarian, F., and Rezvani, M. (2015). Effects of some chemical factors, prechilling treatments and interactions on the seed dormancy-breaking of two Papaver species. *Weed Biology and Management* 15: 11–19.

Gómez-Maqueo, X. and Gamboa-deBuen, A. (2016). The dynamics of plant cell wall in muro modifications and its physiological implications on seed germination. In: *New*

Challenges in Seed Biology-basic and Translational Research Driving Seed Technology (ed. S. Araujo and A. Balestrazzi), 155–177. London: InTech. https://doi.org/10.5772/64085.

Govindasamy, P., Sarangi, D., Provin, T. et al. (2020). No-tillage altered weed species dynamics in a long-term (36-years) grain sorghum experiment in southeast Texas. *Weed Science* 68: 1–33.

Graglia, E., Melander, B., and Jensen, R.K. (2006). Mechanical and cultural strategies to control Cirsium arvense in organic arable cropping systems. *Weed Research* 46: 304–312.

Gruber, S. and Claupein, W. (2009). Effect of tillage intensity on weed infestation in organic farming. *Soil and Tillage Research* 105: 104–111.

Grundy, A.C. and Mead, A. (2000). Modeling weed emergence as a function of meteorological records. *Weed Science* 48: 594–603.

Grünwald, N.J., Hu, S., and van Bruggen, A.H.C. (2000). Short-term cover crop decomposition in organic and conventional soils: characterization of soil C, N, microbial and plant pathogen dynamics. *European Journal of Plant Pathology* 106: 37–50.

Guillemin, J.-P., Gardarin, A., Granger, S. et al. (2013). Assessing potential germination period of weeds with base temperatures and base water potentials. *Weed Research* 53: 76–87.

Gul, B., Marwat, K.B., Hassan, G. et al. (2009). Impact of tillage, plant population and mulches on biological yield of maize. *Pakistan Journal of Botany* 41: 2243–2249.

Håkansson, I., Stenberg, M., and Rydberg, T. (1998). Long-term experiments with different depths of mouldboard ploughing in Sweden. *Soil and Tillage Research* 46: 209–223.

Håkansson, S. (2003). *Weeds and Weed Management on Arable Land: An Ecological Approach*. CABI Publishing.

Haramoto, E.R. and Gallandt, E.R. (2004). Brassica cover cropping for weed management: a review. *Renewable Agriculture and Food Systems* 19: 187–198.

Haramoto, E.R. and Gallandt, E.R. (2005). Brassica cover cropping: I. Effects on weed and crop establishment. *Weed Science* 53: 695–701.

Haring, S.C. and Flessner, M.L. (2018). Improving soil seed bank management. *Pest Management Science* 74: 2412–2418.

Harker, K., Clayton, G., Blackshaw, R. et al. (2003). Seeding rate, herbicide timing and competitive hybrids contribute to integrated weed management in canola (Brassica napus). *Canadian Journal of Plant Science* 83: 433–440.

Hasan, M., Ahmad-Hamdani, M.S., Rosli, A.M., and Hamdan, H. (2021). Bioherbicides: an eco-friendly tool for sustainable weed management. *Plants* 10: 1212.

Hatcher, P. and Melander, B. (2003). Combining physical, cultural and biological methods: prospects for integrated non-chemical weed management strategies. *Weed Research* 43: 303–322.

Heap, I. (2020). The international herbicide-resistant weed database. Online. Monday, May 16, 2022. www.weedscience.org.

Heisel, T., Schou, J., Andreasen, C., and Christensen, S. (2002). Using laser to measure stem thickness and cut weed stems. *Weed Research* 42: 242–248.

Hofmeijer, M.A., Krauss, M., Berner, A. et al. (2019). Effects of reduced tillage on weed pressure, nitrogen availability and winter wheat yields under organic management. *Agronomy* 9: 180.

Hussain, M., Farooq, S., Merfield, C., and Jabran, K. (2018). Mechanical weed control. In: *Non-Chemical Weed Control* (ed. K. Jabran and B.S. Chauhan), 133–155. Elsevier.

Jabran, K. (2019). *Role of Mulching in Pest Management and Agricultural Sustainability*. Springer.

Jabran, K. and Chauhan, B.S. (2018). Weed control using ground cover systems. In: *Non-Chemical Weed Control*, 61–71. Elsevier.

Jha, P., Kumar, V., Godara, R.K., and Chauhan, B.S. (2017). Weed management using crop competition in the United States: a review. *Crop Protection* 95: 31–37.

Johnson, G.A., Defelice, M.S., and Helsel, Z.R. (1993). Cover crop management and weed control in corn (Zea mays). *Weed Technology* 7: 425–430.

Júnior, F.W.R., Scariot, M.A., Forte, C.T. et al. (2019). New perspectives for weeds control using autochthonous fungi with selective bioherbicide potential. *Heliyon* 5: e01676.

Kakabouki, I., Karkanis, A., Travlos, I.S. et al. (2015). Weed flora and seed yield in quinoa crop (Chenopodium quinoa Willd.) as affected by tillage systems and fertilization practices. *International Journal of Pest Management* 61: 228–234.

Kanatas, P., Travlos, I., Gazoulis, J. et al. (2020). The combined effects of false seedbed technique, post-emergence chemical control and cultivar on weed management and yield of barley in Greece. *Phytoparasitica* 48: 131–143.

Kennedy, A. (1999). Soil microorganisms for weed management. *Journal of Crop Production* 2: 123–138.

Khamare, Y., Marble, S.C., and Chandler, A. (2020). Fertilizer placement effects on eclipta (Eclipta prostrata) growth and competition with container-grown ornamentals. *Weed Science* 68: 1–28.

Khan, M.B., Khan, M., Hussain, M. et al. (2012). Bio-economic assessment of different wheat-canola intercropping systems. *International Journal of Agriculture and Biology* 14.

King, A. (2017). The future of agriculture. *Nature* 544: S21–S23.

Knezevic, S.Z., Horak, M.J., and Vanderlip, R.L. (1997). Relative time of redroot pigweed (Amaranthus retroflexus

L.) emergence is critical in pigweed-sorghum [Sorghum bicolor (L.) Moench] competition. *Weed Science* 45: 502–508.

Kolb, L. and Gallandt, E. (2012). Weed management in organic cereals: advances and opportunities. *Organic Agriculture* 2: 23–42.

Kolb, L., Gallandt, E., and Molloy, T. (2010). Improving weed management in organic spring barley: physical weed control vs. interspecific competition. *Weed Research* 50: 597–605.

Koocheki, A., Nassiri, M., Alimoradi, L., and Ghorbani, R. (2009). Effect of cropping systems and crop rotations on weeds. *Agronomy for Sustainable Development* 29: 401–408.

Korres, N.E., Norsworthy, J.K., Mauromoustakos, A., and Williams, M.M. (2020). Soybean density and Palmer amaranth (Amaranthus palmeri) establishment time: effects on weed biology, crop yield and economic returns. *Weed Science* 68: 1–27.

Kremer, R.J. and Li, J. (2003). Developing weed-suppressive soils through improved soil quality management. *Soil and Tillage Research* 72: 193–202.

Kruidhof, H.M., Bastiaans, L., and Kropff, M.J. (2009). Cover crop residue management for optimizing weed control. *Plant and Soil* 318: 169–184.

Kuang, W. and Nelson, S. (1998). Low-frequency dielectric properties of biological tissues: a review with some new insights. *Transactions of t.he ASAE* 41 (1): 173–184.

Kumar, V., Obour, A., Jha, P. et al. (2020). Integrating cover crops for weed management in the semiarid US Great Plains: opportunities and challenges. *Weed Science* 68: 311–323.

Kunz, C., Sturm, D.J., Varnholt, D. et al. (2016). Allelopathic effects and weed suppressive ability of cover crops. *Plant, Soil and Environment* 62: 60–66.

Kurstjens, D. and Kropff, M. (2001). The impact of uprooting and soil-covering on the effectiveness of weed harrowing. *Weed Research* 41: 211–228.

Kurstjens, D. and Perdok, U. (2000). The selective soil covering mechanism of weed harrows on sandy soil. *Soil and Tillage Research* 55: 193–206.

Lament, W.J. (1993). Plastic mulches for the production of vegetable crops. *HortTechnology* 3: 35–39.

Lawley, Y.E., Teasdale, J.R., and Weil, R.R. (2012). The mechanism for weed suppression by a forage radish cover crop. *Agronomy Journal* 104: 205–214.

Leblanc, M.L., Cloutier, D.C., and Stewart, K.A. (2006). Rotary hoe cultivation in sweet corn. *HortTechnology* 16: 583–589.

Lee, S.-M., Lee, B.-M., Lee, Y. et al. (2012). Effect of green-manure crop rotation on weed control. *Korean Journal of Organic Agriculture* 20: 201–209.

Lee, W.S., Slaughter, D., and Giles, D. (1999). Robotic weed control system for tomatoes. *Precision Agriculture* 1: 95–113.

Lemerle, D., Gill, G., Murphy, C. et al. (2001). Genetic improvement and agronomy for enhanced wheat competitiveness with weeds. *Australian Journal of Agricultural Research* 52: 527–548.

Lemerle, D., Luckett, D., Koetz, E., and Wu, H. (2011). Canola (Brassica napus) competition for weed management. *23 rd Asian-Pacific Weed Science Society Conference*, 278.

Leon, R.G. and Ferreira, D.T. (2008). Interspecific differences in weed susceptibility to steam injury. *Weed Technology* 22: 719–723.

Li, J., Huang, L., Zhang, J. et al. (2019). Diversifying crop rotation improves system robustness. *Agronomy for Sustainable Development* 39: 38.

Li, Y., Sun, Z., Zhuang, X. et al. (2003). Research progress on microbial herbicides. *Crop Protection* 22: 247–252.

Liebman, M. and Davis, A. (2000). Integration of soil, crop and weed management in low-external-input farming systems. *Weed Research-Oxford-* 40: 27–48.

Liebman, M. and Dyck, E. (1993). Crop rotation and intercropping strategies for weed management. *Ecological Applications* 3: 92–122.

Liebman, M., Mohler, C.L., and Staver, C.P. (2001). *Ecological Management of Agricultural Weeds*. Cambridge University Press.

Little, N.G., Mohler, C.L., Ketterings, Q.M., and DiTommaso, A. (2015). Effects of organic nutrient amendments on weed and crop growth. *Weed Science* 63: 710–722.

Liu, Q., Xu, R., Yan, Z. et al. (2013). Phytotoxic allelochemicals from roots and root exudates of Trifolium pratense. *Journal of Agricultural and Food Chemistry* 61: 6321–6327.

Loddo, D., Scarabel, L., Sattin, M. et al. (2020). Combination of herbicide band application and inter-row cultivation provides sustainable weed control in maize. *Agronomy* 10: 20.

Lovett, J. (1991). Changing perceptions of allelopathy and biological control. *Biological Agriculture & Horticulture* 8: 89–100.

Lu, P., Jiang, B., and Weiner, J. (2020). Crop spatial uniformity, yield and weed suppression. In: *Advances in Agronomy*, 117–178. Elsevier.

Macías, F.A., Mejías, F.J., and Molinillo, J.M. (2019). Recent advances in allelopathy for weed control: from knowledge to applications. *Pest Management Science* 75: 2413–2436.

MacLaren, C., Swanepoel, P., Bennett, J. et al. (2019). Cover crop biomass production is more important than diversity for weed suppression. *Crop Science* 59: 733–748.

Mahajan, G., Sharda, R., Kumar, A., and Singh, K. (2007). Effect of plastic mulch on economizing irrigation water and weed control in baby corn sown by different methods. *African Journal of Agricultural Research* 2: 19–26.

Marín, C. and Weiner, J. (2014). Effects of density and sowing pattern on weed suppression and grain yield in three varieties of maize under high weed pressure. *Weed Research* 54: 467–474.

Mas, M.T. and Verdú, A.M. (2003). Tillage system effects on weed communities in a 4-year crop rotation under Mediterranean dryland conditions. *Soil and Tillage Research* 74: 15–24.

Masiunas, J.B. (1998). Production of vegetables using cover crop and living mulches—a review. *Journal of Vegetable Crop Production* 4: 11–31.

Mathiassen, S.K., Bak, T., Christensen, S., and Kudsk, P. (2006). The effect of laser treatment as a weed control method. *Biosystems Engineering* 95: 497–505.

Matthews, J.M., Llewellyn, R., Powles, S., and Reeves, T. (1996). Integrated weed management for the control of herbicide resistant annual ryegrass. *Proceedings of the 8th Australian Agronomy Conference*, Toowoomba, Queensland, Australia, 417–420.

Melander, B., Holst, N., Rasmussen, I.A., and Hansen, P.K. (2012). Direct control of perennial weeds between crops–Implications for organic farming. *Crop Protection* 40: 36–42.

Melander, B., Rasmussen, I.A., and Bàrberi, P. (2005). Integrating physical and cultural methods of weed control—examples from European research. *Weed Science* 53: 369–381.

Melander, B., Rasmussen, I.A., and Olesen, J.E. (2020). Legacy effects of leguminous green manure crops on the weed seed bank in organic crop rotations. *Agriculture, Ecosystems & Environment* 302: 107078.

Mennan, H., Jabran, K., Zandstra, B.H., and Pala, F. (2020). Non-chemical weed management in vegetables by using cover crops: a review. *Agronomy* 10: 257.

Mennan, H., Ngouajio, M., Isık, D., and Kaya, E. (2009). Effects of alternative winter cover cropping systems on weed suppression in organically grown tomato (Solanum lycopersicum). *Phytoparasitica* 37: 385–396.

Meurer, K.H.E., Haddaway, N.R., Bolinder, M.A., and Kätterer, T. (2018). Tillage intensity affects total SOC stocks in boreo-temperate regions only in the topsoil—a systematic review using an ESM approach. *Earth-Science Reviews* 177: 613–622.

Mikić, A., Mihailović, V., Ćupina, B. et al. (2013). Agronomic characteristics related to grain yield and crude protein content in common vetch (Vicia sativa) accessions of diverse geographic origin. *New Zealand Journal of Agricultural Research* 56: 297–308.

Mirsky, S., Gallandt, E., Mortensen, D. et al. (2010). Reducing the germinable weed seedbank with soil disturbance and cover crops. *Weed Research* 50: 341–352.

Mirsky, S.B., Curran, W.S., Mortensen, D.A. et al. (2009). Control of cereal rye with a roller/crimper as influenced by cover crop phenology. *Agronomy Journal* 101: 1589–1596.

Mirsky, S.B., Ryan, M.R., Curran, W.S. et al. (2012). Conservation tillage issues: cover crop-based organic rotational no-till grain production in the mid-Atlantic region, USA. *Renewable Agriculture and Food Systems* 27: 31–40.

Mirsky, S.B., Ryan, M.R., Teasdale, J.R. et al. (2013). Overcoming weed management challenges in cover crop–based organic rotational no-till soybean production in the eastern United States. *Weed Technology* 27: 193–203.

Mohler, C.L. (1996). Ecological bases for the cultural control of annual weeds. *Journal of Production Agriculture* 9: 468–474.

Monks, C.D., Monks, D.W., Basden, T. et al. (1997). Soil temperature, soil moisture, weed control, and tomato (Lycopersicon esculentum) response to mulching. *Weed Technology* 11: 561–566.

Moonen, A. and Barberi, P. (2004). Size and composition of the weed seedbank after 7 years of different cover-crop-maize management systems. *Weed Research* 44: 163–177.

Morin, L. (2020). Progress in biological control of weeds with plant pathogens. *Annual Review of Phytopathology* 58.

Munkholm, L.J., Heck, R.J., and Deen, B. (2013). Long-term rotation and tillage effects on soil structure and crop yield. *Soil and Tillage Research* 127: 85–91.

Mwendwa, J., Brown, W., Weston, P. et al. (2019). Evaluation of selected commercial canola cultivars for early vigour, weed suppression and yield in southern New South Wales. Mechanisms of Weed Suppression in Wheat (Triticum aestivum L.) and Canola (Brassica napus L.) in Southeastern Australia, 105.

Mwendwa, J.M. (2019). Mechanisms of weed suppression in wheat (Triticum aestivum L.) and canola (Brassica napus L.) in southeastern Australia. Charles Sturt University, Wagga Wagga, Australia.

Mwendwa, J.M., Brown, W.B., Weidenhamer, J.D. et al. (2020). Evaluation of commercial wheat cultivars for canopy architecture, early vigour, weed suppression, and yield. *Agronomy* 10: 983.

Narwal, S. (2000). Weed management in rice: wheat rotation by allelopathy. *Critical Reviews in Plant Sciences* 19: 249–266.

Nelson, E.B. (2018). The seed microbiome: origins, interactions, and impacts. *Plant and Soil* 422: 7–34.

Ngouajio, M. and Ernest, J. (2004). Light transmission through colored polyethylene mulches affects weed populations. *HortScience* 39: 1302–1304.

Nichols, V., Verhulst, N., Cox, R., and Govaerts, B. (2015). Weed dynamics and conservation agriculture principles: a review. *Field Crops Research* 183: 56–68.

Nikolić, N., Squartini, A., Concheri, G. et al. (2020). Weed seed decay in no-till field and planted riparian buffer zone. *Plants* 9: 293.

Njøs, A. and Ekeberg, E. (1980). Forsok med ploying til to dybder host og vAr pA morenejord i Stange i Arene 1969–1975. Forskning og forsok i landbruket.= Research in Norwegian agriculture.

Norsworthy, J.K., Ward, S.M., Shaw, D.R. et al. (2012). Reducing the risks of herbicide resistance: best management practices and recommendations. *Weed Science* 60: 31–62.

Nosratti, I., Almaleki, S., and Chauhan, B.S. (2019). Seed germination ecology of Soldier Thistle (*Picnomon acarna*): an invasive weed of rainfed crops in Iran. *Weed Science* 67: 261–266, 266.

Olsen, J., Kristensen, L., and Weiner, J. (2005). Effects of density and spatial pattern of winter wheat on suppression of different weed species. *Weed Science* 53: 690–694.

Olsen, J.M., Griepentrog, H.-W., Nielsen, J., and Weiner, J. (2012). How important are crop spatial pattern and density for weed suppression by spring wheat? *Weed Science* 60: 501–509.

Ona, A., Grazina, K., and Birute, J. (2018). The influence of crop density and sowing delay on weed germination in winter wheat. *Julius-Kühn-Archiv* 458: 307–310.

Osipitan, O.A., Dille, J.A., Assefa, Y., and Knezevic, S.Z. (2018). Cover crop for early season weed suppression in crops: systematic review and meta-analysis. *Agronomy Journal* 110: 2211–2221.

Pallottino, F., Menesatti, P., Figorilli, S. et al. (2018). Machine vision retrofit system for mechanical weed control in precision agriculture applications. *Sustainability* 10: 2209.

Pardo, G., Cirujeda, A., Perea, F. et al. (2019). Effects of reduced and conventional tillage on weed communities: results of a long-term experiment in southwestern Spain. *Planta Daninha* 37.

Pedersen, T.R. and Gustavsson, A.-M.D. (2003). *Rotogräs: Råd i praktiken*. Jordbruksverket.

Peigné, J., Ball, B.C., Roger-Estrade, J., and David, C. (2007). Is conservation tillage suitable for organic farming? A review. *Soil Use and Management* 23: 129–144.

Peruzzi, A., Martelloni, L., Frasconi, C. et al. (2017). Machines for non-chemical intra-row weed control in narrow and wide-row crops: a review. *Journal of Agricultural Engineering* 48 (2): 57–70.

Place, G.T., Reberg-Horton, S.C., and Jordan, D.L. (2010). Interaction of cultivar, planting pattern, and weed management tactics in peanut. *Weed Science* 58: 442–448.

Plaza, E.H., Kozak, M., Navarrete, L., and Gonzalez-Andujar, J.L. (2011). Tillage system did not affect weed diversity in a 23-year experiment in Mediterranean dryland. *Agriculture, Ecosystems & Environment* 140: 102–105.

Pouryousef, M., Yousefi, A.R., Oveisi, M., and Asadi, F. (2015). Intercropping of fenugreek as living mulch at different densities for weed suppression in coriander. *Crop Protection* 69: 60–64.

Prill, S. (2017). *Biodegradable Plant Mulch Device*. Google Patents.

Rad, S.V., Valadabadi, S.A.R., Pouryousef, M. et al. (2020). Quantitative and qualitative evaluation of Sorghum bicolor L. under intercropping with legumes and different weed control methods. *Horticulturae* 6: 78.

Raffaelli, M., Martelloni, L., Frasconi, C. et al. (2013). Development of machines for flaming weed control on hard surfaces. *Applied Engineering in Agriculture* 29: 663–673.

Rask, A.M., Larsen, S., Andreasen, C., and Kristoffersen, P. (2013). Determining treatment frequency for controlling weeds on traffic islands using chemical and non-chemical weed control. *Weed Research* 53: 249–258.

Rasmussen, J. (1991). A model for prediction of yield response in weed harrowing. *Weed Research* 31: 401–408.

Rasmussen, J., Bibby, B.M., and Schou, A.P. (2008). Investigating the selectivity of weed harrowing with new methods. *Weed Research* 48: 523–532.

Reberg-Horton, S.C., Grossman, J.M., Kornecki, T.S. et al. (2012). Utilizing cover crop mulches to reduce tillage in organic systems in the southeastern USA. *Renewable Agriculture and Food Systems* 27: 41–48.

Rebetzke, G., Bruce, S., and Kirkegaard, J. (2005). Longer coleoptiles improve emergence through crop residues to increase seedling number and biomass in wheat (Triticum aestivum L.). *Plant and Soil* 272: 87–100.

Robert, S.G. and John, C. (1998). The effect of light environment during tillage on the recruitment of various summer annuals. *Weed Science* 46: 214–216.

Rose, M. (2019). Hand weeding really isn't effective for most crops. *New Scientist* 243: 26.

Ross, S.M., King, J.R., Izaurralde, R.C., and O'Donovan, J.T. (2001). Weed suppression by seven clover species. *Agronomy Journal* 93: 820–827.

Roth, C.M., Shroyer, J.P., and Paulsen, G.M. (2000). Allelopathy of sorghum on wheat under several tillage systems. *Agronomy Journal* 92: 855–860.

Rueda-Ayala, V., Rasmussen, J., and Gerhards, R. (2010). Mechanical weed control. 279–294.

Sahin, H. (2021). Comparative analysis of microwave and electric arc as a new method in weed control. *European Journal of Engineering and Technology Research* 6: 132–135.

Sahin, H. and Yalınkılıc, M. (2017). Using electric current as a weed control method. *European Journal of Engineering and Technology Research* 2: 59–64.

Sainju, U.M. and Singh, B.P. (2008). Nitrogen storage with cover crops and nitrogen fertilization in tilled and nontilled soils. *Agronomy Journal* 100: 619–627.

Samarajeewa, K., Horiuchi, T., and Oba, S. (2006). Finger millet (Eleucine corocana L. Gaertn.) as a cover crop on weed control, growth and yield of soybean under different tillage systems. *Soil and Tillage Research* 90: 93–99.

Samtani, J.B., Derr, J., Conway, M.A., and Flanagan, R.D. (2017). Evaluating soil solarization for weed control and strawberry (Fragaria xananassa) yield in annual plasticulture production. *Weed Technology* 31: 455–463.

Sans, F.X., Berner, A., Armengot, L., and Mäder, P. (2011). Tillage effects on weed communities in an organic winter wheat–sunflower–spelt cropping sequence. *Weed Research* 51: 413–421.

Santín-Montanyá, M. and Sombrero Sacristán, A. (2019). The effects of soil tillage techniques on weed flora in high input barley systems in northern Spain. *Canadian Journal of Plant Science* 100.

Sarabi, V. (2019). Factors that influence the level of weed seed predation: a review. *Weed Biology and Management* 19: 61–74.

Sardana, V., Mahajan, G., Jabran, K., and Chauhan, B.S. (2017). Role of competition in managing weeds: an introduction to the special issue. *Crop Protection* 95: 1–7.

Sartorato, I., Zanin, G., Baldoin, C., and De Zanche, C. (2006). Observations on the potential of microwaves for weed control. *Weed Research* 46: 1–9.

Sato, S., Isobe, S., Asamizu, E. et al. (2005). Comprehensive structural analysis of the genome of red clover (Trifolium pratense L.). *DNA Research* 12: 301–364.

Scavo, A., Abbate, C., and Mauromicale, G. (2019). Plant allelochemicals: agronomic, nutritional and ecological relevance in the soil system. *Plant and Soil* 442: 23–48.

Scavo, A. and Mauromicale, G. (2021). Crop allelopathy for sustainable weed management in agroecosystems: knowing the present with a view to the future. *Agronomy* 11: 2104.

Scavo, A., Restuccia, A., Lombardo, S. et al. (2020). Improving soil health, weed management and nitrogen dynamics by Trifolium subterraneum cover cropping. *Agronomy for Sustainable Development* 40.

Schipanski, M.E., Barbercheck, M., Douglas, M.R. et al. (2014). A framework for evaluating ecosystem services provided by cover crops in agroecosystems. *Agricultural Systems* 125: 12–22.

Schulz, M., Marocco, A., Tabaglio, V. et al. (2013). Benzoxazinoids in rye allelopathy-from discovery to application in sustainable weed control and organic farming. *Journal of Chemical Ecology* 39: 154–174.

Schutte, B., Tomasek, B., Davis, A. et al. (2014). An investigation to enhance understanding of the stimulation of weed seedling emergence by soil disturbance. *Weed Research* 54: 1–12.

Schwarzländer, M., Hinz, H.L., Winston, R.L., and Day, M.D. (2018). Biological control of weeds: an analysis of introductions, rates of establishment and estimates of success, worldwide. *BioControl* 63: 319–331.

Scopel, A., Ballaré, C., and Radosevich, S. (1994). Photostimulation of seed germination during soil tillage. *New Phytologist* 126: 145–152.

Shahzad, M., Hussain, M., Farooq, M. et al. (2017). Economic assessment of conventional and conservation tillage practices in different wheat-based cropping systems of Punjab, Pakistan. *Environmental Science and Pollution Research* 24: 24634–24643.

Shearin, A., Reberg-Horton, S., and Gallandt, E. (2014). Direct effects of tillage on the activity density of ground beetle (Coleoptera: Carabidae) weed seed predators. *Environmental Entomology* 36: 1140–1146.

Simberloff, D. and Stiling, P. (1996). How risky is biological control? *Ecology* 77: 1965–1974.

Singh, S., Nouri, A., Singh, S. et al. (2020). Soil organic carbon and aggregation in response to thirty-nine years of tillage management in the southeastern US. *Soil and Tillage Research* 197: 104523.

Skuodienė, R., Karčauskienė, D., Repšienė, R., and Šiaudinis, G. (2018). Changes in the weed communities as affected by different primary soil tillage and deep loosening. *Acta Agriculturae Scandinavica, Section B—Soil & Plant Science* 68: 643–648.

Sorkheh, M., Zaefarian, F., and Gharineh, M.h. (2020). Effect of green manure under different conditions of tillage on weed characteristics and corn (Zea mays L.) yield. *Journal of Plant Productions* 43: 281–294.

Sosnoskie, L.M., Herms, C.P., and Cardina, J. (2006). Weed seedbank community composition in a 35-yr-old tillage and rotation experiment. *Weed Science* 54 (263–273): 211.

Srinivasan, A. (2006). *Handbook of Precision Agriculture: Principles and Applications.* CRC press.

Stapleton, J.J. (2000). Soil solarization in various agricultural production systems. *Crop Protection* 19: 837–841.

Steinmetz, Z., Wollmann, C., Schaefer, M. et al. (2016). Plastic mulching in agriculture. Trading short-term agronomic benefits for long-term soil degradation? *Science of the Total Environment* 550: 690–705.

Sturm, D., Peteinatos, G., and Gerhards, R. (2018). Contribution of allelopathic effects to the overall weed suppression by different cover crops. *Weed Research* 58: 331–337.

Swanton, C.J., Shrestha, A., Roy, R.C. et al. (1999). Effect of tillage systems, N, and cover crop on the composition of weed flora. *Weed Science* 47: 454–461.

Tabaglio, V., Marocco, A., and Schulz, M. (2013). Allelopathic cover crop of rye for integrated weed control in sustainable agroecosystems. *Italian Journal of Agronomy* 8 (35): 3535–40.

Tang, L., Cheng, C., Wan, K. et al. (2014). Impact of fertilizing pattern on the biodiversity of a weed community and wheat growth. *Plos One* 9: e84370.

Tate, M. and Enneking, D. (2006). Vetch: from feed to food? *Grain Legume* 47: 14–15.

Teasdale, J., Brandsaeter, L., Calegari, A. et al. (2007). Cover crops and weed management. In: *Non Chemical Weed Management Principles. Concepts and Technology* (ed. M.K. Upadhyaya and R.E. Blackshaw), 49–64. Wallingford, UK: CABI.

Teasdale, J., Hatfield, J., Buhler, D., and Stewart, B. (1998). Cover crops, smother plants, and weed management.

In: *Integrated Weed and Soil Management*, (ed. J.L. Hatfield, D.D. Buhler, and B.A. Stewart), 247–270. Chelsea, MI: Ann Arbor Press.

Teasdale, J., Mirsky, S., and Cavigelli, M. (2018a). Meteorological and management factors influencing weed abundance during 18 years of organic crop rotations. *Weed Science* 66: 1–8.

Teasdale, J.R., Devine, T.E., Mosjidis, J.A. et al. (2004). Growth and development of hairy vetch cultivars in the northeastern United States as influenced by planting and harvesting date. *Agronomy Journal* 96: 1266–1271.

Teasdale, J.R., Mirsky, S.B., and Cavigelli, M.A. (2018b). Meteorological and management factors influencing weed abundance during 18 years of organic crop rotations. *Weed Science* 66: 477–484.

Teasdale, J.R. and Mohler, C.L. (2000). The quantitative relationship between weed emergence and the physical properties of mulches. *Weed Science* 48: 385–392.

Testa, G., Reyneri, A., and Blandino, M. (2016). Maize grain yield enhancement through high plant density cultivation with different inter-row and intra-row spacings. *European Journal of Agronomy* 72: 28–37.

Thomsen, M.G., Brandsæter, L.-O., and Fykse, H. (2013). Regeneration of Canada thistle (Cirsium arvense) from intact roots and root fragments at different soil depths. *Weed Science* 61: 277–282.

Tilman, D., Isbell, F., and Cowles, J.M. (2014). Biodiversity and ecosystem functioning. *Annual Review of Ecology, Evolution, and Systematics* 45: 471–493.

Tursun, N., Işık, D., Demir, Z., and Jabran, K. (2018). Use of living, mowed, and soil-incorporated cover crops for weed control in apricot orchards. *Agronomy* 8: 150.

Ulloa, S.M., Datta, A., Bruening, C. et al. (2012). Weed control and crop tolerance to propane flaming as influenced by the time of day. *Crop Protection* 31: 1–7.

Upadhyaya, M.K. and Blackshaw, R.E. (2007). *Non-chemical Weed Management: Principles, Concepts and Technology*. Cabi.

Van der Weide, R., Bleeker, P., Achten, V. et al. (2008). Innovation in mechanical weed control in crop rows. *Weed Research* 48: 215–224.

Van der Weide, R., Bleeker, P., Molema, G. et al. (2005). Innovation in mechanical weed control in row crops. *13th EWRS Symposium*, Bari.

van Elsen, T. (2000). Species diversity as a task for organic agriculture in Europe. *Agriculture, Ecosystems & Environment* 77: 101–109.

Vidotto, F., Fogliatto, S., Milan, M., and Ferrero, A. (2016). Weed communities in Italian maize fields as affected by pedo-climatic traits and sowing time. *European Journal of Agronomy* 74: 38–46.

Vleeshouwers, L. (1997). Modelling the effect of temperature, soil penetration resistance, burial depth and seed weight on pre-emergence growth of weeds. *Annals of Botany* 79: 553–563.

Vrignon-Brenas, S., Celette, F., Piquet-Pissaloux, A. et al. (2018). Intercropping strategies of white clover with organic wheat to improve the trade-off between wheat yield, protein content and the provision of ecological services by white clover. *Field Crops Research* 224: 160–169.

Wallace, J.M., Curran, W.S., and Mortensen, D.A. (2019). Cover crop effects on horseweed (Erigeron canadensis) density and size inequality at the time of herbicide exposure. *Weed Science* 67: 327–338.

Ward, M.J., Ryan, M.R., Curran, W.S. et al. (2011). Cover crops and disturbance influence activity-density of weed seed predators Amara aenea and Harpalus pensylvanicus (Coleoptera: Carabidae). *Weed Science* 59: 76–81.

Weiner, J., Griepentrog, H.-W., and Kristensen, L. (2001). Suppression of weeds by spring wheat Triticum aestivum increases with crop density and spatial uniformity. *Journal of Applied Ecology* 38: 784–790.

Weston, L.A., Alsaadawi, I.S., and Baerson, S.R. (2013). Sorghum allelopathy—from ecosystem to molecule. *Journal of Chemical Ecology* 39: 142–153.

Woźniak, A. (2018). Effect of tillage system on the structure of weed infestation of winter wheat. *Spanish Journal of Agricultural Research* 16: 14.

Woźniak, A. and Rachoń, L. (2019). Effect of tillage systems on pea crop infestation with weeds. *Archives of Agronomy and Soil Science* 65: 877–885.

Xiao, W., Li, J., Zhang, Y. et al. (2022). A fungal Bipolaris bicolor strain as a potential bioherbicide for goosegrass (Eleusine indica) control. *Pest Management Science* 78: 1251–1264.

Xiong, Y., Ge, Y., Liang, Y., and Blackmore, S. (2017). Development of a prototype robot and fast path-planning algorithm for static laser weeding. *Computers and Electronics in Agriculture* 142: 494–503.

Yamada, S. and Nemoto, M. (2020). Effects of weed abundance and frequency of hand weeding on the establishment of transplanted Imperata cylindrica. *Landscape and Ecological Engineering* 16: 1–9.

Yang, Q., Wang, X., Shen, Y., and Philp, J. (2013). Functional diversity of soil microbial communities in response to tillage and crop residue retention in an eroded Loess soil. *Soil Science and Plant Nutrition* 59: 311–321.

Young, F. and Thorne, M. (2004). Weed-species dynamics and management in no-till and reduced-till fallow cropping systems for the semi-arid agricultural region of the Pacific Northwest, USA. *Crop Protection* 23: 1097–1110.

Zhang, Y., Yang, X., Zhu, Y. et al. (2019). Biological control of Solidago canadensis using a bioherbicide isolate of Sclerotium rolfsii SC64 increased the biodiversity in invaded habitats. *Biological Control* 139: 104093.

5

The Underestimated Role of Cultural Practices in Ecologically Based Weed Management Approaches

Ilias Travlos[1,*], *Ioannis Gazoulis*[1], *Milena Simić*[2], *and Panagiotis Kanatas*[3]

[1] *Agricultural University of Athens, Athens, Greece*
[2] *Maize Research Institute "Zemun Polje", Belgrade–Zemun, Serbia*
[3] *University of Patras (Teaching Staff P.D. 407/80), Mesolonghi, Greece*
[*] *Corresponding author*

5.0 Introduction

Weeds are the most important obstacle in agricultural production systems, with the highest potential of crop yield loss in comparison to insect pests and pathogens (Oerke and Dehne 2004). Sharing the same ground with the crop, weeds compete strongly with cultivated plants for the available resources (water, nutrients) and light (Bastiaans et al. 2008). Weeds are adaptable to all environments and if left uncontrolled, they can cause significant yield losses for any crop (Chauhan 2020). The use of herbicides has been the most widely used tool for weed control worldwide since the late 1960s. Herbicides are effective and their use is associated with less soil erosion and a lower carbon footprint compared to mechanical weed control (Giannesi et al. 2013).

The reduction of herbicide inputs in agriculture has emerged as a new research topic in the sector of weed science during the last two decades (Rydahl 2003; Jørgensen et al. 2007; Sønderskov et al. 2015; Montull et al. 2020). There are several reasons for investigating ways to reduce herbicide use and search for alternative, nonchemical weed management practices. Herbicides can be responsible for soil, groundwater, and surface water pollution (Ighalo et al. 2021; Myers et al. 2022; Wołejko et al. 2022). They can be harmful to non-target plants and toxic to other non-target organisms such as birds, fish, and beneficial insects (Aktar et al. 2009). Moreover, there are growing concerns regarding the possible effects of herbicide toxicity on human health (Hasanuzzaman et al. 2020). Issues related to public health and sustainability cannot be ignored and make imperative the need for a more environmentally friendly weed management. An additional reason to investigate non-chemical weed management practices lies on the fact that herbicides might have been a once-in-a-century option to control weeds in agricultural fields (Davis and Frisvold 2017). No novel herbicide modes of action (MOAs) exist from 1980s and the consecutive use of active ingredients with the same mode of action has inevitably led to the selection of resistant weed populations (Hasanuzzaman et al. 2020; Perotti et al. 2020). At present, there are 509 unique cases of herbicide-resistant weeds globally, with 153 dicotyledonous and 113 monocotyledonous weed species having evolved resistance to 164 different herbicides (Heap 2022). It should be noted that herbicide efficacy is an exhaustible resource that can be depleted by use over time.

Therefore, there is an urgent need to develop alternative, ecologically based weed-management systems based on cultural practices that reduce the emergence and growth of weeds in the field (Korres 2018; Perotti et al. 2020). It should be here noted that farmers often underestimate cultural practices, for several reasons:

- Farmers are accustomed to the rapid action of herbicides. They usually see results within days or weeks. The psychological response of instant gratification from herbicides is very strong compared to the lack of visual cues typical of most cultural and nonchemical methods (Moss 2019).
- Cultural practices are often more expensive initially. The economic costs associated with adopting cultural practices are immediate, but the benefits are only seen in the long term (Hurley and Frisvold 2016). Using cultural practices may be more expensive than herbicides for the level of weed control achieved (Sardana et al. 2017; Vincent-Caboud et al. 2017).

Ecologically Based Weed Management: Concepts, Challenges, and Limitations, First Edition. Edited by Nicholas E. Korres, Ilias S. Travlos, and Thomas K. Gitsopoulos.

- Farmers may lack the appropriate equipment needed to implement certain cultural practices.
- Their practical application under real field conditions can be time-consuming (Moss 2019).

However, recent evidence suggests that the role of cultural practices in the development of sustainable weed management systems should not be further underestimated. Selection of an appropriate tillage system to reduce weed emergence in the field, use of crop diversification practices, utilization of crop competition against weeds, selection of the optimal sowing date in relation to weed emergence, and optimal resource management should be considered as important components of ecologically based weed management systems in field crops (Hussain et al. 2021). The role of tillage systems and their relationship with ecologically based weed management approaches is discussed in Chapter 14.

This chapter focuses first on cultural practices that contribute to weed management by increasing crop diversity in the field. These practices include crop rotation, intercropping, and cover cropping (Hufnagel et al. 2020; Sharma et al. 2021). Cultural practices that utilize crop competition to suppress weeds are also analyzed. This category includes the use of weed-suppressive cultivars and hybrids, increased seeding rates and establishment of crops at narrow row spacing (Mhlanga et al. 2016; Datta et al. 2017; Manalil et al. 2017; van der Meulen and Chauhan 2017). The selection of sowing date and optimal management of resources (fertilization and irrigation) are also important factors to consider when developing ecologically based weed management systems (Jha and Norsworthy 2009; Kaur et al. 2018; Korres 2018).

The main objective of this review was to summarize research data and information highlighting the potential of the above-mentioned cultural practices to contribute to the development of ecologically based weed management systems in cereals, legumes, and industrial crops, based on the results of field trials conducted during the last two decades.

5.1 Role of Crop Diversification in Ecologically Based Weed Management

5.1.1 Role of Crop Rotation in Ecologically Based Weed Management

Crop monocultures or short rotations of crops with similar characteristics involve the use of similar cultivation practices in successive years in the same field. Weed species that are better adapted to these practices (repeated every year) gradually dominate the field and compete very strongly with the cultivated crops (Blackshaw et al. 2007). In addition, continuous application of herbicides with the same mode of action significantly increases the risk of developing herbicide-resistant weed populations (Heap 2022). In contrast to monoculture, crop rotation is an alternative, sustainable cropping system that involves growing different crops in a recurring sequence for a specified period of time on the same land (Sharma et al. 2021). It aims to achieve efficient economic production at the lowest possible cost without compromising soil fertility. Apart from other advantages, crop rotation is also one of the most valuable cultural practices used for weed management. This is because troublesome weed species cannot adapt and are less competitive and persistent in the field in the context of a diverse crop rotation that includes species with different characteristics and whose cultivation requires the use of different management practices each year (Knezevic 2014; Kanatas 2020).

Crop selection must follow specific criteria to facilitate successful weed management. As a first recommendation, it is advisable to rotate cereals with broadleaf crops (e.g. legumes and crucifers) to allow the use of different selective herbicides with different modes of action (Korres 2018). In the experiments of MacLaren et al. (2021), seed production of rigid ryegrass (*Lolium rigidum* Gaud.) was found to decrease by 31% and 44%, respectively, when wheat–annual medic–wheat–annual medic and wheat–canola–wheat–lupine were selected as crop rotations compared to wheat monoculture. These authors attributed these differences to the use of herbicides (e.g. propyzamide and tepraloxydim) that were selective for broadleaf crops included in the rotations and controlled rigid ryegrass, reducing its seed production. Rotation of winter and spring crops involves soil disturbance at different times between years and is another practice that reduces adaptability, emergence, and seed production of troublesome weed species (Anderson 2007; Korres 2018). A recent example is that rotations of maize with wheat and spring barley with wheat reduced the number of blackgrass (*Alopecurus myosuroides*) panicles by up to 59% compared to wheat monoculture in the study by Zeller et al. (2021). These authors explained that the introduction of spring crops in the rotation reduced the density and seed production of blackgrass because this species prefers winter crops and germinates predominantly in the fall when winter cereals and oilseed rape are usually sown in western and central Europe.

Growing crops whose cultivation is associated with tillage practices may also facilitate weed suppression. The 10-year study by Brankov et al. (2021) also showed that rotating maize with wheat reduced total biomass and total weed density by 29% and 43%, respectively, compared to maize monoculture. Crop rotation also increased maize grain yield by 26%. The same researchers also found that the maize–wheat rotation reduced biomass of perennial weeds and, in particular, johnsongrass [*Sorghum halepense* (L.) Pers.], Canada thistle [*Cirsium arvense* (L.) Scop.], and field bindweed (*Convolvulus arvensis* L.) by 41%, 60%, and 70%, respectively, compared to the maize monoculture. These results were attributed to the tillage practices implemented during the years of crop rotation. Wheat harvest was followed by shallow plowing (0.1 m depth) and deep plowing in the second half of October. During these tillage operations, the underground vegetative organs of the perennial weeds were cut off and then brought to the soil surface so that they were exposed to the low winter temperatures, which had a draining effect on them.

Rotation of crops grown in different growing seasons (e.g. wheat–maize) poses also an agroecological approach to reduce emergence and seed production of annual weeds. Different sowing and harvest dates for these crops with different life cycles provide growers with more opportunities to prevent both weed emergence and seed production. According to the example of Anderson (2007), green foxtail (*Setaria viridis* L.) emerges between late spring to mid-summer and then begins flowering in early August. Since wheat is harvested in early July, growers can control green foxtail before it flowers and produces seed. A similar opportunity exists with cool–season weeds; they are easily controlled prior to sowing summer crops to prevent seed production (Anderson 2004). Rotating cool–season crops with different sowing dates also reduces weed emergence. Planting dry pea (*Pisum sativum* L.) in the first year of a rotation with wheat provides an opportunity to control annual weeds that emerge over the winter because dry pea is seeded in late March. The result is lower weed seed production and lower weed density in the wheat crop that follows the next year of the rotation. Similar are the benefits for rotating summer crops with different sowing dates. Specifically, the crop that is sown earlier should lead the rotation and the later sown crop should follow in the next year. As Anderson (2005) explained, sowing maize in early May in the first year of the rotation results in a flush of weeds that are controlled three to four weeks later, so they do not produce seed. As a result, weed density is 35–50% lower in the following crop (e.g. proso millet), that is sown in the second year of the rotation in early June. Thus, an optimal rotation plan may include growing two cool-season crops with different sowing dates and then growing two warm-season crops, also with different sowing dates (Anderson 2007).

Another recommendation is to rotate tap-rooting crops with crops that have a fibrous rooting pattern for complementary and uniform use of nutrients in the rotation (Korres 2018). It is also beneficial to include perennial forage crops in the rotation to suppress weed seed production. For alfalfa, in particular, it has been found that its participation in a four- or five-year rotation with maize and wheat can reduce weed seed production by 40% compared to a monoculture of maize (Cardina et al. 2002). In addition to alfalfa's competitive ability, alfalfa cultivation results in a series of five to seven harvests per year. At each harvest, several summer weeds are mowed along with the harvested hay. This reduces weed seed production and deposition in the weed seedbank to some degree (Anderson 2010; Kanatas et al. 2021). Anderson (2010) recommend alfalfa in a nine-year rotation with cereals and legumes because in the third year of alfalfa, weed seed production can be reduced by 90%. The scheme they proposed was alfalfa for three years (maize – soybeans – oats / beans – winter wheat – soybeans – maize) with the aim of improving weed suppression in organic production systems. A future research objective should also be to include weed-suppressive industrial crops such as industrial hemp, kenaf, and sunn hemp in crop rotation schemes to evaluate their performance in terms of weed suppression (Zegada-Lizarazu and Monti 2011; Gazoulis et al. 2021). It should be noted here that a special category of crops, cover crops, are widely used in crop rotations for this purpose.

Cover crops are species used for soil fertilization and weed management purposes and often characterized by rigorous growth and high biomass production. The predominant cover crops are winter-grown cereals, legumes, and species belonging to the botanical family *Brassicaceae* (Osipitan et al. 2018). The most common practice is to plant a cover crop in early fall so that it accumulates significant amounts of biomass during the winter rainy season. In spring, cover crops can be terminated by herbicide application or mechanical methods such as disk–harrowing, flail–mowing, and roller–crimping. In conventional tillage systems, cover crop residues are disk-incorporated into the soil as green manures before sowing the major summer crop. In conservation agriculture and no-till systems, the residue is left on the soil surface to form a dense layer that suppresses weed emergence in the major cash crop (Dhima et al. 2006; Davis 2010). The most widely used cover crop species is undoubtedly cereal rye. Cereal rye can typically produce more than 10000 kg ha^{-1} of biomass under favorable conditions, and its residues have allelopathic effects on weeds; roller-crimped cereal rye cover crops have been reported to suppress weed growth in soybeans by more than 80% early in the season (Mischler et al. 2010; Wells et al. 2014). Concerning other winter cereal cover crops, barley and oat also provide sufficient reductions in weed emergence. In maize grown under a conventional tillage system, barley residue was incorporated into the soil as green manure and reduced the stem number of bristly foxtail [*Setaria verticillata* (L.) P. Beauv.], barnyardgrass [*Echinochloa crus-galli* (L.) P. Beauv.], and large crabgrass [*Digitaria sanguinalis* (L.) Scop.] by 59%, 72%, and 78%, respectively, and increased grain yield by 45% (Dhima et al. 2006). An oat cover crop that was flail-mowed prior processing tomato transplanting reduced weed emergence by 93% (Campiglia et al. 2010). In the same study, a hairy vetch cover crop also reduced weed density by up to 79% and increased marketable yield by up to 28%. Austrian winter pea and clover species are also important legume cover crops that can both suppress weed growth and enrich the soil with high significant amounts of nitrogen (Osipitan et al. 2018).

Species from the *Brassicaceae* family form another group of cover crops that provide a high degree of weed suppression. White mustard, turnip, forage radish, and other cruciferous cover crops are characterized by high growth rates, adequate biomass production, and significant allelopathic potential against weeds (Haramoto and Gallandt 2004). For example, a forage radish cover crop sown in late summer and frost-killed in late winter reduced weed ground coverage by up to 73% at the time of maize planting (Lawley et al. 2011). In addition, the same authors observed that forage radish rapidly covered the soil and outcompeted winter weeds, resulting in very low weed ground coverage.

In this context, it should be noted that winter cover crops suppress the growth of winter annual weeds. Regarding the competitiveness of cereal and legume cover crops, Hayden et al. (2012) observed that hairy vetch and cereal rye reduced winter weed biomass by 71–91% and 95–98%, respectively, during their active growth period. Cover crop mixtures also hold great promise for further increasing plant diversity in the field, balancing nitrogen fixation and retention, and promoting resilience to unpredictable weather conditions (Baraibar et al. 2018). However, the mechanism of weed suppression in a cover crop mixture is still not well explained and should be further investigated in spatially and temporarily replicated field trials. Another target that is likely to receive more attention in the upcoming years is the introduction of summer cover crops into the rotation. Sorghum–sudangrass and tropical legumes such as velvet bean and sunn hemp are characterized by high biomass production and remarkable allelopathic effects on weeds (Correia et al. 2014; Soti and Racelis 2020). After harvesting the main summer crops, these species can either be used to cover the ground and reduce seed production of late-emerging summer annual weeds or mowed down so that their mulch suppresses weed growth in a subsequent winter crop (Soti and Racelis 2020).

5.1.2 Role of Intercropping in Ecologically Based Weed Management

Intercropping is an agricultural system including the cultivation of two or more crop species for a significant period of time in a growing season, targeted to improve crop productivity through increased efficiency in the use of resources (Dowling et al. 2021). Relay intercropping include the establishment of a companion crop before the major crop is mature; mixed intercropping is referred to the spontaneous cultivation of two or more crop species in the same row; row intercropping is the cultivation of two or more crops in alternate rows; strip intercropping includes the spontaneous cultivation of two or more crops in alternate strips of uniform width (Dowling et al. 2021; Sharma et al. 2021). Intercropping leads to spatial diversification of crops, which can improve crop competition with weeds. Intercrops may also shade the soil surface more effectively than monocultures, reducing, thus weed emergence and growth or may have allelopathic effects on weeds (Blackshaw et al. 2007).

There is much evidence proving the potential of intercropping for improved weed suppression and higher yields. In particular, row intercropping with nude oat (*Avena nuda* L.) as companion crop and lentil (*Lens culinaris* L.) as primary crop, reduced grass weed density and biomass by 22 and 30%, respectively, in comparison to the case where lentil was grown as a sole crop (Kraska et al. 2020). Intercropping also slightly improved 1000 seed weight of lentil. Mixed intercropping between lentil and spring-naked barley at the seeding rate of 240 seeds m^{-2} and with a cropping ratio of 3:1 (lentil:barley) resulted in low values of weed biomass (≤ 50 g m^{-2}) when combined with early sowing dates and the use of the most competitive lentil cultivar ("Anita") by Wang et al. (2013). It should be noted that these researchers did not compare the weed-suppressive potential and yield of their intercrops to the corresponding of lentil as a sole crop. However, the above-mentioned intercrop resulted in higher yields compared to that obtained for lentil when grown as a sole crop, which was mechanically weeded multiple times (Alba et al. 2020). In another study where the same mixed intercrop was evaluated study and the dominant weeds were catchweed bedstraw (*Galium aparine* L.), brittlestem hempnettle (*Galeopsis tetrahit* L.), and blackgrass, lentil-barley intercropping resulted in 40–73% lower weed biomass and increased grain yield by 16–22%, compared to lentil monoculture (Wang et al. 2012). Corre-Hellou et al. (2011) recorded 55% and 72% lower weed dry weight per unit area at the beginning of flowering and maturity of field pea (*Pisum sativum* L.) when it was grown simultaneously with barley (row intercropping) compared to growing field pea as a sole crop. The intercrop also yielded up to 27% more biomass compared to monocrops of both species. Weed suppression and overall productivity were higher (34 and 19%, respectively) under a chickpea (*Cicer arietinum* L.) wheat row intercropping system as compared to monocrops of either species in the five-year field trials of Banik et al.

(2006). Saucke and Ackermann (2006) reported that pea–false flax (*Camelina sativa* L.) intercropping suppressed the emergence and growth of black bindweed [*Fallopia convolvulus* (L.) Á. Löve], common sowthistle (*Sonchus oleraceus* L.), and wild chamomile (*Matricaria recutita* L.) when preferred over monocropping systems. Field pea was the primary crop and false flax the companion crop (row intercropping). In addition to weed suppression, this system resulted in an overall 30% higher yield, compared to field pea monoculture.

In another study, maize–soybean row intercropping, at a row spacing of 10 cm and for a given medium-maturity maize hybrid, reduced weed biomass by 24% and increased overall grain yield by 18% compared to maize monocropping. Such findings highlight that intercropping not only suppresses weed growth, but is also a valuable practice to improve land use efficiency. Strip intercropping systems between maize and sweet sorghum provided 48–52% control of purple nutsedge (*Cyperus rotundus* L.) compared to growing maize as a sole crop (Mahmood et al. 2013). It should be noted that this practice resulted in yield loss for maize.

However, maize yield loss was compensated for by the yield of sorghum. Other recent research findings indicate that intercropping is beneficial for generating high biomass yields in the first year of cultivation of industrial crops characterized by slow early growth rates. In particular, von Cossel et al. (2019) showed that maize biomass could compensate for giant miscanthus (*Miscanthus* × *Giganteus* et. Deuter) lack of yield in the first two years of cultivation; maize also reduced weed presence in the field. It should be noted that in this study, the two crops grew in the same field only for the first years of cultivation of the primary crop (e.g. giant miscanthus is a perennial species). In other words, these authors evaluated the process of relay intercropping. As for more information on relay, Amossé et al. (2013) interseeded forage crops in a wheat field in late March and when the crop was at the end of tillering. After wheat harvest and until the next spring, all interseeded species remained in the field to cover the soil. In early December, results showed that weed biomass was 139 times lower in plots where red clover (*Trifolium pretense* L.) had been interseeded compared to values obtained in control plots. This suggests that since fewer weed seedlings emerged when a cover crop was present in autumn, fewer weed seeds were produced, resulting in less seed inputs to the weed seedbank (Mirsky et al. 2010). Moreover, successful relay intercropping schemes between maize and tropical legumes have been proposed. For example, interseeding velvetbean (*Mucuna pruriens* (L.) DC var. *utilis*) in maize, three to five weeks after maize sowing, results in excellent weed suppression (≥90%) and 58% higher grain yield for maize (Correia et al. 2014). In general, relay intercropping between maize with tropical legumes is a beneficial cultural practice, and this can become apparent after maize harvest.

In any case, species selection is critical to develop a win–win intercrop that both suppresses weeds and produces higher yields. Species grown as intercrops should be characterized by complementary resource use. To achieve that, the component species of an intercrop should use different resources or use the same resources at different places or different times (Liebman and Staver 2001). Intercropping with species that differ in nutrient uptake, photosynthetic response to different light intensities, rooting depth etc., may minimize resource competition among cultivated species, and allow intercrops to capture a greater range and quantity of resources that are consequently limited to weeds (Pakeman et al. 2020).

5.2 Role of Crop Competition in Ecologically Based Weed Management

5.2.1 Role of Competitive Cultivars and Competitive Hybrids in Ecologically Based Weed Management

The selection of modern-day cultivars focused to achieve higher grain yields is related to morphological traits often associated with lower competitive ability against weeds. However, the selection of competitive genotypes should be minded as one of the most important components for the design of effective and environmentally friendly methods to control weeds in arable crops. The competitive ability of different cultivars and hybrids can be attributed to various traits such as plant height, early vigor, and canopy architecture (Andrew et al. 2015; Mhlanga et al. 2016; Jabran 2017). According to Andrew et al. (2015), there are two main aspects of crop competitiveness against weeds – namely, tolerance and suppressive ability. The former is the ability of the selected genotype to withstand weed competition and resist to yield losses under weedy conditions, whereas the latter term refers to the ability of a particular genotype to reduce the fitness of weeds infesting the cultivated area. Competitive cultivars that can outcompete weeds can contribute to sustainable weed management in the long term.

In the experiments of Korres and Froud-Williams (2002) where chickweed (*Stellaria media* L.), annual meadow grass (*Poa annua* L.), *Matricaria* spp. and *Lamium* spp. were the dominant weed species, the tallest wheat cultivar Maris Huntsman reduced weed dry weight (40–62%) and seed production (41–62%) compared to the shorter cultivars Flame and Fresco. Weed density decreased by 25% and grain yield increased by 15% in the presence of a medium–tall wheat cultivar with very good tillering ability, namely Alliance, compared to the value obtained in the presence of a short cultivar with lower tillering capacity, namely TAM 107 (Wicks et al. 2004). In the study by Mwendwa et al. (2020), the early maturing wheat cultivars Federation and Condo showed early vigor, faster biomass accumulation, and canopy closure, suppressing weed growth by 73–83%, compared to a slow-maturing cultivar, namely Gregory.

Another factor of major importance determining cultivar selection in winter cereals is tillering capacity. Old wheat cultivars with increased tillering capacity were reported to suppress weeds more effectively than modern-day cultivars in China at low crop densities (200 plants m^{-2}) (Wu et al. 2010). However, at higher crop densities (600 plants m^{-2}), modern-day cultivars were found to compete well with weeds and reach a harvest index that surpassed 0.35. Therefore, the selection of a cultivar with low tillering capacity should be accompanied with the adoption of increased seeding rates so as to increase the number of wheat stems per unit area. Similar results were reported by Korres and Froud-Williams (2002), who also found that the negative effects of wheat tillering capacity on weed biomass and fecundity were evident for Maris Widgeon, Riband, and Rialto. Similar findings indicating the importance for selecting genotypes with higher tillering potential have been also reported in barley. Zhana cultivar reduced sterile oat (*Avena sterilis* L. subsp. *ludoviciana*) dry weight per unit area by up to 19% lower compared to Grace cultivar and reached 18% higher grain yield (Kanatas et al. 2020). These authors partially attributed the weed suppressive ability of the superior cultivar to its higher tillering capacity and early vigor.

Regarding spring cereals, Simić et al. (2009) also noticed that weed biomass was by 22% and 47% lower in the presence of ZPSC 735 maize hybrid compared to the presence of ZPSC 434 and ZPSC 578 hybrids, respectively, in the evaluation performed before maize harvest. These authors attributed the competitive ability of ZPSC 735 hybrid to its increased leaf area and canopy architecture that enabled maize plants to shade the ground and smother weeds grown under crop canopy (Simić et al. 2009). Canopy architecture as well as rapid vegetative growth resulted in lower weed biomass (29–44%) under the canopies of leafy reduced stature (LRS) and conventional Pioneer (P3979) hybrids compared to the the late-maturing, big leaf (LMBL) hybrid (Begna et al. 2001). Maize hybrids with horizontal leaf architecture (Pioneer 3260), reduced the biomass and seed production of fall panicum (*Panicum dichotomiflorum* Michx.), giant foxtail (*Setaria faberi* Herrm.), and large crabgrass [*Digitaria sanguinalis* (L.) Scop.] by 90–94% compared to hybrids with an upright leaf architecture (Pioneer 3394) in the study of Sankula et al. (2004). The weed-suppressive hybrid also reduced seed production of grass weeds by 59–93%.

However, it should be noted that the selection of this competitive hybrid was combined with pre-emergence herbicide application. In addition, the reduction in weed biomass did not result in higher grain yield when maize was grown under rainfed conditions. Indeed, selecting hybrids with horizontal leaf orientation for weed management purposes is a risky decision. On the one hand, these hybrids shade soil surface and reduce weed emergence. On the other hand, hybrids with planophilic leaves tend to reduce photosynthesis on the leaves closest to the ear, the organs that contribute most to the supply of assimilates to the ear (Hammer et al. 2009).

In sweet corn, GH2547 hybrid exhibited higher growth rates, LAI values at crop anthesis, and subsequently intercepted more light to weeds compared to Spirit hybrid, in the two-year field experiments of Williams et al. (2007). The competitive hybrid reduced proso millet (*Panicum miliaceum* L.) shoot biomass by 25–31% under irrigated conditions and by 70–91% more suppressive under rainfed conditions (Williams et al. 2007). This hybrid also reduced proso millet seed production by up to 25% under irrigated conditions and by up to 90% under rainfed conditions compared to Spirit. In rice, the cultivation of competitive genotypes is recommended for the successful management of noxious species belonging to *Echinochloa* spp. In a rice field flooded with 15 cm of water, Ronaldo was the most competitive cultivar that reduced early barnyardgrass [*Echinochloa oryzoides* (Ard.) Fritsch.] number of tillers by more than 30% and reached 20% higher yield compared to Nembo due to its earlier season vigor (Kaya-Altop et al. 2019). The cultivar Koral was reported to suppress barnyardgrass number of tillers by 16–30% and was highlighted as the most competitive genotype in a study conducted in Turkey; this cultivar achieved higher grain yield compared to other cultivars under weed-infested conditions (Mennan et al. 2012). The explanation lies in the fact that this cultivar exhibited higher values of tillering capacity, early growth crop biomass, and plant height compared to Osmancık, Kızılırmak, and Karadeniz cultivars (Mennan et al. 2012).

Concerning legumes, competitive cultivars are valuable as a weed control option in soybean (Datta et al. 2017). Nordby et al. (2007) suggested that late-maturing cultivars tended to withstand yield loss due to weed competition more effectively than early-maturing cultivars. The presence of SC00-883 reduced weed dry weight per unit area up to 82% compared to SC00-892, whose cultivation resulted in less ground coverage (Norsworthy and Shipe 2006). The results of this study indicated that fast-growing cultivars planted in narrow rows can shade the soil surface and suppress weed growth at some point. In wider row spacings, supplementary between-row mechanical operations can potentially lead to sufficient weed control in the area between crop rows. This is also encouraged from the findings of Radicetti et al. (2012a, 2012b) in chickpea who found that selection of competitive cultivars along with two inter-row mechanical weed control is a beneficial practice to reduce weed pressure and achieve higher seed yield. Even under weed-infested conditions, Alto Lazio exhibited faster biomass accumulation and ground coverage during the early growth stages (25 days after crop emergence). As a result, Alto Lazio reduced weed biomass by approximately 25–35% compared to C133, C134, C6150, and Sultano cultivars and did also withstand lower seed yield loss due to weed interference (Radicetti et al. 2012a). In another study conducted by the same researchers using the same chickpea cultivars, it was found that in the presence of 32 plants of prostrate knotweed (*Polygonum aviculare* L.), weed biomass was significantly lower in plots where Alto Lazio was grown (Radicetti et al. 2012b).

Cultivar selection is critical to reduce weed competition also in industrial crops. In a study where the presence of plant of common cocklebur (*Xanthium strumarium* L.) per meter of cotton row caused a lint yield loss of 31%, Rezakhanlou et al. (2013) denoted that the traditional cotton cultivars Syland and Sindoz were able to compete strongly with two common cocklebur plants per meter of cotton row and achieve high lint yields. On the contrary, modern-day cotton cultivars Mehr and Ariya were the less competitive cultivars. Such findings indicate that the modern cultivars, although superior in yield are less competitive due to short height, reduced vigor, and low light interception (Manalil et al. 2017). In sorghum, increased plant height, high leaf area index, and early vigor of Bonus MR reduced Japanese barnyard millet [*Echinochloa esculenta* (A. Braun) H. Scholz.] biomass, panicle number, and seed production by 30%, 26%, and 31%, respectively, compared to the genotype 86G87 (Wu et al. 2010). Bonus MR also achieved 15–17% higher biomass yield compared to the less-competitive cultivar. These authors also suggested that the use of competitive sorghum cultivars should be combined with increased plant populations (7.5 plants m^{-2} instead of 7.5 plants m^{-2}) to further improve weed suppression and biomass yield. Similar results are reported in canola. Specifically, the InVigor 2273 hybrid caused a two-fold decrease in wild oat (*Avena fatua* L.) biomass compared to the open-pollinated cultivar (Exceed) (Harker et al. 2003). In addition, combining InVigor cultivation with the highest seeding rate (200 seeds ha^{-1}), and the earliest time of weed removal (at the two-leaf growth stage of canola) led to a 41% yield increase compared with the combination of Exceed with the lowest seeding rate (100 seeds ha^{-1}) and the latest time of weed removal (at the six-leaf growth stage of canola). In cardoon (*Cynara cardunculus* L.), the presence of C12 cultivar resulted in significantly lower (22%) sterile oat populations and higher biomass yield (13%) compared to the presence of Bianco Avorio cultivar; this outcome was related to higher growth rates and faster canopy closure by C12 cultivar (Vasilakoglou and Dhima 2014). In a study by An et al. (2013) conducted in switchgrass, "Cave-in-Rock," an intermediate ecotype, was the only cultivar able to suppress weed growth during the whole cultivation period. The explanation for this result lies in the higher photosynthetic rates, increased biomass production, and longer growing period of this cultivar compared to others that were investigated.

5.2.2 Role of Increased Seeding Rates in Ecologically Based Weed Management

Agronomic practices that accelerate the development of a dense crop canopy that can shade the soil surface and suppress weed emergence play a prominent role in the development of effective ecologically based weed management systems (Datta et al. 2017; van der Meulen and Chauhan 2017). There is evidence that increased seeding rates can improve the competitiveness of cereal crops against weeds. In a two-year field trial conducted at two wheat-growing regions in the Mediterranean region, Mennan and Zandstra (2005) reported that the seeding rate of 250 kg seed ha^{-1} reduced catchweed bedstraw shoot dry weight and seed production per plant by more than 30% (for both parameters) compared to the seeding rate of 150 kg seed ha^{-1} in plots where weed density was set at 80 plants m^{-2} and the most competitive wheat cultivar was sown. At the same weed density and for the same wheat cultivar, the higher seeding rate (250 kg seed ha^{-1}) resulted in significantly higher ($\geq$ 25%) grain yield compared to the lower seeding rate (150 kg seed ha^{-1}) in the region where climatic conditions were optimal for wheat cultivation. Similarly, increased seeding density of a competitive oat (*Avena sativa* L.) cultivar from 250 to 500 seeds m^{-2} resulted in 63% weed biomass reduction and 11% higher yield (Benaragama and Shirtliffe 2013). In barley, seeding at 200 kg of seeds ha^{-1} reduced wild oat shoot dry weight and seed production by approximately 65% (for both parameters) when preferred over 85 kg of seeds ha^{-1} (O'Donovan et al. 2000). Concerning spring cereals, Tharp and Kells (2001) found that a maize population of 83,900 plants ha^{-1} reduced common lambsquarters (*Chenopodium album* L.) biomass and seed production by 58% and 65%, respectively, compared to a population of 59,300 plants ha^{-1}. However, it should be noted that this study was conducted in glufosinate-resistant maize and herbicides were applied in all plots. As a result, these findings do not indicate that increased seeding rates achieve sufficient levels of weed suppression without supplementary weed control treatments. Yield increased slightly (6%), but this may be due to the fact that these researchers collected yield data only from plots that were kept weed-free by herbicide application along with hand weeding. Mohammadi et al. (2012) proposed a more ecologically based approach and combined increased seeding rates with narrow row spacing. They found that a maize population of 99999 plants ha^{-1} along with planting at a row spacing of 50 cm resulted in 37% lower weed biomass and 40% higher grain yield compared to a population of 66,666 plants ha^{-1} along with planting at a row spacing of 75 cm. In aerobic rice, the quadratic model of Chauhan et al. (2011) predicted that seeding rates of 95–125 kg seed ha^{-1} and 83–92 kg seed ha^{-1} for the inbred cultivars and hybrids, respectively, were needed to enable rice to compete with weeds and avoid severe yield losses.

Concerning legume crops, Place et al. (2009) obtained a significant, negative linear regression between weed ground coverage and soybean seeding rate across three different locations in their two-year field trials conducted in organically grown soybean. At the site infested mainly with *Amaranthus* spp., weed density was around 75% lower and seed yield was by 21% higher in plots where soybean was sown at 586,000 seeds ha^{-1} than in plots where the crop was sown at 185,000 seeds ha^{-1}. However, it should be noted that this practice was integrated with three inter-row mechanical weed control treatments. This practice is also recommended for weed control in organically grown pulse crops. In lentils, seeding to achieve a density of 260 plants m^{-2} instead of 130 plants m^{-2} reduced weed biomass by 16% and increased seed yield by 23% when data were averaged over weed control treatments (Alba et al. 2020). Combining increased seeding rates with three mechanical weed control treatments resulted in an even greater reduction in weed biomass. As a result, seed yield increased by approximately 40% compared to seed yield of lentil in plots where weeds were left uncontrolled and seeding was performed at the recommended rates. In chickpea, a 1.5-fold increase in seeding rate reduced weed biomass by 72% and increased seed yield by 56% when combined with two mechanical weed control treatments, compared with values observed in plots that were not mechanically treated and conventional seeding rates were used (180 kg seed ha^{-1}) (Kanatas and Gazoulis 2022). Except for conventional food crops, manipulations in seeding rates and plant populations can be also utilized to suppress weeds in crops grown for either industrial or energy purposes (Gazoulis et al. 2021). For instance, an increase from 100 to 200 plants m^{-2} markedly reduced weed weight by 72% in industrial hemp in the study of Hall et al. (2014). Further reductions in weed biomass ($\geq$ 95%) were observed at 300 and 400 plants m^{-2}. In the field trials of Webster (2007), cotton was sown to achieve crop densities between 49,000 and 148,000 plants ha^{-1}. The results showed that sicklepod presence was around 70% lower at the maximum seeding rate. In a recent two-year study conducted in organically grown sunflower (*Helianthus annuus* L.), the higher seeding rate of 8.6 seeds m^{-2} resulted in 35% lower weed dry weight per unit area compared to the lower seeding rate of 6.1 seeds m^{-2} (Mouillon et al. 2020). Furthermore, these researchers observed cause–effect relationships between weed biomass and seeding rates but not between seed yield and seeding rate (Mouillon et al. 2020). Wu et al. (2010) increased sorghum (*Sorghum bicolor* L.) seeding rates by 40% to limit the growth, biomass, and seed production of Japanese barnyard millet by 22%, 27%, and 32%, respectively. In a recent study conducted in cardoon, Zenobi et al. (2021) noticed that sowing at 1 m distance between rows and at 0.5 m distance within rows reduced weed biomass by 65% compared to seeding at 1 m inter- and intrarow distances.

It is well established that increased seeding rates can contribute to ecologically based weed management. However, there is a large gap between the theoretical benefits of high crop densities and actual practice because farmers are reluctant to adopt higher seeding rates due to the higher economic costs associated with the use of more seed (Moss 2019). Moreover, dense crop populations in the row increase the risk of disease development in wet conditions such as at sites with high precipitation or in no-till systems where the crop may be directly through cover crop mulches (Blackshaw et al. 2007). In addition, at higher intrarow populations crop plants tend to grow taller increasing the risk for lodging (Liebert and Ryan 2017). Yield reductions may be also obtained due to intraspecific competition between cultivated plants; yield may begin to plateau as density increases or declines above an optimum density (Weiner et al. 2001). It can be hypothesized that this cultural practice is applicable for weed management at sites where resources are not limited so as to avoid the impact of intraspecific competition on crop yield, at sites where climatic conditions do not favor disease development, in organic agriculture and especially in crops that are poor competitors against weeds. Seeding rate also depends on seed size of a particular crop or cultivar (Mohammed et al. 2020). In general, there are several agronomic factors to consider before deciding to increase the seeding rate. In any case, it should be noted that this cultural practice needs to be combined with other cultural practices or mechanical weed control to achieve a sufficient level of weed suppression as shown in most cases reported in the literature.

5.2.3 Role of Narrow Row Spacing in Ecologically Based Weed Management

Narrow row spacing is another agronomic practice improving weed suppression weeds in field crops. The long-term experiments of Borger et al. (2016) in wheat, revealed that selecting 9 cm row spacing can reduce rigid ryegrass seed production by 26%, 74%, and 83% compared to 18, 27, and 36 cm row spacings, respectively. Significant reductions on plant height and biomass were reported for catchweed bedstraw when wheat was established in the row spacing of 11 cm than in cases where sowing was performed in the wider row spacings of 15 and 23 cm (Fahad et al. 2015). Regarding the seed production of catchweed bedstraw, when the weed emerged at the same time as the crop, the row spacing of 11 cm resulted in the production of 900 less seeds per plant compared to the row spacing of 23 cm; wheat yield increased by at approximately 15%. Drews et al. (2009) also observed that weed biomass at wheat flowering tended to be lower in 12 cm than in 24 cm spaced rows. Similar were the findings of Kolb et al. (2010) regarding the comparison of 11.4 and 22.8 cm row spacings in the barley crop. A variety of studies emphasizes the role of row spacing manipulations for weed management purposes, also in spring cereals. Planting maize at 75- and 60-cm spaced rows reduced weed biomass by 41–61% and 60–65% compared to wider row spacing (90 cm) as observed across two soil types in semiarid environments (Mashingaidze et al. 2009). In addition, the row spacing of 60 cm increased ear weight and the number of kernels per ear by in 16 and 19, respectively. In case where perennial weeds such as bermudagrass [*Cynodon dactylon* (L.) Pers.] are infesting a maize field along with troublesome annual species, even narrower row spacings might contribute to weed management (Acciares and Zuluaga 2006). In particular, 35 cm between-row distances reduced weed dry weight by up to 68% and increased grain yield by 30–33% when preferred over 35 cm between-row distances (Acciares and Zuluaga 2006). Row spacing of 15 cm resulted in 24–26% lower weed biomass and 27–32% higher grain yield compared to 30 cm in a direct–seeded rice field where junglerice [*Echinochloa colona* (L.) Link] and goosegrass [*Eleusine indica* (L.) Gaertn.] were among the dominant weeds for two years (Chauhan and Johnson 2011. Narrow row spacing is essential to reduce the seed proliferation of competitive, mimicry weeds of rice. Chauhan and Johnson (2010) observed that rice row spacing of 20 cm reduced barnyardgrass [*Echinochloa crus–galli* (L.) P. Beauv.] and junglerice seed production per plant by 27% and 29%, respectively, when weeds emerged at the same time with the crop.

In legumes, the results of a recent study by Daramola et al. (2019) in soybeans showed that weed density was by 74% and 76% lower at 50 cm row spacing than the corresponding values recorded at 75 cm and 100 cm row spacings, respectively. The corresponding increases in seed yield reached 28% and 31%. Moreover, Rosset and Gulden (2020) demonstrated that adopting narrow row spacings (19 cm) is an agronomic practice to shorten the critical period for weed control in soybean by 85 to 94 growing degree days (GDD) to avoid yield loss of 2.5% and by 77 to 156 GDD to avoid yield loss of 5%, in comparison to the wide row spacing of 76 cm. In two-year field trials carried out in the cowpea (*Vigna unguiculata* L.) crop under semi-arid conditions, the use of 60 cm row spacing contributed to 18–39% and 17–27% reductions in weed density and weed biomass, respectively, compared to 75 and 90 cm row spacings (Adigun et al. 2020). Narrow row spacing also increased seed yield by 27–56% compared to wider row spacings. Field pea managed to partially suppress rigid ryegrass and avoid severe yield losses when crop establishment was performed at 36 cm than at 18 cm row spacing in the research of Lemerle et al. (2006). The results of Dusabumuremyi et al. (2014) showed that weed biomass decreased by up to 56% and yield increased by 30–70% across two growing seasons when common bean establishment was carried out at 30 cm instead of 60 cm spaced rows. When it comes to nonchemical weed management in industrial crops and specifically in cotton, row spacing manipulations are required to minimize the effects of weed competition on crop productivity. Although typically planted in wide row spacings ranging between 70 and 102 cm, narrow row and ultranarrow row spacings provide cotton with a competitive advantage over weeds improving crop's growth and yield performance (Manalil et al. 2017). As also highlighted for soybean, narrow row configurations shorten the critical period for weed and reduce the costs related to weed control operations (Tursun et al. 2016). Regarding quantitative data, the recent findings of Iqbal et al. (2020) revealed that narrow row spacing of 50 cm suppressed weed growth by 55% and achieved higher lint yields by 26% compared to the double row spacing (100 cm). The differences become even more obvious when ultra-narrow row spacings (≤ 25 cm) are adopted. Averaged across four growing seasons, 25 cm row spacing resulted in 60% lower weight to height ratio for prickly sida (*Sida spinosa* L.) in comparison to 100 cm row spacing (Molin et al. 2004). Moreover, prickly sida branch weight decreased by 92% while the number of seed capsules per plant was by 82% lower in 25 cm compared to 100 cm spaced cotton rows.

Considering all these findings, row spacing manipulations can be valuable in ecologically based weed management systems. Although promising results are reported in the literature, there are also some risks associated with crop establishment at narrow row spacings (Blackshaw et al. 2007). First, plant population increases (as when higher seeding rates are used), increasing seed costs and risks associated with yield loss due fungal infection in a dense crop stand. Another disadvantage is that without additional control treatments, weeds between rows are not eliminated and continue to produce seeds. In addition, interrow mechanical weed control treatments may become very risky, as they can cause unacceptable levels of crop damage (Rueda-Ayala et al. 2010).

5.3 Role of Sowing Timing in Ecologically Based Weed Management

In ecologically based weed management systems, sowing time manipulation is another cultural practice to reduce weed pressure and achieve higher yields. In wheat and particularly in no-till systems where previous crop's residues are retained as mulches on soil surface facilitating weed suppression, early sowing can improve crop competitiveness and further reduce weed populations in the field. For example, early sowing of wheat into rice residues was reported to reduce littleseed canarygrass (*Phalaris minor* Retz.) density by 68–80% compared to delaying sowing for one month (Kumar et al. 2013). Similar results were reported in zero-tillage forage barley. Specifically, sowing in May improved crop growth resulting in 95% higher biomass yield and 72% lower weed biomass compared to sowing in late June (Lenssen 2008). In addition, early sowing reduced the seed production of green foxtail and redroot pigweed (*Amaranthus retroflexus* L.) by 96% and 100%, respectively, compared to late sowing. In canola, early sowing accompanied with the selection of a competitive cultivars reduced rigid ryegrass density by 72% and increased seed yield by 76% compared to four-week delayed sowing and selection of a less competitive cultivar (Asaduzzaman et al. 2014). There is also evidence suggesting that early sowing can improve the competitiveness of intercrops against weeds. In particular, Wang et al. (2013) found that early sowing (early April) of lentil–barley intercrops can reduce weed biomass by more than 50% and increase the total yield of the intercrop by 18–55%, compared to late sowing (in May) depending on lentil cultivar and climatic conditions.

In soybean, Jha and Norsworthy (2009) also observed that sowing on early April resulted in significantly lower Palmer amaranth (*Amaranthus palmeri* S. Wats.) density compared to sowing on mid-May. This was due to the fact that early sowing enabled the formation of a dense canopy shading soil surface before the cumulative emergence of the weed. According to these authors, this also explains why Palmer amaranth is less problematic in maize in the United States because crop establishment usually occurs four weeks before initial Palmer amaranth emergence. Early sowing may be useful to reduce the impact of weed competition on cotton productivity. Webster et al. (2009) studied the effects of Benghal dayflower (*Commelina benghalensis* L.) interference (at densities of 11–37 plants m^{-2}) on cotton yield. They observed that a yield loss of 40–60% due to corresponded to the late sowing date but a significantly lower yield loss (21–30%) corresponded to the earlier sowing date (May). In general, late sowing should be avoided; because the damp and moist weather in the month of July favors the emergence and growth of troublesome grass and drought-tolerant broadleaf weeds. In addition to the increase in weed density in cotton fields, late sowing has an additional disadvantage of poor canopy development (Tariq et al. 2020). A general fact for summer crops is that sowing early may provide an advantage over species that usually emerge later in the growing season (e.g. *Amaranthus* spp.) while delayed sowing may allow to control weed species such as common lambsquarters or velvetleaf (*Abutilon theophrasti* Medik.) that tend to emerge earlier in the season (Knezevic 2014; Vidotto et al. 2016; Tariq et al. 2020).

As already mentioned, delaying sowing for a short period after seedbed preparation allows for control of emerging weed seedlings by shallow tillage, a principle known as false seedbed (Travlos et al. 2020). In the meantime, after seedbed preparation and crop sowing, irrigations or rainfall can stimulate further emergence of weeds that are controlled by shallow (10 cm depth) tillage before sowing. A great example is found in the study by Mouillon et al. (2020) who conducted three field trials in the sunflower crop. In their third trial, sowing was delayed because of rain and wet soil conditions that occurred immediately after the initial tillage operations (moldboard plowing and disking). As a result, a false seedbed was created by accident, allowing more weeds to emerge. To prepare the seedbed, the final tillage operation was performed after 20 days and controlled these weeds right before sowing. This resulted in 89% lower weed biomass and 27% higher seed yield in the third trial compared to the corresponding values observed in the second trial. In their second experimental year, Raffaelli et al. (2002) created a false seedbed to control weeds before sowing in maize. In this way, they reduced weed biomass by approximately 86%. In barley, Kanatas et al. (2020) observed that false seedbed reduced sterile oat biomass by 66–75% and increased grain yield by 29–30%. In winter wheat, Shahzad et al. (2021) delayed sowing for one week and controlled weeds by shallow tillage in a false seedbed. Compared to the untreated control, weed biomass decreased by 84%. Akhter et al. (2021) indicated that preparing a seedbed but delaying sowing time under concept of false seedbed may reduce the density and seedbank dynamics of rattail fescue [*Vulpia myuros* (L.) C.C. Gmel.] in wheat. In grass forage crops, Jensen (2019) also found that a false seedbed could reduce the density of rattail fescue and other winter grass weeds by up to 80%. Gürsoy et al. (2014) showed that delaying sowing to control weeds by shallow tillage before lentil establishment reduced weed density by 79%

and increased seed yield by 67% compared to conventional practices.

The selection of the optimum sowing time related to weed emergence can be quite complex; both early and delayed sowing are associated with risks. Sowing too early may expose spring crops to frost damage or result in poor establishment due to inadequate soil moisture and temperature. Early sowing needs to be performed after frost periods have passed and when the soil is warm enough; this is a prerequisite to obtain uniform and vigorous crop growth and increase crop competitiveness against weeds that are present in the field (Korres 2018; Gazoulis et al. 2021). Concerning false seedbed, although this cultural practice has a great potential for weed management purposes, delayed sowing may expose the crop to detrimental weather conditions or may be the cause of canceling crop seeding due to unfavorable weather conditions. In addition, delayed sowing shortens the growth cycle of a crop and may reduce the tillering ability of winter cereals so that increased seeding may be a requirement to overcome reductions in tillering and avoid yield losses (Kolb and Gallandt 2012). In any case, further research is required to optimize sowing time manipulations for weed management purposes.

5.4 Role of Irrigation and Fertilization Management in Weed Management

Crops share the same aboveground and underground resources with weeds. As a result, fertilizer applications and irrigation water affect the competitive interactions between crops and weeds (Kaur et al. 2018). Banding fertilizers under the crop rows provide crops a competitive advantage over weeds by giving cultivated plants optimal access to nutrients (Gazoulis et al. 2021). Because most annual weeds germinate in the upper soil layers, broadcast fertilizer incorporation boosts their emergence and growth. In contrast, precise application of fertilizers at a certain depth reduces weed seed germination and also limits the access of germinated seedlings to the added nutrients (Kaur et al. 2018). Blackshaw et al. (2004) noticed also that weed biomass decreased with subsurface banded or point-injected nitrogen compared surface broadcast nitrogen applications while weed seed production decreased by 25–63% with point-injected nitrogen compared to broadcast fertilization practices (Blackshaw et al. 2004). In addition, nitrogen fertilizer banded near seeds instead of broadcasting resulted in 60% lower biomass for foxtail malleted [Setaria italica (L.) P. Beauv.], when combined with increased seeding rates in maize grown under semiarid conditions (Anderson 2000). However, the impact of fertilizer placement on crop–weed interactions may depend on weed species. Fertilizer application to the 0–15 cm soil profile has little influence on weeds with a tap root but has a pronounced effect on weeds with fibrous root systems (Blackshaw et al. 2004; Kaur et al. 2018; Gazoulis et al. 2021).

Concerning methods that lead to optimal irrigation management, drip and subsurface drip irrigation provide crops with localized supplies of water (Coolong 2013). Compared to conventional surface irrigation systems, de Freitas Souza et al. (2020) recorded 62–97% lower levels of weed infestation in onion across three consecutive growing seasons when they applied drip irrigation compared to conventional practices. Subsurface drip irrigation is another system that not only keeps the soil surface dry, but also improves the growth of the rooting system of the crop improving the competitiveness of cultivated plants against weeds (Coolong 2013). Such findings have been validated under real field conditions in the study of Shrestha et al. (2007) conducted in processing tomato (Lycopersicon esculentum Mill.). Subsurface drip irrigation resulted in 71–96% and 90–99% lower weed density on beds and furrows, respectively, in comparison to furrow irrigation in a field infested with black nightshade, Amaranthus spp. and Echinochloa spp. (Shrestha et al. 2007). Although such systems demand higher installation costs they offer the opportunity for precise irrigation and are expected to have a prominent role in the design of weed management strategies in organic and integrated farming. Moreover, in flooded crops such as rice, flooding can be an important tool for cultural weed management. For example, winter flooding was reported to reduce the number of viable weedy rice (Oryza sativa L.) seeds on the soil surface by 95% in comparison to letting fields dry between rice growing seasons (Fogliatto et al. 2010). Another recent finding related to irrigation timing was reported recently in established furrow-irrigated alfalfa. In their two-year field trials, Kanatas et al. (2021) noticed that irrigating the stand one week prior to harvest or two weeks after harvest resulted in lower levels of weed pressure and higher forage yields of higher quality in the subsequent harvest. In contrast, these authors reported that irrigations right after harvest should be avoided because they stimulate weed emergence at a time where alfalfa canopy has not recovered from the harvest operations.

5.5 Conclusions

The current chapter highlights that the role of cultural practices in developing of sustainable ecologically based weed management systems should not be further underestimated. To make weeds less adaptable, development of suitable crop rotation systems is a priority. Crop selection must follow specific criteria to facilitate successful weed management. Including a cover crop in the rotation is another beneficial practice to further suppress weed emergence and seed production. Moreover, intercropping leads to complementary use of resources from the component species of the intercrop crop mixtures and limits the amounts of resources that are available to weeds. Competitive cultivars and hybrids have much to offer in ecologically based weed management systems. The same is noted for the use of increased seeding rates and planting in narrow row spacings, which are cultural practices that increase crop populations and improve crop competitiveness against weeds. Optimal sowing dates in relation to weed emergence and optimal resource management should also be considered as important components of ecologically based weed management systems. Research shows that combinations among different cultural practices may facilitate sufficient reductions to weed emergence and weed seed production and lead to higher crop yields.

References

Acciares, H.A. and Zuluaga, M.S. (2006). Effect of plant row spacing and herbicide use on weed aboveground biomass and corn grain yield. *Planta Daninha* 24: 287–293. https://doi.org/10.1590/S0100-83582006000200011.

Adigun, J.A., Adeyemi, O.R., Daramola, O.S., et al. (2020). Response of cowpea (*Vigna unguiculata*, L., Walp) to inter-row spacing and weed competition. *Agricultura Tropica et Subtropica* 53: 73–79. https://doi.org/10.2478/ats-2020-0008.

Akhter, M.J., Kudsk, P., Mathiassen, S.K., et al. (2021). Rattail fescue (*Vulpia myuros*) interference and seed production as affected by sowing time and crop density in winter wheat. *Weed Science* 69: 52–61. https://doi.org/10.1017/wsc.2020.84.

Aktar, M.W., Sengupta, D., and Chowdhury, A. (2009). Impact of pesticides use in agriculture: their benefits and hazards. *Interdisciplinary Toxicology* 2: 1–12. https://doi.org/10.2478/v10102-009-0001-7.

Alba, O.S., Syrovy, L.D., Duddu, H.S., et al. (2020). Increased seeding rate and multiple methods of mechanical weed control reduce weed biomass in a poorly competitive organic crop. *Field Crops Research* 245: 107648. https://doi.org/10.1016/j.fcr.2019.107648.

Aluko, O.A. and Olasoji, J.O. (2017). Influence of nitrogen fertilizer on kenaf performance and weed suppression. *Journal of Agriculture and Veterinary Sciences* 10: 54–60. http://iosrjournals.org/iosr–javs/papers/Vol10–issue8/Version–1/K1008015460.pdf (accessed 14 May 2021).

Amossé, C., Jeuffroy, M.H., Celette, F., et al. (2013). Relay–intercropped forage legumes help to control weeds in organic grain production. *European Journal of Agronomy* 49: 158–167. https://doi.org/10.1016/j.eja.2013.04.002.

An, Y., Ma, Y., and Shui, J. (2013). Switchgrass (*Panicum virgatum* L.) plants and switchgrass residue reduce the biomass and density of associated weeds. *Acta Agriculturae Scandinavica, Section B — Soil & Plant Science* 63: 107–113. https://doi.org/10.1080/09064710.2012.729605.

Anderson, R.L. (2000). Cultural systems to aid weed management in semiarid corn (*Zea mays*). *Weed Technology* 14: 630–634. https://doi.org/10.1614/0890-037X(2000)014[0630:CSTAWM]2.0.CO;2.

Anderson, R.L. (2004). Sequencing crops to minimize selection pressure for weeds in the Central Great Plains. *Weed Technology* 18: 157–164. https://doi.org/10.1614/WT-03-090R.

Anderson, R.L. (2005). A multi–tactic approach to manage weed population dynamics in crop rotations. *Agronomy Journal* 97: 1579–1583. https://doi.org/10.2134/agronj2005.0194.

Anderson, R.L. (2007). Managing weeds with a dualistic approach of prevention and control. A review. *Agronomy for Sustainable Development* 27: 13–18. https://doi.org/10.1051/agro:2006027.

Anderson, R.L. (2010). A rotation design to reduce weed density in organic farming. *Renewable Agriculture and Food Systems* 25: 189–195. https://doi.org/10.1017/S1742170510000256.

Andrew, I.K.S., Storkey, J., and Sparkes, D.L. (2015). A review of the potential for competitive cereal cultivars as a tool in integrated weed management. *Weed Research* 55: 239–248. https://doi.org/10.1111/wre.12137.

Asaduzzaman, M., Luckett, D.J., Cowley, R.B. et al. (2014). Canola cultivar performance in weed–infested field plots confirms allelopathy ranking from *in vitro* testing. *Biocontrol Science and Technology* 24: 1394–1411. https://doi.org/10.1080/09583157.2014.942596.

Banik, P., Midya, A., Sarkar, B.K., et al. (2006). Wheat and chickpea intercropping systems in an additive series experiment: advantages and weed smothering. *European Journal of Agronomy* 24: 325–332. https://doi.org/10.1016/j.eja.2005.10.010.

Baraibar, B., Hunter, M.C., Schipanski, M.E. et al. (2018). Weed suppression in cover crop monocultures and mixtures. *Weed Science* 66: 121–133. https://doi.org/10.1017/wsc.2017.59.

Bastiaans, L., Paolini, R., and Baumann, D.T. (2008). Focus on ecological weed management: what is hindering adoption? *Weed Research* 48: 481–491. https://doi.org/10.1111/j.1365-3180.2008.00662.x.

Begna, S.H., Hamilton, R.I., Dwyer, L.M. et al. (2001). Weed biomass production response to plant spacing and corn

(*Zea mays*) hybrids differing in canopy architecture. *Weed Technology* 15: 647–653. https://doi.org/10.1614/0890-037X(2001)015[0647:WBPRTP]2.0.CO;2.

Benaragama, D. and Shirtliffe, S.J. (2013). Integrating cultural and mechanical methods for additive weed control in organic systems. *Agronomy Journal* 105: 1728–1734. https://doi.org/10.2134/agronj2013.0007.

Blackshaw, R.E., Anderson, R.L., and Lemerle, D.E. (2007). Cultural weed management. In: *Non-Chemical Weed Management: Principles, Concepts and Technology* (ed. M.K. Upandhyaya and R.E. Blackshaw), 35–48. Wallingford: CAB International. ISBN: 978-1-84593-291-6.

Blackshaw, R.E., Molnar, L.J., and Janzen, H.H. (2004). Nitrogen fertilizer timing and application method affect weed growth and competition with spring wheat. *Weed Science* 52: 614–622. https://doi.org/10.1614/WS-03-104R.

Boerboom, C.M. and Young, F.L. (1995). Effect of postplant tillage and crop density on broadleaf weed control in dry pea (*Pisum sativum*) and lentil (*Lens culinaris*). *Weed Technology* 9: 99–106. https://doi.org/10.1017/S0890037X00023022.

Borger, C.P.D., Riethmuller, G., and D'Antuono, M. (2016). Eleven years of integrated weed management: long-term impacts of row spacing and harvest weed seed destruction on Lolium rigidum control. *Weed Research* 56: 359–366. https://doi.org/10.1111/wre.12220.

Bowles, T.M., Mooshammer, M., Socolar, Y. et al. (2020). Long–term evidence shows that crop–rotation diversification increases agricultural resilience to adverse growing conditions in North America. *One Earth* 2: 284–293. https://doi.org/10.1016/j.oneear.2020.02.007.

Brankov, M.M., Simić, M., and Dragičević, V. (2021). The influence of maize–winter wheat rotation and pre–emergence herbicides on weeds and maize productivity. *Crop Protection* 143: 105558. https://doi.org/10.1016/j.cropro.2021.105558.

Campiglia, E., Mancinelli, R., Radicetti, E., et al. (2010). Effect of cover crops and mulches on weed control and nitrogen fertilization in tomato (*Lycopersicon esculentum* Mill.). *Crop Protection* 29: 354–363. https://doi.org/10.1016/j.cropro.2009.12.001.

Cardina, J., Herms, C.P., and Doohan, D.J. (2002). Crop rotation and tillage system effects on weed seedbanks. *Weed Science* 50: 448–460. https://doi.org/10.1614/0043-1745(2002)050[0448:CRATSE]2.0.CO;2.

Chauhan, B.S. (2020). Grand challenges in weed management. *Frontiers in Agronomy* 1: 3. https://doi.org/10.3389/fagro.2019.00003.

Chauhan, B.S. and Johnson, D.E. (2010). Implications of narrow crop row spacing and delayed *Echinochloa colona* and *Echinochloa crus–galli* emergence for weed growth and crop yield loss in aerobic rice. *Field Crops Research* 117: 177–182. https://doi.org/10.1016/j.fcr.2010.02.014.

Chauhan, B.S., and Johnson, D.E. (2011). Row spacing and weed control timing affect yield of aerobic rice. *Field Crops Research* 121: 226–231. https://doi.org/10.1016/j.fcr.2010.12.008.

Chauhan, B.S., Singh, V.P., Kumar, A., et al. (2011). Relations of rice seeding rates to crop and weed growth in aerobic rice. *Field Crops Research* 121: 105–115. https://doi.org/10.1016/j.fcr.2010.11.019.

Coolong, T. (2013). Using irrigation to manage weeds: a focus on drip irrigation. In: *Weed and Pest Control–Conventional and New Challenges* (ed. S. Soloneski and M. Larramendy), 161–179. Rijeka: IntechOpen. https://doi.org/10.5772/50276.

Corre-Hellou, G., Dibet, A., Hauggaard-Nielsen, H. et al. (2011). The competitive ability of pea–barley intercrops against weeds and the interactions with crop productivity and soil N availability. *Field Crops Research* 122: 264–272. https://doi.org/10.1016/j.fcr.2011.04.004.

Correia, M.V., Pereira, L.C., De Almeida, L. et al. (2014). Maize–mucuna (*Mucuna pruriens* (L.) DC) relay intercropping in the lowland tropics of Timor–Leste. *Field Crops Research* 156: 272–280. https://doi.org/10.1016/j.fcr.2013.10.011.

Daramola, O.S., Adeyemi, O.R., Adigun, J.A., et al. (2019). Weed interference and control in soybean, as affected by row spacing, in the transition zone of South West Nigeria. *Journal of Crop Improvement* 34: 103–121. https://doi.org/10.1080/15427528.2019.1674759.

Datta, A., Ullah, H., Tursun, N. et al. (2017). Managing weeds using crop competition in soybean [*Glycine max* (L.) Merr.]. *Crop Protection* 95: 60–68. https://doi.org/10.1016/j.cropro.2016.09.005.

Davis, A.S. (2010). Cover–crop roller–crimper contributes to weed management in no–till soybean. *Weed Science* 58: 300–309. https://doi.org/10.1614/WS-D-09-00040.1.

Davis, A.S. and Frisvold, G.B. (2017). Are herbicides a once in a century method of weed control? *Pest Management Science* 73: 2209–2220. https://doi.org/10.1002/ps.4643.

de Freitas Souza, M., Lins, H.A., de Mesquita, H.C. et al. (2020). Can irrigation systems alter the critical period for weed control in onion cropping? *Crop Protection* 105457. https://doi.org/10.1016/j.cropro.2020.105457.

Dhima, K.V., Vasilakoglou, I.B., Eleftherohorinos, I.G., and Lithourgidis, A.S. (2006). Allelopathic potential of winter cereals and their cover crop mulch effect on grass weed suppression and corn development. *Crop Science* 46: 345–352. https://doi.org/10.2135/cropsci2005-0186.

Dowling, A., Sadras, V.O., Roberts, P. et al. (2021). Legume–oilseed intercropping in mechanised broadacre agriculture–a review. *Field Crops Research* 260: 107980. https://doi.org/10.1016/j.fcr.2020.107980.

Drews, S., Neuhoff, D., and Köpke, U. (2009). Weed suppression ability of three winter wheat varieties at different row spacing under organic farming conditions. *Weed Research* 49: 526–533. https://doi.org/10.1111/j.1365-3180.2009.00720.x.

Dusabumuremyi, P., Niyibigira, C., and Mashingaidze, A.B. (2014). Narrow row planting increases yield and suppresses weeds in common bean (*Phaseolus vulgaris* L.) in a semi–arid agro–ecology of Nyagatare, Rwanda. *Crop Protection* 64: 13–18. https://doi.org/10.1016/j.cropro.2014.05.021.

Fahad, S., Hussain, S., Chauhan, B.S. et al. (2015). Weed growth and crop yield loss in wheat as influenced by row spacing and weed emergence times. *Crop Protection* 71: 101–108. https://doi.org/10.1016/j.cropro.2015.02.005.

Fogliatto, S., Vidotto, F., and Ferrero, A. (2010). Effects of winter flooding on weedy rice (*Oryza sativa* L.). *Crop Protection* 29: 1232–1240. https://doi.org/10.1016/j.cropro.2010.07.007.

Forcella, F. (1984). Wheat and ryegrass competition for pulses of mineral nitrogen. *Australian Journal of Experimental Agriculture* 24: 421–425. https://doi.org/10.1071/EA9840421.

Fracchiolla, M., Renna, M., D'Imperio, M. et al. (2020). Living mulch and organic fertilization to improve weed management, yield and quality of broccoli raab in organic farming. *Plants* 9: 177. https://doi.org/10.3390/plants9020177.

Gai, Z., Zhang, J., and Li, C. (2017). Effects of starter nitrogen fertilizer on soybean root activity, leaf photosynthesis and grain yield. *PLoS One* 12: e0174841. https://doi.org/10.1371/journal.pone.0174841.

Gazoulis, I. Kanatas, P., Papastylianou, P. et al.. (2021). Weed management practices to improve establishment of selected lignocellulosic crops. *Energies* 14: 2478. https://doi.org/10.3390/en14092478.

Gianessi, L.P. (2013). The increasing importance of herbicides in worldwide crop production. *Pest Management Science* 69: 1099–1105. https://doi.org/10.1002/ps.3598.

Gürsoy, S., Özaslan, C., Urğun, M. et al. (2014). The effect of sowing time, tillage system and herbicides on weed species density, weed biomass and yield of lentil within a lentil–wheat sequence. *Agriculture & Forestry/Poljoprivreda i Sumarstvo* 60: 75–83. https://www.cabdirect.org/cabdirect/abstract/20143356381 (accessed 13 May 2021).

Hall, J., Bhattarai, S.P., and Midmore, D.J. (2014). Effect of industrial hemp (*Cannabis sativa* L.) planting density on weed suppression, crop growth, physiological responses, and fibre yield in the subtropics. *Renewable Bioresources* 2: 1–7. https://doi.org/10.7243/2052-6237-2-1.

Hammer, G.L., Dong, Z., McLean, G. et al. (2009). Can changes in canopy and/or root system architecture explain historical maize yield trends in the US Corn Belt? *Crop Science* 49: 299–312. https://doi.org/10.2135/cropsci2008.03.0152.

Haramoto, E.R. and Gallandt, E.R. (2004). Brassica cover cropping for weed management: a review. *Renewable Agriculture and Food Systems* 19: 187–198. https://doi.org/10.1079/RAFS200490.

Harker, K.N., Clayton, G.W., Blackshaw, R.E. et al. (2003). Seeding rate, herbicide timing and competitive hybrids contribute to integrated weed management in canola (*Brassica napus*). *Canadian Journal of Plant Science* 83: 433–440. https://doi.org/10.4141/P02-073.

Hasanuzzaman, M., Mohsin, S.M., Bhuyan, M.H.M.B. et al. (2020). Phytotoxicity, environmental and health hazards of herbicides: challenges and ways forward. In: *Agrochemicals Detection, Treatment and Remediation. Pesticides and Fertilizers* (ed. M.N.V. Prasad), 55–99. Amsterdam: Elsevier. https://doi.org/10.1016/b978-0-08-103017-2.00003-9.

Hayden, Z.D., Brainard, D.C., Henshaw, B., et al.(2012). Winter annual weed suppression in rye–vetch cover crop mixtures. *Weed Technology* 26: 818–825. https://doi.org/10.1614/WT-D-12-00084.1.

Heap, I. (2022). *International herbicide resistant weed database*. http://www.weedscience.org (accessed 08 March 2022).

Hufnagel, J., Reckling, M., and Ewert, F. (2020). Diverse approaches to crop diversification in agricultural research. A review. *Agronomy for Sustainable Development* 40: 1–17. https://doi.org/10.1007/s13593-020-00617-4.

Hurley, T.M. and Frisvold, G. (2016). Economic barriers to herbicide–resistance management. *Weed Science* 64: 585–594. https://doi.org/10.1614/WS-D-15-00046.

Hussain, M.I., Abideen, Z., Danish, S. et al. (2021). Integrated weed management for sustainable agriculture. In: *Sustainable Agriculture Reviews 52* (ed. E. Lichtfouse), 367–393. Cham: Springer. https://doi.org/10.1007/978-3-030-73245-5_11.

Ighalo, J.O., Adeniyi, A.G., and Adelodun, A.A. (2021). Recent advances on the adsorption of herbicides and pesticides from polluted waters: performance evaluation via physical attributes. *Journal of Industrial and Engineering Chemistry* 93: 117–137. https://doi.org/10.1016/j.jiec.2020.10.011.

Iqbal, N., Manalil, S., Chauhan, B.S., and Adkins, S.W. (2020). Effect of narrow row–spacing and weed crop competition duration on cotton productivity. *Archives of Agronomy and Soil Science* 1–13. https://doi.org/10.1080/03650340.2020.1836344.

Jabran, K. (2017). Sorghum allelopathy for weed control. In: *Manipulation of Allelopathic Crops for Weed Control* (ed. K. Jabran), 65–75. Cham: Springer. https://doi.org/10.1007/978-3-319-53186-1_8.

Jensen, P.K. (2019). Use of integrated weed management tools in crop rotations with grass seed production. *Acta*

Agriculturae Scandinavica, Section B—Soil & Plant Science 69: 209–218. https://doi.org/10.1080/09064710.2018.1530295.

Jha, P. and Norsworthy, J.K. (2009). Soybean canopy and tillage effects on emergence of Palmer amaranth (*Amaranthus palmeri*) from a natural seed bank. *Weed Science* 57: 644–651. https://doi.org/10.1614/WS-09-074.1.

Jørgensen, L.N., Noe, E., Langvad, A.M. et al. (2007). Decision support systems: barriers and farmers' need for support. *EPPO Bulletin* 37: 374–377. https://doi.org/10.1111/j.1365-2338.2007.01145.x.

Kanatas, P. (2020). Mini–Review: the role of crop rotation, intercropping, sowing dates and increased crop density towards a sustainable crop and weed management in arable crops. *Agraarteadus* 31: 22–27. https://doi.org/10.15159/jas.20.11.

Kanatas, P., Gazoulis, I., and Travlos, I. (2021). Irrigation timing as a practice of effective weed management in established alfalfa (*Medicago sativa* L.) crop. *Agronomy* 11: 550. https://doi.org/10.3390/agronomy11030550.

Kanatas, P.J. and Gazoulis, I. (2022). The integration of increased seeding rates, mechanical weed control and herbicide application for weed management in chickpea (*Cicer arietinum* L.). *Phytoparasitica* 50: 255–267. https://doi.org/10.1007/s12600-021-00955-3.

Kanatas, P.J., Travlos, I.S., Gazoulis, J. et al. (2020). The combined effects of false seedbed technique, post–emergence chemical control and cultivar on weed management and yield of barley in Greece. *Phytoparasitica* 48: 131–143. https://doi.org/10.1007/s12600-020-00783-x.

Kaur, S., Kaur, R., and Chauhan, B.S. (2018). Understanding crop–weed–fertilizer–water interactions and their implications for weed management in agricultural systems. *Crop Protection* 103: 65–72. https://doi.org/10.1016/j.cropro.2017.09.011.

Kaya-Altop, E., Şahin, M., Jabran, K. et al. (2019). Effect of different water management strategies on competitive ability of semi–dwarf rice cultivars with *Echinochloa oryzoides*. *Crop Protection* 116: 33–42. https://doi.org/10.1016/j.cropro.2018.10.009.

Knezevic, S.Z. (2014). Integrated weed management in soybean. In: *Recent Advances in Weed Management* (ed. B. Chauhan and G. Mahajan), 223–237. New York: Springer. https://doi.org/10.1007/978-1-4939-1019-9_10.

Kolb, L.N. and Gallandt, E.R. (2012). Weed management in organic cereals: advances and opportunities. *Organic Agriculture* 2: 23–42. https://doi.org/10.1007/s13165-012-0022-y.

Kolb, L.N., Gallandt, E.R., and Molloy, T. (2010). Improving weed management in organic spring barley: physical weed control vs. interspecific competition. *Weed Research* 50: 597–605. https://doi.org/10.1111/j.1365-3180.2010.00818.x.

Korres, N.E. (2018). Agronomic weed control: a trustworthy approach for sustainable weed management. In: *Non-chemical Weed Control* (ed. K. Jabar and B.S. Chauhan), 97–114. Cambridge: Academic Press. https://doi.org/10.1016/B978-0-12-809881-3.00006-1.

Korres, N.E. and Froud-Williams, R.J. (2002). Effects of winter wheat cultivars and seed rate on the biological characteristics of naturally occurring weed flora. *Weed Research* 42: 417–428. https://doi.org/10.1046/j.1365-3180.2002.00302.x.

Kraska, P., Andruszczak, S., Kwiecińska-Poppe, E. et al. (2020). Supporting crop and different row spacing as factors influencing weed infestation in lentil crop and seed yield under organic farming conditions. *Agronomy* 10: 9. https://doi.org/10.3390/agronomy10010009.

Krishnan, G., Holshouser, D.L., and Nissen, S.J. (1998). Weed control in soybean (*Glycine max*) with green manure crops. *Weed Technology* 12: 97–102. https://doi.org/10.1017/S0890037X00042639.

Kuchinda, N.C., Ndahi, W.B., Lagoke, S.T.O., et al. (2001). The effects of nitrogen and period of weed interference on the fibre yield of kenaf (*Hisbiscus cannabinus* L.) in the northern Guinea Savanna of Nigeria. *Crop Protection* 20: 229–235. https://doi.org/10.1016/S0261-2194(00)00138-1.

Kumar, V., Singh, S., Chhokar, R.S. et al. (2013). Weed management strategies to reduce herbicide use in zero–till rice–wheat cropping systems of the Indo–Gangetic Plains. *Weed Technology* 27: 241–254. https://doi.org/10.1614/WT-D-12-00069.1.

Lawley, Y.E., Weil, R.R., and Teasdale, J.R. (2011). Forage radish cover crop suppresses winter annual weeds in fall and before corn planting. *Agronomy Journal* 103: 137–144. https://doi.org/10.2134/agronj2010.0187.

Lemerle, D., Verbeek, B., and Diffey, S. (2006). Influences of field pea (*Pisum sativum*) density on grain yield and competitiveness with annual ryegrass (*Lolium rigidum*) in south–eastern Australia. *Australian Journal of Experimental Agriculture* 46: 1465–1472. https://doi.org/10.1071/EA04233.

Lenssen, A. (2008). Planting date and preplant weed management influence yield, water use, and weed seed production in herbicide–free forage barley. *Weed Technology* 22: 486–492. https://doi.org/10.1614/WT-08-009.1.

Lenssen, A. (2019). Planting date and preplant weed management influence yield, water use, and weed seed production in herbicide–free forage barley. *Weed Technology* 22: 486–492. https://doi.org/10.1614/WT-08-009.1.

Liebert, J.A. and Ryan, M.R. (2017). High planting rates improve weed suppression, yield, and profitability in organically–managed, no–till–planted soybean. *Weed Technology* 31: 536–549. https://doi.org/10.1017/wet.2017.35.

Liebman, M. and Staver, C.P. (2001). Crop diversification for weed management. In: *Ecological Management of*

Agricultural Weeds (ed. M. Liebman, C.L. Mohler, and C.P. Staver), 322–374. Cambridge: Cambridge University Press. ISBN: 0-511-03821-6.

MacLaren, C., Labuschagne, J., and Swanepoel, P.A. (2021). Tillage practices affect weeds differently in monoculture vs. crop rotation. *Soil and Tillage Research* 205: 104795. https://doi.org/10.1016/j.still.2020.104795.

Mahmood, A., Cheema, Z.A., Mushtaq, M.N., et al. (2013). Maize–sorghum intercropping systems for purple nutsedge management. *Archives of Agronomy and Soil Science* 59: 1279–1288. https://doi.org/10.1080/03650340.2012.704547.

Manalil, S., Coast, O., Werth, J., and Chauhan, B.S. (2017). Weed management in cotton (*Gossypium hirsutum* L.) through weed–crop competition: a review. *Crop Protection* 95: 53–59. https://doi.org/10.1016/j.cropro.2016.08.008.

Mashingaidze, A.B., Van Der Werf, W., Lotz, L.A.P. et al. (2009). Narrow rows reduce biomass and seed production of weeds and increase maize yield. *Annals of Applied Biology* 155: 207–218. https://doi.org/10.1111/j.1744-7348.2009.00331.x.

Mennan, H., Ngouajio, M., Sahin, M. et al. (2012). Competitiveness of rice (*Oryza sativa* L.) cultivars against *Echinochloa crus-galli* (L.) Beauv. in water-seeded production systems. *Crop Protection* 41: 1–9. https://doi.org/10.1016/j.cropro.2012.04.027.

Mennan, H. and Zandstra, B.H. (2005). Effect of wheat (*Triticum aestivum*) cultivars and seeding rate on yield loss from *Galium aparine* (cleavers). *Crop Protection* 24: 1061–1067. https://doi.org/10.1016/j.cropro.2005.02.012.

Mhlanga, B., Chauhan, B.S., and Thierfelder, C. (2016). Weed management in maize using crop competition: a review. *Crop Protection* 88: 28–36. https://doi.org/10.1016/j.cropro.2016.05.008.

Mirsky, S.B., Gallandt, E.R., Mortensen, D.A. et al. (2010). Reducing the germinable weed seedbank with soil disturbance and cover crops. *Weed Research* 50: 341–352. https://doi.org/10.1111/j.1365-3180.2010.00792.x.

Mischler, R.A., Curran, W.S., Duiker, S.W., et al. (2010). Use of a rolled-rye cover crop for weed suppression in no-till soybeans. *Weed Technology* 24: 253–261. https://doi.org/10.1614/WT-D-09-00004.1.

Mohammadi, G.R. and Amiri, F. (2011). Critical period of weed control in soybean (*Glycine max*) as influenced by starter fertilizer. *Australian Journal of Crop Science* 5: 1350–1355. http://www.cropj.com/mohamadi_5_11_2011_1350_1355.pdf (accessed 14 May 2020).

Mohammadi, G.R., Ghobadi, M.E., and Sheikheh-Poor, S. (2012). Phosphate biofertilizer, row spacing and plant density effects on corn (*Zea mays* L.) yield and weed growth. *American Journal of Plant Sciences* 3: 425–429. https://doi.org/10.4236/ajps.2012.34051.

Mohammed, Y.A., Miller, Z., Hubbel, K., et al. (2020). Variety and weed management effects on organic chickpea stand establishment and seed yield. *Agrosystems, Geosciences & Environment* 3: e20035. https://doi.org/10.1002/agg2.20035.

Molin, W.T., Hugie, J.A., and Hirase, K. (2004). Prickly sida (*Sida spinosa* L.) and spurge (*Euphorbia hyssopifolia* L.) response to wide row and ultra-narrow row cotton (*Gossypium hirsutum* L.) management systems. *Weed Biology and Management* 4: 222–229. https://doi.org/10.1111/j.1445-6664.2004.00145.x.

Montull, J.M., Taberner, A., Bøjer, O., and Rydahl, P. (2020). IPMwise: a decision support system for multispecies weed control in cereal crops. In: *Decision Support Systems for Weed Management* (ed. G. Chantre and J. González-Andújar), 279–298. Cham: Springer. https://doi.org/10.1007/978-3-030-44402-0_13.

Moss, S. (2019). Integrated weed management (IWM): why are farmers reluctant to adopt non-chemical alternatives to herbicides? *Pest Management Science* 75: 1205–1211. https://doi.org/10.1002/ps.5267.

Mouillon, P., Caldwell, B.A., Cordeau, S. et al. (2020). Crop density affects weed suppression in organically managed sunflower. *Agronomy Journal* 112: 450–457. https://doi.org/10.1002/agj2.20059.

Mwendwa, J.M., Brown, W.B., Weidenhamer, J.D. et al. (2020). Evaluation of commercial wheat cultivars for canopy architecture, early vigour, weed suppression, and yield. *Agronomy* 10: 983. https://doi.org/10.3390/agronomy10070983.

Myers, J.H., Rose, G., Odell, E. et al. (2022). Household herbicide use as a source of simazine contamination in urban surface waters. *Environmental Pollution* 299: 118868. https://doi.org/10.1016/j.envpol.2022.118868.

Nordby, D.E., Alderks, D.L., and Nafziger, E.D. (2007). Competitiveness with weeds of soybean cultivars with different maturity and canopy width characteristics. *Weed Technology* 21: 1082–1088. https://doi.org/10.1614/WT-06-190.1.

Norsworthy, J.K. and Shipe, E. (2006). Evaluation of glyphosate-resistant Glycine max genotypes for competitiveness at recommended seeding rates in wide and narrow rows. *Crop Protection* 25: 362–368. https://doi.org/10.1016/j.cropro.2005.06.006.

O' Donovan, J.T., Harker, K.N., Clayton, G.W., et al. (2000). Wild oat (*Avena fatua*) interference in barley (*Hordeum vulgare*) is influenced by barley variety and seeding rate. *Weed Technology* 14: 624–629. https://doi.org/10.1614/0890-037X(2000)014[0624:WOAFII]2.0.CO;2.

Oerke, E.-C. and Dehne, H.-W. (2004). Safeguarding production—losses in major crops and the role of crop protection. *Crop Protection* 23: 275–285. https://doi.org/10.1016/j.cropro.2003.10.001.

Osipitan, O.A., Dille, J.A., Assefa, Y., and Knezevic, S.Z. (2018). Cover crop for early season weed suppression in crops: systematic review and meta-analysis. *Agronomy Journal* 110: 2211–2221. https://doi.org/10.2134/agronj2017.12.0752.

Pakeman, R.J., Brooker, R.W., Karley, A.J. et al. (2020). Increased crop diversity reduces the functional space available for weeds. *Weed Research* 60: 121–131. https://doi.org/10.1111/wre.12393.

Perotti, V.E., Larran, A.S., Palmieri, V.E. et al. (2020). Herbicide resistant weeds: a call to integrate conventional agricultural practices, molecular biology knowledge and new technologies. *Plant Science* 290: 110255. https://doi.org/10.1016/j.plantsci.2019.110255.

Place, G.T., Reberg-Horton, S.C., Dunphy, J.E., et al. (2009). Seeding rate effects on weed control and yield for organic soybean production. *Weed Technology* 23: 497–502. https://doi.org/10.1614/WT-08-134.1.

Radicetti, E., Mancinelli, R., and Campiglia, E. (2012a). Combined effect of genotype and inter–row tillage on yield and weed control of chickpea (*Cicer arietinum* L.) in a rainfed Mediterranean environment. *Field Crops Research* 127: 161–169. https://doi.org/10.1016/j.fcr.2011.11.013.

Radicetti, E., Mancinelli, R., and Campiglia, E. (2012b). The competitive ability of different chickpea (*Cicer arietinum*) genotypes against *Polygonum aviculare* under field conditions. *Crop Protection* 42: 312–319. https://doi.org/10.1016/j.cropro.2012.06.011.

Raffaelli, M., Bàrberi, P., Peruzzi, A., et al. (2002). Options for mechanical weed control in grain maize–effects on weeds. *Proceedings of 5th EWRS Workshop on Physical and Cultural Weed Control*, 147–152.

Rezakhanlou, A., Mirshekari, B., Zand, E. et al. (2013). Evaluation of competitiveness of cotton varieties to cocklebur (*Xanthium strumarium* L.). *Journal of Food, Agriculture & Environment* 11: 308–311. https://doi.org/10.1234/4.2013.4272.

Rosset, J.D. and Gulden, R.H. (2020). Cultural weed management practices shorten the critical weed–free period for soybean grown in the Northern Great Plains. *Weed Science* 68: 79–91. https://doi.org/10.1017/wsc.2019.60.

Rueda-Ayala, V., Rasmussen, J., and Gerhards, R. (2010). Mechanical weed control. In: *Precision Crop Protection–the Challenge and Use of Heterogeneity* (ed. E.C. Oerke, G. Menz, and R. Sikora) 279–294. Dordrecht: Springer. https://doi.org/10.1007/978-90-481-9277-9_17.

Rydahl, P. (2003). A web-based decision support system for integrated management of weeds in cereals and sugarbeet. *EPPO Bulletin* 33: 455–460. https://doi.org/10.1111/j.1365-2338.2003.00679.x.

Sankula, S., VanGessel, M.J., and Mulford, R.R. (2004). Corn leaf architecture as a tool for weed management in two corn production systems. *Weed Science* 52: 1026–1033. https://doi.org/10.1614/P2001-125.

Sardana, V., Mahajan, G., Jabran, K., et al. (2017). Role of competition in managing weeds: an introduction to the special issue. *Crop Protection* 95: 1–7. https://doi.org/10.1016/j.cropro.2016.09.011.

Saucke, H. and Ackermann, K. (2006). Weed suppression in mixed cropped grain peas and false flax (*Camelina sativa*). *Weed Research* 6: 453–461. https://doi.org/10.1111/j.1365-3180.2006.00530.x.

Shahzad, M., Hussain, M., Jabran, K. et al. (2021). The impact of different crop rotations by weed management strategies' interactions on weed infestation and productivity of wheat (*Triticum aestivum* L.). *Agronomy* 11: 2088. https://doi.org/10.3390/agronomy11102088.

Sharma, G., Shrestha, S., Kunwar, S., et al. (2021). Crop diversification for improved weed management: a review. *Agriculture* 11: 461. https://doi.org/10.3390/agriculture11050461.

Shrestha, A., Mitchell, J.P., and Lanini, W.T. (2007). Subsurface drip irrigation as a weed management tool for conventional and conservation tillage tomato (*Lycopersicon esculentum* Mill.) production in semi–arid agroecosystems. *Journal of Sustainable Agriculture* 31: 91–112. https://doi.org/10.1300/J064v31n02_07.

Simić, M., Dolijanović, Ž., Maletić, R. et al. (2009). The genotype role in maize competitive ability. *Genetika* 41: 59–67. https://doi.org/10.2298/GENSR0901059S.

Sønderskov, M., Fritzsche, R., de Mol, F. et al. (2015). DSSHerbicide: weed control in winter wheat with a decision support system in three South Baltic regions–Field experimental results. *Crop Protection* 76: 15–23. https://doi.org/10.1016/j.cropro.2015.06.009.

Soti, P. and Racelis, A. (2020). Cover crops for weed suppression in organic vegetable systems in semiarid subtropical Texas. *Organic Agriculture* 10: 429–436. https://doi.org/10.1007/s13165-020-00285-4.

Tariq, M., Abdullah, K., Ahmad, S. et al. (2020). Weed management in cotton. In: *Cotton Production and Uses* (ed. S. Ahmad and M. Hasanuzzaman), 145–161. Springer: Singapore. https://doi.org/10.1007/978-981-15-1472-2_9.

Tharp, B.E. and Kells, J.J. (2001). Effect of glufosinate–resistant corn (*Zea mays*) population and row spacing on light interception, corn yield, and common lambsquarters (*Chenopodium album*) Growth. *Weed Technology* 15: 413–418. https://doi.org/10.1614/0890-037X(2001)015[0413:EOGRCZ]2.0.CO;2.

Travlos, I., Gazoulis, I., Kanatas, P. et al. (2020). Key factors affecting weed seeds' germination, weed emergence and their possible role for the efficacy of false seedbed technique as weed management practice. *Frontiers in Agronomy* 2: 1. https://doi.org/10.3389/fagro.2020.00001.

Tursun, N., Datta, A., Budak, S. et al. (2016). Row spacing impacts the critical period for weed control in cotton

(*Gossypium hirsutum*). *Phytoparasitica* 44: 139–149. https://doi.org/10.1007/s12600-015-0494-x.

Uchino, H., Iwama, K., Jitsuyama, Y. et al. (2012). Effect of interseeding cover crops and fertilization on weed suppression under an organic and rotational cropping system: 1. Stability of weed suppression over years and main crops of potato, maize and soybean. *Field Crops Research* 127: 9–16. https://doi.org/10.1016/j.fcr.2011.10.007.

van der Meulen, A. and Chauhan, B.S. (2017). A review of weed management in wheat using crop competition. *Crop Protection* 95: 38–44. https://doi.org/10.1016/j.cropro.2016.08.004.

Vasilakoglou, I. and Dhima, K. (2014). Potential of two cardoon varieties to produce biomass and oil under reduced irrigation and weed control inputs. *Biomass Bioenergy* 63: 177–186. https://doi.org/10.1016/j.biombioe.2014.01.052.

Vidotto, F., Fogliatto, S., Milan, M., et al. (2016). Weed communities in Italian maize fields as affected by pedo-climatic traits and sowing time. *European Journal of Agronomy* 74: 38–46. https://doi.org/10.1016/j.eja.2015.11.018.

Vincent-Caboud, L., Peigné, J., Casagrande, M., and Silva, E.M. (2017). Overview of organic cover crop-based no-tillage technique in Europe: farmers' practices and research challenges. *Agriculture* 7: 42. https://doi.org/10.3390/agriculture7050042.

von Cossel, M., Mangold, A., Iqbal, Y. et al. (2019). How to generate yield in the first year—a three-year experiment on miscanthus (*Miscanthus* × *giganteus* (Greef et Deuter)) establishment under maize (*Zea mays* L.). *Agronomy* 9: 237. https://doi.org/10.3390/agronomy9050237.

Wang, L., Gruber, S., and Claupein, W. (2012). Effects of woodchip mulch and barley intercropping on weeds in lentil crops. *Weed Research* 52: 161–168. https://doi.org/10.1111/j.1365-3180.2012.00905.x.

Wang, L., Gruber, S., and Claupein, W. (2013). Effect of sowing date and variety on yield and weed populations in a lentil-barley mixture. *The Journal of Agricultural Science* 151: 672. https://doi.org/10.1017/S0021859612000895.

Webster, T., Grey, T., Flanders, J., et al. (2009). Cotton planting date affects the critical period of benghal dayflower (*Commelina benghalensis*) control. *Weed Science* 57: 81–86. https://doi.org/10.1614/WS-08-118.1.

Webster, T.M. (2007). Cotton row spacing and plant population affect weed seed production. *World Cotton Research Conference-4*, Lubbock, TX, USA (10–14 September 2007), 36–57.

Weiner, J., Griepentrog, J., and Kristensen, L. (2001). Suppression of weed by spring wheat Triticum aestivum increases with crop density and spatial uniformity. *Journal of Applied Ecology* 38: 784–790. https://doi.org/10.1046/j.1365-2664.2001.00634.x.

Wells, M.S., Reberg-Horton, S.C., and Mirsky, S.B. (2014). Cultural strategies for managing weeds and soil moisture in cover crop based no-till soybean production. *Weed Science* 62: 501–511. https://doi.org/10.1614/WS-D-13-00142.1.

Wicks, G.A., Nordquist, P.T., Baenziger, P.S. et al. (2004). Winter wheat cultivar characteristics affect annual weed suppression. *Weed Technology* 18: 988–998. https://doi.org/10.1614/WT-03-158R1.

Williams, M.M., Boydston, R.A., and Davis, A.S. (2007). Wild proso millet (*Panicum miliaceum*) suppressive ability among three sweet corn hybrids. *Weed Science* 55: 245–251. https://doi.org/10.1614/WS-06-123.1.

Wołejko, E., Wydro, U., Odziejewicz, J.I., et al. (2022). Biomonitoring of soil contaminated with herbicides. *Water* 14: 1534. https://doi.org/10.3390/w14101534.

Wu, H., Walker, S.R., Osten, V.A., and Robinson, G. (2010). Competition of sorghum cultivars and densities with Japanese millet (*Echinochloa esculenta*). *Weed Biology and Management* 10: 185–193. https://doi.org/10.1111/j.1445-6664.2010.00383.x.

Wu, T., Xi, N., Weiner, J., et al. (2021). Differences in weed suppression between two modern and two old wheat cultivars at different sowing densities. *Agronomy* 11: 253. https://doi.org/10.3390/agronomy11020253.

Zand, E. and Beckie, H.J. (2002). Competitive ability of hybrid and open-pollinated canola (*Brassica napus*) with wild oat (*Avena fatua*). *Canadian Journal of Plant Science* 82: 473–480. https://doi.org/10.4141/P01-149.

Zegada-Lizarazu, W. and Monti, A. (2011). Energy crops in rotation. A review. *Biomass and Bioenergy* 35: 12–25. https://doi.org/10.1016/j.biombioe.2010.08.001.

Zeller, A.K., Zeller, Y.I., and Gerhards, R. (2021). A long-term study of crop rotations, herbicide strategies and tillage practices: effects on *Alopecurus myosuroides* Huds. Abundance and contribution margins of the cropping systems. *Crop Protection* 145: 105613. https://doi.org/10.1016/j.cropro.2021.105613.

Zenobi, S., Fiorentini, M., Aquilanti, L. et al. (2021). Effect of planting density in two thistle species used for vegetable rennet production in marginal Mediterranean areas. *Agronomy* 11: 135. https://doi.org/10.3390/agronomy11010135.

6

The Role of Agri-Chemical Industry on Ecologically Based Weed Management

Vasileios P. Vasileiadis[1], *Vijay K. Varanasi*[2], *Parminder Chahal*[3], *and Nicholas E. Korres*[4,*]

[1] Syngenta Crop Protection, Athens, Greece
[2] Bayer Crop Science, St. Louis, Missouri, USA
[3] FMC Corporation, University of Nebraska, Lincoln, Nebraska, USA
[4] Department of Agriculture, University of Ioannina, Kostakii, Arta, Greece
*Corresponding author

6.0 Introduction

For more than 60 years, herbicides have been very efficient in controlling annual and perennial weed species worldwide, greatly reducing yield losses and stabilizing weed infestations at acceptable levels (Chikowo et al. 2009). They can be applied as either soil-active (pre-sowing, preemergence to a crop) or foliar-active herbicides (i.e. postemergence) depending on herbicide, soil and climatic conditions, weed composition and infestation levels, and farmers' perception of chemical weed management (Vasileiadis et al. 2017). Besides their high efficacy, the rapid adoption of herbicides has also been driven by their relative low cost compared to other methods.

6.1 Herbicide Resistance

The extensive use of herbicides over time, often as the sole control method, has resulted in the evolution of herbicide-resistant weeds in all major agricultural systems (Peterson et al. 2018). Herbicide resistance has been and will remain one of the biggest problems for the next few decades. Globally, it is becoming increasingly difficult to manage resistance in troublesome weeds such as *Lolium rigidum* from Australia, *Alopecurus myosuroides* from northwestern Europe and *Amaranthus palmeri/tuberculatus* from corn and soybean production areas in the USA, to name but a few (Pannell et al. 2016). Due to resistance issues, the cost of crop production is increasing, making it less profitable for the farmers.

Herbicide resistance in weeds is categorized into two major groups: target-site resistance (TSR) and nontarget-site resistance mechanism (NTSR). The more common TSR mechanism involves a single amino acid change in the target enzyme, preventing herbicide binding (Tranel and Wright 2002; Powles and Yu 2010) or due to target-site gene amplification (Gaines et al. 2010). In plants, NTSR can be due to decreased herbicide uptake; reduced herbicide translocation; increased herbicide detoxification (metabolism) due to cytochrome P450 monooxygenases (P450s), glutathione S-transferases (GSTs), or glucosyltransferases; or sequestration into vacuoles (Yuan et al. 2007; Powles and Yu 2010). Because of its complex nature, NTSR is poorly understood and is considered a greater threat to herbicide sustainability. According to the International Herbicide-resistant Weed Database, globally there are 512 unique cases of herbicide resistant weeds in 152 dicots and 110 monocot weeds. Weeds have evolved resistance to 23 of the 26 known herbicide sites of action in 92 crops (Heap 2020). The evolution of herbicide-resistant weeds has become a global issue, suggesting that the adoption of integrated weed management (IWM) practices is critical to control resistant weeds for global food security (Norsworthy et al. 2012; Chahal and Jhala 2019).

There are at least six ways to ease the pressure of herbicide resistance development. Norsworthy et al. (2012) recommended some best management practices (BMPs) as part of IWM to mitigate the evolution of herbicide resistance:

1) *Understand biology of weeds and their life cycles.* Knowing how weeds germinate, their developmental stages, seed production and viability, and their dispersal mechanisms would help in devising sustainable weed control strategies.
2) *Use diversified weed management tactics.* These include cultural (i.e. crop rotation, cover crops), mechanical (i.e.

Ecologically Based Weed Management: Concepts, Challenges, and Limitations, First Edition. Edited by Nicholas E. Korres, Ilias S. Travlos, and Thomas K. Gitsopoulos.

tillage, harvest weed seed control), biological (natural enemies) along with chemical (herbicides) approaches to decrease the amount of seed that is being added to the soil seed bank each year.

3) *Scout fields regularly for weed escapes.* The purpose of field scouting is to examine what weed species are in the field (richness), the location of these species (where they are) and the weed density (number of weeds per unit area). This information is important to design a management program that fits the specific needs of each field.
4) *Plant weed-free seed at the beginning of the season.* It is very important to use certified seed and clean of weed infestation. This preventive measure is one of the most important actions producers can employ to keep their fields free from weeds.
5) *Use multiple sites of action.* Herbicides with different sites of action can be applied either sequentially or as tank mix partners.
6) *Follow herbicide label.* It is important to use the herbicides as intended, in the dosage and strength recommended.

6.2 Climate Change

Climate change is another major challenge we need to stand up to. Scientists need to come up with innovative solutions to counter changes in climatic patterns affecting crop production systems. Climate change refers to a long-term shift in weather events such as temperature, precipitation, and wind, associated with corresponding increases in greenhouse gases such as carbon dioxide, methane, and nitrous oxide. It is predicted that by the year 2100, Earth's surface temperature will increase by 2–4 °C (Pachari and Meyer 2014; Korres et al. 2016). This would impact all agriculture systems, including weed management, in a significant fashion (Dukes and Mooney 1999; Ziska and Goins 2006; Singer et al. 2013). The efficacy of many of the commonly used herbicides for weed control is determined by various climatic factors such as light, temperature, relative humidity, and wind velocity. Therefore, it is imperative that any change in climate is going to have an impact on the performance of herbicides. Additionally, any long-term change in climate could result in shifts in weed species in terms of physiology and growth habitats (Jugulam et al. 2019). It is therefore important to devise a more holistic weed control approach incorporating both chemical and nonchemical methods for sustainable crop production.

Weed management is at the crossroads of use of some traditional chemical and some newly emerging weed control practices (Korres et al. 2019). The challenge is to meet the food and nutritional requirements of ever-increasing population in a sustainable and environmentally friendly way. For various stakeholders (growers, consumers, industry, academia) involved, it is going to be a balancing act between increasing food production and reducing environmental impact. The current and the next-generation weed scientists, whether in academia or industry, must embrace and accept new challenges that include herbicide resistance, climate change, and environmental footprint. This is especially important for achieving a sustainable future.

Major agricultural companies have realized the need for more sustainable practices for optimized weed control and yield protection, and in this direction have taken several initiatives as discussed in this chapter. The role of industry is essential to support sustainable weed management approaches that satisfy all dimensions of sustainability – economic, environmental, and social.

6.3 Environmentally Sound Weed Control Approaches

Effective traditional and new methods, which are categorized as mechanical/physical, cultural, biological and prevention, have been implemented for weed control (Korres 2005). Integrating cover crops along with PRE and/or POST herbicides, for example, is being recommended as a sustainable weed control approach in soybean, cotton, and corn cropping systems (Wiggins et al. 2015, 2016, 2017; Whalen et al. 2020). Cover crops, due to their high biomass density or allelochemical production, prevent weed emergence through physical suppression and reduced light transmission (Barnes et al. 1987; Chase et al. 1991; Akemo et al. 2000; Saini et al. 2006). Some tested cover crop species include hairy vetch (*Vicia villosa*), cereal rye (*Secale cereale*), crimson clover (*Trifolium incarnatum*), red clover (*T. pratense*), white clover (*T. repens*), peas (*Pisum spp.*), bird's-foot trefoil (*Lotus corniculatus*), common oat (*Avena sativa*), ryegrass (*Lolium spp.*), and fescues (*Festuca* spp.) (Korres 2005). Cover crops usage, especially legumes, has been recommended as an ecologically sound practice for improving soil chemical and physical properties, and for elevating the soil nutrient status (Korres 2005, 2018; Price and Norsworthy 2013; Korres and Norsworthy 2015; Norsworthy et al. 2016). Other tactics include crop seed rate even though the cost, if not justified, can be proved economically infeasible; use of competitive crops even though crop breeding programs do not test for crop competitiveness but only for yield and yield components (Korres and Froud-Williams 2002); crop rotation, an approach that has been proven effective for weed soil seedbank reduction (Schwartz-Lazaro et al. 2022); and mechanical means, although issues about soil compaction, fertility, and erosion should be taken into consideration (Korres 2018). Relatively new preventive tactics, along with new technologies such as narrow windrow burning, weed seed harvesting and destruction, precision agriculture approaches, and data knowledge acquisition, are efficient and environmentally sound approaches for weed management (Korres et al. 2019).

Recently, US EPA, under its reduced risk pesticide program, registered certain new active ingredients and identified some new uses of already registered herbicides (EPA 2022). These new registrations were categorized as green chemistries due to significantly less impact on human health and environment, less groundwater pollution, extremely low application rates, and most importantly, their compatibility with various IWM strategies such as cover crops. Examples of such green chemistries predicted to transform future weed control include florpyrauxifen-benzyl, imazamox, and mesotrione (van Wychen 2020).

6.4 Environmentally Friendly Industry Initiatives

Major agricultural companies have realized the demand for more sustainable practices for optimized weed control and yield protection, and in this direction have taken several initiatives as discussed below.

6.4.1 Syngenta Crop Protection AG

Syngenta is actively promoting the implementation of sustainable integrated pest management (IPM), which Europe has mandated by all professional users of crop protection products since 1 January 2014 (Directive 2009/128/EC; European Parliament 2009). Integrated weed management (IWM), an essential component of IPM, has the potential to suppress weed populations, reduce the environmental impact of individual weed management practices and reduce the selection pressure for weed resistance to herbicides by combining preventive, cultural, mechanical, biological, and chemical control methods making use of available decision-aid methods like decision support systems (DSS) or predictive weed emergence models (Harker and O'Donovan 2013; Vasileiadis et al. 2015, 2016). Syngenta offers growers a comprehensive portfolio of chemical and biological crop protection tools, as well as seeds. Its services include agronomic advice, grower programs, digital agriculture, and stewardship.

Besides the broad portfolio of chemical control solutions for weed management in different crops, Syngenta is promoting the use of diversified crop rotation as the primary IWM tool for disrupting the growing cycle of difficult-to control or herbicide resistant (HR) weeds with the diverse weed management practices implemented in different crops. The use of winter and summer crops, the choice of competitive crop varieties, and use of cover crops within the rotation can increase the range of stress and mortality factors that regulate weed population dynamics and maintain effective weed control (Liebman et al. 2014). Trials at our Barton Blackgrass Innovation Centre and Ryegrass Innovation Centre at Doncaster, UK determined that for the difficult to control Alopecutus myosuroides (black-grass) and Lolium perenne (ryegrass), competitive cropping is one management tool that could sustainably reduce densities. Black-grass and ryegrass (Figure 6.1) were determined to be out competed by Syngenta's hybrid winter barley compared to the conventional variety and winter wheat (Cook and Roche 2018).

The type of primary tillage operations within a crop rotation will largely affect the weed densities and type of weed species emerging, since it affects the weed seed distribution in

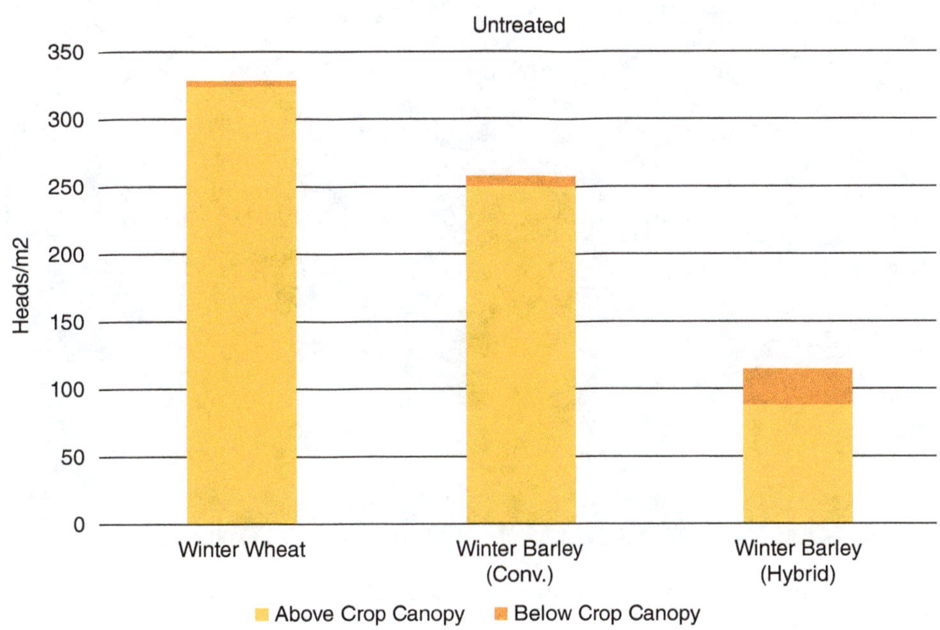

Figure 6.1 Competitive ability of hybrid winter barley against ryegrass in comparison to winter wheat and conventional barley without herbicide treatment (https://www.syngenta.co.uk/ryegrass).

the soil. Syngenta has developed the "Cultivation Insight" tool (https://www.syngenta.co.uk/cultivation-insight-tool) that helps growers understand their seedbank based on recent or planned cultivations (Figure 6.2). Understanding where seed is and how that relates to the time it takes for emergence can help to refine herbicide strategies. Where seed is left on the surface or the top-soil horizon, and factors such as dormancy of freshly shed seed and moisture availability, will determine the weed emergence pattern. If these seeds are most likely to emerge in one rapid flush, then a single herbicide application should deliver effective control. When deeper soil cultivations have been practiced, seed position will play an important role in determining the emergence pattern. The deeper weed seed is in the seedbank, the longer it takes to germinate and reach the surface, thus the longer a grower would need residual herbicides to be active. In trials, Syngenta has seen that noninversion cultivations on the top 15cm spread seeds throughout the soil profile creating a continuous flush of grass weeds as seeds and plants germinate from a range of different depths making it challenging to control.

Specifically, a trial at Barton Blackgrass Innovation Centre in UK, has shown the effect of different cultivation matrix scenarios with herbicide application on blackgrass control in winter wheat. Findings over the last three years suggest that using the same cultivation year after year will not achieve the best margin or black-grass control. Results have shown that the best margin was achieved with a direct-drilling (DD)/Min-till/DD scenario, best control with a DD/Plough/DD scenario, worst margin under a plough/DD/DD scenario and worst control under continuous min-till (Figure 6.3).

Digital tools have the ability to enhance agronomic knowledge, enable better decision-making and resource use efficiency, and generate higher yields. Decision Support Systems (DSS) can assist farmers in making decisions about *if*, *when*, and *how* to treat, and field data can provide the information needed to run the DSS (Berti et al. 2003). Syngenta recently launched Cropwise Protector Scouting, a digital tool that makes agronomic decision easier and faster, supporting the grower with precision monitoring

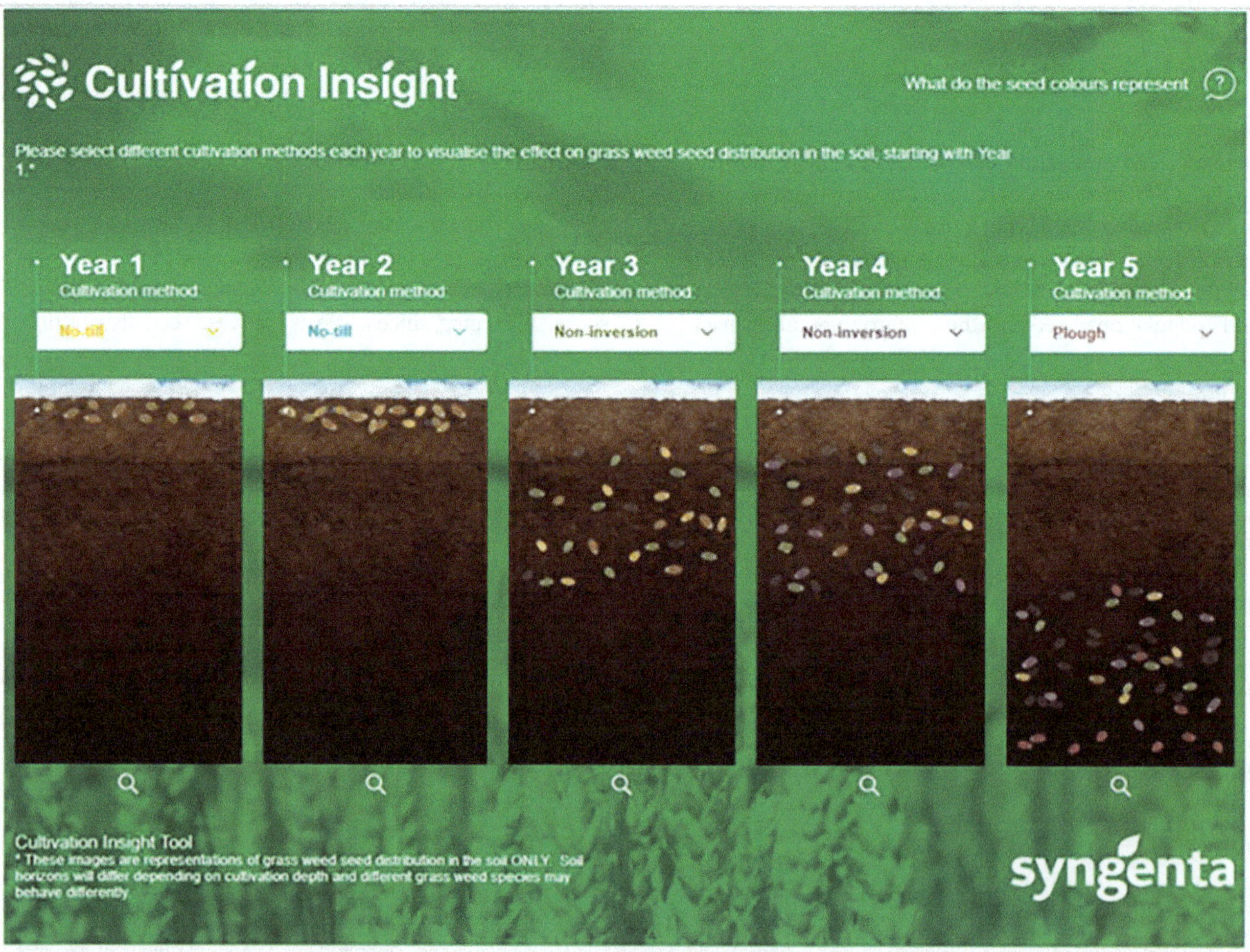

Figure 6.2 Weed seed distribution in the soil profile as affected by a no-till/no-till/ min-till/min-till/plough scenario according to Syngenta's Cultivation Insight tool.

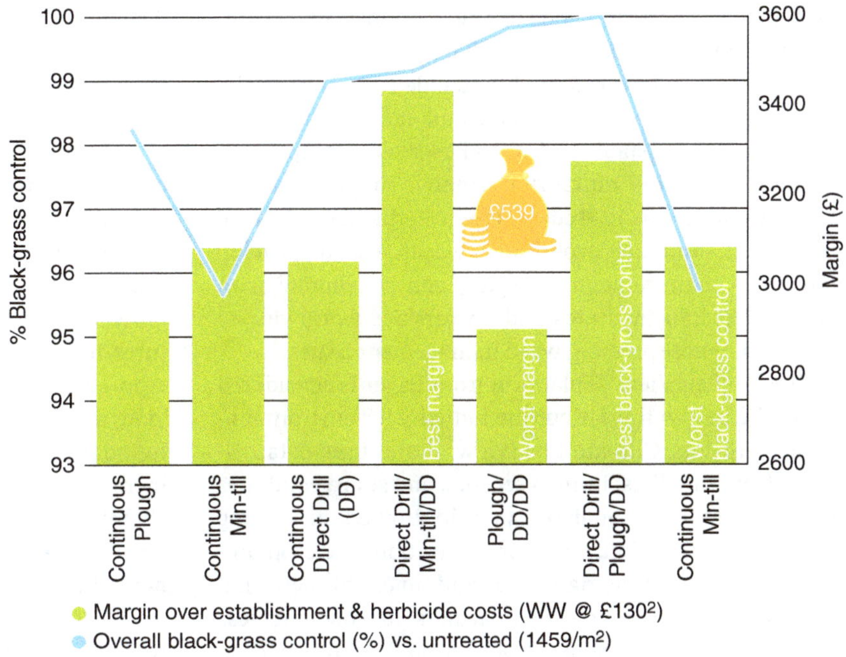

Figure 6.3 Blackgrass control (%) and margin (£) under different cultivation matrix scenarios in herbicide treated winter wheat over three years (https://www.syngenta.co.uk/black-grass).

and analysis of the results. Protector Scouting can be used on tablets and/or cell phones enabling simplified monitoring of the main agronomic data like pests, diseases, weeds, and parameters of quality and evolution of the crop, speeding up the visualization and analysis of the results so that the grower can more closely monitor the real situation of the crop and take action when needed.

Computer-based simulation models have also proved a useful tool to decipher how the diverse biological, genetic, and agroecological parameters interact toward evolution and population dynamics of herbicide resistance (Neve et al. 2011). In Syngenta, we have built generalized individual-based models (IBM) to account for both single and multigene quantitative resistance effects in annual weeds (Liu et al. 2017), as well as to represent the life cycle of perennial weed species characterized by both sexual and asexual reproductive systems (Liu et al. 2019). These models incorporate the whole life history of the weed and analyze how different traits and management scenarios impact the evolution of resistance in the species, generating the density of surviving weeds and the corresponding risk of herbicide resistance evolution (Kaundun 2021).

Essentially, for a successful pest, disease, and weed control a grower needs to follow the best management practices for a sustainable spray application of crop protection products and stewardship. Syngenta has recently launched the digital tool "Spray Assist" a new app created to help sprayer operators to select the most appropriate application techniques (https://www.syngenta.co.uk/spray-assist). The simple to use app links to live local weather data to analyze the factors influencing accurate spraying application and potential risk of spray drift. The app suggests techniques to enable sprayer operators to mitigate risks or alter practices. Users can enter their cropping plans, locations, product use, and application timings. Further personalizing the data with the farm's application equipment enables advanced algorithms to provide guidance on the most appropriate available sprayer setup in any selected situation, for the forecasted weather conditions. With planned spraying operations for specific crops and treatment targets, the app will also suggest the most appropriate nozzle selections and sprayer operation (e.g. for speed and water volume), to deliver the best possible results. For example, in our trials, we have indeed seen that the optimum speed for preemergence herbicides application was 10 km/h, whereas an average 8% increase in grass weed control was achieved when herbicides were sprayed at 200 L/ha than the commonly used 100 L/ha. Moreover, lowering the sprayer boom from 100 to 50 cm above the crop increased preemergence herbicide efficacy from 70% to 87%. Black-grass control was also consistently better when using 90% drift reduction nozzles via a more even spray distribution compared to flat fan nozzles (Kaundun 2021). Syngenta continues to investigate other alternative weed control methods such as the integration of cover crops in the rotation, which ultimately also enables the fight against climate change by helping carbon sequestration and water soil retention among other benefits.

6.4.2 Bayer CropScience

Bayer has taken up various innovative approaches geared toward sustainable, holistic, and scalable agriculture. To enhance agricultural productivity and food security, Bayer promotes best management practices such as IWM programs and herbicide stewardship. Bayer is encouraging farmers to use varied toolbox of practices, instead of a single weed control measure. This will help in reducing selection pressure and evolution of herbicide resistance in the long term. To better understand and manage herbicide resistance, Bayer established Weed Resistance Competence Center (WRCC) in Frankfurt, Germany. The main goal of starting WRCC was to study different herbicide resistance mechanisms and their evolution in weeds, test and develop new weed management strategies, and collaborate with leading research institutes and universities worldwide to address the most pressing weed management issues.

Climate FieldView™ platform from Bayer is considered one of the best in the agriculture industry for environmental stewardship. Climate FieldView™ app uses data collected from satellites, in-field sensors, and smart combines to create predictive analytics to help growers in making decisions on irrigation, fertilizer application, and crop protection products at the right place and time. This new digital farming tool has enabled farmers to consume less inputs, conserve precious resources, and reduce carbon emissions. Climate FieldView™ helps farmers to keep track of carbon sequestration in their soil while achieving yield targets to meet demands of the growing population.

More recently, Bayer has introduced a carbon sequestration initiative (called carbon initiative) to reward farmers adopting ecologically friendly and climate smart agronomic practices such as no-till farming and cover crops. The carbon initiative from Bayer is a part of its commitment to secure carbon-zero future for agriculture by achieving 30% reduction in field greenhouse gas emissions by the year 2030. Farmers participating in this carbon initiative will be rewarded for implementing best management practices on their soils.

In addition to above initiatives, Bayer's ForwardFarming network of independent farmers from across the world showcases on-farm sustainable practices, tailored solutions, digital tools, and stewardship measures to achieve high yields and preserve environment. ForwardFarming platform brings together growers, universities, value-chain partners, etc. to promote dialogue on modern methods of sustainable agriculture.

6.4.3 FMC Corporation

FMC, an agricultural sciences company based in Philadelphia, Pennsylvania, is promoting various sustainable crop protection and environmental safety initiatives to achieve global food production goals. In a recent article, Ganie et al. 2021 highlights FMC's key strategies to support its sustainable weed control initiatives through discovery of new MOA, stewardship, and collaborations with like-minded institutes. As a member of the global and regional Herbicide Resistance Action Committee (HRAC) and Take ACTION herbicide resistance management initiative, FMC is creating awareness about herbicide resistance (HR) and the mitigation of HR using sustainable weed control programs. FMC advocates the use of IWM practices and offers a wide variety of herbicide premixes containing distinct modes of action (MOA) to manage existing herbicide resistant weeds and to delay the evolution of HR. An example of FMC's innovative weed management technology is the PrecisionPac® – an application innovation that offers flexibility to the farmers by creating customized herbicide mixtures/blends with multiple herbicide MOAs specific to the crop and weed spectrum in the field. PrecisionPac® provides an effective strategy to slow down the process of evolutionary HR by reducing selection pressure and helping increase the durability of existing herbicides. FMC is chemistry focused and has taken a lead role in the discovery of novel MOA herbicides to combat hard to control herbicide-resistant weeds.

FMC collaborates with numerous universities and research institutions to research a plethora of IWM practices, such as crop row spacing, use of overlapping residual herbicides, and multiple MOA herbicides to develop viable management programs. FMC provides comprehensive information on weed management practices and in-field agronomic support to growers and crop consultants through its exclusive precision agriculture platform and innovative smart applications. As part of this commitment, FMC Evalio AgriXpert® is one of the digital services that simplifies the process of weed identification and has the capability to select available FMC products required for the management of those weed species. FMC recently launched a new precision agriculture platform known as Arc™ farm intelligence that uses the power of real-time data and predictive modeling to assist with the selection and precise application of products for pest management. This advanced software could reduce unnecessary pesticide applications, optimizing crop yield and grower profitability. FMC's revolutionary 3RIVE 3D® application system delivers a consistent seedling defense directly into the furrow. This application system efficiently covers ground using 90% less water than a traditional liquid system which helps growers save water, fuel, labor, and time.

HR is a major challenge to sustainable crop production throughout the world. It requires strategic partnership between industry, academia, growers, and policy makers to develop innovative weed management solutions for a continued growth in crop productivity to meet the food demands of increasing population. FMC supports its sustainable weed control initiatives through the discovery of new modes of action (MOA), stewardship, and collaborations with private and public research institutes (Ganie et al. 2021).

FMC recommends use of herbicide mixtures (premixes and tank mixes) to manage and delay herbicide resistance in weeds. Development of customizable blend of herbicides such as PrecisionPac®, which are field and weed specific, are a step in the right direction. Using tailormade and customizable herbicide mixtures containing multiple sites of action could be an effective strategy to slow down the process of herbicide resistance evolution by reducing selection pressure and to increase the durability of existing herbicides until viable alternatives are developed, such as green chemistries or nonchemical methods are optimized for weed control. FMC also advocates the use of IWM practices and provides in-field agronomic support to growers and crop consultants through its various digital tools and smart applications such as FMC Evalio AgriXpert® and FMC Evalio AgroSystems®.

References

Akemo, M.C., Regnier, E.E., and Bennett, M.A. (2000). Weed suppression in spring-sown rye (*Secale cereale*) pea (*Pisum sativum*) cover crop mixes. *Weed Technology* 14: 545–549.

Barnes, J.P., Putman, A.R., Burke, B.A., and Aasen, A.J. (1987). Isolation and characterization of allelochemicals in rye herbage. *Phytochemistry* 26: 1385–1390.

Berti, A., Bravin, F., and Zanin, G. (2003). Application of decision-support software for postemergence weed control. *Weed Science* 51 (4): 618–627. https://doi.org/10.1614/0043-1745(2003)051[0618:AODSFP]2.0.CO;2.

Chahal, P.S. and Jhala, A.J. (2019). Integrated management of glyphosate-resistant horseweed (*Erigeron canadensis*) with tillage and herbicides in soybean. *Weed Technology* 33: 859–866.

Chase, W.R., Nair, M.G., and Putman, A.R. (1991). 2,2′-oxo-1,1′-azobenzene–selective toxicity of rye (*Secale cerale* L.) allelochemicals to weed and crop species: II. *Journal of Chemical Ecology* 19: 9–19.

Chikowo, R., Faloya, V., Petit, S., and Munier-Jolain, N.M. (2009). Integrated Weed Management systems allow reduced reliance on herbicides and long-term weed control. *Agriculture, Ecosystems & Environment* 132: 237–242.

Cook, S. and Roche, P. (2018). Enhancing the competitive ability of hybrid barley for the control of blackgrass (Alopecurus myosuroides) in the UK. *Aspects of Applied Biology* 141: 47–55.

Dukes, J.S. and Mooney, H.A. (1999). Does global change increase the success of biological invaders? *Trends in Ecology and Evolution* 14: 135–139.

EPA (US Environmental Protection Agency). (2022). Conventional reduced risk pesticide program. https://www.epa.gov/pesticide-registration/conventional-reduced-risk-pesticide-program (accessed January 2022).

European Parliament. (2009). *Directive 2009/128/EC* of the European parliament and of the council. http://eur-lex.europa.eu/LexUriServ/LexUriServ.do?uri=OJ:L:2009:309:0071:0086:EN:PDF (accessed January 2022).

Gaines, T.A., Zhang, W.L., Wang, D. et al. (2010). Gene amplification confers glyphosate resistance in *Amaranthus palmeri. Proceedings of the National Academy of Sciences of the United States of America* 107: 1029–1034.

Ganie, Z.A., Obrigawitch, T., Kang, I.H. et al. (2021). An outlook of FMC's current and future herbicide-resistance management strategies. *Pest Management Science* 77: 1559–1563.

Harker, K.N. and O'Donovan, J.T. (2013). Recent weed control, weed management and Integrated Weed Management. *Weed Technology* 27: 1–11.

Heap, I. (2020). *The international survey of herbicide resistant weeds*. http://www.weedscience.org (accessed 25 September 2020).

Jugulam, M., Varanasi, A.K., Varanasi, V.K., and Prasad, P.V.V. (2019). Climate change influence on herbicide efficacy and weed management. In: *Food Security and Climate Change*, (ed. S.S. Yadav R.J. Redden, J.L. Hatfield, et al.). 433–442. Hoboken, NJ: John Wiley & Sons Ltd.

Kaundun, S.S. (2021). Syngenta's contribution to herbicide resistance research and management. *Pest Management Science* 77: 1564–1571.

Korres, N.E. (2005). *Encyclopaedic Dictionary of Weed Science: Theory and Digest*, 177–178. Andover, UK: Intercept, Paris: Lavoisier Publishing.

Korres, N.E. (2018). Agronomic weed control: a trustworthy approach for sustainable weed management. In: *Non-chemical Weed Control* (ed. K. Jabran and B.S. Chauhan), 97–114. London, San Diego: Academic Press, Elsevier.

Korres, N.E., Burgos, N.R., Vurro, M. et al. (2019). New directions for integrated weed management: modern technologies, tools and knowledge discovery. *Advances in Agronomy* 155: 243–319.

Korres, N.E. and Froud-Williams, R.J. (2002). Effects of winter wheat cultivars and seed rate on the biological characteristics of naturally occurring weed flora. *Weed Research* 42: 417–428.

Korres, N.E. and Norsworthy, J.K. (2015). Influence of a rye cover crop on the critical period for weed control in cotton. *Weed Science* 631: 346–352.

Korres, N.E., Norsworthy, J.K., Tehranchian, P. et al. (2016). Cultivars to face climate change effects on crops and weeds: a review. *Agronomy for Sustainable Development* 36 (12).

Liebman, M., Miller, Z.J., Williams, C.L. et al. (2014). Fates of Setaria faberi and *Abutilon theophrasti* seeds in three crop rotation systems. *Weed Research* 54: 293–306.

Liu, C., Bridges, M.E., Kaundun, S.S. et al. (2017). A generalised individual-based algorithm for modelling the evolution of quantitative herbicide resistance in arable weed populations. *Pest Management Science* 73: 462–474.

Liu, C., Scursoni, J.A., Moreno, R. et al. (2019). An individual-based model of seed- and rhizome-propagated perennial plant species and sustainable management of *Sorghum halepense* in soybean production systems in Argentina. *Ecological Evolution* 9: 10017–10028.

Neve, P., Norsworthy, J., Smith, K., and Zelaya, I.A. (2011). Modelling evolution and management of glyphosate resistance in Amaranthus palmeri. *Weed Research* 51: 99–112.

Norsworthy, J.K., Korres, N.E., Walsh, M.J., and Powles, S.B. (2016). Integrating herbicide programs with harvest weed seed control and other fall management practices for the control of glyphosate-resistant Palmer amaranth. *Weed Science* 64: 540–550.

Norsworthy, J.K., Ward, S.M., Shaw, D.R. et al. (2012). Reducing the risks of herbicide resistance: best management practices and recommendations. *Weed Science* 60: 31–62.

Pachauri, R.K., Meyer, L.A. (eds.) (2014). Climate Change 2014: Synthesis Report. *Contribution of Working Groups I, II and III to the Fifth Assessment Report of the Intergovernmental Panel on Climate Change*. Geneva, Switzerland: IPCC. 151.

Pannell, D., Tillie, P., Rodríguez-Cerezo, E. et al. (2016). Herbicide resistance: economic and environmental challenges. *AgBio-Forum* 19: 136–155.

Peterson, M.A., Collavo, A., Ovejero, R. et al. (2018). The challenge of herbicide resistance around the world:a current summary. *Pest Management Science* 74: 2246–2259.

Powles, S.B. and Yu, Q. (2010). Evolution in action: plants resistant to herbicides. *Annual Review of Plant Biology* 61: 317–347.

Price, A.J. and Norsworthy, J.K. (2013). Cover crops for weed management in southern reduced tillage vegetable cropping systems. *Weed Technology* 27: 212–217.

Saini, M., Price, A.J., and Santen, E.V. (2006). Cover crop residue effects on early-season weed establishment in a conservation-tillage corn-cotton rotation. *Proceedings of the 28th Southern Conservation Systems Conference*. Bushland, TX: USDA-ARS Conservation and Production Research Laboratory Report No. 06-1, 175–178.

Schwartz-Lazaro, L.M., Korres, N.E., Bararpour, T. et al. (2022). Population dynamics of naturally occurring weed flora in response to crop rotation and HPPD-inhibiting herbicide-based treatments. *Weed Technology* 36: 426–435.

Singer, A., Travis, J.M.J. Johst, K. (2013). Interspecific interactions affect species and community responses to climate shifts. *Oikos* 122: 358–366.

Tranel, P.J. and Wright, T.R. (2002). Resistance of weeds to ALS-inhibiting herbicides: what have we learned? *Weed Science* 50: 700–712.

van Wychen, L. (2020). Scientists highlight improved green chemistries for weed management. *Weed Science Society of America*. https://wssa.net/2020/07/scientists-highlight-improved-green-chemistries-for-weed-management (accessed 10 July 2023).

Vasileiadis, V.P., Kudsk, P., and Sattin, M. (2017). Weed management of maize grown under temperate conditions: the case of Europe and the United States. In: *Achieving Sustainable Cultivation of Maize: Cultivation Techniques, Pest and Disease Control*, vol. 2, 1e (ed. D. Watson), 415–437. London: Taylor & Francis.

Vasileiadis, V.P., Otto, S., van Dijk, W. et al. (2015). On-farm evaluation of integrated weed management tools for maize production in three different agro-environments in Europe: agronomic efficacy, herbicide use reduction and economic sustainability. *European Journal of Agronomy* 63: 71–78.

Vasileiadis, V.P., van Dijk, W., Verschwele, A. et al. (2016). Farm-scale evaluation of herbicide band application integrated with inter-row mechanical weeding for maize production in four European regions. *Weed Research* 56: 313–322.

Whalen, D.M., Shergill, L.S., Kinne, L.P. et al. (2020). Integration of residual herbicides with cover crop termination in soybean. *Weed Technology* 34: 11–18.

Wiggins, M.S., Hayes, R.M., Nichols, R.L., and Steckel, L.E. (2017). Cover crop and postemergence herbicide integration for Palmer amaranth control in cotton. *Weed Technology* 31: 348–355.

Wiggins, M.S., Hayes, R.M., and Steckel, L.E. (2016). Evaluating cover crops and herbicides for glyphosate-resistant Palmer amaranth (*Amaranthus palmeri*) control in cotton. *Weed Technology* 30: 415–422.

Wiggins, M.S., McClure, M.A., Hayes, R.M., and Steckel, L.E. (2015). Integrating cover crops and POST herbicides for glyphosate-resistant Palmer amaranth (*Amaranthus palmeri*) control in corn. *Weed Technology* 29: 412–418.

Yuan, J.S., Tranel, P.J., and Stewart, C.N., Jr. (2007). Non-target-site herbicide resistance: a family business. *Trends in Plant Science* 12: 6–13.

Ziska, L.H., and Goins, E.W. (2006). Elevated atmospheric carbon dioxide and weed populations in glyphosate treated soybean. *Crop Science* 46: 1354–1359.

7

Ecologically Based Weed Management to Support Pollination and Biological Pest Control

Vaya Kati[1,2,*] *and Filitsa Karamaouna*[2,*]

[1] Laboratory of Agronomy, School of Agriculture, Faculty of Agriculture, Forestry and Natural Environment, Aristotle University of Thessaloniki, Thessaloniki, Greece
[2] Scientific Directorate of Pesticides Control and Phytopharmacy, Benaki Phytopathological Institute, Kifissia, Greece
*Corresponding authors

7.0 Introduction

A weed is "*a plant whose virtues have not yet been discovered,*" as Ralph Waldo Emerson famously said. Indeed, weeds have traditionally been associated with adverse impacts on economic, environmental or aesthetic aspects of any system and even on human health, and on the approaches used to limit this (Blaix et al. 2018). When it comes to crop production, the yield loss sustained by weeds surpasses that caused by other biotic stressors like pests and pathogens (Oerke 2006). Also, unlike pests and pathogens whose infestations fluctuate annually depending on environmental and other conditions, weeds are expected to occur every year in a field. Therefore, weed management has always been the main crop protection investment a farmer has to make, throughout the cropping season. With the intensification of agriculture, weed control measures evolved towards the fast and indiscriminate eradication of weed populations, and invariably relied on the use of herbicides. This reductionist approach was adopted in all aspects of crop protection, driven by economic reasons, the lack of an easy alternative, and the false assumption that there is an ample homeostasis within agricultural ecosystems, allowing them to sustain a considerable amount of abuse with no serious side-effects (Altieri et al. 1983). Thus, intensive farming measures that helped to increase productivity in the short term, have been scrutinised in recent years as potentially responsible for a series of side-effects, which threaten the sustainability of farming and undermine the very same goals they were employed to safeguard. Indeed, yield-gap analysis (theoretical *vs*. actual) indicated that the yields obtained from high-input intensive farming systems may be heading toward stagnation (van Ittersum et al. 2013), raising questions regarding their sustainability.

Intensive farming practices have led to simplified crop rotations, decreased spatial and temporal diversity in weed management, and increased agricultural landscape homogeneity, with palpable consequences for the weed communities and weed evolution (Neve et al. 2009; Storkey et al. 2012) and eventually for the diversity of associated pollinating insects and natural enemies of pests (Kremen et al. 2002; Garibaldi et al. 2016). Weeds are known for their ability to adapt (Vigueira et al. 2013). However, the adaptation to pressures exerted by intensive farming practices is variable among weed species. Depending on the farming practices applied, the consequences are shifts in species abundance and diversity (Fried et al. 2019), evolution of resistance to herbicides (Délye et al. 2013; Kati et al. 2019; Heap 2021), or rarefication and even extinction of weed species with lower adaptation skills (Gibson et al. 2006; Kolářová et al. 2013). For example, a 50-year survey in central and northern Germany from the 1950s/1960s to 2009, which analyzed changes in the arable flora of nearly 400 fields from different study areas, revealed a significant loss of diversity up to 65% on average (Meyer et al. 2013). Examples of arable weeds that have been reported as rare or in decline include *Anchusa arvensis* (L.) Bieb., *Myosotis arvensis* (L.) Hill. (Boraginaceae); *Legousia hybrida* (L.) Delarbre, *L. speculum-veneris* L. (Campanulaceae); *Agrostemma githago* L., *Silene conica* L., *Silene noctiflora* L. (Caryophyllaceae); *Lamium amplexicaule* L. (Lamiaceae); *Ornithogalum umbellatum* L. (Hyacinthaceae); *Papaver argemone* L. (Papaveraceae); *Consolida regalis* L., *Nigella arvensis* L., *Ranunculus arvensis* L. (Ranunculaceae); *Viola arvensis* Murray, *V. tricolor* L. (Violaceae) (Rollin et al. 2016).

The decrease of weed species at the intra-field scale (a diversity) and the shift of weed communities toward species with a greater intrinsic capacity to adapt to weed control strategies have been correlated with the wider loss of cropping systems' resilience (Storkey and Neve 2018; Storkey et al. 2019). Indeed, the reduction in weed species diversity is considered among the key drivers behind the

Ecologically Based Weed Management: Concepts, Challenges, and Limitations, First Edition. Edited by Nicholas E. Korres, Ilias S. Travlos, and Thomas K. Gitsopoulos.

ongoing rapid decline of pollinators (Potts et al. 2010). Biella et al. (2019) elaborated on the reasons behind this relationship, arguing that the loss of a small subset of functionally important plant species induces an immediate reduction of resources available to the pollinator guild and thus has a disproportionate impact on the vital ecosystem service of pollination. The decline in pollinators is threatening food security, with serious economic implications.

Worldwide, 75% of plant species cultivated for food production rely primarily on insects for pollination (Klein et al. 2007), while pollination services to crops have an estimated annual value of €153bn (Gallai et al. 2009). Habitat loss is also destabilizing food webs of natural enemies and herbivorous insects, thus having a negative impact on the biological control of pests, another valuable ecosystem service. In a recent meta-analysis, Naranjo et al. (2019) estimated the value of natural pest control services to crop production at $74 per hectare on average, while earlier studies reported an estimated global value of around €40 bn to €74 bn (Naylor and Ehrlich 1997). Even if not entirely lost, the degraded habitats become inefficient in supporting the needs of insect communities, rendering them susceptible to more harmful stressors, such as diseases and pesticides (Goulson et al. 2015; Maia et al. 2019). On the other hand, experimental evidence has suggested that a high diversity of weeds (i.e. noncrop plants) is associated with lower population densities of crop pests and higher densities of natural enemies (Altieri and Letourneau 1982; Schellhorn and Sork 1997). Reversing the side effects of agricultural intensification on habitat loss or degradation, as expressed by the decreasing arable weed flora diversity and the associated decline in pollinators and natural enemies of pests, while also maintaining crop yields, is a multitask challenge. In the following sections, the interactions between weeds, pollinators, and natural enemies are reviewed, and management practices to exploit these interactions for the benefit of functional biodiversity are discussed, along with practical considerations for their implementation.

7.1 Weed-Insect Interactions

Weeds are generalist plants, hosting a wide range of insects and playing an important and probably overlooked role in the conservation of species-rich ecosystems and in the support of ecosystem services (Rollin et al. 2016; Biella et al. 2019). Predicting the functional role of plants including weeds in the agro-ecosystems, should be based on the set of traits expressed by the plants growing there, as affected by genetics, abiotic factors, and biological interactions – so far, however, it is not clear whether any particular traits consistently favour higher biodiversity (MacLaren et al. 2020). The identified knowledge gap on the functional interaction between plants including weeds, and arthropods (Balzan et al. 2016) led several studies to look into the traits and population dynamics affecting the functional diversity of plants within cultivated systems (Storkey et al. 2015; Cresswell et al. 2019). Global trait databases, such as the TRY database of plant traits developed by Kattge et al. (2011) and updated by Kattge et al. (2020), along with a better understanding of the relationships between traits and multiple ecosystem services (de Bello et al. 2010) are important tools toward this initial screen of candidate species for a range of systems (Storkey et al. 2015). In the agroecosystem, this knowledge is required to assess the ecological value of existing weed vegetation and to determine the types of weeds (or selected plants sown for this purpose) that promote insect biodiversity (MacLaren et al. 2020). One of the more explored and obvious functional relationships between weeds and insects is the one between floral resources and pollinators (Gaba et al. 2017). Hawes et al. (2009) investigated the invertebrate community composition within the arable food web and identified two main interaction groups: one between omnivores, generalist predators and detritivores, which are positively associated with monocots; and one between omnivores, parasitoids, sap feeders and leaf chewers, which have a stronger association with dicots. More work is needed, however, to determine which particular traits consistently favor higher biodiversity (MacLaren et al. 2020).

7.1.1 Weeds and Pollinating Insects

The population dynamics of weeds and pollinators in agricultural landscapes are regulated by their bilateral interactions and governed through the mutually beneficial function of pollination: weeds need pollination to produce seeds, and pollinators need the pollen and nectar offered by weed vegetation (Figure 7.1).

Depending on the diversity of flowering species in the agroecosystem, weeds can provide a continuous supply of pollen and nectar to pollinators, which is vital when the main crop is not in bloom (Hannon and Sisk 2009). A consistent floral supply is especially critical early and late in the year (Morrison 2018) when flowering of most crops is not available. During these periods, floral supplies by weeds are the main food source for pollinating insects preparing to hibernate or emerging from it in early spring. Moreover, weeds help to create the diverse microhabitats needed for egg-laying and larval-development of pollinating insects (Nicholls and Altieri 2013). On the other hand, pollination contributes to weed propagation and abundance in agricultural landscapes. Most entomophilous weeds are visited for pollen and/or nectar by a large

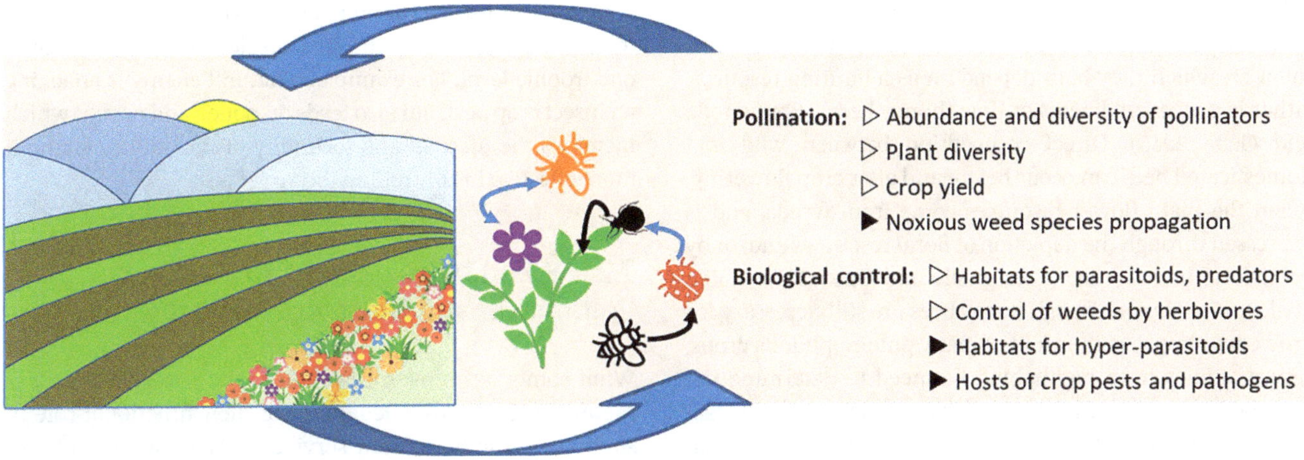

Figure 7.1 Graphic representation of the field margin vegetation (weeds, sown plant mixtures) functional role on the ecosystem services of pollination and biological control, as habitats for pollinating insects, natural enemies of pests, and associated services (▷) and dis-services (▶).

number of generalist insects from the orders of Hymenoptera, Diptera, Lepidoptera and Coleoptera, forming a largely mutualistic relationship, normally based on "attract and reward" mechanisms (Petanidou and Lamborn 2005; Rollin et al. 2016). Of these visitors, weed pollination is served mainly by Hymenoptera and especially wild bees, followed by honey bees *Apis mellifera* L. (Biesmeijer et al. 2006; Potts et al. 2010). Bees are considered the most important pollinators because they actively seek pollen to feed their larvae (Lee-Mader et al. 2019). Non-bee insects are generally considered less efficient pollinators, but what they lack in efficiency they compensate by the larger number of visits they make, thus resulting in valuable pollination services, often comparable to those provided by bees (Rader et al. 2016). Wild bees include solitary bees (e.g. in the Mediterranean arable lands: Andrenidae, Apidae, Colletidae, Halictidae, Megachilidae, Melitidae) and eusocial bumble bees, *Bombus* spp. (in Rollin et al. (2016) and references therein). Gibson et al. (2006) reported that wild bees are also the main pollinating insects of rare weed species, and are thus helping considerably to conserve floral biodiversity in the agricultural landscape (Biesmeijer et al. 2006).

Key plant traits associated with attracting pollinators include floral display size, flower color, pollen content, and nectar content and quality (Cresswell et al. 2018, 2019). A larger floral display is more attractive to pollinators who probably associate it with a larger reward, though this is not always the case (Makino and Sakai 2007). An extensive literature review by Rollin et al. (2016) is presenting the association of weed species and pollinating insects, in relation to the reward involved, i.e. pollen and/or nectar. Several of these weed species reward insects only with pollen. For e.g. common poppy *Papaver rhoeas* L. (Papaveraceae), a completely self-incompatible species that relies only on insect pollination for its fertilization, is a prolific pollen producer that can account for the majority of pollen resources in extensive cereal crops during late spring (Odoux et al. 2012). Pollen availability depends of course on the weed population abundance after weeding (Cirujeda et al. 2003). In other species, nectar is also offered as a reward, for e.g., in *Consolida regalis* Gray, *Stellaria media* (L.) Vill., *Xanthium strumarium* L. etc., where it is the only reward, or in *Centaurea cyanus* L., *Borago officinalis* L., *Echium vulgare* L., etc., where it is offered in addition to pollen. Nectar composition and secretion rate are important parameters that affect insect visits on weed flowers (Corbet 1978) and are considered among the key floral traits that can shape pollinator communities (Potts et al. 2003). In particular, nectar quality and quantity is important for social bees because it affects the energetics of transporting the load to the hives and concentrating or diluting it (Corbet 1978).

The effectiveness or efficiency of insects to transfer pollen depends on qualitative and quantitative parameters, such as the amount of compatible pollen transferred on each visit, and the frequency of visits on the weed flowers (Gibson et al. 2006). Solitary bees carry a larger amount of dry pollen that is available for pollination compared to honey bees and bumble bees, which moisten the collected pollen grains with nectar and compress them into dense pellets in their hind leg baskets (*tibial corbicula*) (Free 1993; Martins et al. 2015). The higher performance of solitary bees over bumble bees and honey bees renders them indispensable in the delivery of the pollination ecosystem service and highlights the need to conserve their populations with targeted and cost-effective measures (Eeraerts et al. 2019b).

Wild bees and honey bees interact directly and indirectly with each other through the weed and crop plant compartment on which they both depend on as a limiting resource, either in space (wild bees) or time (honey bees) (Bretagnolle and Gaba 2015). Direct competition between wild and domesticated bees can occur before and after crop flowering, when the main flower resources come from weeds, and is expressed through the depletion of floral resources and/or by behavioral interference (Bretagnolle and Gaba 2015; Raine et al. 2017). Thus, although honey bees are still necessary for growers to improve the yield of many entomophilous crops, recent studies have highlighted the need to determine the optimal honey bee density for specific crops to avoid displacing wild pollinators and therefore reducing their indispensable role in the total pollination services (Eeraerts et al. 2019a). Furthermore, the negative consequences of competition between pollinators can be counteracted with the provision of abundant flowering resources, especially by species that flower sequentially. These can support an increased number of pollinator visits, thus facilitating the pollination of more abundant or rarer plant species throughout the year in a form of *effective mutualism* (Gibson et al. 2006).

7.1.2 Weeds and Natural Enemies

The trophic interrelations between weeds and natural enemies (i.e. predators and parasitoids of insect pests) are regulated by spatiotemporal parameters, depending on whether the weeds serving as alternative hosts for crop-damaging arthropods, occur within the field or are external to the crop. Norris and Kogan (2000) reviewed these parameters and discussed the effect of each case on pest population dynamics and impact on adjacent crops, through herbivory but also as vectors of viruses and other pathogens. Weed-natural enemy interaction is governed by the mutually beneficial function of pollination, as discussed above for pollinators (weeds offer pollen and nectar to insects, and may get pollinated in return). Nevertheless, this interaction is more complex through the provision of niches to parasitoids, predators, hyperparasitoids, and intraguild predators (weeds are hosts for phytophagous insects, which can attack the crop or even the weeds, and are in turn hosts for parasitoids or prey for predators, which can also be parasitized or preyed upon by other species) (Figure 7.1). Most of the predatory and parasitoid insects either need or benefit from alternative food sources, like nectar or pollen, during at least one stage of their life, in addition to their host or prey. Consequently, like pollinators, many of the natural pest enemies also benefit from floral resources.

Depending on the trophic levels involved, tritrophic or polytrophic relationships can be formed. At the level of tritrophic interaction, weeds serve as a food source for herbivorous insects (including crop pests), which in turn are attacked by carnivorous insects (natural enemies). At the polytrophic interaction, the same arthropod species feeds on more than one trophic level. For example: a natural enemy is attacking an insect crop pest, but also feeds on pollen and nectar, which increases the lifespan and fecundity of the adult, making it more effective in controlling the target pest.

7.2 Weed Management to Support Pollination and Biological Control

What comes as an overdue credit to the virtues of weeds is the increasing acknowledgment of their functional role in supporting the ecosystem services of pollination and biological control in agricultural land. Rollin et al. (2016) stated that stability of weed-insect pollinator networks is essential to maintaining biodiversity in the arable landscapes. Increasing plant diversity within agricultural landscapes increases the resilience of associated organisms under the pressures posed by climate change, helping to mitigate the negative effects of rising temperatures on wild bee species richness and total abundance (Papanikolaou et al. 2017). Also, a more diverse weed community is likely to be less competitive for the crop and less prone to be dominated by a few widespread, highly adapted, or herbicide-resistant weed species, and hence it represents a good indicator of sustainable cropping systems (Storkey and Neve 2018).

Crop plants, weeds, and insects interact in a dynamic and changeable way over time. Detailed knowledge on these interactions is required throughout the life cycle of plants and arthropods to ensure safe and efficient management practices for weeds and pests alike. Understanding both the trophic interactions between plants (cultivated or weeds) and insects (pollinators, natural enemies, and pests) and knowledge on these species' abundance and diversity are required to assess the associated services and disservices. A schematic outline of these relationships is presented in Figure 7.1. Crop pests can be specialists, preadapted to feed only on some plant species often from the same family, or polyphagous but usually with a distinct preference hierarchy, feeding more widely only when preferred hosts are unavailable (Capinera 2005). Therefore, weed species taxonomically related to the crop may host common insect pests. This means that weeds should be carefully managed to minimize the risk of the insect pests migrating in search of new feeding grounds and attacking the crop in the event of untimely weed removal (Norris and Kogan 2000). Also, timing of weed removal will also affect the associated natural enemies. In other words, weed and pest management measures, which may be appropriate at one stage or cropping system, could be detrimental on a different occasion (Norris and Kogan 2005). Obviously, maintaining weed communities to support desirable ecosystem services may conflict with the goal of maximizing crop yield. To navigate through this conflict, it is necessary to set economic thresholds and understand the factors affecting crop-weed balance within a crop season, (e.g., weed biology, ecology and density, crop and weed competitive capacity, timing of weed emergence and management), so that crop yields are not economically reduced (Buchanan 1977; Zimdahl 1980; Altieri and Letourneau 1982). When shifting this balance, the question is which and how many of the weeds to leave for an economically acceptable but also ecologically significant output (Storkey 2006). More robust tools are required to support such decisions (Storkey 2006) and assess whether more weeds of selected species could be tolerated without increasing the competition with crops (MacLaren et al. 2020). Weed species that are relatively short, shade-tolerant and produce a relatively small number of seeds per unit biomass are likely to benefit biodiversity, while not threatening crop production (Storkey 2006). Other desirable weed traits include capturing resources at a slow rate, that allows them to be resilient and survive competition from the crop plants, but not to the expense of the crop (MacLaren et al. 2020).

In a study of arable weeds in the UK, Marshall et al. (2003) list several species of high value for invertebrates and low competitive index (e.g. *Polygonum aviculare* L., *Senecio vulgaris* L., *Sonchus oleraceus* L. and *Poa annua* L.) and separates them from those that are competitive to the crop and associated with insect pests (e.g. *Galium aparine* L., *Sinapis arvensis* L.), or noncompetitive to the crop but of low biodiversity value for arthropods (e.g. *Viola arvensis* L.). Of course, the temporal and spatial variability of weed phenotypes means that such ranking of weed species should be defined on a pedoclimatic and crop-scenario level. Weed management practices that encourage the establishment of less competitive weed communities could help resolve the conflict between removing weeds to prevent competition with crops, and the need to retain them for biodiversity and ecosystem function (Storkey 2006; MacLaren et al. 2020; Smith et al. 2020). In order to benefit the overall functional diversity of the farm ecosystem, weed populations adapted in both the cultivated body of the field and the field margins should be sustained with careful management, since these sites tend to have different plant species (Storkey 2006). By simply relaxing weed control either rotationally or in conservation headlands, the most fecund and competitive weed species are likely to benefit, indicating the need of a new approach to weed management, with the explicit aim of maintaining specific weed assemblages (Marshall et al. 2003). Such management practices could include the establishment of selected plant species along field margins or as cover crops. The following sections discuss weed management in field margins and through cover cropping, as practices to create habitats and offer food sources for beneficial insects (i.e. pollinators and natural enemies of pests) by establishing desirable species from the existing weed flora and/or by sowing selected plant mixtures.

7.2.1 Field Margins

Field margins in arable and perennial cropping systems are described as typified forms of boundary structures, usually associated with herbaceous vegetation adjacent to the crop (Greaves and Marshall 1987; Marshall 2002). These structures create semi-natural habitats, which are key ecological features in agroecosystems (Storkey et al. 2019; Blaix and Moonen 2020). Marshall and Moonen (2002) described three principal components in a field margin: the preexisting boundary of the field (like a hedge, fence, or wall), which is not always present; the field margin strip established between the boundary and the crop; and the crop edge (i.e. the outer few meters of the crop). A particular type of field margins are the *conservation headlands*, which are low-input cereal crop headlands, managed to allow populations of rare, broad-leaved weeds and the associated insect fauna to develop (Holland et al. 2016). These components maintain a high diversity of vegetation typologies offering a wide range of ecological benefits. Mkenda et al. (2019) grouped the values of these benefits under the three major ecological roles of the field margins reported by Smith et al. (2008): namely, enhancing biodiversity (biodiversity value), providing habitat refuges for rare and endangered species (conservation value), and promoting ecosystem services like natural pest regulation, pollination, and nutrient cycling (functional value).

Since the pioneering works of Dambach (1948) and Marcovitch (1935), field margins have been studied extensively for their role to support pollination and biological control of pests (Altieri and Letourneau 1982; Balzan et al. 2016; Blaix and Moonen 2020; Kati et al. 2021). Management of field margins to support these ecosystem services should be based on a sound understanding of their ecology (Marshall 2002). The target is to offset the side effects of agricultural intensification and compensate for the vegetational simplification associated with the expansion of monocultures, and specifically the elimination of natural plant communities, low within-field biodiversity (Altieri and Letourneau 1982; Holland et al. 2014; Storkey et al. 2019; Egan et al. 2020), and dominance of a few competitive, highly adapted, and widely distributed weed species (Storkey and Neve 2018). Today more than ever, field margins are required to deliver multiple ecosystem services concurrently, to meet the increasing pressure on the available agricultural land (Cresswell et al. 2019). This should be considered from the early stages of management, since margins designed to deliver only a specific ecosystem service are likely to fail in supporting other ecological roles and will require adjustments in the plant species composition to achieve this; for example, plant strips established for the conservation of a bird species provided limited benefits in terms of enhancing the biological control of pests in adjacent cotton crops (Olson and Wäckers 2006).

Field margin size is crucial for the attracted arthropod fauna. However, in many cases, these buffer zones have been reduced to a 1m-wide grass strip (Ricou et al. 2014). A study in Finland that revised the species-area relationship ($S = cA^z$, where S is the number of species, A is the area and c, z are fitted constants) developed by Arrhenius (1921) and Preston (1962), reported that patch width was the most effective spatial indicator for species richness in the margins (Ma et al. 2002). In a similar study, the width of field margins significantly affected the total density of bumblebees and cuckoo bumblebees, but not the species number or diversity (Bäckman and Tiainen 2002). Ma et al. (2002) reported that more species per unit area increment could be obtained by widening the buffer zones rather than by lengthening them and concluded that designing wider buffer zones may be an easy and cost-effective management for conservation in agricultural landscape.

7.2.2 Weed Management

Maintaining a diverse flora along the field margin that contains a mixture of desirable weeds and/or purposely sown plant species, is needed to create a multifunctional boundary vegetation that caters for the pollen and nectar requirements and meets other habitat needs of pollinators and natural enemies. A common concern by farmers is whether this practice can cause weed infestation into the crop (Mante and Gerowitt 2009). There are several factors that can influence this possibility, such as the type of field margin, plant composition, and distance from the field crop (Blumenthal and Jordan 2001; Mkenda et al. 2019). Weed communities in field margins are normally affected by adjacent land use (Marshall and Arnold 1995). Reports on the risk of these weed species infesting adjacent crops are contradictory. Reberg-Horton et al. (2010) reported that the presence of weeds, including noxious species, such as *Amaranthus retroflexus* L., *Cyperus esculentus* L., *Urochloa platyphylla* (Munro ex C. Wright) R.D. Webster, *Ipomoea* sp., *Sida spinosa* L. and *Senna obtusifolia* (L.) H.S. Irwin & Barneby, was higher near the crop edge, that is, toward the margin of maize or peanut fields in USA, than around the field center. Taking into account the variable functional traits of weed dispersal, they argued that this occurrence is probably due to the different management that takes place at the crop edges compared to the rest of the crop, rather than to a contamination effect from the field margins. Also, a review of the field margin functions across different farms in the UK presented evidence that sown grass and/or wildflower strips along arable fields not only increased plant species richness and floral diversity, but also reduced the abundance of pernicious weeds, such as *Anisantha sterilis* (L.) Nevski, *Galium aparine* L., *Poa trivialis* L. and *Urtica dioica* L. (Moonen and Marshall 2001). In another study in Essex, UK, the margin flora and the arable weed flora at 5 and 50 m into the studied fields indicated no relationship between the plant communities (Marshall and Arnold 1995), and hence implied no weed movement from the margins. Recently, however, Blaix and Moonen (2020) reported that field margins can harbor weed species that could potentially invade the cropped fields, and Egan et al. (2020) urged researchers and advisors to pay more attention on this subject, which might have been underestimated. They emphasized the importance of critical management actions like mowing and termination of weeds before going to seed. Several other studies elaborated on the need for proper vegetation management to prevent weeds in field margins from spreading into the field (Boutin et al. 2001; Bokenstrand et al. 2004; De Cauwer et al. 2008), even if they are attractive to beneficial insects. For example, in a UK study of 120 field margins, Pywell et al. (2005) reported that natural regeneration of field margins resulted in habitats of questionable value because, although they provided good foraging for bumblebee species, most of the key forage plants were noxious weeds, such as *Cirsium* spp. Such competitive weeds require control to ensure the safety of the crop and to facilitate a more diverse flora. Field margin management with sown mixtures of selected plant species could help to smother noxious weeds as reported by Moonen and Marshall (2001) and could also upgrade the floral value of these sites, to make them more attractive for pollinating insects and natural enemies. For example (Figure 7.2), this was observed in the margins of a processing tomato crop (Kati et al. 2021) and a grapevine orchard (unpublished data). Another concern is whether weed flora along field margins can also harbor diseases that can harm the crop (Wisler and Norris 2005). Although it is beyond our scope here, it is worth noting that management practices encouraging weed maintenance or sown plant mixtures in field margins should consider this risk.

7.2.3 Pollinators and Field Margins

Pollinators are more attracted to field margins with flowering plants rich in nectar and pollen (Bäckman and Tiainen 2002; Carvell et al. 2006; Ricou et al. 2014; Barbir et al. 2015; Mkenda et al. 2019). Ricou et al. (2014) developed an indicator that can be used at the field margin to predict the pollination value of floral diversity provided by plants, whether existing weeds or purposely sown species. They

Figure 7.2 Field margins management with purposely sown mixtures of plants, selected to attract pollinating insects and natural enemies of pests, in a processing tomato crop (left), and a grapevine orchard (right), in Greece. (Photos, V. Kati, F. Karamaouna).

based their model on criteria such as visual attraction, flower accessibility, and floral reward. A general rule is that the plant mixtures sown along a field margin should provide a species-rich habitat (De Cauwer et al. 2006) with a high proportion of different dicotyledonous species (Bäckman and Tiainen 2002) and a long-lasting flowering period (Dramstad and Fry 1995), to cover the needs of a diverse range of insect groups. Seed mixtures developed to fulfill these preconditions, and also meet farmers' preferences for low-cost, competitive mixtures that can smother noxious weeds, are likely to be efficient from a nature conservation and agronomic point of view (Mante and Gerowitt 2009).

Several plant mixtures developed to provide resources for pollinators tend to be dominated by Fabaceae (Campbell et al. 2012); however, preferences for floral resources and foraging distance differ among species. For example, honey bees have been firmly associated with mass-flowering crops while wild solitary bees with semi-natural habitats and bumble bees with an intermediate behavior as habitat generalists (Rollin et al. 2013), while butterflies benefit from grassy field margins (Delattre et al. 2010). In addition, different margin characteristics influence the value of pollination service to the crop yield. For instance, linear habitats such as field margins and road verges appear to be less effective in providing quality flower-based ecosystem services when compared with permanent grasslands, because of their low plant species richness (Denisow and Wrzesień 2015; Kütt et al. 2016). According to Denisow and Wrzesień (2015), bee species richness and abundance in the field margin benefit if there are natural habitats, like meadows and forests, at a distance of less than 1000 m, or if the field area is less than 10 ha. Availability of floral resources close to cropped land enhances pollinator abundance, with associated increased pollination service. The type of field margin, whether cropped or uncropped, may also influence the insect population in such habitats because of the differences in plant species composition. For example, uncropped field margins with naturally regenerated rich flowering weed flora harbored more bumblebees and honey bees as compared with cropped margins (Kells et al. 2001). In addition, Knapp et al. (2019) highlighted the important interactive effect of the abundance and species richness of flowering weeds on different groups of pollinators, both within the crop field and the seminatural habitats (woodlands and uncultivated areas) in a 500–2000 m radius from the field boundaries. They concluded that species-rich flora in field margins was an effective way of boosting the abundance of bumblebees and honey bees, which were important crop flower visitors, and reported no competition with the crop flowers. They also observed that increasing floral resources at the field scale favored the abundance and species richness of solitary bees, thereby benefiting pollinator conservation, especially in sites with fewer seminatural habitats.

7.2.4 Natural Enemies and Field Margins

Strips and borders along the field edge with flowering vegetation, either from the weed flora or purposely sown plants, can increase the abundance and diversity of arthropod natural enemies (Mkenda et al. (2019) and references therein). The plant species in the seed mixtures selected to attract natural enemies in the field margins tend to include Apiaceae and other species with exposed floral nectaries (Campbell et al. 2012). The resulting habitats function as a reservoir of beneficial insects that can recolonize the adjacent crop field. For example, they serve as diversity hotspots for hoverflies, ground beetles, etc, or as nectar resources for parasitoids. Even a very small area (a few square meters) of noncrop habitat can have a significant positive effect on the population of ground-dwelling spiders and carabid beetles. Other components of field margins, like trees and shrubs, can be refuge sites for an increased population of predatory insects. Field margins can also act as refuge areas during pesticide application in conventionally managed fields (Mkenda et al. 2019). However, studies on the abundance of weed seed predators, which are associated with ecologically based weed management, reported no site interaction regarding field margins and in-field populations (Chauhan et al. 2010).

The most studied natural enemies in managed field margins include spiders and ground beetles (Carabidae) since these organisms are regarded as biological indicators in biodiversity and conservation assessments as well as indicators of change in terrestrial ecosystems (Perner and Malt 2003), but also ladybirds (Coccinellidae), hover flies (Syrphidae), tachinid flies (Tachinidae), predatory bugs (including Miridae, Reduviidae), parasitoid species of various families (Chalcidoidea, Ichneumonoidea, Chrysidoidea and Proctotrupoidea), Neuroptera, and ants (Formicidae) (Anderson et al. 2013; Balzan et al. 2014; Bowie et al. 2014).

Natural enemy populations in field margins are related to timely availability of three key resources: prey, floral resources, and overwintering sites in case of disturbances in the field crop (Ramsden et al. 2015). Some invertebrates move from the field margin to the field crop during the growing season when there are abundant food resources and later back to the margin when the resources are scarce, or due to agronomic disturbances (Mkenda et al. 2019). D'Alberto et al. (2012) underpin crop characteristics (annuals vs. perennials due to the periodic field disturbance in the former) and regional differences, as factors that can cause a weaker influence of noncrop vegetation on arthropod populations. Noordijk et al. (2010) report the effect of the margin age, i.e., decrease of predators' population as the field margin age increases, because plant species numbers and species evenness are reduced. Finally, van Rijn et al. (2016) conclude that the main driver determining flower resource suitability, flower choice and abundance of hoverflies in arable field margins, is nectar and not pollen accessibility. They suggest that preserving the right flowers in and around agricultural fields could enhance local insect populations and the services of pest control and pollination they provide.

7.2.5 Cover Crops

Cover cropping is one of the most recognizable nature-based solutions. It enhances diversity and provides various ecosystem-services in annual and perennial crops (Daryanto et al. 2018; Keesstra et al. 2018; Couëdel et al. 2019). Cover crop services span a wide range of functions that include protection of soil from erosion, soil fertilization, tilth (aeration and ease of tillage), increased water infiltration (drainage), reduced runoff, weed management, and provision of habitats for pollinators and natural enemies of pests (Altieri and Letourneau 1982; Nicholls et al. 2000; Snapp et al. 2005; Kruidhof et al. 2008; Storkey et al. 2015; Damien et al. 2017; Wittwer et al. 2017). Lee-Mader et al. (2019) described the function of flowering field margins as the *skeleton* of habitat provision for pollinators and natural enemies at the field-level scale, while extensive cover crops act as the *muscle* that implements the scale-up of this function.

A cover crop sown either before or with a main crop (i.e. undersown and maintained as a ground cover throughout the growing season) is also known as *living mulch* or *companion crop* (Hartwig and Ammon 2002). Living mulches are often perennial legume species. Cover crop mulching on the other hand (i.e. incorporating the biomass of the cover crop species into the field prior to sowing the cash crop) has been applied to smother weeds, utilizing the competitive and allelopathic capacity of the cover crops (Dhima et al. 2006). Cover crops may also serve as *green manure* when used primarily as a soil amendment or a nutrient source for subsequent crops, and normally include fast-growing legumes (Cherr et al. 2006). However, Norsworthy et al. (2005) used a brassicaceous mix as green manure to control weeds. When a cover crop is used to scavenge nutrients following the economic crop and prevent nutrient leaching during the winter, it is called catch crop. As pointed out by Maughan and Drost (2016), the terms *cover crop, green manure*, and *catch crop* are often used interchangeably, which may be confusing. Here we will use the term *cover crops* to refer to a single or a mixture of plant species established within a field as an ecological weed management tool that can also provide habitats for beneficial insect populations.

7.2.5.1 Current Status

Cover crops in the United States are currently planted on more than 15 million acres, according to the latest Sustainable Agriculture Research and Education annual report (SARE 2020). Significant drivers behind this wide adoption of cover cropping by US farmers include improving soil structure or health, managing herbicide-resistant weeds, harboring beneficial insects, and improving insect or disease control. These functions have led to increased profitability, depending on the cash crop. In Europe, the establishment of cover crops with noncrop species to support biological control is in its infancy, and hence the related benefits are not as well documented compared to those from sown flowering strips along the field margins (Holland et al. 2016).

7.2.5.2 Sowing and Termination

A cover crop can be sown or created by encouraging the establishment of suitable species from the weed flora. Timing of cover crop sowing and termination will affect the efficacy of all associated ecosystem services. Possible periods for cover crop establishment include during the winter fallow, during a summer fallow between cash crops, during a small-grain rotation, or during a full year of improved fallow (Lee-Mader et al. 2019), depending on the cash crop, the selected cover crop species and the expected benefits.

Termination of cover crops before seedset will help to avoid volunteer cover crop plants from emerging in the following season (Kruidhof et al. 2008; Keene et al. 2017). On the other hand, delaying termination up until the end of cover crop flowering can extend the window of floral resources and habitat provision to beneficial insects and can maximize ecosystem services such as soil fertilization. Obviously, the decision of when to terminate a cover crop is further influenced by other conditions, such as allowing soil water to recharge, which would require terminating a cover crop well in advance of establishing the cash crop. Methods of terminating cover crops include tillage, herbicide spraying, mowing, allowing the cover crops to winter kill, using plastic or tarps to block light and moisture, or using a roller/crimper which is often preferred in cover crop–based no-till systems (Liebman et al. 2018; Lee-Mader et al. 2019). In our opinion, chemical input to terminate cover crop should be avoided as it counteracts the benefits provided by cover crops.

7.2.5.3 Impact on Weeds, Pollinators, and Natural Enemies

Some common cover crop species include the following:

- Grasses (Poaceae): cereal rye (*Secale cereale* L.), winter wheat (*Triticum aestivum* L.), oats (*Avena sativa* L.), Italian ryegrass (*Lolium multiflorum* Lam.), tall fescue (*Festuca arundinacea* Scherb.), grain sorghum (*Sorghum bicolor* (L.) Moench), and sudangrass (*Sorghum sudanense* (Piper) Stapf.)
- Legumes (Fabaceae): hairy vetch (*Vicia villosa* R.), red clover (*Trifolium pratense* L.), and sweet clover (*Melilotus officinalis* L.)
- Cruciferous forages (Brassicaceae): turnips (*Brassica rapa* L.), rapeseed (*Brassica napus* L.), and radishes (*Raphanus sativus* L.); and other forbs, like buckwheat (*Fagopyrum esculentum* Moench, Polygonaceae) and lacy phacelia (*Phacelia tanacetifolia* Benth. Boraginaceae)

An expected advantage of cover crops is a more even distribution pattern of natural enemies across fields compared to field margins, where densities tend to be higher along the edge (Holland et al. 2016). However, a single species cover crop may have a weaker effect on the diversity of attracted beneficial insects compared to field margins, because the latter have a richer plant diversity and create complex habitats that benefit both biological control agents and biodiversity in general (Ditner et al. 2013). Also, single species cover crops typically offer a feast-or-famine situation for bees, since most dicotyledonous species used have a short flowering period (Lee-Mader et al. 2019). With regard to weed suppression, although certain species such as cereal rye are effective, it is preferable to include them in cover crop mixtures in low seeding rates to allow for other, purposely selected species to deliver additional ecosystem services (Baraibar et al. 2017). Although traditionally farmers have been using single species cover crops, there is an increasing interest by farmer groups and the private sector in multi-mix cover crops (Chapagain et al. 2020). This interest indicates a requirement by cover crops to deliver multiple services, which also applies for field margins. Recently, Couëdel et al. (2019) reviewed the benefits of selected crucifer-legume mixtures and discussed the potential benefits of multi-service cover crops (MSCC). They concluded that MSCC are a promising tool in the quest for the sustainable intensification of agriculture. Since the concept is new, the existing evidence for the services of cover crops remain focused on a single target – from our perspective here, only weed management or biological control, or more rarely, pollination.

Weed suppression by cover crops is regulated by various intertwined mechanisms, that depend on the selected species competitive, allelopathic, and environment-modifying capacity, regarding light, soil moisture, nitrogen, and other nutritional components (Teasdale 1996; Justes 2017). Kruidhof et al. (2008) reported various levels of weed suppression with cover crops established between the main crop periods with species from the families Brassicaceae (*Brassica napus* L., *Raphanus sativus* L.), Poaceae (*Secale cereale* L., *Lolium multiflorum* Lam.), and Fabaceae (*Medicago sativa* L., *Lupinus albus* L.). The results depended on the species sown and the timing of sowing and termination. Overall, delayed autumn sowing in combination with a stale seedbed resulted in significant reduction of weed pressure in spring, while species-wise, winter oilseed rape, fodder radish and winter rye had a strong competitive ability. Brassicaceae and Fabaceae species are also attractive to pollinators, which prefer dicotyledonous flowering species as opposed to grasses, because the latter do not provide nectar and their pollen has lower protein content than broadleaf plants (Lee-Mader et al. 2019). However, grasses can be valuable cover crop species as they attract alternate prey or hosts for predators and parasitoids of crop pests (Lee-Mader et al. 2019).

Holland et al. (2016) reviewed European studies on the impact of cover cropping on the conservation of natural enemies of pests in several crops, such as cereals (Vickerman 1978), vegetables (Ditner et al. 2013), vineyards (Altieri and Schmidt 1985) and orchards, like citrus (Monzó et al. 2008, 2011; Silva et al. 2010; Aguilar-Fenollosa et al. 2011a, 2011b, 2012, 2016), olive tree (Markó and Keresztes 2014), and pome fruit (Rieux et al. 1999; Markó and Keresztes 2014). They reported that several taxa of in-field natural enemy populations were enhanced in abundance and sometimes diversity by cover cropping in these cash crops, including Araneae, Opiliones, Hymenoptera parasitoids, Coleoptera, Carabidae ground beetle especially, and Diptera. A modified cover crop system was studied in an *Oxalis pes-caprae* L. infested olive grove, where ground-cover mixtures of selected plants from Brassicaceae, Fabaceae, Apiaceae, Asteraceae, and Boraginaceae, were sown in 3 m^2 patches along the tree line (as opposed to covering the entire ground) and attracted higher numbers of honey bees and wild bees compared to similar patches of weed vegetation, as well as high number of parasitoids and predators associated with olive pests (Karamaouna et al. 2019). This system of sowing the plant mixtures in patches facilitated cultivation practices and minimized competition with the crop, since the sown patches were maintained up until early summer. In other cases, cover crops supported beneficial insect populations even when they did not reach flowering. Some predators and parasitoids do not feed on nectar and pollen, but, rather, need a continuous supply of prey/host insects to maintain their local populations at an effective level. Hence, when the main crops are absent, nonflowering cover crops can support pests as a stable food source for beneficial insects outside the cropping period as well as a shelter for overwintering and breeding sites. For example, unmowed areas or "beetle banks" of perennial grasses outside crop fields can maintain ground beetles, which are generalist predators of slugs, caterpillars, and grasshopper eggs. Similarly, even if prey insects found in cover crops are not pests of the main crops, they can still be an important food resource for predator and parasitoid insects that will switch their prey/host preference once the crop pests become available (Lee-Mader et al. 2019).

7.2.5.4 Cover Cropping Effect on Insect Pests

The impact of cover crops on within-field pest populations is context specific (Holland et al. 2016), since not all pests respond similarly to greater vegetational diversity (Dunbar et al. 2016) while their population will be regulated by the associated natural enemies attracted by the cover crop plants. For example, spring-sown cereals with undersown mixtures of grasses and legumes had lower abundance of aphid populations than cereals without the cover crop, which was attributed to increased predation (Vickerman 1978). Similarly, a *Festuca arundinacea* Schreb. cover crop in citrus orchards increased predation on the Mediterranean fruit fly *Ceratitis capitata* Wiedemann (Monzó et al. 2008, 2011) and the two-spotted spider mite *Tetranychus urticae* Koch (Aguilar-Fenollosa et al. 2011c). On the other hand, ground cover with annual and perennial flowering herbs or grasses, while greatly increasing spider abundance in the canopy of apple trees, had no impact on the control of major pests in the studied apple orchard system (Markó and Keresztes 2014). A similar increase in spider densities was observed on grapevines in plots with a mixture of vetch and oat cover crop, but this was deemed not enough to affect the abundance of the leaf hopper *Erythroneura variabilis* Beamer (Hanna et al. 2003). An additional aspect that farmers should consider regarding the impact of cover crops on pests, is that some polyphagous insects, such as true armyworm (*Mythimna unipuncta* Haworth), black cutworm (*Agrotis ipsilon* Hufnagel), and common stalk borer (*Papaipema nebris* Guenee), whose host range includes rye, have the potential to cause injury to crops following a cereal rye cover crop (Dunbar et al. 2016).

7.3 Challenges for Implementation of Ecological Weed Management in Practice

Ecological weed management approaches such as those discussed earlier, are facing several challenges that should be acknowledged and addressed, to ensure a successful implementation in practice toward a more sustainable agriculture, through the support of pollination and biological control ecosystem services. Based on the current knowledge and experience, these are summarized as follows:

- Increase research effort to cover knowledge gaps in assessing the multiple benefits arising from ecologically based weed management, such as the impact on pollinators and natural enemies of pests, and develop decision support tools that will help in enhancing the associated ecosystem services for sustainability in agriculture.
- Understand the trade-offs between abiotic and biotic services to maximize ecosystem services and avoid disservices of weed management interactions with associated insects, by taking a wider-systems approach, beyond pest management (Kruidhof et al. 2008; Lee-Mader et al.

2019). This is vital to ensure that weeds are sufficiently abundant for pollinators and natural enemies, but not too abundant to reduce crop yield (Bretagnolle and Gaba 2015).

- Prioritize cases that need tailoring of floral and habitat composition, which, as Egan et al. (2020) suggested, are:
  1) Certain crop systems or landscapes without sufficient quantity or quality of resources for pollinators or natural enemies (e.g. forage, nutrition, nesting and shelter sites)
  2) Crops that require phenological complementarity in floral resource availability
  3) Crops requiring more specialised pollinators or natural enemies
  4) When existing or introduced floral species are either noxious weeds, or weak competitors against agronomic weeds, or are co-hosts to specialized pest insects or diseases
- Encourage the involvement of various stakeholders (who may vary depending on country) such as farmers, local authorities, researchers, policy makers, nongovernmental organizations, charities and land or estate owners, in the discovery of the scientific knowledge for easy adoption. Understanding of their personal, social and economic dynamics in the context of innovation adoption is essential.
- Raise awareness among the farmers on the multiple services and benefits of ecological weed management to help alleviate their concern that the new technologies might interfere with their usual farming practices or have a financial cost (Mkenda et al. 2019). As Blaix et al. (2018) suggested, it is important to provide farmers with concrete data regarding the benefits they can obtain from mixed farming, reduced herbicide use, inclusion of semi-natural habitats in their farms and the use of cover crops.
- Provide incentives and rewards toward ecologically based weed practices for agriculturalists and farmers, who are the principal managers of the most productive lands on earth (Tilman et al. 2002). According to Bokenstrand et al. (2004), the establishment and maintenance of field margins is costly to farmers, and such measures are unlikely to succeed if not supported financially. EU legislation, from the Directive 2009/128/EC on the sustainable use of pesticides to the 2014–2020 CAP reform, including the recent Green Deal (European Commission 2019), is setting out an ambitious agenda to preserve and restore ecosystems and biodiversity and could provide valuable funds on this direction. Such agri-environment schemes have been an increasingly important tool for the maintenance and restoration of farmland biodiversity in Europe (Kleijn et al. 2006).
- Exploit native or well adapted plant species from the local flora, for use in field margins or cover crops. This capacity varies between countries which have different levels of readiness in this direction. The disadvantages of using non-native species include low establishment capacity in new environments, genetic erosion of the same plant species in local populations, highly variable attractiveness to wild native bees.
- Combine ecological weed management approaches with other integrated pest management tactics to maximize the benefits for pollinators and natural enemies.

7.4 Conclusions

Today, weeds remain one of the most important crop production hazards. However, the management of weed vegetation is more than ever driven toward an integrated approach that encompasses methods to minimize the negative impacts of weeds but also to utilize their function as components of the ecosystem services in the agricultural environment for a more sustainable crop protection (Storkey et al. 2019). Innovative, interdisciplinary research is needed on functional biodiversity-related subjects, that will increase knowledge on the drivers of weed-arthropod interactions in agroecosystems across different spatial and temporal scales, as well as on a crop-specific basis (Barberi et al. 2010). This knowledge would be of utmost importance for fine-tuning management actions and designing improved agri-environment schemes aimed at enhancing the agroecological services related to pest and weed management.

References

Aguilar-Fenollosa, E., Ibáñez-Gual, M.V., Pascual-Ruiz, S. et al. (2011a). Effect of ground-cover management on spider mites and their phytoseiid natural enemies in clementine mandarin orchards (I): bottom-up regulation mechanisms. *Biological Control* 59 (2): 158–170. https://doi.org/10.1016/j.biocontrol.2011.06.013.

Aguilar-Fenollosa, E., Ibáñez-Gual, M.V., Pascual-Ruiz, S. et al. (2011b). Effect of ground-cover management on spider mites and their phytoseiid natural enemies in clementine mandarin orchards (II): top-down regulation mechanisms. *Biological Control* 59 (2): 171–179. https://doi.org/10.1016/j.biocontrol.2011.06.012.

Aguilar-Fenollosa, E., Pascual-Ruiz, S., Hurtado, M.A., and Jacas, J.A. (2011c). Efficacy and economics of ground cover management as a conservation biological control strategy against Tetranychus urticae in clementine mandarin orchards. *Crop Protection* 30 (10): 1328–1333. https://doi.org/10.1016/j.cropro.2011.05.011.

Aguilar-Fenollosa, E., Pina, T., Gómez-Martínez, M.A. et al. (2012). Does host adaptation of Tetranychus urticae populations in clementine orchards with a Festuca arundinacea cover contribute to a better natural regulation of this pest mite? *Entomologia Experimentalis et Applicata* 144 (2): 181–190. https://doi.org/10.1111/j.1570-7458.2012.01276.x.

Aguilar-Fenollosa, E., Rey-Caballero, J., Blasco, J.M. et al. (2016). Patterns of ambulatory dispersal in Tetranychus urticae can be associated with host plant specialization. *Experimental and Applied Acarology* 68 (1): 1–20. https://doi.org/10.1007/s10493-015-9969-1.

Altieri, M.A. and Letourneau, D.K. (1982). Vegetation management and biological control in agroecosystems. *Crop Protection* 1 (4): 405–430. https://doi.org/https://doi.org/10.1016/0261-2194(82)90023-0.

Altieri, M.A., Martin, P.B., and Lewis, W.J. (1983). A quest for ecologically based pest management systems. *Environmental Management* 7 (1): 91–99. https://doi.org/10.1007/BF01867047.

Altieri, M.A. and Schmidt, L.L. (1985). Cover crop manipulation in northern California orchards and vineyards: effects on arthropod communities. *Biological Agriculture & Horticulture* 3 (1): 1–24. https://doi.org/10.1080/01448765.1985.9754453.

Anderson, A., Carnus, T.I.M., Helden, A.J. et al. (2013). The influence of conservation field margins in intensively managed grazing land on communities of five arthropod trophic groups. *Insect Conservation and Diversity* 6 (3): 201–211. https://doi.org/10.1111/j.1752-4598.2012.00203.x.

Arrhenius, O. (1921). Species and area. *Journal of Ecology* 9 (1): 95–99. https://doi.org/10.2307/2255763.

Bäckman, J.-P.C. and Tiainen, J. (2002). Habitat quality of field margins in a Finnish farmland area for bumblebees (Hymenoptera: *Bombus* and *Psithyrus*). *Agriculture, Ecosystems & Environment* 89 (1): 53–68. https://doi.org/10.1016/S0167-8809(01)00318-8.

Balzan, M.V., Bocci, G., and Moonen, A.-C. (2014). Augmenting flower trait diversity in wildflower strips to optimise the conservation of arthropod functional groups for multiple agroecosystem services. *Journal of Insect Conservation* 18 (4): 713–728. https://doi.org/10.1007/s10841-014-9680-2.

Balzan, M.V., Bocci, G., and Moonen, A.-C. (2016). Utilisation of plant functional diversity in wildflower strips for the delivery of multiple agroecosystem services. *Entomologia Experimentalis et Applicata* 158 (3): 304–319. https://doi.org/10.1111/eea.12403.

Baraibar, B., Hunter, M.C., Schipanski, M.E. et al. (2017). Weed suppression in cover crop monocultures and mixtures. *Weed Science* 66 (1): 121–133. https://doi.org/10.1017/wsc.2017.59.

Barberi, P., Burgio, G., Dinelli, G. et al. (2010). Functional biodiversity in the agricultural landscape: relationships between weeds and arthropod fauna. *Weed Research* 50 (5): 388–401. https://doi.org/10.1111/j.1365-3180.2010.00798.x.

Barbir, J., Badenes-Pérez, F.R., Fernández-Quintanilla, C., et al. (2015). The attractiveness of flowering herbaceous plants to bees (Hymenoptera: Apoidea) and hoverflies (Diptera: Syrphidae) in agro-ecosystems of Central Spain. *Agricultural and Forest Entomology* 17 (1): 20–28. https://doi.org/10.1111/afe.12076.

Biella, P., Akter, A., Ollerton, J. et al. (2019). Experimental loss of generalist plants reveals alterations in plant-pollinator interactions and a constrained flexibility of foraging. *Scientific Reports* 9 (1): 7376. https://doi.org/10.1038/s41598-019-43553-4.

Biesmeijer, J.C., Roberts, S.P., Reemer, M. et al. (2006). Parallel declines in pollinators and insect-pollinated plants in Britain and the Netherlands. *Science* 313 (5785): 351–354. https://doi.org/10.1126/science.1127863.

Blaix, C. and Moonen, A.C. (2020). Structural field margin characteristics affect the functional traits of herbaceous vegetation. *PLoS One* 15 (9): e0238916. https://doi.org/10.1371/journal.pone.0238916.

Blaix, C., Moonen, A.C., Dostatny, D.F. et al. (2018). Quantification of regulating ecosystem services provided by weeds in annual cropping systems using a systematic map approach. *Weed Research* 58 (3): 151–164. https://doi.org/10.1111/wre.12303.

Blumenthal, D. and Jordan, N. (2001). Weeds in field margins: a spatially explicit simulation analysis of Canada thistle population dynamics. *Weed Science* 49 (4): 509–519. https://doi.org/10.1614/0043-1745(2001)049[0509:WIFMAS]2.0.CO;2.

Bokenstrand, A., Lagerlöf, J., and Torstensson, P.R. (2004). Establishment of vegetation in broadened field boundaries in agricultural landscapes. *Agriculture, Ecosystems & Environment* 101 (1): 21–29. https://doi.org/10.1016/s0167-8809(03)00275-5.

Boutin, C., Jobin, B., Bélanger, L., and Choinière, L. (2001). Comparing weed composition in natural and planted hedgerows and in herbaceous field margins adjacent to crop fields. *Canadian Journal of Plant Science* 81 (2): 313–324. https://doi.org/10.4141/P00-048.

Bowie, M.H., Klimaszewski, J., Vink, C.J. et al. (2014). Effect of boundary type and season on predatory arthropods associated with field margins on New Zealand farmland. *New Zealand Journal of Zoology* 41 (4): 268–284. https://doi.org/10.1080/03014223.2014.953552.

Bretagnolle, V. and Gaba, S. (2015). Weeds for bees? A review. *Agronomy for Sustainable Development* 35 (3): 891–909. https://doi.org/10.1007/s13593-015-0302-5.

Buchanan, G.A. (1977). Weed biology and competition. In: *Research Methods in Weed Science*, 2e (ed. B. Truelove), 25–41. Southern Weed Science Society.

Campbell, A.J., Biesmeijer, J.C., Varma, V., et al. (2012). Realising multiple ecosystem services based on the response of three beneficial insect groups to floral traits and trait diversity. *Basic and Applied Ecology* 13 (4): 363–370. https://doi.org/10.1016/j.baae.2012.04.003.

Capinera, J.L. (2005). Relationships between insect pests and weeds: an evolutionary perspective. *Weed Science* 53 (6): 892–901. https://doi.org/10.1614/WS-04-049R.1.

Carvell, C., Meek, W.R., Pywell, R.F. et al. (2006). Comparing the efficacy of agri-environment schemes to enhance bumble bee abundance and diversity on arable field margins. *Journal of Applied Ecology* 44 (1): 29–40. https://doi.org/10.1111/j.1365-2664.2006.01249.x.

Chapagain, T., Lee, E.A., and Raizada, M.N. (2020). The potential of multi-species mixtures to diversify cover crop benefits. *Sustainability* 12 (5). https://doi.org/10.3390/su12052058.

Chauhan, B.S., Migo, T., Westerman, P.R., and Johnson, D.E. (2010). Post-dispersal predation of weed seeds in rice fields. *Weed Research* 50 (6): 553–560. https://doi.org/10.1111/j.1365-3180.2010.00807.x.

Cherr, C.M., Scholberg, J.M.S., and McSorley, R. (2006). Green manure approaches to crop production: a synthesis. *Agronomy Journal* 98 (2): 302–319. https://doi.org/10.2134/agronj2005.0035.

Cirujeda, A., Melander, B., Rasmussen, K., et al. (2003). Relationship between speed, soil movement into the cereal row and intra-row weed control efficacy by weed harrowing. *Weed Research* 43 (4): 285–296. https://doi.org/10.1046/j.1365-3180.2003.00345.x.

European Commission. (2019). *Communication from the Commission to the European Parliament, the European Council, the Council, the European Economic and Social Committee and the Committee of the Regions, the European Green Deal COM/2019/640 final.*

Corbet, S.A. (1978). Bee visits and the nectar of Echium vulgare L. and Sinapis alba L. *Ecological Entomology* 3 (1): 25–37. https://doi.org/10.1111/j.1365-2311.1978.tb00900.x.

Couëdel, A., Kirkegaard, J., Alletto, L., et al. (2019). *Crucifer-legume cover crop mixtures for biocontrol: toward a new multi-service paradigm.* 55–139. https://doi.org/10.1016/bs.agron.2019.05.003.

Cresswell, C.J., Cunningham, H.M., Wilcox, A., et al. (2018). What specific plant traits support ecosystem services such as pollination, bio-control and water quality protection in temperate climates? A systematic map. *Environmental Evidence* 7 (1). https://doi.org/10.1186/s13750-018-0120-8.

Cresswell, C.J., Cunningham, H.M., Wilcox, A., et al. (2019). A trait-based approach to plant species selection to increase functionality of farmland vegetative strips. *Ecology and Evolution* 9 (8): 4532–4543. https://doi.org/10.1002/ece3.5047.

D'Alberto, C.F., Hoffmann, A.A., and Thomson, L.J. (2012). Limited benefits of non-crop vegetation on spiders in Australian vineyards: regional or crop differences? *BioControl* 57 (4): 541–552. https://doi.org/10.1007/s10526-011-9435-x.

Dambach, C.A. (1948). *Ecology of Crop Field Borders.* Columbus, OH: Ohio State University Press.

Damien, M., Le Lann, C., Desneux, N. et al. (2017). Flowering cover crops in winter increase pest control but not trophic link diversity. *Agriculture, Ecosystems & Environment* 247: 418–425. https://doi.org/10.1016/j.agee.2017.07.015.

Daryanto, S., Fu, B., Wang, L. et al. (2018). Quantitative synthesis on the ecosystem services of cover crops. *Earth-Science Reviews* 185: 357–373. https://doi.org/10.1016/j.earscirev.2018.06.013.

de Bello, F., Lavorel, S., Díaz, S. et al. (2010). Towards an assessment of multiple ecosystem processes and services via functional traits. *Biodiversity and Conservation* 19 (10): 2873–2893. https://doi.org/10.1007/s10531-010-9850-9.

De Cauwer, B., Reheul, D., De Laethauwer, S. et al. (2006). Effect of light and botanical species richness on insect diversity. *Agronomy for Sustainable Development* 26 (1): 35–43. https://doi.org/10.1051/agro:2005058.

De Cauwer, B., Reheul, D., Nijs, I., et al. (2008). Management of newly established field margins on nutrient-rich soil to reduce weed spread and seed rain into adjacent crops. *Weed Research* 48 (2): 102–112. https://doi.org/10.1111/j.1365-3180.2007.00607.x.

Delattre, T., Pichancourt, J.-B., Burel, F., et al. (2010). Grassy field margins as potential corridors for butterflies in agricultural landscapes: a simulation study. *Ecological Modelling* 221 (2): 370–377. https://doi.org/10.1016/j.ecolmodel.2009.10.010.

Délye, C., Jasieniuk, M., and Le Corre, V. (2013). Deciphering the evolution of herbicide resistance in weeds. *Trends in Genetics* 29 (11): 649–658. https://doi.org/10.1016/j.tig.2013.06.001.

Denisow, B. and Wrzesień, M. (2015). The importance of field-margin location for maintenance of food niches for pollinators. *Journal of Apicultural Science* 59 (1): 27–37. https://doi.org/10.1515/jas-2015-0002.

Dhima, K.V., Vasilakoglou, I.B., Eleftherohorinos, I.G., et al. (2006). Allelopathic potential of winter cereals and their cover crop mulch effect on grass weed suppression and corn development. *Crop Science* 46 (1): 345–352. https://doi.org/10.2135/cropsci2005-0186.

Ditner, N., Balmer, O., Beck, J. et al. (2013). Effects of experimentally planting non-crop flowers into cabbage fields on the abundance and diversity of predators. *Biodiversity and Conservation* 22 (4): 1049–1061. https://doi.org/10.1007/s10531-013-0469-5.

Dramstad, W. and Fry, G. (1995). Foraging activity of bumblebees (Bombus) in relation to flower resources on arable land. *Agriculture, Ecosystems & Environment* 53 (2): 123–135. https://doi.org/10.1016/0167-8809(94)00561-R.

Dunbar, M.W., O'Neal, M.E., and Gassmann, A.J. (2016). Increased risk of insect injury to corn following rye cover crop. *Journal of Economic Entomology* 109 (4): 1691–1697. https://doi.org/10.1093/jee/tow101.

Eeraerts, M., Smagghe, G., and Meeus, I. (2019a). Pollinator diversity, floral resources and semi-natural habitat, instead of honey bees and intensive agriculture, enhance pollination service to sweet cherry. *Agriculture, Ecosystems & Environment* 284. https://doi.org/10.1016/j.agee.2019.106586.

Eeraerts, M., Vanderhaegen, R., Smagghe, G., et al. (2019b). Pollination efficiency and foraging behaviour of honey bees and non-Apis bees to sweet cherry. *Agricultural and Forest Entomology* 22 (1): 75–82. https://doi.org/10.1111/afe.12363.

Egan, P.A., Dicks, L.V., Hokkanen, H.M.T., et al. (2020). Delivering integrated pest and pollinator management (IPPM). *Trends in Plant Science* 25 (6): 577–589. https://doi.org/10.1016/j.tplants.2020.01.006.

Free, J.B. (1993). *Crop Pollination by Insects*, 2e. Academic Press.

Fried, G., Cordeau, S., Metay, A., et al. (2019). Relative importance of environmental factors and farming practices in shaping weed communities structure and composition in French vineyards. *Agriculture, Ecosystems & Environment* 275: 1–13. https://doi.org/10.1016/j.agee.2019.01.006.

Gaba, S., Perronne, R., Fried, G. et al. (2017). Response and effect traits of arable weeds in agro-ecosystems: a review of current knowledge. *Weed Research* 57 (3): 123–147. https://doi.org/10.1111/wre.12245.

Gallai, N., Salles, J.-M., Settele, J., et al. (2009). Economic valuation of the vulnerability of world agriculture confronted with pollinator decline. *Ecological Economics* 68 (3): 810–821. https://doi.org/10.1016/j.ecolecon.2008.06.014.

Garibaldi, L.A., Carvalheiro, L.G., Vaissière, B.E. et al. (2016). Mutually beneficial pollinator diversity and crop yield outcomes in small and large farms. *Science* 351 (6271): 388. https://doi.org/10.1126/science.aac7287.

Gibson, R.H., Nelson, I.L., Hopkins, G.W. et al. (2006). Pollinator webs, plant communities and the conservation of rare plants: arable weeds as a case study. *Journal of Applied Ecology* 43 (2): 246–257. https://doi.org/10.1111/j.1365-2664.2006.01130.x.

Goulson, D., Nicholls, E., Botias, C., et al. (2015). Bee declines driven by combined stress from parasites, pesticides, and lack of flowers. *Science* 347 (6229): 1255957. https://doi.org/10.1126/science.1255957.

Greaves, M. and Marshall, E. (1987). Field margins: definitions and statistics. *Monographs - British Crop Protection Council (BCPC)* 35.

Hanna, R., Zalom, F.G., and Roltsch, W.J. (2003). Relative impact of spider predation and cover crop on population dynamics of Erythroneura variabilis in a raisin grape vineyard. *Entomologia Experimentalis et Applicata* 107 (3): 177–191. https://doi.org/10.1046/j.1570-7458.2003.00051.x.

Hannon, L.E. and Sisk, T.D. (2009). Hedgerows in an agri-natural landscape: potential habitat value for native bees. *Biological Conservation* 142 (10): 2140–2154. https://doi.org/10.1016/j.biocon.2009.04.014.

Hartwig, N.L. and Ammon, H.U. (2002). Cover crops and living mulches. *Weed Science* 50 (6): 688–699. https://doi.org/10.1614/0043-1745(2002)050[0688:AIACCA]2.0.CO;2.

Hawes, C., Haughton, A.J., Bohan, D.A., et al. (2009). Functional approaches for assessing plant and invertebrate abundance patterns in arable systems. *Basic and Applied Ecology* 10 (1): 34–42. https://doi.org/10.1016/j.baae.2007.11.007.

Heap, V. (2021). *The international herbicide-resistant weed database*. Online. Thursday, July 8, 2021. www.weedscience.org Copyright © 1993-2021 WeedScience.org.

Holland, J.M., Bianchi, F.J., Entling, M.H. et al. (2016). Structure, function and management of semi-natural habitats for conservation biological control: a review of European studies. *Pest Management Science* 72 (9): 1638–1651. https://doi.org/10.1002/ps.4318.

Holland, J.M., Storkey, J., Lutman, P.J.W. et al. (2014). Utilisation of agri-environment scheme habitats to enhance invertebrate ecosystem service providers. *Agriculture, Ecosystems & Environment* 183: 103–109. https://doi.org/10.1016/j.agee.2013.10.025.

Justes, É. (2017). *Cover Crops for Sustainable Farming*. Springer. https://doi.org/10.1007/978-94-024-0986-4.

Karamaouna, F., Kati, V., Volakakis, N. et al. (2019). Ground cover management with mixtures of flowering plants to enhance insect pollinators and natural enemies of pests in olive groves. *Agriculture, Ecosystems & Environment* 274: 76–89. https://doi.org/10.1016/j.agee.2019.01.004.

Kati, V., Karamaouna, F., Economou, L. et al. (2021). Sown wildflowers enhance habitats of pollinators and beneficial arthropods in a tomato field margin. *Plants (Basel)* 10 (5). https://doi.org/10.3390/plants10051003.

Kati, V., Scarabel, L., Thiery-Lanfranchi, D. et al. (2019). Multiple resistance of *Papaver rhoeas* L. to 2,4-D and acetolactate synthase inhibitors in four European countries. *Weed Research* 59 (5): 367–376. https://doi.org/10.1111/wre.12376.

Kattge, J., Bonisch, G., Diaz, S. et al. (2020). TRY plant trait database – enhanced coverage and open access. *Global Change Biology* 26 (1): 119–188. https://doi.org/10.1111/gcb.14904.

Kattge, J., Diaz, S., Lavorel, S. et al. (2011). TRY – a global database of plant traits. *Global Change Biology* 17 (9): 2905–2935. https://doi.org/10.1111/j.1365-2486.2011.02451.x.

Keene, C.L., Curran, W.S., Wallace, J.M. et al. (2017). Cover crop termination timing is critical in organic rotational no-till systems. *Agronomy Journal* 109 (1): 272–282. https://doi.org/10.2134/agronj2016.05.0266.

Keesstra, S., Nunes, J., Novara, A. et al. (2018). The superior effect of nature based solutions in land management for enhancing ecosystem services. *Science of the Total Environment* 610–611: 997–1009. https://doi.org/10.1016/j.scitotenv.2017.08.077.

Kells, A.R., Holland, J.M., and Goulson, D. (2001). The value of uncropped field margins for foraging bumblebees. *Journal of Insect Conservation* 5 (4): 283–291. https://doi.org/10.1023/A:1013307822575.

Kleijn, D., Baquero, R.A., Clough, Y. et al. (2006). Mixed biodiversity benefits of agri-environment schemes in five European countries. *Ecology Letters* 9 (3): 243–254. https://doi.org/10.1111/j.1461-0248.2005.00869.x.

Klein, A.M., Vaissiere, B.E., Cane, J.H. et al. (2007). Importance of pollinators in changing landscapes for world crops. *Proceedings of the Royal Society B: Biological Sciences* 274 (1608): 303–313. https://doi.org/10.1098/rspb.2006.3721.

Knapp, J.L., Shaw, R.F., and Osborne, J.L. (2019). Pollinator visitation to mass-flowering courgette and co-flowering wild flowers: implications for pollination and bee conservation on farms. *Basic and Applied Ecology* 34: 85–94. https://doi.org/10.1016/j.baae.2018.09.003.

Kolářová, M., Tyšer, L., Soukup, J., et al. (2013). Impact of site conditions and farming practices on the occurrence of rare and endangered weeds on arable land in the Czech Republic. *Weed Research* 53 (6): 489–498. https://doi.org/10.1111/wre.12045.

Kremen, C., Williams, N.M., and Thorp, R.W. (2002). Crop pollination from native bees at risk from agricultural intensification. *Proceedings of the National Academy of Sciences* 99 (26): 16812. https://doi.org/10.1073/pnas.262413599.

Kruidhof, H.M., Bastiaans, L., and Kropff, M.J. (2008). Ecological weed management by cover cropping: effects on weed growth in autumn and weed establishment in spring. *Weed Research* 48 (6): 492–502. https://doi.org/10.1111/j.1365-3180.2008.00665.x.

Kütt, L., Lõhmus, K., Rammi, I.-J. et al. (2016). The quality of flower-based ecosystem services in field margins and road verges from human and insect pollinator perspectives. *Ecological Indicators* 70: 409–419. https://doi.org/10.1016/j.ecolind.2016.06.009.

Lee-Mader, E., Anne Stine, A., Jarrod Fowler, J. et al. (2019). *Opportunities in agriculture. Cover cropping for pollinators and beneficial insects. Xerces Society for Invertebrate Conservation; Sustainable Agriculture Research and Education (SARE) project in collaboration with: USDA* Natural Resources *Conservation* Service *(NRCS); National Institute of Food and Agriculture (NIFA), USDA award number 2014-38640-22173*, 16.

Liebman, A.M., Grossman, J., Brown, M. et al. (2018). Legume cover crops and tillage impact nitrogen dynamics in organic corn production. *Agronomy Journal* 110 (3): 1046–1057. https://doi.org/10.2134/agronj2017.08.0474.

Ma, M., Tarmi, S., and Helenius, J. (2002). Revisiting the species–area relationship in a semi-natural habitat: floral richness in agricultural buffer zones in Finland. *Agriculture, Ecosystems & Environment* 89 (1): 137–148. https://doi.org/10.1016/S0167-8809(01)00325-5.

MacLaren, C., Storkey, J., Menegat, A. et al. (2020). An ecological future for weed science to sustain crop production and the environment. A review. *Agronomy for Sustainable Development* 40 (4). https://doi.org/10.1007/s13593-020-00631-6.

Maia, K.P., Vaughan, I.P., and Memmott, J. (2019). Plant species roles in pollination networks: an experimental approach. *Oikos* 128 (10): 1446–1457. https://doi.org/10.1111/oik.06183.

Makino, T.T. and Sakai, S. (2007). Experience changes pollinator responses to floral display size: from size-based to reward-based foraging. *Functional Ecology* 21 (5): 854–863. https://doi.org/10.1111/j.1365-2435.2007.01293.x.

Mante, J. and Gerowitt, B. (2009). Learning from farmers' needs: identifying obstacles to the successful implementation of field margin measures in intensive arable regions. *Landscape and Urban Planning* 93 (3–4): 229–237. https://doi.org/10.1016/j.landurbplan.2009.07.010.

Marcovitch, S. (1935). Experimental evidence on the value of strip farming as a method for the natural control of injurious insects with special reference to plant lice. *Journal of Economic Entomology* 28 (1): 62–70. https://doi.org/10.1093/jee/28.1.62.

Markó, V. and Keresztes, B. (2014). Flowers for better pest control? Ground cover plants enhance apple orchard spiders (Araneae), but not necessarily their impact on pests. *Biocontrol Science and Technology* 24 (5): 574–596. https://doi.org/10.1080/09583157.2014.881981.

Marshall, E.J.P. (2002). Introducing field margin ecology in Europe. *Agriculture, Ecosystems & Environment* 89 (1): 1–4. https://doi.org/10.1016/S0167-8809(01)00314-0.

Marshall, E.J.P. and Arnold, G.M. (1995). Factors affecting field weed and field margin flora on a farm in Essex, UK.

Landscape and Urban Planning 31 (1): 205–216. https://doi.org/10.1016/0169-2046(94)01047-C.

Marshall, E.J.P., Brown, V.K., Boatman, N.D. et al. (2003). The role of weeds in supporting biological diversity within crop fields. *Weed Research* 43 (2): 77–89. https://doi.org/10.1046/j.1365-3180.2003.00326.x.

Marshall, E.J.P. and Moonen, A.C. (2002). Field margins in northern Europe: their functions and interactions with agriculture. *Agriculture, Ecosystems & Environment* 89 (1): 5–21. https://doi.org/10.1016/S0167-8809(01)00315-2.

Martins, K.T., Gonzalez, A., and Lechowicz, M.J. (2015). Pollination services are mediated by bee functional diversity and landscape context. *Agriculture, Ecosystems & Environment* 200: 12–20. https://doi.org/10.1016/j.agee.2014.10.018.

Maughan, T. and Drost, D. (2016). Introduction to cover crops for vegetable production in Utah. In *All Current Publications* (Paper 1587 ed.).

Meyer, S., Wesche, K., Krause, B. et al. (2013). Dramatic losses of specialist arable plants in Central Germany since the 1950s/60s – a cross-regional analysis. *Diversity and Distributions* 19 (9): 1175–1187. https://doi.org/10.1111/ddi.12102.

Mkenda, P.A., Ndakidemi, P.A., Mbega, E. et al. (2019). Multiple ecosystem services from field margin vegetation for ecological sustainability in agriculture: scientific evidence and knowledge gaps. *PeerJ* 7: e8091. https://doi.org/10.7717/peerj.8091.

Monzó, C., Mollá, Ó., Castañera, P., et al. (2008). Activity-density of Pardosa cribata in Spanish citrus orchards and its predatory capacity on Ceratitis capitata and Myzus persicae. *BioControl* 54 (3): 393–402. https://doi.org/10.1007/s10526-008-9199-0.

Monzó, C., Sabater-Muñoz, B., Urbaneja, A., et al. (2011). The ground beetle Pseudophonus rufipes revealed as predator of Ceratitis capitata in citrus orchards. *Biological Control* 56 (1): 17–21. https://doi.org/10.1016/j.biocontrol.2010.09.004.

Moonen, A.C. and Marshall, E.J.P. (2001). The influence of sown margin strips, management and boundary structure on herbaceous field margin vegetation in two neighbouring farms in southern England. *Agriculture, Ecosystems & Environment* 86 (2): 187–202. https://doi.org/10.1016/S0167-8809(00)00283-8.

Morrison, J. (2018). *Flowering native weeds for the conservation of wild pollinators in agroecosystems.* Universitat Politècnica de Catalunya.

Naranjo, S.E., Frisvold, G.B., and Ellsworth, P.C. (2019). Economic value of arthropod biological control. In: *The Economics of Integrated Pest Management of Insects*, 49–85. https://doi.org/10.1079/9781786393678.0049.

Naylor, R. and Ehrlich, P.R. (1997). Natural pest control services and agriculture. In: *Nature's Services: Societal Dependence on Natural Ecosystems* (ed. G.C. Daily). Island Press. http://www.loc.gov/catdir/enhancements/fy0666/96040401-d.html.

Neve, P., Vila-Aiub, M., and Roux, F. (2009). Evolutionary-thinking in agricultural weed management. *New Phytologist* 184 (4): 783–793. https://doi.org/10.1111/j.1469-8137.2009.03034.x.

Nicholls, C.I. and Altieri, M.A. (2013). Plant biodiversity enhances bees and other insect pollinators in agroecosystems. A review. *Agronomy for Sustainable Development* 33 (2): 257–274. https://doi.org/10.1007/s13593-012-0092-y.

Nicholls, C.I., Parrella, M.P., and Altieri, M.A. (2000). Reducing the abundance of leafhoppers and thrips in a northern California organic vineyard through maintenance of full season floral diversity with summer cover crops. *Agricultural and Forest Entomology* 2 (2): 107–113. https://doi.org/10.1046/j.1461-9563.2000.00054.x.

Noordijk, J., Musters, C.J.M., van Dijk, J., and de Snoo, G.R. (2010). Invertebrates in field margins: taxonomic group diversity and functional group abundance in relation to age. *Biodiversity and Conservation* 19 (11): 3255–3268. https://doi.org/10.1007/s10531-010-9890-1.

Norris, R.F. and Kogan, M. (2000). Interactions between weeds, arthropod pests, and their natural enemies in managed ecosystems. *Weed Science* 48 (1): 94–158. https://doi.org/10.1614/0043-1745(2000)048[0094:Ibwapa]2.0.Co;2.

Norris, R.F. and Kogan, M. (2005). Ecology of interactions between weeds and arthropods. *Annual Review of Entomology* 50: 479–503. https://doi.org/10.1146/annurev.ento.49.061802.123218.

Norsworthy, J.K., Brandenberger, L., Burgos, N.R., et al. M. (2005). Weed suppression in Vigna unguiculata with a spring-seeded brassicaceae green manure. *Crop Protection* 24 (5): 441–447. https://doi.org/10.1016/j.cropro.2004.09.015.

Odoux, J.-F., Feuillet, D., Aupinel, P. et al. (2012). Territorial biodiversity and consequences on physico-chemical characteristics of pollen collected by honey bee colonies. *Apidologie* 43 (5): 561–575. https://doi.org/10.1007/s13592-012-0125-1.

Oerke, E.C. (2006). Crop losses to pests. *The Journal of Agricultural Science* 144 (1): 31–43. https://doi.org/10.1017/S0021859605005708.

Olson, D.M. and Wäckers, F.L. (2006). Management of field margins to maximize multiple ecological services. *Journal of Applied Ecology* 44 (1): 13–21. https://doi.org/10.1111/j.1365-2664.2006.01241.x.

Papanikolaou, A.D., Kühn, I., Frenzel, M. et al. (2017). Semi-natural habitats mitigate the effects of temperature rise on wild bees. *Journal of Applied Ecology* 54 (2): 527–536. https://doi.org/10.1111/1365-2664.12763.

Parker, A.J., Tran, J.L., Ison, J.L. et al. (2015). Pollen packing affects the function of pollen on corbiculate bees but not non-corbiculate bees. *Arthropod-Plant Interactions* 9 (2): 197–203. https://doi.org/10.1007/s11829-015-9358-z.

Perner, J. and Malt, S. (2003). Assessment of changing agricultural land use: response of vegetation, ground-dwelling spiders and beetles to the conversion of arable land into grassland. *Agriculture, Ecosystems and Environment* 98 (1–3): 169–181. https://doi.org/10.1016/S0167-8809(03)00079-3.

Petanidou, T. and Lamborn, E. (2005). A land for flowers and bees: studying pollination ecology in Mediterranean communities. *Plant Biosystems – An International Journal Dealing with All Aspects of Plant Biology* 139 (3): 279–294. https://doi.org/10.1080/11263500500333941.

Potts, S.G., Biesmeijer, J.C., Kremen, C. et al. (2010). Global pollinator declines: trends, impacts and drivers. *Trends in Ecology & Evolution* 25 (6): 345–353. https://doi.org/10.1016/j.tree.2010.01.007.

Potts, S.G., Vulliamy, B., Dafni, A. et al. (2003). Linking bees and flowers: how do floral communities structure pollinator communities? *Ecology* 84 (10): 2628–2642. https://doi.org/10.1890/02-0136.

Preston, F.W. (1962). The canonical distribution of commonness and rarity: part I. *Ecology* 43 (2): 185–215. https://doi.org/10.2307/1931976.

Pywell, R.F., Warman, E.A., Carvell, C. et al. (2005). Providing foraging resources for bumblebees in intensively farmed landscapes. *Biological Conservation* 121 (4): 479–494. https://doi.org/10.1016/j.biocon.2004.05.020.

Rader, R., Bartomeus, I., Garibaldi, L.A. et al. (2016). Non-bee insects are important contributors to global crop pollination. *Proceedings of the National Academy of Sciences of the United States of America* 113 (1): 146–151. https://doi.org/10.1073/pnas.1517092112.

Raine, N.E., Mallinger, R.E., Gaines-Day, H.R., et al. (2017). Do managed bees have negative effects on wild bees?: a systematic review of the literature. *PLoS One* 12 (12). https://doi.org/10.1371/journal.pone.0189268.

Ramsden, M.W., Menéndez, R., Leather, S.R., et al. (2015). Optimizing field margins for biocontrol services: the relative role of aphid abundance, annual floral resources, and overwinter habitat in enhancing aphid natural enemies. *Agriculture, Ecosystems & Environment* 199: 94–104. https://doi.org/10.1016/j.agee.2014.08.024.

Reberg-Horton, S.C., Mueller, J.P., Mellage, S.J. et al. (2010). Influence of field margin type on weed species richness and abundance in conventional crop fields. *Renewable Agriculture and Food Systems* 26 (2): 127–136. https://doi.org/10.1017/s1742170510000451.

Ricou, C., Schneller, C., Amiaud, B. et al. (2014). A vegetation-based indicator to assess the pollination value of field margin flora. *Ecological Indicators* 45: 320–331. https://doi.org/10.1016/j.ecolind.2014.03.022.

Rieux, R., Simon, S., and Defrance, H. (1999). Role of hedgerows and ground cover management on arthropod populations in pear orchards. *Agriculture, Ecosystems & Environment* 73 (2): 119–127. https://doi.org/10.1016/S0167-8809(99)00021-3.

Rollin, O., Benelli, G., Benvenuti, S. et al. (2016). Weed-insect pollinator networks as bio-indicators of ecological sustainability in agriculture. A review. *Agronomy for Sustainable Development* 36 (1). https://doi.org/10.1007/s13593-015-0342-x.

Rollin, O., Bretagnolle, V., Decourtye, A. et al. (2013). Differences of floral resource use between honey bees and wild bees in an intensive farming system. *Agriculture, Ecosystems & Environment* 179: 78–86. https://doi.org/10.1016/j.agee.2013.07.007.

SARE. (2020). *Annual report 2019–2020 – national cover crop survey August 2020* (Sustainable Agriculture Research and Education, Issue).

Schellhorn, N.A. and Sork, V.L. (1997). The impact of weed diversity on insect population dynamics and crop yield in collards, Brassica oleraceae (Brassicaceae). *Oecologia* 111 (2): 233–240. https://doi.org/10.1007/s004420050230.

Silva, E.B., Franco, J.C., Vasconcelos, T., et al. (2010). Effect of ground cover vegetation on the abundance and diversity of beneficial arthropods in citrus orchards. *Bulletin of Entomological Research* 100 (4): 489–499. https://doi.org/10.1017/S0007485309990526.

Smith, B.M., Aebischer, N.J., Ewald, J. et al. (2020). The potential of arable weeds to reverse invertebrate declines and associated ecosystem services in cereal crops. *Frontiers in Sustainable Food Systems* 3. https://doi.org/10.3389/fsufs.2019.00118.

Smith, J., Potts, S.G., Woodcock, B.A., et al. (2008). Can arable field margins be managed to enhance their biodiversity, conservation and functional value for soil macrofauna? *Journal of Applied Ecology* 45 (1): 269–278. https://doi.org/10.1111/j.1365-2664.2007.01433.x.

Snapp, S.S., Swinton, S.M., Labarta, R. et al. (2005). Evaluating cover crops for benefits, costs and performance within cropping system niches. *Agronomy Journal* 97 (1): 322–332. https://doi.org/10.2134/agronj2005.0322a.

Storkey, J. (2006). A functional group approach to the management of UK arable weeds to support biological diversity. *Weed Research* 46 (6): 513–522. https://doi.org/10.1111/j.1365-3180.2006.00528.x.

Storkey, J., Bruce, T.J.A., McMillan, V.E., et al. (2019). The future of sustainable crop protection relies on increased diversity of cropping systems and landscapes. In: *Agroecosystem Diversity*, 199–209. https://doi.org/10.1016/b978-0-12-811050-8.00012-1.

Storkey, J., Döring, T., Baddeley, J. et al. (2015). Engineering a plant community to deliver multiple ecosystem services. *Ecological Applications* 25 (4): 1034–1043. https://doi.org/10.1890/14-1605.1.

Storkey, J., Meyer, S., Still, K.S., et al. (2012). The impact of agricultural intensification and land-use change on the European arable flora. *Proceedings: Biological Sciences* 279 (1732): 1421–1429. https://doi.org/10.1098/rspb.2011.1686.

Storkey, J. and Neve, P. (2018). What good is weed diversity? *Weed Research* 58 (4): 239–243. https://doi.org/10.1111/wre.12310.

Teasdale, J.R. (1996). Contribution of cover crops to weed management in sustainable agricultural systems. *Journal of Production Agriculture* 9 (4): 475–479. https://doi.org/10.2134/jpa1996.0475.

Tilman, D., Cassman, K.G., Matson, P.A. et al. (2002). Agricultural sustainability and intensive production practices. *Nature* 418 (6898): 671–677. https://doi.org/10.1038/nature01014.

van Ittersum, M.K., Cassman, K.G., Grassini, P. et al. (2013). Yield gap analysis with local to global relevance—a review. *Field Crops Research* 143: 4–17. https://doi.org/10.1016/j.fcr.2012.09.009.

van Rijn, P.C.J., Wäckers, F.L., and Cadotte, M. (2016). Nectar accessibility determines fitness, flower choice and abundance of hoverflies that provide natural pest control. *Journal of Applied Ecology* 53 (3): 925–933. https://doi.org/10.1111/1365-2664.12605.

Vickerman, G.P. (1978). The arthropod fauna of undersown grass and cereal fields. *Scientific Proceedings of the Royal Dublin Society, Series A* 6: 273–283. https://www.gwct.org.uk/research/scientific-publications/1970-79/1978/vickerman1978.

Vigueira, C.C., Olsen, K.M., and Caicedo, A.L. (2013). The red queen in the corn: agricultural weeds as models of rapid adaptive evolution. *Heredity (Edinb)* 110 (4): 303–311. https://doi.org/10.1038/hdy.2012.104.

Wisler, G.C. and Norris, R.F. (2005). Interactions between weeds and cultivated plants as related to management of plant pathogens. *Weed Science* 53 (6): 914–917. https://doi.org/10.1614/WS-04-051R.1.

Wittwer, R.A., Dorn, B., Jossi, W., et al. (2017). Cover crops support ecological intensification of arable cropping systems. *Scientific Reports* 7: 41911. https://doi.org/10.1038/srep41911.

Zimdahl, R.L. (1980). *Weed–Crop Competition – A Review.* International Plant Protection Center.

8

Use of Arthropods for Ecologically Based Weed Management in Agriculture

Michael D. Day[1,*], *Arne B. R. Witt*[2], *and Rachel L. Winston*[3]

[1] *Department of Agriculture and Fisheries, Brisbane, Queensland, Australia*
[2] *CABI, Nairobi, Kenya*
[3] *MIA Consulting, Shelley, Idaho, USA*
[*] *Corresponding author*

8.0 Introduction

Invasive alien plants affect both agricultural and natural ecosystems, impacting on productivity and biodiversity. Within agricultural ecosystems, invasive alien plants can affect croplands, grazing lands, and plantations, and can impact on these systems directly and indirectly (US Congress 1993; Colautti et al. 2006). For example, aquatic and semiaquatic weeds can directly affect rice production, decrease suitable habitat in fisheries, and damage fishing gear. Aquatic weeds indirectly impact on agriculture by blocking irrigation canals, damaging infrastructure used to store water, and preventing access to water for livestock. In addition, aquatic weeds can indirectly affect human health by providing habitats for vectors of diseases such as malaria (Ding et al. 2001).

Invasive alien plants, referred to as weeds hereafter, unless stated otherwise, cost the agricultural sector billions of dollars annually, both in lost production and through control costs, and if left uncontrolled, can result in 100% yield loss. Annual economic losses by weeds to crops and pastures in five countries are estimated at US$35 billion (USA) (Pimentel et al. 2005), US$38.7 billion (India), US$17 billion (Brazil), US$1.5 billion (South Africa), and US$1.4 billion (United Kingdom) (Pimentel et al. 2001). All values here and elsewhere in this chapter are as per original source and have been not adjusted to account for currency fluctuations or inflation. Weeds cost the agricultural sector in Australia A$4 billion per annum (p.a.) (Sinden et al. 2004) and NZ$1.66 billion p.a. in New Zealand (Saunders et al. 2017). In Australia, the overall cost of weeds (non-native and native) to Australian grain growers has been estimated at A$3.3 billion annually (Llewellyn et al. 2016), while weeds cause A$70–80 million p.a. due to poisoning of stock alone (Culvenor 1985). Weeds have invaded over 10 million hectares (ha) in South Africa, impacting on cropping and grazing as well as water availability by using over 3.3 billion m^3 of water above that of native vegetation (Le Maitre et al. 1996). It is estimated that 71% of natural grazing land would be lost in South Africa if invasive alien plants were not managed (van Wilgen et al. 2008). The cost of all alien invasive species (plants and animals) to agriculture in Southeast Asia is estimated between US$23 billion and US$34 billion p.a. (Nghiem et al. 2013), while for China alone, the losses are estimated at US$14.5 billion p.a. (Xu et al. 2006). In Canada, alien plants and animals are estimated to cost the agricultural sector between US$13 billion and US$34.5 billion p.a. (Colautti et al. 2006). In general, weeds cause an average of 10% yield loss in developed countries but 25% in developing countries (Akobundu 1987). For example, in the USA, the average yield loss due to weeds is 12% (Pimentel et al. 2005), while in India, it is 30% or US$90 billion p.a. (Singh 1996).

Weeds (native and non-native) account for a total loss of approximately 200 million tons of grain globally (Chauhan 2020), accounting for a loss of approximately 56 million metric tons, valued at US$12 billion in rice production alone (Smith 1983). Globally, between 46% and 95% of cassava production is lost to weeds (Melifonwu 1994 and references therein). An analysis by Oerke (2006) of crop losses in wheat, rice, maize, potatoes, soybeans, and cotton, at a global level, found that weeds (native and non-native) accounted for 34% of losses, compared to animal pests (18%) and pathogens (16%). Weeds, both native and non-native, if unmanaged, could reduce yields in maize, rice, soybeans, cotton, potatoes, and wheat by 40, 37, 37, 36, 30, and 23%, respectively (Oerke 2006).

Managing weeds is costly and varies across regions. In the USA, the cost of weed control can vary from US$100 million p.a. to control aquatic weeds (US Congress 1993) to

Ecologically Based Weed Management: Concepts, Challenges, and Limitations, First Edition. Edited by Nicholas E. Korres, Ilias S. Travlos, and Thomas K. Gitsopoulos.

US$2 billion p.a. in rangelands (DiTomaso 2000), while in South Africa, approximately US$457 million was spent over 15 years to clear 1.6 million ha of weeds from natural ecosystems (van Wilgen et al. 2012). In North America, agriculture is very mechanized and relies heavily on herbicides, with US$3 billion spent annually on herbicides in cropping systems, and US$5 billion p.a. for control in pastures and rangelands (Pimentel et al. 2005). In Africa, however, which has some of the poorest countries on Earth, weeding is mainly conducted by hand and is, therefore, very time-consuming. In fact, about 100 million people in Africa spend approximately 20 billion hours weeding per year (Gianessi 2009). A recent analysis found that the average number of days spent annually on weeding 1 ha of cereal (excluding maize), legumes, maize, root crops, and vegetables in Africa was estimated at 50, 54.2, 54.6, 92, and 82.7 days, respectively (Eschen et al. 2021). Eschen et al. (2021) estimated that the combined annual weeding costs of non-native weeds in five crops in Africa amount to US$36.34 billion. In agro-forestry areas in India, weeds are the number one constraint in cultivation, with weeding accounting for 37% of all production costs (Muraleedharan and Anitha 2001).

Worldwide, some of the largest impacts to agriculture can be attributed to just a few weed species. For example, *Lantana camara* L. (Verbenaceae) has invaded over 60 countries globally (Day et al. 2003). In Australia, *L. camara* has invaded approximately 5 million ha (Bhagwat et al. 2012), costing A$121 million p.a. in lost production and control (AECgroup 2007). *Lantana camara* has invaded over 13 million ha in India, costing the economy US$924 million p.a. (Pimentel et al. 2001), and 2 million ha in South Africa (Bhagwat et al. 2012). About 40% of respondents at one locality in Uganda reported that *L. camara* reduced forage for stock by 50%, while 33% reported a reduction in crop yields of up to 50%. Overall, *L. camara* costs individual households about US$400–$500 p.a. (Shackleton et al. 2017). Considering that *L. camara* is widespread throughout Africa, and the majority of people on the continent are directly involved in agriculture, the cost of *L. camara* to African households is likely to be millions of dollars annually.

Mikania micrantha Kunth (Asteraceae) (mile-a-minute) has invaded about 40 countries (CABI 2021), impacting on annual crops, plantations, orchards, and pastures (Day et al. 2016; Han et al. 2017). Economic losses due to *M. micrantha* in the Pearl River Delta of China are estimated at up to an equivalent of US$1.3 billion p.a. (Peng et al. 2013 in Han et al. 2017), while economic losses of up to an equivalent of US$1.2 million are estimated for Neilingding Island, China (Zhong et al. 2004). In India, losses of 42% in tea production have been reported (Puzari et al. 2010), and the cost of maintaining teak plantations free of *M. micrantha* for the first two years after planting are estimated at an equivalent of US$96/ha (Muraleedharan and Anitha 2000). In Nepal, the average number of cattle held by each household halved following invasion by *M. micrantha*, and forest product collection trips made by locals decreased by 75% due to *M. micrantha* invasion (Rai and Scarborough 2014). In the Pacific region, several crops and trees, including taro, oil palms, coconuts, and cocoa, have been seriously affected by *M. micrantha* (Day et al. 2012a, 2016). In Papua New Guinea, 40% of respondents to a nationwide survey believed *M. micrantha* caused >30% of crop losses (Day et al. 2012b) and 35% of respondents indicated they must weed weekly to keep *M. micrantha* from damaging crops (Day et al. 2012a). In Fiji, root crop farmers spend up to A$31/ha to control *M. micrantha* (Macanawai et al. 2012).

Parthenium hysterophorus L. (Asteraceae) has invaded over 40 countries worldwide (CABI 2021). Annual losses in Australia (both reduced forage production and control costs) are approximately A$16.5 million (Chippendale and Panetta 1994). *Parthenium hysterophorus* reduced sorghum yields by 97% in field trials in Ethiopia (Tamado et al. 2002) and pasture carrying capacity by up to 90% in India (Jayachandra 1971). In Pakistan, crops most affected by parthenium are potato, sugarcane, and maize (Bajwa et al. 2019). Costs of managing parthenium in these and other crops amounted to approximately US$205 per household per year (Bajwa et al. 2019). Each farmer surveyed in Pakistan lost approximately US$935 annually due to the impacts of weed infestations on fodder collection sites while a significant proportion of farmers also reported negative effects of the weed on animal (22–36%) and human health (14–24%) (Bajwa et al. 2019). Farmers who have developed serious allergies as a result of *P. hysterophorus* invasions, have been known to abandon their land (McFadyen 1992).

Pontederia crassipes Mart. (Pontederiaceae) (water hyacinth) is a major aquatic weed invading many countries in Africa, Asia, and Oceania as well as the USA (CABI 2021). As with other water weeds, *P. crassipes* impacts on livelihoods by reducing access to water for people and livestock, depleting fish populations, impeding water transport, blocking irrigation canals, hampering hydroelectricity generation, and providing breeding sites for vectors of human and animal diseases (Coetzee et al. 2009). In southern Benin, a water hyacinth invasion was found to have reduced the annual income of 200,000 people by about US$84 million (de Groote et al. 2003). China spends $12 million p.a. on *P. crassipes* control alone (Ding et al. 2001), and the control costs annually have been as high as US$35 million in the southern USA (CABI 2021). On the Panama Canal in Central America, on the Nile in Sudan, and in the Democratic Republic of Congo, costs of control or removal of *P. crassipes* have amounted to millions of dollars p.a. (CABI 2021).

The high costs of conventional interventions, such as chemical and physical control are not sustainable or feasible for many countries (Culliney 2005). In many rangelands overrun with weeds, the benefits of weed control are significant. However, the costs of control are often higher than the actual value of the land, negating any potential financial benefits to landowners (Chuk 2010). The negative impacts of some conventional control methods, especially chemical control, also decrease their application in some settings. Physical control disturbs the soil and may cause erosion; some herbicides pose risks to crops, native plant species, wildlife and human health, and incorrect use has led to resistance in numerous weed species (Culliney 2005). In fact, more than 500 unique cases of herbicide-resistant weeds have been reported globally, with the highest number of resistant weeds reported in wheat, maize, rice, soybean, spring barley, canola, and cotton (Heap 2019). Opportunities for the development of herbicides with different modes of action are limited, with no new developments in this area in the past three decades (Heap 2019).

Due to the drawbacks associated with conventional weed control methods, classical biological control, using co-evolved host specific natural enemies (mostly insects, mites, and pathogens) of the target weeds is increasingly being implemented as a long-term, environmentally friendly, and sustainable method of controlling many of the world's worst weeds. The first attempt at biological control of weeds was in the 1830s. Since then, over 500 biological control agents have been deliberately released against over 200 weed species worldwide (Winston et al. 2021). The main countries undertaking weed biological control research are Australia, Canada, New Zealand, South Africa, and the USA, with CABI also undertaking biological control research from many of their research centres, especially in the United Kingdom and Switzerland. Many countries have benefited from this research, with over 90 countries worldwide having deliberately introduced at least one biological control agent (Schwarzländer et al. 2018).

Biological control of weeds can benefit both agricultural and natural ecosystems alike, and there have been some spectacular successes where weeds have been completely controlled, such that no other control measures need to be undertaken (McFadyen 1998; Winston et al. 2021). Biological control of weeds has resulted in approximately 66% of all weed species targeted being under at least partial control – that is, have biological control agents causing medium to high impacts on the target species in at least some areas. The vast majority of weed species (>85%) targeted for biological control are of agricultural and other concern; only 28 species targeted to date are considered to be primarily environmental weeds (Schwarzländer et al. 2018; Winston et al. 2021).

This chapter discusses the role and benefits of weed biological control using co-evolved host-specific arthropods in agricultural systems. Weed biological control using pathogens is discussed in Chapter 9. In particular, this chapter focuses on three broad agricultural systems: cropping, grazing lands, and plantations and agroforestry. This chapter also discusses biological control of weeds of aquatic systems, as aquatic weeds can have direct and indirect impacts on agricultural production. For each system, a case study is presented to highlight successful biological control using arthropods.

8.1 Weed Biological Control in Agriculture

Weed biological control has been successfully deployed in all three broad agricultural systems viz cropping, grazing lands, and plantations and agroforestry. Biological control has also been successfully deployed against numerous aquatic weeds (Winston et al. 2021). In general, biological control is more suitable in perennial systems such as grazing lands and plantations as there is usually less disturbance than in intensive cropping systems, especially annual cropping systems, where there is regular disturbance through harvesting and ploughing. However, the authors are of the opinion that conservation agriculture, or zero-till cropping systems, could enhance the efficacy of weed biological control agents, by reducing disturbance in these systems.

Most (132) of the 200+ weeds targeted for biological control are found in grazing lands, which includes managed pastures as well as natural rangelands. There are 36 weed species targeted for biological control that occur in cropping systems and 35 weed species are found in plantations and agroforestry (Figure 8.1). Many of these weeds may impact on more than one agricultural system, e.g. *Chromolaena odorata* (L.) R.M. King and H. Rob. (Asteraceae) and *L. camara* can impact on grazing lands, as well as plantations and agroforestry (Day et al. 2003; Zachariades et al. 2009) and are recorded as such.

Within the agricultural systems in which weeds have been targeted for biological control, aquatic weeds have had the highest success, with over 94% of weed species under at least partial control, meaning their biological control agents provide medium to high impacts. Just over 68% of weed species in plantations and in grazing systems that have been targeted for biological control have been deemed under at least partial control. Cropping has had the lowest percentage of weed species successfully controlled using biological control agents (56%) (Figure 8.1).

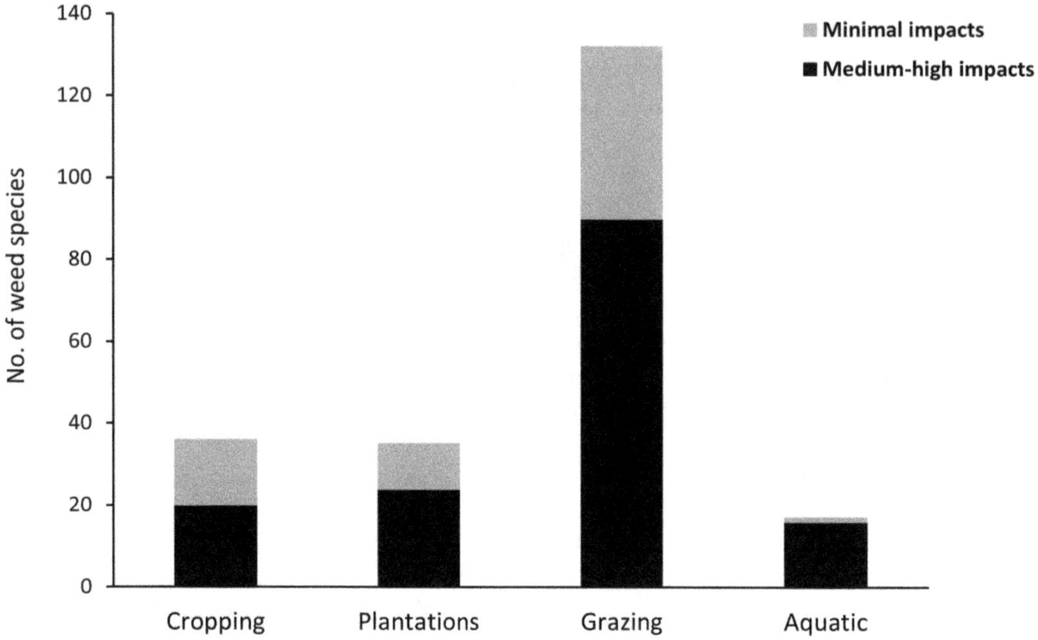

Figure 8.1 The total number of weed species targeted for biological control in the different agricultural systems and the number of weeds deemed to have biological control agents causing medium to high levels of impact (Winston et al. 2021).

8.2 Biological Control in Cropping Systems

Cropping systems here are defined as annual crops usually grown in managed areas. They include large-scale crop production areas as found in Australia, Canada, the USA, and elsewhere, and small-scale subsistence farms as found in many low- and middle-income countries. In large-scale cropping systems, some of the most common and difficult to control weeds include *Amaranthus* spp. and *Bassia scoparia* (L.) A.J. Scott (both Amaranthaceae), *Bromus* spp. (Poaceae), *Brassica* spp. (Brassicaceae), and *Cirsium arvense* (L.) Scop. and *Taraxacum officinale* (L.) Weber ex F.H. Wigg (both Asteraceae) (Van Wychen 2019). In small-scale subsistence farming typical of the Pacific region, crops such as taro, yam, sweet potato, and green leafy vegetables are often overrun with weeds such as *C. odorata* and *M. micrantha* (Day et al. 2012a, 2013a). In Africa, small-scale farmers growing the staple foods maize and cassava are heavily impacted by, among others, *C. odorata*, *P. hysterophorus*, *Solanum elaeagnifolium* Cav. (Solanaceae), *Tithonia* spp. (Asteraceae), and *Xanthium strumarium* L. (Asteraceae).

Biological control is generally not widely used in many cropping systems, with only 16% of weeds targeted for biological control considered problematic in cropping systems. In large-scale farming operations, this is because the farming practices are often highly mechanized and the land is highly disturbed as a result of regular ploughing and other interventions. Frequent or extensive disturbance, including the removal of host plants, interferes with the life cycles of many biological control agents and can prevent the buildup of large populations, thus inhibiting the control of target weeds (Coombs et al. 2004).

Despite these constraints, there have been some outstanding successes of biological control in cropping systems (Table 8.1), particularly in subsistence farming systems. In these systems, there is often limited soil disturbance, with crops generally being tended manually, and weeds being either hand-pulled or slashed. The lack of disturbance and the fact that weeds are rarely completely removed between harvests ensures there are always weeds present to sustain populations of the biological control agents. In addition, subsistence cropping areas are limited in size and are often surrounded by extensive tracts of unmanaged land in which biological control agents and their target species can persist, with extensive spillover of agents onto target weeds growing within the croplands themselves.

Biological control in subsistence farming is further aided by the fact that many farmers cannot afford herbicides, or have no access to herbicides, resulting in no potential negative impacts of chemical use on any agents present. Even on large-scale commercial farms, the presence of target weeds and their associated agents in areas adjacent to croplands means that agents can move into croplands to feed on any emerging weeds. We postulate that reductions in weed propagules in surrounding lands, due to biological control, directly result in fewer propagules invading fields, contributing to lower physical and chemical weed control costs.

Table 8.1 A sample of weed species occurring in cropping systems that have been controlled or partially controlled by arthropods. The general impact caused by each agent and the number of countries in which each occurs is also indicated.

Weed Family	Weed Species	Agent Family	Agent Species	General Impact	Countries
Asteraceae	*Centaurea solstitialis* L.	Curculionidae	*Eustenopus villosus* (Boheman)	Medium	1
Asteraceae	*Tripleurospermum inodorum* (L.) Sch.Bip.	Brentidae	*Omphalapion hookerorum* (Kirby)	Medium	1
Boraginaceae	*Cordia curassavica* (Jacq.) Roem. & Schult.	Chrysomelidae	*Metrogaleruca obscura* (DeGeer)	High	3
Convolvulaceae	*Convolvulus arvensis* L.	Eriophyidae	*Aceria malherbae* Nuzzaci	Variable	1
Plantaginaceae	*Linaria vulgaris* Mill.	Curculionidae	*Rhinusa linariae* (Panzer)	High	1
Polygonaceae	*Rumex crispus* L. *R. obtusifolius* L.	Sesiidae	*Pyropteron doryliformis* (Ochsenheimer)	High	1
Rosaceae	*Rubus argutus* Link	Tortricidae	*Croesia zimmermani* Clarke	Variable	1
Solanaceae	*Solanum elaeagnifolium* Cav.	Chrysomelidae	*Leptinotarsa texana* Schaeffer	High	1
Solanaceae	*Solanum sisymbriifolium* Lam.	Chrysomelidae	*Gratiana spadicea* (Klug)	Medium	1
Zygophyllaceae	*Tribulus cistoides* L. *T. terrestris* L.	Curculionidae	*Microlarinus lareynii* (Jacquelin du Val) *M. lypriformis* (Wollaston)	Variable–High	3

Examples of successful biological control in cropping systems include black sage, *Cordia curassavica* (Jacq.) Roem. & Schult. (Boraginaceae), which has been well controlled by the wasp *Eurytoma attiva* Burks (Hymenoptera: Eurytomidae) and the leaf-feeding beetle *Metrogaleruca obscura* (DeGeer) (Coleoptera: Chrysomelidae) in Mauritius, Malaysia, and Sri Lanka (Greathead 1971; Simmonds 1981; Anwar 1996; Fowler et al. 2000; Winston et al. 2021). Together these biological control agents have reduced weed populations to such an extent that *C. curassavica* is no longer considered a problem in any of these countries (Winston et al. 2021). Two puncture vines, *Tribulus cistoides* L. and *T. terrestris* L. (Zygophyllaceae), are considered weeds in North and South America, the Caribbean, and the Pacific. Two beetles, *Microlarinus lareynii* (Jacquelin du Val) and *M. lypriformis* (Wollaston) (Coleoptera: Curculionidae), have been released in most regions with great success. Control appears to be better in the Pacific, particularly on Hawai'i, where *M. lypriformis* is the most important agent. Control also appears successful in the Federation of St Kitts and Nevis in the Caribbean and some mainland states of the USA (Winston et al. 2021).

8.2.1 Case Study: Biological Control of *Solanum elaeagnifolium* Cav. (Solanaceae)

Silverleaf nightshade is native to the Americas, with its centre of origin in the southwestern USA or northern Mexico and possibly Argentina (Boyd et al. 1984). In Texas and Oklahoma, USA, more than 814,000 ha of cropland have been invaded (Abernathy and Keeling 1979). In Texas, it is ranked as being among the top 10 weed species, causing about 75% of all losses in cotton (Brandon 2005). In some counties in California, USA, invasions can reduce peanut yields by up to 65% (Hackett and Murray 1982). In Morocco, *S. elaeagnifolium* invasions have resulted in crop losses of 64% in maize and 78% in cotton, while in Australia crop losses of 12–50% in wheat, 4–10% in sorghum, and 5–14% in cotton have been recorded (EPPO 2021). Silverleaf nightshade invasions cause losses to Australian agriculture of approximately A$80 million each year (Heap and Wu 2018). It has also been recorded as naturalized or invasive in many African, Asian, and European countries (Mekki 2007; EPPO 2021).

Solanum elaeagnifolium was probably inadvertently introduced to South Africa as a contaminant of pig fodder around 1905 (Wassermann et al. 1988) and/or hay during the 1940s or 1950s. It was first recorded as a problem in 1952 and declared a weed in 1966. An eradication campaign was initiated in 1968 but was unsuccessful, and the weed continued to spread (Wassermann et al. 1988). Subsequent research indicated that the weed is very difficult to control with herbicides (Wassermann et al. 1988; Heap et al. 1997); at the time, chemical costs in South Africa were estimated to be more than US$286/ha (Marais et al. 2004). As a result, a biological control program was initiated, resulting in the introduction of two leaf-feeding beetles, *Leptinotarsa defecta* (Stål) and *L. texana* Schaeffer (Coleoptera: Chrysomelidae) in 1992 (Hoffmann et al. 1998). *Leptinotarsa texana* established and spread rapidly,

Figure 8.2 (a) *Leptinotarsa texana* (Texas potato beetle) adult. *Solanum elaeagnifolium* in Free State Province, South Africa, before (b) and after (c) the release of *L. texana*. (Credits: beetle – John Brush, iNaturalist CC BY-NC 4.0; before and after – Helmuth Zimmermann).

while *L. defecta* remains localized (Hoffmann et al. 1998). Both agents attack the leaves, flowers, young fruits, and stems of *S. elaeagnifolium* during their larval and adult stages, causing considerable defoliation (Figure 8.2).

The utilization of *L. texana* for the control of *S. elaeagnifolium* by some farmers and the continued use of herbicides (e.g., Roundup®) by others, provided the ideal situation to compare the effectiveness of these two management interventions. Comparative studies between two farms in South Africa found that densities of the weed were higher in the chemical treated plots compared to the biological control plots (Pitso 2010). However, plant height was significantly higher in the biological control plot while stem diameter and levels of fruiting were not significantly different. Pitso (2010) concluded that biological and chemical control were having a substantial impact on the weed and that the two situations did not exhibit many differences except for the negative impact of herbicides on the agent. Since both treatments were beneficial compared to a no-control scenario, a comparative economic analysis was conducted on the benefits derived from both control methods and different effectiveness levels. Even at a low level of effectiveness, biological control provided higher financial returns than chemical control due to the lower costs involved (Pitso 2010).

8.3 Biological Control in Grazing Lands

Approximately 60% of all weeds targeted for biological control are also found in grazing lands (Figure 8.1). Here, we define grazing lands as any land on which livestock graze. These include pastures and open rangelands as well as vacant blocks and roadsides and other lands where livestock graze.

Biological control is the most feasible and sustainable solution to manage many weeds in these systems, as invasions can be large and beyond the financial reach of most livestock owners (Sheley et al. 2011). Some of the more common weeds of rangelands include various species of cacti namely *Opuntia* spp., *Cylindropuntia* spp. and *Harrisia* spp. (Cactaceae), especially in Australia and South Africa, *L. camara* in Africa, Asia, Australia and Oceania, *P. hysterophorus* in Africa, Asia, and Australia, and *Jatropha gossypiifolia* L. (Euphorbiaceae), particularly in Australia and parts of Asia (CABI 2021) (Table 8.2). In pastures, *Carduus* spp. (Asteraceae), *Hypericum perforatum* L. (Hypericaceae), *Onopordum* spp. (Asteraceae), and *Jacobaea vulgaris* Gaertn. (Asteraceae) are some of the most widespread weeds causing the most impacts (DiTomaso et al. 2013).

There have been some notable successes in the biological control of weeds in grazing systems, including *Opuntia* spp. by *Cactoblastis cactorum* (Berg) (Lepidoptera: Pyralidae) and *Dactylopius* spp. (Cockerell) (Hemiptera: Dactylopiidae) in numerous countries, including Australia and South Africa (Paterson et al. 2011; Hoffmann et al. 2020; Winston et al. 2021) and *Cylindropuntia fulgida* (Engelm.) F.M. Knuth var. *fulgida* (Cactaceae) by *Dactylopius tomentosus* (Lamark) in South Africa (Paterson et al. 2011) and more recently Australia (Jones et al. 2023).

In Australia, Vanuatu, and a few other countries in the Pacific, *Sida acuta* Burm. f. and *S. rhombifolia* L. (Malvaceae) were successfully controlled by the leaf-feeding beetle *Calligrapha pantherina* Stål (Coleoptera: Chrysomelidae) (Heard and Day 2012; Day and Winston 2016). Following the success in Australia, *C. pantherina* was subsequently released in Vanuatu in 2005 and established on 11 islands. By 2015, it had spread to 21 islands where large stands were reduced to a few plants/m^2 within 12 months, resulting in a 50% increase in productivity (Day and Bule 2016).

Table 8.2 A sample of weed species occurring in grazing systems that have been controlled or partially controlled by arthropods. The general impact caused by each agent and the number of countries in which each occurs is also indicated.

Weed Family	Weed Species	Agent Family	Agent Species	General Impact	Countries
Asteraceae	*Cirsium vulgare* (Savi) Ten.	Tephritidae	*Urophora stylata* (Fabricius)	Variable–Medium	4
Asteraceae	*Jacobaea vulgaris* Gaertn.	Chrysomelidae	*Longitarsus jacobaeae* (Waterhouse)	Variable–High	4
Asteraceae	*Parthenium hysterophorus* L.	Chrysomelidae	*Zygogramma bicolorata* Pallister	Variable	3
Cactaceae	*Cylindropuntia fulgida* var. *mamillata* (A. Schott ex Engelm.) Backeb. *C. imbricata* (Haw.) F.M. Knuth	Dactylopiidae	*Dactylopius tomentosus* (Lamark)	High	3
Cactaceae	*Opuntia* spp. (several)	Dactylopiidae	*Dactylopius opuntiae* (Cockerell)	Variable–High	14
Euphorbiaceae	*Euphorbia esula* L.	Chrysomelidae	*Aphthona lacertosa* Rosenhauer	High	2
Fabaceae	*Mimosa diplotricha* C.Wright ex Sauvalle	Psyllidae	*Heteropsylla spinulosa* Muddiman, Hodkinson & Hollis	Variable–High	11
Hypericaceae	*Hypericum perforatum* L.	Chrysomelidae	*Chrysolina quadrigemina* (Suffrian) *C. hyperici* (Forster)	Variable–High	7
Malvaceae	*Sida acuta* Burm.f.; *S. rhombifolia* L.	Chrysomelidae	*Calligrapha pantherina* Stål	High	4
Plantaginaceae	*Linaria dalmatica* (L.) Mill.	Curculionidae	*Mecinus janthiniformis* Toševski & Caldara	High	2

In managed pastoral systems, *J. vulgaris* is a weed common in Australia, New Zealand, and North America. Ten biological control agents have been deliberately released across these regions, the most effective of which has been *Longitarsus jacobaeae* (Waterhouse) (Coleoptera: Chrysomelidae) (Winston et al. 2021). It has established on *J. vulgaris* in all four countries in which it has been released, namely Australia, Canada, New Zealand, and mainland USA. In the USA, populations of *J. vulgaris* have decreased by up to 90%, while in Australia, where numerous agents have been released, the effects of all agents combined have resulted in substantial reductions of the weed in most areas (Ireson and McLaren 2012; Winston et al. 2021).

8.3.1 Case Study: *Hypericum perforatum* L. (Hypericaceae)

Hypericum perforatum L. is a perennial forb native to Europe, northern Africa, and Asia. It has a long history of use in herbal remedies for mild to moderate depression, chronic fatigue syndrome, healing wounds, suppressing coughs and as an antiviral agent (Cech 1997; Upton 1997). As such, it was intentionally introduced into numerous countries on multiple occasions by European settlers interested in the plant's medicinal properties (Mitich 1994). In many countries where it was introduced, the plant escaped cultivation and became weedy. It is especially problematic in the USA, Canada, Australia, New Zealand, Chile, and South Africa (Briese and Cullen 2012).

Hypericum perforatum is a vigorous competitor in pastures, rangelands, and natural areas, displacing desirable forage as well as native species. It is also toxic to livestock, producing the phototoxin hypericin that causes photosensitization and liver damage (Bourke 1997). Affected animals often become emaciated from starvation and dehydration, and photosensitive sheep produce poor-quality wool (Kingsbury 1964; Mitich 1994). At its peak abundance in California, USA, *H. perforatum* was considered the leading cause of economic loss, attributed both to the loss of pastures and rangelands and to stock fatalities (Turner and Szczawinski 1991; Mitich 1994). In New South Wales, Australia, losses in animal production were estimated at A$22.5 million p.a. (Bourke 1997), compounded by further losses through competition with desirable pasture.

Biological control of *H. perforatum* began in 1930, with the introduction of three beetles into Australia: *Chrysolina brunsvicensis* (Gravenhorst), *C. hyperici* (Forster), and *C. varians* (Schaller) (Coleoptera: Chrysomelidae) (Briese and Cullen 2012). Since then, 11 species have been intentionally released into six countries (Winston et al. 2021). Of these 11 species, the most successful agent is *Chrysolina quadrigemina* (Suffrian). The larvae and adults of this

Figure 8.3 *Hypericum perforatum* in California USA: (a) before the release of *Chrysolina* spp. and (b) after the release of *Chrysolina* spp. (c) Monument in honor of *Chrysolina* spp. (Credits: USDA ARS European Biological Control Laboratory, www.bugwood.org, CC BY 3.0 US).

beetle feed on *H. perforatum* foliage, and repeated defoliations kill plants (Piper 2004). It is established and having medium to very high impacts in all six countries in which it was intentionally released, and it is also present in Argentina, where it spread naturally following its introduction to neighboring Chile. *Chrysolina quadrigemina* often occurs in mixed populations with another leaf-feeding beetle, *C. hyperici*, the second-most effective agent (Winston et al. 2021). *Chrysolina hyperici* established in five of the six countries in which it was released and is having medium to high impacts in all five.

Both beetles established readily in New Zealand and Chile, and the *H. perforatum* biological control programs in both countries are considered a success (Villaneuva and Faure 1959; Winston et al. 2021 and references therein). In South Africa and Hawaii, USA, *C. quadrigemina* provides widespread and excellent control of *H. perforatum*, in conjunction with the medium to high impacts of the gall midge *Zeuxidiplosis giardi* (Kieffer) (Diptera: Cecidomyiidae) (Gordon and Kluge 1991; Conant et al. 2013).

In the continental USA, Canada, and Australia, both *C. quadrigemina* and *C. hyperici* in combination have greatly contributed to the control of *H. perforatum* in open sunny areas and can cause extensive defoliation over wide areas (Jensen et al. 2002; Piper 2004; Briese and Cullen 2012). In California, USA, these beetles were so effective at reducing *H. perforatum* infestations that this program became one of the most famous examples of a classical biological control success story (Figure 8.3a, b). Grateful landowners even erected a monument in honor of the beetles that saved their rangelands (Figure 8.3c), and the weed is still considered under complete control in California, 75 years later (Pitcairn 2018).

In all three countries (the USA, Canada, and Australia) however, the beetles are not well adapted to shaded habitats or areas with high summer rainfall, where *H. perforatum* plants frequently recover from defoliation (Jensen et al. 2002; Piper 2004; Briese and Cullen 2012). While the largest invasions in the continental USA and Canada were controlled within 10 years of the beetles' introduction, invasions at cooler, more moist sites have persisted, although their overall extent is significantly less than historical levels (Winston et al. 2021 and references therein). In Australia, despite the established agents providing only partial control, the annual benefit to agriculture has been reported at A$20.6 million (Page and Lacey 2006).

8.4 Biological Control in Plantations and Agroforestry Systems

Just over 16% of weeds targeted for biological control are also problematic in plantations and agro-forestry systems (Figure 8.1). These systems include perennial crops such as, among others, citrus, papaya, bananas, tea, coffee, cocoa, coconuts, and oil palm, as well as species grown for timber such as teak, eucalypts, pines, and acacias. Often these species are grown as commercial enterprises; however, subsistence farmers in Oceania, Asia, and Africa may also maintain small holdings for local markets.

In these systems, weeds may be most damaging during the establishment phase of the plantation crop (Muraleedharan and Anitha 2000). However, fast-growing vines such as *M. micrantha*, may smother mature crops such as bananas, papaya, and cocoa, reducing flowering and subsequent yields (Day et al. 2016). Other vines that cause significant impacts in plantations are *Coccinia grandis* (L.) Voigt (Cucurbitaceae) (Muniappan et al. 2009), *Cardiospermum* spp. (Sapindaceae), *Convolvulus arvensis* L.

(Convolvulaceae) and *Dioscorea bulbifera* L. (Dioscoreaceae) (Rayamajhi et al. 2016; CABI 2021).

Two of the most widespread and important weeds of plantations are *C. odorata* (see case study 8.4.1) and *L. camara* (Day et al. 2003; Zachariades et al. 2009). *Lantana camara* impacts on both agricultural and natural ecosystems. In plantations, it can reduce growth of young plantation species, as well as increase the risk of fire as it can take flames into the canopy (see Day et al. 2003 and references therein). Other important weeds of plantations are *Clidemia hirta* (L.) D. Don (Melastomataceae), which can be a particular problem under coffee or in plantations in cooler, higher-elevation areas in Asia, and *Baccharis halimifolia* L. (Asteraceae) in pine plantations in Queensland, Australia (Palmer and Sims-Chilton 2012) (Table 8.3).

Biological control of weeds in plantations and agroforestry areas has been particularly successful, with approximately 69% of weeds under at least partial control. *Coccinia grandis* was successfully controlled by two biological control agents, *Acythopeus cocciniae* O'Brien & Pakaluk (Coleoptera: Curculionidae) and *Melittia oedipus* Oberthür (Lepidoptera: Sesiidae), in Guam, the Commonwealth of the Northern Mariana Islands, and Hawai'i. An additional benefit of the biological control program in Guam was that the melon fly, *Zeugodacus cucurbitae* (Coquillett) (Diptera: Tephritidae), which is a pest of fruits and uses *C. grandis* as a host, also decreased following the control of *C. grandis* (Muniappan et al. 2009).

Biological control of *L. camara* commenced in 1902 and since then, 44 biological control insects, mites, and pathogens have been deliberately released in 33 countries. Of these, 28 agents have established and are providing variable levels of control, depending on country and the varieties of *L. camara* present (Day et al. 2003; Zalucki et al. 2007; Winston et al. 2021). Overall, damage appears to be seasonal and mainly caused by *Teleonemia scrupulosa* Stål (Hemiptera: Tingidae) and *Uroplata girardi* Pic (Coleoptera: Chrysomelidae), with extensive dieback and defoliation occurring in some countries and regions. More recently, area-wide damage by *Ophiomyia camarae* Spencer (Diptera: Agromyzidae) has been reported in north Queensland, Australia. In addition, the budmite *Aceria lantanae* (Cook) (Acari: Eriophyidae) has destroyed up to 95% of flowers on some varieties of *L. camara* in Zambia and Malawi (Winston et al. 2021). Studies on how these biological agents have resulted in increased yield and production in agricultural systems are still to be conducted. Day et al. (2003) provides a good account of the factors affecting successful biological control of *L. camara*.

8.4.1 Case Study: Biological Control of *Chromolaena odorata* R.M. King & H. Robinson (Asteraceae)

Chromolaena odorata is a perennial shrub native to tropical America, from the southern USA to northern Argentina.

Table 8.3 A sample of weed species occurring in plantations and agroforestry systems that have been controlled or partially controlled by arthropods. The general impact caused by each agent and the number of countries in which each occurs is also indicated.

Weed Family	Weed Species	Agent Family	Agent Species	General Impact	Countries
Asteraceae	*Baccharis halimifolia* L.	Cerambycidae	*Megacyllene mellyi* (Chevrolat)	High	1
Asteraceae	*Chromolaena odorata* (L.) R.M. King & H. Rob.	Tephritidae	*Cecidochares connexa* (Macquart)	Variable–High	8
Asteraceae	*Mikania micrantha* Kunth	Nymphalidae	*Actinote anteas* (Doubleday) *A. thalia pyrrha* (Fabricius)	High	1
Bignoniaceae	*Dolichandra unguis-cati* (L.) L. G. Lohmann	Tingidae	*Carvalhotingis visenda* Drake & Hambleton	Variable–Medium	2
Cucurbitaceae	*Coccinia grandis* (L.) Voigt	Curculionidae	*Acythopeus cocciniae* O'Brien & Pakaluk	Variable–High	3
Dioscoreaceae	*Dioscorea bulbifera* L.	Chrysomelidae	*Lilioceris cheni* Gressitt & Kimoto	High	1
Melastomataceae	*Clidemia hirta* (L.) D. Don	Phlaeothripidae	*Liothrips urichi* Karny	Variable–High	4
Rosaceae	*Rubus alceifolius* Poir.	Argidae	*Cibdela janthina* (Klug)	High	1
Scrophulariaceae	*Buddleja davidii* Franch.	Curculionidae	*Cleopus japonicus* Wingelmüller	High	1
Verbenaceae	*Lantana camara* L.	Chrysomelidae	*Uroplata girardi* Pic	Variable–High	14

However, it is now a major weed in over 60 countries in Africa, Asia, and Oceania, where it impacts on agriculture and natural ecosystems (Zachariades et al. 2009). In agricultural systems, it can invade grazing lands, plantations, and subsistence farms, reducing productivity and yield (Day et al. 2013a). In plantations in particular, *C. odorata* can quickly form a dense understory, thus interfering with the establishment and growth of young plants and impeding the ability to harvest fallen coconuts and oil palm nuts (Orapa et al. 2002).

There have been several dedicated attempts at biological control of *C. odorata* since the 1960s. This has culminated in 13 biological control agents being deliberately released in 19 countries across the three regions. Two agents, the stem-galling fly *Cecidochares connexa* (Macquart) (Diptera: Tephritidae) and the leaf-feeding moth *Pareuchaetes pseudoinsulata* Rego Barros (Lepidoptera: Erebidae), are the most widespread and damaging of the agents that have established (Winston et al. 2021).

Pareuchaetes pseudoinsulata was the first agent to have established on *C. odorata* in many countries (Zachariades et al. 2009). The moth has now been deliberately released into 15 countries but has established in only 10. However, through natural dispersal, it is now reported in 14 countries across the three regions (Winston et al. 2021). Larvae feed on the leaves of *C. odorata* and can cause seasonal defoliation of plants over wide areas (McFadyen 1999; Day et al. 2013a). In several countries, *P. pseudoinsulata* was able to reduce *C. odorata* cover to about 30% cover (Zachariades et al. 2009; Aigbedion-Atalor et al. 2019).

Cecidochares connexa has been deliberately released into 13 countries, and through natural dispersal, it is now reported in 19 countries (Winston et al. 2021). Galling on the stems results in a reduction of flowering and seed set, and if gall numbers are high enough, the stems, branches, and plants can also die, significantly reducing plant density (McFadyen 1999; Cruz et al. 2006; Day et al. 2013a, 2013b) (Figure 8.4).

Together, these two biological control agents have contributed to the control or near control of *C. odorata* in 12 countries across all three regions (Winston et al. 2021). In most countries, formal assessments on the benefits of the biological control agents to agriculture and the environment have not been conducted. However, reports in Indonesia have found that biological control of *C. odorata* reduced the use of herbicides as well as the number of times plants need to be slashed each year, resulting in a reduction in the cost of weeding by 75% (Desmier de Chenon et al. 2002). In Ghana, control of *C. odorata* by *P. pseudoinsulata* resulted in the reduction of time spent weeding by local farmers (Timbilla 1996). However, in Papua New Guinea, *P. pseudoinsulata* is very localized. Despite it being seasonally abundant and causing total defoliation of *C. odorata*, it is found in only a few areas (Day and Bofeng 2007; Day et al. 2013a).

In one of only two detailed studies on the benefits of the biological control of *C. odorata*, Braimah et al. (2013) reported that over 72% of farmers in Ghana thought there was an overall reduction of *C. odorata* over the last 10 years; the remaining respondents stated they were not sure. None of the farmers thought *C. odorata* had increased during this time. Over 83% of farmers interviewed thought the reduction of *C. odorata* due to biological control was beneficial, with many stating that time spent weeding had been reduced (Braimah et al. 2013).

In Papua New Guinea, almost 300 farmers from 10 provinces were interviewed regarding their perceptions of the biological control program in relation to control of *C. odorata* and productivity. Over 86% of respondents believe there is now less *C. odorata* than before the biological control agents were introduced; 40% reported that less than 25% of the *C. odorata* remains (Day et al. 2013b). Approximately 50% of respondents no longer need to control the weed, while most of the remaining respondents reported that

(a) (b) (c)

Figure 8.4 (a) *Cecidochares connexa* adult. (b) *Chromolaena odorata* in New Ireland, Papua New Guinea before the release of the gall fly in 2003 and (c) in 2011 after the release of the gall fly. (Credits: Queensland Department of Agriculture and Fisheries).

control times and costs have been significantly reduced. Over 60% of respondents believe there have been moderate to major benefits since the introduction of the biological control agents, with over 78% of respondents reporting an increase in yield and income (Day et al. 2013b).

In addition to these direct benefits to agriculture, there were numerous indirect benefits as a result of the biological control programme in Papua New Guinea. A reduction in *C. odorata* enabled farmers to increase the size of their farmed plots because there was less time spent on weeding and more land became available due to less *C. odorata*, thus increasing productivity and income. There was more time available to do other activities such as maintain houses, fences, and fishing nets. There were also considerable health benefits, such as reduced cuts to skin, given that the need to slash *C. odorata* was reduced. The reduction in *C. odorata* infestations also resulted in fewer areas to harbor snakes and wild pigs, and less *C. odorata* along roadsides, resulting in greater visibility and less need to walk on roads (Day et al. 2013b).

8.5 Biological Control in Aquatic Systems

Overall, we classified 17 weed biological control targets as aquatic species (Table 8.4). The most common of the floating aquatic species, include *Pistia stratiotes* L. (Araceae), *P. crassipes* and *Salvinia molesta* D.S. Mitch. (Salviniaceae). These each have a global distribution, being found throughout Africa, Asia, North America, and Oceania (CABI 2021). Much has been written on these three species and their impacts to agriculture, biodiversity, human health, and livelihoods (see reviews by Coetzee et al. 2009; Julien et al. 2009; Neuenschwander et al. 2009). Aquatic weeds such as *Alternanthera philoxeroides* (Mart.) Griseb. (Amaranthaceae), *Azolla filiculoides* Lam. (Salviniaceae), *Hydrilla verticillata* (L.f.) Royle (Hydrocharitaceae), and *Myriophyllum aquaticum* (Vell.) Verdc. (Haloragaceae) also cause significant impacts to agriculture but are not as widespread and are not as important globally (CABI 2021).

We also categorized *Arundo donax* L. (Poaceae), *Crassula helmsii* (Kirk) Cockayne (Crassulaceae), *Melaleuca quinquenervia* (Cav.) S. T. Blake (Myrtaceae), *Mimosa pigra* L. (Fabaceae), and *Tamarix* spp. (Tamaricaceae) as aquatic weed species due to their direct and indirect impacts on aquatic ecosystems and agriculture (e.g. wetland areas, grazing lands, and water availability) (DiTomaso et al. 2013; CABI 2021).

Collectively, all of the aquatic weeds included in Table 8.4 affect agriculture directly through competition with crops or by limiting the availability of water, interfering with fisheries, blocking irrigation canals, and reducing access to water for humans and livestock. Some also impact indirectly on agriculture by interfering with transport to

Table 8.4 A sample of weed species occurring in aquatic systems that have been controlled or partially controlled by arthropods. The general impact caused by each agent and the number of countries in which each occurs are also indicated.

Weed Family	Weed Species	Agent Family	Agent Species	General Impact	Countries
Amaranthaceae	*Alternanthera philoxeroides* (Mart.) Griseb.	Chrysomelidae	*Agasicles hygrophila* Selman & Vogt	Variable–High	6
Araceae	*Pistia stratiotes* L.	Curculionidae	*Neohydronomus affinis* Hustache	Variable–High	17
Azollaceae	*Azolla filiculoides* Lam.	Curculionidae	*Stenopelmus rufinasus* Gyllenhal	High	2
Fabaceae	*Mimosa pigra* L.	Gracillariidae	*Neurostrota gunniella* (Busck)	High	1
Haloragaceae	*Myriophyllum aquaticum* (Vell.) Verdc.	Chrysomelidae	*Lysathia* sp.	High	1
Hydrocharitaceae	*Hydrilla verticillata* (L.f.) Royle	Ephydridae	*Hydrellia pakistanae* Deonier	Variable	1
Poaceae	*Arundo donax* L.	Eurytomidae	*Tetramesa romana* (Walker)	Medium	2
Pontederiaceae	*Pontederia crassipes* Mart.	Curculionidae	*Neochetina bruchi* Hustache *N. eichhorniae* Warner	Variable–High	24
Salviniaceae	*Salvinia molesta* D.S. Mitch.	Curculionidae	*Cyrtobagous salviniae* Calder & Sands	Variable–High	18
Tamaricaceae	*Tamarix* spp.	Chrysomelidae	*Diorhabda* spp.	Variable	2

markets and hospitals, hydroelectricity generation, or providing a habitat for vectors of human disease (Coetzee et al. 2009; Julien et al. 2009).

Alternanthera philoxeroides was the first of the aquatic species to be targeted for biological control, with the 1967 release of the flea beetle *Agasicles hygrophila* Selman & Vogt (Coleoptera: Chrysomelidae) in the USA. Since then, the beetle has established in six more countries, providing good control of the weed in situations in which *A. philoxeroides* forms floating populations. Control, however, does not appear as effective where the weed is found in terrestrial or semiaquatic areas (Winston et al. 2021).

Pontederia crassipes has probably been the most studied aquatic weed in terms of biological control, with 10 biological control agents released in 44 countries. The two most common agents to be released, are the beetles *Neochetina bruchi* Hustache and *N. eichhorniae* Warner (Coleoptera: Curculionidae), which have been deliberately released in 40 and 42 countries, respectively. Together, they are contributing to the control of *P. crassipes* in 25 countries, including Papua New Guinea and on Lake Victoria in Africa (Wilson et al. 2007; Coetzee et al. 2009; Winston et al. 2021). However, control is less successful in areas where there is high eutrophication and plants may outgrow the damage caused by the agents (Coetzee and Hill 2011).

The weevil *Neohydronomus affinis* Hustache (Coleoptera: Curculionidae) has been released in 19 countries against *P. stratiotes*, resulting in the control of the weed in most countries in which the weevil has established (Winston et al. 2021). Despite its success in Africa and Oceania, the weevil has not been introduced into any Asian country (Winston et al. 2021) where the weed has proven to be a major problem.

In Vanuatu, both *P. stratiotes* and *P. crassipes* were controlled within four years in most areas where their respective biological control agents had established (Day and Bule 2016). *Pistia stratiotes* has also been controlled in Australia (Harley et al. 1990) and the Republic of Congo (Mbati and Neuenschwander 2005).

8.5.1 Case Study: *Salvinia molesta* D.S. Mitchell (Salviniaceae)

Salvinia molesta D.S. Mitchell (Salviniaceae) is a floating fern native to Brazil. However, it is now a major weed in many tropical and subtropical countries in Africa, Asia, North America, and Oceania (Julien et al. 2009). *Salvinia molesta* is particularly fast growing, doubling its biomass in three days, and can completely cover lakes, dams, and rivers where it interferes with fishing, transport, water supplies, and access to water by stock (Barrett 1989; Chikwenhere and Keswani 1997; Julien et al. 2009). In Asia, it can block irrigation canals and drainage ditches of rice paddies, and invades rice paddies where it competes for space, water, and nutrients (Doeleman 1989). *Salvinia molesta* reduced the supply of rice in Senegal, resulting in an increase in the price of rice due to lower yields and higher production costs (Diop and Hill 2009). Dense mats of *S. molesta* can also harbor vectors for disease such as rural encephalitis, malaria, and dengue fever (Barrett 1989; Doeleman 1989; Julien et al. 2009 and references therein).

The *S. molesta* biological control program began in 1961, and four agents have since been deliberately released. Of these, the most widely used and successful agent is *Cyrtobagous salviniae* Calder & Sands (Coleoptera: Curculionidae) (Julien et al. 2009). *Cyrtobagous salviniae* adults feed on young leaves, and larvae feed within the rhizomes (Sands et al. 1983), causing plants to become waterlogged and the plants to sink (Julien et al. 1987).

Cyrtobagous salviniae was first released in Australia in 1980 (Figure 8.5); following its success there, it was released

(a) (b) (c)

Figure 8.5 (a) *Cyrtobagous salviniae* adult. *Salvinia molesta* in New South Wales, Australia, (b) before the release of the beetle in the 1980s and (c) three years after the release of the beetle. (Credits: beetle – David Comben, Queensland Department of Agriculture and Fisheries; before and after – Mic Julien, CSIRO).

in a further 22 countries in Africa, Asia, and Oceania as well as the USA, establishing in all but one country. In most countries, *S. molesta* is now considered under control or at least partial control in most areas where *C. salviniae* has established (Winston et al. 2021).

There have been numerous studies documenting the benefits of releasing *C. salviniae* (Doeleman 1989; Chikwenhere and Keswani 1997; Page and Lacey 2006). In Zimbabwe, control of *S. molesta* in two impoundments on a commercial farm was achieved within two years, with an overall benefit:cost ratio of 10.6:1. While some benefits, such as increased water for irrigation and stock, couldn't be quantified, there were significant savings in the costs of manual removal and use of herbicides (Chikwenhere and Keswani 1997). *Salvinia molesta* is reported to be under control in all areas in Zimbabwe where the beetle is present (G. Chikwenhere pers. comm. 2012).

An assessment of the benefits of the *S. molesta* biological control program in Sri Lanka found that the return of investment was 53:1 (Doeleman 1989). The benefits are derived from increased yield in rice paddies due to less *S. molesta*, reduced time weeding or setting up barricades to prevent *S. molesta* entering rice paddies, and the reestablishment of fishing areas. There was also a significant decrease in human health issues such as insect-borne diseases due to the reduction in suitable breeding sites for malaria vectors (Doeleman 1989).

In the Congo Basin in Central Africa, the three water weeds *S. molesta*, *P. stratiotes* and *P. crassipes* invaded waterways, impeding fishing and transport as well as harboring disease vectors (Mbati and Neuenschwander 2005). Within three years following the release and establishment of biological control agents against all three weeds, fishing resumed, and river transport was no longer impeded. Fishing ponds once covered by the weeds were now clear. Channels that were used as transport by villagers were cleared of the weed, and families reclaimed villages once abandoned due to lack of access (Mbati and Neuenschwander 2005).

In Papua New Guinea, *S. molesta* was first reported on the Sepik River in 1971; it quickly spread to occupy 32 km of waterways and lagoons. The impacts of the weed were significant as local villagers relied on the waterways to fish, harvest sago, and access markets, schools, and hospitals (Thomas and Room 1986). *Cyrtobagous salviniae* was introduced into the system in 1982 (Room and Thomas 1985). By 1983, beetle numbers were such that they could be harvested and released in other lagoons along the Sepik River. Control was so quick and complete that by June 1985, there was only 0.5% of the water surface occupied by *S. molesta*. Villages that had been deserted were reoccupied, and people were able to navigate waterways, harvest sago, and fish (Thomas and Room 1986).

8.6 Benefits of Biological Control

Approximately 66% of weeds targeted for biological control have been successfully or partially controlled (Schwarzländer et al. 2018; Winston et al. 2021). Overall, biological control agents have reduced plant size, seed production, and plant density (Clewley et al. 2012), resulting in huge benefits to the agricultural sector through increased production and reduced weed management costs. Consequently, biological control has resulted in massive benefit:cost ratios. In a study assessing 36 weed biological control projects in Australia, the average benefit:cost ratio for all projects assessed was 23:1, and it was estimated that benefits of A$78.1 million p.a. would flow to the agricultural sector given current investments in weed biological control. The benefit:cost ratio for cactus species alone was over 300:1, partly due to the control of *O. stricta* for the past 85+ years (Page and Lacey 2006).

In New Zealand, it is suggested that the net benefit of weed biological control programs returned between NZ$11 million and NZ$217 million p.a. through control of pasture weeds such as *Carduus nutans* L. (Asteraceae), *J. vulgaris*, and *H. perforatum* (Suckling 2013).

In South Africa, weed biological control projects returned benefit:cost ratios ranging from 8:1 to 3,726:1 through increased productivity and reduced management costs (van Wilgen and de Lange 2011). For recent projects such as *A. filiculoides*, the returns on investment are low, about 2.5:1, while weeds that have been controlled over a longer period, have far greater returns on investment. Benefits due to the biological control of invasive *Opuntia* spp. are estimated at 2,731:1 (van Wilgen and de Lange 2011).

There are numerous studies on the benefits of controlling various weed species, particularly those affecting grazing lands or aquatic ecosystems. Many have been conducted in just one country or region within a country (e.g. Doeleman 1989; Nordblom et al. 2001; Witt et al. 2020). There have also been studies comparing the cost of controlling weeds using biological control with the cost of conventional control methods (e.g. Pitso 2010; Maluleke 2019). Other studies have documented the economic benefits to agencies funding the implementation of biological control (e.g. Lubulwa and McMeniman 1998).

In Australia, *L. camara* costs the grazing sector over A$100 million p.a. in lost production and increased management costs (AECgroup 2007). In some cases, there are huge incentives to clear *L. camara*, but in others the costs of control are similar to the benefits that accrue from doing so, a significant disincentive. In these situations, the benefits of biological control become apparent. Even a 5% reduction of *L. camara* is expected to produce benefits of 9:1 (AECgroup 2007).

In Oregon, USA, biological control of *J. vulgaris* has resulted in a benefit:cost ratio of 13:1, with an annual benefit of US$5 million p.a. These savings comprise a reduction in stock losses of over US$3 million p.a. and increased productivity through improved pastures of approximately US$1.3 million p.a. There are also savings from reduced herbicide use of nearly US$1 million p.a. (Coombs et al. 1996). The overall savings across the USA and Canada due to the successful control of *J. vulgaris* is an order of magnitude higher considering the number of states and provinces initially infested with the weed.

The biological control of the pasture weeds *Echium* spp. (Boraginaceae), which at one point had invaded over 30 million ha in Australia, also returned a high benefit:cost ratio of 14:1, rising to a predicted 47:1 by 2050, with annual benefits of around A$73 million (Nordblom et al. 2001). *Echium* spp. are highly toxic, particularly to stock, so there have been significant savings in terms of reduced deaths, along with increased returns due to improved pastures. The benefits quoted do not include savings to reduced herbicide use, which are likely to increase as biological control agents are distributed and spread to new areas and the benefits are seen over more of southern Australia (Nordblom et al. 2001).

The successful biological control of numerous weed species on a broad-scale has demonstrated that biological control has been more efficient and produced greater benefits, especially in the long term, than control using either herbicides or physical means. Both herbicide and physical control need continued applications, partly because these methods rarely kill all plants and partly because of the large soil seed banks. Once established, biological control agents will continue to attack new plants arising.

In one of only a few studies that compares chemical control costs with the cost of biological control, Maluleke (2019) found the cost of controlling four aquatic weeds using biological control was ZAR7,843,205, compared to chemical control of the four weeds using a boat, a pickup, or knapsack of ZAR149,580,142, ZAR268,264,838, and ZAR881,711,738, respectively. These figures do not include any follow-up interventions. This equates to significant benefit:cost ratios of 90.24:1, 164.97:1, and 557.99:1 in implementing biological control, respectively.

In addition, conventional control methods are not feasible to control large invasions covering thousands of ha (Culliney 2005; Sheley et al. 2011). In Kenya, the cost of physically clearing 1,000 ha of *O. stricta* was estimated at US$540,000. In contrast, the cost of implementing biological control and introducing the cochineal *D. opuntiae* was estimated at US$35,000 (Witt et al. 2020). *Dactylopius opuntiae* has successfully controlled or contributed to the control of *O. stricta* in numerous countries, including Australia, India, South Africa, and Sri Lanka (Winston et al. 2021).

Invasions at study sites in Kenya were reduced by 80% and are expected to decrease further as the cochineal continues to build up numbers and spread to new areas. Already the initial investment in biological control has yielded significant savings in control costs. In addition to these savings, productivity is expected to increase as populations of *O. stricta* decrease, allowing for the reestablishment of preferred pasture species (Witt et al. 2020). Biological control agents also reduce the spread of the weed, thus alleviating future control costs in new areas.

8.7 Conclusions

Biological control of weeds has now been practiced in over 90 countries worldwide. Over 66% of the weeds targeted for biological control have been at least partially controlled, offering significant savings through reduced conventional control costs as well as increased productivity and income through improved pastures or crop production (Nordblom et al. 2001; Suckling 2013; Schwarzländer et al. 2018). The savings in control costs encompass reduced labor and herbicide usage, the latter having additional benefits such as reduced risks to the environment and livestock and human health. Added to this, reduced time spent weeding, particularly by subsistence farmers, means that more land can be cultivated, thus increasing productivity and income for families. Reduced weeding times can also free up time to spend on other activities such as maintenance of houses and fences, etc. (Day et al. 2013b). Many children, especially in low and middle income countries (LMIC), are taken out of school so that they can assist with weeding during peak weeding periods. More effective weed control could reduce the number of days learners are absent from school.

Reducing weed populations reduces competition for resources such as space, water, light, and nutrients. In grazing lands, this means improved pastures and more forage for livestock, especially in cases where the weeds are unpalatable. In rangelands, reducing weed populations can also provide greater opportunities for preferred species to grow. In cropping and plantation systems, reductions in invasive vine cover can increase flowering and subsequent yield of cocoa and oil palm (Day et al. 2016); reducing competition from *P. hysterophorus*, *C. curassavica*, and *S. elaeagnifolium* has resulted in increased yields in affected crops (Greathead 1971; Fowler et al. 2000; Pitso 2010).

While a typical biological control program may cost upward of US$1 million, returns on investment indicate that biological control is extremely cost effective, with returns on investment upward of 10:1. In addition, supplying effective biological control agents to LMIC with the same weeds, costs a fraction of the initial costs of developing a biological control agent because the best candidate agents have been identified and host specificity testing, to confirm that the agent poses no risk to crops or native plant species, have already been conducted (Julien et al. 2007; Day and Witt 2019; Day et al. 2020). Some biological control agents have been released into other countries for less than A$10,000. In Vanuatu, three biological control agents were introduced and released under an Australian government-funded project, costing approximately A$225,000 over three years (M. Day, pers. comm.).

The economic benefits of some of these aid programs have been assessed (e.g. Lubulwa and McMeniman 1998). The successful biological control of *S. molesta* returned benefits to the numerous partner countries of over A$27 million, while biological control of *Mimosa pigra* L. (Fabaceae) returned benefits of over A$22 million (Lubulwa and McMeniman 1998). Globally, however, the returns on investment are considerably higher given that the main biological control agent controlling *S. molesta*, *C. salviniae*, is now found in 22 countries in Africa, Asia, and Oceania (Winston et al. 2021).

Considering the significant benefits of utilizing biological control to manage many of agriculture's most important weeds, and that many of the most successful and damaging agents are present in only a fraction of the countries in which the weed occurs, there is tremendous scope to introduce effective and host specific biological control agents into other countries (Day and Winston 2016; Day et al. 2018, 2020; Day and Witt 2019). For example, biological control agents for *C. odorata* and *P. hysterophorus* are found in only a fraction of the countries where the weeds are considered a problem. In addition, no biological agent for *Mimosa diplotricha* C. Wright ex Sauvalle (Fabaceae) has been released in Africa, and despite *N. affinis* controlling *P. stratiotes* in many regions of the world, the beetle has never been released in Asia (Winston et al. 2021).

Some of the reasons that many host specific and highly effective biological control agents have not been released more widely, is the reluctance of governments and funding agencies to support biological control due to perceived risks of nontarget damage, ineffectiveness of agents or that biological control agents will not control the target weeds in all places where the weed is a problem (Day and Witt 2019).

Given the long history of weed biological control, its high rate of success (Schwarzländer et al. 2018; Winston et al. 2021), and low risk of nontarget damage (Suckling and Sforza 2014; Hinz et al. 2019), governments and funding agencies concerned with increasing food security, especially in LMIC, could do well to work with other countries with experience in weed biological control to increase knowledge and awareness of biological control and to fund weed biological control projects where the return on investments is highly significant. In the meantime, biological control agents will continue to suppress weeds, offering relief to many farmers, and spread to neighboring countries providing unintended benefits across the agricultural sector.

References

Abernathy, J.R. and Keeling, J.W. (1979). Silverleaf nightshade control in cotton with glyphosate. *Proceedings of the 32nd Annual Meeting of the Southern Weed Science Society*, 380.

AECgroup. (2007). *Economic impact of lantana on the Australian grazing industry*. Department of Natural Resources and Water, Brisbane, 37 p.

Aigbedion-Atalor, P.O., Adom, M., Day, M.D. et al. (2019). Eight decades of invasion by *Chromolaena odorata* (Asteraceae) and its biological control in West Africa: the story so far. *Biocontrol Science and Technology* 29 (12): 1215–1233.

Akobundu, I.O. (1987). *Weed Science in the Tropics. Principles and Practices*. Chicester, UK: Wiley.

Anwar, A.I. (1996). Biological control efforts towards the management of some common weeds in Malaysia. In: *Proceedings of the Symposium on Biology and Management of Weeds and 4th Tropical Weed Science Conference* (ed. S. Tjitrosemito and I. Soerianegara), 149–157. Bogor: SEAMEO BIOTOP.

Bajwa, A.A., Farooq, M., Nawaz, A. et al. (2019). Impact of invasive plant species on the livelihoods of farming households: evidence from *Parthenium hysterophorus* invasion in rural Punjab, Pakistan. *Biological Invasions* 21: 3285–3304. https://doi.org/10.1007/s10530-019-02047-0.

Barrett, S.C. (1989). Waterweed invasions. *Scientific American* 261: 90–97.

Bhagwat, S.A., Breman, E., Thekaekara, T. et al. (2012). A battle lost? Report on two centuries of invasion and management of *Lantana camara* L. in Australia, India and South Africa. *PloS One* 7: e32407.

Bourke, C.A. (1997). Effects of *Hypericum* spp. (St. John's wort) on animal health and production. *Plant Protection Quarterly* 12: 91–93.

Boyd, J.W., Murray, D.S., and Tyrl, R.J. (1984). Silverleaf nightshade, *Solanum elaeagnifolium*, origin, distribution, and relation to man. *Economic Botany* 38: 210–216.

Braimah, H., Ekyem, S.O., Issah, U.S., et al. (2013). Social perceptions and ecological impacts following biological control of *Chromolaena odorata* by *Pareuchaetes pseudoinsulata* in the forest region of Ghana. In: *Proceedings of the Eighth International Workshop on Biological Control and Management of Chromolaena odorata and Other Eupatorieae* (ed. C. Zachariades, L.W.

Strathie, M.D. Day, and R. Muniappan), 110–116. Pretoria: ARC-PPRI.

Brandon, L.J. (2005). *Weed management in Roundup Ready Flex cotton*. MSc Thesis in crop science, Graduate Faculty of Texas Tech University, Lubbock.

Briese, D.T. and Cullen, J.M. (2012). *Hypericum perforatum* L. -St John's wort. In: *Biological Control of Weeds in Australia* (ed. M. Julien, R. McFadyen, and J. Cullen), 299–307. Melbourne: CSIRO Publishing.

CABI. (2021). *CABI invasive species compendium*. https://www.cabi.org/ISC (accessed 30 April 2021).

Cech, R. (1997). *Herb of the Sun: Saint John's Wort*. Williams, Oregon: Horizon Herbs Publication, 21 p.

Chauhan, B.S. (2020). Grand challenges in weed management. *Frontiers in Agronomy* 1: 1–4.

Chikwenhere, G.P. and Keswani, C.L. (1997). Economies of biological control of Kariba weed (*Salvinia molesta* Mitchell) at Tengwe in north-western Zimbabwe – a case study. *International Journal of Pest Management* 43: 109–112.

Chippendale, J.F. and Panetta, F.D. (1994). The cost of parthenium weed to the Queensland cattle industry. *Plant Protection Quarterly* 9: 73–76.

Chuk, M. (2010). Invasive cacti–a threat to the rangelands of Australia. In: *Proceedings of the 16th Biennial Conference of the Australian Rangeland Society* (ed. D.J. Eldridge and C. Waters), 1–5. Perth: Australian Rangeland Society.

Clewley, G.D., Eschen, R., Shaw, et al. (2012). The effectiveness of classical biological control of invasive plants. *Journal of Applied Ecology* 49: 1287–1295.

Coetzee, J.A. and Hill, M.P.et al. (2011). The role of eutrophication in the biological control of water hyacinth, *Eichhornia crassipes*, in South Africa. *BioControl* 57: 247–261.

Coetzee, J.A., Hill, M.P., Julien, M.H. et al. (2009). *Eichhornia crassipes* (Mart.) Solms–Laub. (Pontederiaceae). In: *Biological Control of Tropical Weeds Using Arthropods* (ed. R. Muniappan, G.V.P. Reddy, and A. Raman), 183–210. Cambridge: Cambridge University Press.

Colautti, R.I., Bailey, S.A., Van Overdijk, C.D. et al. (2006). Characterised and projected costs of nonindigenous species in Canada. *Biological Invasions* 8: 45–59.

Conant, P., Garcia, J.N., Johnson, M.T. et al. (2013). Releases of natural enemies made in Hawaii since 1980 for classical biological control of weeds. In: *Proceedings of the XIII International Symposium on Biological Control of Weeds* (ed. Y. Wu, T. Johnson, S. Sing et al.), 230–242. Morgantown, WV, USA: Forest Health Technology Enterprise Team.

Coombs, E.M., Clark, J.K., Piper, G.L., and Cofrancesco, A.F., Jr. (eds.) (2004). *Biological Control of Invasive Plants in the United States*. Corvallis: Oregon State University Press, 467.

Coombs, E.M., Radtke, H., Isaacson, D.L., and Snyder, S.P. (1996). Economic and regional benefits from the biological control of tansy ragwort, *Senecio jacobaea*, in Oregon. In: *Proceedings of the IX International Symposium on Biological Control of Weeds* (ed. V.C. Moran and J.H. Hoffmann), 489–494. South Africa: University of Cape Town.

Cruz, Z.T., Muniappan, R., and Reddy, G.V.P. (2006). Establishment of *Cecidochares connexa* (Diptera: Tephritidae) in Guam and its effect on the growth of *Chromolaena odorata* (Asteraceae). *Annals of the Entomological Society of America* 99: 845–850.

Culliney, T.W. (2005). Benefits of classical biological control for managing invasive plants. *Critical Reviews in Plant Sciences* 24: 131–150.

Culvenor, C.C.J. (1985). Economic loss due to poisonous plants in Australia. In: *Proceedings of the Australia-U.S.A. Poisonous Plants Symposium Queensland Department of Primary Industries* (ed. A.A. Seawright, M.P. Hegarty, L.F. James, and R.F. Keeler), 3–13. Brisbane, Australia.

Day, M., Witt, A., and Winston, R. (2020). Weed biological control in low-income and middle-income countries. *Current Opinion in Insect Science* 38: 92–98.

Day, M.D. and Bofeng, I. (2007). Biocontrol of *Chromolaena odorata* in Papua New Guinea. In: *Proceedings of the 7th International Workshop on the Biological Control and Management of Chromolaena odorata and Mikania micrantha* (ed. P.-Y. Lai, G.V.P. Reddy, and R. Muniappan), 53–67. Taiwan: National Pingtung University of Science and Technology.

Day, M.D., Bofeng, I., and Nabo, I. (2013a). Biocontrol of *Chromolaena odorata* in Papua New Guinea. In: *Proceedings of the Eighth International Workshop on Biological Control and Management of Chromolaena odorata and Other Eupatorieae* (ed. C. Zachariades, L.W. Strathie, M.D. Day, and R. Muniappan), 117–126. Pretoria: ARC-PPRI.

Day, M.D., Bofeng, I., and Nabo, I. (2013b). Successful biological control of *Chromolaena odorata* (Asteraceae) by the gall fly *Cecidochares connexa* (Diptera: Tephritidae) in Papua New Guinea. In: *Proceedings of the XIII International Symposium on Biological Control of Weeds* (ed. Y. Wu, T. Johnson, S. Sing et al.), 400–408. Morgantown, WV, USA: Forest Health Technology Enterprise Team.

Day, M.D. and Bule, S. (2016). The status of weed biological control in Vanuatu. *NeoBiota* 30: 151–166.

Day, M.D., Clements, D.R., Gile, C. et al. (2016). Biology and impacts of Pacific Islands invasive species. 13. *Mikania micrantha* Kunth (Asteraceae). *Pacific Science* 70: 257–285.

Day, M.D., Kawi, A., Kurika, K. et al. (2012a). *Mikania micrantha* Kunth (Asteraceae) (mile-a-minute): its distribution and physical and socioeconomic impacts in Papua New Guinea. *Pacific Science* 66: 213–223.

Day, M.D., Kawi, A., Tunabuna, A. et al. (2012b). Distribution and socio-economic impacts of *Mikania micrantha* in

Papua New Guinea and Fiji and prospects for its biocontrol. *Pakistan Journal of Weed Science Research* 18 (Special Issue): 169–179.

Day, M.D., Shen, S.-C., Xu, G. et al. (2018). Weed biological control in the Greater Mekong Subregion: status and opportunities for the future. *CAB Reviews* 13: 1–12.

Day, M.D., Wiley, C.J., Playford, J., et al. (2003). *Lantana: current management status and future prospects*. Australian Centre for International Agricultural Research, Canberra.

Day, M.D. and Winston, R.L. (2016). Biological control of weeds in the 22 Pacific island countries and territories: current status and future prospects. *NeoBiota* 30: 167–192.

Day, M.D. and Witt, A.B.R. (2019). Weed biological control: challenges and opportunities. *Weeds-Journal of the Asian-Pacific Weed Science Society* 1: 34–44.

de Chenon, R.D., Sipayung, A., and Sudharto, P. (2002). A decade of biological control against *Chromolaena odorata* at the Indonesian Oil Palm Research Institute in Marihat. In: *Proceedings of the Fifth International Workshop on Biological Control and Management of Chromolaena odorata* (ed. C. Zachariades, R. Muniappan, and L.W. Strathie), 46–52. Pretoria: ARC-PPRI.

de Groote, H., Ajuonu, O., Attignon, S. et al. (2003). Economic impact of biological control of water hyacinth in Southern Benin. *Ecological Economics* 45: 105–117.

Ding, J., Wang, R., Fu, W., et al. (2001). Water hyacinth in China: its distribution, problems and control status. In: *Biological and Integrated Control of Water Hyacinth, Eichhornia crassipes* (ed. M.H. Julien, M.P. Hill, T.D. Center, and J. Ding), 102: 29–32. ACIAR.

Diop, O. and Hill, M.P. (2009). Quantitative post-release evaluation of biological control of floating fern, *Salvinia molesta* DS Mitchell (Salviniaceae), with *Cyrtobagous salviniae* Calder and Sands (Coleoptera: Curculionidae) on the Senegal River and Senegal River Delta. *African Entomology* 17: 64–70.

DiTomaso, J.M. (2000). Invasive weeds in rangelands: species, impacts, and management. *Weed Science* 48: 255–265.

DiTomaso, J.M., Kyser, G.B., Oneto, S.R. et al. (2013). *Weed Control in Natural Areas in the Western United States*. Davis: Weed Research and Information Center, University of California, 544 p.

Doeleman, J.A. (1989). Biological control of *Salvinia molesta* in Sri Lanka: an assessment of costs and benefits. *ACIAR Technical Reports* 12: 14 p.

EPPO Global Database. (2021). *Solanum elaeagnifolium*. https://gd.eppo.int/taxon/SOLEL (accessed 12 April 2021).

Eschen, R., Beale, T., Bonnin, J.M. et al. (2021). Towards estimating the economic cost of invasive alien species to African crop and livestock production. *CABI Agriculture and Bioscience* 2: 18. https://doi.org/10.1186/s43170-021-00038-7.

Fowler, S.V., Ganeshan, S., Mauremootoo, J., et al. (2000). Biological control of weeds in Mauritius: past successes revisited and present challenges. In: *Proceedings of the X International Symposium on Biological Control of Weeds* (ed. N. Spencer), 43–50. Bozeman, USA: Montana State University.

Gianessi, L.P. (2009). Solving Africa's weed problem: increasing crop production and improving the lives of women. *Aspects of Applied Biology* 96: 9–23.

Gordon, A.J. and Kluge, R.L. (1991). Biological control of St John's wort, *Hypericum perforatum* (Clusiaceae), in South Africa. *Agriculture, Ecosystems and Environment* 37: 77–90.

Greathead, D.J. (1971). *A review of biological control in the Ethiopian region*. Technical Communication No. 5. Commonwealth Institute of Biological Control, Farnham Royal, England, 162 p.

Hackett, N.M. and Murray, D.S. (1982). Interference of siverleaf nightshade (*Solanum elaeagnifolium*) with Spanish peanuts (*Arachis hypogaea*). *Proceedings of the 35th Annual Meeting of the Southern Weed Science Society*, 66.

Han, S., Li, Z., Xu, Q., et al. (2017). Mile-a-minute weed *Mikania micrantha* Kunth. In: *Biological Invasions and Its Management in China* (ed. F. Wan, M. Jiang, and A. Zhan), 131–141. Singapore: Springer Nature.

Harley, K.L.S., Kassulke, R.C., Sands, D.P.A., et al. (1990). Biological control of water lettuce, *Pistia stratiotes* (Araceae) by *Neohydronomus affinis* (Coleoptera: Curculionidae). *Entomophaga* 35: 363–374.

Heap, I. (2019). *International survey of herbicide resistant weeds*. http://www.weedscience.org (accessed 17 April 2019).

Heap, J., Honan, I., and Smith, E. (1997). *Silverleaf nightshade: a technical handbook for Animal and Plant Control Boards in South Australia*. Primary Industries South Australia, Animal and Plant Control Commission, Naracoorte, Australia.

Heap, J. and Wu, H. (2018). *Silverleaf nightshade: Australian best practice management manual*. Primary Industries and Regions, South Australia and the NSW Department of Primary Industries.

Heard, T.A. and Day, M. (2012). *Sida acuta* Burm.f – spinyhead sida, *Sida rhombifolia* L. – paddy's lucerne, *Sida cordifolia* L. – flannel weed. In: *Biological Control of Weeds in Australia* (ed. M. Julien, R. McFadyen, and J. Cullen), 544–550. Melbourne: CSIRO Publishing.

Hinz, H.L., Winston, R.L., and Schwarzländer, M. (2019). How safe is weed biological control? A global review of direct nontarget attack. *The Quarterly Review of Biology* 94: 1–27.

Hoffmann, J.H., Moran, V.C., and Impson, F.A.C. (1998). Promising results from the first biological control programme against a solanaceous weed (*Solanum elaeagnifolium*). *Agriculture Ecosystems and Environment* 70: 145–150.

Hoffmann, J.H., Moran, V.C., Zimmermann, H.G., et al. (2020). Biocontrol of a prickly pear cactus in South Africa: reinterpreting the analogous, renowned case in Australia. *Journal of Applied Ecology*. https://doi.org/10.1111/1365-2664.13737.

Ireson, J.E. and McLaren, D. (2012). *Jacobaea vulgaris* Gaertn.–ragwort. In: *Biological Control of Weeds in Australia* (ed. M. Julien, R. McFadyen, and J. Cullen), 314–323. Melbourne: CSIRO Publishing.

Jayachandra, J. (1971). Parthenium weed in Mysore State and its control. *Current Science* 40: 568–569.

Jensen, K.I.N., Harris, P., and Sampson, M.G. (2002). *Hypericum perforatum* L., St John's wort (Clusiaceae). In: *Biological Control Programmes in Canada 1981–2000* (ed. P.G. Mason and J.T. Huber), 361–368. Wallingford, UK: CAB International.

Jones, P.K., Day, M.D., and McConnachie, A.J., et al. (2023). The successful biological control of *Cylindropuntia fulgida* var. *mamillata* (A. Schott ex Engelm.) Backeb. (Cactaceae) (coral or boxing glove cactus) at two field sites in Queensland, Australia. *Biological Control* 182. https://doi.org/10.1016/j.biocontrol.2023.105235.

Julien, M.H., Bourne, A.S., and Chan, R.R. (1987). Effects of adult and larval *Cyrtobagous salviniae* on the floating weed *Salvinia molesta*. *Journal of Applied Ecology* 24: 935–944.

Julien, M.H., Hill, M.P., and Tipping, P.W. (2009). *Salvinia molesta* DS Mitchell (Salviniaceae). In: *Biological Control of Tropical Weeds Using Arthropods* (ed. R. Muniappan, G.V.P. Reddy, and A. Raman), 378–407. Cambridge: Cambridge University Press.

Julien, M.H., Scott, J.K., Orapa, W., et al. (2007). History, opportunities and challenges for biological control in Australia, New Zealand and the Pacific islands. *Crop Protection* 26: 255–265.

Kingsbury, J. (1964). *Poisonous Plants of the United States and Canada*. Englewood Cliffs, NJ: Prentice-Hall, Inc, 52–57 and 171–175.

Le Maitre, D.C., van Wilgen, B.W., Chapman, R.A., et al. (1996). Invasive plants and water resources in the Western Cape Province, South Africa: modelling the consequences of a lack of management. *Journal of Applied Ecology* 33: 161–172.

Llewellyn, R.S., Ronning, D., Ouzman, J. et al. (2016). *Impact of weeds on Australian grain production: the cost of weeds to Australian grain growers and the adoption of weed management and tillage practices*. Report for GRDC. CSIRO, Australia.

Lubulwa, G. and McMeniman, S. (1998). ACIAR-supported biological control projects in the South Pacific (1983–1996): an economic assessment. *Biocontrol News and Information* 19: 91N–98N.

Macanawai, A.R., Day, M.D., Tumaneng-Diete, T., et al. (2012). Impact of *Mikania micrantha* on crop production systems in Viti Levu, Fiji. *Pakistan Journal of Weed Science Research* 18 (Special Issue): 357–365.

Maluleke, M. (2019). *Economic evaluation of chemical and biological control methods on four aquatic weeds in South Africa*. MC, Rhodes University, Grahamstown.

Marais, C., Van Wilgen, B.W., and Stevens, D. (2004). The clearing of invasive alien plants in South Africa: a preliminary assessment of costs and progress. *South African Journal of Science* 100: 97–103.

Mbati, G. and Neuenschwander, P. (2005). Biological control of three floating water weeds, *Eichhornia crassipes*, *Pistia stratiotes*, and *Salvinia molesta* in the Republic of Congo. *BioControl* 50: 635–645.

McFadyen, R.C. (1999). Successful biological control of *Chromolaena odorata* weed in Indonesia using new agents from the Americas. In: *Biological Control in the Tropics: Proceedings of the Symposium held at MARDI Training Centre* (ed. L.W. Hong, S.S. Sastroutomo, I.G. Caunter et al), 55–58. Kuala Lumpur: SEARC-CAB International.

McFadyen, R.E. (1992). Biological control against parthenium weed in Australia. *Crop Protection* 11: 400–407.

McFadyen, R.E.C. (1998). Biological control of weeds. *Annual Review of Entomology* 43: 369–393.

Mekki, M. (2007). Biology, distribution and impacts of silverleaf nightshade (*Solanum elaeagnifolium* Cav.). *Bulletin OEPP/EPPO Bulletin* 37: 114–118.

Melifonwu, A.A. (1994). Weeds and their control in cassava. *African Crop Science Journal* 2: 519–530.

Mitich, L. (1994). Common St. Johnswort. *Weed Technology* 8 (3): 658–661.

Muniappan, R., Reddy, G.V.P., and Raman, A. (2009). *Coccinia grandis* (L.) Voigt (Cucurbitaceae). In: *Biological Control of Tropical Weeds Using Arthropods* (ed. R. Muniappan, G.V.P. Reddy, and A. Raman), 173–180. Cambridge: Cambridge University Press.

Muraleedharan, P.K. and Anitha, V. (2000). The economic impact of *Mikania micrantha* on teak plantations in Kerala. *Indian Journal of Forestry* 23: 248–251.

Muraleedharan, P.K. and Anitha, V. (2001). The economic impact of *Mikania micrantha* on the agroforestry production system in the Western Ghats of Kerala. In: *Alien Weeds in Moist Tropical Zones, Banes and Benefits* (ed. K.V. Sankaran, S.T. Murphy, and H.C. Evans), 80–85. Kerala Forestry Research Institute, Peechi, India and CABI Bioscience, UK Centre (Ascot).

Neuenschwander, P., Julien, M.H., Center, T.D., et al. (2009). *Pistia stratiotes* L. (Araceae). In: *Biological Control of Tropical Weeds Using Arthropods* (ed. R. Muniappan, G.V.P. Reddy, and A. Raman), 332–352. Cambridge: Cambridge University Press.

Nghiem, L.T.P., Soliman, T., Yeo, D.C.J. et al. (2013). Economic and environmental impacts of harmful non-indigenous species in Southeast Asia. *PLoS ONE* 8: e71255. https://doi.org10.1371/journal.pone.0071255.

Nordblom, T.L., Smyth, M., Swirepik, A. et al. (2001). Benefit-cost analysis for biological control of echium weed species (Paterson's curse/Salvation Jane). In: *45th Annual Conference of the Australian Agricultural and Resource Economics Society*, 1–11. Adelaide, Australia: Australian Agricultural and Resource Economics Society.

Oerke, E.C. (2006). Crop losses to pests. *Journal of Agricultural Science* 144: 31–43.

Orapa, W., Donnelly, G., and Bofeng, I. (2002). The distribution of Siam weed, *Chromolaena odorata*, in Papua New Guinea. In: *Proceedings of the Fifth International Workshop on Biological Control and Management of Chromolaena odorata* (ed. C. Zachariades, R. Muniappan, and L.W. Strathie), 19–25. Pretoria: ARC-PPRI.

Page, A.R. and Lacey, K.L. (2006). *Economic impact assessment of Australian weed biological control*. CRC for Australian Weed Management, Technical Series 10, 165 p.

Palmer, W. and Sims-Chilton, N. (2012). *Baccharis halimifolia* L.–groundsel bush. In: *Biological Control of Weeds in Australia* (ed. M. Julien, R. McFadyen, and J. Cullen), 86–95. Melbourne: CSIRO Publishing.

Paterson, I.D., Hoffmann, J.H., Klein, H. et al. (2011). Biological control of Cactaceae in South Africa. *African Entomology* 19 (sp): 230–246.

Pimentel, D., McNair, S., Janecka, J. et al. (2001). Economic and environmental threats of alien plant, animal, and microbe invasions. *Agriculture, Ecosystems and Environment* 84: 1–20.

Pimentel, D., Zuniga, R., and Morrison, D. (2005). Update on the environmental and economic costs associated with alien-invasive species in the United States. *Ecological Economics* 52: 273–288.

Piper, G.L. (2004). St. Johnswort, *Hypericum perforatum*. In: *Biological Control of Invasive Plants in the United States* (ed. E.M. Coombs, J.K. Clark, G.L. Piper, and A.F. Cofrancesco), 322–334. Corvallis: Oregon State University Press.

Pitcairn, M.J. (2018). Weed biological control in California, USA: review of the past and prospects for the future. *BioControl* 63: 349–359.

Pitso, D.C. (2010). A biophysical and economic evaluation of biological and chemical control methods for *Solanum elaeagnifolium* (silverleaf nightshade) in the Limpopo Province, South Africa. PhD Thesis, University of Cape Town, South Africa.

Puzari, K.C., Bhuyan, R.P., Pranab, D., et al. (2010). Distribution of *Mikania* and its economic impact on tea ecosystem of Assam. *Indian Journal of Forestry* 33: 71–76.

Rai, R.K. and Scarborough, H. (2014). Invasive plants and rural livelihoods: an assessment of the effects of the invasion of *Mikania micrantha* in rural livelihoods in Nepal. In: *Proceedings of Invasive Alien Species Management* (ed. G.J. Thapa, N. Subedi, M.R. Pandey et al.), 39–51. Nepal: National Trust for Nature Conservation.

Rayamajhi, M.B., Pratt, P.D., Tipping, P.W. et al. (2016). Seasonal growth, biomass allocation, and invasive attributes manifested by *Dioscorea bulbifera* L. (air-potato) plants generated from bulbils in Florida. *Invasive Plant Science and Management* 9: 195–204.

Room, P.M. and Thomas, P.A. (1985). Nitrogen and establishment of a beetle for biological control of the floating weed Salvinia in Papua New Guinea. *Journal of Applied Ecology* 22: 139–156.

Sands, D.P.A., Schotz, M., and Bourne, A.S. (1983). The feeding characteristics and development of larvae of a salvinia weevil *Cyrtobagous* sp. *Entomologia Experimentalis et Applicata* 34: 291–296.

Saunders, J.T., Greer, G., Bourdôt, G. et al. (2017). The economic costs of weeds on productive land in New Zealand. *International Journal of Agricultural Sustainability* 15: 380–392.

Schwarzländer, M., Hinz, H.L., Winston, R.L., et al. (2018). Biological control of weeds: an analysis of introductions, rates of establishment and estimates of success, worldwide. *BioControl* 63: 313–331.

Shackleton, R.T., Witt, A.B.R., Aool, W., and Pratt, C.F. (2017). Distribution of the invasive alien weed, *Lantana camara*, and its ecological and livelihood impacts in eastern Africa. *African Journal of Range and Forage Science* 34: 1–11.

Sheley, R.L., James, J.J., Rinella, M.J. et al. (2011). Invasive plant management on anticipated conservation benefits: a scientific assessment. In: *Conservation Benefits of Rangeland Practices: Assessment, Recommendations, and Knowledge Gaps* (ed. D.D. Briske), 291–336. Washington, DC: Natural Resources Conservation Service, U.S. Department of Agriculture.

Simmonds, F.J. (1981). Control of *Cordia curassavica* in Sri Lanka. *Biocontrol News and Information* 2: 2.

Sinden, J., Jones, R., Hester, S. et al. (2004). *The economic impact of weeds in Australia*. CRC for Australian Weed Management, Technical Series No. 8, 65 p.

Singh, S.P. (1996). Biological control. In: *50 Years of Crop Science Research in India* (ed. R.S. Paroda and K.L. Chadha), 88–116. New Delhi: Indian Council of Agricultural Research.

Smith, R.J., Jr. (1983). Weeds of major economic importance in rice and yield losses due to weed competition. In: *Proceedings of the Conference on Weed Control in Rice*, 19–36. Manila, Philippines: International Rice Research Institute.

Suckling, D.M. (2013). Benefits from biological control of weeds in New Zealand range from negligible to massive: a retrospective analysis. *Biological Control* 66: 27–32.

Suckling, D.M. and Sforza, R.F.H. (2014). What magnitude are observed non-target impacts from weed biocontrol? *PloS ONE* 9: e84847.

Tamado, T., Ohlander, L., and Milberg, P. (2002). Interference by the weed *Parthenium hysterophorus* L. with grain sorghum: influence of weed density and duration of competition. *International Journal of Pest Management* 48: 183–188.

Thomas, P.A. and Room, P.M. (1986). Successful control of the floating weed *Salvinia molesta* in Papua New Guinea: a useful biological invasion neutralizes a disastrous one. *Environmental Conservation* 13: 242–248.

Timbilla, J.A. (1996). Status of *Chromolaena odorata* biological control using *Pareuchaetes pseudoinsulata*, in Ghana. In: *Proceedings of the IX International Symposium on Biological Control of Weeds* (ed. V.C. Moran and J.H. Hoffmann), 327–332. South Africa: University of Cape Town.

Turner, N. and Szczawinski, A. (1991). *Common Poisonous Plants and Mushrooms of North America*. Portland, Oregon: Timber Press, 24, 136–137.

U.S. Congress, Office of Technology Assessment. (1993 September). *Harmful non-indigenous species in the United States, OTA-F-565*. Washington, DC, U.S. Government Printing Office.

Upton, R. (1997). St. John's Wort *Hypericum perforatum*. Quality Control, Analytical and Therapeutic Monograph. *Herbalgram* 40: 32 p.

van Wilgen, B.W. and de Lange, W.J. (2011). The costs and benefits of biological control of invasive alien plants in South Africa. *African Entomology* 19: 504–514.

van Wilgen, B.W., Forsyth, G.G., Le Maitre, D.C. et al. (2012). An assessment of the effectiveness of a large, national-scale invasive alien plant control strategy in South Africa. *Biological Conservation* 148: 28–38.

van Wilgen, B.W., Reyers, B., Le Maitre, D.C. et al. (2008). A biome-scale assessment of the impact of invasive alien plants on ecosystem services in South Africa. *Journal of Environmental Management* 89: 336–349.

Van Wychen, L. (2019). Survey of the most common and troublesome weeds in broadleaf crops, fruits & vegetables in the United States and Canada. *Weed Science Society of America National Weed Survey Dataset*. https://wssa.net/wp-content/uploads/2019-Weed-Survey_broadleaf-crops.xlsx.

Villaneuva, H.L. and Faure, G.O. (1959). Biological control of St. John's-wort in Chile. *FAO Plant Protection Bulletin* 7: 144–146.

Wassermann, V.D., Zimmermann, H.G., and Neser, S. (1988). The weed silverleaf bitter apple ("Satansbos") (*Solanum elaeagnifolium* Cav.) with special reference to its status in South Africa. Plant Protection Research Institute, Pretoria.

Wilson, J.R., Ajuonu, O., Center, T.D. et al. (2007). The decline of water hyacinth on Lake Victoria was due to biological control by *Neochetina* spp. *Aquatic Botany* 87: 90–93.

Winston, R.L., Schwarzländer, M., Hinz, H.L. et al. (eds.) (2021). *Biological Control of Weeds: A World Catalogue of Agents and Their Target Weeds*. https://www.ibiocontrol.org/catalog (accessed 21 May 2021).

Witt, A.B.R., Nunda, W., Makale, F., et al. (2020). A preliminary analysis of the costs and benefits of the biological control agent *Dactylopius opuntiae* on *Opuntia stricta* in Laikipia County, Kenya. *BioControl* 65: 515–523.

Xu, H., Ding, H., Li, M. et al. (2006). The distribution and economic losses of alien species invasion to China. *Biological Invasions* 8: 1495–1500.

Zachariades, C., Day, M., Muniappan, R., et al. (2009). *Chromolaena odorata* (L.) King and Robinson (Asteraceae). In: *Biological Control of Tropical Weeds Using Arthropods* (ed. R. Muniappan, G.V.P. Reddy, and A. Raman), 130–162. Cambridge: Cambridge University Press.

Zalucki, M.P., Day, M.D., and Playford, J. (2007). Will biological control of *Lantana camara* ever succeed? Patterns, processes and prospects. *Biological Control* 42: 251–261.

Zhong, X., Huang, Z., Si, H., and Zan, Q. (2004). Analysis of ecological-economic loss caused by weed *Mikania micrantha* on Neilingding Island, Shenzhen, China. *Journal of Tropical and Subtropical Botany* 12: 167–170. (in Chinese).

9

Ecologically Based Weed Management

Bioherbicides, Nanotechnology, Heat, and Microbially Mediated Soil Disinfestation

Raghavan Charudattan[1,], Susan M. Boyetchko[2], Erin N. Rosskopf[3], Kaydene T. Williams[4], Andrea Monroy Borrego[5], and Nicole F. Steinmetz[5,6]*

[1] *President & CEO, BioProdex, Inc., Emeritus Professor, University of Florida, Gainesville, FL, USA*
[2] *Deceased. Formerly Research Scientist, Saskatoon Research and Development Centre, Agriculture and Agri-Food Canada, Canada*
[3] *Research Microbiologist, USDA-ARS, US Horticultural Research Laboratory, Fort Pierce, FL, USA*
[4] *Post-doctoral Associate, USDA-ARS, US Horticultural Research Laboratory, Fort Pierce, FL, and University of Florida-Gulf Coast Research and Education Center, Wimauma, FL, USA*
[5] *University of California-San Diego, La Jolla, CA, USA*
[6] *Moores Cancer Center, University of California-San Diego, La Jolla, CA, USA*
[*] *Corresponding author*

This chapter is dedicated to our erstwhile colleague, the late Susan Boyetchko, a dedicated and accomplished scientist, an able research leader, and a dear friend.

9.0 Biological Control of Weeds by Using Plant Pathogens

There has been a sustained level of interest in using plant pathogens as biological control agents to manage weeds in agricultural and environmental sites. By reviewing published and unpublished reports from the year 1970 to 2000, I reported that approximately 100 different pathogens had been studied and many tested and or used in the field as biological control agents for about 86 problematic weeds in row crops, citrus, flood plains, forestry, golf courses, natural areas, rangelands, tree plantations, turf, and water bodies (lakes, ponds, waterways) (Charudattan 2001). Recently, Winston et al. (2014, with an online update, 2021) have published a worldwide catalog of biological control agents for weeds using insects and pathogens and is an indispensable source of information in the field. Listed in this catalog are 74 weed species that have been targets for control attempts by 138 fungal, bacterial, and viral pathogens. Herein, the readers can find reliable, up-to-date worldwide information on weed targets, biocontrol agents (insects, nematodes, mites, and plant pathogens), weed's origin and taxonomic information, agent's source, taxonomic information, and release details, level of weed control (agent's impact) achieved, current status of control, key publications for reference, and other useful information.

Two distinct strategies are used in biological control of weeds (by insects or pathogens). The first is the classical biocontrol tactic, aka inoculative biocontrol, which is used against invasive weeds and is done typically by importing a pathogen or a suite of pathogens (theoretically feasible) from the weed's native country or geographic range into a country where the weed has been targeted for control as an undesirable exotic invasive species. The second strategy is the bioherbicide tactic – inundative or augmentative tactic – which typically utilizes a native pathogen or pathogens against native or naturalized weeds. Whereas, in the classical tactic, a pathogen, once determined to be safe to nontarget plants and permitted for use under appropriate regulations, is "released" into the new country by manually infecting a small number of individual weeds and allowing (hoping, assuming) it to become established on the weed population and spread widely, eliciting high levels of disease. The pathogen is not expected to kill the weed upon release but reduce the weed population and thereby its invasiveness over a few years, by a gradual suppression of weed growth, reproduction, and spread. A pathogen to be

Ecologically Based Weed Management: Concepts, Challenges, and Limitations, First Edition. Edited by Nicholas E. Korres, Ilias S. Travlos, and Thomas K. Gitsopoulos.

used as a bioherbicide is normally native to the country or region where it is intended to be used. It is tested for proof of weed-control efficacy and safety to nontarget plants and often patented as a bioherbicide agent to encourage commercial interest. To transition a pathogen from a "potential bioherbicide" to a registered bioherbicide product, ways should be found to mass-produce it by microbial fermentation techniques, formulate it, as needed, to assure storage stability, ease of application, and herbicidal efficacy, and register the pathogen as the herbicide active ingredient (a.i.) and the formulated product as the biopesticide (aka bioherbicide or mycoherbicide if the a.i. is a fungus). In most countries, governmental approval would be required to import foreign pathogens and prescribe foreign or native pathogens for use as weed control agents.

In the United States, according to the US Environmental Protection Agency (EPA) that regulates pesticides under the Federal Insecticide, Fungicide and Rodenticide Act (FIFRA) (US Environmental Protection Agency 2021), pathogens proposed to be used as weed biocontrol agents, either in the classical or the inundative biocontrol tactic, are considered biopesticides, and they must be registered by this agency before public use would be legally permissible. In the case of an exotic pathogen, its importation, subsequent testing under quarantine, and eventual release into the United States are regulated and approved by the US Department of Agriculture, Animal and Plant Health Inspection Service, Plant Protection and Quarantine (USDA-APHIS-PPQ). Subsequently, these approved pathogens would also be registered by the EPA. Similar regulations exist in most other countries.

The subject of biological control of weeds with plant pathogens has been plentifully reviewed since the 1980s. Rather than reiterating the previously published information, we recommend to the readers of this chapter the most recent *Annual Review of Phytopathology* article by Louise Morin (2020) as an excellent source of up-to-date, comprehensive, account of the history, practice, and progress in biological control of weeds with plant pathogens. The two biocontrol tactics mentioned above are defined and described in greater detail in this review.

Also dealt with are the various aspects of discovery, testing, and release (use) of classical biocontrol pathogens: (i) the process and aids for searching and selecting suitable pathogens for evaluations, (ii) the importance of matching the pathogen strain intended for use with the weed genotype targeted for control to achieve the intended outcome, (iii) the methods and protocols generally used in the proof-of-efficacy and nontarget plant safety determinations, (iv) importation of the pathogen into a quarantine facility in the receiving country for further testing, and (v) field release methods and strategies for widespread establishment of the pathogen on the target weed.

Important international and national regulatory frameworks in place to oversee collection of biological control agents from a source country (i.e. code of conduct for collecting and sharing of global bioresources) and their proper introduction and release into new countries are described. Also discussed are the process of discovery, testing, and development of pathogens for use as bioherbicides, technological adjuncts that are a necessary part of insuring bioherbicide effectiveness, commercial development, and registration. The feasibility of small-scale, not-for-profit production and distribution systems as an alternative to the commercial pesticide model and how such a system might be suitable for low-income countries as well as for weed targets that lack commercial incentives is also covered. Two examples of fungi that are produced and distributed free of charge in the not-for-profit model are *Cylindrobasidium laeve* (product name: Stumpout) for preventing regrowth from cut stumps of *Acacia mearnsii* and *A. pycnantha* (black and golden wattles, respectively) and *Colletotrichum acutatum* f. sp. *hakeae* (Hakatak) to control *Hakea sericea* (silky hakea), both in South Africa (Morris et al. 1999; Winston et al. 2014, 2021).

For a broader coverage of bioherbicides, the reviews by Weaver et al. (2007), Bailey (2014), Watson (2017), Boyetchko (1999), and Rosskopf et al. (2020) are highly recommended. Described in these reviews is the process of bioherbicide development from the discovery stage to the commercial end step, including the proof-of-concept research, technology development, transfer to the commercial phase, and registration. The review by Bailey (2014) is also noteworthy for succinct descriptions of 11 registered bioherbicides around the world (as of 2014), including several previously registered that are currently out of registration.

9.1 A Critical Assessment of the Role of Plant Pathogens in Weed Management

Undeniably, several pathogens used as classical biocontrol agents have provided satisfactory levels of control of their target weeds (all invasive species) with no need for additional controls (examples in Winston et al. 2014, 2021). While the classical biocontrol strategy has been effective for invasive weeds in natural areas, aka environmental weeds, it has not been shown to be effective for agricultural weeds in croplands. The latter – due to the frequent disturbance from activities such as tillage, cropping sequences, short growing cycles, pesticide use, and other actions – are not conducive for the classical biocontrol tactic that requires relatively undisturbed conditions for the pathogens to multiply and cause high levels of disease. Morin (2020, Supplemental Table 9.1) has listed 36 exotic plant pathogenic fungi that were introduced and used in classical biocontrol against 32 invasive weeds. Countries that have led in classical biocontrol introductions are Australia (with 14 introductions), USA (11, of which 3 into the mainland, 7 into Hawaii, and 1 into Guam), New Zealand and South Africa (7 each), and Canada (5). These countries are followed by Cook Islands (3) and Chile (2). Completing the list are Argentina, China, Fiji, French Polynesia, Great Britain (England and Wales only), India, Palau, Papua New Guinea, and Taiwan, each with one introduction (with some of the pathogens released in more than one country). An analysis of the listings in Winston et al. (2021) indicates that 11 of the fungi have inflicted heavy impacts, implying a substantial level of weed control, on their weed targets in at least one of the counties where they were released. Impacts from others range from variable (13), medium (6), slight (5), none (failed to establish 4), and unknown (2). So, while not all fungal introductions have yielded expected results, more than a quarter (27%) have achieved their purpose. There is no ready or easy answer to explain why some pathogen introductions are successful while others fail. It is possible that the host and the pathogen in these cases have evolved into a steady state of balanced interrelationship (homeostasis) where the pathogen damages the host minimally and in return it gets to parasitize and survive on the host. It is also possible that the released strain is not the most highly pathogenic of the stains present and such a strain was missed during surveys.

Pathogens used in classical biological control have been primarily rust fungi (Pucciniomycetes, 23), Dothideomycetes (7), and smut fungi (Exobasidiomycetes, 2). These are biotrophic fungi that are host-specific at the host species or genus level. By reviewing the postrelease and postestablishment safety of these agents, namely whether any unanticipated pathogenic attacks had occurred on nontarget plants since their establishment in new regions, Jane Barton has concluded that there is clear evidence that it is safe to use these pathogens (Barton 2004, 2011). This commendable record of safety is assured by a risk-assessment process based on rigorous host-range testing on nontarget plant species, as well as a good understanding of the taxonomy, biology, and ecology of the pathogen, the target weed, and nontarget species (Berner and Bruckart 2005). The few instances of infections on nontarget species (six reported in Barton 2011) were all predicted during the nontarget host-range studies (Barton 2011) and therefore not surprising.

The following two examples help highlight classical biocontrol of exotic invasive weeds by using pathogens. The first is the control of *Acacia saligna* (Port Jackson willow), a medium-sized, troublesome invasive alien tree in South Africa by the gall-forming rust fungus *Uromycladium morrisii* (formerly placed under *U. tepperianum*) imported from Australia. The introduction was made in 1987 in the Western Cape Province, and by 1995, the fungus had spread naturally by wind-dispersed spores and infected the weed wherever it occurred in the province. By 1997, the mean number of rust galls per tree ranged from five to approximately 1500, and the tree density had declined by 80% at all sites (Morris 1997). Through annual monitoring of five tree populations from 1991 to 2005, Wood and Morris (2007) established that the tree density had declined between 87% and 98% with an annual mortality rate of 18% ±2%. The reduced canopy mass of infected trees and the presence of relatively younger trees at 19 monitored sites further confirmed the effectiveness of *U. morrisii* in biologically suppressing *A. saligna*. Overall, the incidence of the rust disease has increased and stayed at high levels, with mature trees having approximately 2,000 galls per tree (Wood 2012). Winston et al. (2021) rated the impact of this rust fungus as "medium" but my observations (RCs in 2007) in Western Cape Province prompt me to rate the impact as "heavy" to "very heavy."

The *U. morrisii-A. saligna* system illustrates the importance of host-pathogen matching as a prelude to selecting and importing a pathogen for classical weed control. When the *A. saligna* biocontrol program was initiated, it was known from published accounts that the rust, then called *U. tepperianum*, had a wide host range and it was suggested that it occurs as several genotypes adapted to particular host species (see Morris 1991). With care and precision, Mike Morris, then with the Plant Protection Research Institute, Stellenbosch, inoculated several *Acacia* spp, including *A. saligna*. When teliospores from the different host species were inoculated onto a range of *Acacia* species, they formed typical galls only on the host species from which they were collected. Further tests confirmed the presence of a highly pathogenic *A. saligna*-specific *U. tepperianum* genotype that was suitably host-specific for use as a biological control agent in South Africa. This host-pathogen matching study laid the foundation for the assured success in the *A. saligna* biocontrol program. In 2018, following a major revision of *U. tepperianum sensu lato*, Doungsa-ard et al. (2018) renamed the fungus used for biocontrol of *A. saligna* in South Africa as *U. morrisii*, a new species, after Mike Morris who discovered and promoted its use for biocontrol in South Africa.

The second example is the control of *Ageratina riparia* (creeping croftonweed, mist flower, or Hamakua Pamakani in Hawaiian) by *Entyloma ageratinae* (invalid past name: *Cercosporella ageratina;* synonyms: *Entyloma compositarum, E. compositarum* f.sp. *ageratinae*, Exobasidiomycetes), imported into Hawaii from Jamaica by Eduardo Trujillo (Trujillo 2005). The fungus, a leaf smut (aka white smut), was released in 1975 and by the late 1970s had produced "spectacular disease epidemics ... wiping out the weed" (Trujillo 2005). The pathogen is reported to have cleared the weed from more than 200,000 ha of infested rangeland, and kikuyu grass (*Pennisetum clandestinum*) pastures were rehabilitated in less than six years after the pathogen's introduction, saving the ranchers millions of dollars (Trujillo 2005). Generally, an agent's (a pathogen or insect) success in one country prompts its use in other countries where the same problematic weed is prevalent. This was the case with *E. ageratinae*. Its remarkable record in Hawaii (Trujillo 2005) was a consideration in its importation from Hawaii and release into South Africa against *A. riparia* in 1989. By the following year, the fungus had well established on the target weed and by 2009 *A. riparia* was reported to have been rarely observed in the field, while *E. ageratinae* was present over most of the weed's range. So, by this circumstantial evidence, the weed is considered to have been brought under control by *E. ageratinae* (Heystek et al. 2011). The fungus was also released into New Zealand in 1998 to control *A. riparia* (Barton et al. 2007). It established on the weed and spread rapidly, crossing 80 km of sea to the Great Barrier Island within two years. The impact of the fungus was monitored annually at up to 51 sites in the North Island but only in association with a gall fly (*Procecidochares alani*, Diptera: Tephritidae), released in 2001. By 2004, the mean percentage of live leaves infected with the fungus was nearly 60% and the maximum plant height declined significantly. Under heavy infestations, the mean percentage cover of *A. riparia* declined from 81% to 1.5%. The fly galls were only recorded toward the end of the impact study, and only at low mean numbers, suggesting that the observed weed decline was most likely due the pathogen. As the weed declined, the species richness and mean percentage cover of native plants were found to have increased. The many plant species colonizing the plots were important native mid- or late-successional shrubs or trees. These, along with other data (in Barton et al. 2007) suggest this rapid, well-monitored weed biocontrol program has been extremely successful. Winston et al. (2021) rate the impact of *E. ageratinae* as "heavy" and the fungus is reported to be widespread throughout the weed's range in New Zealand.

Two years after its introduction into New Zealand, *E. ageratinae* was discovered in Queensland, Australia, but the pathway of introduction is unknown (Morin et al. 2012). Field surveys by Morin et al. confirmed that the fungus was widespread in southeastern Queensland and north coast of New South Wales (NSW) and was reported "present" in the mid-north coast of NSW but not further south in NSW. Host-specificity testing of the fungus on plant species closely related to *A. riparia* that had not been tested previously indicated that *Ageratina adenophora* (croftonweed) was susceptible but to a much lesser extent than *A. riparia*. The fungus was released at noninfected *A. riparia* sites in NSW in May 2011 and five to six months later, heavy defoliation of *A. riparia* was observed at the release sites. Observations in other *A. riparia* infested sites in NSW revealed that the fungus was spreading and causing severe defoliation (Morin et al. 2012). These results suggest an emerging success scenario in Australia. Winston et al. (2021) have rated the impact of this fungus on *A. riparia* as "heavy" in all countries where it has been introduced.

Unlike these examples of classical biological control, there is not a single case of a bioherbicide that can be rated as "successful" on a pragmatic measure of extent (area) of use or volume of sales. While gaining EPA registration is indeed a major step toward success, and registered bioherbicides can indeed provide high levels of weed kill or weed suppression, still no bioherbicide has been used as a stand-alone control measure or a significant component of integrated management of a weed problem. Nearly all registered commercial bioherbicides, with the exception of the most recently registered Di-Bak Parkinsonia (Morin 2020, Table 9.1) to control *Parkinsonia aculeata* (Jerusalem thorn, Mexican palo-verde), have been used on extremely small areas of a few hundreds to a few thousands of hectares at best. Di-Bak Parkinsonia, registered in 2018 and now commercially available, has not been in the market sufficiently long to rate the extent of its use. Nonetheless, measuring bioherbicide success on the basis of use and sales figures would cause us to overlook examples such as Hakatak, SolviNix, and Stumpout that are serving client needs on an on-demand basis. Also, it cannot be ruled out that the next bioherbicide to come along might be a big winner serving a large market, generating a sizable revenue.

Of the 17 bioherbicides that have been registered worldwide since 1981 (Morin 2020, Table 9.1, Supplemental Material B), 8 are reported to be still registered and 2 (Di-Bak and SolviNix) are commercially available. Not included in this list is a product containing *Pseudomonas fluorescens*, strain D7, a deleterious rhizobacterium (DRB) developed from Ann Kennedy's work (Kennedy 2018). It has been registered and is available as "D7" from Verdesian Life Sciences, Cary, North Carolina. It is registered to suppress downy brome (aka cheatgrass, *Bromus tectorum*), an invasive grass in cereal crops, alfalfa, rangeland, and non-croplands (Anonymous 2018). The United States has led in the number of registrations (9), followed by Canada (5), South Africa (2), Australia (1), and Japan (1). Chontrol (*Chondrostereum purpureum*, Agaricomycetes, used for preventing regrowth from cut stumps of *Populus* spp. *Alnus* spp., and other broadleaved tree and shrub species) has been registered jointly in USA and Canada. Registrations of nine of the bioherbicides have lapsed or presumed to have lapsed for different reasons of which the primary are: competition from newer chemical herbicides or alternative methods (Collego, DeVine, BioMal), unsatisfactory level of control based on premature and unfair user perceptions

Table 9.1 Bioherbicide claims that were abandoned and the reasons for abandonment.

Weed	Bioherbicide agent	Reason abandoned	A key reference
Amaranthus spp. (pigweeds/ amaranths)	*Phomopsis amaranthicola*	No commercial interest (NCI)	Rosskopf et al. 2000
Cuscuta spp, (dodders)	*Alternaria destruens*	NCI	Cook et al. 2009
Cyperus rotundus (purple nutsedge)	*Dactylaria higginsii*	Unable to mass-produce	Kadir and Charudattan 2000
Hydrilla verticillata	*Fusarium culmorum*	Aquatic risks	Smither-Kopperl et al. 1998
Imperata cylindrica (cogongrass)	*Bipolaris sacchari*	Lacked field efficacy	Yandoc et al. 2004
Pontederia (Eichhornia) crassipes (waterhyacinth)	*Cercospora piaropi (C. rodmanii)* *Alternaria eichhorniae*	NCI NCI	Shabana et al. 1995; Tessmann et al. 2008
Senna obtusifolia (Sicklepod)	*Alternaria cassiae*	NCI[a]	Charudattan et al. 1986; Bannon 1988
Several weedy grasses	*Drechslera gigantea, Exserohilum longirostratum*, and *E. rostratum* – fungal cocktail	NCI	Chandramohan et al. 2002

[a]The company Mycogen developed and tested a product named CASST, but commercial development was abandoned due to the product's low efficacy in the field and competition from newly introduced chemical herbicides that were more effective and cheaper to use.

(Smolder), and difficulty in production (Dr. BioSedge and Woad Warrior). The last two contain rust fungi that can only be produced on living host plants – a slow, tedious, low-yielding process. The registration status of Camperico, *Xanthomonas campestris* pv. *poae*, developed and registered by Japan Tobacco Inc. to control *Poa annua* (annual bluegrass) in golf courses in Japan (Imaizumi et al. 1997), is unknown and is presumed to have lapsed.

There are several reasons for the extremely limited use of bioherbicides, following, in order of importance here and then described in detail:

1) Cost to the bioherbicide users
2) Inconsistent performance or incomplete weed control
3) Difficulty in gaining acceptance of the still unfamiliar products by users who are accustomed to chemical herbicides
4) Limited understanding and capabilities in formulating bioherbicides
5) Bioherbicide literature describing pathogens that are useful as bioherbicides

The cost of bioherbicides to users is an inherent drawback. Unlike classical biocontrol (with insects or pathogens) that are discovered, developed, and used for the common good with public funds, usually paid for by research grants through universities, regional governments, and national governments, the bioherbicide follows two distinct steps. The first is the discovery to proof-of-bioherbicide-potential research that is typically carried out in academic and or governmental laboratories with public funds as research grants. The second is the development of a bioherbicide product, its registration, and commercialization, which are expected to be borne by commercial enterprises. The latter phase requires investment in an agent-specific technology for mass production of the pathogen, formulation of inoculum biomass into a product form, rigorous testing of the product to assure its performance, stability, storability, and registration. Generally, for registration some new or supplemental data beyond those generated with public funds during the initial phase would be required. Important among such data are the nontarget plant safety and toxicology generated at the Tier I level of stringency. Generally, toxicology of the end product (formulated product) may be required, and it should be done by certified commercial testing laboratories. The mandatory pesticide (bioherbicide) label is written with the help of field efficacy data, application specifications, application methods, and other information gained from actual trials with the bioherbicide product. Besides, outlays for product promotion, marketing, liability insurance, and other product-related expenses would add further recurrent costs to the registrant (the commercial entity).

All of these costs should be weighed against the potential returns from the bioherbicide's sales. The unit cost of the bioherbicide to the buyer (cost per hectare of weed controlled) should be comparable to or better than those of competing products and methods. A bioherbicide that cannot be produced and sold at an affordable price due to inherent high costs is unlikely to be pursued. As we have learned over the years, unless a bioherbicide promises significant returns for the investment, medium and large companies are unable to justify their entry into the bioherbicide market. Small companies, particularly lifestyle businesses, are a possible alternative, and this approach should be explored vigorously, not only for bioherbicides but also for biopesticides in general.

The second reason is the problem of inconsistent performance or incomplete weed control. Inconsistent performance can be a true problem with a product, but incomplete weed control could be often a perceptual issue stemming from comparisons of the bioherbicide's level of weed control with a competing chemical herbicide. Inconsistency often results from the pathogen having a precise set of ambient conditions for infection and disease development that are difficult to achieve consistently in nature. In this regard, foliar fungal pathogens are more likely to fail to perform in the field compared to vector-aided, root-infecting, and wound-invading fungal pathogens, bacteria, and viruses that are generally unaffected by small differences in ambient conditions during infection. Weeds that have strong vegetative reproductive capability (e.g. aquatic weeds, vines, plants with stolons or resilient roots, grasses and nutsedges with central, well-shieled apical meristems) are difficult to kill completely by pathogens and will tend to regrow even after sustaining a severe disease-inflicted loss of photosynthetic tissues. Consequently, the bioherbicide's effect would appear impressive at the beginning only to turn increasingly disappointing as the new growth replaces older diseased and dead tissues. Consequently, the product gets "ineffective" rating from unsatisfied users (Charudattan 2005).

However, not all bioherbicides, experimental or registered, are prone to perform inconsistently. From our experience, pathogens that completely kill their weed targets fairly quickly as the outcome of the disease they cause (e.g. wilt, systemic necrosis) as opposed to those that cause nonlethal diseases expressed as transient or chronic stress (e.g. foliar leaf spots and lesions) are the most effective bioherbicide agents. For example, *C. gloeosporioides* f.sp. *aeschynomene* (first registered as Collego) can kill northern jointvetch (*Aeschynomene virginica*), an annual weed in rice and irrigated soybean in Arkansas, Mississippi, and Louisiana in the United States. While the pathogen does not kill the weed after an initial application, it causes lesions on stems, petioles, and leaves. These lesions expand rapidly, allowing the fungus to colonize tissues around, beneath, and above the lesions along the stem. The expanding lesions girdle the stem and cause plant parts above the lesions to be killed. The fungus produces secondary inoculum in each lesion, greatly multiplying the number of spores in the field. The spores are dispersed from the lesions by multiple means: by water, wind, contact between infected and healthy plants, and by grasshoppers and green tree frogs that move about on the stem. The secondary dispersal of spores causes an exponential increase in the level of disease on each plant, typically hundreds of lesions per plant, as well as throughout the weed population in the field. This disease increase shortens the time needed for field-wide control of the weed. The fungus can persist from year to year on infected northern jointvetch stubble and infected seeds, but not in soil or irrigation water for more than a few months. Seedlings emerging from infected seeds become diseased. To supplement this natural survival and reinfection of the weed following an initial application, annual repeat applications of Collego are made to provide acceptable levels of northern jointvetch control (National Research Council 2011, p. 34).

Another example of a killer pathogen is the virus *Tobacco mild green mosaic tobamovirus* (Tobacco mild green mosaic virus, TMGMV). It kills the exotic noxious weed *Solanum viarum* (tropical soda apple) by eliciting systemic necrosis resulting from a specific, genetically programed interaction between the virus and *S. viarum*. As such, once the weed is infected, the virus never fails to kill it completely, including the root system. About eight days after manual inoculation, *S. viarum* develops necrotic foliar local lesions in inoculated leaves only – a manifestation of a resistance reaction to the virus. Typically, local lesions to a virus infection confine the virus to the lesions, halting its further spread, and the rest of the plant remains healthy. In *S. viarum*, the virus "escapes" from the lesions and spreads asymptomatically throughout the plant until the plant begins to wilt and then is completely killed with the onset of rapid systemic necrosis (Charudattan and Hiebert 2007; Charudattan et al. 2020) (Figure 9.1). TMGMV was registered as the bioherbicide SolviNix LC in 2014 (Charudattan and Hiebert 2015), the first worldwide example of an herbicide containing a plant virus as the a.i. and SolviNix as the first virus-containing bioherbicide. SolviNix is used for eradicating small, scattered infestations of the tropical soda apple in small- to medium-sized ranches, woodlands, and natural areas. In nearly 15 years of research use followed by 6 years of commercial use, TMGMV has not failed to kill *S. viarum*, and there is no evidence of natural host resistance to TMGMV among *S. viarum* acquisitions from the United States or New South Wales, Australia (Charudattan, unpublished).

Other notable examples include *C. purpureum* (Chontrol, registered for the control of *Alnus* and *Populus* spp.) and a combination of three pathogens, *Lasiodiplodia pseudotheobromae, Neoscytalidium novaehollandiae*, and *Macrophomina phaseolina* (all Dothideomycetes) (Di-Bak Parkinsonia, mentioned previously). The former is a ubiquitous wound-invading fungus occurring in temperate regions. It naturally invades wounded and weakened dicot trees and shrubs, infects the cambial layer, and gradually colonizes and kills the plant. When applied to cut stumps of weedy broadleaved trees and shrubs, it causes the regrowth to wilt and eventually kills the stump (Hintz

Figure 9.1 Sequence of disease and symptoms development in TMGMV U2-infected tropical soda apple plants in the field. A healthy, uninfected plant (A); foliar necrotic local lesions (B); chlorosis of the plant with epinasty in shoot apices and onset of wilting (C); wilting of the entire plant (D); advanced wilting and drying of the plant with partially green branches (E); fully wilted, dried plants (F); and the effects on fruit: top row, right to left: unaffected mature green fruit and mature, ripe, yellow fruit; fruit viewed from adaxial (stalk) and abaxial sides; bottom row, right to left: mature green fruit with rotting around the stalk, rotting of mature yellow fruit, and two stages of dead, dried fruit (G) Reproduced with permission from Dr. Nimal Chandrasena, Editor-in-Chief, WEEDS (Official Journal of the Asian-Pacific Weed Science Society, APWSS (https://weeds-apwss.scholasticahq.com).

2007). The ability to treat and suppress regrowth of a wide range of dicot tree and shrub species in several families makes Chontrol a desirable product potentially for worldwide use in the temperate regions. Di-Bak Parkinsonia is novel, for it is the world's first registered multiple-pathogen bioherbicide containing three pathogens (Galea and Goulter 2013 cited in Morin 2020). As the product's name implies, the fungi cause cankers and die-back in woody plants. The fungi are applied simultaneously to parkinsonia stem with a specially designed applicator, Injecta 800 stem injector (https://www.bioherbicides.com.au/product/1000-capsules-plug-included).

Verticillium nonalfalfae, a fungal pathogen that kills *Ailanthus altissima* (tree-of-heaven) was first discovered in Pennsylvania and later in Ohio and Virginia in the United States and in Austria (Schall and Davis 2009a; Rebbeck et al. 2013; Snyder et al. 2013; Maschek and Halmschlager 2015). A vascular wilt-causing pathogen, *V. nonalfalfae* kills *A. altissima* of all ages, from seedlings to mature trees. The disease spreads naturally in the field from infected to healthy *A. altissima* trees through root grafts, thereby increasing the number of trees that are killed in addition to the primary infected trees. Several published papers confirm the bioherbicide potential of this fungus (Schall and Davis 2009a, 2009b; Kasson et al. 2014, 2015; O'Neal and Davis 2015; Brooks et al. 2020a, 2020b). It has been documented that this native pathogen has naturally killed several thousand mature (canopy) trees of *A. altissima* in the mid-Atlantic US states.

The third reason is the difficulty in gaining acceptance of the still unfamiliar products by users who are accustomed to chemical herbicides. Bioherbicides do indeed require some specific usage conditions such proper handling and storage of the product and timing and method of application – requirements that appear to deter user acceptance. Consequently, there is greater need for instructional material and outreach to the users from the seller.

The fourth is our limited understanding and capabilities thus far in formulation of bioherbicides (and of microbial pesticides in general). Formulation in this case means putting together of certain components in an appropriate relationship or structure, according to a set formula (recipe) to assure, improve, prolong, etc. of performance of a bioherbicide. In general, formulation is more important for fungal pathogens of aerial plant surfaces that require conducive ambient temperature and free moisture for growth and infection and protection from solar irradiation. So, attempts are made to protect the fungal inoculum from unfavorable temperature and irradiation while providing moisture. However, not all registered bioherbicides have needed formulation: The first bioherbicide registered in the United States, DeVine (*Phytophthora palmivora*) to control stranglervine (*Morrenia odorata*) consisted simply of an aqueous suspension of chlamydospores (Kenney 1986). Collego consisted of asexual spores produced by submerged fermentation that were slow-dried and packaged without additives. They were rehydrated *slowly* before application (Churchill 1982) using an osmotic regulator such as sorbitol. SolviNix is composed of an aqueous concentrate of purified TMGMV without additives. It is applied to *S. viarum* foliage with a high-pressure backpack sprayer without any abrasive or surfactant that are unnecessary and prohibited. Until now, the published attempts at formulating bioherbicide pathogens have been largely aimed at providing an optimum moisture level to aid fungal infection and leaf penetration with the help of adjuvants (Boyette et al. 1996). Haynes and Boyetchko (2006) and Ash (2010) have aptly pointed out the need for a concerted effort to develop formulations by involving a multidisciplinary team of experts to optimize bioherbicide yield, efficacy, storage stability, and delivery. An aspect of formulation that is of considerable interest is the application of nanotechnology (Kremer 2019; Vurro et al. 2019). Of course, most bioherbicide pathogens, now and in the past, are fungi with spores and cells that fall far outside the nanoscale range of between 1 and 100 nm in at least one dimension. Kremer (2019) and Vurro et al. (2019) have proposed examples of microbial metabolites with herbicidal activity (biorational herbicides) that could be nano-formulated using nontoxic biological material (e.g., chitosan, cyclodextrin, lignin, zein, others). Nano-formulations using these natural materials can improve efficacy, reduce effective dosage, enable controlled-release, protect against biodegradation, and increase persistence of biorational herbicides. Unlike synthetic molecules and metals, biomolecules would be nontoxic without the problem of environmental buildup. In addition, with newer genetic engineering capabilities such as CRSPR gene editing, it is possible to generate novel biorational herbicides for nano-delivery. Plant viruses are nature's ready-made nanomaterials that could be used to formulate conventional herbicides, biorational herbicides, and botanical extracts, a topic discussed in the section on nanotechnology.

Finally, the bioherbicide literature is replete with reports of pathogens claimed as potential bioherbicides, or in other words, pathogens that appeared to have usefulness as bioherbicides. Before a pathogen could be considered a true bioherbicide candidate, it should have been demonstrated to inflict an acceptable level of weed-control activity in scientifically valid field trials before it could be termed a "potential bioherbicide." Even among promising candidates, most fail to advance beyond the initial claims. The majority of these initial claims are based on university research with no follow-up. The reasons for the lack of follow-up are never published as there is no incentive to publish such information. In Table 9.1 we provide a few examples of bioherbicide agents from our work that had the attributes to justify their further development (i.e., they were "potential bioherbicides") but were abandoned and the principal reason for the abandonment. The most common cause for abandonment is the lack of commercial interest. *Alternaria destruens* and *C. piaropi* (as *C. rodmanii)* were in fact commercially developed to an extent, and the former registered, but were given up by the companies on perceptions of insufficient efficacy and doubtful commercial prospects. There was considerable interest in developing *D. higginsii* for purple nutsedge control but the fungus, an extremely slow grower, could not be mass produced on solid or liquid substrates. *Bipolaris sacchari* by itself was insufficiently damaging to be capable of suppressing cogongrass.

F. culmorum isolated from diseased *Stratiotes aloides* (water soldiers or water pineapple, in the same family as hydrilla, Hydrocharitaceae) killed hydrilla shoots when incubated in a suspension of 10^5 spores per ml. Hydrilla in mesocosms of about 1,900 liters (500 gallons) could be killed but the logistics of producing and inoculating natural waters with such a massive inoculum load (microbial load) and the risk of introducing *Fusarium* mycotoxins halted further interest in this project. Although a *C. piaropi* (*C. rodmanii*) bioherbicide did not materialize, this pathogen is widely prevalent in many tropical and subtropical countries where it causes periodic epidemics on the invasive weed waterhyacinth (Martyn 1985; Morris 1990; Jimenez and Charudattan 1998; Charudattan 2001; de Ávila and Pitelli 2004; Tessmann et al. 2008; Firehun et al. 2013; Mutebi et al. 2021). It has been also intentionally introduced and established in Mexico and South Africa as a classical biocontrol agent for waterhyacinth (Winston 2021).

9.2 Expectations for the Future of Bioherbicides

Undoubtedly, biological control will have its place in weed management in the coming decades. It is a reasonable expectation that classical biological control of weeds with pathogens would remain relevant, given that the science and practice of this approach is well established and there is opportunity to build on the past experiences. Worldwide, the incursion of invasive weeds appears to be never abating, and one of the best ways to manage them is with classical biological control (with insects and pathogens). The returns from investment in successful classical biological control of weeds would continue to add over years with little or no additional expenditure of funds. To maximize the returns from investment in research, it would be prudent to have both entomologists and plant pathologists collaborate in every classical biological control project from the discovery and agent-evaluation phases onwards as is done in Australia, New Zealand, and South Africa. For bioherbicides to remain relevant, researchers should continue to discover prospective pathogens, understand the molecular basis of pathogenicity, host-specificity, and efficacy in weed-pathogen systems, improve methods of mass-production, formulation, and application including application tools, and find newer ways to bring bioherbicides to the market at price levels that are attractive to users. In addition, understanding how certain pathogens kill plant cells, tissues, or entire plants should be a high priority research objective to discover genetically based weed control methods. To have wider acceptance in the coming decades, bioherbicides of the future should be capable of controlling multiple rather than single weed targets. Their efficacy should be based on certain stable gene functions or metabolites that can be regulated (tweaked) to assure and improve the agent's efficacy independent of minor fluctuations in environmental conditions. Westwood et al. (2018) in their forecast of weed management in 2050 have stressed the key role that genetic engineering is expected to play in creating new approaches to weed management while advances in understanding plant pathogen interactions would contribute to developing new biological control agents. We strongly agree with this prognosis.

9.2.1 Deleterious Rhizobacteria to Suppress Weed Growth and Competition

The role of microorganisms that secrete plant growth–inhibitory metabolites has been well studied and, in this regard, certain bacteria in the rhizosphere, the rhizobacteria, have been explored as possible bioherbicides. Rhizobacteria (aka root-colonizing bacteria) that have an inhibitory effect on plant growth are often referred to as deleterious rhizobacteria (DRB) (Suslow and Schroth 1982; Kremer and Kennedy 1996). They are characterized as nonparasitic (exopathogenic) microbes. Despite their ubiquitous nature in soil, this group of bacteria have often been ignored because their symptoms are subtle and not obvious in the field. DRB are selected for their ability to attack a target weed when they are at their most vulnerable state: at the seed germination stage and/or the early seedling stage. Some of the key factors to consider when selecting DRB as weed biocontrol agents is their root-colonizing ability, chemotactic attraction to colonize roots that are mediated by root exudates, ability to produce secondary metabolites that affect the target weed, and ability to be rhizosphere competent (Kremer et al. 1990; Kremer and Sarwar 1995).

A number of projects in the United States and Canada using DRB for biological control have been summarized (Kremer and Kennedy 1996; Boyetchko 1999; Rosskopf et al. 1999a). Some of the first reports of root-colonizing *Pseudomonas* spp. on plants were described on wheat, potato, and sugarbeet (Kremer et al. 1990; Elliott and Lynch 1985; Fredrickson and Elliott 1985). The potential of DRB being pursued as biological weed control agents has been reported for 18 weed species, including velvetleaf (*Abutilon theophrasti*), lambsquarters (*Chenopodium album*), pigweed (*Amaranthus retroflexus*), smartweed (*Polygonum* spp.), cocklebur (*Xanthium* spp.), downy brome (*Bromus tectorum*), wild oat (*Avena fatua*), green foxtail (*Setaria viridis*), and jointed goatgrass (*Aegilops cylindrica*) (Cherrington and Elliott 1987; Kremer et al. 1990; Kennedy et al. 1992; Boyetchko and Mortensen 1993; Kremer and Kennedy 1996; Boyetchko 1999). The majority of the projects targeted annual weeds in cereal and row crops, and several of these projects focused on preliminary screening of bacteria in laboratory experiments.

Kremer (1987) found specific bacterial associations with five annual weed species where phytopathogenicity was observed in approximately 15% of all bacterial isolates evaluated, with the majority of the inhibitory bacteria residing in the rhizosphere belonging to Gram-negative bacteria. Many of the bacterial isolates belong to the genus Pseudomonas, but additional bacteria identified as deleterious to plants include *Enterobacteria*, *Klebsiella*, *Citrobacter*, *Arthrobacter*, and *Achromobacter* (Suslow and Schroth 1982). Gram-negative bacteria are found to readily colonize *the* rhizosphere because they utilize a variety of carbon sources in the root exudates, thereby producing various secondary metabolites such as hydrogen cyanide, extracellular enzymes, siderophores (iron-chelating compounds), and other antibiotics that compete and suppress other microbial groups (Kremer 1987, 1993; Sarwar and Kremer 1995a, 1995b; Kremer and Souissi 2001; Dar et al. 2020). Seedling vigor and the nonviable portion of seeds in soil were reduced, while seedling emergence was decreased by 55% with the soil application of two bacterial isolates.

Dar et al. (2020) reported several weed-suppressive bacteria in canary grass (*Phalaris minor*) and wild oat that were HCN-producing while they also produced siderophores, tested positive for oxidase and catalase activity, and caused plant-growth promotion in wheat through phosphorus solubilization. Screening of several bacterial strains against annual grass weeds – such as downy brome, green foxtail, and wild oat – have been conducted in Canada (Boyetchko and Mortensen 1993; Kennedy et al. 1991; Boyetchko 1997). In many instances, the weed-suppressive activity of DRB has been attributed to the production of secondary metabolites and phytotoxins (Kremer and Kennedy 1996; Boyetchko 1999). Partial purification of a phytotoxin from the *P. fluorescens* D7 strain caused root discoloration and reduced root elongation (Tranel et al. 1993b), and this toxin was found to be complex, consisting of at least two polypeptides, fatty acid esters, and lipopolysaccharides (Gurusiddaiah et al. 1992). Attempts to further purify the compound resulted in near complete loss of phytotoxicity, suggesting that either the purification process damaged the phytotoxin or that more than one component in the complex is necessary for activity. Further investigation into the mode of action revealed that the active fraction from the D7 strain affected lipid synthesis and membrane integrity while there was no effect on cell division, respiration, or protein synthesis (Tranel et al. 1993a). Other researchers have also demonstrated that the overproduction of auxin (indole-3-acetic acid [IAA]) of some bacterial strains, when L-tryptophan as the precursor is added to the media, results in greater root length inhibition of field bindweed (Kremer et al. 1990; Sarwar and Kremer 1995a, 1995b). The production of cyclic lipopeptides is speculated to be one of the major modes of action for suppressing weeds and can often be dose-dependent (Tranel et al. 1993a, 1993b; Kremer and Kennedy 1996).

Banowetz et al. (2008) reported a germination-arrest factor (GAF) in filtrates of *Pseudomonas fluorescens* cultures. This GAF is comprised of a naturally occurring herbicide that is found in seedlots containing annual blue grass (*Poa annua*) seeds that suppress the germination of several weeds, particularly those that are graminaceous. The GAF irreversibly arrests or prevents the germination of weeds, but the chemical nature of GAF is unknown.

Evaluation of the bacteria on other grass weed species such as foxtail barley (*Hordeum jubatum*), crabgrass (*Digitaria* spp.), annual ryegrass (*Lolium multiflorum*), barnyardgrass (*Echinochloa crus-galli*), yellow foxtail (*Setaria pumila*), and Italian ryegrass (*Lolium perenne*) have indicated that some bacterial strains have broad-spectrum activity, while others have a narrow host range (Boyetchko and Mortensen 1993; Boyetchko 1997). Many deleterious rhizobacteria are known to have a wide spectrum of activity on various plant species and have often been criticized as being non-host specific. However, several researchers have discovered that there is a degree of host-specificity amongst bacterial strains (Elliott and Lynch 1985; Cherrington and Elliott 1987 ; Kennedy et al. 1991). Boyetchko and Holmström-Ruddick (1996) reported that specific bacterial strains and their metabolites that exhibited suppressive properties to downy brome did not inhibit growth of spring and winter cereal crops. In some instances, plant-growth promotion was observed in these cereal crops and the plant response was dependent on the bacterial strain and crop cultivar. However, Kremer and Kennedy (1996) suggested that researchers should recognize the risk of utilizing DRB with a wide host-range because of the potential detrimental effects on crop plants. The ability of DRB to be weed suppressive in the presence of crop plants demonstrates the importance of this group of bacteria as weed biocontrol agents (Kremer and Kennedy 1996).

Preemergent soil application of three DRB strains resulted in significant reductions in growth of two grass weeds (Sawchyn et al. 2000). Application of a bacterial strain in a granular formulation in the field resulted in reductions in aboveground biomass and weed emergence of wild oat by 57% and 64%, respectively (Boyetchko, unpublished data). Improvements in formulations resulted in significant increases in weed suppression by the DRB. Patent applications were obtained for the screening and application of bacterial agents for biological control of downy brome and jointed goatgrass (Elliott and Kennedy 1991; Kennedy et al. 1992). Spray application of bacterial strains *P. fluorescens* D7 and 2V19 to the soil surface at a rate of 108 cells/m^2 in 1 L water provided strong weed suppressive activity against natural populations of downy brome. The inhibition of the weed led to increases of winter wheat yields by 18% to 35%, and the increases in wheat yield were attributed to reductions in the competition by downy brome (Fredrickson and Elliott 1985). Johnson et al. (1993) showed that soil factors play a significant role on the efficacy of DRB. Temperatures of 10–15 °C resulted in greater colonization and inhibition of downy brome roots, and moist, cool soils were found to be more conducive to root colonization of *P. fluorescens*. When specific DRB strains were applied to soil in combination with reduced rates of herbicides (such as metribuzin and diclofop), significant increases in weed suppressive activity were observed.

Canadian researchers reported a *P. fluorescens* strain BRG100 as a preemergent bioherbicide for control of green foxtail (Boyetchko and Mortensen 1993; Boyetchko 1997). This bacterium has further exhibited its ability to control herbicide-resistant green foxtail populations: Group 1 (accase inhibitor) and Group 3 (trifluralin resistant) (Boyetchko, unpublished results). The secondary metabolites produced by *P. fluorescens* strains BRG100 and 189 were characterized as a complex mixture of cyclic peptides and X-ray crystallography determined the chemical structure of two compounds with both herbicidal and fungicidal activity, named pseudophomin A and pseudophomin B (Quail et al. 2002; Pedras et al. 2003). They are complex cyclic peptides with nine amino acid residues and a lipophilic portion with water soluble peptides. *P. fluorescens* BRG100 was transformed using the green fluorescent protein in order to visualize root colonization and environmental fate of the bacterium (Hanson 2008; Caldwell et al. 2012).

Pseudomonas fluorescens BRG100 was found to be an effective root colonizer and disperses from a granule formulation based on an oat flour matrix (called "pesta") that is delivered as a soil-applied bioherbicide (Daigle et al. 2002). Factors such as soil moisture, temperature, and soil type have an effect on the dispersal of the bacteria from the pesta into the soil (Daigle et al. 2002; Hanson 2008; Hynes and Boyetchko 2011). Amendments to the pesta formulation were made and a 16-month shelf life at 8.5 log10 colony-forming units (cfu) per gram of pesta was reported (Hynes and Boyetchko 2011). A dryer water activity (moisture content) of 0.2 aw of the pesta was required compared to 0.8 aw where the bacterial population decreased to 7.3 log10 cfu/g over only six months. Release, and thus dispersal of the bacteria from the pesta granule is critical to establish the bioherbicide activity of *P. fluorescens* BRG100. Amendment of the pesta with various starches improved dispersal of the bacterium from the granule where pea>potato>corn>rice starches increased disintegration characteristics of the pesta granule. However, it could not be demonstrated that the bioherbicidal properties of the pesta-formulated *P. fluorescens* was also improved. Formulation is a technology for this soil-applied, preemergent bioherbicide that remains unresolved before these bacteria can be considered an effective weed biocontrol. However, it shows promise as a bioherbicide to control green foxtail and herbicide-resistant weeds.

Zdor et al. (2005) also reported that formulations influence the inhibitory effects of DRB toward green foxtail and velvetleaf seedlings. Semolina flour, as a formulation, was compatible with DRB, but the presence of antimicrobial properties in corn gluten meal negatively impacted the activity of the *P. fluorescens*. In addition, soil type, such as soils containing clay, influenced the activity of the DRB. Soils such as a fine sandy loam contain less clay content which interferes with the phytotoxic substances produced by the DRB that affect the weed-seedling growth.

9.2.2 Weed Seedbank Management

Weed seedbanks represent a major cause of weed infestations in agricultural production (Kremer 1993, 1998). It has been suggested that weed seeds entering and surviving in the soil for at least a year comprise the persistent portion of the seedbank. Weed-associated microbes can either indirectly attack seeds under specific agricultural cropping systems or can directly attack seeds as soil-applied biocontrol agents. Kremer (1993) therefore proposed that soil microorganisms can be exploited to deplete the weed seedbank. Although quantitative evidence on the mortality of seeds by soil microbes is limited, the impact of microbial activity on seedbank mortality has been indirectly documented through buried seed studies. Kennedy and Smith (1995) provided supporting evidence that suggested only a fraction (<1%) of the 1 million to ten billion bacteria contained in 1 g of soil are known or have been partially characterized. Yet, the soil represents an excellent resource of genetic information and biological activity.

Aldrich (1984) was one of the first to suggest that weed seed numbers could be reduced in the soil by using microorganisms to remove part of the seed coat that provides protection against seed deterioration. Seedborne fungi associated with velvetleaf seeds were capable of attacking the seeds via seed coat openings (Kremer 1993). If the seed coat of velvetleaf was punctured, more than 50% of the seeds and developing seedlings were attacked by the fungi. The isolation of several fungal and bacterial microbes from the seeds of giant foxtail and velvetleaf resulted in a reduction of root growth by up to 85% in lettuce seedling bioassays (Kremer 1998). Production of phytotoxins by seedborne bacteria may be responsible for affecting seed viability and seedling vigor (Fredrickson and Elliott 1985). On the other hand, fungi infesting wild oat seeds were not found to be pathogenic in seed bioassays, suggesting that poor seed viability could not be solely attributed to seedborne fungi.

Depletion of the weed seedbank with soil microbes has tremendous potential; however, we have very little understanding of the impact of these groups of organisms on seed deterioration and decay. Kremer (1993) proposed that the ability of these seedborne microbes to chemotactically respond to exudates diffused into the soil by intact or germinating weed seeds, their ability to rapidly colonize the seed and produce toxins or other weed-inhibitory secondary metabolites were key characteristics for these soil microbes to manage weed seedbanks. However, it is critical to understand the environmental factors in the soil that affect seed decomposition and weed growth and development and how to overcome variable and inconsistent field performance (Kremer and Kennedy 1996).

9.2.3 Future Developments

DRB provide a very promising option as a bioherbicide weed control tool, particularly where application of a foliar pathogen is not feasible, but soil applied bacteria are more practical and achievable. However, more effort into research on the formulation and application technology is imperative. Preliminary testing of DRB by researchers often entailed the application of the bacteria as a soil drench, which does not examine the realism of the weed management tool in crop production. How the DRB are introduced into the field, either as a seed treatment to the crop, a soil drench, a granular formulation, or other formulation will certainly resolve how the user will apply the bioherbicide in order to reach the target weed. If applied as a granule, how are the bacteria released from the granule in order to target the weed and how does soil moisture affect release? Also, the water activity of the granule may affect how the bacteria reach their target. If the granule is too dry, the DRB may not be as readily available and released from the formulation. Along with the formulation issue, a simple application of the formulation to the soil surface will not suffice, but examination of the principles for incorporation of the formulated DRB to achieve effective weed control will be necessary. This means comparing broadcast application, side-row banding, and mid-row banding of the DRB at seeding time when the weed seeds start to emerge, thereby reducing their competitive ability to the crop and minimizing crop yield losses. This entails greater intellectual input from agronomists and weed ecologists who understand the principles of critical period (which involves understanding weed and crop competition) in order to resolve this issue, which is lacking in the literature. Agronomists will understand how to incorporate the DRB into a particular crop rotation and how farm management will affect the application of the bioherbicide. For example, does application of fertilizers or other chemicals have a detrimental effect on the DRB and how can we resolve this? These are practical questions that must be answered before this is a useful weed management tool for farmers. We have much to understand, but recognizing the practicality of introducing DRB for weed management will become a greater possibility.

9.3 Nonchemical Soil Disinfestation

Some multi-kingdom soil pest control measures that have been investigated originated through attempts to develop nonchemical, integrated pest management tactics that could replace the use of broad-spectrum fumigants. These include anaerobic soil disinfestation (ASD), steam, soil solarization, and biosolarization, all of which alter the soil microbiome to varying degrees and result in weed and disease suppression through an array of complex mechanisms, both physical and biological. Improved microbial community analysis techniques, particularly DNA sequencing approaches that enable whole-community characterizations at multiple taxonomic levels, are leading to a significantly more in-depth understanding of the biologically derived mechanisms associated with pest control efficacy of these methods and the role that microbes play in these pathways (Hewavitharana et al. 2014; Strauss and Kluepfel 2015). A recent review of the impacts of these approaches on plant pathogens and nematodes is available by Rosskopf et al. (2020).

9.3.1 Anaerobic Soil Disinfestation (ASD)

Anaerobic soil disinfestation (ASD) is one of the nonchemical, preplant soil treatments that is useful in an integrated approach to weed management. ASD, also referred to as "reductive soil disinfestation" (Shinmura et al. 1999) and "biological soil disinfestation" (Blok et al. 2000; Goud et al. 2004), is a more recently developed technique that has evolved from work conducted independently in both Japan (Shinmura et al. 1999) and the Netherlands (Blok et al. 2000) adapted for both protected and open field production systems, with a primary focus on soilborne plant-pathogenic fungi and bacteria. Various approaches to this method, adapted to regional resources and production practices, have been adopted by small-scale growers internationally.

This soil treatment method includes the incorporation of organic amendments that contain readily decomposable carbon (C), irrigation of the soil to saturation, and then coverage with a gas-impermeable mulch. ASD treatment duration can range from three weeks to several months, depending on the cropping system and the target pests (Butler et al. 2012a; Rosskopf et al. 2015; Di Gioia et al. 2016; Guo et al. 2017; Browne et al. 2018). During the ASD process, the soil microbial composition shifts to facultative anaerobes that utilize the amended carbon source and deplete the soil of oxygen. In addition, the soil pH is lowered and short-chain fatty acids such as acetic, butyric, lactic, citric, isovaleric, and propionic acids are produced. In addition, there are increased quantities of aldehydes, alcohols, metal ions (primarily Mn^{2+} and Fe^{2+}), and other volatile organic compounds (VOC) that are toxic and or suppressive against several soilborne pathogens, phytoparasitic nematodes, and weeds are released (Momma et al. 2006; Momma 2008; Oka 2010; van Agtmaal et al. 2015; Guo et al. 2018).

This method was originally developed with control of soilborne plant pathogenic fungi as the target, but the observed, wider spectrum of activity that resulted from ASD application has led to the evaluation of the approach for control of various weeds as well as plant parasitic nematodes. Rosskopf et al. (2010) reported that ASD treatments using composted poultry litter and feed-grade blackstrap molasses as substrate applied under clear film, provided good control of nutsedge and excellent control of grass and broadleaf weeds. In addition, the control of grass and broadleaf weeds with all combinations of litter and molasses was excellent when compared to the untreated check and was equivalent to, or better than methyl bromide. Rosskopf et al. (2010) further noted that most ASD treatments provided nutsedge control that was significantly better than solarization alone but was not equivalent to methyl bromide. Butler et al. (2012b) found that, in greenhouse trials, the use of molasses as the C source resulted in the reduction of yellow nutsedge germination to 31% and use of the cover crops sorghum-Sudangrass, pearl millet, cowpea, and sunn hemp resulted in germination between 49% and 61%. Di Gioia et al. (2016) found that ASD treatments applied using an opaque, totally impermeable film, were successful in suppressing weeds such as nutsedges by 85%, which was enough to prevent an impact on yield. Work reported by Khadka et al. (2021a) further confirmed high levels of grass weed control resulting from ASD applications using various C sources. Liu et al. (2016) demonstrated weed suppression in greenhouse trials using a variety of carbon sources and interestingly, further redroot pigweed germination with the addition of ethanol to spent brewer's grain. Ethanol use for ASD in the United States may not be permitted, but the authors were able to obtain similar results by adding yeast with several C sources, which improved weed control. Shrestha et al. (2018) reported that ASD treatments amended with dried molasses and wheat bran mixtures at four carbon-to-nitrogen (C: N) ratios resulted in significantly reduced tuber germination and biomass accumulation of yellow nutsedge (*Cyperus esculentus)* compared to those in the non-amended control when initial tubers were placed at 15 cm. In addition, *C. esculentus* above- and below-ground biomass was significantly reduced when wheat bran-based amendments were compared to plant growth in the molasses treatments. Although a meta-analysis (Shrestha et al. 2016) led to the conclusion that liquid C sources resulted in improved weed control (Figure 9.2), it may be that the composition of the amendment and availability of the C source for microbial utilization plays a more critical role than previously identified.

9.3.2 Steam

Steam has been used for the management of soilborne pests for many years. In fact, steaming is one of the oldest methods of soil disinfestation used in nurseries and greenhouse crop production systems to control soilborne pathogens and pests including weeds (Baker and Roistacher 1957). Steam weeding involves the application of steam either to weed foliage or to soil at temperatures of approximately 100 ºC for a specific duration of time, depending on the target.

Direct contact with leaves causes intracellular water to expand and rupture the cell membrane thus causing weed dehydration and mortality. Steaming not only enables more rapid transfer of lethal heat into the cell structure of a weed but it will penetrate deep into the crown of the plant and can destroy the growth point (Kanissery and Ampatzidis 2020). Kolberg and Wiles (2002) evaluated the effectiveness of foliar steam application compared with herbicides for the control of cropland weeds under field conditions. They found that steaming (3,200 kg/ha, energy dosage equivalent to 890 kJ/m^2, speed of 0.8 m/s) and glyphosate (560 g ai/ha) treatments resulted in similar control (> 90%) of the common lambsquarters (*Chenopodium album*) and redroot pigweed (*Amaranthus retroflexus*) seedlings. The steam applied at the higher rate significantly reduced weed biomass for up to 9 weeks, while steam applied at a rate of 1,600 kg/ha (1.6 m/s) was not significantly different from the control.

Using steam application as a burndown approach may not provide complete control. Kanissery and Ampatzidis (2020) have observed regrowth of weeds a few weeks after steaming and suggested that steam treatment may require repetition to maintain control throughout the production season.

Success of steam application to foliage also depends on thorough canopy and stem penetration. Thus, steam application is relatively ineffective on weeds like pusley due to their prostrate (creeping) growth nature, which prevents adequate delivery of steam on the entire plant (Kanissery and Ampatzidis 2020). Moreover, the amount of steam applied, weed species, and growth stage are key factors in achieving the desired level of control (Kolberg and Wiles 2002). Although limited application of steam directly to foliage does occur, other approaches to thermal weed control, such as flaming and soil steaming have become more common (Ascard et al. 2007).

It is well-established that many plant pathogens, insects, and weeds are killed when moist soils are heated to maintain temperatures of 65 °C or greater for 30 minutes (Baker and Roistacher 1957). Others have reported that steam application must raise soil temperatures to 60 °C in the 0–15 cm layer to control weeds in vegetable crops and much deeper for longer season crops, such as strawberry (Baker 1962; Fennimore et al. 2014).

Weed seeds can be killed when steam is directly injected into the soil so that the soil temperatures reach >60 °C and

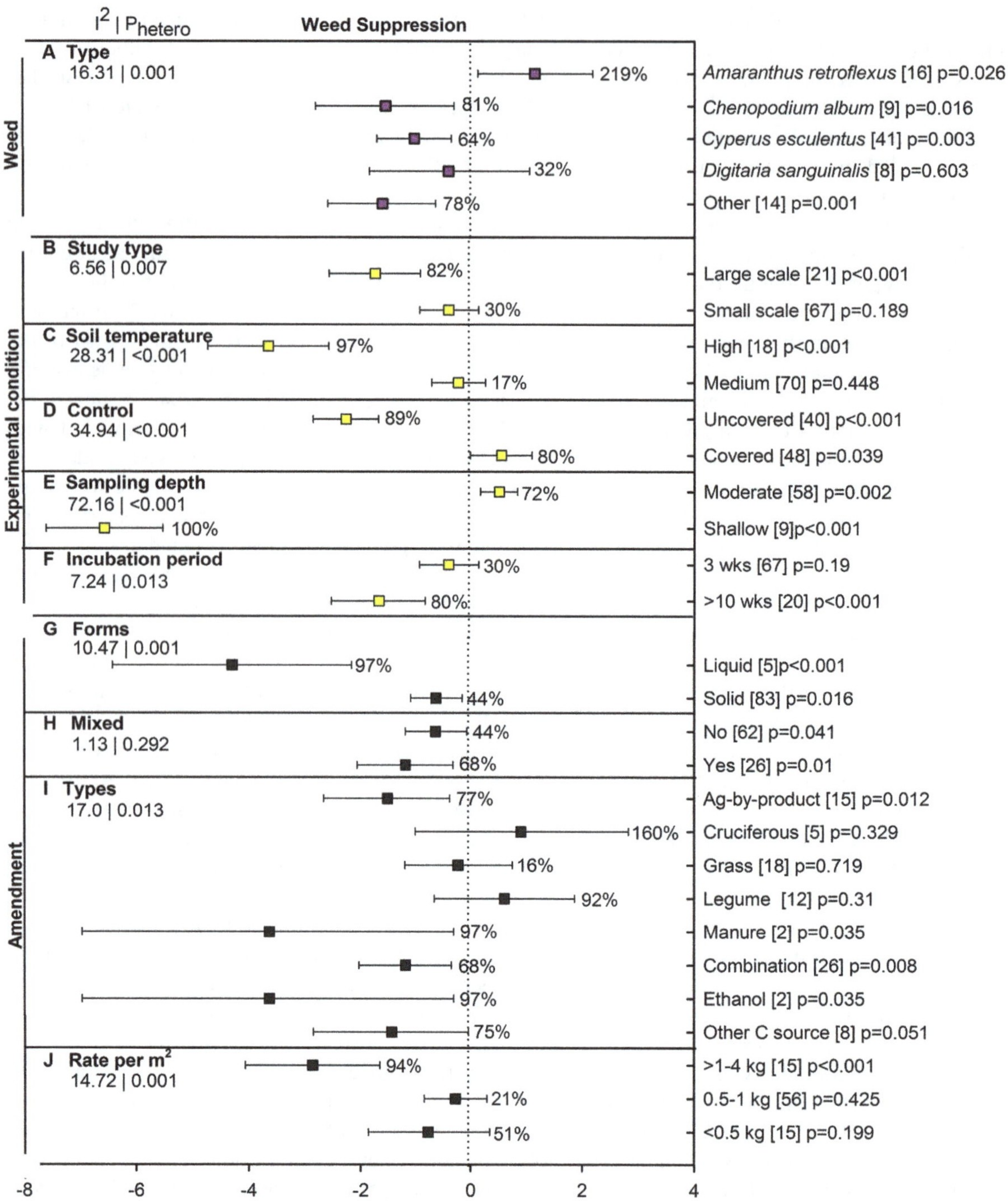

Figure 9.2 Weighted summary effect sizes (lnR) and 95% CIs for ASD effect on weed suppression. Comparisons among levels of (A) weed type, (B) study type, (C) soil temperature, (D) control, (E) sampling depth, (F) incubation period, (G) forms, (H) mixed, (I) types, and (J) rate per m^2. For each level of moderator, values to the right of the CI line with negative effective size are percent weed suppression and with positive effect size are percent of promotion. Number of studies reporting data for each level of moderator is given in parentheses. The moderator level was significantly different from zero if p-value ⩽ 0.05. Values below panel titles to the left are I2 (percentage of heterogeneity due to true variation among moderator levels) and Phetero (test of the null hypothesis, that all studies share a common effect size if Phetero > 0.1) for each moderator. Copyright © 2016 Shrestha, Augé and Butler.

are maintained for 15–20 minutes (Guerra and Fennimore 2020). In this study, weed control in the treated area persisted for several weeks or months, similar to the effects of a preemergence herbicide. Guerra and Fennimore (2020) steam-treated soil with and without hydrogen peroxide, an organically approved compound that helps improve heat distribution. They reported that in one trial, steam and steam + peroxide resulted in 90% and 92% weed control relative to survival of seeds in soil that was not steamed. In another trial, steam and steam + peroxide resulted in 66%

and 84% weed control, respectively. Hairy nightshade (*Solanum physalifolium*), shepherd's purse (*Capsella bursa-pastoris*), and little mallow (*Malva parviflora*) were among the weeds that were controlled. Steam applied alone reduced hand-weeding times 22% to 35% and steam + peroxide reduced hand-weeding times 36% to 40% compared to the nontreated. Fennimore et al. (2014) also observed that most weed seed samples placed in the fields prior to steaming were controlled by steam, with the exception of little mallow. Small-seeded weeds, including common chickweed (*Stellaria media*), burning nettle (*Urtica urens*), annual bluegrass, common purslane (*Portulaca oleracea*), as well as *C. esculentus* were effectively controlled using injected steam.

There are some limitations on the use of soil steaming that, if resolved, could significantly increase the use of the practice from organic or greenhouse production system to more widespread open-field use. Fuel for steam generation is costly, and mobile steam units are also expensive, with limited commercial availability in the United States. Treating large areas can take days, but steam can be easily combined with other disinfestation practices to improve efficacy and potentially reduce costs through increased efficiency (Michuda et al. 2021). Kim et al. (2021) found that steam and steam + mustard seed meal treatments reduced the survival of purslane seeds and nutsedge tubers. In addition, the application of steam with the soil biofumigant, allylisothiocyanate, also increased the efficacy of steam for the control of pigweed (*A. retroflexus*) and burning nettle, especially when seeds were placed farther away from the point of biofumigant and steam injection.

9.3.3 Soil Solarization

Soil solarization is a method of pest control using high temperatures produced by capturing radiant energy from the sun (Katan et al. 1976). The process involves covering soil with clear plastic for a minimum of four to six weeks when the soil will receive the most direct sunlight. The plastic tarps trap the sun's radiant energy at the soil surface, and when done properly, can result in temperatures up to as high 60 °C depending on the geographic location (Stapleton et al. 2000). The effect of solarization is greatest at the surface of the soil, with temperatures lethal to many soilborne pests, including weeds, fungal and bacterial plant pathogens, plant parasitic nematodes, and insects. These lethal temperatures are typically achieved from the soil surface to approximately 25 cm, but temperatures can easily drop as much as 10–15 °C from the surface to this depth (Chellemi et al. 1997).

To initiate soil solarization, plastic sheeting can be applied by hand or machine either for complete coverage of the planting area, or strip coverage, where either a raised bed or designated planting area of the field is covered. Elmore et al. (1997), recommend complete coverage if the soil is heavily infested with pathogens, nematodes, or perennial weeds. Treating the full field results in minimizes pests in nonplanted areas as well and reduces the likelihood of infested soil being moved into planting areas as a result of the movement of machinery, and cultivation and irrigation practices. Strip or bed applications must be coupled with other weed control measures to manage weeds in the nontreated areas but can be extremely useful for high value crops. In addition, Chellemi et al. (1997) found that strip solarization on a raised bed resulted in greater temperatures due to the heating of soil on the bed edges, where temperatures at both 5 and 15 cm were greater than those at the same depths at the center of the bed.

Weed control resulting from solarization is optimized when it can be applied during the hottest part of the year. Temperatures can be reached that will kill many weed seeds and seedlings (Elmore et al. 1997; Rubin and Gamliel 2018). Typically, soil temperatures under solarization conditions in the field range from 40–60 °C at a depth of 5 cm and from 30–40 °C at 45 cm Elmore et al. (1997). The increased soil temperatures may result in suicidal germination of weed seeds or there may be more rapid microbial degradation arising from increased microbial and chemical activity in solarized soil (Cohen et al. 2008). (Elmore 1991a) noted that although soil solarization is restricted to summer months, it can significantly reduce the emergence of species that would be problems for the production of fall or spring crops.

Research has been widely conducted to determine the effect of soil solarization on weed seeds (Elmore 1991a, 1991b; Al-Masoom et al. 1993; Saghir 1997; Yaduraju and Mishra 2004). The effectiveness of the treatment depends on precipitation events, solarization period, application parameters, as well as the weed species present (Roe et al. 2004). Since soil solarization efficacy depends on the degree to which the soil temperature is increased, the best results were reported from countries with hot climates (Al-Masoom et al. 1993; Saghir 1997) and when it is applied in warm summer months when radiation is high.

Numerous mechanisms have been proposed to explain the effect of soil solarization on weed populations. Solarization may result in foliar scorching of weeds under the plastic mulch, seed dormancy interruption by high temperatures, microbial parasitism on weed seeds weakened by sublethal temperatures (Egley 1990; Katan and DeVay 1991), and changes in cell metabolism and ultrastructure (Singla et al. 1997). In addition, imbalance of oxygen and carbon dioxide or release of acetaldehyde, ethylene, and other volatile toxic compounds may also play a role in weed mortality (Rubin and Benjamin 1984; Gamliel et al. 2000) under solarized conditions.

Different weed species vary in their sensitivity to soil heating as well (Economou et al. 1997; Elmore et al. 1997; Elmore 1998). Restuccia et al. (1994) classified weed species based on heat sensitivity as sensitive, resistant, or not well defined. Winter annual species, which are known to germinate under short-day and low-temperature conditions, are more temperature-sensitive than summer annual species, thus requiring smaller temperature increases to be effectively controlled. According to Katan and DeVay (1991), one week of solarization was enough to control many temperature-sensitive winter annuals such as *Poa, annua*, *Montia perfoiata*, and *Senecio vulgaris*. However, summer annual species studied by Egley (1990) required higher solarization temperatures with longer duration. Although summer annual weeds may be generally less temperature-sensitive than winter annuals, Elmore et al. (1997) concludes that most can be effectively managed using solarization. However, weeds such as purslane (*Portulaca oleracea*) and crabgrass (*Digitaria sanguinalis*) are often more difficult to control. Even under successful solarization treatments, these species are commonly found in planting holes once the solarization period is terminated. Multiple annual applications of soil solarization resulted in poor control of Bermudagrass (*Cynodon dactylon*), a warm-weather, creeping perennial, by the second season (Ozores-Hampton et al. 2002). When solarization can be applied during the summer months, management of even difficult-to-control weeds is achieved to some extent and the timing is ideal for use in fall-planted crops. Nevertheless, there are exceptions; white sweet clover (*Melilotus alba*), for example, is one of the few winter annuals that is poorly controlled by solarization (Elmore et al. 1997). Good soil moisture, tight-fitting plastic tarps, and high solar radiation (Roe et al. 2004) are required to control weeds for which heat-resistant propagules present a challenge to control by solarization.

Soil solarization has also been reported to be more effective in controlling annual weeds than perennials (Elmore et al. 1997; D'Addabbo et al. 2010). Masabni and Franco (2016) found that 147 days after plastic cover was removed, solarized plots had only perennial weeds. Perennial weeds may be more difficult to control due underground vegetative reproductive structures. Deeply buried roots, tubers, and rhizomes may re-sprout (Elmore et al. 1997), occurring at soil depths not exposed to lethal temperatures (Rubin and Benjamin 1984). Thermal death of seeds of perennials such as *Sorghum halepense* (Johnsongrass) and *Convolvulus arvensis*. (field bindweed) occurred only in the upper 3–4 cm of solarized soil (Standifer et al. 1984; Katan and DeVay 1991). Elmore et al. (1997) found that even though solarization controlled the seeds of Bermudagrass, Johnsongrass, and field bindweed, the rhizomes of Bermudagrass and Johnsongrass were only controlled through solarization when they occurred near the soil surface and solarization did not effectively control the rhizomes of field bindweed.

Purple and yellow nutsedges (*C. rotundus* and *C. esculentus*), subtropical to tropical creeping perennials, are especially resilient. These weeds have deeply buried rhizomes as well as extensive tuber production. Several researchers have reported sprouting of tubers of *C. rotundus*. during solarization (Egley 1983; Kumar et al. 1993; Elmore et al. 1997; Miles et al. 2002; Roe et al. 2004). In some studies, nutsedge emergence was increased as a result pronounced diurnal temperature fluctuations during solarization and was also influenced by the characteristics of the polyethylene mulch used (Miles et al. 1996; Webster 2005). While the majority of studies reported that solarization fails to control these types of perennial weeds (Duranti and Cuocolo 1988; Rosskopf et al. 1999a; Stapleton et al. 2005; Candido et al. 2008), extending the solarization period up to 8–10 weeks did improve the control of *C. rotundus* and other perennial weeds (Rubin and Benjamin 1984; Chase et al. 1998) with solarization, while in the study by Ricci et al. (2000) a 90-day solarization provided a significant reduction of the weed population which included *C. rotundus*. The application of heat-retentive and UV-stabilized films coupled with extending the solarization period further improved the control of *Cyperus* spp. due to a greater proportion of emerged weeds succumbing to foliar scorching (Chase et al. 1998, 1999a). Patterson (1998) also observed that purple nutsedge shoot emergence was reduced or eliminated under translucent polyethylene film mulches compared to a conventional, opaque, polyethylene film. Translucent mulches reduced purple nutsedge shoot biomass, tuber and rhizome number, and tuber biomass by 85–99%. Under field conditions, mixed populations of purple and yellow nutsedge were significantly reduced using clear film (Chellemi et al. 1997). Other authors have reported only a partial reduction in nutsedge emergence (Herrera and Ramirez 1996; Kamra and Gaur 1998), which may be followed by rapid recovery of *C. rotundus* following the removal of the plastic film (Lira-Saldivar et al. 2004).

Weed control efficacy using soil solarization can be improved by increasing soil moisture. Wetting soil results in transfer of heat deeper into the soil, potentially increasing control through reaching more deeply buried seeds and other reproductive structures. Egley (1990) tested multiple weed species for their susceptibility to high temperature under two moisture regimes, 2% and 19%. The majority of the tested species were killed more rapidly under the higher moisture level. This may occur through seed heat-inactivation but may also result in killing of a germinating seed whose dormancy has been broken by the changes in soil temperature. Since the heat generated from solarization can also reduce beneficial organisms, Masabni and

Franco (2016) suggest that compost should be added after solarization to replace beneficial populations.

9.3.4 Biosolarization

When solarization can be optimized, it provides significant suppression of some weed populations, but it is possible to increase the efficacy of solarization by incorporating organic materials such as compost, cover crops, manure, and crop residues into the soil prior to solarization. Treatments that combine aspects of solarization and organic material decomposition are often termed *biosolarization*. This process involves the incorporation of organic amendment into the soil followed by irrigating the soil and covering it with clear solarization film. When the sun heats the soil to temperatures greater than typical field conditions, the hot, moist soil leads to rapid microbial decomposition of the organic amendments, resulting in the production of herbicidal compounds. Gamliel and Stapleton (1993a, 1993b) documented an increase in the production of compounds such as dimethyl disulfide when cabbage residues were incorporated into soil prior to heating by soil solarization. These compounds, combined with the temperature increase, have a negative impact on soilborne diseases and weed seeds without the addition of fungicide or herbicide. Nunez and Hooks (2019) highlighted that another advantage of biosolarization is that it is suitable in climates that would otherwise be suboptimal for solarization. Unlike solarization, biosolarization is effective over a greater portion of the growing season and not just during the warmest periods of the year. Furthermore, the required treatment period is significantly less than that required for soil solarization without organic amendment, which generally takes weeks to months to be effective. For example, biosolarization effectively killed weed seeds of black mustard (*Brassica nigra*) and black nightshade (*Solanum nigrum*) within eight days after initiation. The estimated treatment time for biosolarization to be effective is 1–1.5 weeks. This may provide growers greater flexibility in planning treatment dates and early season plantings.

Jacobs (2019) tested the effects of biosolarization using Sudan grass residues on weeds and soilborne pathogens. The results showed that biosolarization significantly reduce weed biomass during the first 3.5 month after tarp removal over non-solarized plots. Overall, a combination of solarization and mulch residues resulted in the best levels of weed control and prevented the regrowth of Sudan grass. From this it was determined that biosolarization can be used in California central coast organic strawberry production to reduce hand-weeding. Similarly, in a direct comparison, Díaz-Hernández et al. (2017) reported that solarization provided no control of *Oxalis corniculata*, *Cyperus rotundus*, *Chenopodium murale*, and *Seteria verticillata*, but when goat manure was incorporated prior to solarizing, weed control was significantly improved over solarization and the application of an MITC generator.

Numerous soil amendments have herbicidal effects when applied alone. Some mustard seed meals are derived from high-glucosinulate varieties and can be phytotoxic when applied preemergent and preplant-incorporated as organic herbicides for controlling germinating and emerging weed seedlings. Webber et al. (2017) found that applications of mustard seed meal provided broadleaf, grass, and total weed control. Wang et al. (2015) investigated the effect of mustard seed meal on various weed species. They found that herbicidal efficacy varied with weed species, application rate, and method employed. Stephenson (2019) found that the effects of the organic amendments such as grape and olive-based composts appeared to alter weed species composition but not overall biomass. However, Cayuela et al. (2008) also found that olive mill waste has a high potential to be used as a biobased-pesticide against several species of weeds. While many of these types of amendments provide only partial weed control when applied alone, there is great potential in combining these materials with solarization for improving weed control as has been accomplished for several plant pathogenic fungi (Gamliel and Stapleton 1993a, 1993b, 1997).

9.3.5 Potential Mechanisms of Soil Disinfestation

Each of these nonchemical weed control approaches may have multiple mechanisms at work, but generally, steam and solarization are highly dependent on reaching lethal temperatures for each weed species. Steam weeding is weather-independent and can be applied under windy conditions without regard for precipitation events. Solarization is temperature-dependent, and its use is limited geographically and is highly weather-dependent, as pooling of water on top of solarization plastic results in weed emergence in the affected areas. Where the climate is characterized by extended periods of solar radiation with limited precipitation during the optimal application period, this approach provides an excellent option, particularly for production systems in which crops are produced after the plastic is removed. While thermal inactivation is the principal mechanism associated with solarization, there are changes in the composition of the microbial community under this approach that could also result in a biological component. For example, Gamliel and Stapleton (1993b) reported an increase in *Pseudomonas* spp. as a result of solarization, which may contribute to disease suppression, but as members of this genus are also deleterious rhizobacteria used for weed suppression (Kremer et al. 1990), they may potentially play a role in suppression of weeds with solarization.

Biosolarization, like soil solarization, uses clear plastic mulch to promote passive solar heating and the decomposition of organic amendments. This approach is characterized by the production of organic compounds similar to those that are produced during ASD, which, coupled with soil heating, provides more consistent weed control across species and regions than solarization used alone. Khadka et al. (2021b) summarize that ASD weed suppression could result from several components of a relatively complex system, including increased biological degradation of seeds as a result of imbibing with water during the saturation phase, oxygen deprivation, rapid changes in soil pH, as well as the wide range of biocidal compounds produced during the microbial degradation of the amendments including organic acids and sulfur compounds such as the soil fumigant dimethyl disulfide (Hewavitharana et al. 2014).

Application of ASD results in microbial shifts that may also favor deleterious rhizobacteria as well, although isolates derived from this approach this have yet not been directly investigated. ASD typically results in an increase in members of the Firmicutes, many of which are associated with pathogen suppression, but microbial shifts have also been documented in which populations of *Pseudomonas* have significantly increased, particularly with the use of rice bran as the carbon source (Hewavitharana et al. 2019, 2021). Similarly, members of the Enterobacteria and Proteobacteria have also been shown to increase, which contain genera bacteria that are associated with weed-suppressive (Suslow and Schroth 1982; Liu et al. 2016; Hewavitharana et al. 2021). These potential mechanisms of weed suppression take place when ASD is employed under tropical, subtropical, and temperate conditions, allowing the utilization of ASD over multiple regions and for cropping systems in which solarization is difficult to employ. Additional research on the characterization of potentially weed-suppressive bacteria associated with ASD could allow for the identification of specific amendments that could naturally increase weed suppressive populations or serve as a substrate and ideal environment for inoculation with selected organisms.

9.3.6 Future Research Addressing Sustainability

While these systems provide nonchemical weed control options that have the potential to reduce the use of herbicides and broad-spectrum fumigants, their use must also consider the potential for other types of environmental impacts. As Zhao et al. (2021) suggest, there is potential for nutrient loss during ASD treatments, as there are from solarization and biosolarization applications. While some approaches to ASD incorporate a nitrogen source with a carbon source, such as is the current practice in Florida (Di Gioia et al. 2017), a distinction must be made concerning systems in which composted animal manure is used as a nitrogen source versus those in which fresh animal waste are used as amendments. While the use of fresh animal manure has the potential to significantly increase toxic compounds that can aid in weed control, saturation of soil to which fresh animal manure is added without a labile C source has could significantly increase the release of nutrients, particularly nitrogen to the environment. Studies conducted in Florida in which composted poultry litter was used as a nitrogen source combined with molasses for ASD resulted in no greater risk of nitrogen loss through leaching of nitrate or the production of nitrous oxide than that resulting from soil fumigation (Di Gioia et al. 2017). In this study, there were significant differences between potential nutrient loss based on location, which may be attributed to differences in soil temperature and moisture.

Establishing optimal practices for soil water application in each of these approaches is also important. ASD typically utilizes soil saturation as a critical step, with 5–10 cm of water applied at initiation, while soil moisture characteristics under solarization or biosolarization are reported as field capacity, while no additional irrigation is added or reported in some applications. For example, Chellemi et al. (1997) report soil moisture at solarization initiation being between 6% and 10% over multiple locations, while others report no soil moisture status at treatment. Standardization of these practices would allow for a more thorough understanding of potential mechanisms of activity as well as allow for more accurate comparison of potential environmental impacts. Field validation of greenhouse or laboratory results regarding generation of greenhouse gases and potential nutrient releases is also critically important in order to have realistic assessments of potential detrimental environmental impacts of approaches like these that are considered more sustainable than pesticide applications. Similarly, there is a great deal of concern regarding the use of polyethylene mulch in crop production systems. This concern is not limited to the methods of weed control suggested here, but the utilization of clear mulch for solarization or biosolarization followed by the application of an opaque mulch for crop production that rely on a plasticulture system typically used in many vegetable and strawberry production regions could limit adoption. Current replacements under investigation for use in ASD include biodegradable mulches as well as silage tarps that can be used repeatedly.

ASD also has the potential to serve as a soil remediation treatment for a variety of issues that result in degraded soils. Application of this system has been used to rejuvenate soils that have been continuously monocultured and have yield-limiting pathogen populations (Huang et al. 2016; Zhou 2017) as well as for soils for which continuous cropping has depleted the microbial community or resulted in soil acidification or salinization (Li et al. 2020; Zhao et al. 2020a). In addition, ASD also has the potential as a remediation tool for herbicide contamination (Li et al. 2020).

9.4 Use of Nanocarriers to Deliver Active Ingredients (a.i)

In the previous sections of this chapter, the use of microorganisms as a means to control weeds was discussed. In this section the focus is on engineered nanoparticles, including those derived from microorganisms, as carriers for delivery of active ingredients such as pesticides and herbicides. While pesticides/herbicides are effective at removing pests and weeds, lack of specificity leads to overuse, therefore increasing costs and resulting in environmental pollution. Thus, there is a need to revolutionize pest and weed control. One approach toward this goal is the application of nanocarriers to enhance stability of formulations, enable targeting to release pesticides/herbicides at a lower overall dose but with higher efficiency and efficacy in a controlled manner. The discussed materials are biopolymers, which are polysaccharides and proteins, including nanoparticles obtained from plant viruses. A schematic of this section can be seen in Figure 9.3.

Several common pesticides, including some herbicides, represent a hazard to the environment because of their nonspecificity and broad use. More than 90% of applied pesticides are lost in the environment; their long-term use leads to accumulation and buildup, and ineffective dosing can lead to pest-resistance (Adisa et al. 2019; Wani et al. 2019; Gahukar and Das 2020; Saini et al. 2020). Nanoparticles and biomaterials can allow for greater control of agrochemical delivery, along with reduced phytotoxicity, lower losses of the active ingredient (a.i.) in the environment, less volatilization, and diminished soil degradation (Camara et al. 2019). This delivery can be by passive diffusion or tailored for specific environmental conditions such as salinity, pH, temperature, pathogens, or the presence of a disease. The application of nanotechnology in agriculture is coined as *precision farming* because it uses nano-size particles for agricultural purposes. For example, many nanoparticles are already being used by the industry for other purposes such as food products, cosmetics, and pharmaceuticals, and these can be repurposed for agricultural practices.

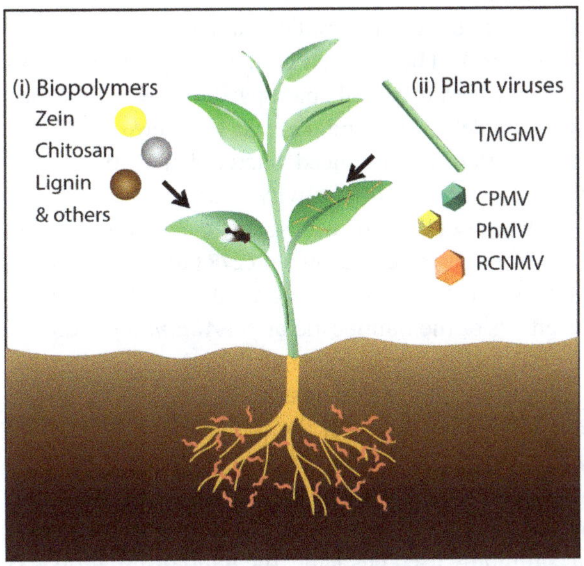

Figure 9.3 Schematics of the nanopesticides discussed in this book chapter. (i) Biopolymers zein, chitosan, liginin, and others; (ii) plant viruses: tobacco mild green mosaic virus (TMGMV), cowpea mosaic virus (CPMV), physalis mosaic virus (PhMV), and red clover necrotic virus (RCNMV) (courtesy Nicole Steinmetz).

Nanoparticle delivery of pesticides, can be defined as a formulation transporting engineered nanomaterials with biocidal properties, including microbicides, herbicides, and insecticides (Adisa et al. 2019). These nanoparticles are produced and assembled from organic and/or inorganic molecules providing new characteristics to the pesticide with enhanced biocompatibility, biodegradability, efficiency, stability, versatility, and slow and control release (Gahukar and Das 2020; Saini et al. 2020). In this chapter, only organic and hybrid nanoparticles will be discussed; we refer the reader to the following review articles on a summary and discussion of inorganic nanoparticles (Saini et al. 2020; Zhao et al. 2020b). The main characteristic of nanoparticles are their dimensions – the nanoscale size allows transport inside organisms and interactions with tissues, cells, and cellular compartments. For example, in plants, transport through the vascular system has been reported, and even cellular transport can be achieved (Raliya et al. 2018).

In comparison to traditional pesticide applications, nanodelivery of the a.i. can be designed to protect and control the release of the bioactive compound, analogous to what is the goal for nanomedicines targeting human health (Chariou et al. 2020). The following can be regulated: concentration, controlled release over time, target specificity – the nanocarrier can be tailored to respond to stimuli such as temperature, and/or pH, and other environmental conditions (Guha et al. 2020). We discuss various biopolymers and plant viruses as nanocarriers; these organic nanoparticles and hybrid versions thereof (organic with inorganic components) have been explored for the delivery of pesticides and essential oils that have pesticidal activity. Fertilizers could be added to nanoformulations in order to provide a complete product to the industry (Kah et al. 2019).

There is tremendous commercial interest and recognition of nanotechnology approaches toward agriculture; for example, India has agrochemical companies exclusively dedicated to precision farming, while companies such as BASF, Bayer Crop Science, Monsanto, and Syngenta have started production lines for nanoagrobiotechnologies (Guha et al. 2020). Another indicator of the interest in nanotechnology, is the exponential growth of patents filed and/or granted for nano-formulations for agricultural purposes (Kah et al. 2019). The versatility of these mechanisms has demonstrated enhanced capacities for agriculture, but there are still multiple unknowns about this technology that need to be addressed (Adisa et al. 2019; Nuruzzaman et al. 2016).

The development pipeline of nano-agrochemicals is moving rapidly and more research is needed to assess its safety for the environment and nontarget organisms; the following questions must be thoroughly addressed: the metabolic processes in which nanoparticles can interfere in animals and humans through nano-biointeractions and bioaccumulation through the food chain (transgenerational studies), the effect of using fertilizers with pesticides, its stability for long term storage and under environmental conditions considering chemical and thermal factors (Nuruzzaman et al. 2016; Prasad et al. 2017; Raliya et al. 2018; Adisa et al. 2019; Kah et al. 2019; Wani et al. 2019; Chariou et al. 2020; de Oliveira et al. 2020; Hofman et al. 2020). The future goals must be to control and minimize pesticide exposure and accumulation in the environment, i.e. we must design formulations that are resistant to external factors such as rain (which can lead to wash-out effects and environmental accumulation); that are targeted as well as specific toward organisms, and that do not promote pest-adaptation to pesticides; in addition, such formulations shall not be toxic toward plants and mammals (including humans) (Nuruzzaman et al. 2016).

While the development pipeline of novel nanocarriers for use in agriculture continues to grow rapidly, regulatory concerns present a critical factor and commercial implementation and translation of these precision farming approaches has been slow (Camara et al. 2019). The lack of public policy has diminished confidence around nanotechnology products, which limits the investments and commercial interest the field has received (Kah et al. 2019). Government and regulatory agencies have the authority to determine the future direction and implementation of nanopesticides. These agencies guide the future in the industry, ultimately shaping the products to which farmers have access to (Saini et al. 2020). At the present time, there are nanopesticide products on the market – however, while at the nanoscale, these products are not labeled as "nano," in an attempt to avoid issues with regulation and public opinion (Chariou et al. 2020). While it has grown out of its infancy, precision framing, and nanotechnology in agriculture is a young and emerging discipline with opportunity to gain acceptance of the general public and create a new chapter for agriculture, in which environmental sustainability and accessible technologies are offered to the farmers in need.

Biopolymers are an attractive material for the formulations of nanopesticides. The biopolymers that have been explored as nanocarriers are mainly polysaccharides and proteins (including the capsids of plant viruses). These biomaterials offer low cost, biodegradability, and low toxicity rendering them biocompatible (Luo et al. 2020). Versatile chemistries and encapsulation strategies have been developed enabling encapsulation and delivery of various a.i. Biopolymer/nanoparticle delivery can enhance the activity of the a.i. as well as to protect it from external factors (UV, oxidation, degradation, evaporation) and even control its mobility in the environment or target organism. Many of these materials have been developed for nanomedicine applications and a large body of data is available to allow pivoting the technology toward agricultural applications (Guha et al. 2020; Saini et al. 2020). While most approaches make use of nanoencapsulation (Nuruzzaman et al. 2016), other techniques used to achieve pesticide loading are conjugation (Chariou et al. 2019), and nanophytosomes (Nazari et al. 2019).

9.4.1 Zein

Zein is a corn protein conformed by hydrophobic amino acids with the capacity to self-assemble and nanoencapsulate different substances because of its high coating capacity. Another positive attribute of zein is that it has the GRAS (generally recognized as safe by the US Food and Drug Administration (FDA)) designation (Pérez-Guzmán and Castro-Muñoz 2020). This protein has been implemented for the nanoencapsulation of several essential oils (or derivates) because it protects the a.i. from oxidative stress enhancing formulation stability. Merino et al. developed a formulation of zein with *Thymbra capitata* essential oil yielding nanoparticles smaller than 180 nm; the main a.i. is carvacrol. The antibacterial activity of the zein-encapsulated essential oil nanoparticles was assessed against *Escherichia coli* and *Listeria monocytogenes*. The results indicated an enhanced bactericidal activity, and when heat was applied this activity was improved (Merino et al. 2019). Pascoli et al. used zein to nanoencapsulate neem oil from *Azadirachta indica* (278±61.5 nm), with a focus on the effect on nontarget organisms. After evaluating the effects of the nanopesticide on *Allium cepa* (on soil microbiota) and *Caenorhabditis elegans*, the nanoencapsulated oil showed decreased toxicity for these organisms; however, the effectiveness of the nanoparticle against the target pest was not tested (Pascoli et al. 2019). Oliveira et al. nanoencapsulated citronella's essential oil (geraniol and R-citronellal) into zein nanoparticles. Citronella's essential oil is a commonly used pesticide, the nanoformulation was tested against *Tetranychus urticae*. The zein nanoparticle formulation protected the oil from photodegradation, therefore providing a higher degree of stability; at the same time the nanoformulation conferred lower cytotoxicity and lower phytotoxicity (Oliveira et al. 2018b). The same group also used the zein nanoparticles to encapsulate a mixture of geraniol, eugenol and cinnamaldehyde; the assembled nanopesticides measured between 90–500 nm (de Oliveira et al. 2019). This new complex nanopesticide was responsive to insects (*Tetranychus urticae* and *Chrysodeixis includes*) and changes in temperature, with enhanced stability for storage conditions, lower toxicity against nontarget organisms (tested in mouse cell lines); higher repellency and insecticidal activity were also achieved (de Oliveira et al. 2019). The authors concluded that the mechanism for zein nanoparticles consists of the bioactive compound passing through the zein protein chain matrix in a decreasing rate release (diffusion) (de Oliveira et al. 2019). In another research study, zein was used as a nanocarrier for gold nanoparticles measuring 50–80 nm, and the metallic nanoparticles were tested against Gram-positive bacteria (*Bacillus pumilus* and *Bacillus subtilis)*, Gram-negative bacteria (*Shigella sonnei* and *Pseudomonas aeruginosa)* and *Aedes aegypti*; antibacterial and larvicidal activity were reported. The use of zein with gold allowed for this inorganic, metallic nanoparticles to be released with control (Suganya et al. 2017). In a different study, also using zein as the platform technology, Hao et al. encapsulated a broad range pesticide, avermectin (AVM). The AVM-zein nanoparticle formulation was conjugated with ethylene glycol diglycidyl ether and then grafted with a sodium carboxymethyl cellulose (CMC) linked to methyl methacrylate and 2-hydroxyethyl methacrylate (180–254 nm) (Hao et al. 2020). The surface coatings achieved pH-controlled release of the AVM: AVM is released at acidic pH but remains encapsulated at neutral and alkaline conditions. The various polymers that were conjugated triggered the release based on their pKa and resulting electrostatic interactions (Camara et al. 2019). The polymer-coated, AVM-loaded zein nanoparticles were tested in cucumber leaves, and the data indicate stronger adhesive properties that may be explained by intermolecular forces between the polymers and the leaf surface. This is an important finding and may be a strategy to avoid the loss of the pesticides by irrigation. UV-protective qualities were also observed. CMC is likely responsible for this property because it is known to absorb UV light. Overall, the polymer-coated zein nanoparticles provide favorable properties for avermectin delivery (Hao et al. 2020).

9.4.2 Chitosan

Chitosan has also been explored for nanoencapsulation, this biopolymer comes from chitin that is extracted from crustaceans and is the second most abundant polysaccharide (after cellulose) (Divya and Jisha 2018). As with zein, several nanoencapsulations of essential oils have been developed and tested. Hasheminejad et al. nanoencapsulated *Syzygium aromaticum* essential oil into a 40–100 nm-sized chitosan nanoformulation. The resulting nanoparticle showed pH sensitive qualities and was used effectively against *Aspergillus niger*, with enhanced efficacy compared to the direct use of the essential oil or oil-free chitosan nanoparticles. A considerable disadvantage of the nanoformulation were its organoleptic effects which could potentially limit its food applications for its undesirable taste (Hasheminejad et al. 2019). Das et al. used the essential oil of *Coriandrum sativum* against fungi in *Oryza sativa seeds*, the nanoparticle (57–80 nm) resulted in improved antifungal, anti-aflatoxigenic, and antioxidant activity (Das et al. 2019). Along the same lines, Dwivedy et al. nanoencapsulated *Illicium verum* essential oil using chitosan and the inorganic compound tripolyphosphate (TPP) resulting in 1–2 micron-sized formulations (so technically not a nanoformulation, but worth mentioning here). The addition of TPP promoted the formulation of spherical particles because of the crosslinking between chitosan and TPP creating a network that promotes the spherical shape. The oil chitosan/TPP formulation was tested against *Aspergillus flavus* and increased antifungal, anti-aflatoxigenic activity and antioxidant potential compared to the direct application of the oil (Dwivedy et al. 2018). The same chitosan/TPP particle formulation was also used to nanoencapsulate *Satureja hortensis* essential oil yielding nanoparticles smaller than 200 nm, which were tested against *T. urticae*. Here prolonged oil release was observed as well as enhanced toxicity and photodegradation protection (Ahmadi et al. 2018). Other polymeric mixtures have also been formulated with chitosan. A chitosan/pectin nanoformulation was used to transport the pesticide carbendazim (70–90 nm), and when tested in *Cucumis sativa*, *Zea mays*, and *Lycopersicum esculentum* against *Fusarium oxysporum* and *Aspergillus parasiticus*, prolonged pesticide release with higher efficacy and lower phytotoxicity were achieved (Kumar et al. 2017). Chitosan/gum arabic nanoformulations have also been explored; for example, the nanoencapsulation of geraniol was achieved and used against *Bemisia tabaci*. Higher attraction to the organism and enhanced stability and protection against UV radiation for geraniol were observed (de Oliveira et al. 2018a). More complex configurations of chitosan have also been developed. Campos et al. functionalized chitosan with cyclodextrin for the nanoencapsulation of carvacol (175.2 nm) and linalool (245.8 nm), and when the nanopesticides were tested against *T. urticae*, these two nanoformulations showed increased thermal stability, repellency, acaricidal activity and decreased oviposition (Campos et al. 2018). Another noteworthy example is the research by Xiang et al., who synthesized a nanoparticle with a nucleus of a bioactive substance (glyphosate or cypermethrin) encapsulated by diatomite (providing magnetic properties) followed by coating with chitosan. The resulting nanoformulation measured ~40 nm and was pH responsive with controlled release and magnetic collectability – nevertheless, while these favorable attributes were achieved, there was no apparent enhancement in efficiency of the pesticides against the target organism (Xiang et al. 2017).

9.4.3 Lignin

Lignin is a biopolymer extracted mainly from wood, present in high abundance in nature (Katahira et al. 2018). Lignin was only recently explored in nano-agriculture, and Gazzurelli et al. used it to nanoencapsulate copper, yielding nanoparticles with distinct morphologies: 2–20 nm spheres and 5–15 nm rods. The spherical conformation was achieved with a wet procedure (stirring of both solutions), while the rods were made by mechanochemistry; the size of the nanoparticle depends on the concentration of copper. The nanoformulations demonstrated slow release of copper with rods outperforming spheres. The copper slow-release formulations demonstrated protection against *B. cinerea*, *P. syringe tomato*, *X. campestris*, and *X. arboricola fragari* in *Fragaria sp.* and *Solanum lycopersicum* (Gazzurelli et al. 2020). Also, Zhang et al. synthesized lignin nanoparticles using a lignin-modified epoxy resin loaded with abamectin (with a size of 150 nm). It was tested against *Meloidogyne incognita*, and increased mobility in the soil, enhanced nematicide effect, and higher stability were observed (Zhang et al. 2020a).

9.4.4 Viral Nanoparticles

Viruses have evolved to become one of nature's cargo delivery system, and mammalian viruses have already made headway in medicine. Their protein-based capsids come in many shapes and sizes and can be explored in nanotechnology. Mammalian viruses, plant viruses, and phages (i.e. viruses that infect bacteria) – all have gained attention as materials and potential applications in various settings are being explored (Wen and Steinmetz 2016). As with other biopolymers, viruses have been explored for nanomedical applications and the design concepts can be pivoted for formulations with agrochemicals. Plant viruses are particularly intriguing for applications in plant health, because plant viruses naturally occur in plants and the environment. An EPA-approved formulation is SolviNix, a bioherbicide based on tobacco mild green mosaic virus (TMGMV); it is commercially produced by BioProdex, Inc. and used in the state of Florida (Charudattan and Hiebert 2007; Ferrell et al. 2008; Charudattan et al. 2009). Extending on this application, we have also developed TMGMV as a pesticide carrier. We developed electrostatic drug-loading methods as well as chemical bioconjugation approaches to load TMGMV with a.i.; in our early work we used crystal violet or fluorophores as model compounds. TMGMV measures 300 × 18 nm and has a 4-nm wide central channel; therefore, a.i. can be loaded on the exterior or interior surfaces. For bioconjugation we make use of the various amino acids that are exposed both on the interior and exterior TMGMV capsid surface; for example, Tyr side chains are readily modified using diazonium-based chemistries. Each 300 × 18 nm TMGMV is self-assembled from 2,130 identical coat protein units; therefore, any covalent modification results in theory in at least 2,130 a.i. presented per particle. Higher loading could be achieved by targeting more than one amino acid (same or different functionality) per coat protein. For nanoencapsulation, electrostatic interactions are utilized to load the a.i. into the central channel. We demonstrated this principle using positively charged compounds, such as crystal violet that self-assembles inside TMGMV's channel that is lined with Glu acid aide chains and hence is negatively charged. Loading with up to 1,500 crystal violet molecules per TMGMV carrier was achieved (Chariou and Steinmetz 2017). Efficacy of the crystal violet-loaded TMGMV formulation was tested against *C. elegans* showing nematicide activity and homogeneous distribution of the nanopesticide inside the organism, with high prevalence in the gastrointestinal system (Chariou and Steinmetz 2017). In a follow-up study, the soil diffusion properties of plant viral vs. synthetic nanoparticles were dissected using computational and experimental approaches: data indicate enhanced soil mobility of TMGMV vs. cowpea mosaic virus (CPMV, a 30 nm-sized icosahedron), virus-like nanoparticles from Physalis mosaic virus (PhMV, a 30 nm-sized icosahedron), mesoporous silica nanoparticles (MSNP), and poly (lactic-co-glycolic acid) (PLGA). Data indicate that TMGMV shows good soil mobility and can reach up to 30 cm depth; this makes it an excellent platform to deliver pesticides to the roots of crops (Chariou et al. 2019). Similarly, others have developed red clover necrotic virus (RCNMV) for agriculture applications: Cao et al. used RCNMV loaded with abamectin (the RCNMV capsid forms a 37 nm-sized icosahedron) and demonstrated the formulations to be effective against *C. elegans*. The system was also proven effective to protect tomato seedlings from attack by *Meloidogyne hapla*. This makes the RCNMV platform an interesting candidate for agricultural applications: RCNMV in soil and has high bioavailability, because it persists in soil. RCNMV is highly host-specific and therefore offers safety to nontarget crops. Its capsid allows for gating, i.e., encapsulation and release of active ingredient based on pH and salt concentrations (Stephen et al. 2018). RCNMV nanopesticide formulation has enhanced stability and mobility in soil, and the treatment for tomato seedlings is superior in efficacy and bioavailability compared to the free pesticide, no viral infection (Cao et al. 2015).

9.4.5 Other Nanoformulations

Other interesting applications of biomaterials for nanoencapsulation are:

1) Casein to camouflage the pesticide eugenol. These nanoformulations measured around 180 nm and were tested against *Artemia salina* sp. Data indicate that the casein nanoformulation enabled controlled release resulting in enhanced cytotoxicity against the target organism (Sahlan et al. 2021).
2) Angum gum and maltodextrin were used for the transport of the essential oil of *Zataria multiflora* and *Eucalyptus globulus*. The nanoparticle formulations measured 60–130 nm and improved toxicity against *Ephestia kuehniella* by prolonged release (Emamjomeh et al. 2021).
3) Nanophytosomes, which are lipid-based vesicular delivery systems, were made with *Allium sativum* and soy phosphatidylcholine for food applications to prevent *E. coli* and *Staphylococcus aureus* infections. The ~100 nm-sized formulations enhanced stability of the active ingredients resulting in enhanced antimicrobial and antioxidant activity with no organoleptic effects (Nazari et al. 2019).
4) Solid lipid nanoparticles have also been explored loaded with pyrethrum extract, an insecticide. The solid nanoparticles measured ~ 260 nm and avoided uptake by honeybees, an off-target organism (Oliveira et al. 2019). However, efficacy of the nanopesticide formulation has not yet been reported.
5) Various polymer mixtures derived from castor oil were used for the transport of azadirachtin protecting rice from brown planthoppers. The nanoformulations consisted of a cationic and anionic polyurethane shell (measuring 30–200 nm), which protected the pesticide and allowed for prolonged release, rain fastness, and thermal stability (Zhang et al. 2020b).

9.5 Summary

Novel technologies that contribute to sustainable crop production and land management are critically needed. Biological control using insects is a proven, practical means of ecologically managing weeds, particularly for invasive weeds in noncrop situations. The use of insects for weed control is reviewed in Chapter 8. In this chapter, we have reviewed the use of plant pathogens and allelopathic microorganisms (deleterious rhizobacteria) as bioherbicides along with applications of nanotechnology to deliver weed- and pest-control compounds through plant viruses and viral capsids employed as nanocarriers. These approaches utilize direct applications of microorganisms or specific manipulations of a specific microbe. More ecologically based alternatives to chemical soil disinfestation, such as soil heating through steam or solarization have direct impacts on weeds, but also on soil microbial communities that can contribute to weed suppression. The relatively new method of anaerobic soil disinfestation (ASD), in which anaerobic microorganisms are judiciously manipulated to kill weed seeds or suppress their germination, utilizes a less directed but microbially dependent form of weed management. ASD can be a natural companion for certain abiotic soil disinfestation methods such as fire and soil solarization, and also has the potential to be combined with other biologically based weed approaches including biological control agents and microbially based nanotechnology. Continued discovery, development, optimization, and use of ecologically based sustainable approaches that complement one another will lead to a shift away from the dominant chemical-based weed control paradigm.

Acknowledgments

The work of N.F.S. was funded in part by a grant from the National Institute of Food and Agriculture NIFA-2020-67021-31255 (to N.F.S.) and the UC San Diego Materials Research Science and Engineering Center (UCSD MRSEC), DMR-2011924 (to N.F.S.).

References

Adisa, I.O., Pullagurala, V.L.R., Peralta-Videa, J.R. et al. (2019). Recent advances in nano-enabled fertilizers and pesticides: a critical review of mechanisms of action. *Environmental Science: Nano* 6: 2002–2030.

Ahmadi, Z., Saber, M., Akbari, A., and Mahdavinia, G.R. (2018). Encapsulation of *Satureja hortensis* L. Lamiaceae in chitosan/TPP nanoparticles with enhanced acaricide activity against Tetranychus urticae Koch Acari: Tetranychidae. *Ecotoxicology and Environmental Safety* 61: 111–119.

Aldrich, R.J. (1984). *Weed-Crop Ecology: Principles in Weed Management*. North Scituate, Massachusetts, USA: Breton.

Al-Masoom, A.A., Saghir, A.R., and Itani, S. (1993). Soil solarization for weed management in the UAE. *Weed Technology* 2: 507–510.

Anonymous. (2018). *Verdesian life sciences announces new biological herbicide*. December 3, 2014, Updated December 13, 2018. https://www.capitalpress.com/archives/verdesian-life-sciences-announces-new-biological-herbicide/article_d29e1f0c-2559-54f4-807d-f6e4080d6af7.html (accessed 6 May 2022).

Ascard, J., Hatcher, P.E., Melander, B. et al. (2007). Thermal weed control. Chapter 10. In: *Non-chemical Weed Management: Principles, Concepts and Technology* (ed. M.K. Upadhyaya and R.E. Blackshaw), 155–175. OX10 8DE, UK: CAB International, Wallingford, Oxfordshire, UK. 239 pages, ePDF 9781845932916.

Ash, G.J. (2010). The science, art and business of successful bioherbicides. *Biological Control* 52: 230–240.

Bailey, K.L. (2014). The bioherbicide approach to weed control using plant pathogens. In: *Integrated Pest Management* (ed. D.P. Abrol), 245–266. San Diego, CA, USA: Academic Press – Elsevier. ISBN: 978-0-12-398529-3.

Baker, K.F. (1962). Principles of heat treatment of soils and planting material. *Journal of the Australasian Institute of Agricultural Science* 28: 118–126.

Baker, K.F., and Roistacher, C.N. (1957). *The U.C. System for Producing Healthy Container-grown Plants* (ed. K.F. Baker). California Agricultural Experiment Station Extension Service Manual 23, University of California, Berkeley, CA.

Bannon, J.S. (1988). CASST™ herbicide (*Alternaria cassiae*); a case history of a mycoherbicide. *American Journal of Alternative Agriculture* 3: 73–76.

Banowetz, G.M., Azevedo, M.D., Armstrong, D.J. et al. (2008). Germination-arrest factor (GAF): biological properties of a novel, naturally-occurring herbicide produced by selected isolates of rhizosphere bacteria. *Biological Control* 46: 380–390.

Barton, J. (née Fröhlich). (2004). How good are we at predicting the field host-range of fungal pathogens used for classical biological control of weeds? *Biological Control* 32: 99–122. https://doi.org/10.1016/j.biocontrol.2004.04.008.

Barton, J. (2011). Predictability of pathogen host range in classical biological control of weeds: an update. *BioControl* 57: 289–305. https://doi.org/10.1007/s10526-011-9401-7.

Barton, J., Fowler, S.V., Gianotti, A.F. et al. (2007). Successful biological control of mist flower *(Ageratina riparia)* in New Zealand: agent establishment, impact, and benefits to the native flora. *Biological Control* 40: 370–385. https://doi.org/10.1016/j.biocontrol.2006.09.010.

Begonia, M.F.T. (1989). *Characterization of attraction of rhizobacteria to weed seeds and seedlings*. Ph.D. Dissertation, University of Missouri, Columbia, Missouri, USA, 165 Pages.

Begonia, M.F.T., Kremer, R.J., Stanley, L., and Jamshedi, A. (1990). Association of bacteria with velvetleaf roots. *Transactions Missouri Academy of Science* 24: 17–26.

Berner, D.K., and Bruckart, W.L. (2005). A decision tree for evaluation of exotic plant pathogens for classical biological control of introduced exotic weeds. *Biological Control* 34: 222–232.

Blok, W.J., Lamers, J.G., Termorshuizen, A.J., and Bollen, G.J. (2000). Control of soilborne plant pathogens by incorporating fresh organic amendments followed by tarping. *Phytopathology* 90: 253–259.

Boyetchko, S.M. (1997). Efficacy of rhizobacteria as biological control agents of grassy weeds. *Proceedings of Soils and Crops* Workshop '97, Saskatoon, Saskatchewan (20–21 February), 460–465.

Boyetchko, S.M. (1999). Innovative applications of microbial agents for biological weed control. In: *Biotechnological Approaches in Biocontrol of Plant Pathogens* (ed. K.G. Mukerji, B.P. Chamola, and K. Upadhyay), 73–97. London: Kluwer Academic/Plenum Publishers.

Boyetchko, S.M. (2017). *The use of microorganisms in integrated weed management*. E-chapter 19. Burleigh Dodds Publishing. www.bdspublishing.com.

Boyetchko, S.M. and Holmstrom-Ruddick, B. (1996). Host-range of rhizobacteria effective as biocontrol agents of downy brome. *Canadian Journal of Plant Pathology* 18: 86–87. [Abstr].

Boyetchko, S.M., and Mortensen, K. (1993). Use of rhizobacteria as biological control agents of downy brome. *Proceedings of Soils and Crops Workshop '93*, Saskatoon, Saskatchewan (25–26 February), 443–448.

Boyette, C.D., Quimby, P.C., Jr., Caesar, A.J. et al. (1996). Adjuvants, formulations, and spraying systems for improvement of mycoherbicides. *Weed Technology* 10: 637–644.

Brooks, R.K., Baudoin, A., and Salom, S.M. (2020a). The natural persistence and distribution of the proposed biological control agent *Verticillium nonalfalfae* on *Ailanthus altissima* in Virginia, USA. https://doi.org/10.1111/efp.12639.

Brooks, R.K., Wickert, K.L., Baudoin, A. et al. (2020b). Field-inoculated *Ailanthus altissima* stands reveal the biological control potential of *Verticillium nonalfalfae* in the mid-Atlantic region of the United States. *Biological Control* 148. https://doi.org/10.1016/j.biocontrol.2020.104298.

Browne, G., Ott, N., Poret-Peterson, A. et al. (2018). Efficacy of anaerobic soil disinfestation for control of prunus replant disease. *Plant Disease* 102: 209–219.

Butler, D.M., Kokalis-Burelle, N., Muramoto, J. et al. (2012a). Impact of anaerobic soil disinfestation combined with soil solarization on plant–parasitic nematodes and introduced inoculum of soilborne plant pathogens in raised-bed vegetable production. *Crop Protection* 39: 33–40.

Butler, D.M., Rosskopf, E.N., Kokalis-Burelle, N. et al. (2012b). Exploring warm-season cover crops as carbon sources for anaerobic soil disinfestation (ASD). *Plant and Soil* 355: 149–165.

Caldwell, C.J., Hynes, R.K., Boyetchko, S.M., and Korber, D. (2012). Colonization and bioherbicidal activity on green foxtail by *Pseudomonas fluorescens* BRG100 in a pesta formulation. *Canadian Journal of Microbiology* 58: 1–9.

Camara, M.C., Campos, E.V.R., Monteiro, A.R. et al. (2019). Development of stimuli-responsive nano-based pesticides: emerging opportunities for agriculture. *Journal of Nanobiotechnology* 17: 1–19.

Campos, E.V.R., Proença, P.L.F., Oliveira, J.L. et al. (2018). Chitosan nanoparticles functionalized with β-cyclodextrin: a promising carrier for botanical pesticides. *Scientific Reports* 8: 2067.

Candido, V., D'Addabbo, T., Basile, M. et al. (2008). Greenhouse soil solarization: effect of weeds, nematodes and yield of tomato and melon. *Agronomy for Sustainable Development* 28: 221–230.

Cao, J., Guenther, R.H., Sit, T.L. et al. (2015). Development of abamectin loaded plant virus nanoparticles for efficacious plant parasitic nematode control. *ACS Applied Materials and Interfaces* 7: 9546–9553.

Cayuela, M.L., Millner, D., Meyer, S.L.F., and Roig, A. (2008). Potential of olive mill waste and compost as biobased pesticides against weeds, fungi, and nematodes. *Science of the Total Environment* 399: 11–18.

Chandramohan, S., Charudattan, R., Sonoda, R.M., and Singh, M. (2002). Field evaluation of a fungal pathogen mixture for the control of several weedy grasses. *Weed Science* 50: 204–213.

Chariou, P.L., Dogan, A.B., Welsh, A.G. et al. (2019). Soil mobility of synthetic and virus-based model nanopesticides. *Nature Nanotechnology* 14: 712–718.

Chariou, P.L., Ortega-Rivera, O.A., and Steinmetz, N.F. (2020). Nanocarriers for the delivery of medical, veterinary, and agricultural active ingredients. *ACS Nano* 14: 2678–2701.

Chariou, P.L. and Steinmetz, N.F. (2017). Delivery of pesticides to plant parasitic nematodes using tobacco mild green mosaic virus as a nanocarrier. *ACS Nano* 11: 4719–4730.

Charudattan, R. (2001). Biological control of weeds by means of plant pathogens: significance for integrated weed management in modern agro-ecology. *BioControl* 46: 229–260.

Charudattan, R. (2005). Ecological, practical, and political inputs into selection of weed targets: what makes a good biological control target? *Biological Control* 35: 183–196. https://doi.org/10.1016/j.biocontrol.2005.07.009.

Charudattan, R. and Hiebert, E. (2007). A plant virus as a bioherbicide for tropical soda apple, *Solanum viarum*. *Outlooks on Pest Management* 18 (4): 167–171. https://doi.org/10.1564/18aug07.

Charudattan, R. and Hiebert, E. (2015). *SolviNix*® LC: a plant virus based bioherbicide. *IR-4 Project Newsletter* 46 (1): 1–3. www.ir4.rutgers.net.

Charudattan, R., Hiebert, E., Currey, W. et al. (2020). Design and testing of field application tools for a bioherbicide with a plant virus as active ingredient. *Weeds – Journal of Asian-Pacific Weed Science Society* 2 (2): 34–45. https://weeds-apwss.scholasticahq.com/article/18619-design-and-testing-of-field-application-tools-for-a-bioherbicide-with-a-plant-virus-as-active-ingredient (accessed 12 June 2021).

Charudattan, R., Pettersen, M.S., and Hiebert, E. (2009 February 24). *Use of tobacco mild green mosaic virus (TMGMV) mediated lethal hypersensitive response (HR) as a novel method of weed control*. United States Patent No. US 7,494,955,B2.

Charudattan, R., Walker, H.L., Boyette, C.D. et al. (1986). Evaluation of *Alternaria cassiae* as a mycoherbicide for sicklepod (*Cassia obtusifolia*) in regional field tests. *Southern Cooperative Series Bull.* 317. Alabama Agricultural Experiment Station, Auburn University, AL, 19 p.

Chase, C.A., Sinclair, T.R., Chellemi, D.O. et al. (1999a). Heat-retentive films for increasing temperatures during solarization in a humid cloudy environment. *HortSci* 34: 1089–1095.

Chase, C.A., Sinclair, T.R., and Locascio, S.J. (1999b). Effects of soil temperature and tuber depth on *Cyperus* spp. control. *Weed Science* 47: 467–472.

Chase, C.A., Sinclair, T.R., Shilling, D.G. et al. (1998). Light effects on rhizome morphogenesis in nutsedges (*Cyperus* spp.): implications for control by soil solarization. *Weed Science* 46: 575–580.

Chellemi, D.O., Olson, S.M., Mitchell, D.J. et al. (1997). Adaptation of soil solarization to the integrated management of soilborne pests of tomato under humid conditions. *Phytopathology* 87: 250–258.

Cherrington, C.A. and Elliott, L.F. (1987). Incidence of inhibitory pseudomonads in the Pacific Northwest. *Plant and Soil* 101: 159–165.

Churchill, B.W. (1982). Mass production of microorganisms for biological control. In: *Biological Control of Weeds with Plant Pathogens* (ed. R. Charudattan and H.L. Walker), 139–156. John Wiley & Sons, Inc. ISBN 0-471-05898-7.

Cohen, O., Riov, J., Katan, J. et al. (2008). Reducing persistent seed banks of invasive plants by soil solarization—the case of *Acacia saligna*. *Weed Science* 56: 860–865.

Cook, J.C., Charudattan, R., Zimmerman, T.W. et al. (2009). Effects of *Alternaria destruens*, glyphosate, and ammonium sulfate individually and integrated for control of dodder (*Cuscuta pentagona*). *Weed Technology* 23: 550–555.

D'Addabbo, T., Miccolis, V., Basile, M., and Candido, V. (2010). Soil solarization and sustainable agriculture. In: *Sociology, Organic Farming, Climate Change and Soil Science* (ed. E. Lichtfouse), 217–274. Dordrecht: Springer.

Daigle, D.J., Connick, W.J., Jr., and Boyetchko, S.M. (2002). Formulating a weed-suppressive bacterium in 'pesta'. *Weed Technology* 16: 407–413.

Dar, A., Zahir, A.A., Asghar, H.N., and Ahmed, R. (2020). Preliminary screening of rhizobacteria for biocontrol of little seed canary grass (*Phalaris minor* Retz.) and wild oat

(*Avena fatua* L.) in wheat. *Canadian Journal of Microbiology* 66: 368–376.

Das, S., Singh, V.P., Dwivedy, A.K. et al. (2019). Encapsulation in chitosan-based nanomatrix as an efficient green technology to boost the antimicrobial, antioxidant and in situ efficacy of *Coriandrum sativum* essential oil. *International Journal of Biological Macromolecules* 133: 294–305.

de Ávila, Z.R. and Pitelli, R.A. (2004). Crescimento, esporulação e virulência do inóculo de *Cercospora piaropi*, agente de biocontrole do aguapé (Growth, sporulation and virulence of *Cercospora piaropi* inoculum, biocontrol agent of water hyacinth). *Fitopatologia Brasileira* 29: 189–192.

de Oliveira, C., Mulinari, J., Reichert, F., and Júnior, A. (2020). *Nano-delivery systems of pesticides active agents for agriculture applications – an overview*. https://doi.org/10.31692/iciagro.2020.0051.

de Oliveira, J.L., Campos, E.V.R., Germano-Costa, T. et al. (2019). Association of zein nanoparticles with botanical compounds for effective pest control systems. *Pest Management Science* 75: 1855–1865.

de Oliveira, J.L., Campos, E.V.R., Pereira, A.E.S. et al. (2018a). Geraniol encapsulated in chitosan/gum arabic nanoparticles: a promising system for pest management in sustainable agriculture. *Journal of Agriculture and Food Chemistry* 66: 5325–5334.

de Oliveira, J.L., Campos, E.V.R., Pereira, A.E.S. et al. (2018b). Zein nanoparticles as eco-friendly carrier systems for botanical repellents aiming sustainable agriculture. *Journal of Agriculture and Food Chemistry* 66: 1330–1340.

Di Gioia, F., Ozores-Hampton, M., Hong, J. et al. (2016). The effects of anaerobic soil disinfestation on weed and nematode control, fruit yield, and quality of Florida fresh-market tomato. *HortScience* 51: 703–771.

Di Gioia, F., Ozores-Hampton, M., Zhao, X. et al. (2017). Anaerobic soil disinfestation impact on soil nutrients dynamics and nitrous oxide emissions in fresh-market tomato. *Agriculture, Ecosystems & Environment* 240: 194–205.

Díaz-Hernández, S., Gallo-Llobet, L., Domínguez-Correa, P., and Rodríguez, A. (2017). Effect of repeated cycles of soil solarization and biosolarization on corky root, weeds and fruit yield in screen-house tomatoes under subtropical climate conditions in the Canary Islands. *Crop Protection* 94: 20–27.

Divya, K. and Jisha, M.S. (2018). Chitosan nanoparticles preparation and applications. *Environmental Chemistry Letters* 16: 101–112.

Doungsa-ard, C., McTaggart, A.R., Geering, A.D.W., and Shivas, R.G. (2018). Diversity of gall-forming rusts (*Uromycladium, Pucciniales*) on *Acacia* in Australia. *Persoonia* 40: 221–238. ISSN (Online) 1878-9080. https://doi.org/10.3767/persoonia.2018.40.09.

Duranti, A. and Cuocolo, L. (1988). Solarization in weed control for onion (*Allium cepa* L). *Advances in Horticultural Science* 2: 104–108.

Dwivedy, A.K., Singh, V.K., Prakash, B., and Dubey, N.K. (2018). Nanoencapsulated *Illicium verum* Hook.f. essential oil as an effective novel plant-based preservative against aflatoxin B1 production and free radical generation. *Food and Chemical Toxicology* 111: 102–113.

Economou, G., Mavrogiannopoulos, G., and Paspatis, E.A. (1997). Weed seed responsiveness to thermal degree hours under laboratory conditions and soil solarization in greenhouse. *Proceedings of the Second Conference on Soil Solarization*. FAO Plant Production and Protection Paper 147, Rome, Italy, 246–263.

Egley, G.H. (1983). Weed seed and seedling reduction by soil solarization with transparent polyethylene sheets. *Weed Science* 31: 404–409.

Egley, G.H. (1990). High-temperature effects on germination and survival of weed seeds in soil. *Weed Science* 38: 429–435.

Elliott, L.F. and Kennedy, A.C. (1991). *Method for screening bacteria and application thereof for field control of the weed downy brome*. U.S. Patent no. 5,030,562.

Elliott, L.F. and Lynch, J.M. (1985). Plant growth-inhibiting pseudomonads colonizing winter wheat (*Triticum aestivum* L.) roots. *Plant and Soil* 84: 57–65.

Elmore, C., Stapleton, J.J., Bell, C.E., and DeVay, J.E. (1997). *Soil solarization, A nonpesticidal method for controlling diseases, nematodes and weeds*. University of California Division of Agriculture and Natural Resources, 17.

Elmore, C.L. (1991a). *Effect of soil solarization on weeds*. Soil Solarization FAO Plant Production and Protection Paper 109. Rome Editors James E DeVay, James J. Stapleton and Clyde Elmore.

Elmore, C.L. (1991b). Weed control by solarization. In: *Soil Solarization* (ed. J. Katan and J.E. DeVay), 61–72. Boca Raton, FL, USA: CRC.

Elmore, C.L. (1998). Sensitivity of pest organisms to soil solarization. In: *Soil Solarization and Integrated Management of Soil Pests: Proceedings of the Second Conference on Soil Solarization* (ed. J.J. Stapleton, J.E. DeVay, and C.L. Elmore), 450–462. Aleppo, Syria. FAO Plant Production and Protection Paper 147, Rome, Italy.

Emamjomeh, L., Imani, S., Talebi Jahromi, K., and Moharramipour, S. (2021). Nanoencapsulation enhances the contact toxicity of *Eucalyptus globulus* Labill and *Zataria multiflora* Boiss essential oils against the third instar larvae of *Ephestia kuehniella* Lepidoptera: pyralidae. *International Journal of Pest Management* 1–9. https://doi.org/10.1080/09670874.2020.1871529.

Fennimore, S.A., Martin, F.N., Miller, T.C. et al. (2014). Evaluation of a mobile steam applicator for soil disinfestation in California strawberry. *HortScience* 49: 1542–1549.

Fernandez-Bayo, J.D., Shea, E.A., Parr, A.E. et al. (2020). Almond processing residues as a source of organic acid biopesticides during biosolarization. *Waste Management* 101: 74–82.

Ferrell, J., Charudattan, R., Elliott, M., and Hiebert, E. (2008). Effects of selected herbicides on the efficacy of tobacco mild green mosaic virus to control tropical soda apple *Solanum viarum*. *Weed Science* 56: 128–132.

Firehun, Y., Struik, P.C., Lantinga, E.A., and Taye, T. (2013). Joint use of insects and fungal pathogens in the management of waterhyacinth (*Eichhornia crassipes*): perspectives for Ethiopia. *Journal of Aquatic Plant Management* 51: 109–121.

Fredrickson, J.K., and Elliott, L.F. (1985). Colonization of winter wheat roots by inhibitory rhizobacteria. *Soil Science Society of America Journal* 49: 1172–1177.

Gahukar, R.T. and Das, R.K. (2020). Plant-derived nanopesticides for agricultural pest control: challenges and prospects. *Nanotechnology and Environmental Engineerinmg* 5: 1–9.

Galea, V. and Goulter, K. (2013). Field evaluation of a bioherbicide for control of parkinsonia (*Parkinsonia aculeata*) in Australia. *Proceedings of the 19th Australasian Plant Pathology Conference*, 43 (Abstr.).

Gamliel, A., Austerweil, M., and Kritzman, G. (2000). Non-chemical approach to soilborne pest management-organic amendments. *Crop Protection* 19: 847–853.

Gamliel, A. and Stapleton, J.J. (1993a). Characterization of antifungal volatile compounds evolved from solarized soil amended with cabbage residues. *Phytopathology* 83: 899–905.

Gamliel, A. and Stapleton, J.J. (1993b). Effect of chicken compost or ammonium phosphate and solarization on pathogen control, rhizosphere microorganisms, and lettuce growth. *Plant Disease* 77: 886–891.

Gamliel, A. and Stapleton, J.J. (1997). Improvement of soil solarization with volatile compounds generated from organic amendments. *Phytoparasitica* 25: S31–S38.

Gazzurelli, C., Migliori, A., Mazzeo, P.P. et al. (2020). Making agriculture more sustainable: an environmentally friendly approach to the synthesis of lignin@Cu pesticides. *ACS Sustainable Chemical Engineering* 14886–14895. https://doi.org/10.1021/acssuschemeng.0c04645.

Goud, J.K.C., Termorshuizen, A.J., Blok, W.J., and van Bruggen, A.H.C. (2004). Long-term effect of biological soil disinfestation on Verticillium wilt. *Plant Disease* 88: 688–694.

Guerra, N. and Fennimore, S. (2020). Evaluation of banded steam for weed and disease control in lettuce. *UC Weed Science*. UCANR retrieved February 21, 2021, from https://ucanr.edu/blogs/blogcore/postdetail.cfm?postnum=4413.

Guha, T., Gopal, G., Kundu, R., and Mukherjee, A. (2020). Nanocomposites for delivering agrochemicals: a comprehensive review. *Journal of Agriculture and Food Chemistry* 68: 3691–3702.

Guo, H., Di Gioia, F., Zhao, X. et al. (2017). Optimizing anaerobic soil disinfestation for fresh market tomato production: nematode and weed control, yield, and fruit quality. *Scientia Horticulturae* 218: 105–116.

Guo, H., Zhao, X., Rosskopf, E.N. et al. (2018). Impacts of anaerobic soil disinfestation and chemical fumigation on soil microbial communities in field tomato production system. *Applied Soil Ecology* 126: 165–173.

Gurusiddaiah, S., Gealy, D.R., Kennedy, A.C., and Ogg, A.G., Jr. (1992). Production, isolation, and characterization of phytotoxic and fungistatic compounds for biocontrol of downy brome (*Bromus tectorum* L.) and plant pathogenic fungi. *Weed Science Society of America Abstracts* 32: 84.

Hanson, C. (2008). Root colonization and environmental fate of the bioherbicide *Pseudomonas fluorescens* BRG100. M.Sc. thesis, University of Saskatchewan, Saskatoon, Saskatchewan.

Hao, L., Lin, G., Wang, H. et al. (2020). Preparation and characterization of zein-based nanoparticles via ring-opening reaction and self-assembly as aqueous nanocarriers for pesticides. *Journal of Agriculture and Food Chemistry* 68: 9624–9635.

Hasheminejad, N., Khodaiyan, F., and Safari, M. (2019). Improving the antifungal activity of clove essential oil encapsulated by chitosan nanoparticles. *Food Chemistry* 275: 113–122.

Herrera, F. and Ramirez, C. (1996). Soil solarization and chicken manure additions on propagule survival of *Cyperus rotundus*, *Rottboellia cochinchinensis* and *Bidens pilosa*. *Agronomia Mesoamericana* 7: 1–8. http://www.mag.go.cr/rev_meso/v07n01_001.pdf.

Hewavitharana, S.S., Klarer, E., Muramoto, J. et al. (2021). Analysis of environmental variables and carbon input on soil microbiome, metabolome and disease control efficacy in strawberry attributable to anaerobic soil disinfestation. *Microorganisms* 9: 1638. https://doi.org/10.3390/microorganisms9081638.

Hewavitharana, S.S., Klarer, E., Reed, A.J. et al. (2019). Temporal dynamics of the soil metabolome and microbiome during simulated anaerobic soil disinfestation. *Frontiers in Microbiology* 10: 2365. https://doi.org/10.3389/fmicb.2019.02365.

Hewavitharana, S.S., Ruddell, D., and Mazzola, M. (2014). Carbon source-dependent antifungal and nematicidal volatiles derived during anaerobic soil disinfestation. *European Journal of Plant Pathology* 140: 39–52.

Heystek, F., Wood, A.R., Neser, S., and Kistensamy, Y. (2011). Biological control of two *Ageratina* species (Asteraceae: Eupatorieae) in South Africa. *African Entomology* 19 (2): 208–216.

Hintz, W. (2007). Development of *Chondrostereum purpureum* as a mycoherbicide for deciduous brush control. In: *Biological Control: A Global Perspective* (ed. C. Vincent, M.S. Goettel, and G. Lazarovits), 285–290. CAB International, Wallingford, OX10 8DE, UK.

Hofmann, T., Lowry, G.V., Ghoshal, S., and Tufenkji, N. (2020). Technology readiness and overcoming barriers to sustainably implement nanotechnology-enabled plant agriculture. *Nature Food* 1: 416–425.

Huang, X., Liu, L., Wen, T. et al. (2016). Reductive soil disinfestations combined or not with Trichoderma for the treatment of a degraded and Rhizoctonia solani infested greenhouse soil. *Scientia Horticulturae* 206: 51–61.

Hynes, R.K. and Boyetchko, S.M. (2006). Research initiatives in the art and science of biopesticide formulations. *Soil Biology & Biochemistry* 38: 845–849. https://doi.org/10.1016/j.soilbio.2005.07.003.

Hynes, R.K. and Boyetchko, S.M. (2011). Improvements in the pesta formulation to promote survival of *Pseudomonas fluorescens* BRG100, green foxtail bioherbicide. *Pest Technology* 5: (Special Issue 1): 80–87.

Imaizumi, S., Nishino, T., Miyabe, K. et al. (1997). Biological control of annual bluegrass (*Poa annua* L.) with a Japanese isolate of *Xanthomonas campestris* pv. *poae* (JT-P482). *Biological Control* 8: 7–14.

Jacobs, T. (2019). *Effects of sudangrass cover crop and soil solarization on weed and pathogen management in organic strawberry production*. Master's Thesis. California Polytechnic State University, San Luis Obispo, 110 p. https://doi.org/10.15368/theses.2019.84.

Johnson, B.N., Kennedy, A.C., and Ogg, A.G., Jr. (1993). Suppression of downy brome growth by a rhizobacterium in controlled environments *Soil Science Society of America Journal* 57: 73.

Kadir, J. and Charudattan, R. (2000). *Dactylaria higginsii*, a fungal bioherbicide agent for purple nutsedge (*Cyperus rotundus*). *Biological Control* 17: 113–124.

Kah, M., Tufenkji, N., and White, J.C. (2019). Nano-enabled strategies to enhance crop nutrition and protection. *Nature Nanotechnology* 14: 532–540.

Kamra, A. and Gaur, H.S. (1998). Control of nematodes, fungi and weeds in nursery beds by soil solarization. *International Journal of Nematology* 8: 46–52.

Kanissery, R. and Ampatzidis, Y. (2020). Using steam for weed management. *Vegetable Specialty Crop News*. Retrieved March 1, 2021, from https://swfrec.ifas.ufl.edu/docs/pdf/weed-science/2020-03-Kanissery-Ampatzidis_VegetableSpecialtyCropNews.pdf.

Kassaby, F.Y. (1985). Solar-heating soil for control of damping-off diseases. *Soil Biology and Biochemistry* 17: 429–434.

Kasson, M.T., O'Neal, E.S., and Davis, D.D. (2015). Expanded host range testing for *Verticillium nonalfalfae*: potential biocontrol agent against the invasive *Ailanthus altissima*. *Plant Disease* 99: 823–835.

Kasson, M.T., Short, D.P.G., O'Neal, E.S. et al. (2014). Comparative pathogenicity, biocontrol efficacy, and multilocus sequence typing of *Verticillium nonalfalfae* from the invasive *Ailanthus altissima* and other hosts. *Phytopathology* 104: 282–292.

Katahira, R., Elder, T.J., and Beckham, G.T. (2018). Chapter 1: A brief introduction to lignin structure. *RSC Energy and Environment Series 2018-Janua, 1–20.*

Katan, J. and DeVay, J.E. (eds.) (1991). *Solarization*. Boca Raton, FL, USA: CRC, 26.

Katan, J., Greenburger, A., Alon, H., and Grinstein, A. (1976). Solar heating by polyethylene mulching for control of diseases caused by soil-borne pathogens. *Phytopathology* 66: 683–688.

Kennedy, A.C. (2018). Selective soil bacteria to manage downy brome, jointed goatgrass, and medusahead and do no harm to other biota. *Biological Control* 123: 18–27.

Kennedy, A.C., Ogg, A.G., Jr., and Yound, F.L. (1992). *Biocontrol of jointed goatgrass*. U.S. Patent no. 5,163,991.

Kennedy, A.C. and Smith, K.L. (1995). Soil microbial diversity and the sustainability of agricultural soils. *Plant and Soil* 170: 75–86.

Kenney, D.S. (1986). DeVine® – the way it was developed – an industrialist's view. *Weed Science* 34 (Suppl. 1): 15–16.

Khadka, R.B., Cardina, J., and Miller, S.A. (2021a). Perspectives on anaerobic soil disinfestation for weed management. *Journal of Integrated Pest Management* 12: 32. https://doi.org/10.1093/jipm/pmab027.

Khadka, R.B. and Miller, S.A. (2021b). Synergy of anaerobic soil disinfestation and *Trichoderma* spp. in rhizoctonia root rot suppression. *Frontiers in Sustainable Food Systems* 5: 76. https://doi.org/10.3389/fsufs.2021.645736.

Kim, D.S., Kim, S.B., and Fennimore, S.A. (2021). *Evaluation of broadcast steam application with mustard seed meal in fruiting strawberry*. https://doi.org/10.21273/HORTSCI15669-20.

Kolberg, R. and Wiles, L. (2002). Effect of steam application on cropland weed 1. *Weed Technology* 16: 44–49.

Kremer, R.J. (1987). Identity and properties of bacteria inhabiting seeds of selected broadleaf weed species. *Microbial Ecology* 14: 29–37.

Kremer, R.J. (1993). Management of weed seed banks with microorganisms. *Ecological Applications* 3: 42–52.

Kremer, R.J. (1998). Microbial interactions with weed seeds and seedlings and its potential for weed management. In: *Integrated Weed and Soil Management* (ed. J.L. Hatfield, D.D. Buhler, and B.A. Stwart), 191–179. Chelsea, MI: Ann Arbor Press.

Kremer, R.J. (2019). Bioherbicides and nanotechnology: current status and future trends. Chapter 15. In: *Nano-biopesticides Today and Future Perspectives* (ed. K. Opender), 353–366. Burlington: Academic Press. ISBN: 978-0-12-815829-6.

Kremer, R.J., Begonia, M.F.T., Stanley, L., and Lanham, E.T. (1990). Characterization of rhizobacteria associated with weed seedlings. *Applied and Environmental Microbiology* 56: 1649–1655.

Kremer, R.J. and Kennedy, A.C. (1996). Rhizobacteria as biocontrol agents of weeds. *Weed Technology* 10: 601–609.

Kremer, R.J. and Sarwar, M. (1995). Microbial metabolites with potential applications in weed management. *Proceedings of the Plant Growth Regulation Society of America* 21: 48.

Kremer, R.J. and Souissi, T. (2001). Cyanide production by rhizobacteria and potential suppression of weed seedling growth. *Current Microbiology* 43: 182–186.

Kumar, B., Yaduraju, N.T., Ahuja, K.N., and Prasad, D. (1993). Effect of soil solarization on weeds and nematodes under tropical Indian conditions. *Weed Research* 33: 423–429.

Kumar, S., Kumar, D., and Dilbaghi, N. (2017). Preparation, characterization, and bio-efficacy evaluation of controlled release carbendazim-loaded polymeric nanoparticles. *Environmental Science and Pollution Research International* 24: 926–937.

Li, Z., Di Gioia, F., Hwang, J.I. et al. (2020). Dissipation of fomesafen in fumigated, anaerobic soil disinfestation-treated, and organic-amended soil in Florida tomato production systems. *Pest Management Science* 76: 628–635.

Lira-Saldivar, R.H., Cruz, J., Beltran, F., and Jimenez, F. (2004). Effect of biofumigation with solarization and *Larrea tridentata* extract on soil-borne pathogens of pepper plants. *Biological Agriculture and Horticulture* 22: 21–29.

Liu, L., Kong, J., Cui, H. et al. (2016). Relationships of decomposability and C/N ratio in different types of organic matter with suppression of *Fusarium oxysporum* and microbial communities during reductive soil disinfestation. *Biological Control* 101: 103–113.

Luo, Y., Wang, Q., and Zhang, Y. (2020). Biopolymer-based nanotechnology approaches to deliver bioactive compounds for food applications: a perspective on the past, present, and future. *Journal of Agriculture and Food Chemistry* 68: 12993–13000. Management in organic strawberry production. Master's Thesis. California Polytechnic State.

Martinez Jimenez, M. and Charudattan, R. (1998). Survey and evaluation of Mexican native fungi for biocontrol of waterhyacinth. *Journal of Aquatic Plant Management* 36: 145–148.

Martyn, R.D. (1985). Waterhyacinth decline in Texas caused by *Cercospora piaropi*. *Journal of Aquatic Plant Management* 23: 29–32.

Masabni, J. and Franco, J.G. (2016). *Soil Solarization*. Texas A&M AgriLife Extension Service. EHT-087.

Maschek, O. and Halmschlager, E. (2015). First report of Verticillium wilt on *Ailanthus altissima* in Europe caused by *Verticillium nonalfalfae*. *Plant Disease* 100: 529. https://doi.org/10.1094/PDIS-07-15-0733-PDN.

Merinoa, N., Berdejo, D., Bentoa, R. et al. (2019). Antimicrobial efficacy of *Thymbra capitata* L. Cav. essential oil loaded in self-assembled zein nanoparticles in combination with heat. *Industrial Crops and Products* 133: 98–104.

Michuda, A., Goodhue, R.E., Hoffmann, M., and Fennimore, S.A. (2021). Predicting net returns of organic and conventional strawberry following soil disinfestation with steam or steam plus additives. *Agronomy* 11: 149. https://doi.org/10.3390/agronomy11010149.

Miles, J.E., Kawabata, O., and Nishimoto, R.K. (2002). Modeling purple nutsedge sprouting undersoil solarization. *Weed Science* 50: 64–71.

Miles, J.E., Nishimoto, R.K., and Kawabata, O. (1996). Diurnally alternating temperatures stimulate sprouting of purple nutsedge (*Cyperus rotundus*) tubers. *Weed Science* 44: 122–125.

Momma, N. (2008). Biological soil disinfestations (BSD) of soil-borne pathogens and its possible mechanisms. *Japan Agricultural Research Quarterly Journal* 42: 7–12.

Momma, N., Yamamoto, K., Simandi, P., and Shishido, M. (2006). Role of organic acids in the mechanisms of biological soil disinfestations (BSD). *Journal of General Plant Pathology* 72: 247–252.

Morin, L. (2020). Progress in biological control of weeds with plant pathogens. *Annual Review of Phytopathology* 58: 201–223. https://doi.org/10.1146/annurev-phyto-010820-012823.

Morin, L., Piper, M., White, A., and Schooler, S. (2012). Spread, specificity and initial impact of the white-smut fungus *Entyloma ageratinae* on mistflower in Australia. *Proceedings Eighteenth Australasian Weed Conference*, 88–90. Peer-reviewed proceedings edited by Valerie Eldershaw, Weed Society of Victoria Inc., ISBN 978-0-646-58670-0. http://caws.org.nz/old-site/awc_contents.php?yr=2012 (accessed 12 June 2021).

Morris, M.J. (1990). *Cercospora piaropi* recorded on the aquatic weed, *Eichhornia crassipes*, in South Africa. *Phytophylactica* 22: 255–256.

Morris, M.J. (1991). The use of plant pathogens for biological weed control in South Africa. *Agriculture, Ecosystems and Environment* 37: 239–255.

Morris, M.J. (1997). Impact of the gall-forming rust fungus *Uromycladium tepperianum* on the invasive tree *Acacia saligna* in South Africa. *Biological Control* 10: 75–82.

Morris, M.J., Wood, A.R., and den Breeÿen, A. (1999). Plant pathogens and biological control of weeds in South Africa: a review of projects and progress during the last decade. *African Entomology Memoir* 1: 129–137.

Mutebi, C.M., Musyimi, D.M., and Opande, G. (2021 May). Bio control of water hyacinth with *Cercospora piaropi* and *Myrothecium roridum* corn oil formulations in the greenhouse for enhanced water resources management and conservation. *East African Journal of Science, Technology and Innovation* 2 (Special issue).

National Research Council, National Academy of Sciences. (2011). Feasibility of using mycoherbicides for controlling illicit drug crops. *Committee on Mycoherbicides for Eradicating Illicit Drug Crops, R. Charudattan, Chair.* The National Academies Press, Washington, DC. See p. 34 for the narrative on Collego written by D. O. TeBeest.

Nazari, M., Ghanbarzadeh, B., Samadi Kafil, H. et al. (2019). Garlic essential oil nanophytosomes as a natural food preservative: its application in yogurt as food model. *Colloids Interface Science Communications* 30: 100176.

Nunez, D. and Hooks, C.R.R. (2019). Biosolarization as an alternative to chemical fumigation *Vegetable and Fruit Headline News*. Anne Arundel County Agriculture Newsletter. Summer Edition, 4–5 Retrieved March 1,2021 from https://extension.umd.edu/sites/extension.umd.edu/files/docs/Biosolarization2019.pdf.

O'Neal, E.S. and Davis, D.D. (2015). Intraspecific root grafts and clonal growth within *Ailanthus altissima* stands influence *Verticillium nonalfalfae* transmission. *Plant Disease* 99: 1070–1077.

Oka, Y. (2010). Mechanisms of nematode suppression by organic soil amendments–a review. *Applied Soil Ecology* 44: 101–115.

Oliveira, C.R., Domingues, C.E.C., de Melo, N.F.S. et al. (2019). Nanopesticide based on botanical insecticide pyrethrum and its potential effects on honeybees. *Chemosphere* 236: 124282.

Ozores-Hampton, M., McSorley, R., Stansly, P.A. et al. (2002 August). Long-term large-scale soil solarization as a low-input production system for Florida vegetables. *XXVI International Horticultural Congress: Sustainability of Horticultural Systems in the 21st Century*. 638, 177–188.

Pascoli, M., Jacques, M.T., Agarrayua, D.A. et al. (2019). Neem oil based nanopesticide as an environmentally-friendly formulation for applications in sustainable agriculture: an ecotoxicological perspective. *Science of the Total Environment* 677: 57–67.

Patterson, D.T. (1998). Suppression of purple nutsedge (*Cyperus rotundus*) with polyethylene film mulch. *Weed Technology* 12: 275–280.

Paudel, B.R., Di Gioia, F., Zhao, X. et al. (2018). Evaluating anaerobic soil disinfestation and other biological soil management strategies for open-field tomato production in Florida. *Renewable Agriculture and Food Systems* 1–12. https://doi.org/10.1017/S1742170518000571.

Pedras, M.S.C., Ismail, N., Quail, J.W., and Boyetchko, S.M. (2003). Structure, chemistry, and biological activity of pseudophomins A and B, new cyclic lipodepsipeptides isolated from the biocontrol bacterium Pseudomonas fluorescens. *Phytochemistry* 62: 1105–1114.

Pérez-Guzmán, C.J. and Castro-Muñoz, R.A. (2020). Review of zein as a potential biopolymer for tissue engineering and nanotechnological applications. *Processes* 8: 1–21.

Prasad, R., Bhattacharyya, A., and Nguyen, Q.D. (2017). Nanotechnology in sustainable agriculture: recent developments, challenges, and perspectives. *Frontiers in Microbiology* 8: 1–13.

Quail, J.W., Ismail, N., Pedras, M.S.C., and Boyetchko, S.M. (2002). Pseudophomins A and B, a class of cyclic lipodepsipeptides isolated from a *Pseudomonas* species. *Acta Crystallographica* C58: o268–o271.

Raliya, R., Saharan, V., Dimkpa, C., and Biswas, P. (2018). Nanofertilizer for precision and sustainable agriculture: current state and future perspectives. *Journal of Agriculture and Food Chemistry* 66: 6487–6503.

Rebbeck, J., Malone, M.A., Short, D.P.G. et al. (2013). First report of Verticillium wilt caused by *Verticillium nonalfalfae* on tree-of-heaven (*Ailanthus altissima*) in Ohio. *Plant Disease* 97: 999.

Restuccia, G., Marchese, M., and Mauromicale, G. (1994). Solarizzazione e lotta contro le infestanti. *Rivista di Agronomia* 28: 21–30.

Ricci, M.S., De Almeida, D.L., Ribeiro, R.D. et al. (2000). *Cyperus rotundus* control by solarization. *Biological Agriculture and Horticulture* 17: 151–157.

Rodríguez-Eugenio, N., McLaughlin, M., and Pennock, D. (2018). *Soil Pollution*. FAO. https://scirp.org/reference/referencespapers.aspx?referenceid=3251104.

Roe, N., Ozores-Hampton, M., and Stansly, P.A. (2004). Solarization effects on weed populations in warm climates. *Acta Horticulturae* 638: 197–200.

Rosskopf, E., Chellemi, D.O., and Kokalis-Burelle, N. (1999a). Alternative soil disinfestations treatments for weed control. *Proceedings of the Annual International Research Conference on Methyl Bromide Alternatives and Emissions Reductions*, San Diego, CA (1–4 November 1999).

Rosskopf, E., Di Gioia, F., Hong, J.C. et al. (2020). Organic amendments for pathogen and nematode control. *Annual Review of Phytopathology* 58: 277–311.

Rosskopf, E.N., Charudattan, R., and Bruckart, W. (2017). Bioherbicides: an overview. In: *Integrated Weed Management for Sustainable Agriculture* (ed. R.L. Zimdahl). E-chapter 18. Burleigh Dodds Publishing. www.bdspublishing.com.

Rosskopf, E.N., Charudattan, R., DeValerio, J.T., and Stall, W.M. (2000). Field evaluation of *Phomopsis amaranthicola*, a biological control agent of *Amaranthus* spp. *Plant Disease* 84: 1225–1230.

Rosskopf, E.N., Charudattan, R., and Kadir, J.B. (1999b). Use of plant pathogens in weed control. In: *Handbook of Biological Control* (ed. T.S. Bellows and T.W. Fisher), 891–918. Academic Press.

Rosskopf, E.N., Kokalis-Burelle, N., Butler, D. et al. (2010). Development of anaerobic soil disinfestation for Florida vegetable and flower production. *Proceedings of Annual International Research Conference on Methyl Bromide Alternatives and Emissions Reductions*, Orlando, 84-1/84-2.

Rosskopf, E.N., Serrano-Pérez, P., Hong, J. et al. (2015). Anaerobic soil disinfestation and soil-borne pest management. In: *Organic Amendments and Soil Suppressiveness in Plant Disease Management* (ed. M.K. Meghavansi and A. Varma), 277–305. Cham, Switzerland: Springer International Publishing.

Rubin, B. and Benjamin, J. (1984). Solar heating of the soil: involvement of environmental factors in the weed control process. *Weed Science* 32: 138–142.

Rubin, B. and Gamliel, A. (2018). Soil solarization: a sustainable method for weed management. In: *Integrated Weed Management for Sustainable Agriculture* (ed. R.L. Zimdahl), 303–318. Cambridge, UK: Burleigh Dodds.

Saghir, A.R. (1997). Soil solarization: an alternative technique for weed management in hot climates. In: *Proceedings of the Second International Conference on Soil Solarization and Integrated Management of Soil-borne Pests* (ed. J.J. Stapleton, J.E. DeVay, and C.L. Elmore), Aleppo, Syrian Arab Republic (16–21 March 1997), 206–211. FAO Plant Protection and Production Paper 147, FAO, Rome, Italy.

Sahlan, M., Tejamaya, M., Arbianti, R. et al. (2021). The effects of nano-casein encapsulation and productions of a controlled-release on eugenol containing bio-pesticide toxicity. In: *Key Engineering Materials*, vol. 874 (ed. Y.M. Syah, A. Patah, E. Hermawati, and M.Y. Azis), 115–127. Frontiers in Molecular Sciences – Selected peer-reviewed full text paper from the International Conference on the Frontiers in Molecular Sciences, ICFMS 2020. Trans Tech Publications Ltd.

Saini, R.K., Patel, S., Bajpai, J., and Bajpai, A.K. (2020). Advanced controlled nanopesticide delivery systems for managing insect pests. *Controlled Release of Pesticides for Sustainable Agriculture* 155–184. https://doi.org/10.1007/978-3-030-23396-9_7.

Sarwar, M. and Kremer, R.J. (1995a). Determination of bacterially derived auxins using a microplate method. *Letters in Applied Microbiology* 20: 282–285.

Sarwar, M. and Kremer, R.J. (1995b). Enhanced suppression of plant growth through production of L-tryptophan-derived compounds by deleterious rhizobacteria. *Plant and Soil* 172: 261–269.

Sawchyn, K., Nelson, T., Gibson, L. et al. (2000). Development of rhizobacteria for biological control of grass weeds. *Proceedings of the Third International Weed Science Congress*, Foz do Iguassu, Brazil (6–11 June 2000), 183. [Abstr].

Schall, M.J. and Davis, D.D. (2009a). *Ailanthus altissima* wilt and mortality: etiology. *Plant Disease* 93: 747–751.

Schall, M.J. and Davis, D.D. (2009b). Verticillium wilt of *Ailanthus altissima*: susceptibility of associated tree species. *Plant Disease* 93: 1158–1162.

Shabana, Y.M., Charudattan, R., and Elwakil, M.A. (1995). Identification, pathogenicity, and safety of *Alternaria eichhorniae* from Egypt as a bioherbicide agent for waterhyacinth. *Biological Control* 5: 123–135.

Shinmura, A., Sakamoto, N., and Abe, H. (1999). Control of Fusarium root rot of Welsh onion by soil reduction (abstract in Japanese). *Japan Journal of Phytopathology* 65: 352–353.

Shrestha, U., Augé, R.M., and Butler, D.M. (2016). A meta-analysis of the impact of anaerobic soil disinfestation on pest suppression and yield of horticultural crops. *Frontiers in Plant Science* 7: 1254. https://doi.org/10.3389/fpls.2016.01254.

Shrestha, U., Rosskopf, E.N., and Butler, D.M. (2018). Effect of anaerobic soil disinfestation amendment type and C:N ratio on *Cyperus esculentus* tuber sprouting, growth and reproduction. *European Weed Research Society* 58: 379–388.

Singla, S.L., Pareek, A., and Grover, A. (1997). High temperature. In: *Plant Ecophysiology* (ed. M.N.V. Prasad), 101–127. New York, USA: Wiley.

Smither-Kopperl, M.L., Charudattan, R., and Berger, R.D. (1998). Dispersal of spores of *Fusarium culmorum* in aquatic systems. *Phytopathology* 88: 382–388.

Snyder, A.L., Kasson, M.T., Salom, S.M. et al. (2013). First report of Verticillium wilt of *Ailanthus altissima* in Virginia caused by *Verticillium nonalfalfae*. *Plant Disease* 97: 837.

Standifer, L.C., Wilson, W., and Porche-Sorbet, R. (1984). Effects of solarization on soil weed populations. *Weed Science* 32: 569–573.

Stapleton, J., Elmore, C., and DeVay, J. (2000). Solarization and biofumigation help disinfest soil. *California Agriculture* 54: 42–45.

Stapleton, J.J. (2000). Soil solarization in various agricultural production systems. *Crop Protection* 19 (8–10): 837–841.

Stapleton, J.J., Molinar, R.H., Lynn-Patterson, K. et al. (2005). Soil solarization provides weed control for limited-resource and organic growers in warmer climates. *California Agriculture* 59: 84–89.

Stephen, S.L., Beales, L., Peyret, H. et al. (2018). Recombinant expression of tandom_HBc virus-like particles. In: *Virus-Derived Nanoparticles for Advanced Technologies: Methods and Protocols, Methods in Molecular Biology*, 1e, vol. 1776 (ed. C. Wege and G.P. Lomonossoff), 97–123. Springer: Imprint: Humana, NY, USA.

Stephenson, G.T. (2019). *The effects of agricultural waste-based compost amendments in organic pest management.* Master's thesis California Polytechnic State University. https://digitalcommons.calpoly.edu/theses/2042GoogleScholar.

Strauss, S.L., and Kluepfel, D.A. (2015). Anaerobic soil disinfestation: a chemical-independent approach to pre-plant control of plant pathogens. *Journal of Integrative Agriculture* 14: 2309–2318.

Suganya, P., Vaseeharan, B., Vijayakumar, S. et al. (2017). Biopolymer zein-coated gold nanoparticles: synthesis, antibacterial potential, toxicity and histopathological effects against the Zika virus vector *Aedes aegypti*. *Journal of Photochemistry and Photobiology. B, Biology* 173: 404–411.

Suslow, T.V. and Schroth, M.N. (1982). Role of deleterious rhizobacteria as minor pathogens in reducing crop growth. *Phytopathology* 72: 111–115.

Tessmann, D.J., Charudattan, R., and Preston, J.F. (2008). Variability in aggressiveness, cultural characteristics, cercosporin production and fatty acid profile of *Cercospora piaropi*, a biocontrol agent of water hyacinth. *Plant Pathology* 57: 957–966.

Tranel, P.J., Gealy, D.R., and Irzyk, G.P. (1993a). Physiological responses of downy brome (*Bromus tectorum*) roots to *Pseudomonas fluorescens* strain D7 phytotoxin. *Weed Science* 41: 483–489.

Tranel, P.J., Gealy, D.R., and Kennedy, A.C. (1993b). Inhibition of downy brome (*Bromus tectorum*) root growth by a phytotoxin from *Pseudomonas fluorescens* strain D7. *Weed Technology* 7: 134–139.

Trujillo, E.E. (2005). History and success of plant pathogens for biological control of introduced weeds in Hawaii. *Biological Control* 33: 113–122. https://doi.org/10.1016/j.biocontrol.2004.11.008.

U.S. Environmental Protection Agency. (2021). *Summary of the federal insecticide, fungicide, and rodenticide act.* 7 U.S.C. §136 et seq. (1996). https://www.epa.gov/laws-regulations/summary-federal-insecticide-fungicide-and-rodenticide-act (accessed 16 June 2021).

van Agtmaal, M., van Os, G.J., Hol, W.G. et al. (2015). Legacy effects of anaerobic soil disinfestation on soil bacterial community composition and production of pathogen-suppressing volatiles. *Frontiers in Microbiology* 6: 1–12.

Vurro, M., Miguel-Rojas, C., and Pérez-de-Lugue, A. (2019). Safe nanotechnologies for increasing the effectiveness of environmentally friendly natural agrochemicals. *Pest Management Science* 75: 2403–2412. https://doi.org/10.1002/ps.5348.

Wang, X., Gu, M., Niu, G., and Baumann, P.A. (2015). Herbicidal activity of mustard seed meal (*Sinapis alba* 'IdaGold' and *Brassica juncea* 'Pacific Gold') on weed emergence. *Industrial Crops and Products* 77: 1004–1013.

Wani, T.A., Masoodi, F.A., NabiBaba, W. et al. (2019). Nanoencapsulation of agrochemicals, fertilizers, and pesticides for improved plant production. *Advances in Phytonanotechnology*. Elsevier Inc. https://doi.org/10.1016/b978-0-12-815322-2.00012-2.

Watson, A.K. (2017). Biocontrol and weed management in rice of asian-pacific region. In: *Weed Management in Rice in the Asian-Pacific Region* (ed. A.N. Rao and H. Matsumoto), 155 to 176. India: Asian-Pacific Weed Science Society (APWSS); The Weed Science Society of Japan, Japan and Indian Society of Weed Science.

Weaver, M.A., Lyn, M.E., Boyette, C.D., and Hoagland, R.E. (2007). Bioherbicides for weed control. Chapter 7. In: *Non-chemical Weed Management* (ed. M.K. Upadhyaya and R.E. Blackshaw), 93–110. CAB International. Wallingford, OX10 8DE, UK.

Webber, C., Boydston, R., White, P.M., and Shrefler, J.W. (2017). Impact of mustard seed meal application on direct seeded cucurbits and weed control. *Journal of Agricultural Science* 9: 81–90.

Webster, T.M. (2005). Patch expansion of purple nutsedge (*Cyperus rotundus*) and yellow nutsedge (*Cyperus esculentus*) with and without polyethylene mulch. *Weed Science* 53: 839–845.

Wen, A.M. and Steinmetz, N.F. (2016). Design of virus-based nanomaterials for medicine, biotechnology, and energy. *Chemical Society Reviews* 45: 4074–4126.2007.

Westwood, J.H., Charudattan, R., Duke, S.O. et al. (2018). Weed management in 2050: perspectives on the future of weed science. *Weed Science* 66 (3): 275–285.

Winston, R.L., Schwarzländer, M., Hinz, H.L. et al. (eds.) (2014). *Biological Control of Weeds: A World Catalogue of Agents and Their Target Weeds*, 5e. Morgantown, West Virginia: USDA Forest Service, Forest Health Technology Enterprise Team. FHTET-2014-04, 838 pages.

Winston, R.L., Schwarzländer, M., Hinz, H.L. et al. (eds.) (2021). *Biological Control of Weeds: A World Catalogue of Agents and Their Target Weeds.* Based on FHTET-2014-04, USDA Forest Service, Forest Health Technology Enterprise Team. Updated 2021. https://www.ibiocontrol.org/catalog (accessed 11 June 2021).

Wood, A.R. (2012). *Uromycladium tepperianum* (a gall-forming rust fungus) causes a sustained epidemic on the weed *Acacia saligna* in South Africa. *Australasian Plant Pathology* 41: 255–261.

Wood, A.R. and Morris, M.J. (2007). Impact of the gall-forming rust fungus *Uromycladium tepperianum* on the invasive tree *Acacia saligna* in South Africa: 15 years of monitoring. *Biological Control* 41: 68–77.

Xiang, Y., Zhang, G., Chi, Y. et al. (2017). Fabrication of a controllable nanopesticide system with magnetic collectability. *Chemican Engineering Journal* 328: 320–330.

Yaduraju, N.T. and Mishra, J.S. (2004). Soil solarization. In: *Weed Biology and Management* (ed. Inderjit), 345–362. Dordrecht, The Netherlands: Springer.

Yandoc, C.B., Charudattan, R., and Shilling, D.G. (2004). Suppression of cogongrass (*Imperata cylindrica*) by a bioherbicidal fungus and plant competition. *Weed Science* 52: 649–653.

Zdor, R.E., Alexander, C.M., and Kremer, R.J. (2005). Weed suppression by deleterious rhizobacteria is affected by formulation and soil properties. *Communications in Soil Science and Plant Analysis* 36: 1289–1299.

Zhang, D.-X., Liu, G., Jing, T.-F. et al. (2020a). Lignin-modified electronegative epoxy resin nanocarriers effectively deliver pesticides against plant root-knot nematodes *Meloidogyne Incognita*. *Journal of Agriculture and Food Chemisty* 68: 13562–13572.

Zhang, Y., Liu, B., Huang, K. et al. (2020b). Eco-friendly castor oil-based delivery system with sustained pesticide release and enhanced retention. *ACS Applied Materialsand Interfaces* 12: 37607–37618.

Zhao, J., Liu, S., Zhou, X. et al. (2020a). Reductive soil disinfestation incorporated with organic residue combination significantly improves soil microbial activity and functional diversity than sole residue incorporation. *Applied Microbiology and Biotechnology* 104: 7573–7588.

Zhao, L., Lu, L., Wang, A. et al. (2020b). Nano-biotechnology in agriculture: use of nanomaterials to promote plant growth and stress tolerance. *Journal of Agriccuture and Food Chemistry* 68: 1935–1947.

Zhao, Y., Lin, S., Wan, L. et al. (2021). Anaerobic soil disinfestation with incorporation of straw and manure significantly increases greenhouse gases emission and reduces nitrate leaching while increasing leaching of dissolved organic N. *Science of the Total Environment* 785: 147307. https://doi.org/10.1016/j.scitotenv.2021.147307.

Zhou, K. (2017). Remediation of watermelon continuous cropping soil by anaerobic soil disinfestation. *Acta Agriculturae Zhejiangensis* 29: 1179–1188.

10

Mechanisms of Weed Suppression by Cover Crops, Intercrops, and Mulches

Richard G. Smith, Natalie P. Lounsbury, and Samuel A. Palmer*

Department of Natural Resources and the Environment, University of New Hampshire, Durham, NH, USA
** Corresponding author*

10.0 Introduction

The overarching objective of weed control and weed management is to minimize the negative effects that weeds have on agricultural production. Current predominant approaches to weed control – herbicides and tillage – focus almost exclusively on killing weeds, usually soon after they emerge from the soil. However, multiple factors are conspiring to make sole reliance on these practices untenable. Chief among these are a growing recognition that herbicides and tillage have deleterious effects on ecosystem and human health (Montgomery 2007; Mortensen et al. 2012; Ramesh et al. 2017). In the case of herbicides, pressure is mounting for stricter regulations on their use, and as resistant genotypes continue to evolve, the durability of herbicide-based weed control is increasingly uncertain (Davis and Frisvold 2017; Storck et al. 2017). At the same time, climate change is exacerbating the challenge of weed control in several ways. On the one hand, more frequent and prolonged droughts heighten the need for practices that reduce the need for tillage and that protect and build resilience-enhancing soil organic matter (Williams et al. 2016). On the other hand, tillage is increasingly being recommended and employed to manage weeds that have evolved resistance to herbicides (Beckie and Harker 2017; Van Deynze et al. 2021). Concomitantly, more extreme precipitation events will accelerate runoff and soil losses when tillage is employed to manage these weeds (Zhang 2012; Van Deynze et al. 2021). Lastly, changes in temperature and the timing and quantity of precipitation may also affect access to fields at times critical for weed management (Tomasek et al. 2017).

While the situation may appear dire, there is also hope. Cover cropping, intercropping, and mulching hold tremendous potential to attenuate the threats to and from our predominant weed management tactics while providing other agroecosystem benefits that will help agriculture adapt to a changing climate (Blanco-Canqui et al. 2015; Kaye and Quemada 2017). Despite strong phenomenological evidence that cover cropping, intercropping, and mulching can suppress weeds (Liebman and Dyck 1993; Osipitan et al. 2018, 2019), we know surprisingly little about the exact mechanisms by which these practices regulate weed dynamics at different stages of the weed life cycle. As will become apparent through the remainder of this chapter, our relative inability to differentiate between the multiple mechanisms that ultimately contribute to weed suppression makes it difficult to rank their relative importance or to make definitive statements about how specific practices are likely to impact specific mechanisms in specific situations.

Cover crops, intercrops, and mulches fall into the broad category of ecologically based weed management (EBWM), which is in many ways the antithesis of the conventional approach to weed control. It is knowledge rather than input-intensive, and proactive instead of reactive. It is also not as simple; it often requires that multiple strategies be integrated in complementary ways to be effective (Liebman and Gallandt 1997; Bàrberi 2002; O'Donovan et al. 2013). Unlike herbicides and tillage, which primarily target the weed-seedling stage, EBWM practices tend to impact factors that stress and/or kill weeds at all stages of their life cycle, from the seed and seedling stage to reproduction. These include influencing the activity of seed-decaying soil microorganisms and seed-predating fauna as well as altering light, moisture, temperature, nutrient, and atmospheric conditions. Employing EBWM is not merely input substitution; it often involves fundamentally changing the cropping system (Liebman 2001). This is especially true with cover crops, intercrops, and mulches because they can change multiple aspects of the cropping system, from the timing of crop management to the dynamics of nutrient

Ecologically Based Weed Management: Concepts, Challenges, and Limitations, First Edition. Edited by Nicholas E. Korres, Ilias S. Travlos, and Thomas K. Gitsopoulos.

and water cycling and non-weed pest populations. Weed management with these practices must therefore be considered in conjunction with other crop management and ecosystem service objectives (Weil and Kremen 2007; Schipanski et al. 2014).

Advancing our mechanistic understanding of how cover crops, intercrops, and mulches affect weed dynamics at all stages of the weed life cycle is critical for both researchers and farmers to implement practices effectively, develop new strategies, and integrate multiple strategies concurrently within the unique contexts of their cropping systems. Rather than review what is known about the practices of cover cropping, intercropping, and mulching individually, we have organized the following chapter by mechanisms, or pathways of weed management that these practices most strongly influence. We will begin by first outlining the traditional view of the weed seedbank and then an alternative view of the seedbank as a subcycle within the larger overall life cycle of an annual weed population. It is through this weed life cycle lens that we will attempt to illuminate what is known and what might be achievable through the use of cover cropping, intercropping, and mulching to enhance weed management. While we acknowledge the important role perennial weeds play in many cropping systems, particularly no-till systems, our discussion will largely center on annual weeds and temperate annual cropping systems.

Examining the pathways by which weeds enter, reside in, and eventually exit the soil seedbank will help to illuminate the mechanisms behind the variable efficacy of different EBWM practices and provide insight into the conditions that are likely to enable them to work most effectively, or conversely, undermine their efficacy (Smith and Mortensen 2017). We also hope this approach to thinking about the mechanisms underpinning weed survival or death will highlight how emerging practices such as "planting green" and interseeding cover crops into standing crops may more effectively reduce the need for weed management at the seedling stage, which should help to slow the evolution toward biotypes capable of avoiding or resisting future control practices. Lastly, we hope this approach will demonstrate how little we actually know about the magnitudes and mechanisms of what may be important pathways determining seed fates in the soil and their responses to different management practices, and thereby reveal areas ripe for future research.

10.1 Traditional View of the Weed Seedbank

The life cycle of an annual weed begins with the seed in the soil and proceeds through the stages of germination, emergence, growth and maturation, flowering and reproduction, and ends with the shedding of seeds back to the soil (i.e. "seed rain"). The seedbank, when considered as a distinct stage in the weed life cycle (e.g. Nichols et al. 2015), is often thought of as a compartment or "bank" in which seeds are stored until they are ready to germinate and emerge (Figure 10.1; Menalled and Schonbeck 2013; Kumar et al. 2019). As such, the traditional mantra for managing the soil weed seedbank often evokes an analogy to an actual bank: "minimize deposits and maximize withdrawals."

Deposits, in this case, are additions of weed seeds to the soil, either via seeds shed from plants in the field or immigration of seeds dispersed from outside the field. *Withdrawals* include any process that results in a decrease in the numbers of viable seeds remaining in the soil, including seed germination and the transition from the seed to seedling life stage. Stimulation of seed germination leading to seedling emergence, after which the seedling is killed in some way, is considered the most important mechanism for "debiting" the seedbank (Menalled and Schonbeck 2013). The practices of creating false or stale seedbeds or chemical fallows are based on this paradigm (Wicks and Smika 1973; Gallandt 2006). Under this conceptualization, seed dormancy is considered an impediment to "withdrawing" seeds from the seedbank, and therefore, a limitation to effective weed control (Mohler 2001; Boyd et al. 2006; Menalled and Schonbeck 2013). Other processes considered under the *withdrawals* category include seed predation, seed decay, and fatal germination (Swanton and Booth 2004). While the important roles these biologically mediated mechanisms play in reducing the weed seedbank are often acknowledged, they are sometimes thought of as somewhat passive rather than active agents of seed mortality that can

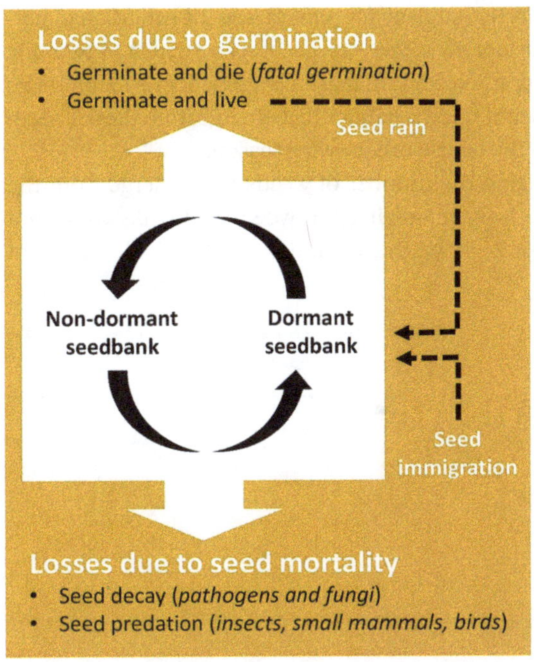

Figure 10.1 A common depiction of the weed seedbank as a compartment. Modified from Menalled and Schonbeck 2013.

and should be managed in ways that increase their efficacy, especially in conjunction with other mechanisms (but see e.g. Kremer 1993; Swanton and Booth 2004; Gallandt 2006). Similarly, abiotic controls on seed dormancy, germination, and seedling emergence are acknowledged to play critical roles in seedbank dynamics (e.g. Benech-Arnold et al. 2000) but are less often considered from the perspective of factors that may be actively managed in order to achieve specific weed management objectives (but see Nichols et al. 2015).

10.2 Alternative View of the Weed Seedbank and the Fate of Weeds

In many ways, managing weeds is a numbers game. The more weed seeds there are in the soil, the more weed seedlings that will emerge and potentially survive to become problematic from the perspective of crop production. This numbers game has implications for both weed control efficacy (Taylor and Hartzler 2000) as well as the durability of weed management practices (Menalled et al. 2016). For example, a paradox of herbicide-based weed control is that the more individuals that are exposed to an herbicide, the greater the probability is that at least one of those individuals possesses a mutation conferring resistance to the herbicide (Figure 10.2; Jasieniuk et al. 1996). In other words, one of the best ways to *avoid* selecting for herbicide resistant weed populations is to ensure that an herbicide is applied to as few weeds as possible. This highlights the important role that the weed seedbank plays in determining the initial size of the weed population that is subsequently exposed to a weed control practice. It also suggests that an effective strategy for minimizing the evolution of herbicide-resistant genotypes (or genotypes capable of resisting or avoiding any specific management practice) is to minimize the number of weed seedlings that emerge from the soil seedbank over the course of a growing season.

Minimizing the number of weeds that emerge from the soil by actively managing the weed seedbank would not only help to slow the evolution of herbicide resistant biotypes but would also help to achieve the overarching goals of weed management, namely, to minimize the negative effects of weeds on crop production. Further, demographic modeling studies have shown that seed survival is the most important stage influencing population growth of many annual weeds (Firbank and Watkinson 1986; Jordan et al. 1995; Davis et al. 2004). But to what degree might we expect the seedbank can be managed to reduce the number of weeds that emerge? And what are the ecological mechanisms governing the fate of weeds that may be amenable to modification through our cropping practices?

To explore these questions, we must first better understand where the weed seedbank fits within the larger life cycle of an annual weed and the nature of the biotic and abiotic factors operating on seeds once they enter the soil. We contend that the soil seedbank should be thought of as a sub-cycle within the larger overall cycle of an annual weed (Figure 10.3). Just as emerged weeds grow and (potentially) reproduce during the growing season, a cohort of weed seeds in the seedbank subcycle experiences major transition phases (i.e. overwintering and growing season phases), during which seeds are differentially susceptible to mortality factors like seed predation, seed decay, and fatal germination. The potential exists for seeds to remain within the seedbank subcycle for multiple overwintering and growing season phases, which presents additional opportunities for encouraging weed seed mortality. As will be discussed in this chapter, cover cropping, intercropping, and mulch may change weed dynamics via multiple mechanisms at all points of the annual weed lifecycle; however, the greatest proactive role of these practices may be to manage underappreciated *withdrawal* pathways via seed mortality that do not rely on additional seedling-stage management.

For this chapter, we will consider cover cropping to mean any crop plant grown, whether as a monoculture or mixture, during a period when a cash crop is not actively growing. We consider intercropping to include any situation in which a crop plant (including a cover crop) is grown at the same time as another cash crop, including living

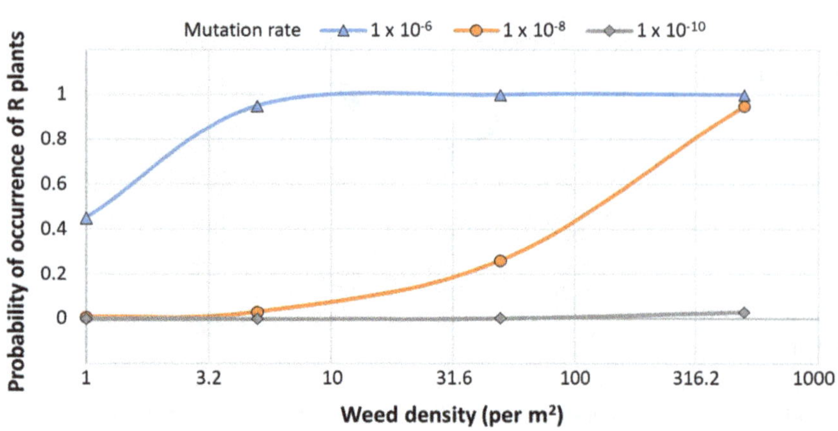

Figure 10.2 Probability of encountering at least one herbicide resistant weed in 30 ha fields as a function of weed density and mutation rate assuming a dominant mutation and random mating. Data are from Table 1 of Jasieniuk et al. (1996). Note: Weed density is presented on a Log_{10} scale.

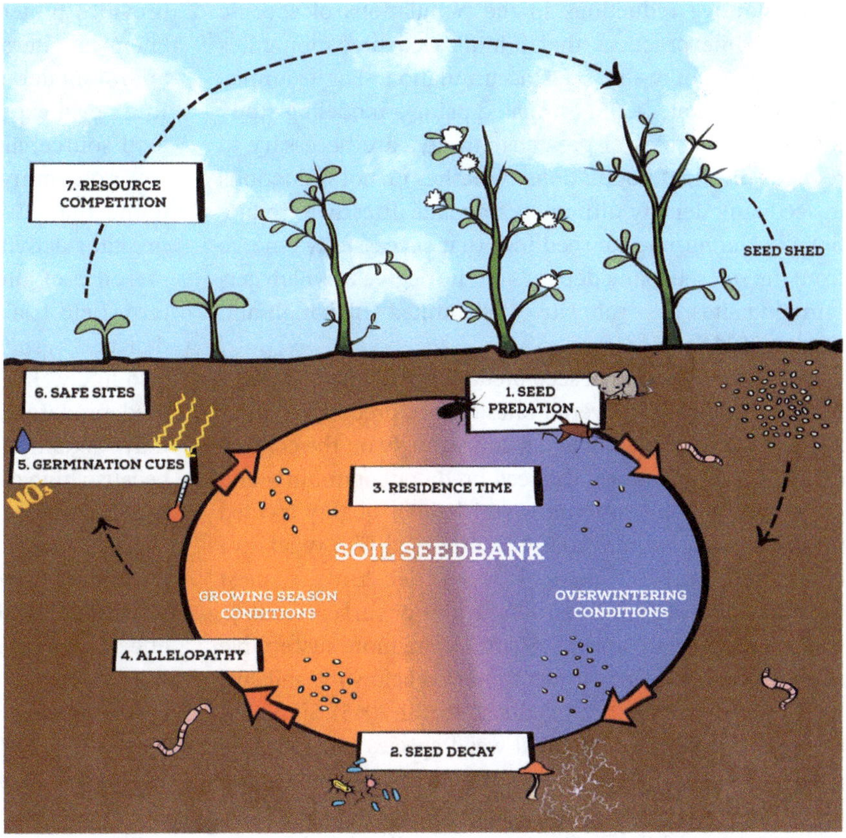

Figure 10.3 An alternative conceptualization of the weed seedbank as a sub-cycle within the overall life cycle of an annual weed. Boxes indicate primary mechanisms by which cover crops, intercrops, and mulches may mediate the fate of weeds. Note that in some cases, placement of specific mechanisms along the time course of the cycle does not necessarily reflect the period of time a particular mechanism may be expected to operate.

mulches and interseeded cover crops (Hartwig and Ammon 2002). We restrict our consideration of mulch here to the residues of dead cover crop plants that are present on the soil surface during or after cash crop production and supplemental intact crop, forage, or cover crop residues (e.g. cut-and-carry mulch) rather than other common mulches such as plastic, wood chips, or composted materials. The remainder of this chapter will discuss how cover crops, intercropping, and mulch are likely to mediate the fate of seeds at each critical phase of the weed life cycle through their effects on both biotic and abiotic mediators of seed and seedling death or survival, growth, and reproduction. We will start our discussion where the life of an annual weed may be thought to begin, entry to the soil seedbank following seed shed (Figure 10.3). This stage is especially critical because recently shed seed can make up a large fraction of the readily germinable portion of the seedbank (Smith and Gross 2006).

10.3 Mechanisms of Weed Suppression by Cover Crops, Intercrops, and Mulches

10.3.1 Seed Predation

Seeds shed to the soil surface are susceptible to seed predation (Figure 10.3). Insects and other animals that live in and utilize agricultural habitats can be considered seed predators if, through their feeding activities, they damage, destroy, or otherwise alter the survival of weed seeds. Common seed predators include field mice, numerous species of carabid beetles, field crickets, ants, birds, slugs, and even earthworms (Smith et al. 2005; Meiss et al. 2010; Sarabi 2019); however, in many studies small mammals and insects, particularly carabid beetles and crickets, appear to be more important in terms of their contribution to overall seed loss compared to birds (Sarabi 2019). Contributions of small mammal and insect predator guilds can sometimes be complementary, with each guild dominating at different portions of the growing or nongrowing season (Williams et al. 2009; Davis and Raghu 2010).

The process of seed predation targets seeds directly and therefore can reduce the number of seeds residing within the soil seedbank (Bohan et al. 2011). Reported rates of seed predation in various agroecosystems and under different management practices have been estimated to range from 10% to 90% depending on crop, location, and time period observed (Alignier et al. 2008). Davis et al. (2013) quantified mean annual rates of weed seed predation attributable to both vertebrate and invertebrate seed predators in corn-soybean-wheat cropping systems in the US Midwest to be 31%, 37%, and 53% for giant ragweed (*Ambrosia trifida* L.), velvetleaf (*Abutilon theophrasti* Medik), and giant foxtail (*Setaria faberi* Herrm), respectively. In theory, increasing these rates through management practices that encourage seed predation could result in substantive reductions in the populations of specific weeds, while practices that inhibit seed predation rates could result in weed seed accumulation. For example, Westerman et al. (2005) estimated using a modeling study that seed loss rates of greater than 40% are necessary to reduce velvetleaf populations. Whether or not reductions in seedbank density ultimately translate directly to reductions in the number of seedlings that successfully emerge from the soil seedbank depends on the degree to which germination sites, i.e. "safe sites," are limited (Blubaugh and Kaplan 2016).

Because the process of seed predation is mediated by animals, the nature and overall rates of seed predation within a given agroecosystem will depend strongly on the composition and abundance of the seed predator community and the environmental conditions and habitat quality created by the crop plant and how it is managed (i.e. vegetative cover and disturbance regime) (Meiss et al. 2010). As most seed predators are surface dwelling and/or surface feeding, weed seeds present on the soil surface are more susceptible to being predated than are seeds buried below the soil surface (Kulkarni et al. 2015; van der Laat et al. 2015). Similarly, the susceptibility of individual weed species to being predated varies and depends on the composition and abundance of the seed predator community as well as the characteristics of the seed itself in terms of its size, defenses, and nutritional quality (Brust and House 1988; Heggenstaller et al. 2006; Berl et al. 2017; Gaba et al. 2019). For example, in field crop agroecosystems of the US Midwest, invertebrates, which tend to target smaller weed seeds, appear to be the dominant guild of seed predators during the growing season (O'Rourke et al. 2006), while in other regions small mammals, which tend to preferentially consume larger weed seeds (Berl et al. 2017), contribute to a greater proportion of the seed predation (Westerman et al. 2003). Likewise, seed predation during the over-winter period by small mammals can also be substantial (Williams et al. 2009; Berl et al. 2017). For these reasons, it can be difficult to generalize across cropping systems and agroclimatic zones; however, there do appear to be certain general principles, especially in regard to the effects that crop plant manipulations, such as cover cropping, intercropping, and mulch, may be expected have on seed predators and seed predation, and how these can be managed in order to maximize potential rates of seed loss.

Much of the temporal and spatial variability of seed predation appears to be due to crop-specific phenology and management that affects the level of vegetative cover (Gallandt et al. 2005; Heggenstaller et al. 2006; Meiss et al. 2010; Davis et al. 2013). In general, rates of seed predation are often positively associated with living vegetative cover (but see Davis et al. 2013). This trend seems to apply most strongly when insects and small mammals are the dominant seed predator guilds, likely because vegetative cover provides protection to seed predators from their natural enemies, alternative food sources, overwintering habitat, and/or moderation of the microenvironment (Sarabi 2019). In a novel experiment separating the effects of weed seed food source, alternative prey food source, and vegetative cover on omnivorous predator diets, Blubaugh et al. (2016) found that cover crops (vegetative cover) increased seed predator activity density and seed predation, even in the presence of alternative food sources. Similarly, in a long-term field trial, cover crops increased macrofauna abundance 93%, suggesting greater rates of seed predation were also likely (Kelly et al. 2021). When birds are the primary seed predators, however, high levels of vegetative cover may actually limit seed predation rates, presumably because high cover makes it difficult for birds to locate and access seeds on the soil surface (Navntoft et al. 2009).

There is scant evidence that the specific species of crop or cover crop strongly influence seed predator activity-densities and seed predation beyond their species-specific effects on production of cover itself (Buchanan and Hooks 2018; but see Shearin et al. 2008). Similarly, whether cover crops are grown as monocultures or mixtures does not appear to influence seed predator activity density (Carabajal-Capitán et al. 2021) or seed predation rates (Buchanan and Hooks 2018).

Effects of intercropping on seed predation should be similar to cover cropping if the overall effect of the intercrop is to increase the level of vegetative cover. If the intercrop is a cash crop or living mulch that increases overall cover, then the potential exists to enhance seed predator activity densities, and therefore presumably seed predation, over the cash crop growing season by increasing the overall level of soil cover. If the intercrop or living mulch persists after cash crop harvest, then the potential exists to extend the period of time that weed seeds are susceptible to seed predation losses beyond the cropping season, again by extending the duration of time that a living cover is present.

There have been comparatively fewer studies of seed predation under the mulch resulting from terminated cover crops, and those that have been conducted provide little evidence that cover crop mulches enhance either seed predator activity-densities or rates of seed predation compared to situations where mulch is not present (Ward et al. 2011; Quinn et al. 2016; Buchanan and Hooks 2018; Lewis et al. 2020). Lewis et al. (2020) postulated that killed cover crops may provide fewer alternative resources for seed predators and may also impede seed predator movement and foraging activities compared to living vegetative cover. Compared to black plastic mulch, however, Pullaro et al. (2006) found greater weed seed predation over the summer period in a cover crop mulch-based system. However, the primary seed predator in their system was the red fire ant, which may be better adapted to navigating the microenvironment under a dense cover crop mulch than other, larger predator species. While the effects of killed cover crop mulch on weed seed loss via predation may be equivocal, the effects of mulch on other mechanisms of seed and seedling mortality, including "safe sites" for germination and seedling emergence, are likely more substantive, as will be discussed later in this chapter.

What does this brief review of the seed predation literature tell us about this important mechanism of seed loss with respect to the use of cover crops, intercropping, and mulch? Given the importance of living vegetative cover for attracting and maintaining seed predators and the fact that seeds on the soil surface are most susceptible to predation, we should expect the highest rates of seed predation to occur in cropping systems that maintain living vegetative cover throughout both the growing and non-growing season and that minimize the use of soil tillage, particularly inversion tillage due, in part, to its potential effect on soil-dwelling seed predator populations (e.g. Menalled et al. 2007), but even more so because it buries weed seeds, making them less susceptible to seed predation. In the context of viewing EBWM practices as part of the overall cropping system with multiple goals, it should be noted that maintaining year-round living cover also prevents erosion and is postulated to be one of the main drivers for increasing soil organic matter (King and Blesh 2018). Management practices that are likely to undermine the effects of cover crops, intercropping, and mulch on predation of weed seeds are those that reduce the quantity and duration of vegetative cover and/or reduce the abundance, activity, and diversity of the seed predator community.

10.3.2 Microbial Seed Decay

Weed seeds in the soil that are not destroyed by seed predators are susceptible to a myriad of seed-attacking microorganisms (Figures 10.3 and 10.4). What is more, damage inflicted to weed seed coats by seed predators or other factors that may not initially kill the seed may facilitate entry and subsequent infection of seeds by soil microbes (Kremer and Spencer 1989). A wide variety of soil bacteria and fungi are capable of colonizing weed seeds and some of these associations may be weed-species specific (Chee-Sanford 2008). Further, a number of studies have demonstrated that weed species can have strongly varying levels of susceptibility to decay (Chee-Sanford et al. 2006; Davis et al. 2006, 2008; Davis 2007). Not surprisingly, then, given its tremendous potential to reduce weed seedbank populations, there has been much interest in identifying cropping practices that might enhance weed seed loss due to microbially mediated decay processes (e.g. Kremer 1993; Wagner and Mitschunas 2008; Müller-Stöver et al. 2016). Unfortunately, seed decay is likely the most complex, least understood, and most difficult to manage component of the mechanisms we will discuss.

Distinguishing between the agents of seed decay remains difficult, especially the delineation between those considered pathogenic versus those that are nonpathogenic but able to use seeds as a carbon (C) substrate (Müller-Stöver et al. 2016). Traditional functional groupings used by microbial ecologists (Nguyen et al. 2016), when applied to the pathways of weed seed mortality, are insufficient because there is no recognized term to describe the nonpathogenic decay of living seeds; by definition, *saprotrophic* (or the outdated term *saprophytic*) means organisms that utilize nonliving C-substrates. Hence, it is likely that microorganisms classified as saprotrophic are indeed responsible for nonpathogenic weed seed decay, but this lack of clarity confuses the discussion of pathways of weed seed decay. As a consequence, many of the most common seedbank viability techniques (e.g. crush test, tetrazolium

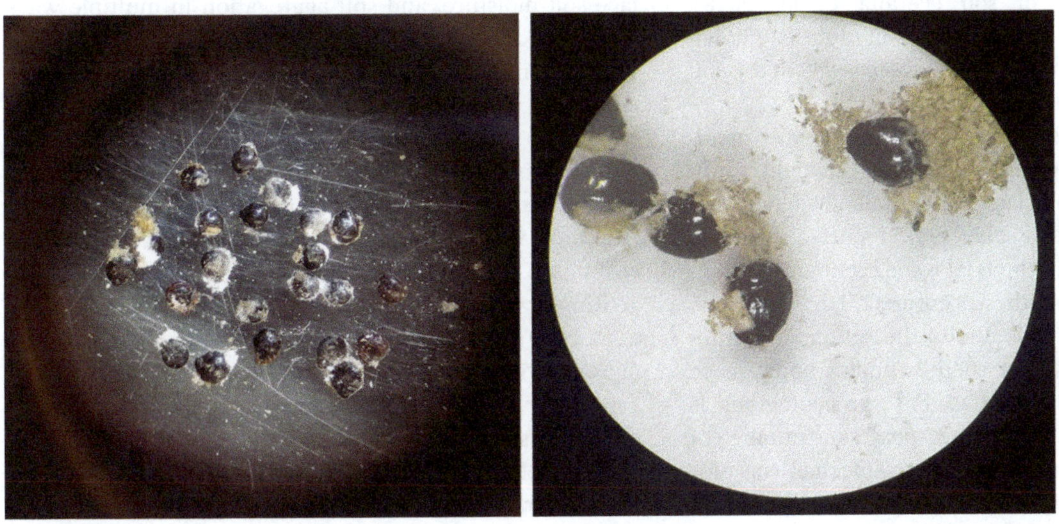

Figure 10.4 Decayed redroot pigweed seeds after placement under a roll-crimped cereal rye/hairy vetch mulch for several weeks in mid-summer. Photo credit: Julia Hobbie.

test, etc.) cannot distinguish between seeds lost to decomposition versus seeds that are destroyed by pathogens early in the germination process. In fact, there is strong evidence that some of the observed increases in weed seed decay as a result of cover cropping and crop rotation are from an increase in pathogenic activity on germinating weed seeds (Davis et al. 2006; Mohler et al. 2012). Lastly, many studies (for a review see Wagner and Mitschunas 2008) have used fungicides to determine the extent of fungal decay of weed seeds, but this technique is indiscriminate with respect to functional guilds and therefore does not elucidate important differences in potential pathways (pathogenic versus nonpathogenic) of weed seed loss from the seedbank.

Clarification of these seed mortality pathways is not merely a matter of semantics; cover crops, intercrops, and mulches have the potential to change microbial community composition and function, as well as the edaphic factors that influence microbial activity in multiple ways. These have potential implications for seed decay mechanisms. For example, cover crops have been shown to alter microbial community composition and biomass while they are actively growing (Finney et al. 2017a) and during their decomposition (Nevins et al. 2018; Thapa et al. 2021). Short-term use of certain cover crops can increase arbuscular mycorrhizal populations (Lehman et al. 2012) and arbuscular mycorrhizae, in turn, can confer greater protection from plant pathogens (Bharadwaj et al. 2012; Veresoglou and Rillig 2012). Whether higher mycorrhizal populations confer increased resistance in weed seeds and seedlings to pathogens is a question that, to the best of our knowledge, has not been investigated. Similarly, long-term (>15 year) cover crop use has been shown to increase soil microbial abundance and shift the community toward a wider metabolic capacity and a wider ability to utilize a range of C inputs (Schmidt et al. 2018), and to increase disease-suppressiveness of soil (Peralta et al. 2018). Whether this translates to a greater ability to break down seed defenses and decompose viable weed seeds in the soil has not been investigated.

Reliable estimates of seed loss due to microbially mediated decay are difficult to derive under realistic field conditions due to methodological constraints (Wagner and Mitschunas 2008). Forcella et al. (1992) reported that dead seeds made up 50–90% of the total weed seedbank in samples collected from across the US cornbelt. However, how much of this dead seed was due to the actions of pathogenic microbes and whether such estimates apply more broadly across cropping systems and agroecoregions is unclear. What is more, weed seeds possess a number of physical, chemical, and biochemical mechanisms for resisting pathogen attack and these may be more prevalent in species capable of prolonged dormancy in the soil (Davis et al. 2016; Pollard 2018). Several studies have attempted to directly quantify weed seed losses in field soils due to decay and their results have been quite variable. Gallandt et al. (2004) quantified decay of dormant wild oat seeds placed in soils in no-till and tilled wheat cropping systems and reported losses due to decay of 4–9% after a year regardless of tillage system. Gomez et al. (2014) observed decay losses of giant foxtail (*Setaria faberi*) seed as high as 31% after 22 weeks of burial but only minimal decay of velvetleaf seeds over that time. The researchers also observed that decay rates appeared to be higher when seeds were placed at shallower soil depths. Degradation of weed seeds buried in mesh bags in a no-till soil increased over time and ranged from 50% to 100% after 643 days, depending on species (Nikolić et al. 2020). Mohler et al. (2012) attributed reduced weed seedling emergence in pots following incorporation of cover crop residues to increases in soil pathogens (primarily *Fusarium* spp.); however, the exact mechanisms remain unclear. Ullrich et al. (2011) found no relationship between microbial abundance and rates of seed decay of smooth pigweed and lambsquarters in contrasting cropping systems.

A broad review of studies that have attempted to quantify seed decay by means of reducing seed exposure to soil fungi, usually through the use of fungicides applied directly to the seeds or to the soil in which the seeds were placed, concluded that soil moisture levels likely play an important role in mediating the levels of seed mortality due to fungal decay (Wagner and Mitschunas 2008). Soil moisture has been shown to be a primary driver of soil decomposer (including bacteria and fungi) activity across soil types and biomes (Manzoni et al. 2012). As such, we can surmise that cover crops, intercrops, and mulches may indirectly alter microbially mediated decay of weed seeds through their effects on soil moisture levels. These effects are complex, as living and dead cover crops affect soil water balance, surface soil moisture, and soil aggregation in multiple ways (Unger and Vigil 1998; Liu et al. 2005). Higher infiltration resulting from cover crops can lead to greater water storage compared to bare soil (Joyce et al. 2002), but a growing cover crop (or intercrop) can also reduce soil moisture levels through higher evapotranspiration relative to situations in which cover crops or intercrops are not actively growing (Kahimba et al. 2008). Once terminated, however, a dead cover crop mulch will likely conserve soil moisture by preventing evaporative losses (Unger and Vigil 1998; Acharya et al. 2019), thereby possibly facilitating weed seed degradation during the period of time the mulch remains on the soil, either by changes in moisture dynamics or as a result of changes to substrate quantity and quality from the mulch itself.

Cover crops can also increase soil organic matter levels over relatively short periods (Poeplau and Don 2015; McClelland et al. 2021). However, clear relationships between microbial community composition, activity, and edaphic factors such as soil organic matter have not been established. In fact, results are conflicting. For example, Kremer and Li (2003) found that agricultural soils with higher organic matter exhibited higher "weed-suppressive activity" in seedling assays. Similarly, Davis et al. (2006) found greater seed decay in cropping systems with higher soil organic matter. In contrast, Nikolić et al. (2020) found lower seed losses in a riparian buffer with higher soil organic matter compared to an adjacent no-till field. Similarly, using organic amendments of variable C:N ratio, De Cauwer et al. (2011) found that compost with the lowest C:N ratio led to the greatest weed seed decay. The lack of an understanding of the relationship between soil organic matter levels and rates of microbial seed loss may be because multiple pathways are at play.

Cover crops also affect soil temperatures (Blanco-Canqui and Ruis 2020) and this too will affect soil microbial activity and likely the decomposition of weed seeds. In general, soil microbial activity increases with increasing soil temperature, providing moisture levels are sufficient (Kirschbaum 1995). Further, cover crops and cover crop residues, through their insulating effect on soils, can delay the onset of soil freezing in the fall and reduce the depth of soil freezing in winter leading to earlier thawing of the soil in the spring (Kahimba et al. 2008; Blanco-Canqui et al. 2015), thereby presumably expanding the period of time that the soil decomposer community is active. During the growing season, living cover crops or intercrops will likely reduce soil temperatures by limiting solar radiation reaching the soil surface. When coupled with a drying effect on soils, this could limit weed seed decay rates. However, the water dynamics of intercrops are complex, and there are documented cases of deep-rooted intercrops providing water to cash crops via hydraulic lift and leveraging of mycorrhizal networks (Brooker et al. 2015). Conversely, a surface layer of mulch over the growing season, while also reducing soil temperature, could help to conserve soil moisture and soil microbial activity (Tu et al. 2006). Obviously, soil temperature and soil moisture have interactive effects on the activities of the decomposer community (Wildung et al. 1975).

The type of cover crop or intercrop (e.g. legume versus nonlegume or residue C:N) may have a strong influence on seed decay; however, evidence for this is indirect, coming from studies that have examined how plants influence the decay of soil organic matter more broadly (Barel et al. 2019). For example, research by Dhakal and Islam (2018) demonstrated that legume monocultures and grass-legume mixtures "cultivated" greater total, bacterial, and actinomycetes microbial biomass in soil compared to grass monocultures and this resulted in higher levels of soil C mineralization. Similarly, in experimental grasslands, the presence of legumes increased decomposition of wood and cotton substrates while the presence of a grass decreased decomposition of both substrates (Scherer-Lorenzen 2008). Soils with greater microbial abundance and activity, as well as higher rates of C mineralization, would be expected to be more favorable to the decomposition of weed seeds compared to soils with more depauperate microbial communities (Kremer 1993; Gomez et al. 2014). We are not aware of studies that have explicitly tested this hypothesis in the context of weed seeds.

As should be apparent from our brief review of the weed seed decay literature, additional research in this area will be necessary to improve our fundamental understanding of how cover crops and other cropping practices alter the decay dynamics of weed seeds in the soil. Research to describe the primary drivers of pregermination seed loss is necessary to translate the known effects of EBWM on microbial communities to rates of weed seed decay. Recent advances in soil ecology and molecular techniques for analyzing microbial communities have made this a possibility (Müller-Stöver et al. 2016). Enzyme-based analysis, where the enzymes necessary to decompose a substrate of interest are measured as a proxy for decomposition activity have allowed researchers in other fields to better understand factors driving organic matter decay in their systems. In the case of weed seedbanks, identifying the enzymes required to decay common seed defense chemicals or structures would likely provide enhanced insight into the effects of EBWM on seed decay.

Despite the variability and uncertainty of the effects of cover crops, intercrops, and mulches on microbial seed decay, the potential to leverage this pathway as a tool remains promising as new research reveals the mechanisms of these seed losses. Additionally, practitioners of EBWM can likely increase the number of weed seeds destroyed by microorganisms by leveraging the better understood germination-inhibiting effects of these practices (described below). By keeping seeds in the soil longer – regardless of the specific rate of decay – more weed seeds will be lost through these mechanisms and thus not manifest as seedlings which subsequently need to be exposed to herbicides or tillage.

10.3.3 Residence Time

As alluded to above, weed seeds in the soil do not become problems unless they germinate and the resulting seedlings successfully establish. Seeds that remain dormant in the soil remain susceptible to stress and mortality factors (i.e. seed predation and decay) and exposure to these mortality factors increases with residence time (Figure 10.3). Evidence for this comes from numerous studies that have attempted to quantify seed longevity for a variety of annual weed species, usually by placing seeds in jars or other containers that protect them from soil organisms and periodically assessing them for viability. Such studies indicate that seeds of many weed species can remain viable for decades or centuries (Chancellor 1986; Telewski and Zeevaart 2002 and references therein). In contrast, when seeds are left on the soil surface or buried in the soil, without protection from natural soil biotic and abiotic factors, seed longevity is often much lower, and in cultivated systems most weed seeds germinate or die in the first few years after entering the soil seedbank (Cavers and Benoit 1989; Lutman et al. 2002; Teasdale et al. 2004; Korres et al. 2018). Even when seeds are buried in the soil, where they are likely protected from seed predators, annual declines can be as high as high as 36% for some species (Chancellor 1986) and seed half-lives have been measured to range from as low as 3 months (*Bassia scoparia*) to 2.2 years (*Abutilon theophrasti*) (Davis et al. 2016). This all raises the intriguing prospect that enhancing the residence time of weed seeds in the soil, by delaying or inhibiting their germination, could be a mechanism for enhancing the effectiveness of EBWM. Certainly, the longer a seed remains in the seedbank the higher the probability that it will succumb to predation, decay, or natural embryonic death (Dell'Aquila 1994). It is also the case that cover crops, intercrops, and mulches have the potential to prevent or delay germination in some situations (see next sections) and therefore, potentially, to deplete the viable weed seedbank via the quasi mechanism of residence time. This said, until we better understand the mechanisms of microbial seed decay, as well as the myriad factors that interact to affect seed survival in the soil, this potential mechanism remains somewhat theoretical, despite promising emerging frameworks aimed at disentangling these factors (e.g. Dalling et al. 2011).

10.3.4 Allelopathy and Biochemical Inhibition

Allelopathy is a mechanism of interference by which one plant inhibits another through the production of plant-produced chemicals and secondary metabolites (Weston 1996). Perhaps the most well-known example of allelopathy is the black walnut tree's ability to suppress the establishment and growth of neighboring plants (Bertin et al. 2003; Weir et al. 2004). Many species of crop and noncrop plants have also been demonstrated to produce a diverse range of different allelochemicals and these can be released to the soil environment through a variety of mechanisms, including volatilization, leaching from tissues, root exudation, and microbial degradation of residues (Cheng and Cheng 2015). These include most common cover crops including cereal rye, hairy vetch, sorghum-sudangrass, wheat, red clover, buckwheat, sunflower, and brassicas, as well as many others (Weston 1996; Burgos and Talbert 2000; Bertin et al. 2003; Haramoto and Gallandt 2004; Schulz et al. 2013; Jabran et al. 2015). As such, there has been much research directed at capitalizing on allelopathic effects of cover crops as a means of ecologically suppressing weeds.

While laboratory studies have documented the existence and chemical profiles of many putative allelopathic chemicals, as well as the nature and magnitude of their inhibitory effects under controlled conditions, it remains challenging under field conditions to separate allelopathic effects from other sources of interference with weed growth, such as pathogens or resource competition (Weston 1996; Inderjit et al. 2005; Duke 2015; Gerhards and Schappert 2020; Koehler-Cole et al. 2020). Hence, while allelopathy is often invoked as one of several potential factors determining the overall levels of weed suppression provided by cover crops in field experiments (e.g. Jabran et al. 2015), it is exceedingly difficult to isolate its unequivocal role. Based on the data available to date, we are not able to provide reliable estimates of the levels of weed suppression that may be attributable solely to allelopathy (Duke 2015). That said, we can assume that allelopathy likely plays some role in weed suppression under most field conditions and there are several general principles that appear robust enough to help guide the management of cover crops, intercrops, and mulch in ways that are likely to maximize their allelopathic effects on weeds.

The most commonly reported inhibitory effects of allelochemicals are on weed seed germination rate and time, length of the radical following germination, and seedling growth. These are the parameters of weed growth most frequently examined in laboratory and greenhouse studies, where the typical approach is to apply allelochemicals extracted or leached from cover crop residues to petri dishes or small pots containing weed seeds (Duke 2015). That said, field studies have also reported reduced and/or delayed emergence following the incorporation of purportedly allelopathic cover crops (e.g. Haramoto and Gallandt 2005). Seed mass appears to be an important determinant of the susceptibility of individual weed species to allelopathic inhibition, potentially due to smaller seeds having a greater surface-to-mass ratio and therefore greater relative exposure compared to larger seeds (Burgos and Talbert 2000; Liebman and Sundberg 2006; Duke 2015; but see Haramoto and Gallandt 2005). Further, evidence suggests that at a given seed size, monocot weeds may be more susceptible to allelopathic inhibition compared to dicot weeds (Liebman and Sundberg 2006), suggesting that strongly allelopathic cropping systems may over time "select" for weed communities dominated by larger-seeded dicot weeds. However, other research has suggested that dicots are more susceptible to allelopathy than monocots (Creamer et al. 1996 and references therein).

Allelopathic effects often appear to be dose-dependent, with the severity of inhibition increasing with concentration (Ohno et al. 2000; Turk and Tawaha 2003; Adler and Chase 2007). Some emerging evidence suggests that some allelochemicals are produced as a stress response, suggesting that allelopathy could be enhanced with certain management practices (Gerhards and Schappert 2020). Concentrations of allelochemicals in the soil appear to be highest immediately after cover crop termination but are short-lived and decline over a period of hours to weeks, depending on the nature of the allelochemical and the physical, chemical, and biological properties of the soil (Rice 1984; Vidal and Bauman 1997; Schulz et al. 2013; Cheng and Cheng 2015). In no-till systems, concentrations of allelochemicals will be highest at the soil surface and the period of release will likely be longer owing to the slower rate of residue decomposition on the surface compared to if residues are incorporated through tillage (Cast et al. 1990; Vidal and Bauman 1997; Roth et al. 2000).

Growing cover crops as mixtures or intercropping cash crops might both be expected to provide a wider spectrum of allelopathic effects because crops produce species-specific allelochemicals (Creamer et al. 1995), however, we are aware of only a few studies that have explicitly examined this. Kunz et al. (2016) compared the effects of extracts of monocultures and mixtures of cover crops and found few differences between mixtures and monocultures on seed germination rate, time, or root length, except in one case where one of their two mixtures increased germination time relative to the other treatments. The researchers also demonstrated that root length inhibition observed in their laboratory study was a strong correlate with weed densities observed in a companion field experiment, suggesting that allelopathy may have played a substantive role in suppressing weeds (Kunz et al. 2016). Similarly, Creamer et al. (1996) found little evidence of enhanced allelopathic effects in a field experiment of cover crop mixtures compared to monocultures. Certainly, evidence that intercrops and mixtures can sometimes be more weed suppressive than monocultures (Liebman and Dyck 1993; Szumigalski and Van Acker 2005; but see Smith et al. 2020) supports the notion of higher allelopathic efficacy; however, as discussed previously, it is difficult to tease allelopathic effects from other sources of weed interference and suppression in field situations (Duke 2015).

For some cover crops, allelopathic compounds may be present, but their effects minimal in comparison to other mechanisms of weed suppression. For example, while there is some evidence that forage and oilseed radish (*Raphanus sativus* L.) has allelopathic effects, changes to red–far red radiation are likely the dominant mechanism for the near-complete weed suppression observed following forage radish (Lawley et al. 2012). This has important management implications because it suggests that seeding date and early canopy growth are critical components of successful implementation of forage radish for optimal weed suppression, and aboveground biomass can be removed after some time (e.g. for human or animal consumption) without affecting weed suppression. This serves as an example of the danger of assuming one pathway for weed suppression is dominant without sufficient evidence.

What does this brief review of the allelopathy literature tell us in respect to its potential for manipulation via the use of cover crops, intercropping, and mulch? Given that most cover crops and many cash crops produce putative allelochemicals, the utilization of these crop plants should be expected to exert at least some level of allelochemical inhibition of weeds. This inhibition might be enhanced with increasing cover crop (or intercrop) biomass production, a longer duration of cover crop or intercrop growth, and/or maintenance of resulting residues as mulch for longer periods of time. Further, because the microbially mediated breakdown of residues appears to be an important driver of allelochemical activity and release to the soil (Inderjit 2005), it might be predicted that for a given quantity of residues, legume residues (or residues with a low C:N) would result in higher concentrations of allelochemicals but for a shorter duration of time compared to more recalcitrant residues such as those from mature grass species. To our knowledge, however, this hypothesis has yet to be explicitly tested. Lastly, because residues will be concentrated at the soil surface and in closer proximity to germinable weed seeds, overall allelopathic effects of residues should be stronger and persist for longer durations in no- or reduced-tillage systems compared to more intensively tilled systems (e.g. Liebman and Davis 2000).

10.3.5 Germination Cues

Weed seeds in the soil that do not succumb to seed predation or pathogenic decay by the end of the growing season or the overwintering period either germinate when subsequent growing season conditions are suitable or remain in the soil in a dormant or nondormant state where they will continue to be susceptible to biotic and abiotic agents of mortality (Figure 10.3). Germination cues are weed species-specific and can include fluctuations in soil moisture, light, temperature, and gases, as well as changes in concentrations of soil nutrients such as nitrate (Benech-Arnold et al. 2000; Mohler 2001). Cover crops, intercrops, and mulches, through their effects on these germination cues, have the potential to either stimulate or inhibit germination or to modify the level of dormancy at the population level (Benech-Arnold et al. 2000). However, because many site-specific and weed-specific factors interact to break dormancy and cue germination (Dalling et al. 2011; Travlos et al. 2020), we do not yet have the ability to generalize the effects that specific cover cropping, intercropping, or mulching practices are likely to have on weed germination ecology. That said, we can generalize to some extent, on how these practices are likely to alter the factors that ultimately influence weed seed germination and seedling emergence.

A living cover crop transpires water and shades the soil from sunlight which will alter soil moisture and temperature regimes, as well as the quantity and quality of light that reaches the soil surface. In general, the more the cover crop grows, the more these regimes will be altered, which could inhibit or stimulate germination depending on the weed species. Living cover crops will also alter concentrations of nitrate in the soil. Legume cover crops will likely increase soil nitrate levels (Thilakarathna et al. 2016), which could stimulate germination in some weed species, while other broadleaf and grass cover crops will generally reduce nitrate levels in the soil. Conversely, a thick cover mulch layer will limit light levels and reduce soil temperatures but could also conserve soil moisture.

The importance of any of these alterations to germination cues will depend on when the cover crop, intercrop, or mulch is present. If cool-season cover crops are planted in the fall after harvest of warm-season cash crops, the effects will be to alter the germination cues of weed species adapted to emerge later in the season (i.e. winter annual and biennial weeds), particularly if tillage is used to establish the cover crops. Overwintering cover crops will further alter soil conditions and subsequent germination cues in the spring through their effects on solar radiation, snow capture and retention, and surface albedo (Kaye and Quemada 2017). Intercropping will have the same effect as a living cover crop, except these effects will manifest during the period cash crops are grown.

Tillage is well known to stimulate the germination of weed seeds and the emergence of seedlings. Tillage warms the soil, brings buried seeds to the soil surface, exposes seeds to light and oxygen, and often results in a flush of nitrate, all of which are cues for germination. Surprisingly, there has been much less work aimed at quantifying the effects of cover crops, intercropping, and mulching on germination cues. Certainly, there are many studies that have quantified the numbers and/or biomass of weeds that emerge in conjunction with these practices; however, differences are often attributed to resource competition (discussed below) and/or allelopathic interference, which as discussed above is challenging to isolate definitively. As described above, allelopathic chemicals from cover crops and mulches can also alter the timing of germination, typically delaying it by a few days, at least in laboratory studies (Duke 2015). Hence, it is highly likely that many field studies purporting to demonstrate allelopathic inhibition of weed seed germination are instead demonstrating changes to germination cues. We are not aware of studies that have effectively parsed out the relative contributions of competitive, pathogenic, or allelopathic interactions from those that are purely acting on the germination cues of weeds, especially under field conditions.

Lastly, the timing of cover crop or intercrop management will likely strongly affect germination cues as well as other factors that contribute to weed suppression, and these may have contrasting weed species-specific effects. For example, as discussed above, the timing of forage radish canopy closure in fall is critical to preventing winter annual weed seeds from receiving the light cues required to initiate germination (Lawley et al. 2012). In a different experiment, the same researchers observed an increase in common lambsquarters (*Chenopodium album*) in the spring, which is responsive to soil nitrate levels, following winter-killed forage radish (Lawley et al. 2011). This highlights that a given cover crop may suppress some weed species via changes to germination cues at one time period via one mechanism, while increasing germination of other weed species at another time period via another mechanism. Altering germination cues appears likely to be an important pathway by which cover crops, intercrops, and mulches suppress weeds; however, the complexity of these relationships warrants further investigation. Suppression of germination could lead to longer residence time for seeds in the seedbank, increasing their susceptibility to other mortality factors, as previously discussed. If this is true, however, relying on germination suppression could lead to a shift toward species that have higher resistance to these factors.

10.3.6 Safe Sites

Successful emergence and establishment of seedlings from germinable weed seeds that escape the mortality factors described above depends on the existence of "safe sites" (Figure 10.3). Safe sites can be thought of as microsites suitable for seedling establishment (Harper et al. 1961; Duncan et al. 2009). One of the primary reasons weeds remain a persistent challenge in cropping systems is because our management of these systems often inadvertently creates an abundance of safe sites for seedling emergence and establishment (Smith 2015). One need only to consider the interrow spaces between crop rows in a typical row crop agroecosystem, where as much as 70% or more of the soil surface remains bare for the majority of the growing season (Figure 10.5). Whereas a conventional approach to weed control may attempt to make these sites less safe through application of postemergence herbicide or cultivation of the interrow space, EBWM practices aim to accomplish this objective through a variety of mechanisms (Liebman and Gallandt 1997), many of which we have already discussed.

Figure 10.5 Bare soil between crop rows encourages weed seed germination and seedling establishment, while unused soil resources are available to support weed growth and seed production. Lack of living cover within these spaces also increases predation risk to granivores. Photo credit: Nick Warren.

One mechanism we have not yet touched upon explicitly is the usurpation of unoccupied space, and therefore reduction in the physical number of potential safe sites. Compared to bare soil, a dense living cover crop, intercrop, or killed *in situ* mulch should be expected to provide much less physical space on the soil for weed seedlings to establish. While this concept is straightforward, evidence to support the notion that cover crops, intercropping, and mulches reduce the physical number of safe sites – and by extension, the number of individual weeds that are able to establish – is less so, for at least two reasons. First, many of the mechanisms we have already described, along with resource competition (discussed below) influence the suitability of the soil microhabitat within which seeds germinate and seedlings emerge; therefore, the safe site concept transcends multiple stress and mortality factors (including seed predation, decay, and resource competition) in addition to the physical space necessary for a seedling to establish. Consequently, observed reductions in the density or biomass of weeds when cover crops, intercrops, or mulches are employed could be due to preemptive occupancy of safe sites by crop plants; however, it could also be due to other factors. Second, even if we restrict the safe site concept to physical space, much of literature on the effects of cover crops, intercropping, and mulches report weed biomass rather than weed densities (Liebman and Dyck 1993; Osipitan et al. 2018), making it difficult to draw definitive conclusions on how these practices influence seedling densities during the cover crop or intercrop growing period. Future research in which both weed density and biomass are measured would help to elucidate whether suppressive effects are due more to reductions in the numbers of individual weed seedlings that emerge and establish (and therefore potentially reductions in the availability of safe sites) versus suppression of individual growth responses, which are more likely to be due to resource competitive effects.

The above caveats aside, of studies that have quantified weed density, many of these indicate that actively growing cover crops and intercrops harbor fewer weed individuals compared to bare fallows or sole crops, respectively (e.g. Bunchek et al. 2020 and references therein; Uchino et al. 2012; Radicetti et al. 2013; Osipitan et al. 2018; Alonso-Ayuso et al. 2018; but see Cho et al. 2015 and Hayden et al. 2012). Working in old-field systems, Reader (1993) demonstrated that removal of ground cover increased seedling emergence more for small seeded species than species with larger seeds, which were more strongly regulated by seed predators, suggesting that the living vegetation is especially suppressive to small-seeded species. Similarly, weed densities in rolled and killed cover crop mulches can be reduced by as much as 80% to 96% compared to a no mulch control (Leavitt et al. 2011). Whether such reductions are fundamentally due to reductions in the actual number of safe sites or instead to changes in the relative suitability of safe sites because of reductions in light or other resources or increases in the activities of biological control organisms or allelopathy is unclear and probably irrelevant. At the end of the day, reducing the abundance of bare soil over periods that weeds would otherwise be able to emerge and grow, either through the use of living cover or intercrops or with killed cover crop mulches should, under most circumstances, be expected to result in a reduction in the number of weeds that are able to emerge and establish.

10.3.7 Resource Competition

Weeds that successfully emerge from the seedbank must acquire sufficient water, nutrients, and light to complete the remainder of their life cycles (Figure 10.3). Actively growing cover crops and intercrops will reduce the availability of these resources to weeds and this should be expected to reduce weed growth and fecundity. The agronomic literature is replete with studies demonstrating that cover crops and intercrops harbor substantially fewer weeds (and/or less weed biomass) compared to when these practices are not implemented (Liebman and Dyck 1993; Osipitan et al. 2018). For example, Hayden et al. (2012) observed that monocultures of cereal rye and vetch suppressed weed biomass at levels ranging from 95% to 98% and 71% to 91%, respectively, compared to no cover crop controls. Similarly, Smith et al. (2020) observed weed suppression levels of 97%, 98%, and 100% compared to weedy fallow controls for monocultures of buckwheat, forage radish, and triticale grown over the summer, fall, and winter/spring time periods, respectively. While a thorough discussion of this extensive literature is beyond the scope of this chapter, readers are encouraged to consult recent meta-analyses by Osipitan and colleagues (Osipitan et al. 2018, 2019) and reviews by Mennan et al. (2020) and Lithourgidis et al. (2011) for more comprehensive summaries of the data reported in these studies.

It is likely fair to say that across most experiments, when weed abundance (or biomass) is reduced in the presence of cover crops or intercrops compared to a control, the most often invoked (and therefore assumed) mechanism to explain the observed response is resource competition. Further, when specific cover crops or intercrops fail to suppress weeds adequately, this is often attributed to a lack of "competitiveness" with weeds (e.g. MacLaren et al. 2019). There is little doubt that competition plays an important role in the suppressive effects these practices exert on weeds; however, as should be clear from our discussion of other potential mechanisms, it is unlikely to be the sole mechanism operating (Gerhards and Schappert 2020). We are aware of only a few studies that have attempted to isolate competitive effects from other potential mechanisms of weed suppression mediated by cover crops and intercrops (e.g. Creamer et al. 1996; Lawley et al. 2012; Gerhards and Schappert 2020), and even fewer that attempt to distinguish among the different types of resources for which crop plants and weeds may be competing.

Given the caveats above, can we at least surmise which resources are most likely driving the competitive effects of cover crops and intercrops on weeds? In many temperate agroecosystems, soil resources are less likely to be limiting due to the addition of fertilizers, and sometimes irrigation, suggesting that relative to other resources, light limitation may be an especially important mechanism by which crop plants exert competitive effects on weeds (Aerts 1999; Page et al. 2010; Gerhards and Schappert 2020 and references therein). The importance of light competition for weed suppression was elegantly demonstrated in several studies conducted by Borger and colleagues (Borger et al. 2010, 2016). Their work demonstrated that merely changing the row orientation (relative to the path of the sun) of wheat and barley, in this case grown for grain rather than as cover crops, increased light interception by the crop canopy and decreased weed biomass by 51% and 37%, respectively, compared to row orientations that permitted greater light penetration to the soil surface (Borger et al. 2010). In a separate study, reductions in weed biomass due to light interception by a crop canopy also reduced overall weed fecundity by nearly 50% (Borger et al. 2016). Of course, changes in light levels reaching the soil surface will also affect other abiotic and biotic properties of the soil environment, which could also have contributed to the observed responses.

If there is a principle with regard to managing cover crops and intercrops for weed suppression via resource competition it is "more is better" (Osipitan et al. 2019). By "more" we mean greater crop biomass or canopy cover. Greater cover crop or intercrop biomass implies greater nutrient uptake, and, perhaps more importantly, more light interception by the crop. The relationship between biomass (or cover) and weed suppression is not linear, however, as weed suppression usually plateaus at or near 100% at very high levels of cover crop biomass (Finney et al. 2016; Florence et al. 2019; Smith et al. 2020). The amount of biomass that a cover crop will produce, and therefore how weed suppressive it will be, depends strongly on the species of cover crop, when and how it is planted, as well as how it is managed (Osipitan et al. 2019). A cover crop planted earlier in the growing season will attain higher biomass than the same cover crop planted later in the season (or in the wrong season). Similarly, drilled cover crops tend to produce more uniform and robust stands compared to cover crops that are broadcast seeded (Brennan and Leap 2014).

Much recent research has focused on understanding whether mixtures of annual cover crops – i.e. "cover crop cocktails" containing three or more species – are more weed suppressive than cover crops sown as monocultures. The rationale for expecting greater resource competition and therefore greater weed suppression in species-diverse mixtures of cover crops follows from ecological niche theory that a more diverse plant community should more effectively use available resources, i.e. the "niche preemption" or "biotic resistance" hypotheses (Tilman et al. 1996; Fargione et al. 2003; Florence et al. 2019). While the theory is solid, in practice few studies have observed consistently greater weed suppression in mixtures compared to their component monocultures. Rather, most studies have found that biomass, rather than species composition or numbers of species, is the primary driver of weed suppression by cover crops (Florence et al. 2019; MacLaren et al. 2019; Florence and McGuire 2020; Smith et al. 2020). Conclusions from these studies tend to support the notion that if weed suppression is the only or primary objective, growers are better off planting the most weed suppressive species of cover crop as a monoculture.

The principle of "more is better" also applies to the level of weed suppression provided by a cover crop mulch, where suppressive effects have been shown to strongly depend on the biomass and thickness of the mulch residue layer (Figure 10.6). A mulch will primarily limit space and light access and act as a physical barrier to weed

Figure 10.6 Weeds have established where a square of roll-crimped cereal rye/hairy vetch mulch was removed several weeks prior. Photo credit: Natalie Lounsbury.

penetration (which ultimately precludes access to light); however, shading from the living cover crop prior to termination may also play an important role in determining weed abundance in the resulting mulch (Liebert et al. 2017). Larger seeded weed species (and perennial weeds) will be better able to overcome the physical barrier that mulch provides (Liebman and Davis 2000). The amount of cover crop biomass required to provide adequate (i.e. greater than 75% inhibition of emergence) weed suppression via the residual mulch alone varies by latitude and growing conditions; in the mid-Atlantic and Southeastern United States, a threshold of 8,000–9,000 kg ha^{-1} biomass is needed, but there is some evidence that requirements are lower in more northern regions (Mirsky et al. 2012; Reberg-Horton et al. 2012; Wallace et al. 2017).

The duration of the suppressive effect of the mulch is strongly related to the duration of time it remains on the soil surface and this will be strongly influenced by the cover crop. As the mulch decomposes, its ability to limit seedling emergence declines. Increasing cover crop residue biomass and C:N were both found to increase the magnitude and duration of weed suppression (Pittman et al. 2020). Mulches of legume cover crops will decompose faster than mulches with higher C:N, such as cereal rye. The nitrogen dynamics of legumes can also sometimes enhance weed growth, particularly those species that are nitrophilic (Hill et al. 2016).

The culminating effects of resource competition and other mechanisms of suppression exerted by cover crops, intercrops, and mulch on weeds, in addition to reducing the impacts of weeds on cash crop production, is to affect the quantity of weed seed that is deposited to the soil seedbank. In general, there is a direct relationship between the size or biomass of a weed at maturity and the quantity of seed it produces (Harrison 1990). Hence, competitive (or other) effects of cover or intercrops on weeds that result in reduced weed growth should reduce the quantity of weed seeds that are subsequently shed and deposited back into the soil seedbank.

10.4 Conclusions and Future Research Directions

Cover crops, intercrops, and mulches suppress weeds via a variety of mechanisms. These mechanisms operate at various stages of the weed life cycle and often interact with one another, making it difficult to isolate the relative importance of specific mechanisms operating at specific times in any given cropping system. Carefully designed experiments that enable the parsing of mechanisms that contribute to weed suppression under realistic field conditions will undoubtedly lead to improvements in the predictability and efficacy of EBWM practices.

Despite our current inability to determine the underlying mechanisms, overall weed suppression often increases with increasing cover crop, intercrop, or mulch cover and/or biomass (Liebman and Dyck 1993; Osipitan et al. 2018). Several general principles can be used to inform the management of cover crops, intercropping systems, and mulch in ways that are likely to maximize their potential to suppress weeds (Figure 10.7). Mechanisms of weed suppression that are strongly mediated by living crop plants, such as weed seed predation, seed decay, safe site limitation, and resource competition, are likely to be most dominant during the periods these crops are actively growing. Hence, cover crops will foster these mechanisms during the "shoulder seasons" after cash crops have been harvested or prior to cash crop planting, whereas intercropping promotes these mechanisms while cash crops are actively growing. Over time, however, this temporal specificity has the potential to shift the composition of weed communities

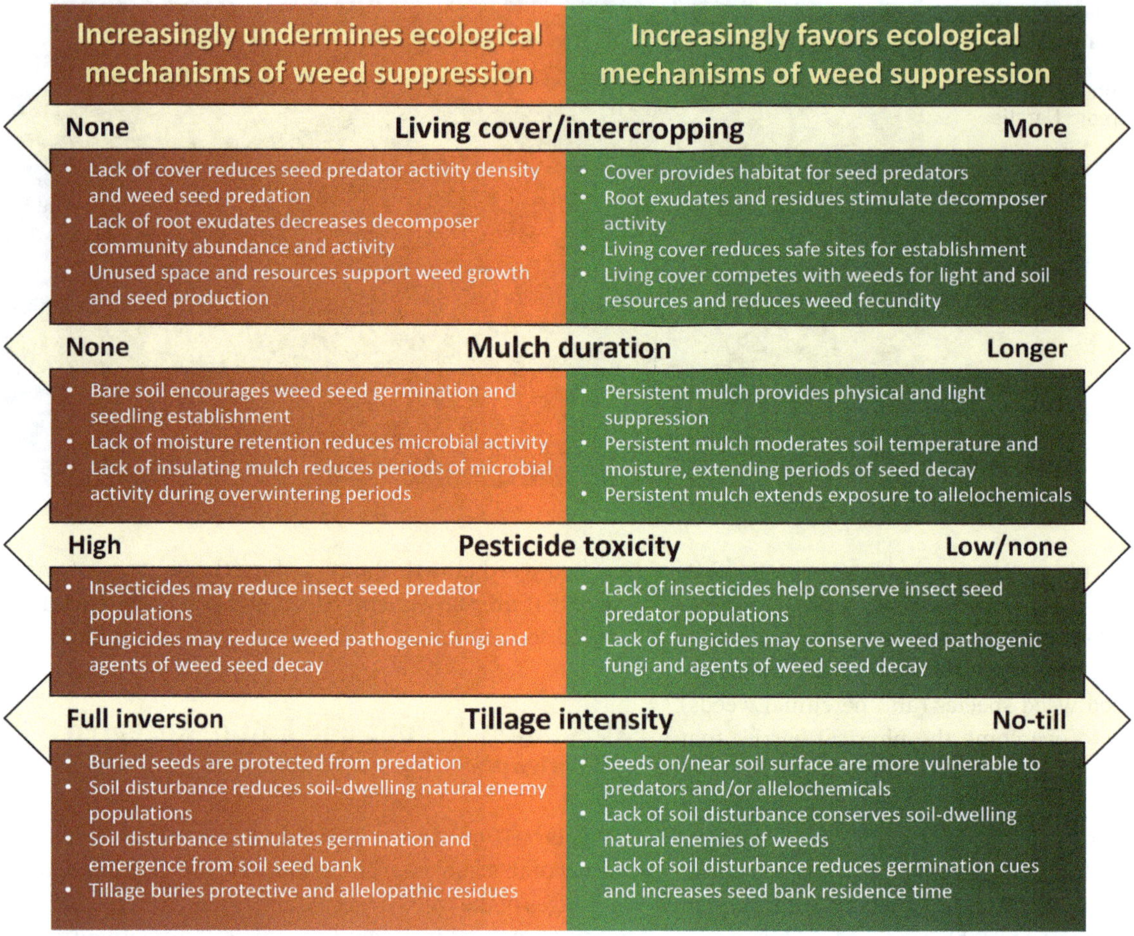

Figure 10.7 General principles for maximizing or undermining weed suppression by cover crops, intercrops, and mulch.

to favor those species or biotypes that either germinate outside the windows of time these mechanisms operate (Cordeau et al. 2017, 2022) or that are otherwise not susceptible to the mechanisms. This argues strongly for identifying crop rotation strategies that make stress and mortality factors associated with these practices less temporally "predictable" from the perspective of weeds and therefore less likely to "select" for weed communities that are more difficult to suppress (Menalled et al. 2016; Weisberger et al. 2019).

Cover crop mulches provide the opportunity to extend some mechanisms of weed suppression longer over the growing and nongrowing seasons. These mechanisms include seed decay, allelopathy, physical suppression/light limitation, moderation of the soil and light environment that stimulates seed germination, and limitation of safe sites. As with living cover and intercrops, the weed-suppressive effects of mulches depend strongly on the biomass/thickness of the mulch and the duration of time the mulch persists on the soil surface (Figure 10.7). Recent research in "planting green," where cash crops are planted into living cover crops that are subsequently terminated with herbicides soon after (Reed et al. 2019), or inter-seeding cover crops into standing cash crops in order get a "head start" on cover crop establishment and growth (e.g. Abdin et al. 1998; Caswell et al. 2019), both represent strategies for extending the period of cover crop growth and maximizing cover crop biomass and the suppressiveness of the resulting mulch layer. Given the relatively low adoption of cover crops and intercropping in many areas owing, in part, to concerns over being able to successfully integrate these crops into existing cropping systems (CTIC 2017; Hamilton et al. 2017), additional research aimed at optimizing these strategies could greatly expand opportunities for EBWM via cover crops and intercropping. Similarly, a fuller accounting and valuation of the other ecosystem services that these practices provide at both the farm-gate and societal levels would likely provide additional incentives for their adoption (Schipanski et al. 2014). Conversely, while maximizing cover crop/mulch biomass is likely to maximize the weed suppression services these practices provide, high rates of residue can sometimes interfere with cash crop production and encourage other crop pests (Mirsky et al. 2012; Finney et al. 2017b). Research aimed at overcoming these potential challenges (i.e. disservices) associated with cover crops and high residue mulches will also help to foster greater adoption of these practices.

The mechanisms of microbial seed decay and allelopathy likely contribute to substantial levels of seed death and weed suppression over both the growing and nongrowing seasons, and living cover and mulch likely strongly influence the timing, duration and overall magnitudes of their effects. Further, the degree to which the weed seedbank can be managed to delay or prevent seed germination could represent a complementary mechanism by which seed decay or other sources of weed seed mortality could be enhanced. Unfortunately, our current understanding of these mechanisms remains limited, thereby precluding our ability to fully capitalize on their potential to suppress weeds. New and emerging molecular techniques (e.g. Müller-Stöver et al. 2016) will likely shed new light on how crop plants can be manipulated to enhance the "weed-suppressiveness of soils" (sensu Kremer and Li 2003).

Because some of the most important mechanisms of weed suppression (i.e. seed predation and seed decay) are mediated by soil dwelling organisms, ancillary management practices that negatively influence their abundance and/or activities will likely also impact their interactions with cover crops, intercrops, and mulch and their efficacy as agents of weed mortality (Figure 10.7). For example, in conventionally managed systems, insecticides and fungicides, applied either as coatings on crop seeds or broadcast, are often used to protect crop seeds and seedlings from insect pests and soilborne fungal pathogens. Evidence suggests some of these pesticides may also indirectly impact weed seedbanks via their effects on seed predators and pathogens (Smith et al. 2016). For example, research has shown that neonicotinoid insecticides, which are a common component of pesticide seed coatings, can have both direct and indirect negative effects on important components of the soil macrofauna community, including ground beetles (Mullin et al. 2005; Douglas et al. 2015), many species of which are omnivorous and consume weed seeds. Additionally, many of the fungal agents of crop disease and seed decay that are the targets of fungicides, including *Pythium ultimum* and *Fusarium* spp., also contribute to disease and decay in seeds and seedlings of agriculturally-important weed species (Mohler et al. 2012; Gomez et al. 2014). The degree to which insecticide and fungicide use might be inadvertently protecting weed seedbanks, particularly those members that are resistant to herbicides, from biological sources of mortality that could be manipulated with cover crops or mulches warrants greater scrutiny.

The importance of living cover and the duration of mulch persistence for maximizing weed suppression highlights the especially important role tillage systems play in facilitating EBWM via these practices, as conservation and no-till enable the persistence of cover crop mulch and its associated mechanisms of weed suppression (Figure 10.7). Further, no-till enables the retention of recently shed weed seeds on the soil surface where they are susceptible to seed predation and where allelochemicals are likely most highly concentrated. This link between tillage and mulch retention comes with important trade-offs, however. In conventional no-till systems, successful cover crop termination often relies on herbicides, which exacerbates selection for herbicide resistant weeds. This argues for strongly weed suppressive cover crops so as not to select for resistance during burn-down applications. In fact, the primary role of cover crops in herbicide resistance management may be in reducing weed seedling exposure to herbicide during the burn-down period (e.g. Wallace et al. 2019; Bunchek et al. 2020).

Trade-offs similarly exist in organic systems, where cover crop termination options are more challenging. If tillage is used to terminate cover crops this will compromise opportunities for extending the period of weed suppression with mulches and will also bury weed seeds and increase safe sites for seed germination and establishment. The predominant no-till strategy for terminating cover crops in organic systems – a roller/crimper – can have variable efficacy, especially if cover crops with differing phenologies are grown in mixture (Mirsky et al. 2012). Research aimed at optimizing termination of high-biomass-producing cover crops for organic systems – e.g. rolling followed by occultating tarps (Lounsbury et al. 2020) – could help alleviate some of these trade-offs, particularly on small-scale operations. Similarly, in cropping systems where it is difficult to achieve high levels of cover crop biomass because of short growing seasons or other factors, cut-and-carry mulch systems, where high biomass-producing crops are grown outside of the cash crop fields in which they will be used (e.g. Erenstein 2003; Youkhana and Idol 2016), may expand opportunities for enhancing the weed suppression services that mulches provide.

Acknowledgments

The authors thank Sydney Smith for illustrating the seedbank/life cycle diagram and Nick Warren and Julia Hobbie for providing photos. This work was funded, in part, by the New Hampshire Agricultural Experiment Station, USDA National Institute of Food and Agriculture Hatch Project 100682, USDA SARE Project LNE18-371, USDA NIFA OREI project 2016-25-8468, USDA NIFA AFRI project

2017-67013-26594, and the NSF Graduate Research Fellowship Program. Any opinions, findings, and conclusions or recommendations expressed in this material are those of the authors and do not necessarily reflect the views of the National Science Foundation.

References

Abdin, O., Coulman, B.E., Cloutier, D. et al. (1998). Yield and yield components of corn interseeded with cover crops. *Agronomy Journal* 90: 63–68.

Acharya, B.S., Dodla, S., Gaston, L.A. et al. (2019). Winter cover crops effect on soil moisture and soybean growth and yield under different tillage systems. *Soil and Tillage Research* 195: 104430.

Adler, M.J. and Chase, C.A. (2007). Comparison of the allelopathic potential of leguminous summer cover crops: cowpea, sunn hemp, and velvetbean. *HortScience* 42: 289–293.

Aerts, R. (1999). Interspecific competition in natural plant communities: mechanisms, trade-offs and plant-soil feedbacks. *Journal of Experimental Botany* 50: 29–37.

Alignier, A., Meiss, H., Petit, S., and Reboud, X. (2008). Variation of post-dispersal weed seed predation according to weed species, space and time. *Journal of Plant Diseases and Protection* 21: 221–226.

Alonso-Ayuso, M., Gabriel, J.L., Garcia-Gonzalez, I. et al. (2018). Weed density and diversity in a long-term cover crop experiment background. *Crop Protection* 112: 103–111.

Bàrberi, P. (2002). Weed management in organic agriculture: are we addressing the right issues? *Weed Research* 42: 177–193.

Barel, J.M., Kuyper, T.W., Paul, J. et al. (2019). Winter cover crop legacy effects on litter decomposition act through litter quality and microbial community changes. *Journal of Applied Ecology* 56: 132–143. https://doi.org/10.1111/1365-2664.13261.

Beckie, H.J. and Harker, K.N. (2017). Our top 10 herbicide-resistant weed management practices. *Pest Management Science* 73: 1045–1052.

Benech-Arnold, R.L., Sánchez, R.A., Forcella, F. et al. (2000). Environmental control of dormancy in weed seed banks in soil. *Field Crops Research* 67: 105–122. https://doi.org/10.1016/S0378-4290(00)00087-3.

Berl, J.L., Johnstone, H.A., Wu, J.Y. et al. (2017). Winter preference for weed seed and waste grain by native mice in row-crop agriculture. *Weed Science* 65: 406–412. https://doi.org/10.1614/WS-D-16-00100.1.

Bertin, C., Yang, X., and Weston, L.A. (2003). The role of root exudates and allelochemicals in the rhizosphere. *Plant and Soil* 256: 67–83.

Bharadwaj, D.P., Alström, S., and Lundquist, P.-O. (2012). Interactions among *Glomus irregulare*, arbuscular mycorrhizal spore-associated bacteria, and plant pathogens under in vitro conditions. *Mycorrhiza* 22: 437–447.

Blanco-Canqui, H. and Ruis, S.J. (2020). Cover crop impacts on soil physical properties: a review. *Soil Science Society of America Journal* 84: 1527–1576.

Blanco-Canqui, H., Shaver, T.M., Lindquist, J.L. et al. (2015). Cover crops and ecosystem services: insights from studies in temperate soils. *Agronomy Journal* 107: 2449–2474.

Blubaugh, C.K., Hagler, J.R., Machtley, S.A., and Kaplan, I. (2016). Cover crops increase foraging activity of omnivorous predators in seed patches and facilitate weed biological control. *Agriculture, Ecosystems & Environment* 231: 264–270. https://doi.org/10.1016/j.agee.2016.06.045.

Blubaugh, C.K. and Kaplan, I. (2016). Invertebrate seed predators reduce weed emergence following seed rain. *Weed Science* 64: 80–86. https://doi.org/10.1614/ws-d-15-00111.1.

Bohan, D.A., Boursault, A., Brooks, D.R., and Petit, S. (2011). National-scale regulation of the weed seedbank by carabid predators. *Journal of Applied Ecology* 48: 888–898.

Borger, C.P.D., Hashem, A., and Pathan, S. (2010). Manipulating crop row orientation to suppress weeds and increase crop yield. *Weed Science* 58: 174–178.

Borger, C.P.D., Hashem, A., and Powles, S.B. (2016). Manipulating crop row orientation and crop density to suppress *Lolium rigidum*. *Weed Research* 56: 22–30.

Boyd, N.S., Brennan, E.B., and Fennimore, S.A. (2006). Stale seedbed techniques for organic vegetable production. *Weed Technology* 20: 1052–1057.

Brennan, E.B. and Leap, J.E. (2014). A comparision of drill and broadcast methods for establishing cover crops on beds. *HortScience* 49: 441–447.

Brooker, R.W., Bennett, A.E., Cong, W.-F. et al. (2015). Improving intercropping: a synthesis of research in agronomy, plant physiology and ecology. *New Phytologist* 206: 107–117. https://doi.org/10.1111/nph.13132.

Brust, G.E. and House, G.J. (1988). Weed seed destruction by arthropods and rodents in low-input soybean agroecosystems. *American Journal of Alternative Agriculture* 3: 19–25.

Buchanan, A.L. and Hooks, C.R.R. (2018). Influence of winter cover crop mulch on arthropods in a reduced tillage cucurbit system. *Environmental Entomology* 47: 292–299.

Bunchek, J.M., Wallace, J.M., Curran, W.S. et al. (2020). Alternative performance targets for integrating cover crops as a proactive herbicide-resistance management tool. *Weed Science* 68: 534–544.

Burgos, N.R. and Talbert, R.E. (2000). Differential activity of allelochemicals from *Secale cereale* in seedling bioassays. *Weed Science* 48: 302–310.

Carabajal-Capitán, S., Kniss, A.R., and Jabbour, R. (2021). Seed predation of interseeded cover crops and resulting impacts on ground beetles. *Environmental Entomology* 50: 832–841.

Cast, K.G., Mcpherson, J.K., Pollard, A.J. et al. (1990). Allelochemicals in soil from no-tillage versus conventional-tillage wheat (*Triticum aestivum*) fields. *Journal of Chemical Ecology* 6: 2277–2289.

Caswell, K., Wallace, J.M., Curran, W.S. et al. (2019). Cover crop species and cultivars for drill-interseeding in mid-Atlantic corn and soybean. *Agronomy Journal* 111: 1060–1067.

Cavers, P.B. and Benoit, D.L. (1989). Seed banks in arable land. In: *Ecology of Soil Seed Banks* (ed. M.A. Leck, V.T. Parker, and R.L. Simpson), 309–329. Academic Press.

Chancellor, R.J. (1986). Decline of arable weed seeds during 20 years in soil under grass and periodicity of seedling emergence after cultivation. *Journal of Applied Ecology* 23: 631–637.

Chee-Sanford, J.C. (2008). Weed seeds as nutritional resources for soil Ascomycota and characterization of specific associations between plant and fungal species. *Biology and Fertility of Soils* 44: 763–771.

Chee-Sanford, J.C., Williams, M.M., Davis, A.S., and Sims, G.K. (2006). Do microorganisms influence seed-bank dynamics? *Weed Science* 54: 575–587.

Cheng, F. and Cheng, Z. (2015). Research progress on the use of plant allelopathy in agriculture and the physicological and ecological mechanisms of allelopathy. *Frontiers in Plant Science* 6: 1020. https://doi.org/10.3389/fpls.2015.01020.

Cho, A.H., Chase, C.A., Treadwell, D.D. et al. (2015). Apical dominance and planting density effects on weed suppression by sunn hemp (*Crotalaria juncea* L.). *HortScience* 50: 263–267.

Cordeau, S., Baudron, A., Busset, H. et al. (2022). Legacy effects of contrasting long-term integrated weed management systems. *Frontiers in Agronomy* 3: 769992. https://doi.org/10.3389/fagro.2021.769992.

Cordeau, S., Smith, R.G., Gallandt, E.R. et al. (2017). Timing of tillage as a driver of weed communities. *Weed Science* 65: 504–514.

Creamer, N.G., Bennett, M.A., Cardina, J., and Regnier, E.E. (1995). Allelopathy in cover crop-based production systems. *HortScience* 30: 819A–819.

Creamer, N.G., Bennett, M.A., Stinner, B.R. et al. (1996). Mechanisms of weed suppression in cover crop-based production systems. *HortScience* 31: 410–413.

[CTIC] Conservation Technology Information Center. (2017). *Report of the 2016–17 national cover crop survey*. Joint Publication of the Conservation Technology Information Center, the North Central Region Sustainable Agriculture Research and Education Program, and the American Seed Trade Association. West Lafayette, IN: CTIC, 46 p.

Dalling, J.W., Davis, A.S., Schutte, B.J., and Arnold, A.E. (2011). Seed survival in the soil: interacting effects of predation, dormancy and the soil microbial community. *Journal of Ecology* 99: 89–95.

Davis, A.S. (2007). Nitrogen fertilizer and crop residue effects on seed mortality and germination of eight annual weed species. *Weed Science* 55: 123–128.

Davis, A.S., Anderson, K.I., Hallett, S.G., and Renner, K.A. (2006). Weed seed mortality in soils with contrasting agricultural management histories. *Weed Science* 54: 291–297.

Davis, A.S., Dixon, P.M., and Liebman, M. (2004). Using matrix models to determine cropping system effects on annual weed demography. *Ecological Applications* 14: 655–668. https://doi.org/10.1890/02-5385.

Davis, A.S. and Frisvold, G.B. (2017). Are herbicides a once in a century method of weed control? *Pest Management Science* 73: 2209–2220.

Davis, A.S., Fu, X., Schutte, B.J. et al. (2016). Interspecific variation in persistence of buried weed seeds follows trade-offs among physiological, chemical, and physical seed defenses. *Ecology and Evolution* 6: 6836–6845.

Davis, A.S. and Raghu, S. (2010). Weighing abiotic and biotic influences on weed seed predation. *Weed Research* 50: 402–412.

Davis, A.S., Schutte, B.J., Iannuzzi, J., and Renner, K.A. (2008). Chemical and physical defense of weed seeds in relation to soil seedbank persistence. *Weed Science* 56: 676–684.

Davis, A.S., Taylor, E.C., Haramoto, E.R., and Renner, K.A. (2013). Annual postdispersal weed seed predation in contrasting field environments. *Weed Science* 61: 296–302.

De Cauwer, B., D'Hose, T., Cougnon, M. et al. (2011). Impact of the quality of organic amendments on size and composition of the weed seed bank. *Weed Research* 51: 250–260.

Dell' Aquila, A. (1994). Wheat seed ageing and embryo protein degradation. *Seed Science Research* 4: 293–298.

Dhakal, D. and Islam, M.A. (2018). Grass-legume mixtures for improved soil health in cultivated agroecosystem. *Sustainability* 10: 2718.

Douglas, M.R., Rohr, J.R., and Tooker, J.F. (2015). Neonicotinoid insecticide travels through a soil food chain, distrupting biological control of non-target pests and decreasing soya bean yield. *Jounral of Applied Ecology* 52: 250–260.

Duke, S.O. (2015). Proving allelopathy in crop–weed interactions. *Weed Science* 63: 121–132.

Duncan, R.P., Diez, J.M., Sullivan, J.J. et al. (2009). Safe sites, seed upply, and the recruitment function in plant populations. *Ecology* 90: 2129–2138.

Erenstein, O. (2003). Smallholder conservation farming in the tropics and sub-tropics: a guide to the development and dissemination of mulching with crop residues and cover crops. *Agriculture, Ecosystems & Environment* 100: 17–37.

Fargione, J., Brown, C.S., and Tilman, D. (2003). Community assembly and invasion: an experimental test of neutral versus niche processes. *Proceedings of the National Academy of Science USA* 100: 8916–8920.

Finney, D.M., Buyer, J.S., and Kaye, J.P. (2017a). Living cover crops have immediate impacts on soil microbial community structure and function. *Journal of Soil and Water Conservation* 72: 361–373.

Finney, D.M., Murrell, E.G., White, C.M. et al. (2017b). Ecosystem services and disservices are bundled in simple and diverse cover cropping systems. *Agricultural & Environmental Letters* 2: 170033. https://doi.org/10.2134/ael2017.09.0033.

Finney, D.M., White, C.M., and Kaye, J.P. (2016). Biomass production and carbon/nitrogen ratio influence ecosystem services from cover crop mixtures. *Agronomy Journal* 108: 39–52.

Firbank, L.G. and Watkinson, A.R. (1986). Modelling the population dynamics of an arable weed and its effects upon crop yield. *Journal of Applied Ecology* 23: 147–159.

Florence, A.M., Higley, L.G., Drijber, R.A. et al. (2019). Cover crop mixture diversity, biomass productivity, weed suppression, and stability. *PLoS ONE* 14 (3): e0206195. https://doi.org/10.1371/journal.pone.0206195.

Florence, A.M. and McGuire, A.M. (2020). Do diverse cover crop mixtures perform better than monocultures? A systematic review. *Agronomy Journal* 112: 3513–3534.

Forcella, F., Wilson, R.G., Renner, K.A. et al. (1992). Weed seedbanks of the US corn belt: magnitude, variation, emergence, and application. *Weed Science* 40: 636–644.

Gaba, S., Deroulers, P., Bretagnolle, F., and Bretagnolle, V. (2019). Lipid content drives weed seed consumption by ground beetles (Coleopterea, Carabidae) within the smallest seeds. *Weed Research* 59: 170–179.

Gallandt, E.R. (2006). How can we target the weed seedbank? *Weed Science* 54: 588–596.

Gallandt, E.R., Fuerst, E.P., and Kennedy, A.C. (2004). Effect of tillage, fungicide seed treatment, and soil fumigation on seed bank dynamics of wild oat (*Avena fatua*). *Weed Science* 52: 597–604.

Gallandt, E.R., Molloy, T., Lynch, R.P., and Drummond, F.A. (2005). Effect of cover-cropping systems on invertebrate seed predation. *Weed Science* 53: 69–76.

Gerhards, R. and Schappert, A. (2020). Advancing cover cropping in temperate integrated weed management. *Pest Management Science* 76: 42–46.

Gomez, R., Liebman, M., and Munkvold, G. (2014). Weed seed decay in conventional and diversified cropping systems. *Weed Research* 54: 13–25.

Hamilton, A.V., Mortensen, D.A., and Kammerer Allen, M. (2017). The state of the cover crop nation and how to set realistic future goals for the popular conservation practice. *Journal of Soil and Water Conservation* 72: 111A–115A. https://doi.org/10.2489/jswc.72.5.111A.

Haramoto, E.R. and Gallandt, E.R. (2004). Brassica cover cropping for weed management: a review. *Renewable Agriculture and Food Systems* 19: 187–198.

Haramoto, E.R. and Gallandt, E.R. (2005). Brassica cover cropping: I. Effects on weed and crop establishment. *Weed Science* 53: 695–701.

Harper, J., Clatworthy, J., McNaughton, I., and Sagar, G. (1961). The evolution and ecology of closely related species living in the same area. *Evolution* 15: 209–227.

Harrison, S.K. (1990). Interference and seed production by common lambsquarters (*Chenopodium album*) in soybeans (*Glycine max*). *Weed Science* 38: 113–118.

Hartwig, N.L. and Ammon, H.U. (2002). Cover crops and living mulches. *Weed Science* 50: 688–699. https://doi.org/10.1614/0043-1745(2002)050(0688:aiacca)2.0.co;2.

Hayden, Z.D., Brainard, D.C., Henshaw, B., and Ngouajio, M. (2012). Winter annual weed suppression in rye-vetch cover crop mixtures. *Weed Technology* 26: 818–825.

Heggenstaller, A.H., Menalled, F.D., Liebman, M., and Westerman, P.R. (2006). Seasonal patterns in post-dispersal seed predation of *Abutilon theophrasti* and *Setaria faberi* in three cropping systems. *Journal of Applied Ecology* 43: 999–1010.

Hill, E.C., Renner, K.A., Sprague, C.L., and Davis, A.S. (2016). Cover crop impact on weed dynamics in an organic dry bean system. *Weed Science* 64: 261–275.

Inderjit. (2005). Soil microorganisms: an important determinant of allelopathic activity. *Plant and Soil* 274: 227–236.

Inderjit, Weston, L.A., and Duke, S.O. (2005). Challenges, achievements and opportunities in allelopathy research. *Journal of Plant Interactions* 1: 69–81.

Jabran, K., Mahajan, G., Sardana, V., and Chauhan, B.S. (2015). Allelopathy for weed control in agricultural systems. *Crop Protection* 72: 57–65.

Jasieniuk, M., Brûlé-Babel, A.L., and Morrison, I.N. (1996). The evolution and genetics of herbicide resistance in weeds. *Weed Science* 44: 176–193.

Jordan, N., Mortensen, D.A., Prenzlow, D.M., and Cox, K.C. (1995). Simulation analysis of crop rotation effects on weed seedbanks. *American Journal of Botany* 82: 390–398.

Joyce, B.A., Wallender, W.W., Mitchell, J.P. et al. (2002). Infiltration and soil water storage under winter cover cropping in California's Sacramento Valley. *Transactions of the ASAE* 45: 315.

Kahimba, F.C., Ranjan, R.S., Froese, J. et al. (2008). Cover crop effects on infiltration, soil temperature, and soil moisture distribution in the Canadian Prairies. *Applied Engineering in Agriculture* 24: 321–333.

Kaye, J.P. and Quemada, M. (2017). Using cover crops to mitigate and adapt to climate change. A review. *Agronomy for Sustainable Development* 37: 4. https://doi.org/10.1007/s13593-016-0410-x.

Kelly, C., Fonte, S.J., Shrestha, A. et al. (2021). Winter cover crops and no-till promote soil macrofauna communities in irrigated, Mediterranean cropland in California, USA. *Applied Soil Ecology* 166: 104068. https://doi.org/10.1016/j.apsoil.2021.104068.

King, A.E. and Blesh, J. (2018). Crop rotations for increased soil carbon: perenniality as a guiding principle. *Ecological Applications* 28: 249–261.

Kirschbaum, M.U.F. (1995). The temperature dependence of soil organic matter decomposition, and the effect of global warming on soil organic C storage. *Soil Biology and Biochemistry* 27: 753–760.

Koehler-Cole, K., Everhart, S.E., Gu, Y. et al. (2020). Is allelopathy from winter cover crops affecting row crops? *Agricultural & Environmental Letters* 5: e20015.

Korres, N.E., Norsworthy, J.K., Young, B.G. et al. (2018). Seedbank persistence of Palmer amaranth (*Amaranthus palmeri*) and waterhemp (*Amaranthus tuberculatus*) across diverse geographical regions in the United States. *Weed Science* 66: 446–456. https://doi.org/10.1017/wsc.2018.27.

Kremer, R.J. (1993). Management of weed seed banks with microorganisms. *Ecological Applications* 3: 42–52.

Kremer, R.J. and Li, J. (2003). Developing weed-suppressive soils through improved soil quality management. *Soil & Tillage Research* 72: 193–202.

Kremer, R.J. and Spencer, N.R. (1989). Interaction of insects, fungi, and burial on velvetleaf (*Abutilon theophrasti*) seed viability. *Weed Technology* 3: 322–328.

Kulkarni, S.S., Dosdall, L.M., Spence, J.R., and Willenborg, C.J. (2015). Depth of seed burial and gender influence weed seed predation by three species of ground beetle (Coleoptera: Carabidae). *Weed Science* 63: 910–915.

Kumar, A., Choudhary, T., Das, S., and Meena, S.K. (2019). Weed seed bank: impacts and management for future crop production. In: *Agronomic Crops* (ed. M. Hasanuzzaman), 207–223. Singapore: Springer. https://doi.org/10.1007/978-981-32-9783-8_12.

Kunz, C., Sturm, D.J., Varnholt, D. et al. (2016). Allelopathic effects and weed suppressive ability of cover crops. *Plant, Soil and Environment* 62: 60–66.

Lawley, Y.E., Teasdale, J.R., and Weil, R.R. (2012). The mechanism for weed suppression by a forage radish cover crop. *Agronomy Journal* 104: 205–214.

Lawley, Y.E., Weil, R.R., and Teasdale, J.R. (2011). Forage radish cover crop suppresses winter annual weeds in fall and before corn planting. *Agronomy Journal* 103: 137–144.

Leavitt, M.J., Sheaffer, C.C., Wyse, D.L., and Allan, D.L. (2011). Rolled winter rye and hairy vetch cover crops lower weed density but reduce vegetable yields in no-tillage organic production. *HortScience* 46: 387–395.

Lehman, R.M., Taheri, W.I., Osborne, S.L. et al. (2012). Fall cover cropping can increase arbuscular mycorrhizae in soils supporting intensive agricultural production. *Applied Soil Ecology* 61: 300–304. https://doi.org/10.1016/j.apsoil.2011.11.008.

Lewis, D.G., Cutulle, M.A., Schmidt-Jeffris, R.A., and Blubaugh, C.K. (2020). Better together? Combining cover crop mulches, organic herbicides, and weed seed biological control in reduced-tillage systems. *Environmental Entomology* 49: 1327–1334. https://doi.org/10.1093/ee/nvaa105.

Liebert, J.A., DiTommaso, A., and Ryan, M.R. (2017). Suppression in cover crop-based organic no-till planted soybean. *Weed Science* 65: 426–439.

Liebman, M. (2001). Weed management: a need for ecological approaches. In: *Ecological Management of Agricultural Weeds* (ed. M. Liebman, C.L. Mohler, and C.P. Staver), Cambridge: Cambridge University Press.

Liebman, M. and Davis, A.S. (2000). Integration of soil, crop and weed management in low-external-input farming systems. *Weed Research* 40: 27–47.

Liebman, M. and Dyck, E. (1993). Crop rotation and intercropping strategies for weed management. *Ecological Applications* 3: 92–122. https://doi.org/10.2307/1941795.

Liebman, M. and Gallandt, E.R. (1997). Many little hammers: ecological management of crop-weed interactions. In: *Ecology in Agriculture* (ed. L.E. Jackson), 291–343. San Diego, CA: Academic Press.

Liebman, M. and Sundberg, D.N. (2006). Seed mass affects the susceptibility of weed and crop species to phytotoxins extracted from red clover shoots. *Weed Science* 54: 340–345. https://doi.org/10.1614/WS-05-54.2.340a.

Lithourgidis, A.S., Dordas, C.A., Damalas, C.A., and Vlachosterigios, D.N. (2011). Annual intercrops: an alternative pathway for sustainable agriculture. *Australian Journal of Crop Science* 5: 396–410.

Liu, A., Ma, B.L., and Bomke, A.A. (2005). Effects of cover crops on soil aggregate stability, total organic carbon, and polysaccharides. *Soil Science Society of America Journal* 69: 2041–2048.

Lounsbury, N.P., Warren, N.D., Wolfe, S.D., and Smith, R.G. (2020). Investigating tarps to facilitiate organic no-till cabbage production with high-residue cover crops. *Renewable Agriculture and Food Systems* 35: 227–233.

Lutman, P.J.W., Cussans, G.W., Wright, K.J. et al. (2002). The persistence of seeds of 16 weed species over six years in two arable fields. *Weed Research* 42: 231–241.

MacLaren, C., Swanepoel, P., Bennett, J. et al. (2019). Cover crop biomass production is more important than diversity for weed suppression. *Crop Science* 59: 1–16.

Manzoni, S., Schimel, J.P., and Porporato, A. (2012). Responses of soil microbial communities to water stress: results from a meta-analysis. *Ecology* 93: 930–938. https://doi.org/10.1890/11-0026.1.

McClelland, S.C., Paustian, K., and Schipanski, M.E. (2021). Management of cover crops in temperate climates influences soil organic carbon stocks: a meta-analysis. *Ecological Applications* 31: e02278. https://doi.org/10.1002/eap.2278.

Meiss, H., Le Lagadec, L., Munier-Jolain, N. et al. (2010). Weed seed predation increases with vegetation cover in perennial forage crops. *Agriculture, Ecosystems & Environment* 138: 10–16.

Menalled, F.D., Peterson, R.K.D., Smith, R.G. et al. (2016). The eco-evolutionary imperative: revisiting weed management in the midst of an herbicide resistance crisis. *Sustainability* 8: 1297.

Menalled, F.D. and Schonbeck, M. (2013). Manage the weed seed bank—minimize "deposits" and maximize "withdrawals." *eOrganic*. https://eorganic.org/node/2806.

Menalled, F.D., Smith, R.G., Dauer, J.T., and Fox, T.B. (2007). Impact of agricultural management on carabid communities and weed seed predation. *Agriculture, Ecosystems & Environment* 118: 49–54.

Mennan, H., Jabran, K., Zandstra, B.H., and Pala, F. (2020). Non-chemical weed management in vegetables by using cover crops: a review. *Agronomy* 10 (2): 257. https://doi.org/10.3390/agronomy10020257.

Mirsky, S.B., Ryan, M.R., Curran, W.S. et al. (2012). Conservation tillage issues: cover crop-based organic rotational no-till grain production in the mid-Atlantic region, USA. *Renewable Agriculture and Food Systems* 27: 31–40. https://doi.org/10.1017/S1742170511000457.

Mohler, C.L. (2001). Weed life history: identifying vulnerabilities. In: *Ecological Management of Agricultural Weeds* (ed. M. Liebman, C.L. Mohler, and C.P. Staver), 40–98. Cambridge: Cambridge University Press.

Mohler, C.L., Dykeman, C., Nelson, E.B., and Ditommaso, A. (2012). Reduction in weed seedling emergence by pathogens following the incorporation of green crop residue. *Weed Research* 52: 467–477.

Montgomery, D.R. (2007). Soil erosion and agricultural sustainability. *Proceedings of the National Academy of Sciences* 104: 13268 LP – 13272. https://doi.org/10.1073/pnas.0611508104.

Mortensen, D.A., Egan, J.F., Maxwell, B.D. et al. (2012). Navigating a critical juncture for sustainable weed management. *BioScience* 62: 75–84.

Müller-Stöver, D., Nybroe, O., Baraibar, B. et al. (2016). Contribution of the seed microbiome to weed management. *Weed Research* 56: 335–339.

Mullin, C.A., Saunders, M.C., II, Leslie, T.W. et al. (2005). Toxic and behavioral effects to Carabidae of seed treatments used on Cry3Bd1- and Cry1Ab/c-protected corn. *Environmental Entomology* 34: 1626–1636.

Navntoft, S., Wratten, S.D., Kristensen, K., and Esbjerg, P. (2009). Weed seed predation in organic and conventional fields. *Biological Control* 49: 11–16.

Nevins, C.J., Nakatsu, C., and Armstrong, S. (2018). Characterization of microbial community response to cover crop residue decomposition. *Soil Biology and Biochemistry* 127: 9–49.

Nguyen, N.H., Song, Z., Bates, S.T. et al. (2016). FUNGuild: an open annotation tool for parsing fungal community datasets by ecological guild. *Fungal Ecology* 20: 241–248. http://doi.org/10.1016/j.funeco.2015.06.006.

Nichols, V., Verhulst, N., Cox, R., and Govaerts, B. (2015). Weed dynamics and conservation agriculture principles: a review. *Field Crops Research* 183: 56–68.

Nikolić, N., Squartini, A., Concheri, G. et al. (2020). Weed seed decay in no-till field and planted riparian buffer zone. *Plants* 9: 293.

O'Donovan, J.T., Harker, K.N., Turkington, T.K., and Clayton, G.W. (2013). Combining cultural practices with herbicides reduces wild oat (*Avena fatua*) seed in the soil seed bank and improves barley yield. *Weed Science* 61: 328–333. https://doi.org/10.1614/WS-D-12-00168.1.

O'Rourke, M.E., Heggenstaller, A.H., Liebman, M., and Rice, M.E. (2006). Post-dispersal weed seed predation by invertebrates in conventional and low-external-input crop rotation systems. *Agriculture, Ecosystems & Environment* 116: 280–288. https://doi.org/10.1016/j.agee.2006.02.018.

Ohno, T., Doolan, K., Zibilske, L.M. et al. (2000). Phytotoxic effects of red clover amended soils on wild mustard seedling growth. *Agriculture, Ecosystems & Environment* 78: 187–192.

Osipitan, O.A., Dille, J.A., Assefa, Y., and Knezevic, S.Z. (2018). Cover crop for early season weed suppression in crops: systematic review and meta-analysis. *Agronomy Journal* 110: 2211–2221. https://doi.org/10.2134/agronj2017.12.0752.

Osipitan, O.A., Dille, J.A., Assefa, Y. et al. (2019). Impact of cover crop management on level of weed suppression: a meta-analysis. *Crop Science* 59: 833–842.

Page, E.R., Tollenaar, M., Lee, E.A. et al. (2010). Shade avoidance: an integral component of crop-weed competition. *Weed Research* 50: 281–288.

Peralta, A.L., Sun, Y., McDaniel, M.D., and Lennon, J.T. (2018). Crop rotational diversity increases disease suppressive capacity of soil microbiomes. *Ecosphere* 9: e02235.

Pittman, K.B., Barney, J.N., and Flessner, M.L. (2020). Cover crop residue components and their effect on summer annual weed suppression in corn and soybean. *Weed Science* 68: 301–310.

Poeplau, C. and Don, A. (2015). Carbon sequestration in agricultural soils via cultivation of cover crops–a meta-analysis. *Agriculture, Ecosystems & Environment* 200: 33–41.

Pollard, A.T. (2018). Seeds vs fungi: an enzymatic battle in the soil seedbank. *Seed Science Research* 28: 197–214.

Pullaro, T.C., Marino, P.C., Jackson, D.M. et al. (2006). Effects of killed cover crop mulch on weeds, weed seeds, and herbivores. *Agriculture, Ecosystems & Environment* 115: 97–104.

Quinn, N.F., Brainard, D.C., and Szendrei, Z. (2016). The effect of conservation tillage and cover crop residue on beneficial arthropods and weed seed predation in acorn squash. *Environmental Entomology* 45: 1543–1551.

Radicetti, E., Mancinelli, R., and Campiglia, E. (2013). Impact of managing cover crop residues on the floristic composition and species diversity of the weed community of pepper crop (*Capsicum annuum* L.). *Crop Protection* 44: 109–119.

Ramesh, K., Matloob, A., Aslam, F. et al. (2017). Weeds in a changing climate: vulnerabilities, consequences, and implications for future weed management. *Frontiers in Plant Science* 8: 95.

Reader, R.J. (1993). Control of seedling emergence by ground cover and seed predation in relation to seed size for some old-field species. *Journal of Ecology* 81: 169–175.

Reberg-Horton, S.C., Grossman, J.M., Kornecki, T.S. et al. (2012). Utilizing cover crop mulches to reduce tillage in organic systems in the southeastern USA. *Renewable Agriculture and Food Systems* 27: 41–48.

Reed, H.K., Karsten, H.D., Curran, W.S. et al. (2019). Planting green effects on corn and soybean production. *Agronomy Journal* 111: 2314–2325.

Rice, E.L. (1984). *Allelopathy* 2e. Academic Press.

Roth, C.M., Shroyer, J.P., and Paulsen, G.M. (2000). Allelopathy of sorghum on wheat under several tillage systems. *Agronomy Journal* 92: 855–860. https://doi.org/10.2134/agronj2000.925855x.

Sarabi, V. (2019). Factors that influence the level of weed seed predation: a review. *Weed Biology and Management* 19: 61–74.

Scherer-Lorenzen, M. (2008). Functional diversity affects decomposition processes in experimental grasslands. *Functional Ecology* 22: 547–555.

Schipanski, M.E., Barbercheck, M., Douglas, M.R. et al. (2014). A framework for evaluating ecosystem services provided by cover crops in agroecosystems. *Agricultural Systems* 125: 12–22.

Schmidt, R., Gravuer, K., Bossange, A.V. et al. (2018). Long-term use of cover crops and no-till shift soil microbial community life strategies in agricultural soil. *PLoS ONE* 13: e0192953.

Schulz, M., Marocco, A., Tabaglio, V. et al. (2013). Benzoxazinoids in rye allelopathy – from discovery to application in sustainable weed control and organic farming. *Journal of Chemical Ecology* 39: 154–174. https://doi.org/10.1007/s10886-013-0235-x.

Shearin, A.F., Chris Reberg-Horton, S., and Gallandt, E.R. (2008). Cover crop effects on the activity-density of the weed seed predator *Harpalus rufipes* (Coleoptera: Carabidae). *Weed Science* 56: 442–450. https://doi.org/10.1614/WS-07-137.1.

Smith, R.G. (2015). A succession-energy framework for reducing non-target impacts of annual crop production. *Agricultural Systems* 133: 14–21.

Smith, R.G., Atwood, L.W., Morris, M.B. et al. (2016). Evidence for indirect effects of pesticide seed treatments on weed seed banks in maize and soybean. *Agricutlure, Ecosystems and Environmment* 216: 269–273.

Smith, R.G. and Gross, K.L. (2006). Rapid change in the germinable fraction of the weed seed bank in crop rotations. *Weed Science* 54: 1094–1100. https://doi.org/10.1614/WS-06-090R.1.

Smith, R.G., Gross, K.L., and Januchowski, S. (2005). Earthworms and weed seed distribution in annual crops. *Agriculture, Ecosystems & Environment* 108: 363–367. https://doi.org/10.1016/j.agee.2005.01.014.

Smith, R.G. and Mortensen, D.A. (2017). A disturbance-based framework for understanding weed community assembly in agroecosystems: challenges and opportunities for agroecological weed management. In: *Agroecological Practices for Sustainable Agriculture: Principles, Applications, and Making the Transition* (ed. A. Wezel), 127–154. London: World Scientific Publishing.

Smith, R.G., Warren, N.D., and Cordeau, S. (2020). Are cover crop mixtures better at suppressing weeds than cover crop monocultures? *Weed Science* 68: 186–194.

Storck, V., Karpouzas, D.G., and Martin-Laurent, F. (2017). Towards a better pesticide policy for the European Union. *Science of the Total Environment* 575: 1027–1033. https://doi.org/10.1016/j.scitotenv.2016.09.167.

Swanton, C.J. and Booth, B.D. (2004). Management of weed seedbanks in the context of populations and communities. *Weed Technology* 18: 1496–1502.

Szumigalski, A. and Van Acker, R. (2005). Weed suppression and crop production in annual intercrops. *Weed Science* 53: 813–825. https://doi.org/10.1614/WS-05-014R.1.

Taylor, K.L. and Hartzler, R.G. (2000). Effect of seed bank augmentation on herbicide efficacy. *Weed Technology* 14: 261–267. https://doi.org/10.1614/0890-037X(2000)014(0261:EOSBAO)2.0.CO;2.

Teasdale, J.R., Mangum, R.W., Radhakrishnan, J., and Cavigelli, M.A. (2004). Weed seedbank dynamics in three organic farming crop rotations. *Agronomy Journal* 96: 1429–1435.

Telewski, F.W. and Zeevaart, J.A.D. (2002). The 120-yr period of Dr. Beal's seed viability experiment. *American Journal of Botany* 89: 1285–1288.

Thapa, V.R., Ghimire, R., Acosta-Martínez, V. et al. (2021). Cover crop biomass and species composition affect soil microbial community structure and enzyme activities in semiarid cropping systems. *Applied Soil Ecology* 157: 103735.

Thilakarathna, M.S., McElroy, M.S., Chapagain, T. et al. (2016). Belowground nitrogen transfer from legumes to non-legumes under managed herbaceous cropping systems. A review. *Agronomy for Sustainable Development* 36: 58. https://doi.org/10.1007/s13593-016-0396-4.

Tilman, D., Wedin, D., and Knops, J.M.H. (1996). Productivity and sustainability influenced by biodiversity in grassland ecosystems. *Nature* 379: 718–720.

Tomasek, B.J., Williams, M.M., II, and Davis, A.S. (2017). Changes in field workability and drought risk from projected climate change drive spatially variable risks in Illinois cropping systems. *PLoS ONE* 12 (2): e0172301. https://doi.org/10.1371/journal.pone.0172301.

Travlos, I., Gazoulis, I., Kanatas, P. et al. (2020). Key factors affecting weed seeds' germination, weed emergence, and their possible role for the efficacy of false seedbed technique as weed management practice. *Frontiers in Agronomy* 2: 1. https://doi.org/10.3389/fagro.2020.00001.

Tu, C., Ristaino, J.B., and Hu, S. (2006). Soil microbial biomass and activity in organic tomato farming systems: effects of organic inputs and straw mulching. *Soil Biology and Biochemistry* 38: 247–255.

Turk, M.A. and Tawaha, A.M. (2003). Allelopathic effect of black mustard (*Brassica nigra* L.) on germination and growth of wild oat (*Avena fatua* L.). *Crop Protection* 22: 673–677. https://doi.org/10.1016/S0261-2194(02)00241-7.

Uchino, H., Iwama, K., Jitsuyama, Y. et al. (2012). Effect of interseeding cover crops and fertilization on weed suppression under an organic and rotational cropping system: 1. Stability of weed suppression over years and main crops of potato, maize and soybean. *Field Crops Research* 127: 9–16.

Ullrich, S.D., Buyer, J.S., Cavigelli, M.A. et al. (2011). Weed seed persistence and microbial abundance in long-term organic and conventional cropping systems. *Weed Science* 59: 202–209.

Unger, P.W. and Vigil, M.F. (1998). Cover crop effects on soil water relationships. *Journal of Soil and Water Conservation* 53: 200–207.

van der Laat, R., Owen, M.D.K., Liebman, M., and Leon, R.G. (2015). Postdispersal weed seed predation and invertebrate activity density in three tillage regimes. *Weed Science* 63: 828–838.

Van Deynze, B., Swinton, S.M., and Hennessy, D.A. (2021). Are glyphosate-resistant weeds a threat to conservation agriculture? Evidence from tillage practices in soybeans. *American Journal of Agricultural Economics*. Early View. https://doi.org/10.1111/ajae.12243.

Veresoglou, S.D. and Rillig, M.C. (2012). Suppression of fungal and nematode plant pathogens through arbuscular mycorrhizal fungi. *Biology Letters* 8: 214–217.

Vidal, R.A. and Bauman, T.T. (1997). Fate of allelochemicals in the soil. *Ciência Rural* 27: 351–357.

Wagner, M. and Mitschunas, N. (2008). Fungal effects on seed bank persistence and potential applications in weed biocontrol: a review. *Basic and Applied Ecology* 9: 191–203. https://doi.org/10.1016/j.baae.2007.02.003.

Wallace, J.M., Curran, W.S., and Mortensen, D.A. (2019). Cover crop effects on horseweed (*Erigeron canadensis*) density and size inequality at the time of herbicide exposure. *Weed Science* 67: 327–338.

Wallace, J.M., Williams, A., Liebert, J.A. et al. (2017). Cover crop-based, organic rotational no-till corn and soybean production systems in the mid-Atlantic United States. *Agriculture* 7: 34.

Ward, M.J., Ryan, M.R., Curran, W.S. et al. (2011). Cover crops and disturbance influence activity-density of weed seed predators *Amara aenea* and *Harpalus pensylvanicus* (Coleoptera: Carabidae). *Weed Science* 59: 76–81.

Weil, R. and Kremen, A. (2007). Thinking across and beyond disciplines to make cover crops pay. *Journal of the Science of Food and Agriculture* 87: 551–557.

Weir, T.L., Park, S.-W., and Vivanco, J.M. (2004). Biochemical and physiological mechanisms mediated by allelochemicals. *Current Opinion in Plant Biology* 7: 472–479. https://doi.org/10.1016/j.pbi.2004.05.007.

Weisberger, D., Nichols, V., and Liebman, M. (2019). Does diversifying crop rotations suppress weeds? A meta-analysis. *PLoS ONE* 14 (7): e0219847. https://doi.org/10.1371/journal.pone.0219847.

Westerman, P.R., Hofman, A., Vet, L.E.M., and van der Werf, W. (2003). Relative importance of vertebrates and invertebrates in epigeaic weed seed predation in organic cereal fields. *Agriculture, Ecosystems & Environment* 95: 417–425. https://doi.org/10.1016/S0167-8809(02)00224-4.

Westerman, P.R., Liebman, M., Menalled, F.D. et al. (2005). Are many little hammers effective? Velvetleaf (*Abutilon theophrasti*) population dynamics in two-and four-year crop rotation systems. *Weed Science* 53: 382–392.

Weston, L.A. (1996). Utilization of allelopathy for weed management in agroecosystems. *Agronomy Journal* 88: 860–866.

Wicks, G.A. and Smika, D.E. (1973). Chemical fallow in a winter wheat-fallow rotation. *Weed Science* 21: 97–102. https://doi.org/10.1017/S0043174500031829.

Wildung, R.E., Garland, T.R., and Buschbom, R.L. (1975). The interdependent effects of soil temperature and water content on soil respiration rate and plant root decomposition in arid grassland soils. *Soil Biology and Biochemistry* 7: 373–378. https://doi.org/10.1016/0038-0717(75)90052-8.

Williams, A., Hunter, M.C., Kammerer, M. et al. (2016). Soil water holding capacity mitigates downside risk and volatility in US rainfed maize: time to invest in soil organic matter? *PLoS ONE* 11 (8): e0160974. https://doi.org/10.1371/journal.pone.0160974.

Williams, C.L., Liebman, M., Westerman, P.R. et al. (2009). Over-winter predation of *Abutilon theophrasti* and *Setaria faberi* seeds in arable land. *Weed Research* 49: 439–447.

Youkhana, A.H. and Idol, T.W. (2016). Leucaena-KX2 mulch additions increase growth, yield and soil C and N in a managed full-sun coffee system in Hawaii. *Agroforestry Systems* 90: 325–337.

Zhang, X.C. (2012). Cropping and tillage systems effects on soil erosion under climate change in Oklahoma. *Soil Science Society of America Journal* 76: 1789–1797. https://doi.org/10.2136/sssaj2012.0085.

11

Soil Seedbank from an Ecological Perspective

Lauren M. Schwartz-Lazaro[1,] and Karla L. Gage[2]*

[1] *Blue River Technology, Sunnyvale, CA, USA*
[2] *School of Agricultural Sciences/School of Biological Sciences, Southern Illinois University, Carbondale, IL, USA*
[*] *Corresponding author*

11.0 Introduction

An ecological understanding of how the soil seedbank (seeds in the soil) functions is vital to ultimately being able to control it if that is the intent. Within an agronomic system, problems associated with simplified weed management have led to the ever-increasing complexity of evolving herbicide-resistant weeds and the corresponding modernization of agricultural technologies. Short-term management of the soil seedbank is no longer a viable option since the trajectory of weed adaptation and evolution operates at extended temporal scales. Long-term management that specifically targets the weed seedbank is critical to the future of agriculture. This can take shape through integrated weed management practices that use the fundamentals of weed biology and ecology to provide a framework for diversification. This chapter focuses on the soil seedbank and its complexities. Further, we explore various avenues of seedbank increase, such as seed dispersal and seed rain, and reduction, through various avenues of seed fate, of the soil seedbank leading to management decisions and the effects those decisions may have. Future weed seedbank management requires an ecological core to foster sustainable changes in how weed management is approached.

11.1 The Soil Seedbank

The seedbank is a collection of dormant and nondormant seeds persisting at varying depths within the soil profile at a particular location. Seeds may be deposited through natural population processes leading to reproduction or through dispersal mechanisms. Seeds may leave the seedbank through germination, decay, predation, or parasitism (Figure 11.1).

Theoretical components of seedbank persistence for a given cohort of seeds include seed losses due to germination, aging, emigration, and mortality (Schafer and Chilcote 1969; Fenner and Thompson 2005). Most viable weed seeds within persistent soil seedbanks are found beneath the soil surface but within the top 30 cm of the soil profile (Reuss et al. 2001; Harrison et al. 2007). Postdispersal seed predation (Heggenstaller et al. 2006; Menalled et al. 2006) causes much of the weed seed rain in arable fields to disappear from the soil surface within days to weeks. Remaining seeds can find their way into deeper layers in the soil profile through various processes, including soil disturbance (Yenish et al. 1992), seed caching by granivores (Pemberton and Irving 1990; Hartke et al. 1998), or seed movement into soil cracks (Thompson et al. 1993). Dormancy cycling helps to ensure that for most weed species, only a small proportion of buried seeds is recruited as seedlings from the soil seedbank in any given year (Baskin and Baskin 2001). Those seeds that remain dormant in the soil seedbank must maintain their viability while withstanding attacks from soil invertebrates and microbes (Chee-Sanford et al. 2006).

Agricultural habitats are characterized by frequent and intense disturbance events. As such, after disturbance, fields are in a constant state of early succession where resources are released and become available for weed reestablishment (Smith 2015). Annual weeds with the strategy of rapid resource uptake, growth, and high fecundity are primary colonizers of agricultural fields after disturbance. The population dynamics of annual weeds are driven by the seedbank, and reducing the seedbank is a primary objective in weed management. Therefore, weed management practices in

Ecologically Based Weed Management: Concepts, Challenges, and Limitations, First Edition. Edited by Nicholas E. Korres, Ilias S. Travlos, and Thomas K. Gitsopoulos.

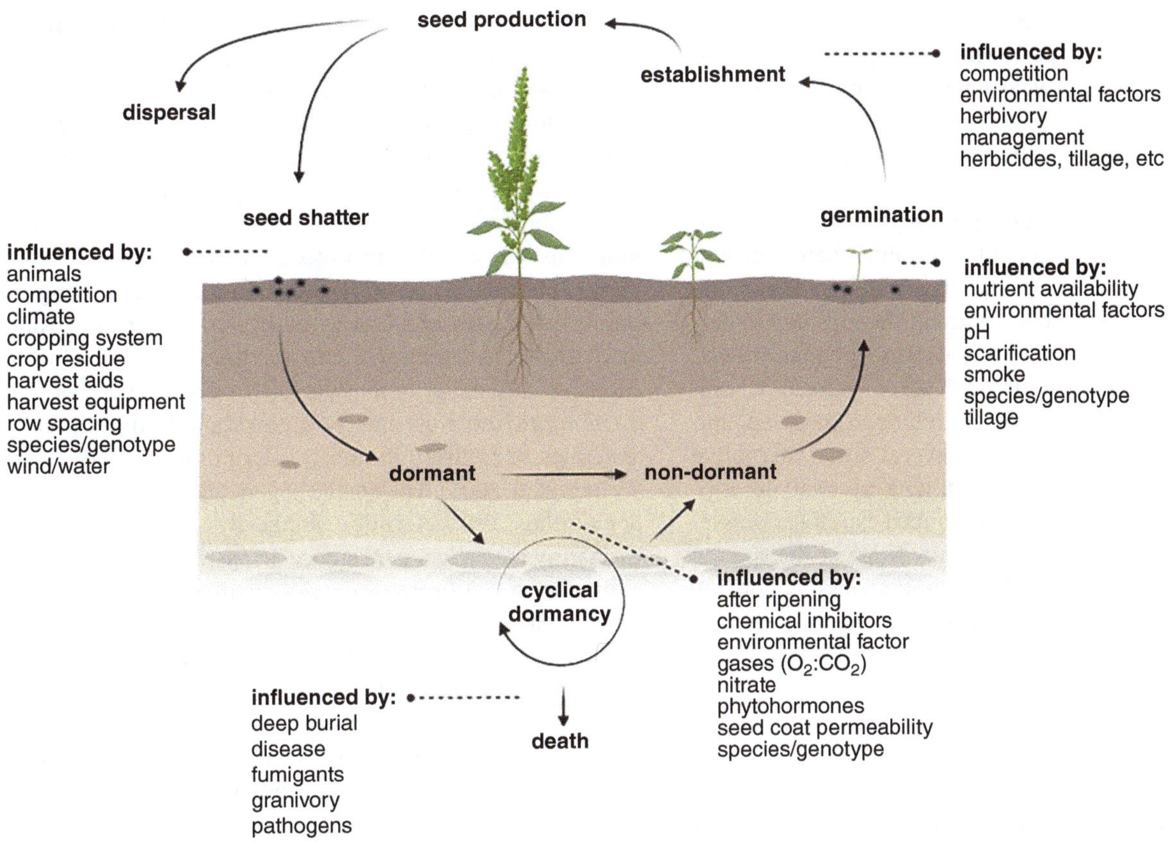

Figure 11.1 Stages of seed movement, from weed seed shatter to seed production, of an annual weed species, including examples of ecological and management influences on specific seed stages. Environmental factors include light, soil moisture, and temperature with additional specific factors within each stage. Created with BioRender.com.

one year will impact the composition of weed communities in subsequent years through the effects on the seedbank (Haring and Flessner 2018; Schwartz-Lazaro and Copes 2019), and weed seedbanks may conserve biodiversity in the agricultural field and present a culmination of weed management history through the deposition and persistence of seeds (Cavers 1995). The persistence of seeds in the seedbank can be influenced by a variety of factors: the physical characteristics of the seeds, spatial distribution/location within the soil profile, the site management activities, and edaphic conditions (Ghersa and Martínez-Ghersa 2000; Chee-Sanford et al. 2006).

11.1.1 Seedbank Types

Seedbanks may be classified according to patterns of transience and persistence (Thompson and Grime 1979). There are generally three types of seedbanks associated with these characteristics (Bakker et al. 1991, 1996; Thompson 1992, 1993): transient, short-term persistence, and long-term persistence. The description of these seedbank types are often incorporated with a seedbank classification proposed by Thompson and Grime (1979), which describes seedbanks based on seasonal variation in germination timing. The Thompson and Grime classification system grouped seedbanks into types I–IV, based on germination and persistence over one year of seasonal sampling but does not describe longevity beyond one year. Therefore, the Thompson and Grime scheme is often combined with other descriptors, which take into account long-term persistence. Transient seedbanks persist for one year or less and consist of type I, fall-germinating species, whose seeds are present during the summer. Type II, spring-germinating species have seeds present mainly in the winter. Seeds of both are often dormant when shattered. Dormancy is broken by a period of low or high temperatures in types I and II, respectively (Fenner and Thompson 2005). Short-term persistent seedbanks are viable for greater than one year but less than five years and have the potential to rescue populations that experience a disruption in one reproductive cycle (Bakker et al. 1996). Long-term persistence seedbanks persist for five or more years, with "five years" being an arbitrary cutoff resulting from the time limitations of persistence studies (Long et al. 2008). Long-term persistence seedbanks may allow community regeneration following a major disturbance event (Bakker et al. 1996).

Persistent seedbanks may be classified as type III, fall-germinating with a small, persistent year-long seedbank, or type IV, large, persistent year-long seedbank (Booth et al. 2003). The same species may behave as a type III or IV at different times and in different places (Cummins and Miller 2002).

An alternative classification proposed by Poschlod and Jackel (1993) takes into account seed rain, where transient seedbanks occur in the upper soil layer with less than one year of persistence, or occur mostly in the topsoil and persist for one to two years; and persistent seedbanks are viable for several years to decades with seeds in the upper and lower layers of soil at great abundance (Poschlod and Jackel 1993; Bakker et al. 1996). Several other alternative classifications have been proposed, with up to 10 descriptive categories; however, these advanced classification systems rely on detailed ecological knowledge of the species included in the seedbank, which is often lacking. For this reason, the classification systems with fewer categories are commonly used (Csontos and Tamás 2003).

11.1.2 Longevity

The persistence of viable seeds in the soil seedbank depends on a wide range of interacting biotic and abiotic factors. These include germination cues, seed dormancy, seed size (Thompson et al. 1994; Hulme 1998; Ooi et al. 2007; Honda 2008), physiological age, predation, and microbial decay, along with environmental conditions, burial depth, and burial duration (Liebman et al. 2001; Davis et al. 2005; Korres et al. 2018). The seeds of most annual weed species are capable of persisting in the soil seedbank for several years or more (Roberts and Feast 1972; Lueschen et al. 1993; Burnside et al. 1996; Buhler and Hartzler 2001; Conn et al. 2006), and thus reinfesting fields for many years after a single seed rain event (Davis et al. 2008).

Many weed species have the characteristic of long-lived seeds with high fecundity (Booth et al. 2003). This longevity of the seedbank is especially important in disturbed habitats and can be described as a bet-hedging strategy to ensure against years of reproductive failure (Cohen 1966). Seedbank persistence may be related to the life history strategy of the contributing weed species; annuals and biennials tend to have longer-lived seeds than perennials (Thompson et al. 1998). While several studies have found correlations between life history strategy and longevity (Chambers 1995; Thompson et al. 1998; Honda 2008), this is not always the case (Stöcklin and Fischer 1999). Seeds with the greatest longevity are not necessarily the most competitive. Prolonged residency in the soil seedbank can be associated with slow germination and growth, negatively impacting competitive ability and persistence in the seedbank (Long et al. 2009).

Longevity appears to have a genetic basis, as populations and species exhibit predictable persistence in various environments and may be influenced by genetically determined dormancy and antimicrobial properties (Walters et al. 2005; Long et al. 2015). Inherent longevity determines persistence and is based on trait expression in processes such as the seed's ability to control and repair damage over time through cellular processes and protection mechanisms (Long et al. 2008). Damage-induced seed aging is influenced by the temperature of the surrounding environment as well as seed water activity; these two factors influence the consistency of the cytosol, the functionality of cell membranes, the movement of antioxidants within the cell, and rates of translation and transcription (Walters 1998; Walters et al. 2005). The degree of genetic variability within populations also determines longevity; a greater genetic variability may allow persistence across diverse spatial and temporal gradients of resource availability.

Seeds can be desiccation-sensitive (recalcitrant); that is, mortality is caused by drying. About 8% of angiosperms are desiccation-sensitive. Alternatively, seeds can be desiccation-tolerant, which describes the remaining 92% of angiosperms, as a result of processes in late seed development leading to the presence of late embryogenesis abundant (LEA) proteins and nonreducing sugars (Long et al. 2015). The presence of LEA and sugars causes a reduction in chemical reactions and cell mobility, possibly for long periods of time (Sallon et al. 2008; Colville and Kranner 2010; Leprince and Buitink 2010). Most desiccation-sensitive species are from tropical regions. Seeds are hydrated and must maintain a minimum water content when dispersed, and these seeds often germinate quickly (Long et al. 2015). Although the longevity of desiccation-sensitive seeds dependents on moisture levels in the environment, these recalcitrant seeds can still form a persistent seedbank (Ferrandis et al. 2011).

Ecological relationships of the maternal plant also have the possibility of influencing seed longevity in the seedbank. Plasticity exhibited by the parental plant may increase the longevity of seeds produced (Kochanek et al. 2011). Longevity, as a trait, is selected for by density-independent mortality (Thompson et al. 1998). Longevity is determined in large part by seed dormancy, which is heavily dependent on maternal investment in chemical and physical seed defenses, to combat granivory, microbial attack, and physical damage by crop management operations (Davis et al. 2016).

11.1.3 Spatial Distribution in the Seedbank

In undisturbed environments, it is assumed that seed age increases with depth, since relatively older seeds have a longer time to become buried; and therefore, burial depth may be an indicator of seed longevity (Bekker et al. 1998). The longevity of seeds in the seedbank may also be related to their vertical distribution in the soil (Bekker et al. 1998; Ghersa and Martínez-Ghersa 2000), but it is difficult to separate the effect of vertical distribution from the relationship between seed mass, seed shape, and longevity (Ghersa and Martínez-Ghersa 2000). Seed mass and seed longevity were inversely related in a survey of more than 500 species (Hodkinson et al. 1998), and the role of seed mass in the vertical distribution in the seedbank is assumed to be partially explanatory for this inverse relationship (Ghersa and Martínez-Ghersa 2000). In undisturbed environments, seeds are mostly found at the surface or shallow soil depths, as opposed to a random distribution with tillage. Sometimes seed can be redistributed after shatter and found in the greatest abundance directly below the soil surface (Espinar et al. 2005), as a result of successional processes (transition of cultivated land to pasture (Chippindale and Milton 1934), buried ruderal seedbanks in late-successional stages (Bekker et al. 1999), natural soil processes, such as cracking of vertic soils (Espinar et al. 2005), and the redistribution of seeds by animals such as earthworms (Willems and Huijsmans 1994; Smith et al. 2005; Regnier et al. 2008)). If seed shatter occurs for a species at a time when a process of redistribution reaches peak activity (e.g. soil cracking), the seedbank of that species may exhibit a bimodal distribution, where seeds are found in abundance both surface and deep layers (Espinar et al. 2005).

In contrast to undisturbed environments, tilled environments may exhibit a random distribution of seeds within the plow layer, and the seed distribution effects of tillage may significantly impact seed longevity. Seeds buried in the soil by tillage have increased longevity for many species, compared to those at shallow depths, but this persistence does not seem to be related to seed mass (Martínez-Ghersa et al. 2000). Rather, rates of loss for the seedbank are high near the soil surface due to seed predators, germination, microbial activity, and advanced aging of seeds (Abdalla and Roberts 1968; Villiers 1973; Ballare et al. 1988; Roberts 1988a; Wagner and Mitschunas 2008). In an evaluation of the effect of tillage system on the rigid ryegrass (*Lolium rigidum*) seedbank, seed decay was higher and seedling recruitment lower in no-till systems as compared to minimum tillage systems (Chauhan et al. 2006a). The impact of tillage and the resulting vertical distribution of the seedbank may be species-specific; no-till may favor species that need light to germinate, but crop residue on the surface may have a negative effect due to physical suppression, changes in soil temperature, or allelopathic properties of residue (Chauhan et al. 2006b). In a study of reduced vs. conventional tillage systems, 85% of all seeds were found at 0–5 cm depth in reduced tillage, compared to 29% in conventional tillage (Pareja et al. 1985).

11.1.4 Seed Morphology

Morphological characteristics may influence seed persistence. Seed coat thickness, for example, can determine longevity and attractiveness to predators (Dalling et al. 2011; Fuerst et al. 2011), as well as seed coat color (Lev-Yadun and Ne'eman 2013). Plants usually exhibit a trade-off between the number of seeds produced and seed size (Booth et al. 2003). Since plants have a finite number of resources for reproduction, large seeds are more costly than small seeds but have more resources and possibly a greater chance of surviving, although there is some debate about the latter in the literature (Long et al. 2015). Large seeds are usually more competitive, can survive for longer in harsh environments, and germinate from greater depths, but large seeds may be targeted by predators and may not be able to disperse as far from the parent plant (Booth et al. 2003). Although large seeds may attract predators, fewer large seeds may satisfy predators than small seeds, may be more difficult to move or pierce the seed coat, and may be more likely to be cached than small seeds (Rodriguez-Perez and Traveset 2007; Tasker et al. 2011; Yang et al. 2012). Small seeds are not as costly to produce; while small seeds may have less probability of survival, a plant's ability to produce more small seeds may overcome the increased mortality when compared to the survival of large seeds. Some studies of the relationship between seed size and longevity found no effect (Leishman and Westoby 1998; Probert et al. 2009), while others found a positive relationship between seed size and longevity (Moles et al. 2003; Hölzel and Otte 2004) or a negative relationship (Bekker et al. 1998).

Based on four main assumptions, it is predicted that seeds of intermediate size in a community will have the highest dispersal and colonization potential. These assumptions are related to the potential trade-off between reproductive investment in seed production quantity and seed size; the unknown potential of seed dispersal and establishment for a species; an assumed positive relationship between seed size and establishment; and the limitations on safe sites for seed establishment. See Eriksson's (2000) detailed explanation of this theoretical model of trade-offs between dispersal and colonization ability of seeds based on seed size.

Both genetics and environment influence seed size and other morphological characteristics within populations and individuals, although some species show morphological consistencies across a wide range of resource availability, such as redroot pigweed (*Amaranthus retroflexus*) (McLachlan et al. 1995; Booth et al. 2003). Several agricultural weeds express stable seed polymorphisms. Size polymorphism in giant ragweed (*Ambrosia trifida*) seeds result in large seeds persisting longer and germinating from greater depths than small seeds (Harrison et al. 2007). Dimorphic seeds of telegraph plant (*Heterotheca grandiflora*) are thought to have contributed to rapid range expansion, where disc achenes disperse greater distances and germinate more readily than ray achenes, affecting both spatial and temporal evolutionary processes (Flint and Palmblad 1978). Seed color polymorphism in ruby dock (*Rumex vesicarius*) (Schatral and Osborne 2006) results in dormancy and germination differences, where light-colored seeds are nondormant and germinate in constant darkness following an after-ripening period, while dark seeds exhibit conditional dormancy, and germination occurs with alternations of light-dark periods. Common lambsquarters (*Chenopodium album*) exhibits a similar seed polymorphism with four seed types: brown-smooth, brown-reticulate, black-smooth, and black reticulate, each differing in germination requirements (Williams and Harper 1965). Therefore, seed polymorphisms may lead to selection and habitat adaptation through differential survival of various morphs across an environmental gradient (Harrison et al. 2007).

11.2 Contributions to the Soil Seedbank

Weed seedbank dynamics, in combination with germination, establishment, growth, reproduction, and dispersal, dictate population changes over time (Davis 2018). The number of seeds a plant produces is dependent on the interaction of various genetic and environmental factors and may be limited by plant traits, such as germination time and plant size at reproduction (Booth et al. 2003). Likewise, seedling recruitment, the number of seeds that germinate and establish, is also dependent on the interaction of various factors, such as the interaction of genetics with resource availability (Muñoz-Rojas et al. 2016), dispersal of seeds to safe sites (Gallandt et al. 1999; Booth et al. 2003), seed retention and shatter timing (Taghizadeh et al. 2012), and crop management operations (Bagavathiannan and Davis 2018). Knowledge of weed recruitment patterns can form an important basis for management actions (Gallandt et al. 1999; Cici and Van Acker 2009). In traditional weed control programs, the seedling recruitment and establishment stages have been the primary target; however, new emerging technologies allow post-reproductive seed control and focus on destroying seed viability through various means, leading to decreased contributions to the soil seedbank.

11.2.1 Seed Dispersal

Plants are sessile organisms with movement only commonly occurring in the pollen or seed phase of the plant life cycle. The ability of seeds to disperse away from the parent plant may increase the likelihood of seed survival (Booth et al. 2003). Dispersal is generally seen as a spatial process, but dormancy is a mechanism that may allow seeds to disperse over time, emerging when conditions are optimal. The full extent of seed dispersal possible for any given population is unknown and difficult to predict (Silvertown and Lovett Doust 1993; Eriksson 2000; Nathan et al. 2002). Some inherent seed traits may allow differences in dispersal distance away from the parent within one cohort. Seed sizes may vary in some species and may directly influence dispersal. Seed dispersal away from the parent reduces competition with the parent and siblings, decreases the chances of inbreeding depression, enhances colonization of new habitats, and may reduce granivore or parasite pressure that may be present in the location of the established parent and siblings (Booth et al. 2003).

Seed mechanisms that promote dispersal may take advantage of seed-derived energy (autochory) or the energy of animals or the environment (allochory). Although wind and water dispersal may not require specialized structures, dispersal distances may be enhanced by the production of specialized appendages, which may be modified by the environment through eco-evolutionary selection processes (Seale and Nakayama 2020). Production of structures that promote dispersal may be costly and may reduce resource allocation to reproduction, representing a trade-off. In a study of seven plant traits (vegetation type, growth form, height, seedbank type, seed mass, life cycle, and taxonomy), the predictive ability of dispersal mechanism was 70% accurate for some floras studied for wind (anemochory), water (hydrochory), ant (myrmecochory), ingestion (endozoochory), attachment (epizoochory), and ballistic (ballochory) dispersal mechanisms (Thompson et al. 2010).

Seeds are an important food source for consumers and provide the base of many agroecological food webs (Pocock et al. 2012), and animals can be effective dispersers (Schupp 1993). Some seeds may need to pass through the digestive tract of an animal to be scarified and promote germination, and animals that can move through multiple, potentially fragmented landscape types provide the greatest dispersal benefits to seeds (Spennemann 2020). The consumption of seeds may lead to destruction in the intestinal tract of animals or may promote dispersal, and one or a few animal species may play a primary role in dispersal of some species, potentially moving seed great distances away from the source population (Rehm et al. 2019). Waterfowl are known to transport weed seeds (Mueller and Van der Valk 2002), including economically damaging agricultural weeds such as Palmer amaranth (*Amaranthus palmeri*), up to 2,900 km (Farmer et al. 2017). Some seeds contain phytochemicals that produce a laxative effect, which may decrease the seed dispersal range but increase the survival of seeds by decreasing gut retention time (Murray et al. 1994).

Humans are also important seed dispersers in agronomic systems (anthropochory), and selection pressure associated with human-mediated dispersal is likely responsible for evolutionary changes in plants (Spengler 2020). Human-aided weed seed dispersal includes vehicles (Pickering and Mount

2010; Taylor et al. 2012; Ansong and Pickering 2013), clothing (Healy 1943) and equipment, horse and donkey fur and dung (Pickering and Mount 2010), global seed trade, and crop management, such as the use of combines in a field or between fields or tillage operations (Wiles and Brodahlb 2004; Benvenuti 2007). Some weed seeds may exhibit complex evolutionary adaptations, such as crop seed mimicry; this may occur through selection associated with mistakes in hand-weeding, or in cases where seed size and morphological traits are selected in harvest or seed cleaning activities (Barrett 1983). Humans, therefore, may be responsible for long distance dispersal of weed seeds at an "unprecedented scale and frequency of dispersal over very broad barriers" (Nathan et al. 2008).

11.2.2 Seed Shatter and Retention

The prolific seed production potential of annual weeds and the formation of persistent seedbanks suggest that management strategies should focus on reducing weed seed inputs rather than solely controlling weed density (Gallandt 2006). Most annual weeds retain a high proportion of the total weed seed produced, which enables its collection and processing at grain crop harvest (Gill and Holmes 1997; Walsh et al. 2018; Schwartz-Lazaro et al. 2021a, 2021b). During a typical grain crop harvest, a high proportion of the total weed seed produced is retained above the harvesting height for many annual weeds, enabling its collection, threshing, and separation from grain to exit the harvester in the chaff fraction, which is evenly redistributed across the field by the harvester's residue-spreading system (Walsh et al. 2013; Broster et al. 2016).

An abundance of case studies surrounding weed seed shattering or retention rates has been made in most major cropping systems (Table 11.1). For example, Palmer amaranth, common waterhemp (*Amaranthus tuberculatus*), giant ragweed (*Ambrosia trifida*), common lambsquarters (*Chenopodium album*), and morningglory species (*Ipomoea* spp.) are the most problematic broadleaf weeds in corn (*Zea mays*) and soybean (*Glycine max*) production systems in the United States (Van Wychen 2015, 2016).

Table 11.1 Seed retention values at crop harvest for major weeds of global cropping systems. Adapted from (Walsh et al. 2018).

Species	Seed retained (%)	Reference
US soybean producing region		
Abutilon theophrasti	50–88	Davis (2008), Schwartz-Lazaro et al. (2021a)
Amaranthus hybridus	98	(Schwartz-Lazaro et al. 2021a)
Amaranthus palmeri	90–99	Green et al. (2016), Schwartz-Lazaro et al. (2017), (Schwartz-Lazaro et al. 2021a)
Amaranthus retroflexus	52–97	Haring et al. (2017), Schwartz-Lazaro et al. (2021a)
Amaranthus tuberculatus	94–99	Schwartz et al. (2016, 2021a)
Amaranthus spp.	<25	Davis (2008)
Ambrosia artemisiifolia	62–66	Haring et al. (2017), Schwartz-Lazaro et al. (2021a)
Ambrosia trifida	60–90	Goplen et al. (2016), Haring et al. (2017), Schwartz-Lazaro et al. (2021a)
Chenopodium album	<25–91	Forcella et al. (1996), Davis (2008), Haring et al. (2017), Schwartz-Lazaro et al. (2021a)
Digitaria sanguinalis	<25–98	Davis (2008), Haring et al. (2017), Schwartz-Lazaro et al. (2021b)
Echinochloa crus-galli	32–41	Schwartz-Lazaro et al. (2017, 2021b)
Ipomoea spp.	75–100	Davis (2008), Schwartz-Lazaro et al. (2021a)
Oryza sativa	13–85	Burgos et al. (2014)
Panicum dichotomiflorum	0	Davis (2008)
Sesbania herbacea	100	Schwartz-Lazaro et al. (2021a)
Setaria faberi	32–60	Davis (2008), Haring et al. (2017), Schwartz-Lazaro et al. (2021b)
Setaria pumila	0	Davis (2008)
Setaria viridis	0	Davis (2008)
Sida spinosa	>50	Davis (2008), Schwartz-Lazaro et al. (2021a)
Speghum halepense	70–100	Schwartz-Lazaro et al. (2021b), Young (2021)
Urochloa platyphylla	50	Schwartz-Lazaro et al. (2021b)
Urochloa texana	68	Schwartz-Lazaro et al. (2021b)

(Continued)

Table 11.1 (Continued)

Species	Seed retained (%)	Reference
Xanthium strumarium	94	Schwartz-Lazaro et al. (2021a)
Great Plains region		
Aegilops cylindrical	76	Soni et al. (2020)
Amaranthus palmeri	91	Schwartz et al. (2021a)
Amaranthus retroflexus	66	Beckie et al. (2017)
Avena fatua	39	Beckie et al. (2017), Burton et al. (2016, 2017), Shirtliffe et al. (2000), Tidemann et al. (2017)
Brassica napus	98	Tidemann et al. (2017)
Bromus tectorum	75	Soni et al. (2020)
Chenopodium album	90	Beckie et al. (2017)
Galium spp.	74–98	Beckie et al. (2017), Burton et al. (2017);
Kochia scoparia	100	Beckie et al. (2017), Burton et al. (2016, 2017), Tidemann et al. (2017)
Lolium perenne ssp. *multiflorum*	58	D. Lyon unpublished data
Malva neglecta	100	Beckie et al. (2017)
Polygonum convolvulus	82	Burton et al. (2017)
Secale cereal	90	Soni et al. (2020)
Setaria viridis	94	Beckie et al. (2017), Burton et al. (2016, 2017)
Sinapis arvensis	70	Forcella et al. (1996), Beckie et al. (2017), Burton et al. (2016, 2017)
Sonchus asper	8	Beckie et al. (2017)
Sorghum halepense	50–100	Youn (2021)å
Pacific Northwest region		
Bromus tectorum	43	San Martin et al. (2021)
Lolium perenne ssp. *multiflorum*	41	San Martin et al. (2021)
Secale cereal	34–70	San Martin et al. (2021)
Vulpia myuros	38	San Martin et al. (2021)
Australia		
Avena spp.	69–84	Walsh and Powles (2014), Widderick et al. (2014)
Bromus spp.	77	Walsh and Powles (2014)
Chloris virgata	47–71	Widderick et al. (2014)
Conyza bonariensis	19–93	Widderick et al. (2014)
Echinochloa colona	9–95	Widderick et al. (2014)
Hibiscus trionum	21–55	Widderick et al. (2014)
Lolium rigidum	85	Walsh and Powles (2014)
Raphanus raphanistrum	99	Walsh and Powles (2014)
Rupistrun rugosum	19–100	Widderick et al. (2014)
Sonchus oleraceus	38–54	Widderick et al. (2014)
Sisymbrium thellungii	100	Widderick et al. (2014)
Europe		
Alopecurus myosuroides	10–50	Walsh et al. (2018)
Avena spp.	4–20	Barroso et al. (2006)
Fallopia convolvulus	44	Bitarafan and Andreasen (2020)
Lolium rigidum	90	Blanco-Moreno et al. (2004)
Secale cereal	90	Blanco-Moreno et al. (2004)
Sinapis arvensis	67	Bitarafan and Andreasen (2020)
Spergula arvensis	45	Bitarafan and Andreasen (2020)
Stellaria media	56	Bitarafan and Andreasen (2020)

A large proportion (>50%) of seeds are retained in these weed species concurrent with the crop harvest window (Davis 2008; Goplen et al. 2016; Schwartz-Lazaro et al. 2016; Schwartz-Lazaro et al. 2017, 2021a; 2021b). Goplen et al. (2016) observed that giant ragweed retained on average 80% of total seeds produced by the time 75% of soybean were already harvested in Minnesota. High seed retention ranging from 95% to 100% in Palmer amaranth and common waterhemp at soybean maturity was reported in a survey across five states in the United States (US) (Schwartz-Lazaro et al. 2016). Davis (2008) reported that ivyleaf morningglory (*Ipomoea hederacea*) retained 75% and 85% of its seed in corn and soybean fields, respectively, in east-central Illinois. Studies examining weed seed retention at crop harvest in Australia and Canada have also reported a high proportion (>70%) of seed retention at crop harvest in broadleaf weeds such as wild radish (*Raphanus raphanistrum*), common sowthistle (*Sonchus oleraceus*), flaxleaf fleabane (*Conyza bonariensis*), African turnip weed (*Sisymbrium thellungii*), cleavers (*Galium* spp.), and wild mustard (*Sinapis arvensis*) (Walsh and Powles 2014; Widderick et al. 2014; Burton et al. 2016). Furthermore, Bitarafan and Andreasen (2020) showed that, on average, 44%, 67%, 45%, and 56% of the seeds were retained on black bindweed (*Fallopia convolvulus*), wild mustard (*Sinapis arvensis*), corn spurry (*Spergula arvensis*), and chickweed (*Stellaria media*), respectively, at spring oat (*Avena sativa*) harvest. Soni et al. (2020) determined seed retention was 75%, 90%, and 76% for downy brome (*Bromus tectorum*), feral rye (*Secale cereale*), and jointed goatgrass (*Aegilops cylindrica*), respectively. However, the level of seed retention in a species is likely to be influenced by agroecological and environmental factors (Shirtliffe et al. 2000; Taghizadeh et al. 2012).

Environmental conditions also have been shown to influence maternal plant phenology and can modify the timing of seed maturity, shatter, and dispersal (Taghizadeh et al. 2012; Maity et al. 2021). Various environmental (i.e., temperature, rainfall, growing season length, etc.) and agronomic (i.e., crop competition, weed population density, etc.) factors influence the inter- and intra-specific variation of seed shatter among weed species (Shirtliffe et al. 2000; Walsh and Powles 2014; Tidemann et al. 2017). Growing degree days (GDD), crop competition, and shattering habits of different species were shown to explain seed shattering differences in wild oat (*Avena fatua*), false cleavers (*Galium spurium*), and volunteer canola (*Brassica napus*) (Tidemann et al. 2017). Accumulated heat units (GDD) were used by Shirtliffe et al. (2000) to predict the seed shatter percentage of wild oat before wheat harvest. Agronomic practices such as row spacing, cropping system, herbicide application timing (harvest aid or postharvest application), and harvest method (swathing vs. direct) have been shown to influence weed-seed shattering patterns (Burton et al. 2016; Beckie et al. 2017; Sun et al. 2021).

Using such seemingly basic ecological data for each economically important weed species, determined by cropping system and region, is invaluable to understanding the life history for each species. Increasing in complexity, each of these case studies is further affected by variables such as crop header height, harvest timing, current cropping system, and rotational cropping system. Any adjustments to these can alter the amount of seed shatter per day, which increases the soil seedbank size.

11.3 Reducing the Soil Seedbank

Weed seeds are not able to survive indefinitely in the seedbank. Although seeds may persist for long periods, seeds are presented with three primary fates: (i) dormancy, (ii) germination, (iii) death. Dormancy is a temporary seed fate, which can be changed under specific scenarios and is typically measured by the absence of germination. Germinating seeds reduce the soil seedbank but also compete with crops in the absence of timely management. Death may occur with inhospitable environmental conditions (soil moisture, pH, temperature), disease and pathogens, granivory, and deep burial. Removing, decreasing, or eliminating the soil seedbank, when unwanted such as in agronomic systems, can be a timely and costly feat. Seedbank reduction applies predominantly to weeds with annual life cycles, but a broader management goal that pertains to all weeds encompasses targeting all life histories of a weed to reduce the population growth rate. However, a loss of 25% to 50% of seeds per year may be sufficient to significantly slow the growth of weed populations (Firbank and Watkinson 1985).

11.3.1 Dormancy

Seed dormancy has been defined by Baskin and Baskin (2004) as a seed that does not have the capacity to germinate in a specified period of time under any combination of normal physical environmental factors that are otherwise favorable for its germination. However, dormancy can only be measured by the absence of germination making definitions complex. It has been stated before that the primary function of seed dormancy is to prevent germination during periods that are unsuitable for germination and establishment. However, dormancy is not a requirement for germination. The imperative function of dormancy is in fact to prevent germination when the conditions for germination are suitable, but the probability of seedling survival and growth is low (Fenner and Thompson 2005).

Dormancy should also be associated with the seed characteristics that determine the conditions required for germination (Vleeshouwers et al. 1995; Thompson 2000; Fenner and Thompson 2005). When dormancy is considered in this way, any environmental cue that alters the conditions required for germination is by definition altering dormancy. Temperature regulates both dormancy and germination and that light regulates germination; however, light can also regulate dormancy (Bewley and Black 1994; Vleeshouwers et al. 1995; Casal and Sanchez 1998; Baskin and Baskin 2004; Fenner and Thompson 2005). Light has been considered both to stimulate germination (Vleeshouwers et al. 1995) and to terminate dormancy (Benech-Arnold et al. 2000; Batlla et al. 2004). Thus, a wide range of factors can alter the dormancy of seeds.

A generally accepted distinction made in dormancy studies is that of primary versus secondary dormancy. Mature dormant seeds are said to have primary dormancy, which can be induced with changes to abscisic acid (ABA) levels during seed maturation (Kucera et al. 2005). Maternal conditions can also influence primary seed dormancy of the progeny; velvetleaf (*Abutilon theophrasti*) grown in competition with a corn crop, rather than under intraspecific competition, had a 30% reduction in the proportion of dormant seeds (Nurse and DiTommaso 2005). Following dispersal, subsequent dormancy release may involve similar factors that are commonly used in a laboratory setting to break dormancy: cold stratification, warm stratification, light, gibberellins and other hormones (Kucera et al. 2005), smoke substances such as butenolide (Krock et al. 2002; Flematti et al. 2004), and compounds such as nitric oxide (Bailly 2004).

Secondary dormancy, on the other hand, is often associated with annual dormancy cycles in the seedbank (Baskin and Baskin 1998, 2004; Fenner and Thompson 2005). Once primary dormancy is lost, due to environmental conditions, secondary dormancy would begin if the conditions required to end dormancy, thus inducing germination, are absent (e.g. light). Secondary dormancy can be cyclical as seasons change until the required germination conditions become available (e.g. through soil disturbances such as tillage). To some extent, the different dormancy types are constrained by the habitats and climates that species possessing them can occur.

Dormancy, as stated above, is considered to be a whole seed characteristic that controls germination. A classification system for seed dormancy has been determined by both morphological and physiological properties of the seed (Nikoleva 1967, 2004). Baskin and Baskin (1998, 2004) have thus proposed a comprehensive classification system that includes five classes of seed dormancy: physiological (PD), morphological (MD), morphophysiological (MPD), physical (PY), and combinational (PY+PD). These five classes can further be divided into levels and types (Finch-Savage and Leubner-Metzger 2006).

Physiological dormancy is the most common form of dormancy and the most prevalent form in temperate climates. PD is the most abundant in weed populations (Finch-Savage and Leubner-Metzger 2006). PD can be divided into three levels: deep, intermediate, and nondeep (Baskin and Baskin 2004; Finch-Savage and Leubner-Metzger 2006).

Morphological dormancy (MD) is evident in seeds with embryos that are underdeveloped but differentiated. These embryos simply need time to grow and germinate as they are not physiologically dormant. Morphophysiological dormancy is also found in seeds with underdeveloped embryos but that additionally have a physiological component to their dormancy (Baskin and Baskin 2004). Therefore, seeds with MPD require assistance in breaking their dormancy; for example, a combination of warm and/or cold stratification. There are eight known levels of MPD (Finch-Savage and Leubner-Metzger 2006).

PY is caused by water-impermeable layers of palisade cells in the seed coat that controls the movement of water (Finch-Savage and Leubner-Metzger 2006). PY dormancy can be broken by chemical or mechanical scarification. Combinational dormancy is found in seeds with simal seed coats as in PY combined with PD (Baskin and Baskin 2004).

This classification system shows a great diversity of morphological and physiological features that have evolved to control dormancy in response to varying environmental conditions (Li and Foley 1997; Baskin and Baskin 2004; Donohue 2005).

11.3.2 Germination

Several interacting environmental factors influence the germination of seeds. The most influential of these factors are temperature, water, and light. The seasonality or predictability of environments may influence the inherent germination cues for species found there. For example, in predictable environments, either without seasonal changes or with strong seasonal changes, seeds will germinate under optimal conditions for survival (Merritt et al. 2007). In situations where environmental conditions fluctuate, there may be considerably more variability in the longevity of the seedbank or the cues for germination within a species, which represents a trade-off in short- and long-term risk (Venable 2007). Seedlings may easily perish following germination; this is perhaps the most vulnerable time for the plant, and the conditions that facilitate germination may not promote establishment. The mortality rate for seedlings is high, but some species may have adaptations that enhance their chances of survival. However, the seed contains limited resources and long-term survival is not possible without establishment.

Temperature is the primary environmental signal regulating germination (Roberts 1988b). Two distinct effects of temperature have been identified, the first acting on dormancy (see above section), and the second determining the rate of progress toward germination in nondormant seeds. The role of temperature is to determine the rate of progress toward completion of germination once a nondormant seed is stimulated to germinate. This can also be complex, as the temperature range permissive for germination widens as dormancy is released and narrows as dormancy is induced (Vegis 1964; Karssen 1982; Benech-Arnold et al. 2000). Thermal emergence cues differ among weed species, and emergence decreases at high and low temperatures around the time of germination (Forcella 1992). Thermal time models for emergence must be tailored to each species, but even then these models may be somewhat simplistic as they only consider weather patterns prior to emergence. For example, WeedCast (Archer et al. 2006) uses a base temperature of 4.4 °C for summer annual weed species, including common cocklebur (*Xanthium strumarium*), wild buckwheat (*Polygonum convolulus*), and giant ragweed (*Ambrosia trifida*), whereas the base temperature reported for yellow foxtail (*Setaria pumila*), barnyardgrass (*Echinochloa crus-galli*), and common sunflower (*Helianthus annuus*) is 10 °C. These data imply that giant ragweed and common cocklebur would emerge earlier in the growing season than common sunflower and barnyardgrass. Thermal time emergence models, however, fail to consider soil surface conditions, such as water availability, water movement, or soil structure.

Water is also essential for seed germination, and its availability is key to gemination timing. To account for the effects of reduced water potential on progress toward germination, Gummerson (1986) proposed the hydrotime concept. Hydrothermal modeling can provide information about potential seed germination response over a wide range of temperature and water potential conditions (Bradford 2002). This method has successfully been used to model seedling emergence of many broadleaf and grass species across many cropping systems (Gummerson 1986; Forcella 1998; Forcella et al. 2000; Roman et al. 2000; Grundy 2003; Grundy et al. 2003; Spokas and Forcella 2009; Gardarin et al. 2010; Masin et al. 2010; Davis et al. 2013; Gonzalez-Andujar et al. 2016). However, these models do not take into account species biological characteristics, such as multiple emergence timings, can be limited by the number of in-season observations, and can be time-consuming to produce over a wide range of temperature and water potential conditions (Clay 2006; Hardegree et al. 2017).

The final germination factor to be discussed is light. In a classic study, Wesson and Wareing (1969) collected soil and screened the soil in the dark and then placed the soil into trays in the greenhouse in either the light or dark. Three experiments took place at varying times of the year. They found that under light conditions, 12 times more dicot and 26 times more monocot seedlings germinated. Several follow-up studies to this yielded similar results (Taylorson 1972; Stroller and Wax 1974). Some species, such as the *Amaranthus* species, require very little light to promote germination (Scopel et al. 1991; Gallagher and Cardina 1998). Seed dimorphisms or polymorphisms may be present in some species (refer to Section 14.1.4). Parental plants produce seeds with different "morphs" or morphologies with different traits, such as optimal conditions for germination. This is a strategy to ensure some germination under various environmental conditions, but may result in a trade-off, as only a small percentage of seeds produced may germinate in optimal conditions. Seeds perceive the quality of light to which they are exposed, and this influences seed dormancy and germination by regulating hormones and signaling pathways. This light response occurs through the perception of red to far-red light ratios by five phytochromes (phyA, phyB, phyC, phyD, and phyE), where red light induces germination and far-red light inhibits germination (Yang et al. 2020). There are two forms of phytochromes: Pr, an inactive form, and Pfr, an active form. Pr absorbs red light and is converted into Pfr as a result, whereas Pfr absorbs far-red light and is converted back to Pr (Li et al. 2011). Pfr is transported to the nucleus ad influences the transcription of genes related to gibberellic acid and abcisic acid (Yang et al. 2020).

Moreover, the natural variability of weed emergence is confounded by difficulties in studying the seedbank. Within a single seedbank quantification study, coefficients of variation for germination typically exceed 50% (Forcella et al. 1992; Webster et al. 2003). Greenhouse studies of seedling emergence have been somewhat more accurate than field studies; however, the artificial nature of the greenhouse challenges study of seed traits such as dormancy or community processes such as self-thinning (Cardina and Sparrow 1996). Even under controlled conditions, some of the mechanisms of seed dormancy are poorly understood. Although germination is not necessary for quantifying the seedbank, it is essential for understanding establishment, highlighting the need for more research into the mechanisms of seed dormancy and other barriers to seed establishment. Improved insights to the natural processes that destroy weed seeds can serve future attempts to manage the seedbank directly.

11.3.3 Death

Factors affecting the rate of seed death in the soil include (i) germination, (ii) seed predators, (iii) decay by microorganisms, and (iv) physiological aging. Technically, germination is the death of seedlings and not seeds, but it is typically treated as a source of seed death. Refer back to section 11.3.2 for more information.

Seed predators, including invertebrates, vertebrates, fungi, and bacteria, consume a significant amount of weed seeds in some agroecosystems. Insects and weeds are part of the same agroecosystem and thus interact in myriad ways (Norris and Kogan 2005). These interactions can be both positive and negative, and understanding these interactions is a key component for the successful implementation of integrated weed management (IWM). Weeds act as important refuges and food for natural enemies, which then feed on insect pests (Barbercheck and Wallace 2021). Weeds are also important sources of pollen and nectar for insects (Barbercheck and Wallace 2021). Flowering weeds provide alternate food sources for pollinators when other resources are scarce. However, in return, many weed species benefit from pollinators through increased seed production (Barbercheck and Wallace 2021). Similarly, pest insects like the soybean looper, *Chrysodeixis includens*, have increased fecundity when adults feed on nectar from weeds like hemp sesbania, *Sesbania herbacea* (Collins and Johnson 1985).

It is not only through their pollen and nectar that weeds provide food for insects. Weed foliage from morningglory (*Ipomoea* spp.), common lambsquarters (*Chenopodium album*), dock (*Rumex* spp.), and Palmer amaranth provides the "green bridge" (bridging hosts) when the main agronomic crop has not emerged (Harding 1976). Once herbicides are applied, pests such as thrips and spider mites in cotton and the fall armyworm, *Spodoptera frugiperda* in soybean leave their weedy hosts and move quickly to their agronomic counterparts. Weed seeds are also an important food source for insects, either pre- or postdispersal. In agroecosystems, carabids (Coleoptera: Carabidae) and ants (Hymenoptera: Formicidae) cause the most postdispersal weed seed predation (Sarabi 2019). So much so that when ants are removed using insecticides, weed seed stands double (Evans and Gleeson 2016).

Further, there are many unknown variables at play in the regulation of seed decay by microorganisms. Susceptibility of weed seeds to microbial pathogens and decay may be species-specific; velvetleaf (*Abutilon theophrasti*) has been cited as having high susceptibility to pathogenic decay, while giant ragweed (*Ambrosia trifida*) has been cited as having low susceptibility (Chee-Sanford et al. 2006). Studies of velvetleaf suggested that diverse bacterial assemblages may be associated with seed decay, while fungal assemblages may be species-specific in association and promotion of decay (Chee-Sanford et al. 2006). Seeds were associated with diverse microbial communities that represented distinct assemblages, depending on weed species, and were not necessarily associated with decay (Chee-Sanford et al. 2006). Additionally, there is evidence that some dormant seeds may be able to respond to pathogenic microorganisms and release biochemical defense enzymes (Fuerst et al. 2018).

Physiological aging of seeds involves various processes such as loss of membrane integrity, organelle deterioration, and damage to DNA (Abdalla and Roberts 1968; Villiers 1973; Roberts 1988b). Under warm conditions with adequate seed moisture (8 to 15%), the aging process of seeds occurs rapidly (Abdalla and Roberts 1968; Villiers and Edgcumbe 1975). Fully imbibed dormant seeds, however, are metabolically active and can repair genetic damage (Villiers and Edgcumbe 1975; Long et al. 2011). Eventually, these seeds will die, possibly due to the accumulation of lethal levels of damaged DNA (Osborne 1980; Villiers 1980). These observations explain why mortality increases toward the soil surface. The conditions near the surface are warmer than deep in the soil profile and periodically can dry out which affects the seeds' imbibed state.

11.4 Managing the Soil Seedbank

Weed control tactics often focus on early emerging weeds, relative to crop growth, with the premise that later emerging weeds do not influence yield loss as much as early emerging weeds. However, early emerging weeds may survive, and late-emerging weeds may still produce some viable seeds. Therefore, shifting the focus from reducing crop yield loss to managing seedbank inputs will need to span multiple years and multiple crop rotations (Gallandt 2006). Decreasing the seedbank density over time will increase the chance of weed management success. Most often, management occurs at the community level and does not target species individually. Management may focus on one or more of the four principles outlined by Swanton and Booth (2004): increasing seedling death after emergence, changing rate or occurrence of seed germination and emergence, destruction of emerging plants, and prevention of seed production. These four principles may be implemented through cultural practices, such as crop rotation, use of mulch or cover crops, solarization of soil; biological practices, such as enhancing habitat for seed predation and seed decay by microbial infection; mechanical practices, such as tillage or harvest weed seed destruction (cultural and mechanical); and chemical control practices (e.g. using effective preemergence or postemergence herbicides) to reduce the number of plants reaching reproductive maturity (Swanton and Booth 2004; Gallandt 2006; Norsworthy et al. 2012; Shaw et al. 2013; Shergill et al. 2020).

Management tactics could be improved through the ability to predict seedling emergence and knowledge of the critical weed-free period for a given crop and weed community. The critical weed-free period is an important concept in crop protection and is defined as the crop growth stage and period of time during which the crop must be kept weed-free in order to prevent yield loss (Zimdahl 1987). The critical weed-free period must be determined for a given crop and associated environmental conditions and cultural management practices (Swanton and Valente 2018). For example, some edaphic factors as well as practices like using narrow row spacing, planting at high seed densities, and planting tall-stature cultivars may decrease the critical weed-free period (Rosset and Gulden 2020). These factors and others, such as phenological models, can be used to predict seedling emergence. Weed emergence prediction models have often been used to understand weed biology and to encourage more effective and timelier weed control. Weed emergence models have been implemented to optimize control of summer annual weeds (Forcella 1993; Masin et al. 2005; Werle et al. 2014; Goplen et al. 2018; Piskackova et al. 2020) but are difficult due to random emergence events.

Soil seedbank management has typically been achieved through short-term victories of targeted aboveground weed management tactics (cultural, mechanical, biological, and chemical). Overreliance on herbicides has led to an increasing problem of herbicide-resistant weeds that in return populate the soil seedbank in agroecosystems. Ecological weed management may help to mitigate this crisis over time.

11.4.1 Cultural Management

Cultural practices may reduce the seedbank directly through physical suppression, decreasing the number of weeds that reach reproductive maturity, or indirectly through various ecological interactions (Schwartz-Lazaro and Copes 2019). Some of these practices include crop rotations, the use of competitive crops, row spacing, and the use of cover crops and intercropping. While many of these management practices, and others, deal with germinated seeds directly, and not the soil seedbank, they are still a critical component of managing the soil seedbank.

11.4.1.1 Crop Rotations

Crop rotation represents a way to provide temporal diversification in crop management practices (Liebman and Dyck 1993), thus allowing more diverse weed control strategies and reducing selection pressures on weed populations. Crop rotation has long been one of the foundational tenets of an effective weed management program. Weeds tend to thrive in crops with similar traits and phenological development timings; sequential planting of crops with diverse traits can reduce weed reproduction and rate of adaptation (Blackshaw et al. 2007). The sequence of crops in rotation can influence the effectiveness of this practice in reducing the seedbank (Zohry and Ouda 2018), and even winter crops, such as the biofuel crop *Brassica carinata*, can lead to reductions in subsequent summer annual weed seedbanks (Tiwari et al. 2021). However, crop rotation in combination with other management tactics may influence the effectiveness of crop rotation in reducing the weed seedbank. While crop rotations generally result in reduced seedbank density, interactions between tillage and crop rotation did not result in reduced weed seedbank density, compared to monoculture wheat, when a no-till system was implemented in four-year rotations of wheat and canola (*Brassica napus*) or lupin (*Lupinus* spp.) (MacLaren et al. 2021). Thus, crop rotation is a factor in weed management that contributes to a manager's ability to implement a diversified system. To lend further support for this premise, data from six 3-year crop rotations with various combinations of alfalfa (*Medicago sativa*), corn, soybean, or wheat, indicated no difference in giant ragweed (*Ambrosia trifida*) seedbank depletion when a zero-weed threshold was implemented (Goplen et al. 2017).

Rotation of crops with green manure or managed fallow systems may be another tool for reducing the weed seedbank. Leguminous green manure crops may build soil fertility, reduce inputs to the weed seedbank, and provide a long-term legacy effect in seedbank reduction (Melander et al. 2020). When leguminous green manure crops were grown for one year, there was a 65% reduction in the weed seedbank after crop termination, and this effect was still detectable three years after termination (Melander et al. 2020). Green manure crops may also contribute to soil phytotoxicity, which may further decrease seedling emergence and establishment but may also negatively impact crop establishment. Reduced corn establishment following a wheat crop under seeded with red clover (*Trifolium pratense*) followed by spring tillage resulted in increased seedbank recruitment of giant foxtail (*Setaria faberi*) when compared to wheat and red clover fall-tilled, or wheat-alone fall- or spring-tilled (Davis and Liebman 2003).

11.4.1.2 Competitive Crops

Many cultural practices, such as crop density, spatial arrangement, planting date, and cultivar affect crop competitive ability with weeds. Most recommendations, however, are based on the assumption that weeds are absent from the field. Many research trials (i.e., fertility, variety, row spacing, etc.) that provide these recommendations are evaluated in weed-free scenarios that are ignoring the effects of weed competition, which can have a direct impact on crop yield.

Ecological theory predicts that as crop density increases, the proportion of available resources, such as light, water, and nutrients, captured by the crop should increase (Mohler 2001a). Research has supported this theory via various measures of weed abundance (e.g., weed biomass) (Lemerle et al. 2004; Mohler 2001a; Singh et al. 2013; Sardana et al. 2017). Mohler (2001a) reviewed 91 cases in the literature across 19 cropping systems and found six cases that failed to show decreasing weediness with increasing crop density. Crop spatial arrangement has been shown to affect weed-crop interactions in some situations. For a given crop population, maintaining desired plant populations through reduced row spacing or increased seeding rate, may increase the ability of a crop to compete with weeds. This effect in the field has been inconsistent. Although most studies have supported that narrow row spacing suppressed weeds, many studies have shown no effect or inconsistent effects. Mohler (2001a) assessed 48 studies, in 16 different cropping systems, examining the effects of narrow rows at constant crop population density on weed density and found weed suppression in 48% of the cases, a neutral effect in 17%, a positive effect in 2%, and a variable response in 33%. In the cases that resulted in fewer or smaller weeds because of narrow rows, the effect is typically seen with increased crop ground cover, leaf area index, dry matter production, and light interception (Teasdale 1995; Mohler 2001a; Drews et al. 2009). Cases in which weed biomass did not decrease with narrow row spacing might be related to soil fertility or rainfall (Kristensen et al. 2008; Oslen et al. 2012).

Narrow row spacing and higher plant density have repeatedly resulted in higher leaf area index (Taylor et al. 1982; Teasdale and Frank 1983). Decreased light penetration with narrow rows, also known as canopy closure, is a result of higher leaf area index (Murphy et al. 1996) and through leaf arrangements (Flenet et al. 1996). Achieving early canopy closure through narrow row planting can potentially reduce weed emergence and decrease reliance on herbicides. For example, light affected Palmer amaranth biomass production, leaf number, partitioning of dry weight to stem tissue, stem elongation, specific leaf area, and photosynthesis (Korres and Norsworthy 2017). The available information to date indicates that under most conditions, narrow row crop spacing will interfere with soil seedbank replenishments of weeds; hence weed population dynamics, a response that might be related to a reduction in light interception at the soil surface (Yelverton and Coble 1991; Bradley 2006; Jha et al. 2008). However, certain weed species, herbicide application timings, weed densities, environmental conditions can potentially reduce the positive effects of row spacing on crop-weed competition (Swanton et al. 2008; Jha and Norsworthy 2009; Sardana et al. 2017). Increasing plant density, or modifying canopy through changed planting geometry, may increase lodging and disease incidence in certain crops and may not be enough to suppress weed populations under commercial production conditions; thus, deteriorating the marketable quality of some crops.

While crop density and spatial arrangement of crops are critical to cultural weed management, planting date also can have substantial effects on weed pressure. Planting date plays a critical role between the crops and the weeds that emerge after crop planting because the crop and weed growth rate both depend on environmental conditions, such as temperature and soil moisture. In the absence of weeds, an earlier crop planting date is ideal because it can allow for successful germination, albeit potentially slower, with more resource availability which leads to a higher yield potential (Gunsolus 1990; Mohler 2001a; Williams 2006). However, weeds may emerge at the same time as the crop, regardless of planting date, and may overtake a crop that is planted earlier due to the slow emergence of the crop and the competitiveness of the weed. This competitive balance depends not only on the growth rate of the individual crop and weed species but also on the crop population as a whole to take up the valuable resources needed to succeed in achieving maximum yields. A common approach to quantify this is through critical weed-free periods. Critical weed-free periods are determined by characterizing functional relationships between two separately measured competition components on crop yield:

1) Yield as a function of the duration of weed interference
2) Yield as a function of the duration of the weed-free period

Basically, focus is on the time during which weeds must be controlled, and not compete with a crop, to prevent yield reductions. Site-specific factors, such as environmental conditions (e.g. fertility, soil moisture, temperature), weed species (e.g. composition and density), and crop production practices (e.g. row spacing, crop variety through enhanced germination, growth rate, plant vigor, and ability to withstand a wider environmental gradient, planting density), can influence the outcome of crop and weed competition that affects yield (Rajcan and Swanton 2001). Additionally, a single event in which weeds are controlled, such as the use of a nonresidual postemergence herbicide, may or may not be enough to constitute a sufficient weed-free period (Weaver 1984; Van Acker et al. 1993; Knezevic et al. 2002; Evans et al. 2003).

11.4.1.3 Cover Crops

Cover crops are an important tool in managing the soil seedbank because they occupy a niche otherwise available to weeds, thus depriving weeds through competition for resources. Inclusion of cover crops in crop rotations provides many ecosystem services (Meisinger et al. 1991; Decker et al. 1994; Kuo et al. 1996; Teasdale 1996):

- Nutrient cycling, adding carbon to build soil organic matter and the associated beneficial soil structural effects
- Preventing loss of sediment and nutrients,
- Altering soil radiation and water exchange with residue retention
- Suppressing weed and other pest populations.

Once terminated, cover crop mulches suppress weeds physically by impeding emergence or attenuating environmental cues that break weed seed dormancy (Teasdale and Mohler 1993, 2000), by releasing phytotoxic compounds (Creamer et al. 1996; Rice et al. 2012; Teasdale et al. 2012), and/or bio-geochemically by immobilizing soil nitrogen (another weed germination cue) in the case of high carbon/nitrogen (C:N) ratio grass mulches (Wells et al. 2013).

Both the amount and composition of cover crop residue are important to maximize weed suppression (Teasdale and Mohler 1993). Most cover crops such as cereal rye and hairy vetch produce biomass levels (without supplemental fertilization) of approximately 5,000 to 7,000 kg ha^{-1} at maturity, which can lower weed germination, emergence, growth rates, and biomass accumulation (Teasdale and Mohler 1993). Teasdale and Mohler (1993) showed that intermediate residue levels can limit light and temperature fluctuation cues that weed seeds often require for germination. The quality of cover crop residue (nitrogen content and C:N ratio) at termination also plays a vital role in physically suppressing weeds. The high biomass levels and residue quality (high C:N ratio, lignin, and hemicellulose content) of cereal rye result in very low inorganic nitrogen released to the soil during residue decomposition and could contribute significantly to weed suppression by rye residue (Mirsky et al. 2013). Maximizing cover crop residue quantity (biomass) at termination is critical for physically suppressing germinating weeds with cover crop residues. Mixtures of grasses and legumes can lead to greater biomass production and nitrogen accumulation due to the complementary ways in which they acquire resources relative to monocultures (Clark et al. 1997; Teasdale and Abdul-Baki 1998; Sainju et al. 2005).

Winter annual cover crops are terminated in spring using herbicides, such as glyphosate, or mechanically using a roller-crimper; then cash crops are no-till planted into the cover crop mulch. Integration of cover crops into field crop production systems has been recommended as a best management practice in herbicide resistance management programs to manage resistant weeds (Norsworthy et al. 2012; Pittman et al. 2020). Wiggins et al. (2017) showed that including cereal rye, cereal rye plus hairy vetch, and wheat with preemergence residual herbicides significantly improved Palmer amaranth control compared to no cover crop usage. The cover crop strategy relies on a combination of effective cover cropping practices and innovations in farm implements, such as high-residue planters, roller-crimpers, and high-residue cultivators (Mirsky et al. 2013).

11.4.1.4 Intercropping

While crop rotation is used to provide temporal diversification in management, intercropping is used to provide spatial diversification in management (Liebman and Dyck 1993). Intercropping is defined as the practice of growing two or more crops simultaneously, even if the sole purpose of additional crops is weed suppression (Vandermeer 1989; Blackshaw et al. 2007). When a cash crop is grown with a noncash crop for weed suppression, this may be referred to as companion plant intercropping (Verret et al. 2017). This practice can improve the consistency of yield over monocultures, since resources are captured more efficiently by crops that can partition niche space (Raseduzzaman and Jensen 2017). Interseeded crops or companion plants compete with weeds for resources and may also compete with the crop, reducing crop yield, if not managed properly. This includes determining the optimal planting and termination time of the interseeded crops to minimize competition with the cash crop. However, in a meta-analysis of interseeded legumes in cereal crops, it was found that weed biomass was decreased by 56% when compared to a nonweeded control and 42% when compared to a weeded control, without decreasing the yield of the cash crop; there was higher cash crop yield in 52% of unweeded control treatments and 36% of weeded control treatments in the reviewed experiments (Verret et al. 2017). Although most measures of weed suppression by interseeded crops were quantified by using weed density or biomass in the field, some studies suggest a concurrent reduction in weed seed density (9 of 12 reviewed studies), as would be expected (Liebman and Dyck 1993). While there is growing global interest in intercropping, the practice is currently more common in Africa, Asia, and Latin America (Blackshaw et al. 2007). If adopted globally, intercropping could be part of a multifaceted solution to support the sustainable intensification of agriculture to feed a growing population (Martin-Guay et al. 2018).

11.4.2 Mechanical Management

Soil cultivation alters the soil's physical properties, often exposing weed seeds to conditions favorable for germination, and thus seedbank reductions (Ball and Miller 1990; Ball 1992). Recently tilled soils tend to be warmer and have high diurnal temperature fluctuations, higher nitrate concentrations, and better aeration, and they prompt seed germination via exposure to light (Sauer and Struik 1964; Dou et al. 1995; Gue´rif et al. 2001). Weed seedbank management is not a primary goal of tillage, but tillage can affect the soil seedbank by burying emerged weeds or uncovering emerging seedlings (Haring and Flessner 2018; Schwartz-Lazaro and Copes 2019). Further, tillage can modify the seedbank composition (Ball 1992) by burying some seeds and exposing others, such that they exit the seedbank through early germination. Tillage timing and depth affect seedbanks depending on weed traits such as germination phenology and seed size, respectively. For example, spring tillage promotes germination of weeds with a summer annual life cycle, whereas fall tillage boosts winter annual weeds (Smith 2006). Shallow tillage can promote germination of recently shattered seeds, but extremely small seeds may be incorporated into the soil seedbank by such tillage (Pekrun and Claupein 2006). While frequent tillage may cause decreases in the seedbank, this practice can also select for a persistent seedbank through the selection for individuals with a high degree of dormancy (Ghersa and Martínez-Ghersa 2000).

Stale seedbed management is an example of tillage explicitly for seedbank management, whereby tillage is used to promote weed germination before another management action stops emergence ahead of crop planting. This practice takes advantage of a shallow seedbank, fast weed germination, and delayed crop planting. The false seedbed method works similarly as the stale seedbed method. In this technique, preparation of the seedbed is followed by one or more shallow tillage, to limit the number of additional seeds brought within emergence depth in the soil, events at one-week intervals prior to planting the cash crop (Mohler 2001b). Germinable seeds at the soil surface will be reduced, directly affecting the weed density in the crop. The false seedbed method is primarily used in organic farming (Wookey 1985; Stopes and Millinton 1991), but Johnson and Mullinix (1995, 1998) found that two shallow passes with a rotary tiller yielded similar results to that of two glyphosate applications in reducing weed populations. There are similar limitations in both stale seedbed and false seedbed methods, including (i) soil conditions must be conducive (i.e. warm and moist) for weed seed germination; (ii) bare and loose soil is prone to an increase in erosion, which goes against best management practices; (iii) there is a potential for yield loss if planting is delayed too much; and (iv) shifts in weed species may occur due to physiological dormancy. Additionally, postharvest tillage is a method that can reduce the soil seedbank by killing weeds before the seeds shatter and has shown to be nearly as effective as postharvest herbicide applications at reducing weed seed production (Kegode et al. 2015).

Harvest weed seed control (HWSC) captures and processes the unshattered seed retained by weeds at harvest time to minimize the number of viable seeds dispersed to the soil. HWSC tactics were first developed and adopted in Australia and include practices such as narrow windrow burning, bale direct, chaff tramlining or lining, and seed impact mills (Walsh and Powles 2007; Walsh et al. 2012, 2018; Schwartz-Lazaro et al. 2017; Norsworthy et al. 2020; Shergill et al. 2020). HWSC has the potential to prevent large proportions of viable seeds that remain on the weed plants at the time of harvest from entering the seedbank, substantially reducing the total seed rain of weeds, including herbicide-resistant weeds. Therefore, HWSC can negatively impact weed population dynamics by limiting seed addition and preventing the buildup of resistant subpopulations in the soil seedbank over time. Reducing viable weed seed additions to the soil seedbank by greater than 40% can reduce weed population growth and increase net returns to weed management (Liebman and Davis 2009). Different HWSC systems aid in effectively managing 60% to 99% seed of various weed species in Australian production systems (Walsh et al. 2013), but the efficacy of these systems depends upon the proportion of weed seed retained on the weed plant at crop harvest, because only those seeds will be captured and removed or processed by the harvest machinery (Gill and Holmes 1997; Schwartz et al. 2021a, 2021b; Walsh et al. 2018).

11.4.3 Biological Management

There are generally four categories of biological management: macroorganisms (seed predators), microorganisms (fungal or bacterial pathogens, viruses), chemical signaling molecules (pheromones), or natural plant- or animal-derived compounds (Cordeau et al. 2016). A diverse assemblage of seed predators, or natural enemies of pests in general, in terms of richness and functional diversity, may have a high efficacy of pest control (Evans 2016), but managing agricultural habitats for biodiversity may pose challenges in some cropping systems (Mortensen et al. 2012). The goal of biological management is to bring pest populations to equilibrium through the use of ecological interactions; usually, eradication is not the goal, nor is it feasible (Cordeau et al. 2016). Biological management of the seedbank may allow diversification of cultural, mechanical, and chemical practices, contributing to an integrated weed management framework (Cordeau et al. 2016).

Various groups of macroorganisms may serve as effective pre- or postdispersal seed predators. Predispersal seed predators are often specialists, such as insects of the orders Diptera, Lepidoptera, Coleoptera, and Hymenoptera (Crawley 2000). Postdispersal seed predators are often generalist species: ants, carabid beetles, and crickets (Orthoptera: Grylloidea) (Cutler et al. 2016; Crawley 2000), birds (Holmes and Froud-Williams 2005), and small mammals (Janzen 1971; Tschumi et al. 2018). The use of seed predators as a management tool may be useful, in that seed predators diversify the types of selection pressure imposed on the seedbank. Predators may preferentially target seeds in a clumped distribution, large seeds, or seeds with soft seed coats (Ghersa and Martínez-Ghersa 2000).

While management of the seedbank with macroorganism seed predators may be possible, there are challenges in using seed predation as a consistent and reliable integrated weed management tactic. Seed predator populations may fluctuate with climate and changes to the surrounding habitat. There may also be incompatibility issues of predator habitat management with intensive crop production system management. For example, tillage disturbance may decrease predator populations and decrease the number, diversity, and/or activity of seed consumption (Landis et al. 2005; Blubaugh and Kaplan 2015). However, efforts are underway to integrate seed predators into weed seedbank management in some production systems. Management of European farmland may include environmental stewardship methodologies such as management of the habitat to promote carabid beetle seed predators. An evaluation of the ability of carabid beetle seedbank regulation at the national level in the UK suggested that carabids provide seedbank regulatory services across various crops and management practices and supported the role of carabids in integrated weed management (Bohan et al. 2011). Another challenge may be an issue of predation timing; while animal grazers may be effective in weed management with appropriately timed grazing, grazers may become seed dispersers if seeds are present; goats and sheep would need to be contained for 96–120 hours after grazing, depending on the weed species, to prevent redistribution of weed seeds (Oveisi et al. 2021).

Microorganisms (fungi, bacteria, and viruses) can be used to target the weed seedbank. Some microorganisms naturally found in the soil may infect weed seeds and cause decay (Kremer and Kennedy 1996; Davis et al. 2006). The possibility of targeting the seedbank using "weed-suppressive soil" has been proposed (Kremer 1993; Kennedy and Kremer 1996; Kennedy 1999; Chee-Sanford et al. 2006), and improvements in soil quality may positively impact soil microbial biodiversity and lead to reductions in the soil seedbank (Gallandt et al. 1999). Isolates of seed-targeting microorganisms can be used as bioherbicides. For example, isolates of *Fusarium solani* and *Epcoccum purpurascens* have been used to inhibit germination of *Striga* and *Echinochloa* spp., respectively (Ahmed et al. 2001; Motlagh 2011; Cordeau et al. 2016). Whether through soil inoculation with seed-pathogenic microorganisms or through management of soils to enhance seed-targeting microorganisms, more work is needed to understand the role of these organisms and how to optimize their use in integrated weed management (Müller-Stöver et al. 2016).

11.4.4 Chemical Management

Herbicide-based weed management systems have historically had a greater probability of reducing weed density and biomass within the crop, and therefore, reducing seed rain and contributions to the weed seedbank, as compared to mechanical weed control (Gallandt 2006). However, the increased occurrence of herbicide-resistant weeds has reduced the reliance on herbicides alone and prompted a shift toward other control tactics, such as mechanical and cultural practices (Bagavathiannan and Davis 2018; Shergill et al. 2020).

Chemical management options that target ungerminated seeds are limited and are associated with risks of off-target impacts on soil microorganisms or arthropods. These consist of fumigants, such as methyl bromide and methyl iodide, which may provide good to excellent control of weed seeds, or metam sodium and metam potassium, which may provide some control of weed seeds, as well as nematodes and diseases (Sande et al. 2011). Fumigants can kill ungerminated seeds by interfering with seed respiration. Methyl bromide was historically the most widely used fumigant in agriculture. However, restrictions on the use of methyl bromide were set by the 1992 Parties of the Montreal Protocol to preserve the stratospheric ozone layer; methyl bromide was classified as a Class I controlled substance with high ozone depletion potential and the use of methyl bromide has been highly restricted to critical usage since 2005 in developed countries and 2015 in developing countries (USDA 2006). Alternatives to methyl bromide with relatively low environmental impact have been sought since this phase-out plan began, but available fumigant alternatives have similar risks (Sande et al. 2011). Although the preplant fumigants, chloropicrin plus 1,3-dichloropropene, may provide partial control and reduced germination of some seeds in some specialty crop production, the loss of control provided by methyl bromide has led to a transition in management practices, such as a greater dependence on mulches, herbicides, and hand weeding (Fennimore and Boyd 2018).

Traditional herbicide programs utilize preemergence herbicides, which act on the germinating seed once the seed coat cracks. Much of the current literature focuses not only on the use of preemergence herbicides, but those with an added residual activity (lay-by), which allows for an extended period of efficacy to control any weed seeds that have germinated prior to crop emergence or to control additional flushes of germinating weed seed. Preharvest herbicides or harvest aids are herbicides that are applied at a time of crop maturity to desiccate weed foliage and to reduce grain moisture at harvest, but many weeds already have mature seed present (Griffin et al. 2010; Hill et al. 2016). These late-season options only control weeds seeds that are still retained on the mother plant. If the seeds have already shattered prior to the use of the harvest aid, then they have entered the seedbank whether they are viable or not. For example, a single glyphosate application to soybean later in the season suppressed seed production of pitted morningglory (*Ipomoea lacunosa*), barnyardgrass (*Echinochloa crus-galli*), Palmer amaranth (*Amaranthus palmeri*), prickly sida (*Sida spinosa*), and sicklepod (*Senna obtusifolia*) by 83–99% (Walker and Oliver 2008). Pitted morningglory seed production was also reduced by a single glufosinate application 14 days before soybean harvest (Bennett and Shaw 2000). Further, eight grass and eleven broadleaf species across the major US soybean cropping regions were evaluated for seed shatter, and at the time of soybean physiological maturity, the grass species had shattered up to 40% of their seeds, whereas the broadleaf weeds had shattered less than 10% (Schwartz-Lazaro et al. 2021a, 2021b).

Late-season or postharvest herbicide treatments are often targeted at the early bloom stage or initiation of seed set to reduce/prevent late-season weed seed rain and seedbank increase (Bennett and Shaw 2000; Mickelson et al. 2004; Jha and Norsworthy 2012). Studies have shown that the residual population may be sufficient to persist for several years after the implementation of weed management programs that are effective in controlling late-season weeds (Schweizer and Zimdahl 1984; Jones and Medd 2005). Reported in other weed species such as Palmer amaranth, glufosinate (0.820 kg ha^{-1}), 2, 4-D (1.06 kg ha^{-1}), or dicamba (0.280 kg ha^{-1}) applied at the inflorescence initiation stage reduced seed production of treated plants by 95%, and seed viability by 39–51% (Jha and Norsworthy 2012). A single late-season application of glyphosate, 2, 4-D, dicamba, or glufosinate in common lambsquarters, redroot pigweed, velvetleaf (*Abutilon theophrasti*), and sicklepod reduced seed production up to 99% (Fawcett and Slife 1978; Biniak and Aldrich 1986; Taylor and Oliver 1997). Additionally, paraquat alone or tank-mixed with several residual herbicides has shown over 90% weed control in several cropping systems and weed species (Crow et al. 2015). In wheat-based crop rotations in the US Great Plains, kochia (*Bassia scoparia*) populations and seed production were reduced 100% with paraquat and metribuzin (Kumar and Jha 2015). Further, in the absence of a postharvest herbicide, uncontrolled kochia plants at a density of 8–10 plants m^{-2} contributed more than 100,000 seeds m^{-2} (Kumar and Jha 2015).

11.5 Seedbank Response to Best Management Practices

Each year, the soil seedbank is the primary source of new annual weed infestations. The composition and diversity of the weed seedbank, and the emerged weeds, determine weed competitiveness with crops. This composition of weeds possesses a set of traits that determines if it can colonize a site and survive the local environmental conditions and biotic interactions (Booth et al. 2003; MacLaren et al. 2020). Dispersal and survivor barriers act as "filters" by removing those weed species that lack the required traits to successfully colonize (Kraft et al. 2015). Strong filters can remove more filters, and soft filters might allow more species to survive. These filters are comparable to Liebman and Gallandt's (1997) "many little hammers" in weed ecology, and the strength of the filter is similar to that of selection pressure. Filters in an agroecological system can consist of disturbances (e.g. chemical or mechanical control), physical conditions (e.g. environmental variables), resource availability, and biotic interactions (e.g. competition, predation, facilitation). For example, the widespread use of herbicides, tillage, and few cultivated crops (strong filters) has led to weed communities being dominated by only a few species (i.e. Palmer amaranth, Italian ryegrass) that are well adapted to these filters.

Utilizing an integrated weed management system or best management practices can promote weed diversity, which is a (not *the*) goal, by decreasing the use of strong filters and by combining multiple weak filters in a system (Neve et al. 2009; Storkey and Neve 2018; Mahaut et al. 2019). For example, varying planting dates and weed management timings between years can be particularly effective (Weisberger et al. 2019). This is particularly true for weeds that have a persistent seedbank that has been adapted to take advantage of favorable growing conditions. A step toward reducing filter strength would be the uptake of precision herbicide applications or precision mechanical weeding. Herbicides or mechanical control could be applied to areas with problematic weed patches while allowing a diversity of weeds to remain in other parts of the field (Metcalfe et al. 2018). The areas where these "strong filters" are required would be expected to decrease as other best management practices were incorporated.

Preventing seeds from contributing to the seedbank is critical for long-term weed management (Norsworthy et al. 2012; Bagavathiannan and Davis 2018; Gage and Schwartz-Lazaro 2019). Targeting weed seed rain in addition to controlling weeds at the seedling stage could aid in further optimization of best management practices that include integrated weed management systems (Davis 2006). Weeds that escape early-season control measures are highly likely to harbor traits for herbicide resistance and, if allowed to set seed, would contribute to the development of herbicide resistance in subsequent seasons via increases to the soil seedbank (Jasieniuk et al. 1996; Gaines et al. 2020). More diverse agroecosystems are typically more resilient, increase crop yield, and are less dependent on strong filters to support the crop (Tittonell and Giller 2013; Altieri et al. 2017; Bowles et al. 2020).

The potential advantages of ecological weed management outlined above raise the question of why existing practices are not more widespread and what can be done to encourage the introduction and adoption of new ecological practices. The research and extension community have long preached that the overreliance on any one weed management tactic will inevitably fail. However, as Bagavathiannan and Davis (2018) point out, proof of concept for several ecological practices offering improvements in the sustainability of weed management has been present in the weed science literature for decades but has not been widely adopted in agriculture. The major barriers to this lack of adoption are still being understood. Kleijn et al. (2019) further emphasize the gap between an abundance of theory on agricultural ecology in the scientific literature and a lack of evidence and recommendations for specific practices that farmers can expect to implement, their effects and the timescales over which these effects will occur. This indicates the need for more on-farm research that directly involves farmers. Although this is not always straightforward, many farmers prefer to adopt new ideas only after having seen them successfully implemented in their own environments and farming systems. Working side by side, farmers and researchers (weed scientists, agronomists, psychologists, economists, and social scientists) can begin to move the needle of adoption in a forward motion (Doohan et al. 2010; Jordan et al. 2016; Liebman et al. 2016; Moss 2019). Including farmers as well as the general public in developing new approaches to agriculture could also help to drive democratic demand for the political and economic shifts required to facilitate sustainable food and farming systems.

11.6 Conclusions

Recently, there has been a strong push to move from the overreliance on herbicides as the sole weed management tool to more ecologically based technologies or approaches (Bagavathiannan and Davis 2018; Gage and Schwartz-Lazaro 2019). However, oversimplified management of weed populations has left the soil seedbank as an afterthought with short-term solutions despite the evidence and need for change. Ecology is once again becoming the core of weed management decisions and will help ensure the sustainability of future weed management programs. To optimize the success of an integrated weed management program, knowledge of seedbank persistence and an understanding of the factors that stimulate or suppress germination is critical (Long et al. 2015). Utilizing a truly diverse approach to reduce the soil seedbank is vital to the success of modern farming.

References

Abdalla, F.H. and Roberts, E.H. (1968). Effects of temperature, moisture, and oxygen on the induction of chromosome damage in seeds of barley, broad beans, and peas during storage. *Annals of Botany, New Series* 32: 119–136.

Ahmed, N., Sugimoto, Y., Babiker, A. et al. (2001). Effects of Fusarium solani isolates and metabolites on Striga germination. *Weed Science* 49: 354–358. doi:10.1614/0043-1745.

Altieri, M.A., Nicholls, C.I., and Montalba, R. (2017). Technological approaches to sustainable agriculture at a crossroads: an agroecological perspective. *Sustainability* 9: 349–362.

Ansong, M. and Pickering, C. (2013). Are weeds hitchhiking a ride on your car? A systematic review of seed dispersal on cars. *PLoS One* 8: e80275.

Archer, D.W., Forcella, F., Korth, A. et al. (2006). *WeedCast.* Available at http://www.ars.usda.gov/services/software/download.htm?softwareid5112 (accessed 22 June 2021).

Bagavathiannan, M.V. and Davis, A.S. (2018). An ecological perspective on managing weeds during the great selection for herbicide resistance. *Pest Management Science* 74: 2277–2286.

Bailly, C. (2004). Active oxygen species and antioxidants in seed biology. *Seed Science Research* 14: 93–107.

Bakker, J.P., Bos, A.F., Hoogveld, J., and Muller, H.J. (1991). The role of the seed bank in restoration management of semi-natural grasslands. In: *Terrestrial and Aquatic Ecosystems: Perturbation and Recovery* (ed. O. Ravera), 449–455. Ellis Horwood.

Bakker, J.P., Poschlod, P., Strykstra, R.J. et al. (1996). Seed banks and seed dispersal: important topics in restoration ecology. *Acta Botanica Neerlandica* 45: 461–490.

Ball, D.A. (1992). Weed seedbank response to tillage, herbicides, and crop rotation sequence. *Weed Science* 40: 654–659.

Ball, D.A. and Miller, S.D. (1990). Weed seed population response to tillage and herbicide use in three irrigated cropping sequences. *Weed Science* 38: 511–517.

Ballare, C.L., Scopel, L., Ghersa, C.M., and Sanchez, R.A. (1988). The fate of Datura ferox seeds in the soil as affected by cultivation, depth of burial and degree of maturity. *Annals of Applied Biology* 112: 337–345.

Barbercheck, M.E. and Wallace, J. (2021). Weed–insect interactions in annual cropping systems. *Annuals of the Entomology Society of America* 114: 276–291.

Barrett, S.H. (1983). Crop mimicry in weeds. *Economic Botany* 37: 255–282.

Barroso, J., Navarrete, L., Sánchez del Arco, M.J. et al. (2006). Dispersal of *Avena fatua* and *Avena sterilis* patches by natural dissemination, soil tillage and combine harvesters. *Weed Res* 46: 118–128.

Baskin, C.C. and Baskin, J.M. (1998). *Seeds – Ecology, Biogeography, and Evolution of Dormancy and Germination.* San Diego, CA, USA: Academic Press.

Baskin, C.C. and Baskin, J.M. (2001). *Seeds: Ecology, Biogeography, and Evolution of Dormancy and Germination*, 2e. San Diego, CA: Academic Press.

Baskin, J.M. and Baskin, C.C. (2004). A classification system for seed dormancy. *Seed Science Research* 14: 1–16.

Batlla, D., Kruk, B.C., and Benech-Arnold, R.L. (2004). Modelling changes in dormancy in weed soil seed banks: implications for the prediction of weed emergence. In: *Handbook of Seed Physiology: Applications to Agriculture*, (ed. R.L. Benech-Arnold and R.A. Sanchez), 245–270. New

York, NY, USA: Food Product Press and the Haworth Reference Press.

Beckie, H., Blackshaw, R., Harker, K.N., and Tidemann, B. (2017). Weed seed shatter in spring wheat in Alberta. *Canadian Journal of Plant Science* 98: 107–114.

Bekker, R.M., Bakker, J.P., Grandin, U. et al. (1998). Seed size, shape and vertical distribution in the soil: indicators of seed longevity. *Functional Ecology* 12: 834–842.

Bekker, R.M., Lammerts, E.J., Schutter, A., and Grootjans, A.P. (1999). Vegetation development in dune slacks: the role of persistent seed banks. *Journal of Vegetation Science* 10: 745–754.

Benech-Arnold, R.L., Sanchez, R.A., Forcella, F. et al. (2000). Environmental control of dormancy in weed seed banks in soil. *Field Crops Research* 67: 105–122.

Bennett, A.C. and Shaw, D.R. (2000). Effect of preharvest desiccants on weed seed production and viability. *Weed Technology* 14: 530–538.

Benvenuti, S. (2007). Weed seed movement and dispersal strategies in the agricultural environment. *Weed Biology and Management* 7: 141–157. DOI: 10.1111/j.1445-6664.2007.00249.x.

Bewley, J.D. and Black, M. (1994). *Seeds – Physiology of Development and Germination*, 2e. New York, NY, USA: Plenum Press.

Biniak, B.M. and Aldrich, R.J. (1986). Reducing velvetleaf (*Abutilon theophrasti*) and giant foxtail (*Setaria faberi*) seed production with stimulated-roller herbicide applications. *Weed Science* 34: 256–259.

Bitarafan, Z. and Andreasen, C. (2020). Seed retention of ten common weed species at oat harvest reveals the potential for harvest weed seed control. *Weed Research* 60: 343–352.

Blackshaw, R.E., Anderson, R.L., and Lemerle, D. (2007). Cultural weed management. In: *Non-chemical Weed Management: Principles, Concepts and Technology* (ed. M.K. Upadhyaya and R.E. Blackshaw), 35–47. CABI.

Blanco-Moreno, J.M., Chamorro, L., Masalles, R.M. et al. (2004). Spatial distribution of *Lolium rigidum* seedlings following seed dispersal by combine harvesters. *Weed Research* 44: 375–387. DOI: 10.1111/j.1365-3180.2004.00412.x.

Blubaugh, C.K. and Kaplan, I. (2015). Tillage compromises weed seed predator activity across developmental stages. *Biological Control* 81: 76–82.

Bohan, D.A., Boursault, A., Brooks, D.R., and Petit, S. (2011). National-scale regulation of the weed seedbank by carabid predators. *Journal of Applied Ecology* 48: 888–898.

Booth, B.D., Murphy, S.D., and Swanton, C.J. (2003). *From Seed to Seedling, in Weed Ecology in Natural and Agricultural Systems*, 81–99. CABI.

Bowles, T.M., Mooshammer, M., Socolar, Y. et al. (2020). Long-term evidence shows that crop-rotation diversification increases agricultural resilience to adverse growing conditions in North America. *One Earth* 2: 1–10.

Bradford, K. (2002). Applications of hydrothermal time to quantifying and modeling seed germination and dormancy. *Weed Science* 50: 248–260.

Bradley, K.W. (2006). A review of the effects of row spacing on weed management in corn and soybean. *Crop Management* 5. https://doi.org/10.1094/CM-2006-0227-02-RV.

Broster, J.C., Walsh, M.J., and Chambers, A.J. (2016). Harvest weed seed control: the influence of harvester set up and speed on efficacy in South-Eastern Australia wheat crops. *20th Australasian Weeds Conference*. Perth, Western Australia, 38–41.

Buhler, D.D. and Hartzler, R.G. (2001). Emergence and persistence of seed of velvetleaf, common waterhemp, woolly cupgrass, and giant foxtail. *Weed Science* 49: 230–235.

Burgos, N.R., Singh, V., Tseng, T.M. et al. (2014). The impact of herbicide-resistant rice technology on phenotypic diversity and population structure of United States weedy rice. *Plant Physiology* 166: 1208–1220. https://doi.org/10.1104/pp.114.242719.

Burnside, O.C., Wilson, R.G., Weisbert, S., and Hubbard, K.G. (1996). Seed longevity of 41 weed species buried 17 years in eastern and western Nebraska. *Weed Science* 44: 74–86.

Burton, N.R., Beckie, H.J., Willenborg, C.J. et al. (2016). Evaluating seed shatter of economically important weed species. *Weed Science* 64: 673–682.

Burton, N.R., Beckie, H.J., Willenborg, C.J. et al. (2017). Seed shatter of six economically important weed species in producer fields in Saskatchewan. *Canadian Journal of Plant Science* 97: 266–276.

Cardina, J. and Sparrow, D.H. (1996). A comparison of methods to predict weed seedling populations from the soil seedbank. *Weed Science* 44: 46–51.

Casal, J.J. and Sanchez, R.A. (1998). Phytochromes and seed germination. *Seed Science Research* 8: 317–329.

Cavers, P.B. (1995). Seed banks: memory in soil. *Canadian Journal of Soil Science* 75: 11–13.

Chambers, J.C. (1995). Disturbance, life history strategies, and seed fates in alpine herb field communities. *American Journal of Botany* 82: 421–433.

Chauhan, B.S., Gill, G., and Preston, C. (2006a). Influence of tillage systems on vertical distribution, seedling recruitment and persistence of rigid ryegrass (*Lolium rigidum*) seed bank. *Weed Science* 54: 669–676.

Chauhan, B.S., Gill, G.S., and Preston, C. (2006b). Tillage system effects on weed ecology, herbicide activity and persistence: a review. *Australian Journal of Experimental Agriculture* 46: 1557–1570.

Chee-Sanford, J.C., Williams, M.M., Davis, A.S., and Sims, G.K. (2006). Do microorganisms influence seed-bank dynamics? *Weed Science* 54: 575–587.

Chippindale, H.G. and Milton, W.E. (1934). On the viable seeds present in the soil beneath pastures. *The Journal of Ecology* 1: 508–531.

Cici, S.Z. and Van Acker, R.C. (2009). A review of the recruitment biology of winter annual weeds in Canada. *Canadian Journal of Plant Science* 89: 575–589.

Clark, A.J., Decker, A.M., Meisinger, J.J., and McIntosh, M.S. (1997). Kill date of vetch, rue, and a vetch rye mixture: I. Cover crop and corn nitrogen. *Agronomy Journal* 89: 427–434.

Clay, S.A. (2006). Developing weed management zones for site specific farming application. *Proceedings of the 8th International Conference of Precision Agriculture*. A to Z Session, Madison, WI: ASA/CSSA/SSSA.

Cohen, D. (1966). Optimising reproduction in a randomly varying environment. *Journal of Theoretical Biology* 12: 119–129.

Collins, F.L. and Johnson, S.J. (1985). Reproductive response of caged adult velvetbean caterpillar and soybean looper to the presence of weeds. *Agriculture, Ecosystems, and Environment* 14: 139–149.

Colville, L. and Kranner, I. (2010). Desiccation tolerant plants as model systems to study redox regulation of protein thiols. *Plant Growth Regulation* 61: 241–255.

Conn, J.S., Beattie, K.L., and Blanchard, A. (2006). Seed viability and dormancy of 17 weed species after 19.7 years of burial in Alaska. *Weed Science* 54: 464–470.

Cordeau, S., Triolet, M., Wayman, S. et al. (2016). Bioherbicides: dead in the water? A review of the existing products for integrated weed management. *Crop Protection* 87: 44–49.

Crawley, M.J. (2000). Seed predators and plant population dynamics. In: *Seeds: The Ecology of Regeneration in Plant Communities*, 2e (ed. M. Fenner), 167–168. CABI Publishing.

Creamer, N.G., Bennett, M.A., Stinner, B.R. et al. (1996). Mechanisms of weed suppression in cover crop-based production systems. *Horticultural Science* 31: 410 413.

Crow, W.D., Steckel, L.E., Hayes, R.M., and Mueller, T.C. (2015). Evaluation of postharvest herbicide applications for seed prevention of glyphosate resistant Palmer amaranth (*Amaranthus palmeri*). *Weed Technology* 29: 405–411.

Csontos, P. and Tamás, J. (2003). Comparisons of soil seed bank classification systems. *Seed Science Research* 13: 101–111.

Cummins, R.P. and Miller, G.R. (2002). Altitudinal gradients in seed dynamics of *Calluna vulgaris* in eastern Scotland. *Journal of Vegetation Science* 13: 859–866.

Cutler, G.C., Astatkie, T., and Chahil, G.S. (2016). Weed seed granivory by carabid beetles and crickets for biological control of weeds in commercial lowbush blueberry fields. *Agricultural and Forest Entomology* 18: 390–397.

Dalling, J.W., Davis, A.S., Schutte, B.J., and Arnold, A.E. (2011). Seed survival in soil: interacting effects of predation, dormancy and the soil microbial community. *Journal of Ecology* 99: 89–95.

Davis, A.S. (2006). When does it make sense to target the weed seed bank? *Weed Science* 54: 558–565.

Davis, A.S. (2008). Weed seed pools concurrent with corn and soybean harvest in Illinois. *Weed Science* 56: 503–508.

Davis, A.S. (2018). Weed ecology and population dynamics. In: *Integrated Weed Management for Sustainable Agriculture*, (ed. R.L. Zimdahl), 3–27. Cambridge, UK: Burleigh Dodds Science Publishing Limited.

Davis, A.S., Anderson, K.I., Hallett, S.G., and Renner, K.A. (2006). Weed seed mortality in soils with contrasting agricultural management histories. *Weed Science* 54: 291–297.

Davis, A.S., Cardina, J., Forcella, F. et al. (2005). Environmental factors affecting seed persistence of annual weeds across the US corn belt. *Weed Science* 53: 860–868.

Davis, A.S., Clay, S., Cardina, J. et al. (2013). Seed burial physical environment explains departures from regional hydrothermal model of giant ragweed (*Ambrosia trifida*) seedling emergence in US Midwest. *Weed Science* 61: 415–421.

Davis, A.S., Dixon, P.M., and Liebman, M. (2004). Using matrix models to determine cropping system effects on annual weed demography. *Ecology Applied* 14: 655–668.

Davis, A.S., Fu, X., Schutte, B.J. et al. (2016). Interspecific variation in persistence of buried weed seeds follows trade-offs among physiological, chemical, and physical seed defenses. *Ecology and Evolution* 6: 6836–6845.

Davis, A.S. and Liebman, M. (2003). Cropping system effects on giant foxtail (*Setaria faberi*) demography: I. Green manure and tillage timing. *Weed Science* 51: 919–929.

Davis, A.S., Schutte, B.J., Iannuzzi, J., and Renner, K.A. (2008). Chemical and physical defense of weed seeds in relation to soil seedbank persistence. *Weed Science* 56: 676–684.

Decker, A.M., Clark, A.J., Meisinger, J.J. et al. (1994). Legume cover crop contributions to no-tillage corn production. *Agronomy Journal* 86: 126–135.

Donohue, K. (2005). Seeds and seasons: interpreting germination timing in the field. *Seed Science Research* 15: 175–187.

Doohan, D., Wilson, R., Canales, E., and Parker, J. (2010). Investigating the human dimension of weed management: new tools of the trade. *Weed Science* 58: 503–510.

Dou, Z., Fox, R.H., and Toth, J.D. (1995). Seasonal soil nitrate dynamics in corn as affected by tillage and nitrogen source. *Soil Science Society of America Journal* 59: 858–864.

Drews, S., Neuhoff, D., and K€opke, U. (2009). Weed suppression ability of three winter wheat varieties at different row spacing under organic farming conditions. *Weed Research* 49: 526–533.

Eriksson, O. (2000). Seed dispersal and colonization ability of plants—assessment and implications for conservation. *Folia Geobotanica* 35: 115–123.

Espinar, J.L., Thompson, K., and García, L.V. (2005). Timing of seed dispersal generates a bimodal seed bank depth distribution. *American Journal of Botany* 92: 1759–1763.

Evans, E.W. (2016). Biodiversity, ecosystem functioning, and classical biological control. *Applied Entomology and Zoology* 51: 173–184.

Evans, S.P., Knezevic, S.Z., Lindquist, J.L. et al. (2003). Nitrogen application influences the critical period for weed control in corn. *Weed Science* 51: 408–417.

Evans, T.A. and Gleeson, P.V. (2016). Direct measurement of ant predation of weed seeds in wheat cropping. *Journal of Applied Ecology* 53: 1177–1185.

Farmer, J.A., Webb, E.B., Pierce, R.A., and Bradley, K.W. (2017). Evaluating the potential for weed seed dispersal based on waterfowl consumption and seed viability. *Pest Management Science* 73: 2592–2603.

Fawcett, R.S. and Slife, F.W. (1978). Effects of 2, 4-D and dalapon on weed seed production and dormancy. *Weed Science* 26: 543–547.

Fenner, M. and Thompson, K. (2005). *The Ecology of Seeds.* Cambridge, England: Cambridge University Press.

Fennimore, S.A. and Boyd, N.S. (2018). Sustainable weed control in strawberry. In: *Weed Control; Sustainability, Hazards, and Risks in Cropping Systems Worldwide* (ed. N.E. Korres, N.R. Burgos, and S.O. Duke), 383–403. CRC Press.

Ferrandis, P., Bonilla, M., and Osorio, L.D. (2011). Germination and soil seed bank traits of *Podocarpus angustifolius* (Podocarpaceae): an endemic tree species from Cuban rain forests. *Revista de Biología Tropical* 59: 1061–1069.

Finch-Savage, W.E. and Leubner-Metzger, G. (2006). Seed dormancy and the control of germination. *New Phytologist* 171: 501–523.

Firbank, L.G. and Watkinson, A.R. (1985). On the analysis of competition within two-species mixtures of plants. *Journal of Applied Ecology* 22: 503–517.

Flematti, G.R., Ghisalberti, E.L., Dixon, K.W., and Trengove, R.D. (2004). A compound from smoke that promotes seed germination. *Science* 305: 977.

Flenet, F., Scenery, J.R., Board, J.E. et al. (1996). Row spacing effects on light extinction coefficients in corn, sorghum, soybean, and sunflower. *Agronomy Journal* 88: 185–190.

Flint, S.D. and Palmblad, I.G. (1978). Germination dimorphism and developmental flexibility in the ruderal weed *Heterotheca grandiflora*. *Oecologia* 36: 33–43. https://doi.org/10.1007/BF00344569.

Forcella, F. (1992). Prediction of weed seedling densities from buried seed reserves. *Weed Research* 32: 29–38.

Forcella, F. (1993). Seedling emergence model for velvetleaf. *Agronomy Journal* 85: 929–933.

Forcella, F. (1998). Real-time assessment of seed dormancy and seedling growth for weed management. *Seed Science Research* 8: 201–209.

Forcella, F., Benech-Arnold, R.L., Sanchez, R.A., and Ghersa, C.M. (2000). Modeling seedling emergence. *Field Crops Research* 67: 123–139.

Forcella, F., Peterson, D., and Barbour, J. (1996). Timing and measurement of weed seed shed in corn (Zea mays). *Weed Technology* 10: 535–543. https://doi.org/10.1017/S0890037X00040409.

Forcella, F., Wilson, R.G., Renner, K.A. et al. (1992). Weed seedbanks of the U.S. Corn Belt: magnitude, variation, emergence, and application. *Weed Science* 40: 636–644.

Fuerst, E.P., Anderson, J.V., Kennedy, A.C., and Gallagher, R.S. (2011). Induction of polyphenol oxidase activity in dormant wild oat (*Avena fatua*) seeds and caryopses: a defense response to seed decay fungi. *Weed Science* 59: 137–144.

Fuerst, E.P., James, M.S., Pollard, A.T., and Okubara, P.A. (2018). Defense enzyme responses in dormant wild oat and wheat caryopses challenged with a seed decay pathogen. *Frontiers in Plant Science* 8. https://doi.org/10.3389/fpls.2017.02259.

Gage, K.L. and Schwartz-Lazaro, L.M. (2019). Shifting the paradigm: an ecological systems approach to weed management. *Agriculture* 9: 179.

Gaines, T.A., Duke, S.O., Morran, S. et al. (2020). Mechanisms of evolved herbicide resistance. *Journal of Biological Chemistry* 295: 10307 10330.

Gallagher, R.S. and Cardina, J. (1998). Ecophysiological aspects of phytochrome-mediated germination in soil seed banks. *Aspects of Applied Biology* 51: 1–8.

Gallandt, E.R. (2006). How can we target the weed seedbank? *Weed Science* 54: 588–596.

Gallandt, E.R., Liebman, M., and Huggins, D.R. (1999). Improving soil quality: implications for weed management. *Journal of Crop Production* 2: 95–121.

Gardarin, A., Guillemin, J.P., Munier-Jolain, N.M., and Colbach, N. (2010). Estimation of key parameters for weed population dynamics models: base temperature and base water potential for germination. *European Journal of Agronomy* 32: 162–168.

Ghersa, C.M. and Martínez-Ghersa, M.A. (2000). Ecological correlates of weed seed size and persistence in the soil under different tilling systems: implications for weed management. *Field Crops Research* 67: 141–148.

Gill, G.S. and Holmes, J.E. (1997). Efficacy of cultural control methods for combating herbicide resistant *Lolium rigidum*. *Pesticide Science* 51: 352–358.

Gonzalez-Andujar, J.L., Chantre, G.R., Morvillo, C. et al. (2016). Predicting field weed emergence with empirical models and soft computing techniques. *Weed Research* 56: 415–423.

Goplen, J.J., Sheaffer, C.C., Becker, R.L. et al. (2016). Giant ragweed (*Ambrosia trifida*) seed production and retention in soybean and field margins. *Weed Technology* 30: 246–253.

Goplen, J.J., Sheaffer, C.C., Becker, R.L. et al (2017). Seedbank depletion and emergence patterns of giant ragweed (*Ambrosia trifida*) in Minnesota cropping systems. *Weed Science* 65: 52–60.

Goplen, J.J., Sheaffer, C.C., Becker, R.L. et al. (2018). Giant ragweed (*Ambrosia trifida)* emergence model performance evaluated in diverse cropping systems. *Weed Science* 66: 36–46.

Green, J.K., Norsworthy, J.K., Palhano, M.G. et al. (2016). Seed retention of Palmer amaranth and barnyardgrass in soybean. San Juan, PR. http://wssaabstracts.com/public/38/proceedings.html (accessed 22 June 2021).

Griffin, J.L., Boudreaux, J.M., and Miller, D.K. (2010). Herbicides as harvest aids. *Weed Science* 58: 355–358.

Grundy, A.C. (2003). Predicting weed emergence: a review of approaches and future challenges. *Weed Research* 43: 1–11.

Grundy, A.C., Peters, N.C.B., Rasmussen, I.A. et al. (2003). Emergence of Chenopodium album and Stellaria media of different origins under different climatic conditions. *Weed Research* 43: 163–176.

Gue´rif, J., Richard, G., Du¨rr, C. et al. (2001). A review of tillage effects on crop residue management, seedbed conditions and seedling establishment. *Soil Tillage Research* 61: 13–32.

Gummerson, R.J. (1986). The effect of constant temperatures and osmotic potential on the germination of sugar beet. *Journal of Experimental Botany* 37: 729–741.

Gunsolus, J.L. (1990). Mechanical and cultural weed control in corn and soybeans. *American Journal of Alternative Agriculture* 5: 114–119.

Hardegree, S.P., Moffet, C.A., Walters, C.T. et al. (2017). Hydrothermal germination models: improving experimental efficiency by limiting data collection to the relevant hydrothermal range. *Crop Science* 57: 2753–2760.

Harding, J.A. (1976). Seasonal occurrence hosts, parasitism and parasites of cabbage and soybean loopers in the lower Rio Grande Valley. *Journal of Environmental Entomology* 5: 672–674.

Haring, S.C. and Flessner, M.L. (2018). Improving soil seed bank management. *Pest Management Science* 74: 2412–2418.

Haring, S.C., Flessner, M.L., Everman, W.J., and Mirsky, S.B. (2017). The role of late-season weather events on seed shattering of redroot pigweed (*Amaranthus retroflexus*) and common ragweed (*Ambrosia artemisiifolia*). *Weed Science Society of America* 5: 175.

Harrison, S.K., Regnier, E.E., Schmoll, J.T., and Harrison, J.M. (2007). Seed size and burial effects on giant ragweed (*Ambrosia trifida*) emergence and seed demise. *Weed Science* 55: 16–22.

Hartke, A., Drummond, F.A., and Liebman, M. (1998). Seed feeding, seed caching, and burrowing behaviors of *Harpalus rufipes* De Geer larvae (Coleoptera: carabidae) in the maine potato agroecosystem. *Biological Control* 13: 91–100.

Healy, A.J. (1943). Seed dispersal by human activity. *Nature* 151: 140.

Heggenstaller, A.H., Menalled, F.D., Liebman, M., and Westerman, P.R. (2006). Seasonal patterns in post dispersal seed predation of Abutilon theophrasti and Setaria faberi in three cropping systems. *Journal of Applied Ecology* 43: 999–1010.

Hill, E.C., Renner, K.A., VanGessel, M.J. et al. (2016). Late-season weed management to stop viable weed seed production. *Weed Science* 64: 112–118.

Hodkinson, D., Askew, A., Thompson, K. et al. (1998). Ecological correlates of seed size in the British flora. *Functional Ecology* 12: 762–766.

Holmes, R.J. and Froud-Williams, R.J. (2005). Post-dispersal weed seed predation by avian and non avian predators. *Agriculture, Ecosystems & Environment* 105: 23–27.

Hölzel, N. and Otte, A. (2004). Assessing soil seed bank persistence in flood-meadows: the search for reliable traits. *Journal of Vegetation Science* 15: 93–100.

Honda, Y. (2008). Ecological correlations between the persistence of the soil seed bank and several plant traits, including seed dormancy. *Plant Ecology* 196: 301–309.

Hulme, P.E. (1998). Post-dispersal seed predation and seed bank persistence. *Seed Science Research* 8: 513–519.

Janzen, D.H. (1971). Seed predation by animals. *Annual Review of Ecology and Systematics* 2: 465–492.

Jasieniuk, M., Brule-Babel, A.L., and Morrison, I.N. (1996). The evolution of genetics of herbicide resistance in weeds. *Weed Science* 44: 176–193.

Jha, P. and Norsworthy, J.K. (2009). Soybean canopy and tillage effects on emergence of Palmer amaranth from a natural seedbank. *Weed Science* 27: 644–651.

Jha, P. and Norsworthy, J.K. (2012). Influence of late-season herbicide applications on control, fecundity, and progeny fitness of glyphosate-resistant palmer amaranth (*Amaranthus palmeri*) biotypes from Arkansas. *Weed Technology* 26: 807–812.

Jha, P., Norsworthy, J.K., Bridges, W., Jr, and Riley, M.B. (2008). Influence of glyphosate timing and row width on Palmer amaranth (*Amaranthus palmeri*) and pusley (*Richardia* spp.) demographics in glyphosate-resistant soybean. *Weed Science* 56: 408–415.

Johnson, W.C., III and Mullinix, B.G., Jr (1995). Weed management in peanut using stale seedbed techniques. *Weed Science* 43: 293–297.

Johnson, W.C., III and Mullinix, B.G., Jr (1998). Stale seedbed weed control in cucumber. *Weed Science* 46: 698–702.

Jones, R.E. and Medd, R.W. (2005). A methodology for evaluating risk and efficacy of weed management technologies. *Weed Science* 53: 505–514.

Jordan, N., Schut, M., Graham, S. et al. (2016). Transdisciplinary weed research: new leverage on challenging weed problems? *Weed Research* 56: 345–358.

Karssen, C.M. (1982). Seasonal patterns in dormancy in weed seeds. In: *The Physiology and Biochemistry of Seed Development, Dormancy and Germination* (ed. A.A. Khan), 243–270. Elsevier Biomedical Press.

Kegode, G.O., Forcella, F., and Durgan, B.R. (2015). Limiting green and yellow foxtail (*Setaria viridis* and *S. glauca*) seed production following spring wheat (*Triticum aestivum*) harvest. *Weed Technology* 13: 43–47.

Kennedy, A.C. (1999). Soil microorganisms for weed management. *Journal Crop Production* 2: 123–138.

Kennedy, A.C. and Kremer, R.J. (1996). Microorganisms in weed control strategies. *Journal Production Agriculture* 9: 480–485.

Kleijn, D., Bommarco, R., Fijen, T.P.M. et al. (2019). Ecological intensification: bridging the gap between science and practice. *Trends in Ecology & Evolution* 34: 154–166.

Knezevic, S.Z., Evans, S.P., Blankenship, E.E. et al. (2002). Critical period for weed control: the concept and data analysis. *Weed Science* 50: 773–786.

Kochanek, J., Steadman, K.J., Probert, R.J., and Adkins, S.W. (2011). Parental effects modulate seed longevity: exploring parental and offspring phenotypes to elucidate pre-zygotic environmental influences. *New Phytologist* 191: 223–233.

Korres, N.E., Norsworth, J.K., Young, B.G. et al. (2018). Seedbank persistence of Palmer amaranth (*Amaranthus palmeri*) and waterhemp (*Amaranthus tuberculatus*) across diverse geographical regions in the United States. *Weed Science* 66: 446–456.

Korres, N.E. and Norsworthy, J.K. (2017). Palmer amaranth (*Amaranthus palmeri*) demographic and biological characteristics in wide-row soybean. *Weed Science* 65: 491–503.

Kraft, N.J.B., Adler, P.B., Godoy, O. et al. (2015). Community assembly, coexistence and the environmental filtering metaphor. *Functional Ecology* 29: 592–599.

Kremer, R.J. (1993). Management of weed seed banks with microorganisms. *Ecology Applied* 3: 42–52.

Kremer, R.J. and Kennedy, A.C. (1996). Rhizobacteria as biocontrol agents of weeds. *Weed Technology* 10: 601–609.

Kristensen, L., Oslen, J., and Weiner, J. (2008). Crop density, sowing pattern, and nitrogen fertilization effects on weed suppression and yield in spring wheat. *Weed Science* 56: 97–102.

Krock, B., Schmidt, S., Hertweck, C., and Baldwin, I.T. (2002). Vegetation-derived abscisic acid and four terpenes enforce dormancy in seeds of the post-fire annual, *Nicotiana attenuata*. *Seed Science Research* 12: 239–252.

Kucera, B., Cohn, M.A., and Leubner-Metzger, G. (2005). Plant hormone interactions during seed dormancy release and germination. *Seed Science Research* 15: 281–307.

Kumar, V. and Jha, P. (2015). Influence of herbicides applied postharvest in wheat stubble on control, fecundity, and progeny fitness of *Kochia scoparia* in the US Great Plains. *Crop Protection* 71: 144–149.

Kuo, S., Sainju, E.M., and Jellum, E.J. (1996). Winter cover cropping influence on nitrogen mineralization and corn yields. *Soil Biology Fertility* 22: 310–317.

Landis, D.A., Menalled, F.D., Costamagna, A.C., and Wilkinson, T.K. (2005). Manipulating plant resources to enhance beneficial arthropods in agricultural landscapes. *Weed Science* 53: 902–908.

Leishman, M. and Westoby, M. (1998). Seed size and shape are not related to persistence in soil in Australia in the same way as in Britain. *Functional Ecology* 12: 480–485.

Lemerle, D., Cousens, R., Gill, G. et al. (2004). Reliability of higher seeding rates of wheat for increased competitiveness with weeds in low rainfall environments. *Journal of Agricultural Science* 142: 395–409.

Leprince, O. and Buitink, J. (2010). Desiccation tolerance: from genomics to the field. *Plant Science* 179: 554–564.

Lev-Yadun, S. and Ne'eman, G. (2013). Bimodal colour pattern of individual Pinus halepensis Mill seeds: a new type of crypsis. *Biological Journal of the Linnean Society* 109: 271–278.

Li, B.L. and Foley, M.E. (1997). Genetic and molecular control of seed dormancy. *Trends in Plant Science* 2: 384–389.

Li, J., Li, G., Wang, H., and Deng, X. (2011). Phytochrome signalling mechanisms. *Arabidopsis Book* 9: e0148.

Liebman, M., Baraibar, B., Buckley, Y. et al. (2016). Ecologically sustainable weed management: how do we get from proof-of-concept to adoption? *Ecological Applications* 26: 1352–1369.

Liebman, M. and Davis, A.S. (2009). Managing weeds in organic farming systems: an ecological approach. In: *Organic Farming: The Ecological System* (ed. C. Francis), 173–196. Madison: American Society of Agronomy.

Liebman, M. and Dyck, E. (1993). Crop rotation and intercropping strategies for weed management. *Ecological Applications* 3: 92–122.

Liebman, M. and Gallandt, E.R. (1997). Many little hammers: ecological approaches for management of crop–weed interactions. In: *Ecology in Agriculture* (ed. L.E. Jackson), 291–343. Academic Press.

Liebman, M., Mohler, C.L., and Staver, C.P. (2001). *Ecological Management of Agricultural Weeds*. Cambridge University Press.

Liebman, M. and Staver, C.P. (2001). Crop diversification for weed management. In: *Ecological Management of Agricultural Weeds* (ed. M. Liebman and C.L. Mohler), 322–374. Staver CP: Cambridge University Press.

Long, R.L., Gorecki, M.J., Renton, M. et al. (2015). The ecophysiology of seed persistence: a mechanistic view of the journey to germination or demise. *Biological Reviews* 90: 31–59.

Long, R.L., Kranner, I., Panetta, F.D. et al. (2011). Wet-dry cycling extends seed persistence by re-instating antioxidant capacity. *Plant and Soil* 388: 511–519.

Long, R.L., Panetta, F.D., Steadman, K.J. et al. (2008). Seed persistence in the field may be predicted by laboratory-controlled aging. *Weed Science* 56: 523–528.

Long, R.L., Steadman, K.J., Panetta, F.D., and Adkins, S.W. (2009). Soil type does not affect seed ageing when soil water potential and temperature are controlled. *Plant and Soil* 320: 131–140.

Lueschen, W.E., Andersen, R.N., Hoverstad, T.R., and Kanne, B.R. (1993). Seventeen years of cropping systems and tillage affect velvetleaf (*Abutilon theophrasti*) seed longevity. *Weed Science* 41: 82–86.

MacLaren, C., Labuschagne, J., and Swanepoel, P.A. (2021). Tillage practices affect weeds differently in monoculture vs. crop rotation. *Soil and Tillage Research* 205: 104795.

MacLaren, C., Storkey, J., Menegat, A. et al. (2020). An ecological future for weed science to sustain crop production and the environment. A review. *Agronomy for Sustainable Development* 40: 24.

Mahaut, L., Gaba, S., and Fried, G. (2019). A functional diversity approach of crop sequences reveals that weed diversity and abundance show different responses to environmental variability. *Journal of Applied Ecology* 56: 1400–1409.

Maity, A., Lamichaney, A., Joshi, D.C. et al. (2021). Seed shattering: a trait of evolutionary importance in plants. *Frontiers in Plant Science* 12: 657773.

Martínez-Ghersa, M.A., Ghersa, C.M., Benech-Arnold, R.L. et al. (2000). Adaptive traits regulating dormancy and germination of invasive species. *Plant Species Biology* 15: 127–137.

Martin-Guay, M.O., Paquette, A., Dupras, J., and Rivest, D. (2018). The new green revolution: sustainable intensification of agriculture by intercropping. *Science of the Total Environment* 615: 767–772.

Masin, R., Loddo, D., Benvenuti, S. et al. (2010). Temperature and water potential as parameters for modeling weed emergence in central-northern Italy. *Weed Science* 58: 216–222.

Masin, R., Zuin, M.C., Archer, D.W. et al. (2005). WEEDTURF: a predictive model to aid control of annual summer weeds in turf. *Weed Science* 53: 193–201.

McLachlan, S.M., Murphy, S.D., Tollenaar, M. et al. (1995). Light limitation of reproduction and variation in the allometric relationship between reproductive and vegetative biomass in *Amaranthus retroflexus* (redroot pigweed). *Journal of Applied Ecology* 1: 157–165.

Meisinger, J.J., Hargrove, W.L., Mikkelsen, R.L., Jr et al. (1991). Effects of cover crops on groundwater quality. In: *Cover Crops for Clean Water* (ed. W.L. Hargrove), 57–68. Soil and Water Conservation Society.

Melander, B., Rasmussen, I.A., and Olesen, J.E. (2020). Legacy effects of leguminous green manure crops on the weed seed bank in organic crop rotations. *Agriculture, Ecosystems & Environment* 302: 107078.

Menalled, F.D., Liebman, M., and Renner, K. (2006). The ecology of weed seed predation in herbaceous crop systems. In: *Handbook of Sustainable Weed Management* (ed. H.P. Singh, D.R. Batish, and R.K. Kohli), 297–327. Haworth.

Merritt, D.J., Turner, S.R., Clarke, S., and Dixon, K.W. (2007). Seed dormancy and germination stimulation syndromes for Australian temperate species. *Australian Journal of Botany* 55: 336–344.

Metcalfe, H., Milne, A.E., Hull, R. et al. (2018). The implications of spatially variable pre-emergence herbicide efficacy for weed management. *Pest Management Science* 74: 755–765. https://doi.org/10.1002/ps.4784.

Mickelson, J.A., Bussan, A.J., Davis, E.S. et al. (2004). Postharvest kochia (*Kochia scoparia*) management with herbicides in small grains. *Weed Technology* 18: 426–431.

Mirsky, S.B., Ryan, M.R., Teasdale, J.R. et al. (2013). Overcoming weed management challenges in cover crop–based organic rotational no-till soybean production in the Eastern United States. 27: 193–203.

Mohler, C.L. (2001a). *Enhancing the Competitive Ability of Crops* (ed. M. Liebman and C.L. Mohler), 269–231. Staver CP: Cambridge University Press.

Mohler, C.L. (2001b). Mechanical management of weeds. In: *Ecological Management of Agricultural Weeds*, (ed. M. Liebman and C.L. Mohler), 192–209. Staver CP: Cambridge University Press.

Moles, A.T., Warton, D.I., and Westoby, M. (2003). Seed size and survival in the soil in arid Australia. *Austral Ecology* 28: 575–585.

Mortensen, D.A., Egan, J.F., Maxwell, B.D. et al. (2012). Navigating a critical juncture for sustainable weed management. *BioScience* 62: 75–84.

Moss, S. (2019). Integrated weed management (IWM): why are farmers reluctant to adopt non-chemical alternatives to herbicides? *Pest Management Science* 75: 1205–1211.

Motlagh, M.R.S. (2011). Evaluation of *Epicoccum purpurascens* as biological control agent of *Echinochloa* spp. in rice fields. *Journal of Food, Agriculture & Environment* 9: 394–397.

Mueller, M.H. and Van der Valk, A.G. (2002). The potential role of ducks in wetland seed dispersal. *Wetlands* 22: 170–178.

Müller-Stöver, D., Nybroe, O., Baraibar, B. et al. (2016). Contribution of the seed microbiome to weed management. *Weed Research* 56: 335–339.

Muñoz-Rojas, M., Erickson, T.E., Martini, D.C. et al. (2016). Climate and soil factors influencing seedling recruitment of plant species used for dryland restoration. *Soil* 28: 287–298.

Murphy, S.D., Yakubu, Y., Weise, S.F., and Swanton, C.J. (1996). Effect of planting patterns and inter row cultivation on competition between corn (*Zea mays*) and late emerging weeds. *Weed Science* 44: 856–870.

Murray, K.G., Russell, S., Picone, C.M. et al. (1994). Fruit laxatives and seed passage rates in frugivores: consequences for plant reproductive success. *Ecology* 75: 989–994.

Nathan, R., Katul, G., Horn, H. et al. (2002). Mechanisms of long-distance dispersal of seeds by wind. *Nature* 418: 409–413.

Nathan, R., Schurr, F.M., Spiegel, O. et al. (2008). Mechanisms of long-distance seed dispersal. *Trends in Ecology & Evolution* 23: 638–647.

Neve, P., Vila-Aiub, M., and Roux, F. (2009). Evolutionary-thinking in agricultural weed management. *New Phytology* 184: 783–793.

Nikolaeva, M.G. (1967). *Physiology of Deep Dormancy in Seeds.* Leningrad, Russia: Izdatel'stvo 'Nauka' (in Russian). [Translated from Russian by Z. Shapiro (1969), National Science Foundation, Washington, DC, USA: 219.].

Nikolaeva, M.G. (2004). On criteria to use in studies of seed evolution. *Seed Science Research* 14: 315–320.

Norris, R.F. and Kogan, M. (2005). Ecology of interactions between weeds and arthropods. *Annual Review Entomology* 50: 479–503.

Norsworthy, J.K., Green, J.K., Barber, T. et al. (2020). Seed destruction of weeds in southern US crops using heat and narrow-windrow burning. *Weed Technology* 34: 589–596.

Norsworthy, J.K., Ward, S.M., Shaw, D.R. et al. (2012). Reducing the risks of herbicide resistance: best management practices and recommendations. *Weed Science* 60: 31–62.

Nurse, R.E. and DiTommaso, A. (2005). Corn competition alters the germinability of velvetleaf (*Abutilon theophrasti*) seeds. *Weed Science* 53: 479–488.

Ooi, M.K.J., Auld, T.D., and Whelan, R.J. (2007). Distinguishing between persistence and dormancy in soil seed banks of three shrub species from fire-prone southeastern Australia. *Journal of Vegetation Science* 18: 405–412.

Osborne, D.J. (1980). Senescence in seeds. In: *Senescence in Plants* (ed. K.V. Thimann), 13–37. CRC Press.

Oslen, J.M., Griepentrog, H.W., Nielsen, J., and Wiener, J. (2012). How important are crop spatial pattern and density for weed suppression by spring wheat? *Weed Science* 60: 501–509.

Oveisi, M., Ojaghi, A., Rahimian Mashhadi, H. et al. (2021). Potential for endozoochorous seed dispersal by sheep and goats: risk of weed seed transport via animal feces. *Weed Research* 61: 1–2.

Pareja, M.R., Staniforth, D.W., and Pareja, G.P. (1985). Distribution of weed seed among soil structural units. *Weed Science* 1: 182–189.

Pekrun, C. and Claupein, W. (2006). The implication of stubble tillage for weed population dynamics in organic farming. *Weed Research* 46: 414–423.

Pemberton, R.W. and Irving, D.W. (1990). Elaiosomes on weed seeds and the potential for myrmecochory in naturalized plants. *Weed Science* 38: 615–619.

Pickering, C. and Mount, A. (2010). Do tourists disperse weed seed? A global review of unintentional human-mediated terrestrial seed dispersal on clothing, vehicles and horses. *Journal of Sustainable Tourism* 18: 239–256.

Piskackova, T.R., Reberg-Horton, S.C., Richardson, R.J. et al. (2020). Incorporating environmental factors to describe wild radish (*Raphanus raphanistrum*) seedling emergence and plant phenology. *Weed Science* 68: 627–638.

Pittman, K., Barney, J., and Flessner, M. (2020). Cover crop residue components and their effect on summer annual weed suppression in corn and soybean. *Weed Science* 68: 301–310.

Pocock, M.J., Evans, D.M., and Memmott, J. (2012). The robustness and restoration of a network of ecological networks. *Science* 335: 973–977.

Poschlod, P. and Jackel, A.K. (1993). Untersuchungen zur Dynamik von generativen diasporenbanken von samenpflanzen in kalkmagerrasen. I. Jahreszeitliche dynamik des diasporenregens und der diasporenbank auf zwei kalkmagerrasenstandorten der schwabischen Alp. *Flora* 188: 49–71.

Probert, R.J., Daws, M.I., and Hay, F.R. (2009). Ecological correlates of ex situ seed longevity: a comparative study on 195 species. *Annals of Botany* 104: 57–69.

Rajcan, I. and Swanton, C.J. (2001). Understanding maize–weed competition: resource competition, light quality and the whole plant. *Field Crops Research* 71: 139–150.

Raseduzzaman, M.D. and Jensen, E.S. (2017). Does intercropping enhance yield stability in arable crop production? A meta-analysis. *European Journal of Agronomy* 91: 25–33.

Regnier, E., Harrison, S.K., Liu, J. et al. (2008). Impact of an exotic earthworm on seed dispersal of an indigenous US weed. *Journal of Applied Ecology* 45: 1621–1629.

Rehm, E., Fricke, E., Bender, J. et al. (2019). Animal movement drives variation in seed dispersal distance in a plant–animal network. *Proceedings of the Royal Society B* 286: 20182007.

Reuss, S.A., Buhler, D.D., and Gonsolus, J.L. (2001). Weed seed associations with soil aggregates: distribution and viability in a silt loam soil. *Applied Soil Ecology* 16: 209–217.

Rice, C.P., Cai, G., and Teasdale, J.R. (2012). Fate of benzoxazinoids in soil treated with rye cover crop. *Journal of Agricultural and Food Chemistry* 60: 4471–4479.

Roberts, E.H. (1988a). Seed aging: the genome and its expression. In: *Senescence and Aging in Plants* (ed. L.D. Nooden and A.C. Lepold), 465–498. Academic Press.

Roberts, E.H. (1988b). Temperature and seed germination. In: *Plants and Temperature* (ed. L. SP and F.I. Woodward), 109–132. Society for Experimental Biology.

Roberts, H.A. and Feast, P.M. (1972). Fate of seeds of some annual weeds in different depths of cultivated and undisturbed soil. *Weed Research* 12: 316–324.

Rodriguez-Perez, J. and Traveset, A. (2007). A multi-scale approach in the study of plant regeneration: finding bottlenecks is not enough. *Perspectives in Plant Ecology Evolution and Systematics* 9: 1–13.

Roman, E.S., Murphy, S.D., and Swanton, C.J. (2000). Simulation of Chenopodium album seedling emergence. *Weed Science* 48: 217–224.

Rosset, J. and Gulden, R. (2020). Cultural weed management practices shorten the critical weed-free period for soybean grown in the Northern Great Plains. *Weed Science* 68: 79–91.

Sainju, U.M., Whitehead, W.F., and Singh, B.P. (2005). Biculture legume–cereal cover crops for enhanced biomass yield and carbon and nitrogen. *Agronomy Journal* 97: 1403–1412.

Sallon, S., Solowey, E., Cohen, Y. et al. (2008). Germination, genetics, and growth of an ancient date seed. *Science* 320: 1464–1464.

Sande, D., Mullen, J., Wetzstein, M., and Houston, J. (2011). Environmental impacts from pesticide use: a case study of soil fumigation in Florida tomato production. *International Journal of Environmental Research and Public Health* 8: 4649–4661.

San Martín, C., Thorne, M., Gourlie, J. et al. (2021). Seed retention of grass weeds at wheat harvest in the Pacific Northwest. *Weed Science* 69: 238–246. https://doi.org/10.1017/wsc.2020.91.

Sarabi, V. (2019). Factors that influence the level of weed seed predation: a review. *Weed Biology Management* 19: 61–74.

Sardana, V., Mahajan, G., Jabran, K., and Chauhan, B.S. (2017). Role of competition in managing weeds: an introduction to the special issue. *Crop Protection* 95: 1–7.

Sauer, J. and Struik, G. (1964). A possible ecological relation between soil disturbance, light-flash, and seed germination. *Ecology* 45: 884–886. https://doi.org/10.2307/1934942.

Schafer, D.E. and Chilcote, D.O. (1969). Factors influencing persistence and depletion in buried seed populations. I. A model for analysis of parameters of buried seed persistence and depletion. *Crop Science* 9: 417–418.

Schatral, A. and Osborne, J.M. (2006). Germination and dormancy states of seeds in the environmental weed *Rumex vesicarius* (Polygonaceae). I. Seed polymorphism and germination of extracted seeds. *Australian Journal of Botany* 54: 773–782.

Schupp, E.W. (1993). Quantity, quality and the effectiveness of seed dispersal by animals. *Vegetation* 107: 15–29.

Schwartz, L.M., Norsworthy, J.K., Young, B.G. et al. (2016). Tall waterhemp (*Amaranthus tuberculatus*) and Palmer amaranth (*Amaranthus palmeri*) seed production and retention at soybean maturity. *Weed Technol* 30: 284–290.

Schwartz-Lazaro, L.M. and Copes, J.T. (2019). A review of the soil seedbank from a weed scientist perspective. *Agronomy* 9: 369.

Schwartz-Lazaro, L.M., Norsworthy, J.K., Walsh, M.J., and Bagavathiannan, M.V. (2017). Efficacy of the integrated Harrington seed destructor on weeds of soybean and rice production systems in the southern United States. *Crop Science* 57: 2812–2818.

Schwartz-Lazaro, L.M., Shergill, L.S., Evans, J.A. et al. (2021a). Seed shattering phenology at soybean harvest of economically important weeds in multiple regions of the United States. Part 1: broadleaf species. *Weed Science* 69: 95–103.

Schwartz-Lazaro, L.M., Shergill, L.S., Evans, J.A. et al. (2021b). Seed shattering phenology at soybean harvest of economically important weeds in multiple regions of the United States. Part 2: grass species. *Weed Science* 69: 104–110.

Schweizer, E.E. and Zimdahl, R.L. (1984). Weed seed decline in irrigated soil after six years of continuous Corn (*Zea mays*) and Herbicides. *Weed Science* 32: 76–83.

Scopel, A.L., Ballare, C.L., and Sanchez, R.A. (1991). Induction of extreme light sensitivity in buried weed seeds and its role in the perception of soil cultivations. *Plant, Cell and Environment* 14: 501–508.

Seale, M. and Nakayama, N. (2020). From passive to informed: mechanical mechanisms of seed dispersal. *New Phytologist* 225: 653–658.

Shaw, D.R., Norsworthy, J.K., and Ward, S.M. (2013). Best management practices to prevent, delay, or manage herbicide resistance. *Outlooks on Pest Management* 24: 132–135.

Shergill, L.S., Schwartz-Lazaro, L.M., Leon, R. et al. (2020). Current outlook and future research needs for harvest weed seed control in North American cropping systems. *Pest Management Science* 76: 3887–3895.

Shirtliffe, S.J., Entz, M.H., and Van Acker, R.C. (2000). *Avena fatua* development and seed shatter as related to thermal time. *Weed Science* 48: 555–560.

Silvertown, J.W. and Lovett, D.J. (1993). *Introduction to Plant Population Biology*. Oxford: Blackwell.

Singh, B., Dhaka, A.K., Pannu, R.K., and Kumar, S. (2013). Integrated weed management-a strategy for sustainable wheat production - a review. *Agricultural Review* 34: 243–255.

Smith, R.G. (2006). Timing of tillage is an important filter on the assembly of weed communities. *Weed Science* 54: 705–712.

Smith, R.G. (2015). A succession-energy framework for reducing non-target impacts of annual crop production. *Agricultural Systems* 133: 14–21.

Smith, R.G., Gross, K.L., and Januchowski, S. (2005). Earthworms and weed seed distribution in annual crops. *Agriculture, Ecosystems & Environment* 108: 363–367.

Soni, N., Nissen, S.J., Westra, P. et al. (2020). Seed retention of winter annual grass weeds at winter wheat harvest maturity shows potential for harvest weed seed control. *Weed Technology* 34: 266–271.

Spengler III, R.N. (2020). Anthropogenic seed dispersal: rethinking the origins of plant domestication. *Trends in Plant Science* 25: 340–348.

Spennemann, D.H. (2020). Frugivory and seed dispersal revisited: codifying the plant-centred net benefit of animal-mediated interactions. *Flora* 263: 151534.

Spokas, K. and Forcella, F. (2009). Software tools for weed seed germination modeling. *Weed Science* 57: 216–227.

Stöcklin, J. and Fischer, M. (1999). Plants with longer-lived seeds have lower local extinction rates in grassland remnants 1950–1985. *Oecologia* 120: 539–543.

Stopes, S. and Millington, S. (1991) Weed control in organic farming systems. *Brighton Crop Protection Conference: Weeds*. Farnham: British Crop Protection Council, 185–192.

Storkey, J. and Neve, P. (2018). What good is weed diversity? *Weed Research* 58: 239–243.

Stroller, E.W. and Wax, L.M. (1974). Dormancy changes and fate of some annual weed seeds in the soil. *Weed Science* 22: 151–155.

Sun, C., Ashworth, M.B., Flower, K. et al. (2021). The adaptive value of flowering time in wild radish (*Raphanus raphanistrum*). *Weed Science* 69: 203–209.

Swanton, C.J. and Booth, B.D. (2004). Management of Weed Seedbanks in the context of populations and communities1. *Weed Technology* 18: 1496–1502.

Swanton, C.J., Mahoney, K.J., Chandler, K., and Gulden, R.H. (2008). Integrated weed management: knowledge-based weed management systems. *Weed Science* 56: 168–172.

Swanton, C.J. and Valente, T. (2018). Key issues and challenges of integrated weed management. In: *Integrated Weed Management for Sustainable Agriculture* (ed. R. Zimdahl), 91–104. Burleigh Dodds Science Publishing.

Taghizadeh, M.S., Nicolas, M.E., and Cousens, R.D. (2012). Effects of relative emergence time and water deficit on the timing of fruit dispersal in *Raphanus raphanistrum* L. *Crop Pasture Science* 63: 1018–1025.

Tasker, E.M., Denham, A.J., Taylor, J.E., and Strevens, T.C. (2011). Post-fire seed predation: does distance to unburnt vegetation matter? *Austral Ecology* 36: 755–766.

Taylor, H., Mason, W., Bennie, A., and Rowse, H. (1982). Responses of soybeans to two row spacings and two soil water levels. I. An analysis of biomass accumulation, canopy development, solar radiation interception and components of seed yield. *Frontiers in Crop Research* 5: 1–14.

Taylor, K., Brummer, T., Taper, M.L. et al. (2012). Human-mediated long-distance dispersal: an empirical evaluation of seed dispersal by vehicles. *Diversity and Distributions* 18: 942–951.

Taylor, S.E. and Oliver, L.R. (1997). Sicklepod (*Senna obtusifolia*) seed production and viability as influenced by late-season postemergence herbicide applications. *Weed Science* 45: 497–501.

Taylorson, R.B. (1972). Phytochrome controlled changes in dormancy and germination of curried weed seeds. *Weed Science* 20: 417–422.

Teasdale, J.R. (1995). Influence of narrow row/high population corn on weed control and light transmittance. *Weed Technology* 9: 113–118.

Teasdale, J.R. (1996). Contribution of cover crops to weed management in sustainable agricultural systems. *Journal of Production Agriculture* 9: 475–479.

Teasdale, J.R. and Abdul-Baki, A.A. (1998). Comparison of mixtures vs monocultures of cover crops for fresh-market tomato production with and without herbicide. *Horticulture Science* 33: 1163–1166.

Teasdale, J.R. and Frank, J.R. (1983). Effect of row spacing on weed competition with snap beans (*Phaseolus vulgaris*). *Weed Science* 31: 81–85.

Teasdale, J.R. and Mohler, C.L. (1993). Light transmittance, soil temperature, and soil moisture under residue of hairy vetch and rye. *Agronomy Journal* 85: 673–680.

Teasdale, J.R. and Mohler, C.L. (2000). The quantitative relationship between weed emergence and the physical properties of mulches. *Weed Science* 48: 385–392.

Teasdale, J.R., Rice, C.P., Cai, G., and Magnum, R.W. (2012). Expression of allelopathy in the soil environment: soil concentration and activity of benzoxazinoid compounds released by rye cover crop residue. *Journal of Plant Ecology* 213: 1893–1905.

Thompson, K. (1992). The functional ecology of seed banks. In: *Seeds: The Ecology of Regeneration in Plant Communities* (ed. M. Fenner), 231–258. CAB International.

Thompson, K. (1993). Seed persistence in soil. In: *Methods in Comparative Plant Ecology* (ed. G.A.F. Hendry and J.P. Grime), 199–202. Chapman & Hall.

Thompson, K. (2000). The functional ecology of soil seed banks. In: *Seeds: The Ecology of Regeneration in Plant Communities* (ed. M. Fenner), 215–235. CAB International.

Thompson, K., Bakker, J.P., Bekker, R.M., and Hodgson, J.G. (1998). Ecological correlates of seed persistence in soil in the north-west European flora. *Journal of Ecology* 86: 163–169.

Thompson, K., Band, S.R., and Hogdson, G. (1993). Seed size and shape predict persistence in soil. *Functional Ecology* 7: 236–241.

Thompson, K., Green, A., and Jewels, A.M. (1994). Seeds in soil and worm casts from a neutral grassland. *Functional Ecology* 1: 29–35.

Thompson, K. and Grime, J.P. (1979). Seasonal variation in the seed banks of herbaceous species in ten contrasting habitats. *The Journal of Ecology* 67: 893–921.

Thomson, F.J., Moles, A.T., Auld, T.D. et al. (2010). Chasing the unknown: predicting seed dispersal mechanisms from plant traits. *Journal of Ecology* 98: 1310–1318.

Tidemann, B.D., Hall, L.M., Harker, K.N. et al. (2017). Suitability of wild oat (*Avena fatua*), false cleavers (*Galium spurium*), and volunteer canola (*Brassica napus*) for harvest weed seed control in Western Canada. *Weed Science* 65: 769–777.

Tittonell, P. and Giller, K.E. (2013). When yield gaps are poverty traps: the paradigm of ecological intensification in African smallholder agriculture. *Forage Crop Research* 143: 76–90.

Tiwari, R., Reinhardt Piskáčková, T.A., Devkota, P. et al. (2021). Growing winter Brassica carinata as part of a diversified crop rotation for integrated weed management. *GCB Bioenergy* 13: 425–435.

Tschumi, M., Ekroos, J., Hjort, C. et al. (2018). Rodents, not birds, dominate predation-related ecosystem services and disservices in vertebrate communities of agricultural landscapes. *Oecologia* 188: 863–873.

USDA-Agricultural Research Service. (2006). National program 308 methyl bromide alternatives action plan. 2007–2012.

Van Acker, R.D., Swanton, C.J., and Weeise, S.F. (1993). The critical period of weed control in soybean [*Glycine max* (L.) Merr.]. *Weed Science* 41: 194–200.

Van Wychen, L. (2015). Baseline survey of the most common and troublesome weeds in the United States and Canada. National Survey Data set. Weed Science Society of America.

Van Wychen, L. (2016). Survey of the most common and troublesome weeds in broadleaf crops, fruits & vegetables in the United States and Canada. National Survey Data set. Weed Science Society of America.

Vandermeer, J. (1989). *The Ecology of Intercropping*. Cambridge UK: Cambridge University Press.

Vegis, A. (1964). Dormancy in higher plants. *Annual Review Plant Physiology* 15: 185–224.

Venable, D.L. (2007). Bet hedging in a guild of desert annuals. *Ecology* 88: 1086–1090.

Verret, V., Gardarin, A., Pelzer, E. et al. (2017). Can legume companion plants control weeds without decreasing crop yield? A meta-analysis. *Field Crops Research* 204: 158–168.

Villiers, T.A. (1973). Ageing and the longevity of seeds in field conditions. In: *Seed Ecology, Proceedings of the 19th Easter School in Agricultural Sciences, University of Nottingham, 1972* (ed. E. Heydecker), 265–288. Pennsylvania State University Press.

Villiers, T.A. (1980). Ultrastructural changes in seed dormancy and senescence. In: *Senescence in Plants* (ed. K.V. Thimann), 39–66. CRC Press.

Villiers, T.A. and Edgcumbe, E.H. (1975). On the cause of seed deterioration in dry storage. *Seed Science and Technology* 3: 761–774.

Vleeshouwers, L.M., Bouwmeester, H.J., and Karssen, C.M. (1995). Redefining seed dormancy: an attempt to integrate physiology and ecology. *Journal of Ecology* 83: 1031–1037.

Wagner, M. and Mitschunas, N. (2008). Fungal effects on seed bank persistence and potential applications in weed biocontrol: a review. *Basic and Applied Ecology* 9: 191–203.

Walker, E.R. and Oliver, L.R. (2008). Weed seed production as influenced by glyphosate applications at flowering across a weed complex. *Weed Technology* 22: 318–325.

Walsh, M., Newman, P., and Powles, S. (2013). Targeting weed seeds in-crop: a new weed control paradigm for global agriculture. *Weed Technology* 27: 431–436.

Walsh, M.J., Broster, J.C., Schwartz-Lazaro, L.M. et al. (2018). Opportunities and challenges for harvest weed seed control in global cropping systems. *Pest Management Science* 74: 2235–2245.

Walsh, M.J., Harrington, R.B., and Powles, S.B. (2012). Harrington seed destructor: a new nonchemical weed control tool for global grain crops. *Crop Science* 52: 1343–1347.

Walsh, M.J. and Powles, S.B. (2007). Management strategies for herbicide-resistant weed populations in Australian dryland crop production systems. *Weed Technology* 21: 332–338.

Walsh, M.J. and Powles, S.B. (2014). High seed retention at maturity of annual weeds infesting crop fields highlights the potential for harvest weed seed control. *Weed Technology* 28: 486–493.

Walters, C. (1998). Understanding the mechanisms and kinetics of seed aging. *Seed Science Research* 8: 223–244.

Walters, C., Wheeler, L.M., and Grotenhuis, J.M. (2005). Longevity of seeds stored in a genebank: species characteristics. *Seed Science Research* 15: 1–20.

Weaver, S.E. (1984). Critical period of weed competition in three vegetable crops in relation to management practices. *Weed Research* 24: 317–325.

Webster, T.M., Cardina, J., and White, A.D. (2003). Weed seed rain, soil seedbanks, and seedling recruitment in no-tillage crop rotations. *Weed Science* 51: 569–575.

Weisberger, D., Nichols, V., and Liebman, M. (2019). Does diversifying crop rotations suppress weeds? A meta-analysis. *PLoS One* 14: 1–12.

Wells, M.S., Reberg-Horton, S.C., Smith, A.N., and Grossman, J.M. (2013). The reduction of plant available nitrogen by cover crop mulches and subsequent effects on soybean performance and weed interference. *Agronomy Journal* 105: 539.

Werle, R., Sandell, L.D., Buhler, D.D. et al. (2014). Predicting emergence of 23 summer annual weed species. *Weed Science* 62: 267–279.

Wesson, G. and Wareing, P.F. (1969). The induction of light sensitivity in weed seeds by burial. *Journal of Experimental Botany* 20: 414–425.

Widderick, M.J., Keenan, M., and Walsh, M.J. (2014). Harvest weed seed control: is there a role in northern region farming systems? In: *19th Australasian Weeds Conference-Science, Community and Food Security: The Weed Challenge* (ed. M. Baker). Tasmanian Weed Society.

Wiggins, M.S., Hayes, R.M., Nichols, R.L., and Steckel, L.E. (2017). Cover crop and postemergence herbicide integration for Palmer amaranth control in cotton. *Weed Technology* 32: 348 355.

Wiles, L. and Brodahl, M. (2004). Exploratory data analysis to identify factors influencing spatial distributions of weed seed banks. *Weed Science* 1: 936–947.

Willems, J.H. and Huijsmans, K.G. (1994). Vertical seed dispersal by earthworms: a quantitative approach. *Ecography* 17: 124–130.

Williams, J.T. and Harper, J.L. (1965). Seed polymorphism and germination, I. The influence of nitrates and low temperatures on the germination of Chenopodium album. *Weed Research* 5: 141–150.

Williams, M.M. (2006). Planting date influences critical period of weed control in sweet corn. *Weed Science* 54: 928–933.

Wookey, C.B. (1985). Weed control practice on an organic arm. *Proceedings of the 1985 British Crop Protection Conference: Weeds, British Crop protection Council*, 577–582.

Yang, L., Liu, S., and Lin, R. (2020). The role of light in regulating seed dormancy and germination. *Journal of Integrative Plant Biology* 62: 1310–1326.

Yang, Y.Q., Yi, X.F., and Niu, K.K. (2012). The effects of kernel mass and nutrition reward on seed dispersal of three tree species by small rodents. *Acta Ethologica* 15: 1–8.

Yelverton, F.H. and Coble, H.D. (1991). Narrow row spacing and canopy formation reduces weed resurgence in soybeans (*Glycine max*). *Weed Technology* 5: 169–174.

Yenish, J.P., Doll, J.D., and Buhler, D.D. (1992). Effects of tillage on vertical distribution and viability of weed seed in soil. *Weed Science* 40: 429–433.

Young, B.L. (2021). Integration of harvest-time and post-harvest tactics for integrated management of johnsongrass (*Sorghum halepense*) in grain sorghum (*Sorghum bicolor*). Master's thesis, Texas A&M University. Available electronically from https://hdl.handle.net/1969.1/195565.

Zimdahl, R.L. (1987). The concept and application of the critical weed free period. In: *Weed Management in Agroecosystems: Ecological Approaches* (ed. M.A. Altieri and M. Liebman), 145–155. CRC Press.

Zohry, A. and Ouda, S. (2018). Crop rotation defeats pests and weeds. In: *Crop Rotation*. 77–88. Cham: Springer.

12

The Role and Relationship of Tillage Systems with Ecologically Based Weed Management Approaches

Thomas Gitsopoulos[1,*] *and Ioannis Vasilakoglou*[2]

[1] *HAO-Demeter, Institute of Plant Breeding and Genetic Resources, Greece*
[2] *University of Thessaly, Dept. of Agriculture – Agrotechnology, Greece*
[*] *Corresponding author*

12.0 Introduction

Weeds compete with crops for water, nutrients, light, and space leading to high crop yield reductions (Oerke 2006). Herbicides, tillage, and soil cultivation are considered the most important weed control practices to control weeds. Herbicides, though, are regarded as the primary method for managing weeds in most countries, globally presenting greater advantage over all the other practices mainly due to their ease and speed of application and their high effectiveness. Herbicides usually save time, money, and labor. However, their continuous use has resulted in 266 herbicide-resistant weed species globally evolving resistance to 21 of the 31 known herbicide sites of action (SoA) and to 165 herbicides (Heap 2022). Moreover, there have been concerns due to herbicide effect on both humans and environment (Bernardes et al. 2015; Mendes et al. 2019).

Herbicide resistance, stricter regulation regarding herbicide registration, and increased interest in organic agriculture call for alternative solutions for weed management with less reliance in herbicide use (Bastiaans et al. 2008). One option, tillage and soil cultivation, are important tools to control weeds by uprooting weed seedlings, exposing underground vegetative propagules of perennial weeds to sun or killing frosts, or by burying weed seeds below optimal emergence zones (Knezevic et al. 2017). Tillage practices are also alternatives to managing herbicide resistance usually in combination with other cultural tactics (Owen et al. 2014). However, they may harm the environment and the natural resources contributing to soil compaction and erosion, soil structure damage, air pollution, water and pesticide runoff and loss of soil organic content. Tillage is regarded as the most energy-intensive operation in cropping systems (Al-Kaisi and Yin 2005; Sharma et al. 2011).

Ecologically based weed management approaches mainly refer to the three principles for reducing yield losses, that is: (i) reducing weed seedling recruitment, (ii) improving crop competitiveness (both applied within the growing season) and (iii) depleting weed density through the gradual depletion of the soil seedbank (as a long-term strategy) (Bastiaans et al. 2008; Bàrberi 2019). Ecologically based weed management approaches encompass proactive noncurative (indirect) methods for long-term weed suppression with at least possible and wise use of reactive or curative (direct) weed control methods (herbicides, thermal, biological, and mechanical weed control) during the crop-growing cycle (Teasdale 2003; Bastiaans et al. 2008; Bàrberi 2019). In this framework, ecological weed management stresses the importance of reducing environmental issues and threats to human health caused by the extensive use of soil cultivation and herbicides to directly control weeds. Tillage systems affect the vertical distribution of weed seeds in the soil surface and determine the presence or absence of crop residues and their bulk density on the soil surface. As a result, weed seed germination, predation and decay, seedling emergence, vegetative perennial propagation, weed-crop interactions, and seed production are significantly affected by each tillage system applied with consequences in the weed seedbank size.

The objective of the present study is to present the role of tillage systems and discuss their effect on the three basic principles of ecological weed management already mentioned. This will provide knowledge and foster awareness regarding the relationship between tillage systems and ecological weed management, helping farmers make decisions regarding the effectiveness of ecological weed management under different tillage system applied.

Ecologically Based Weed Management: Concepts, Challenges, and Limitations, First Edition. Edited by Nicholas E. Korres, Ilias S. Travlos, and Thomas K. Gitsopoulos.

12.1 Tillage Systems

The term *tillage* refers to the mechanical disturbance of soil that is conducted for seedbed preparation. Tillage reduces the soil compaction and mulches the residues of the previous crop. It is distinguished in the main or primary and in the secondary tillage (Greil et al. 2002). The main tillage is applied with heavy machinery, such as the reversible plow (moldboard plow), the chisel plow, the chisel, the disc (heavy offset harrow disc), and the rotary harrow, which move slowly and disturb the ground to a great depth. In most cases, the main tillage inverts the soil. It is distinguished to (i) deep tillage (usually at 25–40 cm according to the crop) (Table 12.1), (ii) subsoiling (at 60–70 cm to break hard pans), and (iii) year-round tillage that is carried out throughout the year for weed growth destruction (soil cultivation) and seedbed preparation. The secondary cultivation follows the main one. It is applied with lighter machinery, such as the rotavator, the tandem harrow disc, and the cultivator, which disturb the soil to a shallow depth (6–10 cm) to break soil clods, uproot weed seedlings, and prepare a firm seedbed.

Both primary and secondary tillage are mainly used for the seedbed preparation. However, at the same time, tillage achieves the destruction of the emerged weeds and the vegetative reproductive organs of perennials. Many of these organs are placed on the soil surface, where they are destroyed by low temperatures, drought, and increased activity of soil microorganisms. Additionally, the soil incorporation of pesticides and fertilizers, as well as the destruction of some ground pests, are achieved by secondary tillage.

The conventional tillage (CT) system, commonly including reversible moldboard plow, tandem harrow disc, and cultivator, is the most common tillage practice in the world to incorporate previous crops' residues, control weeds, break soil clods, and prepare a firm seedbed before sowing (Table 12.2). In general, conventional tillage is characterized by soil inversion with a moldboard plow (primary tillage) followed by disks, field cultivators, and/or harrowing (secondary operations), whereas its application results in less than 15% of the soil covering by crop residues after seeding (Horowitz et al. 2010; Al-Hajaj 2021). However, this tillage system depletes soil moisture and organic matter (Halvorson et al. 2002), increasing the risk of soil erosion and carbon dioxide emission (Kassam et al. 2019).

Conversely, conservation tillage leaves at least 30% of the previous year's crop residue before and after sowing the next crop (Carter 2005) (Table 12.2). This tillage system aims to reduce soil erosion and water runoff and improve soil structure and production efficiency (Griffith et al. 1992; Choudhary et al. 1997; Uri 1999; Carter 2005; Wang et al. 2006). It has been characterized as environmentally friendly because it reduces carbon dioxide, nitrous oxide, and methane emissions from agricultural land, which contribute to the greenhouse effect (Blanco-Canqui and Lal 2008; Horowitz et al. 2010; Kan et al. 2020). In addition, soil temperature, bulk density, and organic matter accumulation are increased by the long-term application of conservation tillage systems (Griffith et al. 1992). As shown in Table 12.2, mulch, ridge, strip, and no-till are types of tillage, while the main forms of conservation tillage are (i) reduced, (ii) minimum, and (iii) shallow tillage (Carter 2005).

Table 12.1 Optimal tillage depth required for the establishment of major field crops.

Crop	Scientific name	Tillage depth (cm)	Reference
Alfalfa	*Medicago sativa*	40–45	Birkàs et al. 2014
Corn	*Zea mays*	25–40	
Cotton	*Gossypium hirsutum*	35–40	Fox et al. 2018
Durum wheat	*Triticum durum*	22–28	Birkàs et al. 2014
Faba bean	*Vicia faba*	25–30	
Garden pea	*Pisum sativum*	20–30	
Grain sorghum	*Sorghum bicolor*	25–30	
Kidney bean	*Phaseolus vulgaris*	20–25	
Oilseed rape	*Brassica napus*	30–35	
Soybean	*Glycine max*	30–40	Botta et al. 2022
Sugar beet	*Beta vulgaris*	30–40	Birkàs et al. 2014
Sunflower	*Helianthus annuus*	25–35	
White mustard	*Sinapis alba*	20–22	
Winter wheat	*Triticum aestivum*	27–30	Botta et al. 2022

Table 12.2 The main characteristics of the tillage systems (modified from Carter 2005).

Tillage systems		Main characteristics of the tillage systems
Conventional tillage system		Primary and secondary tillage operations are applied for optimal seedbed preparation.
Conservation tillage system		
Conservation tillage forms:		
	Reduced tillage	Reduction in total number of primary and secondary tillage operations.
	Minimum tillage	Minimum use of primary of primary and secondary tillage operations to meet to crop production requirements.
	Shallow or non-inversion tillage	Primary tillage is applied to shallow soil depth (<15 cm) without soil inversion.
Conservation tillage types:		
	Mulch tillage	Crops are seeded into shallow cultivated soil covered by high percentage (≥30%) of crop residues.
	Ridge tillage	Crops are planted on raised ridges or shallow beds in rows. Crop residues are left on the surface between ridges.
	Zone or strip tillage	Crops are seeded into cultivated bands (strips) separated by undisturbed soil bands.
	No-tillage or Zero tillage	Crops are seeded directly into undisturbed soil (not any type of primary or secondary tillage is applied).

The minimum and the reduced tillage forms include primary tillage with chisel plow or disk focusing to a minimum number of tillage operations. However, in the mulch tillage type ≥30% of crop residues are left on the soil surface. On the contrary, in the no-tillage type, the crop residues are left on the soil surface and the soil is left undisturbed from harvest to seeding of the next crop (Carter 2005). Strip or zone tillage type combines the benefits of the no-tillage and the conventional tillage, but tillage is confined to 15- to 20-cm strips into which fertilizers and seeds are placed (Carter 2005). Under strip tillage, narrow strips for planting are created, whereas the soil between strips remain untilled maintaining previous crops residues. Conservation tillage as a pillar of conservation agriculture (CA) is related to permanent soil coverage and to diverse crop rotation. The conservation tillage systems have been widely accepted and adopted for more than 180,000 ha worldwide. In fact, they were increased by 70% in the 2015/16 growing season as compared with the 2008/09 season (Kassam et al. 2019).

Tillage affects weed communities' composition as it incorporates weed seeds and vertically distributes them in the soil (Yenish et al. 1992). Variability in the vertical seed distribution pattern depends on soil humidity, soil texture, ploughing depth, and working speed of tillage implements (van Esso et al. 1986). However, the distribution posed to weeds by tillage depends on the type of implement used rather than on the tillage depth. Implements that do not invert the soil, such as chisels (although they can plow at 45 cm depth), contribute to the distribution of most of the weed seeds in the upper soil horizon favoring the abundance of those species that germinate from shallow depth (Gruber and Claupein 2009). In moldboard plowed fields, most of the weed seeds are placed in 10–15 cm depth, whereas in chisel-plowed fields a high percentage (80–90%) of them is distributed in the top 10 cm; in NT systems weed seeds accumulate on or close to the soil surface in the top 0–5 cm of the soil (Cousens and Moss 1990; Yenish et al. 1992; Hoffman et al. 1998; Murphy et al. 2006).

12.2 Ecologically Based Weed Management Approaches and Tillage Systems

12.2.1 Reduced Weed Seedling Recruitment from Weed Seedbank

Weed seedling recruitment is the successful germination and emergence of weeds (Harper 1977). A reduction in weed density within a growing season can be obtained through a reduced weed seed germination and seedling recruitment (Bastiaans et al. 2008). Tillage affects the survival and dispersal of weed seeds, the survival and spread of the underground vegetative propagules of perennial weeds, and the weed seed germination, mainly by the vertical distribution of the seeds/propagules in the soil, the exposure to light, heat, or cold and by altering many other aspects (e.g.

temperature) of the soil environment that influence seed dormancy and germination (Yenish et al. 1992; Scopel et al. 1994; Gallagher and Gardina 1998; Mirsky et al. 2010; Merfield 2013). Conventional tillage with ploughing distributes weed seeds deeper in the soil, compared to reduced- or no-tillage systems that accumulate weed seeds close or on the soil surface, that affects seed viability and seedling recruitment (Chauhan et al. 2006d; Zhang et al. 2019). Cardina et al. (2002) reported that common chickweed (*Stellaria media L. Vill.*) did not tolerate deep plowing with moldboard plow (50 germinable seeds m^{-2}) but survived with absence or moderate soil disturbance (150 and 210 seeds m^{-2} in no-tillage (NT) and chisel plow (CP), respectively). Similarly, barnyardgrass (*Echinochloa crus-galli L. P. Beauv.*) seeds did not survive well at the soil surface in NT (20 seeds m^{-2}) or when buried deep in the soil in moldboard plow (80 seeds m^{-2}); however, they were favored to moderate soil disturbance with chisel plow (180 seeds m^{-2}). The densities of both redroot pigweed (*Amaranthus retroflexus L.*) and corn speedwell (*Veronica arvensis* L.) were lower in moldboard plow (310 and 90 seeds m^{-2}, respectively) compared to NT plots (620 and 560 seeds m^{-2}, respectively), whereas higher densities of *Oxalis stricta* L. (390 seeds m^{-2}) and *Chenopodium album* L. (100 seeds m^{-2}) were recorded in NT. Regarding the effect of seed burial by tillage on weed species emergence it has been revealed that palmer amaranth (*Amaranthus palmeri L. S. Wats.*) seedlings cannot emerge from depths of 5 cm or more; therefore, a method to reduce palmer amaranth densities is to bury seeds buried deeper than 5 cm (Aulakh et al. 2012, 2013). For the same weed species another study reported around 50% control with a moldboard plow that buries its seeds at least 10 cm (Prostko 2012). Reducing seedling recruitment in such way is important for the management of herbicide resistant weeds (Prostko 2012). In other cases, adoption of reduced-tillage fallow was limited for the control of winter annual grass weeds such as for downy brome (*Bromus tectorum* L.) in the western USA, because shallow (or eliminated) tillage operations did not bury its seeds deep enough to prevent seedling emergence (Kettler et al. 2000).

As stated for CA systems most of the weed seeds are on or near the soil surface and this may cause a large accumulation of seeds on the soil that may lead to higher seedling recruitment in the upper soil layers. Therefore, deep inversion tillage operation may be useful to bury weed seeds in depths that are not able to emerge. Kettler et al. (2000) reported that occasional tillage with the moldboard plow in a reduced-or no-tillage system will help control annual grass weeds. Occasional tillage may be also employed for the control of perennial weeds (Armengot et al. 2016), or to disadvantage low-dormant grass species whose seeds are usually unable to emerge from deep soil layers (Bàrberi 2003). A previous study suggested ploughing once every five years systems to manage *Alopecurus myosuroides Huds.* that thrived in the reduced tillage fields (Cussans and Moss 1982). An occasional inversion tillage in autumn has been suggested to greatly reduce palmer amaranth emergence in farms highly infested with glyphosate-resistant palmer amaranth (Aulakh et al. 2013). Moldboard operation though, can bring deeply buried weed seeds to depths suitable for seed germination, although the return of buried seeds has been reported to be much less than the burial of surface seeds by a factor of 2–13 times (van Esso et al. 1986; Mohler et al. 2006). However, repeated moldboard ploughing may lead to continuous reintroduction of seeds. Deeply buried seeds with moldboard ploughing may reduce short-term weed numbers but extend the potential longevity of the seedbank (Zhang et al. 2019). To avoid weed seed lift moldboard may be practiced every 4–8 years. Overall, weed seed response to seed burial is a matter of emergence pattern of seeds buried at different depths.

Regarding the recruitment of perennial weeds, deep tillage in CT is applied as a preventive method that reduces the abundance of these weed species. This is achieved when the underground vegetative propagules of these weeds are exposed to sunlight or cold (Govindasamy et al. 2020). Desiccation of roots and rhizomes by heat in summer fallowing for *Sorghum halepense* (L.) Pers., *Cynodon dactylon* (L.) Pers., and *Cyperus rotundus* L. or by freezing for *Cirsium arvense* (L.) Scop., have been widely used as preventive weed control (reviewed by Schimming and Messersmith 1988; Mohler 2001) after moldboard plowing. However, CT with rotovators or RT at shallower depths with tines or discs may break up the vegetative reproductive organs of perennial weeds and partially scatter them across the field, promoting their further spread and persistence in infested areas (Froud-Williams et al. 1981; Peigné et al. 2007; Armengot et al. 2016). In contrast, RT can prevent and reduce the spread of perennial propagules within a field after tillage operations due to less field traffic and machinery operations compared to conventional tillage (Nichols et al. 2015).

In CT systems, many buried seeds are exposed to light after tillage operation, whereas in NT a large proportion of weed seedbank accumulates on or close to the soil (Bàrberi and Lo Cascio 2001; Govindasamy et al. 2020). Light is a necessary promoter for seed germination of many weed species (Merfield 2013). Therefore, germination of light-sensitive weed seeds may be favored in CT systems compared to NT. Photocontrol (tilling soil in darkness) has been reported as a practice in CT to prevent light flash hitting weed seeds exposed to the soil surface (Hartmann and Nezadal 1990). Tillage performed at night or covering tillage equipment with black plastic or non-light-transmitting cloth to reduce weed emergence have been investigated to reduce weed germination (Hartmann and Nezadal 1990; Ascard 1994; Riemens et al. 2007). Although this preventive ecologically based approach may substantially reduce seedling recruitment, the results have been highly variable between weed species, year, location, and soil type (Riemens et al. 2007). This is mainly due to variable seed dormancy and insensitivity of many weeds to light for seed germination (Juroszek and Gerhards 2004; Bastiaans et al. 2008; Bàrberi 2019). For example, some weed species such as *Avena fatua* L., *Malva parviflora* L. germinate equally in light and dark; seeds of *Gallium tricornutum* Dandy do not germinate in the light; seeds of *Eclipta prostrata* L. and *Cyperus difformis* L. do not germinated in the dark at all, whereas seed germination in *barnyardgrass Sisymbrium orientale* L. is stimulated by light, although this is not an absolute requirement for seed germination (Boyd and Van Acker 2004; Chauhan et al. 2006a, 2006b, 2006c; Chauhan and Johnson 2010).

A significant approach to reducing weed seedling recruitment is mulching, particularly that of plant material remaining after harvest. Mulches spread or laid over the soil surface suppress weed germination and establishment by acting as a physical barrier, altering radiation or through the release of allelopathic compounds, thus making the environment at the soil surface unsuitable for weed emergence (Reddy 2003; Reeves et al. 2005; Marles et al. 2010). Cash crops such as rice (*Oryza sativa* L.), barley (*Hordeum vulgare* L.), wheat (*Triticum aestivum* L.), oat (*Avena sativa* L.), oilseed rape (*Brassica napus* L.), sunflower (*Helinathus annus* L.), or sorghum (*Sorghum bicolor* L.) can suppress weeds with the release of allelopathic compounds when their residues are left on the soil surface after harvest (Boydston and Hang 1995; Ben-Hammouda et al. 2001; Chung et al. 2002; Chon and Kim 2004; Reeves et al. 2005; Dhima et al. 2006b; Kong et al. 2008; Norsworthy et al. 2011).

Other crops defined as cover crops grown in rotations for ecological benefits (e.g. prevention of soil erosion, conservation of soil moisture, enhancing soil health) and not for harvest can change the dynamics of weeds either in the form of living plants or as dead plant residues after their termination (Teasdale et al. 2007). Cover crops compete with weeds for light, space, nutrients, and water and have the potential to reduce weed biomass and weed seedbank (Mirsky et al. 2010; Alonso-Ayuso et al. 2018; MacLaren et al. 2019; Smith et al. 2020). After their termination, cover crop residues remaining on the soil surface as mulches prevent weed seed germination by obstructing light penetration, decreasing soil temperature, affecting soil moisture, and reducing oxygen availability; they also act as physical barriers for weed seedling emergence, and some of them release allelopathic compounds that affect weed seed germination and seedling emergence (Bartley and Frankland 1982; Barnes et al. 1987; Teasdale and Mohler 1993; Mirsky et al. 2011; De Vore et al. 2013; Rawat et al. 2017; Hussain et al. 2021; Nikolić et al. 2021). Cover crops with allelopathic properties may be different species – for example, cereal rye (*Secale cereale* L.) or black oat (*Avena strigosa* Schreb.) from cereals, hairy vetch (*Vicia villosa* Roth), or clovers (*Trifolium* spp.) from legumes, brown mustard (*Brassica juncea* (L.)), or white mustard (*Sinapis alba* (L.)) from brassicas (Teasdale et al. 1991, 2012; Liebl et al. 1992; Johnson et al. 1993; Yenish et al. 1996; Burgos and Talbert 2000; Dhima et al. 2006b; Norsworthy et al. 2011; Buchanan et al. 2016). Crop residues on the soil surface are widely practiced in NT or RT systems as maintenance of permanent soil cover is one of the three principles of CA ("minimum soil disturbance" and "diversification of plant species" the other two) (FAO 2022).

In contrast, cover crops in CT are usually used as green manure and they are soil incorporated by tillage before cash crop planting. In NT cover cropping systems, herbicides are the dominant practice to terminate cover crops prior to cash crop planting. Mowing or rolling down cover crops can replace herbicides to implement ecological weed management and reduce reliance upon herbicides. Mowing is a method employed to terminate cover crops by using flail or rotary mower. This method, though, has several drawbacks; the small pieces of the mowed crop residues are less thick, allowing light to reach germinating weed seedlings, decompose easily, and may be unevenly distributed on the soil surface, whereas regrowth of cover crops could occur. In contrast, roller crimpers or the rolling stalk choppers flatten and crimp cover crops, creating a thick and uniform longer-lasting mat of mulch providing potential for long weed suppression (Creamer and Dabney 2002; Ashford and Reeves 2003; Teasdale 2003; Davis 2010).

The management of cover crop residues is significant for successful weed management. Killing cover crops without herbicides depends on the growth stage and species involved (Reberg-Horton et al. 2012). Different residue management may alter the suppressive effect of the residues and the weed species response. The germination response of weeds depends on weed biology and on the quantity, the position and the allelopathic potential of the crop residues (Chauhan et al. 2006a, 2006b). Chauhan and Abugho (2013) studying the effect of rice residues on seedling emergence of eight weed species in ZT reported that the greatest and most substantial reductions in weed emergence occurred with 6 t ha^{-1}. Mirsky et al. (2012) reported that cover crop biomass must reach more than 8 t ha^{-1} for satisfactory weed control. Residues on the soil surface as in NT decompose at slower extent compared to those incorporated in the soil as in CT (Dou et al. 1995) resulting in slower release rate of allelochemicals but, in longer-lasting release rate of allelochemicals (Kruidhof et al. 2009). In tilled soils, faster degradation of winter rye mulches has been reported due to better contact with soil microbes than when left as mulch on the soil surface (Pinnamaneni et al. 2022). Cover crops effectively reduced early season palmer amaranth emergence in both inversion tillage and noninversion tillage system in cotton (*Gossypium hirsutum* L.) (Aulakh et al. 2013; Price et al. 2016).

In undisturbed soils, though, greater effect on weed seed germination may be achieved, since the majority of weed seeds are accumulated at or on the soil surface in proximity to allelopathic compounds released by the crop residues (Nichols et al. 2015); however, it has been stated that in NT systems allelopathic compounds may not diffuse deeply into the soil profile to reach weed seeds or germinating seedlings (Liebman and Mohler 2001). Tabaglio et al. (2013) studying the allelopathic effect of rye residues on summer weeds in NT and CT in a subsequent maize (*Zea mays* L.) crop they reported greater grass and broadleaf control in NT system.

Regarding the effect of residue placement for winter rye the surface mulch proved a relatively stronger and more consistent inhibitor of emergence of the receptor species tested compared to soil-incorporated residues (Kruidhof et al. 2009). The opposite, however, was recorded for oilseed rape residues for the same receptors, indicating that the effect of placement of the residue material on weed management seems to be cover crop species dependent (Kruidhof et al. 2009). In a study with *V. villosa* and *S. cereale* used as cover crops it was revealed different response of the weed species studied; *A. retroflexus*, *C. album*, *Panicum capillare* L. and *E. crus-galli* showed a monotonic decline in emergence with increasing rates of residues, whereas the emergence of *Rumex crispus* L., *S. media*, and *Taraxacum officinale* Webber in Wiggers increased at low residue rates (Mohler and Teasdale 1993). In the same study, *Abutilon theophrastii* Medik. appeared to be the least sensitive to both residues, either due to its negligible germination response to light (Horowitz and Taylorson 1984), or/and to its large seed size that provide some resistance to allelopathic toxins (Putnam et al. 1983). Small-seeded annuals are more affected by the presence of crop residue on the soil surface (Teasdale and Mohler 1993; Liebman and Davis 2000; Teasdale 2003).

Cover crop termination is critical to maximize weed suppression and to minimize regrowth of cover crop that in turn minimizes volunteer problems in the subsequent crop, as reported for cereal rye that has become an important weed in winter wheat (Keene et al. 2017). Proper timing of cover crop termination should allow restorage of soil moisture depleted by cover crops, and this usually takes place three weeks before the planting of the next crop (Price et al. 2009). Moreover, an adequate time from cover crop termination to planting the next crop is needed to avoid allelopathic effects (Reeves 1994). As mentioned earlier, in CT systems cover crops are soil incorporated, contributing to reduction in seedling recruitment due to allelochemicals. However, soil disturbance may also enhance seedling germination due to seed light exposure or due to the high quantity of soluble nitrogen released in the soil when legumes are used as green manures that stimulate weed seed germination (Ballaré et al. 1992; Blum et al. 1997). A novel approach to enhance the reduction of seedling recruitment is to promote weed seed dormancy with the use of abscisic acid (ABA) hormone combined with the use of cover crops. Both cover crops and ABA may also enhance microbial activity for weed seed decomposition over time. ABA may be combined with herbicides for cover crop termination and/or with PRE herbicides (Oliveira et al. 2020). The impact of ABA on cultivated and noncultivated seeds and its fate and effectiveness under field conditions, especially when combined with PRE herbicides and the presence of cover crop residues should be investigated (Oliveira et al., 2020).

Weed suppression by cover crops is generally affected by cultivar and environmental conditions and management practices (Schomberg et al. 2006; Vasilakoglou et al. 2006). Inclusion of cover crops in a production system has been proved to be an effective method for suppressing weeds that could lead to a lesser need for weed control treatments (Alberts and Neibling 1994; Dabney et al. 2001; Korres and Norsworthy 2015; Weber et al. 2017; Nikolić et al. 2021). Cover crops used in conservation tillage may offer an early season weed suppression that may contribute to the reduction in herbicide use (Teasdale 1996; Yenish et al. 1996; Reeves et al. 2005; Teasdale et al. 2005; Vasilakoglou et al. 2006; Mischler et al. 2010). To achieve such reduction in herbicide use, proper management of cover crop residues is needed. Cover crops may decrease the use of PRE herbicides; however, the same time may increase the use of POST herbicides. This is more evident in NT systems where soil cultivation is limited compared to CT and cover crops cannot successfully control weeds.

12.2.2 Improved Crop Competitiveness

Any management practice that shifts the temporal and/or spatial access to resources in a direction favorable to the crop and unfavorable to weeds enhances crop competitiveness (Bàrberi 2019). Early crop emergence, narrow interrow space, homogenous crop spatial arrangement and adjusted crop planting date delaying weed seedling emergence relative to the crop emergence are regarded as ecologically based approaches that can increase crop competitiveness over weeds (Bastiaans et al. 2008; Bàrberi 2019). However, tillage regimes may interfere with these practices influencing their effect.

Crop residues discussed previously apart from suppressing weeds may suppress the cash crop by interfering with its establishment regarding seed placement, by releasing phytotoxins, by maintaining cooler soils or by enhancing seeding diseases (Teasdale et al. 2007). Inappropriate management of crop residues may lead to poor crop establishment or delays in crop emergence that will adversely impact crop-weed competition leading to yield loss. As stated, the thickness of the mulch layer plays a significant role for effective weed suppression (Teasdale and Mohler 2000). A high biomass of mulch uniformly distributed across the soil surface with long persistence should be achieved for optimum weed suppression. High amount of crop residue (resulting in lower amplitude of temperature fluctuations, lower light level, and increased concentrations of allelopathic compounds) is normally needed to suppress weed germination and growth, although, it is important to determine that this does not hinder crop emergence (Chauhan et al. 2012). Unevenly distributed and/or high bulk density of crop residues may affect the optimum and uniform seeding depth, causing thin stands, poor crop establishment or retardation in vegetative crop growth. Likewise, crop residues may promote wet conditions and lower soil temperature leading to fungal contamination of crop seeds. Crop residues in NT must be adequately handled during seeding, so they will not be buried into the furrow interfering with seed germination and seedling emergence or clump over the seed rows and block seeders (Aikins et al. 2019). Weber et al. (2017) reported a significant reduction in soybean (*Glycine* max L. Merr.) density in NT resulting in lower crop yield due to poor plant emergence associated to thick mulch layer that caused technical problems at soybean seeding.

In NT systems furrow openers for crop seeding should maintain adequate surface residue distribution and regular inter-plant spacing, whereas they should provide accurate and uniform placement of the seeds and fertilizer (Aikins et al. 2020). Seed-fertilizer adequate separation within the furrow in no-tillage farming is required to minimize the effect of fertilizer toxicity and ensure efficient access by plants (Baker et al. 2007; O'Donovan et al. 2008). Seeding machines with no separation between N caused 48% reduction in oilseed rape (*Brassica napus* L.) establishment, whereas as tine openers, either with a single or double seed chute, that separate seed and fertilizer resulted in the best oilseed rape performance (Swanepoel and Labuschagne 2020). The same authors highlighted the risk of applied N in the band at the same position as the seed. Farmers should be aware of the moisture status of fields in the springtime as continued vegetative growth of the cover crop may result in depletion of stored soil moisture, particularly if low-precipitation spring conditions occur (Williams et al. 2000). Luna et al. (2012) reported several crop failures or yield reduction in organic crops in NT systems after the use of roller crimper to kill cover crops. In particular, cooler temperatures under cover crop mulch resulted in delays in cotton emergence, whereas the continued growth of the cover crops caused soil surface to dry resulted in reduced cotton crop germination and emergence; continuous growth of cover crops resulted in delay in transplanting eggplants and tomatoes leading to poor stands and yield reduction; in cowpeas cover crops impeded and delayed crop emergence due to their presence into the slot created by disk openers on the no-till planter and also served as a weed causing cowpea yield loss (Luna et al. 2012). Davis (2010) reported soybean yield loss under NT compared to CT due to reduced soil water content after the cover crop vegetation period. Reddy (2001) studying the effect of different cover crop species also in soybean reported reduction in soybean emergence and height relative to no cover controls due to the physical barrier of cover crop residues and to the plant-derived chemicals released. Moreover, crop residues may immobilize soil nitrogen and cause reducing nitrogen use efficiency to the following crops (Radicetti et al. 2016).

To avoid problems in crop establishment in NT systems due to the presence of crop residue farmers should be aware of the proper time needed for cover crop termination and should align the period of the cash and cover crops (Mirsky et al. 2009). Proper timing of cover crop termination is important to avoid regrowth issues or allelopathic effect on the following cash crop (Vollmer et al. 2020). For example, to avoid the allopathic effect of rye on summer cash crops, rye should be terminated at its reproductive stage (Mirsky et al. 2009; Teasdale et al. 2012). Limited regrowth of both winter wheat and rye mowed by a sickle-bar mower was reported when mowing took place after heading (Wilkins and Bellinder 1996). Both rolling and roll-chopping may be preferred than mowing because apart from depositing the residue uniformly on the soil surface they also orientate the residues to facilitate planting (Creamer and Dabney 2002). Increased crop seeding rates (Oplinger and Philbrook 1992) and use of cultivars with longer coleoptile that can tolerate the presence of residues on soil surface (Rebetzke et al. 2005) have been recommended to avoid crop establishment issues. Vincent-Caboud et al. (2017) and Weber et al. (2017) highlighted the necessity of purchasing specific equipment for managing cover crop residue and for successful seeding in NT.

Modifying row spacing is an effective cultural management practice to increase crop competitiveness due to early and rapid canopy closure and to deplete seedbank over time. Fahad et al. (2015) averted that narrow row spacing of wheat (11 cm instead of 15 cm and 23 cm) reduced the growth of *Galium aparine* L. and *Lepidium sativum* L. leading to higher grain yields, highlighting that wider wheat rows lack an overlapping canopy early in growth, allowing substantial weed growth and consequent lower grain yield. In soybean lower weed density by 74% to 76% was recorded at 50-cm row spacing compared with 75-cm and 100-cm row spacing (Daramola et al. 2020). Higher yields and better weed suppression were achieved when maize was planted at 60 cm row distance than planting at 75 or 90 cm (Mashingaidze et al. 2009). Borger et al. (2016) after long-term impacts on row spacing on *Lolium rigidum* L. control reported that narrow interrow spacing of wheat, barley, *field peas* (*Pisum sativum* L.), and chickpeas (*Cicer arietinum* L.) increased crop competitiveness and reduced *L. rigidum* seed production.

In NT wider row spacing is often employed to reduce plugging with seed openers (Lemerle et al. 2001), adjusting from 18 cm for cereals, to between 30 and 50 cm for appropriate management of greater quantities of retained stubble. However, the effect of row spacing on winter cereal crop yield depends on the growing season rainfall, time of sowing and the yield potential of the crop, with greater negative impact of wide rows on cereal yields when high yield potential is expected (Routley et al. 2011). Field trials in New South Wales, Australia, with yield potential greater than 3.5 t ha^{-1} revealed that 30 cm row spacing produced significantly higher yields compared to 40 cm spacing, whereas in central Queensland for wheat yields up to 2 t ha^{-1} there was no effect on increasing row spacing from 25 to 50 cm. For greater yield potentials (2 to more than 4 t ha^{-1}) various yield penalties between 0.2 to 0.6 t ha^{-1} were observed when row spacing increased from 25 to 37.5 or 50 cm (Routley et al. 2011). In addition, cereals grown on wider row spacing may be taller and have the potential for lodging in high yielding years. In general, if row spacing increases to more than 40 cm in cereals this will induce failure in canopy closure making weed control extremely important (Riethmuller et al. 2011).

Increasing row spacing in canola from 18 to 36 cm have led to an average yield loss of 13.7% ranging from 9% to 27% in eight different trials in Australia (Riethmuller et al. 2011). However, pulses in NT responded positively to wider rows depending on the level of rainfall. More particular in lupins, wider rows (50 to 60 cm) had higher yield potential compared to narrow inter-rows under a warm and dry environment of medium to low rainfall in northern Australia, due to less water use during winter and more water availability to the crop at the end of the season to fill pods (Riethmuller et al. 2011). The same authors reported that narrower rows (<50 cm) were more likely to lead to increased yields in cooler climate. Overall, growers should weigh the risk of a potential yield decrease against the cost savings when they increase row spacing (May et al. 2020). In general, the wider row spacing applied under NT may cause greater weed competition increasing both herbicide use and herbicide rates to control weeds within the retained stubble (Riethmuller et al. 2011; Routley et al. 2011).

Increased row spacing increases the number of seeds per linear meter and this may cause interplant competition particularly under stressful environmental conditions (May et al. 2020). Furthermore, increasing row spacing increases the amount of fertilizer product applied in the sideband. Therefore, monitoring of plant density is required that ensures an adequate density particularly since all seeding equipment may not maintain the separation of seed and fertilizer (May et al. 2020). When separation of seed and fertilizer is not maintained damages to emerging seedlings may occur due to greater and toxic concentrations of N fertilizer as reported for oilseed rape, oat, and spring wheat (Xie et al. 1998; May et al. 2004, 2008).

Lower yields under RT have been reported in crops sown in widely spaced rows such as maize, soybean, or fava bean and reduced tillage concurs with higher weed infestations; the great amount of bare soil between and within rows provide less spacing shading conditions compared to other crops such as cereals, thus promoting the growth of weeds (Armengot et al. 2014).

Adjusting the planting date to delay weed seedling emergence is another nonchemical weed management method to provide crop a competitive advantage over weeds. An example of increased crop competitiveness under NT and early planting is the case of herbicide-resistant canary grass (*Phalaris minor* Retz) management in rice-wheat systems of the Indo-Gangetic plains where adoption of NT allowed wheat crops to be planted one to two weeks earlier and establish before the emergence of the still-dormant canary grass. Early planting gave an advantage to wheat over the herbicide-resistant canary grass and reduced the emergence of the weed due to minimum soil disturbance resulted in less crop-weed competition (Mahajan et al. 1999; Franke at al. 2001; Singh 2007; Chauhan et al. 2012). Adjusting planting date may be easier to apply in NT than in CT, due to less work and time needed for crop planting.

Overall, improving crop competitiveness in terms of early crop emergence or narrow interrow space under NT may be an issue compared to CT, and may increase the need for curative weed control methods such as herbicides.

12.2.3 Reduced Weed Seedbank Size

An important ecological weed management strategy that limits potential weed populations and weed pressure is to promote gradual weed seedbank depletion (Bastiaans et al. 2008), i.e. reduction in the number of weed seeds and vegetative structures such as rhizomes, stolons, tubers, bulbs, or transformed roots through which the perennial weeds propagate themselves. Most of the seedbank deposits (50–90%) are found on the first 0–5 cm of the soil, whereas there is a decrease in seed deposits as soil depth increases (Govindasamy et al. 2020). Weed seedbank may be gradually depleted by maximizing losses (withdrawals) and by minimizing inputs (deposits) of seeds (Forcella et al. 1993). Weed seedbank depletion in the frame of ecologically weed management can occur by enhancing seed predation / decay and seed germination and by preventing production and shedding of new seeds as a result of increased competition through diversified crop rotations (Bastiaans et al. 2008; Bàrberi 2019). Tillage systems strongly influence seed predation/decay and seed germination, and they may interact with crop rotation affecting weed seedbank.

Seedbank losses mostly occur in the upper soil layer by generation of seedlings from nondormant seeds, by seed death or predation (Gallandt 2006). Therefore, seeds at or close to the topsoil surface as in NT are more vulnerable to predation, decay, desiccation, and harsh weather conditions compared to seeds buried in the soil. Seed predators may be beetles, fire ants, crickets, rodents, and carabids (Wang et al. 2022). Seed predators also show preference to surface seeds and usually avoid digging to save energy (Hulme 1994; Gallandt et al. 2005; Puricelli et al. 2005; Chauhan et al. 2010; Blubaugh and Kaplan 2015; Nikolić et al. 2020; Daouti et al. 2022). In contrast, seed burial with tillage increases seed survival by protecting the seeds in the soil from adverse environmental conditions and surface predators (Froud-Williams et al. 1984). Thomas et al. (1986) reported that more than 50% of green foxtail (*Setaria viridis* (L.) P. Beauv.) seeds were alive after two years of burying at 10 cm in soil compared with less than 10% of seeds surviving when remained on the soil surface, due to adverse environmental conditions and predation.

Nikolić et al. (2020), studying the seed longevity and seed degradation of six weed species often classified as long-lived (seeds surviving ≥ 3 years) in an NT soil after 643 days of burial, found out that seeds of *A. myosuroides* and *D. sanguinalis* completely lost their viability, those of *P. oleracea*, *A. retroflexus* and *A. theophrasti* had a residual survival of 10–20%, whereas those of *S. halepense* maintained avalability of 50%. This study revealed high activity of cellulolytic microorganisms that feed on seed coat, mostly made of polyphenolic polymers such cellulose or lignin and scarce cellulotic matter in the non-tilled field, a combination that possibly made microbes more prone to colonize the seeds (Nikolić et al. 2020). A positive relationship between seed coat thickness and seed viability in the soil has been reported, since seed coat preserves seeds against external aggression of abiotic and biotic factors with microbial attack being the most important (Davis et al. 2008; Gardarin et al. 2010; Nikolić et al. 2020).

NT regimes resulted in an increase in seed predation, as reported by Menalled et al. (2007), that revealed a threefold increase in seed predation by carabid species in NT over the CT system. Murphy et al. (2006) revealed that establishment of NT resulted in decrease in the seedbank density and in increased weed species diversity compared to CT over a six-year study in Canada; they reported a decrease in the mean weed seedbank from 41,000 seeds m^{-3} of soil to 8,000 seeds m^{-3} in NT fields compared to 34,000 and 49,000 seeds m^{-3} of soil in chisel-plowed fields (RT) and moldboard-plowed fields (CT), respectively, mainly due to seed destruction rather than prevention of germination.

Tillage apart from moving seeds to depths away from their predators may be responsible for the damage of the nests of harvester ants (Díaz 1991). However, Cromar et al. (1999) reported higher weed seed consumptions rates in both NT and in fields tilled with moldboard compared to chisel-plowed fields. Cardina et al. (1996) also found no difference in seed predation of *A. theophrasti* seeds between NT and conventional-tilled treatments. These contradictory results may suggest that some predators are well adapted to soil disturbance (Westerman et al. 2003; Shearin et al. 2007). Cover crops and crop residue left on the soil surface may improve the quality of the habitat for weed seed foraging (Meiss et al. 2010), whereas sufficient cover crop biomass is important for increasing the activity of seed predators (Shearin et al. 2008; Ward et al. 2011). However, Cromar et al. (1999) mentioned that high amounts of crop residue may reduce insect mobility throughout the field affecting the quantity of seed predation. Sometimes the mulch itself may increase the alternative food supply that the natural enemies utilize, resulting in reduction of their overall activity (Quinn et al. 2016). Overall, seed predation has been reported as an important ecological service by reducing weed seedbank inputs, however, highly variable in agricultural fields as well with uncertain contribution in weed management (van der Laat 2015). The same authors suggested that tillage is more important in determining the number of weed seeds on the soil surface available to predators, rather than directly affecting the predator activity density. Yearly variation in abiotic and biotic factors might exert a great influence on insects (Trichard et al. 2013; Quinn et al. 2016).

Weed seedbank depletion is also accomplished by stimulating weed seed germination and seedling emergence with subsequent control of the weeds that have emerged. As already stated, tillage is considered the most effective promoter of seed germination, particularly for the weed species that require light to germinate and have low initial dormancy. Stimulating seed emergence by tillage and controlling the emerged seedlings is achieved by the false seedbed technique applied in CT system. This technique consists of an early application of tillage to promote weed seed germination and control the emerged weed seedlings before planting by shallow soil cultivation to limit soil disturbance and possible lift of buried seeds to the soil surface, so the crop is planted into clean soil (Travlos et al. 2020).

In NT systems where most of the seeds are placed on the topsoil, weed seed germination is encouraged only by irrigation or rainfall and the emerged seedlings appear after a week (Chauhan et al. 2012). For successful weed seedbank depletion, weed control and time crop sowing should be delayed until the main flush of emergence has passed (Bond and Grundy 2001). This necessitates an adjustment in crop planting time in such a way that allows the crop to get established before the troublesome weed species emerge, providing a significant competitive advantage (Chauhan et al. 2012). In CA systems, greater area of autumn-sown cereals can be planted compared to CT systems because of the more rapid performance of the farming operations under conditions that may not be suitable for traditional ploughing methods (Froud-Williams et al. 1981). In contrast, in CT any delays in seedbed preparation due to the application of false seed bed technique or due to the occurrence of unfavorable weather conditions may postpone the planting date, shortening the growing season with consequences in yield production.

A novel approach to induce weed seed synchronous germination could be the use of germination stimulants such as the gibberellic acid (GA3). This approach integrates cover cropping combined with GA3 hormone, as in the case of ABA discussed earlier, but with the opposite effect (i.e. inducing synchronous weed seed germination and homogenous emergence of weed species) (Oliveira et al. 2020). Cover crop residues on soil surface can provide an unfavorable environment for weed establishment, whereas the presence of GA3 could induce a homogenous emergence of weeds. In case of applying PRE herbicides, the combined effect of cover cropping and GA3 would enhance the effect of PRE herbicides, increasing the likelihood of successful weed control and thus reducing the reliance on POST control treatments (Oliveira et al. 2020). As already stated for ABA, the fate and effectiveness of GA3 under field conditions, particularly when combined with PRE herbicides and the presence of cover crop residues, should be further studied (Oliveira et al. 2020).

Weed seed germination may be also induced with tillage. Seed burial favors the germination of the weed species that require burial to break dormancy and are better adapted to moldboard ploughing (Buhler 1995). For example, approximately 94% of seeds of *S. halepense* falling from panicles are dormant (Valla et al. 1980); if these seeds remain on soil surface and escape predation and decay, approximately 30% become nondormant in spring, whereas if they are buried at a depth of 15 cm and return to the soil surface in spring, the nondormant seed population will be more than 80%. Seed burial can also contribute to seed losses if fatal germination takes place (Fenner and Thompson 2005) – that is, the buried seeds germinate but the seedlings never emergence. This happens when seeds are placed too deep in the soil and they do not have enough reserves in its endosperm for seedling growth to reach for the soil and become autotroph (Bàrberi 2019). Moldboard tillage buries seeds in deeper soil profiles, mainly affecting short-lived weed species that might die before they manage to emerge (Gunsolus et al. 2010). For example, moldboard plowing is an effective method to control downy brome (*Bromus tectorum* L.) in winter-wheat fallow system, since its seeds buried at a soil depth of 8 cm or greater generally germinate but do not emerge (Wicks et al. 1971; Wicks 1984). Grass weeds with a comparatively low dormancy and seed persistence are not favored under soil inversion (Lampkin 2002; Riemens et al. 2007; Gruber and Claupein 2009), whereas large seeds that have more reserves succeed in emerging even from 20 cm deep (Lampkin 2002; Riemens et al. 2007; Gardarin et al. 2009).

Another ecologically based weed long-term strategy to deplete weed seedbank is to prevent production and shedding of new seeds through a well-planned crop rotation (Légère et al. 2011; Bàrberi 2019). Continuous cropping contributes to the dominance of certain species. RT under continuous winter cereal cropping can lead to dominance of grasses with low-dormant seeds after a few years, such as *A. myosuroides* and *Bromus* spp. (Froud-Williams 1983), whereas cultivation of perennial leys under RT will decrease tillage operations and in combination with the repeated mowing of the ley will eliminate annual weeds shifting weed community composition toward biennials and perennials (Bàrberi 2003).

Crop rotation provides greater variability in the type and timing of soil, crop, and weed management practices, thus increasing the opportunities for weed mortality in crop sequences when compared to monoculture (Martin and Felton 1993). Other authors reported that crop rotation creates such conditions that differentially affect emergence, development, and dissemination of weed species (Dorado et al. 1999; Sosnoskie et al. 2006). As already stated, crop rotation is one of three aspects of CA, although RT or NT may be applied without accompanied by crop rotation plans. In CT a reduction in crop rotational diversity due to adoption of herbicides and herbicide-resistant crops is often recorded (Hurley and Frisvold 2016; Ramankutty et al. 2018). The abundance and composition of weed species in the field can be affected by interaction between rotation and tillage, although there have been reports that excelled either crop rotation (Ball 1992) or tillage (Doucet et al. 1999; Bàrberi and Lo Cascio 2001) for their effect on weed density or weed seedbank structure and composition. Crop rotation and tillage system may interact together affecting the abundance and composition of weed species in the field (Murphy et al. 2006). Ruisi et al. (2015) studied the effect of continuous application of CT moldboard ploughing or NT on weed seedbank within wheat monoculture and two crop rotations, reporting interaction effect between tillage and crop rotation. They pointed out that tillage system and crop rotation both act as filters that often interact with each other in determining the composition and abundance of weed species in the soil seedbank and suggested farmers should apply NT within an appropriate crop sequence.

Cardina et al. (2002) characterized crop rotation as more important determinant of weed seedbank than tillage; however, they reported significant interactions between crop rotation and tillage. Studying the effect of crop rotation (maize-soybean, maize-oat-hay, and continuous maize) and tillage systems (NT, CP, MP (moldboad plow)) on weed seedbanks, they revealed the highest total weed density in maize monoculture in NT, whereas they observed lower total weed density in CP- and MP- than in NT-maize-oat-hay by 45% and 60%, respectively. They concluded that tillage and crop rotation act as important environmental filters that quite often interact determining the abundance and composition of seeds in the soil seedbank. A decline in the occurrence of one species might leave resources to other species with better tolerance of micro-environmental conditions or with a greater ability to gain resources provided by different crop rotation and tillage systems. The same authors highlighted that proper manipulation of tillage, rotation and other cultural practices could be combined in such way to allow available niches to be filled by species with the little effect on the crop. Different effect of tillage practices on weed seedbank were observed in wheat monoculture and two crop rotations (MacLaren et al. 2021). In particular, assessing the effects of four tillage practices (soil inversion through ploughing CT, shallow soil loosening MT, direct drilling with disc openers ZT, and direct drilling with tine openers NT), MacLaren et al. reported that direct drilling with disc openers (ZT) counteracted any benefits in weed suppression gained from crop rotation. In contrast, a study conducted over seven years to investigate the effect of different soil tillage intensities and crop rotation on weed species composition, weed density, and biomass in maize revealed no significant interaction between crop rotation and soil tillage intensity (Demjanová et al. 2009). Bàrberi and Lo Cascio (2001) studying the long-term tillage (MP and CP at 45 cm, MT at 15 cm, and NT) and crop rotation (continuous winter wheat and pigeon bean (*Cajanus cajan* (L.) Millsp./ winter wheat two-year rotation) effect on weed seedbank reported that tillage system influenced seedbank size and composition to a greater extent than crop rotation, however, with no significant interaction effects. They reported that modification of the cropping system from plough-based wheat monoculture to noninversion tillage and crop rotation would not result in increased weed problems. The same authors detected some significant interactions between tillage and crop rotation, however with lower levels of significance than the main effects; moreover, they reported that management practices have little effect on the reserve of biodiversity in the soil due to the seed longevity of many weed species (Bàrberi and Lo Cascio 2001).

Diversified weed communities, such those recorded under reduced- or no-tillage, tend to be less competitive with no significant effect on crop yield compared to weed monocultures (Sans et al. 2011; Armengot et al. 2013; Cooper et al. 2016; Storkey and Neve 2018; Adeux et al. 2022). Decreased weed biomass associated with high levels of diversified weed communities has been reported to increase crop productivity, indicating that the increased weed diversity in a field may limit the negative effect of competitive and dominant species (Adeux et al. 2019); however, this could be more detrimental in case this community composes highly competitive species with traits that could increase interference with a given crop or cropping system (Bàrberi et al. 2018). Furthermore, although weed abundance was reported higher under reduced tillage practices as documented by several studies, this trend has been reported to be crop specific and not constant over time (Armengot et al. 2014, 2016). Vakali et al. (2011) revealed that reduced tillage reduced crop growth for barley but not for rye possibly through the promotion of weeds, whereas Légère et al. (2013) reported similar yield for barley and red clover (*Trifolium pratense* L.) but not for both maize and soybean across three tillage treatments (moldboard plow, chisel plow, and no-till system) in low-input cropping systems mainly to poor weed control in a barley-red clover-maize-soybean crop rotation. Moreover, environmental conditions may influence the effect of tillage (Armengot et al. 2014). Dhima et al. (2006a) found that the stem number and total fresh weight of sterile oat (*Avena sterilis* L.) emerged in barley were less in CT (moldboard plow, tandem harrow disc, and cultivator) compared to those in RT (heavy offset harrow disc and cultivator) and minimum tillage (cultivator) in a wet growing season. In contrast, in a dry growing season, sterile oat growth was greater in CT than in reduced or minimum tillage. Weather conditions may have greater impact on weed abundance and seedbank in the soil compared to tillage practices (Alarcón Víllora et al. 2019; Romaneckas et al. 2021).

A recent study with data of seven trials in different European climatic regions studied the effect of conventional and reduced tillage on functional attributes of the weed communities and the consequences in relation to crop-weed competition and revealed that different weed functional trait values favorable to crop production were observed under both tillage systems (Armengot et al. 2016). Weeds in RT were shorter and less competitive because they had less affinity for nutrients, whereas, under CT lower abundance of perennials and weeds with higher seed weight but with lower seed production were observed. However, apart from the relevant important role of tillage system on weed communities, the authors highlighted that crop choice is the main driver of the shifts in the functional composition of the weed communities (Armengot et al. 2016). The different management regimes associated with each crop, such as time and tillage operations, date and pattern of sowing, and weed control management interact together and lead to different compositions of weed community (Fried et al. 2008).

Adoption of reduced- or no-tillage usually results in increased abundance of small-seeded weeds that can germinate and emerge from shallow depths; perennial weeds are also favored since their control is restricted due to limited soil cultivation (Cardina et al. 1991; Zanin et al. 1997; Swanton et al. 1999). In general, it has been reported that long-term RT or NT systems contribute to changes on weed community composition with more abundant weed species, richness, and diversity in favor to small-seeded annuals, grasses, perennials, and wind-dispersed species (Buhler and Daniel 1988; Cardina et al. 1991; Buhler 1995; Mulugeta et al. 2001; Armengot et al. 2014, 2016; Govindasamy et al. 2020). However, other studies have reported higher weed densities in CT (Shrestha et al. 2002) or no effect of weed diversity (Plaza et al. 2011) (Table 12.3). Higher densities of *Sorghum halepense* (L.) Pers., *Chamaesyce humistrata* (Engelm. ex A. Gray) Small, *Amaranthus tuberculatus* (Moq.) Sauer, *Lamium amplexicaule* L., *Elytrigia repens* (L.) Nevski, *Poa annua* L., *S. media*, *Matricaria perforata* Merat, *Cirsium arvense* L. (Scop.), and *Elymus repens* (L.) Gould were observed under NT or RT system compared to CT (Tørresen et al. 2003; Govindasamy et al. 2020).

In another study, perennial species such as *Taraxacum officinale* Weber in Wiggers was strongly associated with ZT, other perennial species such as *C. arvense*, and *Sonchus arvensis* L. with both zero- and reduced-tillage systems, whereas annual species were associated with a range of tillage systems. More specifically, *Salsola iberica* Sennen & Pau was associated with ZT, *Thlaspi arvense* L. with CT, whereas *C. album*, *A. retroflexus*, *Brassica kaber* (DC.) L.C. Wheeler, and *Polygonum convolvulus* L. were equally abundant in conventional-, reduced- and zero-tillage systems (Thomas et al. 2004). Cessation of tillage in autumn in NT fields may favor winter annuals, biennials, and perennial weeds such as *Tragopogon dubius* L. and *Hieracium* spp. as reported by Murphy et al. (2006). Tuesca et al. (2001), in wheat/soybean and maize/soybean rotations over six years found that annual broad-leaved weed species (such as *Portulaca oleracea* L., *Datura ferox* auct. non L., *C. album*, *Amaranthus hybridus* L., and *Capsella bursa-pastoris* (L.) Medik.) emerged in higher populations in wheat cultivated under CT (moldboard-plough, disk, and harrow), compared to annual grasses (*Digitaria sanguinalis* (L.) Scop. and *E. crus-galli*) and perennial (*S. halepense* and *Cyperus esculentus* L.) weed species emerged in lower populations.

In NT or RT more seeds remain close to the soil surface that are expected to germinate compared to buried seeds in CT. Although higher recruitment of seedlings subject to control and higher seed predation are expected in NT and RT that will induce weed seedbank diminishment, nevertheless, large weed seedbanks usually occur in these tillage systems. This is possibly due to rarely successful depletion of weed seedbank by noninversion tillage (Bàrberi 2003) and to higher weed seed production of the emerged plants (Gruber and Claupein 2009).

Weeds are characterized by high genetic diversity and ability to quickly adapt and take advantage of conditions created by crop management systems, such crop rotation and tillage system. Therefore, more diverse crop rotations with different tillage practices should be applied. Bàrberi et al. (2001) reported a reduction in weed density in soybean-winter wheat rotation by altering deep (at 50 cm) moldboard ploughing with rotary harrowing (at 15 cm), minimum tillage compared to chisel ploughing (at 50 cm), minimum tillage, or shallow ploughing (at 25 cm). To foster weed control, it is imperative to rotate with more competitive crops or with crops requiring different tillage operations (Bàrberi 2003); proper manipulations of rotation with tillage that both determine abundance and composition of weed seeds should focus on replacement of a dominant species by more than one species that may have little effect on the crop (Cardina et al. 2002). Different cover crop species should also be included in crop rotation to avoid selection for weed species. It should highlight the need to rotate the pattern of planting of the cover crops otherwise weed species will be adapted to a continuous pattern of tillage (Teasdale 2003). Crop sequence effects cannot be separated from associated cultural practices, particularly from the effects of weed control methods, that provide temporal variation in the chemical and physical microenvironment influencing weed community composition (Légère and Samson 1999; Menalled et al. 2001). Tillage system, environment, crop rotation, fertilization, crop type, type, and timing of weed management are factors that interact each other and strongly affect shifts in weed flora (Légère and Samson 1999; Shrestha et al. 2002). Overall, the effects on weed communities of changing tillage and crop rotation need several years to stabilize (Dick and Daniel 1987).

Table 12.3 Tillage effects on weed density and weed seed bank.

	Weed density		Weed seedbank		Reference
	NT[a] or MT[b]	CT[c]	NT or MT	CT	
Avena sterilis	+[d]	–			Dhima et al. 2006a
Avena sterilis	–	+			
Avana fatua	/	/	/	/	Martin and Felton 1993
Avena sterilis	+	–			Mas and Verdú 2003
Avena sterilis	+	–	+	–	Navarrete and Fernandez-Quintanilla 1996
Portulaca oleracea	–	+			Tuesca et al. 2001
Datura ferox	–	+			
Chenopodium album	–	+			
Amaranthus hybridus	–	+			
Capsella bursa-pastoris	–	+			
Digitaria sanguinalis	+	–			
Echinochloa crus-galli	+	–			
Sorghum halepense	+	–			
Cyperus esculentus	+	–			
Sonchus oleraceus	+	–			
Carduus acanthoides	+	–			
Conyza bonariensis	+	–			
Polygonum aviculare	–	+			Dorado and Lopez-Fando 2006
Phalaris paradoxa	–	+			
Anthemis arvensis	+	–			
Lolium rigidum	+	–			
Hirschfeldia incana	+	–			
Capsella bursa pastoris	/	/			
Torilis nodosa	/	/			
Veronica persica	–	+			Fried et al. 2008
Papaver rhoeas	–	+			
Varonica hederifolia	–	+			
Alopecurus myosuroides	–	+			
Viola arvensis	–	+			
Falopia convolvulus	–	+			
Thlaspi arvense	–	+			
Chenopodium album	–	+			
Amaranthus retroflexus	–	+			
Polygonum persicaria	–	+			
Echinochloa crus-galli	–	+			
Sonchus arvensis	–	+			
Amaranthus spp.	–	+			Zanin et al. 1997
Chenopodium album	–	+			
Echinochloa crus-galli	–	+			
Conyza canadensis	+	–			
Digitaria sanguinalis	+	–			
Kickxia eletine	+	–			

(Continued)

Table 12.3 (Continued)

	Weed density		Weed seedbank		Reference
	NT[a] or MT[b]	CT[c]	NT or MT	CT	
Galium spurium	+	–			Salonen et al. 2012
Lamium spp.	+	–			
Lapsana communis	+	–			
Poa annua	+	–			
Senecio vulgaris	+	–			Froud-Williams et al. 1981
Marticaria spp.	+	–			
Papaver rhoeas	/	/			
Veronica persica	/	/			
Anagalis arvensis	/	/			
Polygonum aviculare	+	–			
Chenopodium album	+	–			
Polygonum aviculare			–	+	Ruisi et al. 2015
Papaver rhoeas			+	–	
Anagalis arvensis			+	–	
Lactuca serriola			+	–	
Lolium spp.			+	–	
Matricaria perforate	+	–			Tørresen et al. 2003
Poa annua	+	–			
Stellaria media	+	–			
Cirsium arvense	+	–			
Elymus repens	+	–			
Conyza canadensis			+	–	Bàrberi and Lo Cascio 2001
Amaranthus retroflexus			+	–	
Anacyclus clavatus			+	–	Dorado et al. 1999
Portulaca oleracea			+	–	
Papaver rhoeas			+	–	
Torilis nodosa			+	–	
Arabidopsis thaliana			+	–	
Draba verna			+	–	
Amaranthus albus			+	–	
Scandix pectin-veneris			+	–	
Veronica triphyllos			+	–	
Conyza canadensis	+	–	–	+	Murphy et al. 2006
Fumaria officinalis	+	–	–	+	
Urtica dioica	+	–	–	+	
Sicyos angulatus	+	–	–	+	
Tragopogon dubius	+	–	–	+	
Pilea pumila	+	–	–	+	
Geranium robertanium	+	–	–	+	

[a] NT: No-till
[b] MT: Minimum tillage
[c] CT: Conventional tillage
[d] +=increase; =decrease; /=no effect

12.3 Herbicide Efficacy, Herbicide Resistance, and Organic Farming

Tillage system may affect the efficacy of herbicides when they used as a curative (re-active) method. In reduced- or no- tillage crop residues remain on the soil surface and may intercept the PRE herbicides, resulting in reduced efficacy as less herbicide can reach the soil surface for weed control (Banks and Robinson 1982; Whalen et al. 2020). Herbicide volatilization and photodegradation may occur for herbicides intercepted by crop residues in CA, depending on herbicide chemical properties and formulation (Chauhan et al. 2012). On the contrary, in CT system PRE herbicides that require soil incorporation to avoid losses due to volatilization or photodegradation can be mechanically soil-incorporated. The efficacy of POST herbicides may be also reduced as well by the presence of crop residues (Wolf et al. 2000). Reduction in the use of PRE herbicides and restricted mechanical control lead to higher reliance to POST herbicides to keep weed populations at low level (Tørresen et al. 2003; Chauhan et al. 2006d).

Adoption of herbicide-resistant crops and high reliance on herbicides have been regarded as the main reasons for the widespread adoption of conservation tillage (Légère et al. 2013), leading to a decreased use of PRE herbicides; however, in an increased use of POST as the sole weed control method. The extensive use of glyphosate has resulted in shifts in problematic weed species and the development of glyphosate resistant weeds in these crops in this cropping system. Weeds such as *Conyza canadensisis* (L.) Cronq. and palmer amaranth have developed herbicide resistance in crop like glyphosate-resistant cotton or soybean cultivated under CA (Mueller et al. 2003; Chahal et al. 2015).

Moldboard plowing has been considered the key strategy to manage weeds, particularly in organic farming that face higher weed infestations (Armengot et al. 2013). For organic farmers the restriction in soil cultivation as a mean for weed management is a major drawback for adopting conservation tillage systems. Up to now, tillage and soil cultivation have been the main solutions for the management of perennial weed species in organic systems (Melander et al. 2013). Therefore, in the frame of ecological weed management, cover crop management, choice of crop sequence and improved crop competitive ability that imply maximum diversification of the cropping systems are regarded as important issues in organic production (Bàrberi 2002).

12.4 Conclusions

Conventional tillage is characterized by soil inversion and incorporates the previous crop residues in the soil, whereas conservation tillage is characterized by less or no soil disturbance maintaining crop residues on soil surface after planting. The vertical weed seed distribution in the soil and the presence or absence of crop residues on the soil surface determined by the tillage system applied can differently affect seedling recruitment, crop competitiveness, and weed seedbank size. There are positive and negative effects of both conventional and conservation tillage systems on ecological weed management that can result in an increase or a decrease in weed species density and weed seedling emergence. Proper selection and management of crop residues is an important issue for successful weed management and crop establishment. The absence of soil cultivation in NT may render ecologically based weed management practices insufficient to keep weed infestations low and may lead to an increased use of herbicides. Considering, the existence of the exceptionally high level of genetic diversity of weeds, their large seedbank, their unique ability to adapt to the any management imposed to them repeatedly, and the different response of weed species to soil disturbances and environmental effcts, tillage systems should be rotated for successful ecological weed management, particularly for the management of herbicide resistant weeds.

References

Adeux, G., Vieren, E., Carlesi, S. et al. (2019). Mitigating crop yield losses through weed diversity. *Nature Sustainability* 2: 1018–1026.

Adeux, G., Yvoz, S., Biju-Duval, L. et al. (2022). Cropping system diversification does not always beget weed diversity. *European Journal of Agronomy* 133: 126438.

Aikins, K.A., Antille, D.L., Jensen, T.A., and Blackwell, J. (2019). Performance comparison of residue management units of no-tillage sowing systems: a review. *Engineering in Agriculture, Environment and Food* 12: 181–190.

Aikins, K.A., Barr, J.B., Ucgul, M. et al. (2020). No-tillage furrow opener performance: a review of tool geometry, settings and interactions with soil and crop residue. *Soil Research* 58: 603–621.

Alarcón Víllora, R., Hernández Plaza, E., Navarrete, L. et al. (2019). Climate and tillage system drive weed communities' functional diversity in a Mediterranean cereal-legume rotation. *Agriculture, Ecosystems & Environment* 283: 1–10.

Alberts, E.E. and Neibling, W.H. (1994). Influence of crop residue on water erosion. In: *Managing Agricultural Residues* (ed. P.W. Unger), 19–44. Boca Raton, Florida: Lewis Publishers. ISBN-13: 9780873717304.

Al-Hajaj, N. (2021). Review of the influence of crop rotation and primary tillage operations on weed seed bank. *International Journal of Research -GRANTHAALAYAH* 9 (7): 407.

Al-Kaisi, M.M. and Yin, X. (2005). Tillage and crop residue effects on soil carbon and carbon dioxide emission in corn-soybean rotations. *Journal of Environmental Quality* 34: 437–445.

Alonso-Ayuso, M., Gabriel, J.L., García-González, I. et al. (2018). Weed density and diversity in a long-term cover crop experiment background. *Crop Protection* 112: 103–111.

Armengot, L., Berner, A., Blanco-Moreno, J.M. et al. (2014). Long-term feasibility of reduced tillage in organic farming. *Agronomy for Sustainable Development* 35: 339–346.

Armengot, L., Blanco-Moreno, J.M., Bàrberi, P. et al. (2016). Tillage as a driver of change in weed communities: a functional perspective. *Agriculture, Ecosystems & Environment* 222: 276–285.

Armengot, L., José-María, L., Chamorro, L., and Sans, F.X. (2013). Weed harrowing in organically grown cereals crops avoids yield losses without reducing weed diversity. *Agronomy for Sustainable Development* 33: 405–411.

Ascard, J. (1994). Soil cultivation in darkness reduced weed emergence. *Acta Horticulturae* 372: 167–177.

Ashford, D.L. and Reeves, D.W. (2003). Use of a mechanical roller-crimper as an alternative kill method for cover crops. *American Journal of Alternative Agriculture* 18 (1): 37–45.

Aulakh, J.S., Price, A.J., Enloe, S.F. et al. (2012). Integrated palmer amaranth management in glufosinate-resistant cotton: I. Soil-inversion, high-residue cover crops and herbicide regimes. *Agron* 2: 295–311.

Aulakh, J.S., Price, A.J., Enloe, S.F. et al. (2013). Integrated palmer amaranth management in glufosinate-resistant cotton: II. Primary, secondary and conservation tillage. *Agron* 3: 28–42.

Baker, C.J., Saxton, K.E., Ritchie, W.R. et al. (2007)., No-tillage seeding in conservation agriculture., CAB International/Food and Agriculture Organization of the United Nations: Oxford, UK.

Ball, D.A. (1992). Weed seedbank response to tillage, herbicides, and crop rotation sequence. *Weed Science* 40: 654–659.

Ballaré, C.L., Scopel, A.L., Sánchez, R.A., and Radosevich, S.R. (1992). Photomorphogenic processes in the agricultural environment. *Photochemistry and Photobiology* 56: 777–788.

Banks, P.A. and Robinson, E.L. (1982). The influence of straw mulch on the soil reception and persistence of metribuzin. *Weed Science* 30: 164–168.

Bàrberi, P. (2002). Weed management in organic agriculture: are we addressing the right issues? *Weed Research* 42: 176–193.

Bàrberi, P. (2003). Preventive and cultural methods for weed management - Chapter 3. Management options and perspectives. In: *Weed Management for Developing Countries. Addendum 1* (ed. R. Labrada). Rome: Food and Agriculture Organization of the United Nations.

Bàrberi, P. (2019). Ecological weed management in Sub-Saharan Africa: prospects and implications on other agroecosystem services. *Advances in Agronomy* 156: 219–264.

Bàrberi, P., Bocci, G., Carlesi, S. et al. (2018). Linking species traits to agroecosystem services: a functional analysis of weed communities. *Weed Research* 58 (2): 76–88.

Bàrberi, P., Bonari, E., Mazzoncini, M. et al. (2001). Weed density and composition in winter wheat as influenced by tillage systems. Conservation agriculture, a worldwide challenge. In: Proceedings of the First World Congress on Conservation Agriculture, Madrid, Spain, 1–5 October, Vol. 2: offered contributions, XUL, 451–455.

Bàrberi, P. and Lo Cascio, B. (2001). Long-term tillage and crop rotation effects on weed seedbank size and composition. *Weed Research* 41: 325–340.

Barnes, J.P., Putnam, A.R., Burke, B.A., and Aasen, A.J. (1987). Isolation and characterization of allelochemicals in rye herbage. *Phytochemistry* 26: 1385–1390.

Bartley, M.R. and Frankland, B. (1982). Analysis of the dual role of phytochrome in the photoinhibition of seed germination. *Nature* 300 (5894): 750–752.

Bastiaans, L., Paolini, R., and Baumann, D.T. (2008). Focus on ecological weed management: what is hindering adoption? *Weed Research* 48: 481–491.

Ben-Hammouda, B.H., Ghorbal, H., Kremer, R.J., and Oueslati, O. (2001). Allelopathic effects of barley extracts on germination and seedlings growth of bread and durum wheats. *Agronomie (Paris)* 21: 65–71.

Bernardes, M.F.F., Pazin, M., Pereira, L.C., and Dorta, D.J. (2015). Impact of pesticides on environmental and human health. In: *Toxicology Studies - Cells, Drugs and Environment* (ed. A.C. Andreazza), and Gustavo, Scola). IntechOpen Books.

Birkàs, M., Jug, D., and Kisić, I. (2014). Tillage systems for the main field crops. In: *Book of Soil Tillage* (ed. M. Birkàs), 219–265. Szent Istvan University Press.

Blanco-Canqui, H. and Lal, R. (2008). No-tillage and soil-profile carbon sequestration: an on-farm assessment. *Soil Science Society of America Journal* 72: 693–701.

Blubaugh, C.K. and Kaplan, I. (2015). Tillage compromises weed seed predator activity across developmental stages. *Biological Control* 81: 76–82.

Blum, U., King, L.D., Gerig, T.M. et al. (1997). Effects of clover and small grain cover crops and tillage techniques on seedling emergence of some dicotyledonous weed species. *American Journal of Alternative Agriculture* 4: 146–161.

Bond, W. and Grundy, A.C. (2001). Non-chemical weed management in organic farming systems. *Weed Research* 41: 383–405.

Borger, C.P.D., Riethmuller, G., and D'Antuono, M. (2016). Eleven years of integrated weed management: long-term impacts of row spacing and harvest weed seed destruction on *Lolium rigidum* control. *Weed Research* 56: 359–366.

Botta, G.F., Nardon, G.F., and Clavijo, R.G. (2022). Soil sustainability: analysis of the soil compaction under heavy agricultural machinery traffic in extensive crops. *Agronomy* 12 (2): 282.

Boyd, N.S. and Van Acker, R.C. (2004). Seed germination of common weed species as affected by oxygen concentration, light, and osmotic potential. *Weed Science* 52: 589–596.

Boydston, R.A. and Hang, A. (1995). Rapeseed (*Brassica napus)* green manure crop suppresses weeds in potato (*Solanum tuberosum*). *Weed Technology* 9: 669–675.

Buchanan, A.L., Kolb, L.N., and Hooks, C.R.R. (2016). Can winter cover crops influence weed density and diversity in a reduced tillage vegetable system? *Crop Protection* 90: 9–16.

Buhler, D.D. (1995). Influence of tillage systems on weed population dynamics and management in corn and soybean in the central USA. *Crop Science* 35: 1247–1258.

Buhler, D.D. and Daniel, T.C. (1988). Influence of tillage systems on giant foxtail, *Setaria faberi*, and velvetleaf, *Abutilon theophrasti*, density and control in corn, *Zea mays*. *Weed Science* 36: 642–647.

Burgos, N.R. and Talbert, R.E. (2000). Differential activity of allelochemicals from *Secale cereale* in seedling bioassays. *Weed Science* 48: 302–310.

Cardina, J., Herms, C.P., and Doohan, D.J. (2002). Crop rotation and tillage system effects on weed seedbanks. *Weed Science* 50: 448–460.

Cardina, J., Norquay, H.M., Stinner, B.R., and McCartney, D.A. (1996). Postdispersal predation of velvetleaf (*Abutilon theophrasti*) seeds. *Weed Science* 44: 534–539.

Cardina, J., Regnier, E., and Harrison, K. (1991). Long-term tillage effects on seed banks in three Ohio soils. *Weed Science* 39: 186–194.

Carter, M.R. (2005). Conservation tillage. In: *Encyclopedia of Soils in the Environment* (ed. D. Hillel), 306–311. USA: Elsevier Ltd.

Chahal, P.S., Aulakh, J.S., Jugulam, M., and Jhala, A.J. (2015). Herbicide-resistant palmer amaranth (*Amaranthus palmeri* S. Wats.) in the United States - Mechanisms of resistance, impact, and management. Herbicides, Agronomic Crops and Weed Biology.

Chauhan, B.S. and Abugho, S.B. (2013). Effect of crop residue on seedling emergence and growth of selected weed species in a sprinkler-irrigated zero-till dry-seeded rice system. *Weed Science* 61: 403–409.

Chauhan, B.S., Gill, G., and Preston, C. (2006a). Factors affecting seed germination of threehorn bedstraw (*Galium tricornutum*) in Australia. *Weed Science* 54: 471–477.

Chauhan, B.S., Gill, G., and Preston, C. (2006b). Factors affecting seed germination of little mallow (*Malva parviflora*) in southern Australia. *Weed Science* 54: 1045–1050.

Chauhan, B.S., Gill, G., and Preston, C. (2006c). Influence of environmental factors on seed germination and seedling emergence of oriental mustard (*Sisymbrium orientale*). *Weed Science* 54: 1025–1031.

Chauhan, B.S., Gill, G., and Preston, C. (2006d). Tillage system effects on weed ecology, herbicide activity and persistence: a review. *Australian Journal of Experimental Agriculture* 46: 1557–1570.

Chauhan, B.S. and Johnson, D.E. (2010). Chapter 6 - The role of seed ecology in improving weed management strategies in the tropics. *Advances in Agronomy* 105 (10): 221–262.

Chauhan, B.S., Migo, T., Westerman, P.R., and Johnson, D.E. (2010). Post-dispersal predation of weed seeds in rice fields. *Weed Research* 50: 553–560.

Chauhan, B.S., Singh, R.G., and Mahajan, G. (2012). Ecology and management of weeds under conservation agriculture: a review. *Crop Protection* 38: 57–65.

Chon, S.U. and Kim, Y.M. (2004). Herbicidal potential and quantification of suspected allelochemicals from four grass crop extracts. *Journal of Agronomy and Crop Science* 190: 145–150.

Choudhary, M.A., Lal, R., and Dick, W.A. (1997). Long-term tillage effect on runoff and soil erosion under simulated rainfall for a central Ohio soil. *Soil and Tillage Research* 42: 175–184.

Chung, I.M., Kim, K.H., Ahn, J.K. et al. (2002). Screening of allelochemicals on barnyardgrass (*Echinochloa crus-galli*) and identification of potentially allelo-pathic compounds from rice (*Oryza sativa*) variety hull extracts. *Crop Protection* 21: 913–920.

Cooper, J., Baranski, M., Stewart, G. et al. (2016). Shallow non-inversion tillage in organic farming maintains crop yields and increases soil C stocks: a meta-analysis. *Agronomy for Sustainable Development* 36: 22.

Cousens, R. and Moss, S.R. (1990). A model of the effects of cultivations on the vertical distribution of weed seeds within the soil. *Weed Research* 30: 61–70.

Creamer, N.G. and Dabney, S.M. (2002). Killing cover crops mechanically: review of recent literature and assessment of new research results. *American Journal of Alternative Agriculture* 17: 32–40.

Cromar, H.E., Murphy, S.D., and Swanton, C.J. (1999). Influence of tillage and crop residue on postdispersal predation of weed seeds. *Weed Science* 47: 184–194.

Cussans, G.W. and Moss, S.R. (1982). Population dynamics of annual grass weeds. In: *Decision Making in the Practice of Crop Protection* (ed. R.B. Austin), 91–98. London: BCPC.

Dabney, S.M., Delgado, J.A., and Reeves, D.W. (2001). Using cover crops to improve soil and water quality. *Communications in Soil Science and Plant Analysis* 32: 1221–1250.

Daouti, E., Jonsson, M., Vico, G., and Menegat, A. (2022). Seed predation is key to preventing population growth of the weed *Alopecurus myosuroides*. *Journal of Applied Ecology* 59: 471–482.

Daramola, O.S., Adeyemi, O.R., Adigun, J.A., and Adejuyigbe, C.O. (2020). Weed interference and control in soybean, as affected by row spacing, in the transition zone of South West Nigeria. *Journal of Crop Improvement* 34: 103–121.

Davis, A.S. (2010). Cover-crop roller crimper contributes to weed management in no-till soybean. *Weed Science* 58: 300–309.

Davis, A.S., Schutte, B.J., Iannuzzi, J., and Renner, K.A. (2008). Chemical and physical defense of weed seeds in relation to soil seedbank persistence. *Weed Science* 56: 676–684.

De Vore, J.D., Norsworthy, J.K., and Brye, K.R. (2013). Influence of deep tillage, a rye cover crop, and various soybean production systems on palmer amaranth emergence in soybean. *Weed Technology* 27: 263–270.

Demjanová, E., Macák, M., Dalovic, I. et al. (2009). Effects of tillage systems and crop rotation on weed density, weed species composition and weed biomass in maize. *Agronomy Research* 7: 785–792.

Dhima, K., Vasilakoglou, I., Lithourgidis, A. et al. (2006a). Tillage system effects on competition between barley and sterile oat. *Agronomy Journal* 98: 1023–1029.

Dhima, K.V., Vasilakoglou, I.B., Eleftherohorinos, I.G., and Lithourgidis, A.S. (2006b). Allelopathic potential of winter cereals and their cover crop mulch effect on grass weed suppression and corn development. *Crop Science* 46: 345–352.

Díaz, M. (1991). Spatial patterns of granivorous and nest abundance and nest site selection in agricultural landscapes of Central Spain. *Insectes Sociaux* 38: 351–363.

Dick, W.A. and Daniel, T.C. (1987). Soil chemical and biological properties as affected by conservation tillage: environmental implications. In: *Effects of Conservation Tillage on Groundwater Quality: Nitrates and Pesticides* (ed. T. Logan, J. Davidson, J. Baker, and M. Overcash), 315–339. Chelsea, MI, USA: Lewis Publishers.

Dorado, J., Delmonte, J.P., and López-Fando, C. (1999). Weed seedbank response to crop rotation and tillage in semi-arid agroecosystems. *Weed Science* 47: 67–73.

Dorado, J. and Lopez-Fando, C. (2006). The effect of tillage system and use of a paraplow on weed flora in a semiarid soil from central Spain. *Weed Research* 46: 424–431.

Dou, Z., Fox, R.H., and Toth, J.D. (1995). Seasonal soil nitrate dynamics in corn as affected by tillage and nitrogen source. *Soil Science Society of America Journal* 59: 858–864.

Doucet, C., Weaver, S.E., Hamill, A.S., and Zhang, J.H. (1999). Separating the effects of crop rotation from weed management on weed density and diversity. *Weed Science* 47: 729–735.

Fahad, S., Hussain, S., Chauhan, B.S. et al. (2015). Weed growth and crop yield loss in wheat as influenced by row spacing and weed emergence times. *Crop Protection* 71: 101–108.

FAO (2022). What is conservation agriculture? (Online). https://www.fao.org/conservation-agriculture/en. (accessed July 2022).

Fenner, M. and Thompson, K. (2005). *The Ecology of Seeds*. Cambridge: Cambridge University Press.

Forcella, F., Eradat-Oskoui, K., and Wagner, S.W. (1993). Application of weed seedbank ecology to low-input crop management. Ecological Applications 3: 74–83. Available online at: http://www.jstor.org/stable/1941793 (verified 23 March 2010).

Fox, J.W., Khalilian, A., Han, Y.J. et al. (2018). Real-time, valuable-depth tillage for managing soil compaction in cotton production. *Open Journal of Soil Science* 8: 147–161.

Franke, A.C., McRoberts, N., Marshall, G. et al. (2001). The contribution of zero tillage for the management of *Phalaris minor* in the Indian rice–wheat system. Brighton Crop Protection Conference-Weeds 2, 901–906.

Fried, G., Norton, L.R., and Reboud, X. (2008). Environmental and management factors determining weed species composition and diversity in France Agr. *Ecosystems & Environment* 128: 68–76.

Froud-Williams, R., Chancellor, R.J., and Drennan, D.S.H. (1984). The effects of seed burial and soil disturbance on emergence and survival of arable weeds in relation tominimal cultivation. *Journal of Applied Ecology* 21: 629–641.

Froud-Williams, R.J. (1983). The influence of straw disposal and cultivation regime on the population dynamics of *Bromus sterilis*. *Annals of Applied Biology* 103: 139–148.

Froud-Williams, R.J., Chancellor, R.J., and Drennan, D.S.H. (1981). Potential changes in weed floras associated with reduced-cultivation systems for cereal production in temperate regions. *Weed Research* 21: 99–109.

Gallagher, R.S. and Cardina, J. (1998). The effect of light environment during tillage on the recruitment of various summer annuals. *Weed Science* 46: 214–216.

Gallandt, E.R. (2006). How can we target the weed seedbank? *Weed Science* 54 (3): 588–596.

Gallandt, E.R., Molloy, T., Lynch, R.P., and Drummond, F.A. (2005). Effect of cover-cropping systems on invertebrate seed predation. *Weed Science* 53 (1): 69–76.

Gardarin, A., Dürr, C., and Colbach, N. (2009). Which model species for weed seedbank and emergence studies? *A Review Weed Research* 49: 117–130.

Gardarin, A., Dürr, C., Mannino, M.R. et al. (2010). Seed mortality in the soil is related to seed coat thickness. *Seed Science Research* 20: 243–256.

Govindasamy, P., Sarangi, D., Provin, T. et al. (2020). No-tillage altered weed species dynamics in a long-term (36-year) grain sorghum experiment in southeast Texas. *Weed Science* 68 (5): 476–484.

Greil, J., Steinmaus, S., and Kempen, H.M. (2002). Mechanical control methods. In: *Principles of Weed Control* (ed. California Weed Science Society), 89–109. Fresno: Thomson Publications.

Griffith, D.R., Moncrief, J.F., Eckert, D.J. et al. (1992). Crop response to tillage systems. In: *Conservation Tillage Systems and Management: Crop Residue Management with No-till, Ridge-till, Mulch-till* (ed. MidWest Plan Serice Committee), 25–33. Ames: Iowa State University.

Gruber, S. and Claupein, W. (2009). Effect of tillage intensity on weed infestation in organic farming. *Soil and Tillage Research* 105: 104–111.

Gunsolus, J., Wyse, D., Moncada, K., and Fernholz, C. (2010). Chapter 6: weed management. In: *Risk Management Guide for Organic Producers* (ed. K.M. Moncada and C.C. Sheaffer), 76–101. University of Minnesota, Dept. of Agronomy and Plant Genetics, MN, USA.

Halvorson, A.D., Peterson, G.A., and Reule, C.A. (2002). Tillage system and crop rotation effect on dryland crop yields and soil carbon in the central Great Plains. *Agronomy Journal* 94: 1429–1436.

Harper, J.L. (1977). *Population Biology of Plants*, 111–147. London: Academic Press.

Hartmann, K.M. and Nezadal, W. (1990). Photocontrol of weeds without herbicides. *Naturwissenschaften* 77: 158–163.

Heap, I. (2022). The international herbicide-resistant weed database. Online. May 21, 2022. Available at www.weedscience.org.

Hoffman, M.L., Owen, M.D., and Buhler, D.D. (1998). Effects of crop and weed management on density and vertical distribution of weed seeds in soil. *Agronomy Journal* 90: 793–799.

Horowitz, J., Ebel, R., and Ueda, K. (2010). No-till farming is a growing practice. A report from the USDA economic research service. Economic Information Bulletin 70. 22.

Horowitz, M. and Taylorson, R.B. (1984). Hard seededness and germinability of velvetleaf (*Abutilon theophrasti*) as affected by temperature and moisture. *Weed Science* 32: 111–115.

Hulme, P.E. (1994). Post-dispersal seed predation in grassland: its magnitude and sources of variation. *Journal of Ecology* 81: 645–652.

Hurley, T.M. and Frisvold, G. (2016). Economic barriers to herbicide-resistance management. *Weed Science* 64: 585–594.

Hussain, M.I., Danish, S., Sánchez-Moreiras, A.M. et al. (2021). Unraveling Sorghum Allelopathy in agriculture: concepts and Implications. *Plants* 10: 1795.

Johnson, G.A., DeFelice, M.S., and Helsel, Z.R. (1993). Cover crop management and weed control in corn (*Zea mays*). *Weed Technology* 7: 425–430.

Juroszek, J. and Gerhards, R. (2004). Photocontrol of weeds. *Journal of Agronomy and Crop Science* 190: 402–415.

Kan, Z.R., Ma, S.T., Liu, Q.Y. et al. (2020). Carbon sequestration and mineralization in soil aggregates under long-term conservation tillage in the North China Plain. *Catena* 188: 104428.

Kassam, A., Friedrich, T., and Derpsch, R. (2019). Global spread of conservation agriculture. *International Journal of Environmental Studies* 76: 29–51.

Keene, C.L., Curran, W.S., Wallace, J.M. et al. (2017). Cover crop termination timing is critical in organic rotational no-till systems. *Agronomy Journal* 109: 272–282.

Kettler, T.A., Lyon, D.J., Doran, J.W. et al. (2000). Soil quality assessment after weed-control tillage in a no-till wheat-fallow cropping system. *Soil Science Society of America Journal* 64: 339–346.

Knezevic, S.Z., Jhala, A., and Datta, A. (2017). Integrated weed management. In: *Encyclopedia of Applied Plant Sciences*, 2e (ed. B. Thomas, B.G. Murray, and D.J. Murphy), 459–462. Academic Press.

Kong, C.H., Hu, F., Wang, P., and Wu, J.L. (2008). Effect of allelopathic rice varieties combined with cultural management options on paddy field weeds. *Pest Management Science* 64 (3): 276–282.

Korres, N.E. and Norsworthy, J.K. (2015). Influence of a rye cover crop on the critical period for weed control in cotton. *Weed Science* 63 (01): 346–352.

Kruidhof, H.M., Bastiaans, L., and Kropff, M.J. (2009). Cover crop residue management for optimizing weed control. *Plant and Soil* 318: 169–184.

Lampkin, P. (2002). *Organic Farming*, 2e. Ipswich, UK: Old Pond Publishing.

Légère, A. and Samson, D.N. (1999). Relative influence of crop rotation, tillage, and weed management on weed associations in spring barley cropping systems. *Weed Science* 47: 112–122.

Légère, A., Stevenson, F., and Benoit, D. (2011). The selective memory of weed seedbanks after 18 years of conservation tillage. *Weed Science* 59 (1): 98–106.

Légère, A., Vanasse, A., and Stevenson, F.C. (2013). Low-input management and mature conservation tillage: agronomic potential in a cool humid climate. *Agronomy Journal* 105: 745–754.

Lemerle, D., Gill, G.S., Murphy, C.E. et al. (2001). Genetic improvement and agronomy for enhanced wheat competitiveness with weeds. *Crop and Pasture Science* 52: 527–548.

Liebl, R., Simmons, F.W., Wax, L.M., and Stoller, E.W. (1992). Effect of rye (*Secale cereale*) mulch on weed control and soil moisture in soybean (*Glycine max*). *Weed Technology* 6: 838–846.

Liebman, M. and Davis, A.S. (2000). Integration of soil, crop, and weed management in low-external-input farming systems. *Weed Research* 40: 27–47.

Liebman, M. and Mohler, C. (2001). Weeds and the soil environment. In: *M. Ecological Management of Agricultural Weeds* (ed. M. Liebman, C. Mohler, and C. Staver), 210–268. Cambridge: Cambridge University Press.

Luna, J.M., Mitchell, J.P., and Shrestha, A. (2012). Conservation tillage for organic agriculture: evolution toward hybrid systems in the western USA. *Renewable Agriculture and Food Systems* 27: 20–31.

MacLaren, C., Labuschagne, J., and Swanepoel, P.A. (2021). Tillage practices affect weeds differently in monoculture vs. crop rotation. *Soil and Tillage Research* 205: 104795.

MacLaren, C., Swanepoel, P., Bennett, J. et al. (2019). Cover crop biomass production is more important than diversity for weed suppression. *Crop Science* 59: 733–748.

Mahajan, G., Brar, L.S., and Sardana, V. (1999). Effect of tillage and time of sowing on the efficacy of herbicides against *Phalaris minor* in wheat. In: The Proceeding of 17th

Asian Pacific Weed Science Society Conference, Bangkok, Thailand, 193–198.

Marles, S.M., Warkentin, T.D., and Holm, F.A. (2010). Field pea seed residue: a potential alternative weed control agent. *Weed Science* 58: 433–441.

Martin, R.J. and Felton, W.L. (1993). Effect of crop rotation, tillage practice, and herbicides on the population dynamics of wild oats in wheat. *Australian Journal of Experimental Agriculture* 33: 159–165.

Mas, M.T. and Verdú, A.M. (2003). Tillage system effects on weed communities in a 4-year crop rotation under Mediterranean dryland conditions. *Soil and Tillage Research* 74: 15–24.

Mashingaidze, A.B., van der Werf, W., Lotz, L.A.P. et al. (2009). Narrow rows reduce biomass and seed production of weeds and increase maize yield. *Annals of Applied Biology* 155: 207–218.

May, W.E., Aldous, L., and Lafond, G.P. (2020). Feasibility of a wider row spacing and recommended nitrogen in no-till wheat. *Agronomy Journal* 112: 4076–4091.

May, W.E., Fernandez, M.R., Holzapfel, C.B., and Lafond, G.P. (2008). Influence of phosphorus, nitrogen, and potassium chloride placement and rate on durum wheat yield and quality. *Agronomy Journal* 100: 1173–1179.

May, W.E., Mohr, R.M., Lafond, G.P. et al. (2004). Effect of nitrogen, seeding date and cultivar on oat quality and yield in the eastern Canadian prairies. *Canadian Journal of Plant Science* 84: 10251036.

Meiss, H., Le Lagadec, L., Munier-Jolain, N. et al. (2010). Weed seed predation increases with vegetation cover in perennial forage crops. *Agriculture, Ecosystems & Environment* 138: 10–16.

Melander, B., Munier-Jolain, N., Charles, R. et al. (2013). European perspectives on the adoption of nonchemical weed management in reduced-tillage systems for arable crops. *Weed Technology* 27: 231–240.

Menalled, D., Smith, R.G., Dauer, J.T., and Fox, T.B. (2007). Impact of agricultural management on carabid communities and weed seed predation. *Agriculture, Ecosystems & Environment* 118: 49–54.

Menalled, F.D., Gross, K.L., and Hammond, M. (2001). Weed aboveground and seedbank community responses to agricultural management systems. *Ecological Applications* 11: 1586–1601.

Mendes, K.F., Régo, A.P.J., Takeshita, V., and Tornisielo, V.L. (2019). Water resource pollution by herbicide residues. In: *Biochemical Toxicology - Heavy Metals and Nanomaterials* (ed. Muharrem Ince, Olcay Kaplan, and Gabrijel Ondrasek). IntechOpen Books.

Merfield, C.N. (2013). False and stale seedbeds: the most effective non-chemical weed management tools for cropping and pasture establishment. Lincoln: The BHU Future Farming Centre. 23.

Mirsky, S.B., Curran, W.S., Mortensen, D.A. et al. (2009). Control of cereal rye with a roller/crimper as influenced by cover crop phenology. *Agronomy Journal* 101: 1589–1596.

Mirsky, S.B., Curran, W.S., Mortenseny, D.M. et al. (2011). Timing of cover-crop management effects on weed suppression in no-till planted soybean using a roller-crimper. *Weed Science* 59 (03): 380–389.

Mirsky, S.B., Gallandt, E.R., Mortensen, D.A. et al. (2010). Reducing the germinable weed seedbank with soil disturbance and cover crops. *Weed Research* 50: 341–352.

Mirsky, S.B., Ryan, M.R., Curran, W.S. et al. (2012). Conservation tillage issues: cover crop-based organic rotational no-till grain production in the mid-Atlantic region, USA. *Renewable Agriculture and Food Systems* 27: 31–40.

Mischler, R., Curran, W., Duiker, S., and Hyde, J. (2010). Use of a rolled-rye cover crop for weed suppression in no-till soybeans. *Weed Technology* 24 (3): 253–261.

Mohler, C. (2001). Mechanical management of weeds. In: *Ecological Management of Agricultural Weeds* (ed. M. Liebman, C.L. Mohler, and C.P. Staver), 139–209. Cambridge: Cambridge University Press.

Mohler, C.L., Frisch, J.C., and McCulloch, C.E. (2006). Vertical movement of weed seed surrogates by tillage implements and natural processes. *Soil and Tillage Research* 86: 110–122.

Mohler, C.L. and Teasdale, J.R. (1993). Response of weed emergence to rate of *Vicia villosa* Roth and *Secale cereale* L. residue. *Weed Research* 33: 487–499.

Mueller, T.C., Massey, J.H., Hayes, R.M. et al. (2003). Shikimate accumulates in both glyphosate-sensitive and glyphosate-resistant horseweed (*Conyza canadensis* L. Cronq.). *Journal of Agricultural and Food Chemistry* 51: 680–684.

Mulugeta, D., Stoltenberg, D.E., and Boerboom, C.M. (2001). Weed species–area relationships as influenced by tillage. *Weed Science* 49: 217–223.

Murphy, S., Clements, D., Belaoussoff, S. et al. (2006). Promotion of weed species diversity and reduction of weed seedbanks with conservation tillage and crop rotation. *Weed Science* 54: 69–77.

Navarrete, L. and Fernandez-Quintanilla, C. (1996). The influence of crop rotation and soil tillage on seed population dynamics of *Avena sterilis* ssp. *ludoviciana*. *Weed Research* 36: 123–131.

Nichols, V., Verhulst, N., Cox, R., and Govaerts, B. (2015). Weed dynamics and conservation agriculture principles: a review. *Field Crops Research* 183: 56–68.

Nikolić, N., Loddo, D., and Masin, R. (2021). Effect of crop residues on weed emergence. *Agronomy* 11 (1): 163.

Nikolić, N., Squartini, A., Concheri, G. et al. (2020). Weed seed decay in no-till field and planted riparian buffer zone. *Plants (Basel)* 9: 293.

Norsworthy, J., McClelland, M., Griffith, G. et al. (2011). Evaluation of cereal and Brassicaceae cover crops in conservation-tillage enhanced glyphosate-resistant cotton. *Weed Technology* 25 (1): 6–13.

O'Donovan, J.T., Clayton, G.W., Grant, C.A. et al. (2008). Effect of nitrogen rate and placement and seeding rate on barley productivity and wild oat fecundity in a zero tillage system. *Crop Science* 48: 1569–1574.

Oerke, E.C. (2006). Crop losses to pests. *Journal of Agricultural Science* 144: 31–43.

Oliveira, M.C., Osipitan, O.A., Begcy, K., and Werle, R. (2020). Cover crops, hormones and herbicides: priming an integrated weed management strategy. *Plant Science* 301: 110550.

Oplinger, E.S. and Philbrook, B.D. (1992). Soybean planting date, row width, and seeding rate response in three tillage systems. *Journal of Production Agriculture* 5: 94–99.

Owen, M.D., Beckie, H.J., Leeson, J.Y. et al. (2014). Integrated pest management and weed management in the United States and Canada. *Pest Management Science* 71: 357–376.

Peigné, J., Ball, B.C., Roger-Estrade, J., and David, C. (2007). Is conservation tillage suitable for organic farming? A review. *Soil Use and Management* 23: 129–144.

Pinnamaneni, S.R., Anapalli, S.S., Molin, W., and Reddy, K.N. (2022). Effect of rye cover crop on weed control, soybean (*Glycine max* L.) yield and profitability. *Frontiers in Agronomy* 4: 907507.

Plaza, E.H., Kozak, M., Navarrete, L., and Gonzalez-Andujar, J.L. (2011). Tillage system did not affect weed diversity in a 23-yr experiment in Mediterranean dryland. *Agriculture, Ecosystems & Environment* 140: 102–105.

Price, A.J., Arriaga, F.J., Raper, R.L. et al. (2009). Comparison of mechanical and chemical winter cereal cover crop termination systems and cotton yield in conservation agriculture. *Cotton Science* 13: 238–245.

Price, A.J., Monks, C.D., Culpepper, A.S. et al. (2016). High-residue cover crops alone or with strategic tillage to manage glyphosate-resistant Palmer amaranth (*Amaranthus palmeri*) in southeastern cotton (*Gossypium hirsutum*). *Journal of Soil and Water Conservation* 71: 1–11.

Prostko, E.P. [Internet]. (2012). Managing herbicide-resistant palmer amaranth (pigweed) in field corn, grain sorghum, peanut and soybean. Available from: http://www.gaweed.com/resistance-2012-tables.pdf (accessed July 2022).

Puricelli, E., Faccini, D., OriolI, G., and Sabbatini, M.R. (2005). Seed survival and predation of *Anoda cristata* in soyabean crops. *Weed Research* 45: 477–482.

Putnam, A.R., Defrank, J., and Barnes, J.P. (1983). Exploitation of allelopathy for weed control in annual and perennial cropping systems. *Journal of Chemical Ecology* 9: 1001–1010.

Quinn, N.F., Brainard, D.C., and Szendrei, Z. (2016). The effect of conservation tillage and cover crop residue on beneficial arthropods and weed seed predation in Acorn Squash. *Environmental Entomology* 45: 1543–1551.

Radicetti, E., Mancinelli, R., Moscetti, R., and Campiglia, E. (2016). Management of winter cover crop residues under different tillage conditions affects nitrogen utilization efficiency and yield of eggplant (*Solanum melanogena* L.) in Mediterranean environment. *Soil and Tillage Research* 155: 329–338.

Ramankutty, N., Mehrabi, Z., Waha, K. et al. (2018). Trends in global agricultural land use: implications for environmental health and food security. *Annual Review of Plant Biology* 69: 789–815.

Rawat, L.S., Maikhuri, R.K., Bahuguna, Y.M. et al. (2017). Sunflower allelopathy for weed control in agriculture systems. *Journal of Crop Science and Biotechnology* 20: 45–60.

Reberg-Horton, S., Grossman, J., Kornecki, T. et al. (2012). Utilizing cover crop mulches to reduce tillage in organic systems in the southeastern USA. *Renewable Agriculture and Food Systems* 27 (1): 41–48.

Rebetzke, G.J., Bruce, S.E., and Kirkegaard, J.A. (2005). Longer coleoptiles improve emergence through crop residues to increase seedling number and biomass in wheat (*Triticum aestivum* L.). *Plant and Soil* 272: 87–100.

Reddy, K.N. (2001). Effects of cereal and legume cover crop residues on weeds, yield, and net return in soybean (*Glycine max*). *Weed Technology* 15: 660–668.

Reddy, K.N. (2003). Impact of rye cover crop and herbicides on weeds, yield, and net return in narrow-row transgenic and conventional soybean (*Glycine max*). *Weed Technology* 17: 28–35.

Reeves, D.W. (1994). Cover crops and rotations. In: *Advances in Soil Science: Crops Residue Management* (ed. J.L. Hatfield and B.A. Stewart), 125–172. Boca Raton, FL: Lewis Publishers, CRC Press Inc.

Reeves, D.W., Price, A.J., and Patterson, M.G. (2005). Evaluation of three winter cereals for weed control in conservation-tillage nontransgenic cotton. *Weed Technology* 19: 731–736.

Riemens, M.M., van der Weide, R.Y., Béeeker, P.O., and Lotz, L.A.P. (2007). Effect of stale seedbed preparations and subsequent weed control in lettuce (cv. Iceboll) on weed densities. *Weed Research* 47: 149–156.

Riethmuller, G., French, B., White, P., and Paynter, B. (2011). Crop placement and row spacing fact sheet. Grains Research and Development Corporation 1-6. Crop-placement-and-row-spacing-western-fact-sheet.pdf (grdc.com.au) (accessed 26 July 2022).

Romaneckas, K., Kimbirauskienė, R., Sinkevičien, A. et al. (2021). Weed diversity, abundance, and seedbank in differently tilled faba bean (*Vicia faba* L.) cultivations. *Agronomy* 11: 529.

Routley, R., Fettell, N., McMullen, G. et al. (2011). Crop placement and row spacing fact sheet. Grains Research and Development Corporation 1-6.

Crop-placement-and-row-spacing-northern-fact-sheet.pdf (grdc.com.au) (accessed 26 July 2022). https://grdc.com.au/__data/assets/pdf_file/0010/210133/crop-placement-and-row-spacing-western-fact-sheet.pdf.

Ruisi, P., Frangipane, B., Amato, G. et al. (2015). Weed seedbank size and composition in a long-term tillage and crop sequence experiment. *Weed Research* 55: 320–328.

Salonen, J., Hyvönen, T., Kaseva, J., and Jalli, H. (2012). Impact of changed cropping practices on weed occurrence in spring cereals in Finland - a comparison of surveys in 1997-1999 and 2007-2009. *Weed Research* 53: 110–120.

Sans, F.X., Berner, A., Armengot, L., and Mäder, P. (2011). Tillage effects on weed communities in an organic winter wheat–sunflower–spelt cropping sequence. *Weed Research* 51: 413–421.

Schimming, W. and Messersmith, C. (1988). Freezing resistance of overwintering buds of four perennial weeds. *Weed Science* 36 (5): 568–573.

Schomberg, H.H., McDaniel, R.G., Mallard, E. et al. (2006). Conservation tillage and cover crop influences on cotton production on a Southeastern U.S. coastal plain soil. *Agronomy Journal* 98 (5): 1247–1256.

Scopel, A., Ballaré, C., and Radosevish, S. (1994). Photostimulation of seed germination during soil tillage. *New Phytologist* 126: 145–152.

Sharma, P., Abrol, V., and Sharma, R.K. (2011). Impact of tillage and mulch management on economics, energy requirement and crop performance in maize–wheat rotation in rainfed subhumid inceptisols, India. *European Journal of Agronomy* 34: 46–51.

Shearin, A.F., Reberg-Horton, S., and Gallandt, E. (2007). Direct effects of tillage on the activity density of ground beetle (Coleoptera: carabidae) weed seed predators. *Environmental Entomology* 36: 1140–1146.

Shearin, A.F., Reberg-Horton, S., and Gallandt, E.R. (2008). Cover crop effects on the activity-density of the weed seed predator *Harpalus rufipes* (Coleoptera: carabidae). *Weed Science* 56: 442–450.

Shrestha, A., Knezevic, S.Z., Roy, R.C. et al. (2002). Effect of tillage, cover crop and crop rotation on the composition of weed flora in a sandy soil. *Weed Research* 42: 76–87.

Singh, S. (2007). Role of management practices on control of isoproturon-resistant littleseed canarygrass (*Phalaris minor*) in India. *Weed Technology* 21: 339–346.

Smith, R.G., Warren, N.D., and Cordeau, S. (2020). Are cover crop mixtures better at suppressing weeds than cover crop monocultures? *Weed Science* 68: 186–194.

Sosnoskie, L.M., Herms, C.P., and Cardina, J. (2006). Weed seedbank community composition in a 35-yr-old tillage and rotation experiment. *Weed Science* 54: 263–273.

Storkey, J. and Neve, P. (2018). What good is weed diversity? *Weed Research* 56: 239–243.

Swanepoel, P.A. and Labuschagne, J. (2020). Canola is injured by in-row nitrogen placement associated with disc openers, but not by tine openers. *Crop Science* 60: 466–474.

Swanton, C.J., Shrestha, A., Roy, R.C. et al. (1999). Effect of tillage systems, N, and cover crop on the composition of weed flora. *Weed Science* 47: 454–461.

Tabaglio, V., Marocco, A., and Schulz, M. (2013). Allelopathic cover crop of rye for integrated weed control in sustainable agroecosystems. *Italian Journal of Agronomy* 8 (1): 35–40.

Teasdale, J.R. (1996). Contribution of cover crops to weed management in sustainable agricultural systems. *Journal of Production Agriculture* 9: 475–479.

Teasdale, J.R. (2003). Principles and practices of using cover crops in weed management systems. Chapter 3. Management options and perspectives. In: *Weed Management for Developing Countries*. Addendum 1 (ed. R. Labrada). Rome: Food and Agriculture Organization of the United Nations.

Teasdale, J.R., Beste, C.E., and Potts, W.E. (1991). Response of weeds to tillage and cover crop residue. *Weed Science* 39: 195–199.

Teasdale, J.R., Brandsaeter, L.O., Calegari, A., and Neto, F.S. (2007). Chapter 4. cover crops and weed management. In: *Non-Chemical Weed Management Principles, Concepts and Technology* (ed. M.K. Upadhyaya and R.E. Blackshaw), 49–64. Wallingford, UK: CABI.

Teasdale, J.R. and Mohler, C.L. (1993). Light transmittance, soil temperature, and soil moisture under residue of hairy vetch and rye. *Agronomy Journal* 85: 673–680.

Teasdale, J.R. and Mohler, C.L. (2000). The quantitative relationship between weed emergence and the physical properties of mulches. *Weed Science* 48: 385–392.

Teasdale, J.R., Pillai, P., and Collins, R.T. (2005). Synergism between cover crop residue and herbicide activity on emergence and early growth of weeds. *Weed Science* 53: 521–527.

Teasdale, J.R., Rice, C.P., Cai, G., and Mangumet, R.W. (2012). Expression of allelopathy in the soil environment: soil concentration and activity of benzoxazinoid compounds released by rye cover crop residue. *Plant Ecology* 213: 1893–1905.

Thomas, A.G., Banting, J.D., and Bowes, G. (1986). Longevity of green foxtail seeds in a Canadina prairie soil. *Canadian Journal of Plant Science* 66: 189–192.

Thomas, A.G., Derksen, D.A., Blackshaw, R.E. et al. (2004). A multistudy approach to understanding weed population shifts in medium- to long-term tillage systems. *Weed Science* 52 (5): 874–880.

Tørresen, K.S., Skuterud, R., Tandsæther, H.J., and Hagemo, M.B. (2003). Long-term experiments with reduced tillage in spring cereals. I. Effects on weed flora, weed seedbank and grain yield. *Crop Protection* 22: 185–200.

Travlos, I., Gazoulis, I., Kanatas, P. et al. (2020). Key factors affecting weed seeds' germination, weed emergence and their possible role for the efficacy of false seedbed technique as weed management practice. *Frontiers in Agronomy* 2: 1.

Trichard, A., Alignier, A., Biju-Duval, L., and Petit, S. (2013). The relative effects of local management and landscape context on weed seed predation and carabid functional groups. *Basic and Applied Ecology* 14: 235–245.

Tuesca, D., Puricelli, E., and Papa, J.C. (2001). A long-term study of weed flora shifts in different tillage systems. *Weed Research* 41: 369–382.

Uri, N.D. (1999). *Conservation Tillage in U.S. Agriculture - Environmental, Economic, and Policy Issues*. New York: Food Products Press, Haworth.

Vakali, C., Zaller, J.G., and Köpke, U. (2011). Reduced tillage effects on soil properties and growth of cereals and associated weeds under organic farming. *Soil and Tillage Research* 111: 113–141.

Valla, L.G., Sanchez, R.A., Soriano, A., and Ghersa, C.M. (1980). Influencia de al gunos faclores internos y externos sobre la germinacion de las semillas de sorgo de alepo. Revisia delnvestigaciones Agropecuarias. INTA, XV, 549–560.

van der Laat, R., Owen, M.D.K., Liebman, M., and Leon, R.G. (2015). Postdispersal weed seed predation and invertebrate activity density in three tillage regimes. *Weed Science* 63: 828–838.

van Esso, M.L., Ghersa, C.M., and Soriano, A. (1986). Cultivation effects on the dynamics of a Johnson grass seed population in the soil profile. *Soil and Tillage Research* 6: 325–335.

Vasilakoglou, I., Dhima, K., Eleftherohorinos, I., and Lithourgidis, A. (2006). Winter cereal cover crop mulches and inter-row cultivation effects on cotton development and grass suppression. *Agronomy Journal* 98: 1290–1297.

Vincent-Caboud, L., Peigné, J., Casagrande, M., and Silva, E.M. (2017). Overview of organic cover crop-based no-tillage technique in Europe: farmers' practices and research challenges. *Agriculture* 7: 42.

Vollmer, K.M., Van Gessel, M.J., Johnson, Q.R., and Scott, B.A. (2020). Influence of cereal rye management on weed control in soybean. *Frontiers in Agronomy* 2: 600568.

Wang, K.-H., Waisen, P., Leslie, A.W. et al. (2022). Relationships between Soil Tillage Systems, Nematode Communities and Weed Seed Predation. *Horticulturae* 8: 425.

Wang, X.-B., Cai, D.-X., Hoogmoed, W.B. et al. (2006). Potential Effect of Conservation Tillage on Sustainable Land Use: a Review of Global Long-Term Studies. *Pedosphere* 16: 587–595.

Ward, M.J., Ryan, M.R., Curran, W.S. et al. (2011). Cover crops and disturbance influence activity-density of weed seed predators *Amara aenea* and *Harpalus pensylvanicus* (Coleoptera: carabidae). *Weed Science* 59: 76–81.

Weber, J.F., Kunz, C., Peteinatos, G.G. et al. (2017). Weed control using conventional tillage, reduced tillage, no-tillage, and cover crops in organic soybean. *Agriculture* 7: 43.

Westerman, P.R., Wes, J.S., Kropff, M.J., and van der Werf, W. (2003). Annual losses of weed seeds due to predation in organic cereal fields. *Journal of Applied Ecology* 40: 824–836.

Whalen, D.M., Shergill, L.S., Kinne, L.P., and Bish, M.D. (2020). Integration of residual herbicides with cover crop termination in soybean. *Weed Technology* 34: 11–18.

Wicks, G.A. (1984). Integrated systems for control of downy brome (*Bromus tectorum*) in cropland. *Weed Science* 32 ((Suppl.) 1): 26–31.

Wicks, G.A., Burnside, O.C., and Fenster, C.R. (1971). Influence of soil type and depth of planting on downy brome seeds. *Weed Science* 19: 82–86.

Wilkins, E.D. and Bellinder, R.R. (1996). Mow-kill regulation of winter cerealsfor spring no-till crop production. *Weed Technology* 10: 247–252.

Williams, M.M., Mortensen, D.A., and Doran, J.W. (2000). No-tillage soybean performance in cover crops for weed management in the western Corn Belt. *Journal of Soil and Water Conservation* 55: 79–84.

Wolf, T.M., Harrison, S.K., Hall, F.R., and Cooper, J. (2000). Optimizing postemergence herbicide deposition and efficacy through application variables in no-till systems. *Weed Science* 48: 761–768.

Xie, H.S., Rourke, D.R.S., and Hargrave, A.P. (1998). Effect of row spacing and seed/fertilizer placement on agronomic performance of wheat and canola in zero tillage systems. *Canadian Journal of Plant Science* 78: 389–394.

Yenish, J.P., Doll, J.D., and Buhler, D.D. (1992). Effects of tillage on vertical distribution and viability of weed seed in soil. *Weed Science* 40: 429–433.

Yenish, J.P., Worsham, A.D., and York, A.C. (1996). Cover crops for herbicide replacement in no-tillage corn (*Zea mays*). *Weed Technology* 10: 815–821.

Zanin, G., Otto, S., Riello, L., and Borin, M. (1997). Ecological interpretation of weed flora dynamics under tillage systems. *Agriculture, Ecosystems & Environment* 66: 177–188.

Zhang, Z., Gao, P., Dai, W. et al. (2019). Effect of tillage and burial depth and density of seed on viability and seedling emergence of weedy rice. *Journal of Integrative Agriculture* 18: 1914–1923.

13

Ecologically Based Weed Management in Vegetable Crops

Matthias Schumacher[1], *Michael Spaeth*[1], *Georg Naruhn*[1], *David Reiser*[2], *Miriam Messelhäuser*[1], *Rosa Witty*[1], *Roland Gerhards*[1], *and Gerassimos Peteinatos*[3,4]

[1] Department of Weed Science Institute of Phytomedicine, University of Hohenheim, Stuttgart, Germany
[2] Department of Technology in Crop Production, Institute of Agricultural Engineering, University of Hohenheim, Stuttgart, Germany
[3] Centre for Automation and Robotics (CSIC), Madrid, Spain
[4] Hellenic Agricultural Organization - DIMITRA, Department of Agricultural Engineering, Soil and Water Research Institute, Athens, Greece

13.0 Introduction

Vegetables play an important role in the well-balanced human diet, because they supply vitamins, minerals, dietary fiber, and phytochemicals (Dias 2012). Through the efforts to achieve ever higher yields, new problems have emerged, like loss of biodiversity as well as natural habitats and pesticide contamination (Wezel et al. 2018). A significant problem is the high reliance on synthetic herbicides, which have increased the negative impact on the environment, leading to residues in the food chain, a higher health risk, and greater weed-resistant populations (Hillocks 2012; Gaba et al. 2016; Özkara et al. 2016). Even if chemical weed control plays a dominant role in current weed management strategies, there is a strong need for alternative measures. Vegetable production systems require careful planning and often constant adjustments of weed management to ensure yield and quality levels of the produced crop (Korres et al. 2019).

Vegetable production can be categorized into three cultivation systems: greenhouse, semi-field, and open field cultivation (Figure 13.1). In greenhouses, controlled environmental conditions are in place, which often differ substantially from the outside. Elevated temperatures in greenhouses promote the growth of weeds as well as vegetables. However, in big production areas, such as Almeria in Spain, the majority of e.g. tomatoes, is grown soilless by utilizing coconut fiber (Boix-Ruiz 2018). The recent ideas of vertical farming, largely depend on hydroponic systems in controlled environmental conditions while they simultaneously produce fish stock (Wirza and Nazir 2021). These systems totally avoid the necessity for weed management. On the other hand, vegetables can also be grown in soil inside greenhouses. There, preventive methods to decrease weed seed intake, the soil seedbank and weed emergence are most important. Air filters can be installed to block the seed flow from outside, whereas sterilized substrates can be used to avoid weed emergence. Additionally, weed plants growing just outside the greenhouse should be removed to eliminate potential sources of infestation (Neal 2015). The soil seedbank is targeted mainly by solarization (Lombardo et al. 2012), but also stale seedbed techniques can be used inside the greenhouse. Mulches (organic and nonorganic) can serve as physical barriers for emerging weeds. Established weed plants are often removed by flaming or hand weeding.

Under semi-field conditions, plastic covered tunnels are used to accelerate the growth of the vegetables outdoors. These tunnels can either be fixed to one location or can be dislocated. Both types of semi-field cultivations are similar to the greenhouse in terms of weed control strategies; however, depending on the tunnel size, small tractors with mechanical implements can be used additionally to remove emerged weeds. In open field cultivation solarization, steaming and flaming are often not economically viable for weed control, whereas options such as crop rotation and mechanical weed control are available.

The two most important vegetables, grown worldwide and in Europe, are onions and tomatoes (Table 13.1). The weed flora of processing tomatoes in Europe was summarized by Tei et al. (2003) and they found a quite diverse community, including dicotyledonous weeds like *Amaranthus* spp., *Polygonum* spp., *Chenopodium album*, *Portulaca oleracea*, but also grass weeds such as *Avena* spp, *Lolium* spp, and *Echinochloa crus-galli*. The most widespread weed in all examined countries was *Solanum nigrum* L. In the cultivation of onions especially *Poa annua* L. and *Polygonum aviculare L.* are widespread, but also *Cyperus* spp., *Convolvulus arvensis* L., *Datura stramonium* L., *Euphorbia* spp. and *Stellaria media* (L.) Vill. Are common weeds in Europe according to a similar study by Tei et al (1999).

Mechanical weed control methods such as harrowing or hoeing, are highly promising alternatives to chemical weed control in vegetable crops. In previous studies, mechanical weed control and herbicide applications showed similar yield and weed control efficacies in vegetable crops; however, these were only a few cases when weather and field

Ecologically Based Weed Management: Concepts, Challenges, and Limitations, First Edition. Edited by Nicholas E. Korres, Ilias S. Travlos, and Thomas K. Gitsopoulos.

Figure 13.1 Two of the three cultivation systems of vegetable crops: (a) greenhouse and (b) open field (Source: M. Spaeth).

Table 13.1 Top five vegetable crops of the European Union and worldwide according to their 2020 harvested area (FAOSTAT 2022).

Region	Vegetables	Harvested area (×1000 ha)
European Union	Tomatoes	233
	Onions, dry	177
	Peas, green	151
	Cabbages and other brassicas	133
	Lettuce and chicory	132
World	Onions, dry	5479
	Tomatoes	5052
	Watermelons	3053
	Peas, green	2532
	Cabbages and other brassicas	2414

conditions were optimal (Wiltshire et al. 2003; Fennimore et al. 2014). Real-time communication of implements with sensor-systems further increases the potential of mechanical weeding and ensures a constant high yield standard, by taking simultaneously several field conditions like weed density, soil structure, and weed species into account (Machleb et al. 2020; Spaeth et al. 2021).

Weed communities in vegetable crops often exhibit high species numbers and vary considerably in their composition according to crop, soil, and weather conditions as well as farming practices (Tei and Panacci 2017). Tei and Panacci (2017) summarized the key weeds in several selected vegetable crops, such as cabbage, onions, and carrots, within several European and Mediterranean countries. They found that weed species that are taxonomically closely related to the vegetables are the most troublesome and dominant (Tei and Pannacci 2017), like *Solanum nigrum* in processing tomatoes (*Solanum lycopersicum* L.) (Tei et al. 2003). Weeds can substantially influence vegetable production, not only by reducing yield due to competition (Oerke 2006), but also through additional reasons like reduction of quality parameters (e.g. size, color, shape) or contamination (e.g. *S. nigrum* fruits in peas) (Knott 2002). Furthermore, weeds can host a range of pests and diseases (e.g. *Stellaria media* can carry the lettuce mosaic virus), increase labor costs and interfere with the harvest procedure (e.g. *Convolvulus arvensis*) (Knott 2002; Tei and Pannacci 2017).

In that sense, weed management, as an important part of the vegetable production, needs to evolve and holistically integrate the interaction of environmental, economic, and societal dimensions of agroecosystems (Korres et al. 2019). It needs to follow the approach of Hill and MacRae (1996) to transform conventional weed management systems into ecological weed management systems. The first step is to increase efficiency by e.g. site-specific field management, and the second step is to substitute chemical inputs with mechanical tools and biological herbicides. These two steps were already intensively studied, and sensor-based/autonomous tools already took these steps and are currently able to achieve similar yield levels as chemical weed control (Home et al. 2002; Wiltshire et al. 2003; Fennimore et al. 2014; Kunz et al. 2015; Machleb et al. 2018; Spaeth et al. 2020, 2021). The last step is called agro-ecosystem redesign, and aims to prevent weed emergence, reduce external inputs and uses ecological processes to manage weed populations. The goal is to strengthen the vegetable crop and utilize the weak points of weeds. Ecological weed management is not a once-determined fixed strategy, but an adaptive system with "many little hammers" (Liebman and Gallandt 1997).

13.1 Cultural Methods

13.1.1 Crop Rotation

To prevent dominance of particular weeds, vegetables crops should be rotated according to their life cycle. Different sowing and harvesting dates stimulate the emergence of different weed species and avoid "problematic" weed communities due to continuous cropping (Graziana et al. 2012; Hosseini et al. 2014; Weisberger et al. 2019). In ecological weed management not only the design of the crop rotation but also the diversity of incorporated crops can substantially contribute to reducing weed densities in the long run (Weisberger et al. 2019).

Tei and Pannacci (2017) and Panacci et al. (2017) describe additional important aspects on how to design a crop rotation for vegetables. One aspect is to grow vegetables with high (pumpkins (*Cucurbita pepo* L.), watermelons (*Citrullus lanatus* (Thunb.) Matsum. and Nakai)) and low (leek (*Allium ampeloprasum* L.), onion, radish (*Raphanus raphanistrum* subsp. *sativus* (L.) Domin)) competitive ability in turn. Alternating crops with different harvesting organs like leaf crops (spinach, lettuce), fruit crops (pumpkin, tomato), bulb crops (onion, garlic (*Allium sativum* L.)), and root as well as tuber crops (carrots, radish) is another aspect. Moreover, they advise inserting cereals whenever possible to reduce weed pressure, because of a higher competitive ability of and different management strategies in cereals. Also avoiding crops originating from the same botanical family like celery (*Apium graveolens* L.) and carrots (Apiaceae), potato (*Solanum tuberosum* L.), tomato (Solanaceae), cucumber (*Cucumis sativus* L.), and melon (Cucurbitaceae) as pre-crops to the planned vegetable is advised (Tei and Pannacci 2017).

Under intensive continuous cropping systems autotoxicity problems may arise from the respective crops. Examples for several vegetables, including cucumber, tomato, and red pepper (*Capsicum annuum* L.), are summarized by John (2010). Another common example is asparagus (Kato-Noguchi et al. 2017); tests to avoid these effects by selecting appropriate succeeding crops are available (Asaduzzaman et al. 2013).

However, crop rotations or crop sequences of vegetable producers are much less "plannable" than those of, e.g. cereal producers. Often crop sequences of particular fields need to be adjusted in the short term, due to the current level of weed infestations or a delay in growth of the crop, so the next planting batch cannot be planted afterward. Moreover, inappropriate soil conditions such as high soil moisture for soil tillage hinder timely sowing or transplantation. Therefore, crop rotation design or choice of crop sequence on a field can be quite spontaneous in vegetable production.

13.1.2 Intercropping

Intercropping is growing two (or more) cash crops simultaneously on the same area of land, often in alternating crop strips. Intercropping can reduce weed densities (MacLaren et al. 2020) within a vegetation period by improving resource use efficiency of the crop plants (Malézieux et al. 2009; Brooker et al. 2015). An example is an intercropping system of leek and celery, where weed biomass of *Senecio vulgaris* was reduced by 58% and total weed density by 98% in comparison to leek alone (Baumann et al. 2000). Cauliflower (*Brassica oleracea* L. var. *botrytis*) intercropped with leaf lettuce (*L. sativa* L. var. *crispa*), onions or snap bean (*Phaseolus vulgaris* L. var. *nanus*), provided land equivalent ratios (LER) higher than 1. This means that from the same area, more vegetables can be harvested in an intercropping system than if they were grown in monoculture (LER = 1). But also, under greenhouse conditions intercropping can be advantageous, although few authors have investigated this. By intercropping eggplant (*Solanum melongena* L.) with leaf lettuce in the greenhouse, Güvenç and Yildirim (2005) also obtained LERs higher than 1.1.

13.1.3 Transplanting

The critical period of weed control determines the optimal timing for weed control measures to prevent yield loss in a crop, often in relation to the crop growth stage (Knezevicz et al. 2002). In cabbage (*Brassica oleracea* Capitata Group) or beet (*Beta vulgaris* subsp. *vulgaris* Conditiva Group), e.g. this is at canopy closure, because weeds emerging afterward will be efficiently suppressed by the crop (Robinson 2014). The starting and end point of the critical period (or the length) can vary considerably, depending on, for example, growth rates, seeding dates, morphology a.s.o. (Martin et al. 2001; Knezevic et al. 2002). A way to shorten the critical period, and to facilitate the crops competitive ability simultaneously, is using transplanting instead of seeding as a planting technique (Weaver 1984). In cabbage, this technique leads to a consistent critical period from transplanting to three weeks post transplanting (Weaver 1984), while the critical period in direct seeded cabbage is fluctuating considerably (Roberts et al. 1976; Miller and Hopen 1991). Transplanting can be performed for almost any vegetable crop, although it is highly uncommon (e.g. in carrots and turnips due to their growth habit below soil). Moreover, economic reasons might stand against transplanting, for example, in tomatoes seeding is much cheaper if their destination is being processed in industry instead of marketed on the fresh market. Transplanting is often combined with the stale seedbed technique to reduce the weed pressure.

13.1.4 False Seedbed

With the technique of false seedbed, the preparation of the soil is done several days or weeks before drilling or transplanting. This stimulates with irrigation or rainfall (Shekhawat et al. 2020) the germination of nondormant weed seeds (Liebman et al. 2001; Rasmussen 2004; Merfield 2015), which is followed by the destruction of emerged weed seedlings. This is either performed thermally by using flames or steam weeders, or mechanically using shallow tillage implements, such as spring tine weeders (harrows), rotary tillers, hoes and specialist false seedbed tillage equipment, such as twin roller false seedbed tillers, and roller undercutter machines (Caldwell and Mohler 2001; Merfield 2015). According to a study of Riemens et al. (2007), the combination of false seedbed preparation followed by weed control measures prior to planting reduced the number of weeds during growth of lettuce (cv. Iceboll) by 43–83%. The effectiveness of a false seedbed technique largely depends on factors affecting weed emergence, in particular weed species, tillage, soil type, and soil physical characteristics, such as temperature and moisture (Yenish et al. 1996; Mohler and Galford 1997; Riemens et al. 2007).

13.1.5 Inorganic and Organic Soil Cover

A way to further strengthen the vegetable crop is to transplant it directly into a black polyethylene sheet that is covering the soil. This is a very widespread way to effectively suppress weeds, e.g. in strawberries (*Fragaria × ananassa* (Duchesne ex Weston) Duchesne ex Rozier), tomatoes and melons (Tei and Pannacci 2017). Black polyethylene sheets are used in greenhouses and in the field. Apart from the weed suppressive effects they offer further advantages such as a clean harvest product. Similar colored sheets are available that selectively transmit light, heating the soil like clear plastic covers (like in solarization) but suppressing weeds by reflecting photosynthetic active radiation (Coolong 2012). This is especially advantageous for temperature-loving vegetable crops like melons and eggplants. Their utilization in vegetable farming, however, is not widespread. An interesting alternative is to use biodegradable sheets made of starch or paper that are more sustainable than synthetic sheets. They last a limited time in the field (two to four months), but often long enough to cover the critical period of weed control (Tei and Pannacci 2017) and can be incorporated into the soil after harvest. As all of the mentioned sheets are rarely used more than once, this is another advantage over synthetic sheets, since they do not need to be removed from the field by hand. Inorganic as well as organic mulches are often not that effective against perennial weeds, such as *Cyperus* spp., *Cirsium arvense* and *Elymus repens*, because these weed species are able to penetrate the mulch layer (Bond and Grundy 2001).

13.2 Preventive Methods

13.2.1 Field Choice

A good part of weed control can arise by carefully choosing the field location for production. Especially low-competitive vegetable crops (e.g. onions (*Allium cepa* L.), carrots (*Daucus carota* subsp. *sativus* (Hoffm.) Arcang)) require fields with a low number of potential weeds to ensure production. Moreover, perennial weeds can become a huge problem, e.g. in asparagus (*Asparagus officinalis* L.) production, if not accounted for during the field choice process. Not only do the historical weed infestations of the field need to be accounted, but also so do the landscape surrounding it. By deliberately manipulating landscape structure, weed communities can be affected, e.g. fields adjacent to bare ground or rivers receive less weed propagules than fields next to grassland, thereby affecting weed density and diversity (Metcalfe et al. 2019). Altered flow of propagules from neighboring habitats might reduce competitive weed species (Metcalfe et al. 2019). Weed mapping (via UAVs) might improve the farmers knowledge about weed distributions to make appropriate field choices (Mink et al. 2018).

13.2.2 Cover Crops

Cover crops are grown in between two cash crops. Cover crops offer several advantages for the producer, such as reduced erosion, increased nutrient availability, and weed suppression (Hartwig and Ammon 2002). In general, cover crops suppress weeds through competition and allelopathy. They interfere with weed emergence or growth from sowing until a certain time period after the next main crop has established (Falquet et al. 2015). However, cover crops also need to be managed properly to avoid negative effects on the concomitant or subsequent vegetable crop.

The use and effects of cover crops in vegetable production systems were recently reviewed by Mennan et al. (2020). The authors argue that cover crops can be a very good option, even or especially for no-till vegetable production (Mennan et al. 2020), as tillage is not available as a weed control option in this system. Also, under reduced tillage cover crops like barley (*Hordeum vulgare* L.) and crimson clover (*Trifolium incarnatum* L.) can lower weed densities by 50% (Buchanan et al. 2016). In tomato, *Lolium multiflorum* Lam. And *Secale cereale* L. were able to reduce weed biomass by more than 75% (Mennan et al. 2009a). Moreover, *Vicia villosa* ROTH. Reduced the density of the commonly occurring weed species *Amaranthus retroflexus*, *Digitaria sanguinalis* and *Portulaca oleracea* on average by 72% in a study by Campiglia et al. (2010) in tomato. The same effects were also observed in cabbage, where the yield of cabbage (*Brassica oleraces* var. acephala DC) was even doubled due to weed control by using *V. villosa* (Mennan et al. 2009b). The extent to which the different cover crop species exert competition and allelopathy is unclear and controversial in the literature (Finney et al. 2016; Baraibar et al. 2018; Gfeller et al. 2018). The use of cover crops as living mulches is not widespread in vegetable production; however, some systems are in place involving mainly legume (e.g. *Trifolium* spp.) species or ryegrass (*Lolium* spp.) as companion crops (Masiunas 1998). Müller-Schärer (1996) found a sufficient weed suppression in leek, when perennial ryegrass (*Lolium perenne* L.) was intercropped after two mechanical weed control treatments with a hoe.

The allelopathic properties of cover crops can be advantageous by suppressing weeds, and might even be enhanced by artificial stress application (Gerhards and Schappert 2020). On the other hand, they can detrimentally affect the subsequent vegetable crops. Crimson clover (*Trifolium incarnatum* L.) and hairy vetch (*Vicia hirsute* (L.) SF. Gray) volatiles hamper germination and development of carrots, onions, and tomato seedlings (Bradow and Connick 1990).

13.3 Direct Methods

13.3.1 Biological Weed Control

Between 1950 and 1970 it was common to utilize herbivores, especially geese, as weeders in vegetable crops such as onion and asparagus (FAO 2002), but also cucumber, tomato and garlic (Popay and Field 1996). They selectively feed on grass weeds (Food and Agriculture Organization of the United Nations 2002). This technique became obsolete after the introduction of herbicides (Food and Agriculture Organization of the United Nations 2002), but might still be valuable for small-scale vegetable production.

13.3.2 Mechanical Weed Control

13.3.2.1 Current Stage

Mechanical weed control in vegetable crops is an established option, which has existed for centuries in farming. However, to achieve an economically valid and efficient weed control level, it requires the combination of different weeding techniques and cultivation strategies. Mechanical weed control is challenging because of the high dependency on weather conditions during a typically narrow application window and the slow working speed of the tools compared to chemical weed control strategies. Despite its complexity, mechanical weed control is a highly promising substitution to the conventional herbicide applications. The main goal for vegetable crops is to find an alternative for effective weeding while avoiding crop damage. Mechanical weed control can be applied intra-row (e.g. finger weeder, torsion weeder) (Table 13.2), between the crop rows (e.g. no-till-sweep, goose feet sweep, side-knife), or the whole field can be cultivated (harrow); see Table 13.3 and Figure 13.2.

Table 13.2 Overview of finger and torsion weeders in specific vegetables and their weed control efficacy.

Working tool	Vegetable	Weed control efficacy	References
Finger weeder	Spinach	45–99%	Tei et al. (2002)
	Onion		Melander et al. (2015)
	Lettuce		Riemens et al. (2007)
	Onion		Pannacci et al. (2007)
	Tomato		Pannacci et al. (2008)
	Tomato		Cirujeda et al. (2015)
Torsion weeder	Onion	45–91%	Melander et al. (2015)
	Leek		Kurstjens et al. (2000)
	Tomato		Cirujeda et al. (2015)
	Carrot		Peruzzi et al. (2004)

Figure 13.2 Overview of conventional and sensor-based hoeing and harrowing systems. (a) Conventional manual steered hoe, company K.U.L.T.-Kress; (b) Sensor-guided hoeing system, company K.U.L.T.-Kress; (c) Conventional manual adjusted harrow, company Einböck; (d) Sensor-based harrowing system, company Hatzenbichler. Source: M. Spaeth.

Different passive and active mechanical working tools are available on the market as shown in Table 13.3. Passive working tools have to be adjusted manually for the current field situation. This can be done by optimizing the working intensity of a harrow or the size and shape of sweeps of the hoe. Machleb et al. (2018) could achieve a higher weed control efficacy (up to 89.1%) with an optimized no-till sweeps. But, Rasmussen et al. (2010), Melander et al. (2015), and Pérez-Ruiz et al. (2015) pointed out, that the biggest problem for vegetable cultivation is caused by intra-row weeds, because vegetable crops are sensitive against contact, and it is not possible to treat the crop row without touching them. In the last decade, several intra-row tools, such as finger weeders or torsion weeders have been developed to achieve a higher intra-row efficacy while keeping the crop damage as low as possible. They need very accurate steering to work as close as possible to the crop, and its use is recommended when crop plants are well rooted to avoid crop damage. In sown spinach, finger weeders achieved a reduction of weed density up to 68% intra-row, while inter-row control efficacy was nearly 90% (Tei et al. 2002). Melander et al. (2015) showed a control efficacy of up to 91% in transplanted onion with torsion weeders. Tillett et al. (2008) could achieve up to 87% weed control inter-row, in a field trial with cabbage.

Flex-tine harrows or rotary harrows allow a whole field cultivation. Normally they are used in field crops like cereals, but over the last years, experiments showed that they can be just as well be used in vegetable crops like carrot, onion, and pea (*Pisum sativum* L.) (Ascard and Bellinder 1996; Raffaelli et al. 2002; Pannacci and Tei 2014). Melander and Hartvig (1995) showed in his results a weed control efficacy of up to 60% by weed harrowing in onions. But, treating planted vegetable crops with a whole field cultivator needs care, because of the high crop damage potential. In order to minimize the crop damage, whole field cultivators like a flex-tine harrow should not be performed before the true four- to six-leaf stage or when the plant is not well established (Lichtenhahn et al. 2005).

To alleviate work-input and to avoid human-based crop damage, sensor technology can be used for steering, tool guidance and crop identification. The active mechanical weeding tools are categorized into half and full autonomous, tractor-pulled implements or robots. The half autonomous, tractor-pulled implements are sensor-based machines, like the Robovator (F. Poulsen Engineering, Denmark), where a driver on the tractor is needed. The full autonomous robots (NAÏO, DINO-Weedrobot) are independent of a driver, or someone who is adjusting the machine. Most of the commercial active implements operate intra-row. Benefits of such active weeding machines are high precision and high weed control efficacy (90% or better), while keeping the crop damage as low as possible (Tillett et al. 2008; Nørremark et al. 2011; Kunz et al. 2015; Spaeth et al. 2020). The main negative aspects are the slow working speed ($< 4\ \text{km h}^{-1}$) and the costs for such a system (half autonomous > 40,000 €, full autonomous (robots) > 100,000 €) (Bakker et al. 2010; Fennimore et al. 2014).

Table 13.3 Overview of the most common, commercial passive and active mechanical weeding tools for intra- and inter-row cultivation.

Tool guidance	Machine for the tool	Weeding tool	Company	Target zone	Working speed km h^{-1}
Passive	Harrow	Flex-tine	All common*	Whole field	8 – 10
Passive		Rotary	All common*	Whole field	> 4
Passive	Hoe	Finger weeder	K.U.L.T. Kress	Intra-row	4
Passive		Torsion weeder	All common*	Intra-row	< 4
Passive		Goose feet	All common*	Inter-row	3 – 4
Passive		Flat hoe blade	All common*	Inter-row	3 – 4
Passive		Down-cut side knives	All common*	Inter-row	3 – 4
Passive		No-till sweep	All common*	Inter-row	3 – 4
Passive		Rotary hoe	All common*	Whole field	20
Passive		Soil ridging		Whole field	
Active	Sensor-based	All hoeing blades, In-row knifes	K.U.L.T. Kress, Robovator	Intra-row	1.5 – 3.5
Active		Camera-hoe, all hoeing blades	All common*	Inter-row	8 – 15
Active		Robocrop in-row weeder	Garford	Intra-row	8 – 15
Active		All hoeing blades, In-row knifes	Drimac, ZIP	Whole field	3 – 4
Active		ILT-rotary how	Inst. für Land-technik Bonn	Intra-row	7
Active		V-shape hoe blades, L-knifes hoe blades, finger weeder	NAÏO, DINO-Weedrobot	Whole field	4
Active	Robot	Five-teeth harrow, spiked-harrow, finger weeder	NAÏO, Oz-Weedrobot	Whole field	1.8
Active		Mechanical chopper	Farming Revolution	Whole field	4

*Agrokraft, Carré, Einböck, Garford, Hatzenbichler, K.U.L.T. Kress, Schmotzer, APV, Thyregod, Maschio Gaspardo and Kongskilde

13.3.2.2 Inter and Intra Row Diversification

Mechanical tools for intra-row cultivation should be aligned as close as possible to the vegetable to increase the efficacy. Bearing this in mind, it is obvious that adjustment by intelligent sensor-technologies is needed for optimization (Machleb et al. 2020). For optimization, the sensors must be able to identify crop rows, structures, weeds, or the single vegetable plant. If there is an error, the crop plants can easily be damaged by the inter- or intra-row tool (Home et al. 2002; Melander et al. 2005). For guiding tools as close as possible to the crop row, precision agriculture requires the utilization of more accurate methods. Typically, RTK-GNSS technology (Global Navigation Satellite System). This technology uses a local reference station (e.g. on the farm) to correct and refine the signal received from the satellites thus offering sub-centimeter level accuracy (Keicher and Seufert 2000; Zhang et al. 2002). Gerhards et al. (2020) used a total station to detect the deviation and could achieve a maximum lateral offset from the row center of 22.53 mm with a 3 m hoe and 18.42 mm with a 6 m hoe. The weed-control efficacies in their study were in the inter-row space up to 96% and intra-row up to 91%; using a camera guided hoe. A further approach to increase the efficacy of weed control is to locate the crops and weeds. There are two common concepts of using georeferencing: first localizing the plants or seeded crops via GNSS and creating a plant map or combining the GNSS navigation with the machine vision (Pérez-Ruiz et al. 2012). The second approach is the more advanced concept (Machleb et al. 2020). In previous studies, using GNSS-referenced crop plants for mechanical weed Nørremark et al. (2008, 2011) could achieve a weed control efficacy of 91% using a cycloid hoe in combination with an intra-row treatment. Langsenkamp et al. (2014) was using a GNSS-based tube stamp and could achieve a weed control efficacy of 94%.

13.3.3 Robotic Weeding

Robots in general have the advantage to be independent of human labor input. This allows new ways for weeding, which can be different to the current conventional farming (Machleb et al. 2020), but right now they are not of interest for farming with conventional machines. The special benefit of using robots is the option to work site-specific even on a "single weed level." This is not of interest in conventional agriculture as it is time consuming and slow. However, selective and energy efficient weeding could have ecological benefits for future farms. The machines could adapt to the plant's individual needs. For example, robots could try to remove just the yield impacting weeds in the field. However, this makes precise detection of single weeds necessary (Peteinatos et al. 2020).

Until now many different sensor setups were used to detect weeds using cameras, structure from motion, light curtains, laser scanners, sonar, and even tactile sensors (Peteinatos et al. 2014). Raja et al. (2020) used a stereo camera with HSV-based weed-crop classification to guide their weeding tool in a lettuce field. Wendel and Underwood (2016) investigated ground-based hyperspectral imaging with NDVI and support vector machines for weed detection in vegetables. A multicamera system combined with sonar sensors to detect and track the weeds for precise treatment with the robotic tool was created by Wu et al. (2020). Andujar et al. (2011, 2013) used a terrestrial lidar and sonar for detecting weeds based on height differences. Ruckelshausen et al. (2009) compared several sensors for weed crop identification, including 3D-Lidar, and light curtains. Also, combination of visual sensors and seed mapping were used in the past (Griepentrog et al. 2004). Camera-based weed detection is possibly the most mentioned application for deep learning in agriculture right now (Kamilaris and Prenafeta-Boldú 2018).

The used mechanical tools and size of the robots vary a lot, starting with research on micro weeding robots (Chen et al. 2003) to automated tractors with advanced implement control (Griepentrog et al. 2006). Bawden et al. (2017) used a flexible robot implement with extendable hoes for onion, carrot, and tomato. Rotary harrows in different sizes were used by Frese et al. (2015) and Reiser et al. (2019) and could achieve a control efficacy of up to 65%. Lee et al. (2014) created a universal robot arm for mechanical weeding mounted to a mobile robot platform. Michaels et al. (2015) performed weeding with a delta robot arm. As tool a tube stamp for "punching" the weeds into the soil was used (Langsenkamp et al. 2014). Even laser weeding was performed to destroy single weeds with an accuracy of more than 97% (Xiong et al. 2017).

So far, selective weeding is not available for robot machines on the market. The main suppliers of robots utilize similar implements as the actual implements available for tractors. For example, Naïo is using conventional tools like hoes and finger weeders, combined with autonomous camera-based row navigation and GNSS navigation. Farmdroid is using the GNSS planting position to guide the weeding tools. Agrointelli tries to be able to attach all conventional tractor implements to the autonomous platform. The named robot machines can already be used in vegetables like cabbage, salat, onion, and leek. Hence, there are a bunch of interesting ideas and solutions for selective weeding in development and possibly will be for sale in the next years.

13.3.4 Thermal Weed Control

Another alternative to mechanical weed control is the thermal weeding, which basically include fire, flaming, hot water, steam, infrared radiation, electrical energy, microwave as well as ultraviolet radiation, lasers, and freezing. But thermal weeding is not so common, because relatively high costs, large energy, and fuel consumption as well as slow working speed are disadvantages (Ascard et al. 2007).

The most common thermal weeding technique is flaming, which has been successfully practiced in many vegetable crops for a long time period. A study by Fontanelli et al. (2015) in organically grown carrot, garlic, and fennel (Foeniculum vulgare Mill.) compared a traditional standard weed management system with different innovative strategies, which combine preventive, cultural, and direct control methods, including also pre-emergence flaming. The results show that the innovative strategies lead to a significantly lower weed dry biomass in all crops and concurrently higher yield for garlic and fennel (Fontanelli et al. 2015). Mainardis et al. (2020) found in their study on different weed control methods in vineyards that controlling weeds by flaming and tillage was more efficient than mowing, while the number of species was substantially unaltered over all treatments. An experiment examining a biomass-fueled flame weeder for intra-row weeding in vineyards showed that flaming resulted in greater control efficacy compared to mowing and tillage in both inter- and intra-row spaces (Pergher et al. 2019). Kerpauskas et al. (2006) showed that more than 98% of weed shoots are destroyed by steaming, which simultaneously resulted in a 9–10% higher onion yield compared to the untreated control plots. After testing a thermal band heater machine for intra-row weed control, Jørgensen et al. (2004) concluded that more than 70 °C is necessary to kill weed seeds. Shrestha et al. (2013) compared different weed control methods in organic vineyards. They stated, that all mechanical treatments were significantly more effective and two to four times more cost-efficient than steaming or the application of an organic herbicide. Therewith, the adoption of this method is challenging and not economically sound.

13.4 Conclusion and Outlook

To sum up, there are several options already available to replace the use of herbicides with ecological weed management methods. The most promising and effective alternative is sensor-based mechanical weed control. However, a disadvantage is the cost factor for the sensor recognition and automation process of these systems. To make the change from conventional to ecological weed management, a more integrated approach should be the starting point: using "many little hammers", including direct methods (mechanical and biological) but also preventive methods such as tillage, crop rotation, and cover crops. At this stage of redesign, the ecological processes of weed population and interaction in the agroecosystem need to be the basis to achieve true ecological weed strategies. The efficiency substitution redesign (ESR) approach of Hill and MacRae (1996) is even extended by Gliessman (2016). He adds two further steps: a more direct connection between producers and consumers for shorter food chains and a new global food system for more sustainable and environmentally sound food production. Most research, however, is carried out for efficiency and substitution, less for redesign. Especially for vegetable crops, marketing aspects like quality and production process are important for the food value chain. If production considers ecological processes this will positivity influence the environment. Although some producers already follow this path, it is still not established on a large-scale due to aspects like hostile policy and market environments (Liebman et al. 2016), but also the lack of willingness to spend more money for ecologically produced vegetables.

Acknowledgments

This project was partially funded by the German federal ministry for agriculture and nutrition and the German federal ministry for research and technology.

References

Andújar, D., Escolà, A., Dorado, J., and Fernández-Quintanilla, C. (2011). Weed discrimination using ultrasonic sensors. *Weed Research* 51 (6): 543–547. https://doi.org/10.1111/j.1365-3180.2011.00876.x.

Andújar, D., Rueda-Ayala, V., Moreno, H. et al. (2013). Discriminating crop, weeds and soil surface with a terrestrial LIDAR sensor. *Sensors (Basel, Switzerland)* 13 (11): 14662–14675. https://doi.org/10.3390/s131114662.

Asaduzzaman, M., Fuad Mondal, M., Ban, T., and Asao, T. (2013). Selection of ideal succeeding crops after asparagus, taro and beans replanting field in seedling growth bioassay. *Allelopathy Journal* 32 (1): 1–22.

Ascard, J. and Bellinder, R.R.B. (1996). Mechanical in-row cultivation in row crops. *2. International Weed Control Congress, Copenhagen (Denmark), 25–28 Jun*. SP.

Ascard, J., Hatcher, P.E., Melander, B., and Upadhyaya, M.K. (2007). Thermal Weed Control. In: *Non-chemical Weed Management. Principles, Concepts and Technology* (ed. M.K. Upadhyaya and R.E. Blackshaw), 155–175. Wallingford, UK: CABI.

Bakker, T., Asselt, K., Bontsema, J. et al. (2010). Systematic design of an autonomous platform for robotic weeding. *Journal of Terramechanics* 47 (2): 63–73. https://doi.org/10.1016/j.jterra.2009.06.002.

Baraibar, B., Hunter, M.C., Schipanski, M.E. et al. (2018). Weed suppression in cover crop monocultures and mixtures. *Weed Science* 66 (1): 121–133. https://doi.org/10.1017/wsc.2017.59.

Baumann, D.T., Kropff, M.J., and Bastiaans, L. (2000). Intercropping leeks to suppress weeds. *Weed Research* 40 (4): 359–374. https://doi.org/10.1046/j.1365-3180.2000.00197.x.

Bawden, O., Kulk, J., Russell, R. et al. (2017). Robot for weed species plant-specific management. *Journal of Field Robotics* 34 (6): 1179–1199. https://doi.org/10.1002/rob.21727.

Boix-Ruiz, A., Gómez-Tenorio, M.A., Ruiz-Olmos, C. et al. (2018). Coconut fiber grow bags as a source of primary inoculum of fusarium oxysporum F. Sp. Radicis-Lycopersici in greenhouse tomato crops in Almeria (Spain). *Acta Horticulturae* 1207: 275–280. https://doi.org/10.17660/ActaHortic.2018.1207.38.

Bond, W. and Grundy, A.C. (2001). Non-chemical weed management in organic farming systems. *Weed Research* 41: 383–405. https://doi.org/10.1046/j.1365-3180.2001.00246.x.

Bradow, J.M. and Connick, W.J. (1990). Volatile seed germination inhibitors from plant residues. *Journal of Chemical Ecology* 16 (3): 645–666. https://doi.org/10.1007/BF01016477.

Brooker, R.W., Bennett, A.E., Cong, W.-F. et al. (2015). Improving intercropping: a synthesis of research in agronomy, plant physiology and ecology. *New Phytologist* 206 (1): 107–117. https://doi.org/10.1111/nph.13132.

Buchanan, A.L., Kolb, L.N., and Hooks, C.R.R. (2016). Can winter cover crops influence weed density and diversity in a reduced tillage vegetable system? *Crop Protection* 90: 9–16. https://doi.org/10.1016/j.cropro.2016.08.006.

Caldwell, B. and Mohler, C.L. (2001). Stale seedbed practices for vegetable production. *HortSci* 36 (4): 703–705. https://doi.org/10.21273/HORTSCI.36.4.703.

Campiglia, E., Caporali, F., Radicetti, E., and Mancinelli, R. (2010). Hairy vetch (*Vicia villosa* Roth.) cover crop residue management for improving weed control and yield in no-tillage tomato (Lycopersicon esculentum Mill.) production. *European Journal of Agronomy* 33 (2): 94–102. https://doi.org/10.1016/j.eja.2010.04.001.

Chen, B., Tojo, S., and Watanabe, K. (2003). Machine vision for a micro weeding robot in a paddy field. *Biosystems Engineering* 85 (4): 393–404. https://doi.org/10.1016/S1537-5110(03)00078-3.

Cirujeda, A., Aibar, J., Moreno, M.M., and Zaragoza, C. (2015). Effective mechanical weed control in processing tomato: seven years of results. *Renewable Agriculture and Food Systems* 30 (3): 223–232. https://doi.org/10.1017/S1742170513000434.

Coolong, T. (2012). Mulches for weed management in vegetable production. In: *Weed Control* (ed. A. Price), 57–74. InTech. http://www.intechopen.com/books/weed-control/mulches-for-weed-management-in-vegetable-production.

Dias, J.S. (2012). Nutritional quality and health benefits of vegetables: a review. *FNS* 03 (10): 1354–1374. https://doi.org/10.4236/fns.2012.310179.

Falquet, B.T., Gfeller, A., Mickaël Pourcelot, L. et al. (2015). Weed suppression by common buckwheat: a review. *ecb* 53 (1): 1–6. https://doi.org/10.2525/ecb.53.1.

Fennimore, S.A., Smith, R.F., Tourte, L. et al. (2014). Evaluation and economics of a rotating cultivator in Bok Choy, Celery, Lettuce, and Radicchio. *Weed Technology* 28 (1): 176–188. https://doi.org/10.1614/WT-D-13-00051.1.

Finney, D.M., White, C.M., and Kaye, J.P. (2016). Biomass production and carbon/nitrogen ratio influence ecosystem services from cover crop mixtures. *Agronomy Journal* 108 (1): 39–52. https://doi.org/10.2134/agronj15.0182.

Fontanelli, M., Frasconi, C., Martelloni, L. et al. (2015). Innovative strategies and machines for physical weed control in organic and integrated vegetable crops. *Chemical Engineering Transactions* 44: 211–216. https://doi.org/10.3303/CET1544036.

Food and Agriculture Organization of the United Nations. 2002. *Goose Production*. Herausgegeben von Roger Buckland und Gerard Guy. *FAO Animal Production and Health Paper 154*. Rome. http://www.fao.org/3/y4359e/y4359e00.htm#Contents.

Frese, C., Frey, C., Meßmer, F. et al. (2015). AgriApps–an app-based solution for field-robot-based agriculture. *Land-Technik AgEng. 73rd International Conference on Agricultural Engineering*.

Gaba, S., Gabriel, E., Chadœuf, J. et al. (2016). Herbicides do not ensure for higher wheat yield, but eliminate rare plant species. *Scientific Reports* 6 (1): 30112. https://doi.org/10.1038/srep30112.

Gerhards, R., Kollenda, B., Machleb, J. et al. (2020). Camera-guided weed hoeing in winter cereals with narrow row distance. *Gesunde Pflanzen* 72 (4): 403–411. https://doi.org/10.1007/s10343-020-00523-5.

Gerhards, R. and Schappert, A. (2020). Advancing cover cropping in temperate integrated weed management. *Pest Management Science* 76 (1): 42–46. https://doi.org/10.1002/ps.5639.

Gfeller, A., Herrera, J.M., Tschuy, F., and Judith, W.I.R.T.H. (2018). Explanations for amaranthus retroflexus growth suppression by cover crops. *Crop Protection* 104: 11–20. https://doi.org/10.1016/j.cropro.2017.10.006.

Gliessman, S. (2016). Transforming food systems with agroecology. *Agroecology and Sustainable Food Systems* 40 (3): 187–189. https://doi.org/10.1080/21683565.2015.1130765.

Graziana, F., Onofri, A., Pannacci, E. et al. (2012). Size and composition of weed seedbank in long-term organic and conventional low-input cropping systems. *European Journal of Agronomy* 39: 52–61. https://doi.org/10.1016/j.eja.2012.01.008.

Griepentrog, H.-W., Nørremark, M., and Nielsen, J. (2006). Autonomous intra-row rotor weeding based on GPS. *Proceedings of the CIGR World Congress Agricultural Engineering for a Better World, Bonn, Germany*, 37:17.

Griepentrog, H.W., Christensen, S., Søgaard, H.T. et al.. (2004). Robotic weeding. *Proceedings of AgEng*.

Güvenç, I. and Yildirim, E. (2005). Intercropping with eggplant for proper utilisation of interspace under greenhouse conditions. *European Journal of Horticultural Science* 70 (6): 300–302.

Hartwig, N.L. and Ammon, H.U. (2002). Cover crops and living mulches. *Weed Science* 50 (6): 688–699. https://doi.org/10.1614/0043-1745(2002)050[0688:AIACCA]2.0.CO;2.

Hill, S.B. and MacRae, R.J. (1996). Conceptual framework for the transition from conventional to sustainable agriculture. *Journal of Sustainable Agriculture* 7 (1): 81–87. https://doi.org/10.1300/J064v07n01_07.

Hillocks, R.J. (2012). Farming with fewer pesticides: EU pesticide review and resulting challenges for UK agriculture. *Crop Protection* 31 (1): 85–93. https://doi.org/10.1016/j.cropro.2011.08.008.

Home, M.C.W., Tillett, N.D., Hague, T., and Godwin, R.J. (2002). An experimental study of lateral positional accuracy achieved during inter-row cultivation. *5th EWRS Workshop on Physical and Cultural Weed Control*. Citeseer.

Hosseini, P., Karimi, H., Babaei, S. et al. (2014). Weed seed bank as affected by crop rotation and disturbance. *Crop Protection* 64: 1–6. https://doi.org/10.1016/j.cropro.2014.05.022.

Jørgensen, M.H., Kristensen, J.K., MELANDER, B. et al. (2004). *Thermal band heating for intra-row weed control*. Not submitted. https://orgprints.org/id/eprint/4822.

Kamilaris, A. and Prenafeta-Boldú, F.X. (2018). Deep learning in agriculture: a survey. *Computers and Electronics in Agriculture* 147: 70–90. https://doi.org/10.1016/j.compag.2018.02.016.

Kato-Noguchi, H., Nakamura, K., Ohno, O. et al. (2017). Asparagus decline: autotoxicity and autotoxic compounds in asparagus rhizomes. *Journal of Plant Physiology* 213: 23–29. https://doi.org/10.1016/j.jplph.2017.02.011.

Keicher, R. and Seufert, H. (2000). Automatic guidance for agricultural vehicles in Europe. *Computers and Electronics in Agriculture* 25 (1–2): 169–194. https://doi.org/10.1016/S0168-1699(99)00062-9.

Kerpauskas, P.A., Sirvydas, P., Lazauskas, P. et al. (2006). Possibilities of weed control by water steam. *Agronomy Research* 4 (Special Issue): 221–225.

Knezevic, S.Z., Evans, S.P., Blankenship, E.E. et al. (2002). Critical period for weed control: the concept and data analysis. *Weed Science* 50 (6): 773–786. https://doi.org/10.1614/0043-1745(2002)050[0773:CPFWCT]2.0.CO;2.

Knott, C.M. (2002). Weed Control in Other Arable and Field Vegetable Crops. In: *Weed Management Handbook* (ed. R.E.L. Naylor), 359–398. Oxford, UK: Blackwell Science Ltd. https://doi.org/10.1002/9780470751039.ch19.

Korres, N.E., Burgos, N.R., Travlos, I. et al. (2019). New directions for integrated weed management: modern technologies, tools and knowledge discovery. 0065-2113 155: 243–319. https://doi.org/10.1016/bs.agron.2019.01.006.

Kunz, C., Weber, J., and Gerhards, R. (2015). Benefits of precision farming technologies for mechanical weed control in soybean and sugar beet—comparison of precision hoeing with conventional mechanical weed control. *Agronomy* 5 (2): 130–142. https://doi.org/10.3390/agronomy5020130.

Kurstjens, D. and Bleeker, P. (2000). Optimising torsion weeders and finger weeders. *Proceedings 4th EWRS Workshop on Physical Weed Control*, Elspeet, The Netherlands, 30–32.

Langsenkamp, F., Sellmann, F., Kohlbrecher, M. et al. (2014). Tube stamp for mechanical intra-row individual plant weed control. *Proceedings of 18th World Congress of CIGR*. Beijing, China: CIGR, 16–29.

Lee, J.J.H., Frey, K., Fitch, R., and Sukkarieh, S. (2014). *Fast path planning for precision weeding*, 1448–2053. https://opus.lib.uts.edu.au/bitstream/10453/106135/1/pap169.pdf.

Lichtenhahn, M., Koller, M., Dierauer, H., and Baumann, D. (2005). *Weed control in organic vegetable cultivation*. Research Institute of Organic Agriculture (FiBL).

Liebman, M., Baraibar, B., Buckley, Y. et al. (2016). Ecologically sustainable weed management: how do we get from proof-of-concept to adoption? *Ecological Applications: A Publication of the Ecological Society of America* 26 (5): 1352–1369. https://doi.org/10.1002/15-0995.

Liebman, M. and Gallandt, E.R. (1997). Many little hammers: ecological management of crop-weed interactions. In: *Ecology in Agriculture and Soil Management* (ed. L.E. Jackson), 291–343. San Diego, CA, USA: Academic Press.

Liebman, M., Mohler, C.L., and Staver, C.P. (2001). *Ecological Management of Agricultural Weeds*. Cambridge University Press.

Lombardo, S., Longo, A.M.G., Lo Monaco, A., and Mauromicale, G. (2012). The effect of soil solarization and fumigation on pests and yields in greenhouse tomatoes. *Crop Protection* 37: 59–64. https://doi.org/10.1016/j.cropro.2012.02.010.

Machleb, J., Kollenda, B., Peteinatos, G., and Gerhards, R. (2018). Adjustment of weed hoeing to narrowly spaced cereals. *Agriculture* 8 (4): 54. https://doi.org/10.3390/agriculture8040054.

Machleb, J., Peteinatos, G.G., Kollenda, B.L. et al. (2020). Sensor-based mechanical weed control: present state and prospects. *Computers and Electronics in Agriculture* 176: 105638. https://doi.org/10.1016/j.compag.2020.105638.

MacLaren, C., Storkey, J., Menegat, A. et al. (2020). An ecological future for weed science to sustain crop production and the environment. A review. *Agronomy Sustainable Development* 40 (4). https://doi.org/10.1007/s13593-020-00631-6.

Mainardis, M., Boscutti, F., del Mar Rubio Cebolla, M., and Pergher, G. (2020). Comparison between flaming, mowing and tillage weed control in the vineyard: effects on plant community, diversity and abundance. *PLoS ONE* 15 (8): e0238396. https://doi.org/10.1371/journal.pone.0238396.

Malézieux, E., Crozat, Y., Dupraz, C. et al. (2009). Mixing Plant Species in Cropping Systems: concepts, Tools and Models: a Review. In: *Sustainable Agriculture* (ed. E. Lichtfouse, M. Navarrete, P. Debaeke et al.), 329–353. Dordrecht: Springer Netherlands.

Martin, S.G., Van Acker, R.C., and Friesen, L.F. (2001). Critical period of weed control in spring canola. *Weed Science* 49 (3): 326–333. https://doi.org/10.1614/0043-1745(2001)049[0326:CPOWCI]2.0.CO;2.

Masiunas, J.B. (1998). Production of vegetables using cover crop and living mulches—a review. *Journal of Vegetable Crop Production* 22.

Melander, B. and Hartvig, P. (1995). Weed harrowing in seeded onions. *9th EWRS (European Weed Research Society) Symposium Budapest 1995: Challenges for Weed Science in a Challenging Europe*, http://orgprints.org/00001764.

Melander, B., Lattanzi, B., and Pannacci, E. (2015). Intelligent versus non-intelligent mechanical intra-row weed control in transplanted onion and cabbage. *Crop Protection* 72: 1–8. https://doi.org/10.1016/j.cropro.2015.02.017.

Melander, B., Rasmussen, I.A., and Paolo, B. (2005). Integrating physical and cultural methods of weed control— examples from European research. *Weed Science* 53 (3): 369–381. https://doi.org/10.1614/WS-04-136R.

Mennan, H., Ngouajio, M., Isık, D., and Kaya, E. (2009a). Effects of alternative winter cover cropping systems on weed suppression in organically grown tomato (Solanum lycopersicum). *Phytoparasitica* 37 (4): 385–396. https://doi.org.10.1007/s12600-009-0048-1.

Mennan, H., Ngouajio, M., Kaya, E., and Isik, D. (2009b). Weed management in organically grown kale using alternative cover cropping systems. *Weed Technology* 23 (1): 81–88. https://doi.org/10.1614/WT-08-119.1.

Mennan, H., Jabran, K., Zandstra, B.H., and Pala, F. (2020). Non-chemical weed management in vegetables by using cover crops: a review. *Agronomy* 10 (2): 257. https://doi.org/10.3390/agronomy10020257.

Merfield, C.N. (2015). False and Stale Seedbeds: the most effective non-chemical weed management tools for cropping and pasture establishment. *The FFC Bulletin* 2015 (V4): 25.

Metcalfe, H., Hassall, K.L., Boinot, S., and Storkey, J. (2019). The contribution of spatial mass effects to plant diversity in arable fields. *Journal of Applied Ecology* 56 (7): 1560–1574. https://doi.org/10.1111/1365-2664.13414.

Michaels, A., Haug, S., and Albert, A. (2015). Vision-based high-speed manipulation for robotic ultra-precise weed control. *2015 IEEE/RSJ International Conference on Intelligent Robots and Systems (IROS)*, 5498–5505: IEEE.

Miller, A.B. and Hopen, H.J. (1991). Critical weed-control period in seeded cabbage (Brassica Oleracea Var Capitata). *Weed Technology* 5 (4): 852–857. https://doi.org/10.1017/S0890037×00033972.

Mink, R., Dutta, A., Peteinatos, G. et al. (2018). Multi-temporal site-specific weed control of cirsium Arvense (L.) Scop. and Rumex Crispus L. in maize and sugar beet using unmanned aerial vehicle based mapping. *Agriculture* 8 (5): 65. https://doi.org/10.3390/agriculture8050065.

Mohler, C.L. and Galford, A.E. (1997). Weed seedling emergence and seed survival: separating the effects of seed position and soil modification by tillage. *Weed Research* 37 (3): 147–155. https://doi.org/10.1046/j.1365-3180.1997.d01-21.x.

Müller-Schärer, H. (1996). Interplanting ryegrass in winter leek: effect on weed control, crop yield and allocation of N-Fertiliser. *Crop Protection* 15 (7): 641–648. https://doi.org/10.1016/0261-2194(96)00033-6.

Neal, J. 2015. Greenhouse weed control. *NC State Extension Publications*. https://content.ces.ncsu.edu/greenhouse-weed-control.

Nørremark, M., Griepentrog, H.W., Nielsen, J., and Søgaard, H.T. (2008). The development and assessment of the accuracy of an autonomous GPS-based system for intra-row mechanical weed control in row crops. *Biosystems Engineering* 101 (4): 396–410. https://doi.org/10.1016/j.biosystemseng.2008.09.007.

Nørremark, M., Griepentrog, H.W., Nielsen, J., and Søgaard, H.T. (2011). Evaluation of an autonomous GPS-based system for intra-row weed control by assessing the tilled area. *Precision Agriculture* 13 (2): 149–162. https://doi.org/10.1007/s11119-011-9234-5.

Oerke, E.-C. (2006). Crop losses to pests. *Journal of Agricultural Science* 144 (1): 31–43. https://doi.org/10.1017/S0021859605005708.

Özkara, A., Akyil, D., and Konuk, M. (2016). Pesticides, Environmental Pollution, and Health. In: *Environmental Health Risk - Hazardous Factors to Living Species* (ed. M. Larramendy and S. Soloneski). InTech.

Pannacci, E., Graziani, F., Guiducci, M., and Tei, F. (2008). Controllo meccanico delle malerbe in colture da seme di cipolla proceedings of research project "Azioni di innovazione e ricerca a supporto del piano sementiero, PRIS2 Progetto di ricerca interregionale, Programma "Sviluppo Rurale" – Sottoprogramma "Innovazione e Ricerca. 265–275.

Pannacci, E., Guiducci, M., and Tei, F. (2007). Mechanical weed control in organic onion seed production. *Proceedings 7th Workshop of the EWRS Working Group: Physical and Cultural Weed Control*, Salem (Mecklenburg-Vorpommern), Germany, 11-14 March, 119–120.

Pannacci, E. and Tei, F. (2014). Effects of mechanical and chemical methods on weed control, weed seed rain and crop yield in maize, sunflower and soyabean. *Crop Protection* 64: 51–59. htpps://doi.org/10.1016/j.cropro.2014.06.001.

Pannacci, E., Lattanzi, B., and Tei, F. (2017). Non-chemical weed management strategies in minor crops: a review. *Crop Protection* 96: 44–58. https://doi.org/10.1016/j.cropro.2017.01.012.

Pérez-Ruiz, M., Gonzalez-de-santos, P., Ribeiro, A. et al. (2015). Highlights and preliminary results for autonomous crop protection. *Computers and Electronics in Agriculture* 110: 150–161. https://doi.org/10.1016/j.compag.2014.11.010.

Pérez-Ruiz, M., Slaughter, D.C., Gliever, C.J., and Upadhyaya, S.K. (2012). Automatic GPS-based intra-row weed knife control system for transplanted row crops. *Computers and Electronics in Agriculture* 80: 41–49. https://doi.org/10.1016/j.compag.2011.10.006.

Pergher, G., Gubiani, R. and Mainardis,M. (2019). Field testing of a biomass-fueled flamer for in-row weed control in the vineyard. *Agriculture* 9 (10): 210. https://doi.org/10.3390/agriculture9100210.

Peruzzi, A., Raffaelli, M., Ginanni, M., and Borelli, M. (2004, March). Physical weed control in organic carrot production. *Proceedings of 6th EWRS Workshop on Physical and Cultural Weed Control*, Lillehammer, Norway. 24–38.

Peteinatos, G.G., Reichel, P., Karouta, J. et al. (2020). Weed identification in maize, sunflower, and potatoes with the aid of convolutional neural networks. *Remote Sensing* 12 (24): 4185. https://doi.org/10.3390/rs12244185.

Peteinatos, G.G., Weis, M., Andújar, D. et al. (2014). Potential use of ground-based sensor technologies for weed detection. *Pest Management Science* 70 (2): 190–199. https://doi.org/10.1002/ps.3677.

Popay, I. and Field, R. (1996). Grazing animals as weed control agents. *Weed Technology* 10 (1): 217–231. https://doi.org/10.1017/S0890037×00045942.

Raffaelli, M., Peruzzi, A., Ginanni, M., and Di Ciolo, S. 2002. *Mechanical weed control in sunflower and soyabean crops using spring-tine harrow: results of two-year trials*".

Raja, R., Nguyen, T.T., Slaughter, D.C., and Fennimore, S.A. (2020). Real-time weed-crop classification and localisation technique for robotic weed control in lettuce. *Biosystems Engineering* 192: 257–274. https://doi.org/10.1016/j.biosystemseng.2020.02.002.

Rasmussen, I.A. (2004). The effect of sowing date, stale seedbed, row width and mechanical weed control on weeds and yields of organic winter wheat. *Weed Research* 44 (1): 12–20. https://doi.org/10.1046/j.1365-3180.2003.00367.x.

Rasmussen, J., Mathiasen, H., and Bibby, B.M. (2010). Timing of post-emergence weed harrowing. *Weed Research* 50 (5): 436–446. https://doi.org/10.1111/j.1365-3180.2010.00799.x.

Reiser, D., Sehsah, E.-S., Bumann, O. et al. (2019). Development of an autonomous electric robot implement for intra-row weeding in vineyards. *Agriculture* 9 (1): 18. https://doi.org/10.3390/agriculture9010018.

Riemens, M.M., van der Weide, R.Y., Bleeker, P.O., and Lotz, L.A.P. (2007). Effect of stale seedbed preparations and subsequent weed control in lettuce (Cv. Iceboll) on weed densities. *Weed Research* 47 (2): 149–156. https://doi.org/10.1111/j.1365-3180.2007.00554.x.

Roberts, H.A., Bond, W., and Hewson, R.T. (1976). Weed competition in drilled summer cabbage. *Ann Applied Biology* 84 (1): 91–95. https://doi.org/10.1111/j.1744-7348.1976.tb01732.x.

Robinson, D.E. (2014). Integrated Weed Management in Horticultural Crops. In: *Recent Advances in Weed Management* (ed. B.S. Chauhan and G. Mahajan), 239–254. New York: Springer New York.

Ruckelshausen, A., Biber, P., Dorna, M. et al. (2009). BoniRob–an autonomous field robot platform for individual plant phenotyping. *Precision Agriculture* 9 (841): 1.

Shekhawat, K., Rathore, S.S., and Chauhan, B.S. (2020). Weed management in dry direct-seeded rice: a review on challenges and opportunities for sustainable rice production. *Agronomy* 10 (9): 1264. https://doi.org/10.3390/agronomy10091264.

Shrestha, A.S., Kaan Kurtural, M.W., Fidelibus, G.D., and Konduru, S. (2013). Efficacy and cost of cultivators, steam, or an organic herbicide for weed control in organic vineyards in the san joaquin valley of california. *hortte* 23 (1): 99–108. https://doi.org/10.21273/HORTTECH.23.1.99.

Spaeth, M., Jannis Machleb, G.G., Peteinatos, M.S., and Gerhards, R. (2020). Smart harrowing—adjusting the treatment intensity based on machine vision to achieve a uniform weed control selectivity under heterogeneous field conditions. *Agronomy* 10 (12): 1925. https://doi.org/10.3390/agronomy10121925.

Spaeth, M., Schumacher, M., and Gerhards, R. (2021). Comparing sensor-based adjustment of weed harrowing intensity with conventional harrowing under heterogeneous field conditions. *Agronomy* 11 (8): 1605. https://doi.org/10.3390/agronomy11081605.

Tei, F., Ascard, J., Baumann, D.T. et al. (1999, June). Weeds and weed management in onion–a review. *11th Symposium EWRS. European Weed Research Society, Basel*, 28, 131.

Tei, F., Montemurro, P., Baumann, D.T. et al. (2003). Weeds and weed management in processing tomatoes. *Acta Horticulturae* 613: 111–121. https://doi.org/10.17660/ActaHortic.2003.613.13.

Tei, F., Stagnari, F., and Granier, A. (2002). Preliminary results on physical weed control in processing spinach. *Proceedings 5 th EWRS Workshop on Physical and Cultural Weed Control*, 164–171.

Tei, F. and Pannacci, E. (2017). Weed Management Systems in Vegetables. In: *Weed Research* (ed. P.E. Hatcher and R.J. Froud-Williams), 355–388. Chichester, UK: John Wiley & Sons, Ltd.

Tillett, N.D., Hague, T., Grundy, A.C., and Dedousis, A.P. (2008). Mechanical within-row weed control for transplanted crops using computer vision. *Biosystems Engineering* 99 (2): 171–178. https://doi.org/10.1016/j.biosystemseng.2007.09.026.

Weaver, S.E. (1984). Critical period of weed competition in three vegetable crops in relation to management practices. *Weed Research* 24 (5): 317–325. https://doi.org/10.1111/j.1365-3180.1984.tb00593.x.

Weisberger, D., Nichols, V., and Liebman, M. (2019). Does diversifying crop rotations suppress weeds? A meta-analysis. *PLoS ONE* 14 (7): e0219847. https://doi.org/10.1371/journal.pone.0219847.

Wendel, A. and Underwood, J. (2016). Self-supervised weed detection in vegetable crops using ground based hyperspectral imaging. *2016 IEEE international conference on robotics and automation (ICRA)*, 5128–5135. IEEE.

Wezel, A., Goris, M., Bruil, J. et al. (2018). Challenges and action points to amplify agroecology in Europe. *Sustainability* 10 (5): 1598. https://doi.org/10.3390/su10051598.

Wiltshire, J.J.J., Tillett, N.D., and Hague, T. (2003). Agronomic evaluation of precise mechanical hoeing and chemical weed control in sugar beet. *Weed Research* 43 (4): 236–244. https://doi.org/10.1046/j.1365-3180.2003.00338.x.

Wirza, R. and Nazir, S. (2021). Urban aquaponics farming and cities- a systematic literature review. *Reviews on Environmental Health* 36 (1): 47–61. https://doi.org/10.1515/reveh-2020-0064.

Wu, X., Aravecchia, S., Lottes, P. et al. (2020). Robotic weed control using automated weed and crop classification. *Journal of Field Robotics* 37 (2): 322–340. https://doi.org/10.1002/rob.21938.

Xiong, Y., Yuanyue, G., Liang, Y., and Blackmore, S. (2017). Development of a prototype robot and fast path-planning algorithm for static laser weeding. *Computers and Electronics in Agriculture* 142: 494–503. https://doi.org/10.1016/j.compag.2017.11.023.

Yenish, J.P., Fry, T.A., Durgan, B.R., and Wyse, D.L. (1996). Tillage effects on seed distribution and common milkweed (Asclepias Syriaca) establishment. *Weed Science* 44 (4): 815–820. https://doi.org/10.1017/S0043174500094765.

Zhang, N., Wang, M., and Wang, N. (2002). Precision agriculture—a worldwide overview. *Computers and Electronics in Agriculture* 36 (2–3): 113–132. https://doi.org/10.1016/S0168-1699(02)00096-0.

14

Ecological Weed Management in Row Crops

Stevan Z. Knezevic

Professor of Integrated Weed Management, Department of Agronomy and Horticulture, University of Nebraska-Lincoln, Lincoln, Nebraska, USA

14.0 Introduction

By a simple definition, a row crop is a crop that is planted in rows, which are usually wide enough to allow access for agricultural machinery designed for the seasonal activities in row crops. The inter-row spaces could also be hoed few times per season and/or the rows are weeded to conserve moisture and improve soil aeration. Examples of row crops include many agronomic crops grown worldwide, including: *(Zea mays*), soybeans (*Glycine max*), sunflower (*Helianthus annus*), canola (*Brassica spp*), cotton (*Gossypium spp*), and sugarbeets (*Betta vulgris).* In 2020, the USA accounted for about 80 million hectares planted with major row crops, including maize, soybean, sorghum, cotton, and sugarbeets, while about 28% of all global land in 2020 was planted with row crops, which is about 14 million square kilometers (1.4 billion hectares).

Regardless of the type of row crops, their growth and production are affected by various types of pests, including weeds. Weed management is widely reported as a significant constraint in crop production worldwide. It is also well known that weeds reduce crop yield through competition for light, nutrients, water, and space. Uncontrolled weed infestations can lead to complete crop loss (Knezevic et al. 2020). Weeds can not only reduce overall yields but also impact the quality of harvested agricultural products. Weeds can also interfere with harvest operations and increase the cost of harvesting in both smallholder and large-scale farms. Finally, the cost of controlling weeds can be high, running as much as 40% of the overall cost of row crop production costs (Stopes and Millington 1991).

There are many kinds of weeds with different life cycles, thus a single control method is not effective. Frankly speaking, controlling weeds with one or two methods provides the weeds a chance to adapt to those practices. Classical example of such an approach was the widespread use of glyphosate-tolerant crops, and overreliance on a single herbicide (e.g. glyphosate), which resulted in weed adaptation in the form of 17 glyphosate resistant weeds in USA (Knezevic et al. 2020). Therefore, instead of using a particular weed control method, there is a need for diversity in agricultural systems.

Diversity in agriculture can be described in terms of the diversity in practices utilized at the farm and the biological diversity (Kristiansen et al. 2001). Diversity of practices refers to types of physical and cultural methods used by producers (e.g. to control weeds), while biological diversity refers to the number and variety of organisms present on a farm, including genetic diversity amongst species (e.g. weeds).

At the farm level, diversity may include the types of activities undertaken at a given time and space, the types of crops and livestock, the variety of invertebrates and microbes present in the soil, the air, and in the existing crop and noncrop vegetation (Kristiansen et al. 2001). These options could act synergistically with other ecosystem functions, e.g. crop diversification promotes diversity within an agroecosystem but also alters the population of dominant (and maybe harmful) weed species. This can all lead toward ecologically based weed management, a sustainable system that can be used long term without harming the environment.

In the context of ecological weed management in row crops, the general belief is that diversity and sustainability at the farm level can be improved by utilizing the concept of integrated weed management (Swanton and Weise 1991).

Ecologically Based Weed Management: Concepts, Challenges, and Limitations, First Edition. Edited by Nicholas E. Korres, Ilias S. Travlos, and Thomas K. Gitsopoulos.

14.1 Integrated Weed Management (IWM) in Row Crops

There are numerous definitions of IWM (Shaw 1982). Knezevic et al. (2017) described IWM as a combination of mutually supportive technologies in order to manage weeds. It is a multidisciplinary approach to weed management utilizing the application of numerous alternative control measures. In practical terms, it means developing a weed management program using a combination of preventive, cultural, mechanical, and chemical practices. It does not mean abandoning chemical weed control but relying on it less.

Based on experiences from the US Midwest, there is a general believe that the development of an IWM program should be based on a few general principles that can be utilized in any row crop production system (Knezevic et al. 2020):

- Use agronomic practices that limit the introduction and spread of weeds (preventing weed problems before they start).
- Help the crop compete with weeds (help "choke out" weeds).
- Use practices that keep weeds "off-balance" (do not allow weeds to adapt).
- Make a weed control decision.
- Document and keep records.

Combining agronomic practices based on the above principles can help design weed control for any farming operation. Also, it is important to understand that an IWM program is not a recipe; it needs to be changed and adjusted to the particular farming operation, and from year to year. The goal is to manage, not eradicate, weeds, as weed eradication is not possible for environmental and economic reasons.

14.1.1 Preventing Weed Problems before They Start

Preventive weed control practices are well covered in the literature (Knezevic et al. 2017). It is important to understand that the best start of any weed management program is to reduce the potential for introducing weed seeds to the field. Preventing practices may include field sanitation, planting certified crop seeds, controlling volunteer weeds and patches of new species or herbicide resistant weeds, cleaning equipment, tarping grain trucks, and use of well composted manure.

Field sanitation involves practices that prevent weeds from entering or spreading across the field. Cleaning equipment, especially combines, before moving from field to field can further reduce the spread of weeds. Tarping grain trucks prevents introduction of weeds on roadsides, which can then invade neighboring fields. Planting certified crop seeds produces vigorous crop seedlings and improves crop emergence and establishment, which is important for improving soybean competition against weeds and overall yields (Knezevic et al. 2020).

Control of volunteer weeds along edges of the field, fence lines, and ditches is useful in preventing the spread of weeds. Patches of newly invading weeds or herbicide resistant weeds should be controlled to prevent their spreading. Manure can be a potential source of weed seeds and introduce new weed species to a field, especially if either the animals or livestock feed was imported from a different region. Therefore, aging or composting manure for at least a year, before spreading in the field will reduce the viability of weed seeds. In general, preventive weed control techniques are usually the least expensive but routinely the most overlooked.

14.1.2 Improve Crop Competition against Weeds

A few simple practices can be utilized to provide crop the advantage over weeds, including adjusting row spacing, increasing planting density, or selecting the most competitive variety. For example, narrowing soybean row spacing has been known to improve crop competitiveness. Soybeans planted at 18 cm and 38 cm are more competitive against weeds than ones planted at 76 cm or 96 cm. It was suggested that narrower rows and higher soybean density significantly reduces biomass of late-emerging weeds. For example, planting soybeans in narrow rows improved early-season crop tolerance to weeds, delayed the critical time for weed removal, and required less intensive weed management programs than in wide rows (Knezevic et al. 2003).

Cultivar selection is also important. For example, certain soybean varieties can be more competitive than others. More specifically, taller soybean varieties close their canopy more completely than shorter types, which helps to shade out weeds. Taller varieties still need to be sprayed for weeds, but the overall weed control should be better due to added crop competition.

14.1.3 Keep Weeds "Off-Balance" – Do Not Let Them Adapt

Developing a farming operation based on a variety of cropping practices that minimize weed establishment is the key in not allowing weeds to adapt to the production system; thus, the goal is to develop a system that "keeps weeds off-balance." Such a goal can be achieved with a variety of practices, of which the most important ones are crop and herbicide rotations, while adjusting planting date, or the use of cover crops can also be helpful. In addition, utilizing various mechanical and alternative methods of weed control, such as flaming, can also help the process.

Crop rotation has multiple benefits to the cropping systems and should be the first step for keeping weeds off balance. Diversified crop rotation will help manage weeds in many different ways and at many different times over the growing season. For example, including forage crops in multiyear rotation allows cutting weeds before they set seeds, which is an important form of weed removal.

Crops also differ in their competitive ability. For example, cereals are generally better competitors than maize and soybean. Rotating crops with different life cycles will also help in preventing weeds from adapting. Annual weeds are more common in annual crops while biennial and perennial weeds are mostly found in perennial crops. Rotating crops with different life cycles will prevent weeds with specific life cycles from adapting and establishing. Rotating crops will also allow rotating herbicide choices. Furthermore, rotating herbicides with different modes of action and application times will help delay weed adaption and reduce a chance for weed resistance.

Finally, selecting herbicides for a particular application window will help keep weeds off balance. Herbicides can be applied at different time intervals, such as before the crop is planted (preplant), after the crop is planted but before emergence (preemergence), and after crop emergence (postemergence). The choice of herbicide application timing depends on many factors and varies from grower to grower and field to field. Much about herbicide use has been written in the literature and is beyond the scope of this chapter. It is important to note that the current cropping system in North America and parts of South America are based on herbicide-tolerant crops (e.g. Roundup-Ready, Liberty-Link, Dicamba-Tolerant, and 2,4-D-tolerant soybeans), which also should be rotated. Technology of herbicide tolerance should be viewed as just another tool for weed control and used accordingly as part of the IWM program (Knezevic 2007).

Crops can also be selected to vary their planting date, planting density, and crop rotation (Liebman and Ohno 1998). Planting dates may be chosen to aid in managing the particular weed species. Planting "early" may provide a better crop start against late-emerging weeds, such as common waterhemp (*Amaranthus rudis* Sauer (=*A.tamariscinus auct. non* Nutt.)). Planting "late" may allow the use of a preplanting (burndown) herbicides or a tillage operation to control early-emerging weeds, such as winter annuals (field pennycress (*Thlaspi arvense* L.), shepherds purse (*Capsella bursa-pastoris* (L.) Medik. [=*Bursa b.* (L.) Britt.]), and tansy mustard (*Descurainia pinnata* Walt.) Britt.). Changing planting dates from year to year will not allow specific weeds to adapt.

Another potential tool for keeping weeds off balance is the use of cover crops. Using cover crops is gaining importance due to numerous benefits including improved soil health, reduced soil erosion, and weed suppression. Weeds are most competitive with crops at early growth stages and a management strategy that ensures early season weed suppression in crops is crucial for crop growth, development, and yield. Osipitan et al. (2018) provided systematic and meta-analytic review, with examples how cover crops can provide satisfactory weed suppression from termination time up to seven weeks after planting the main crop.

The impact of cover crops as a weed control input on main crop yield was also evaluated. A total of 46 relevant studies were evaluated, with all using fall-sown cover crops; 70% of the cover crops were terminated mechanically as compared to using chemical control. Main crops were planted one to three weeks after termination of the cover crops. They concluded that cover crops could provide early weed suppression comparable to those provided by chemical and mechanical weed control methods in many cropping systems. The use of cover crops for early season weed suppression had no effect on main crop grain yields but could increase vegetable crop yields when compared to no cover crop. Decisions about selecting cover crops species type (broadleaf or grass) or number (single or mixtures) were not as important as identifying cover crops with inherently competitive characteristics that suppress weeds, such as high biomass productivity, allelopathic phytotoxicity, and persistent residue (Osipitan et al. 2018).

Mechanical and alternative weed control methods can be also valuable tools for keeping weeds "off balance." They might include a variety of mechanical implements to do various types of disking and cultivation. Tillage and cultivation have proven to be effective tools for controlling emerged weeds or burying their seeds below optimal emergence zones. The primary goal of cultivation is to uproot weed seedlings and for use early in the cropping cycle on very small weeds. Cultivation, however, leaves a strip of uncontrolled weeds within the crop row, directly influencing crop yield. In addition, repeated cultivation causes loss of soil organic matter, destroys soil aggregates, increases the chance for soil erosion, and promotes the emergence of new weed flushes (Wszelaki et al. 2007).

Finally, the alternative weed control methods can also be useful components of IWM. The two methods that are gaining some popularity are the use of abrasive grit, which is more at the experimental stage, and flame weeding, which is gaining popularity, especially in organic crop production.

Carlson et al. (2018) reported that abrasive grit, applied at high pressure and directed at plant base, can control weeds and increase yield. They evaluated fertilizer [pelletized turkey (*Meleagris gallopavo*) litter] and nonfertilizer [walnut (*Juglans regia*) shell] grits for maize and soybean in-row (IR) weed management. They applied grit at V1 and V5 of maize, and V1 and V3 of soybean. Between-row weed cultivation was done alone (BR), or in combination with grit (I/B), after grit application. Small weeds (<4 cm) were controlled after grit treatment, but larger broadleaf weeds, grass weeds (treated when growing points were below ground), and later emerging weeds resulted in IR weed biomass similar between season-long weedy (SLW) and IR treatments by August. In maize, fertilizer and nonfertilizer I/B treatments averaged 44% and 14% greater yields, respectively, than SLW.

14.1.4 Flame Weeding

Alternative weed control method, such as flame weeding, can be also very useful (Ascard 1998), especially in those fields where weed resistance to herbicides has developed, or during wet spring when mechanical means are not possible. Propane-fueled flame weeding is a process of exposing plant tissues to the heat coming from a propane burner (Ascard et al. 2007). Flaming should not be confused with burning, as plant tissue does not ignite but heats rapidly to the point of rupturing cell membranes and tissues, which results in weed death. The heat that comes from the burners can have temperatures of up to 1,900 °C, which raises the temperature of the exposed plant tissues rapidly. An increase of temperature above 50 °C inside plant cells can result in the coagulation (denaturation and aggregation) of membrane proteins leading to loss of membrane integrity. Consequently, flamed weeds die or their competitive ability against the crop is severely reduced.

The efficacy of flame weeding was reported to be influenced by several factors, including the growth stages of the plant, the physical location of the growing point at the time of flaming, the presence of protective layers of hair or wax and lignification, the technique of flaming, the regrowth potential of plant species and the leaf relative water content of plant species (Ulloa et al. 2010a).

The most important factor affecting the success of flame weeding is the life cycle of the weed. In general, perennials are more tolerant to flaming than annuals (Rifai et al. 2002). Flaming has no residual soil activity and the heat cannot penetrate the soil; therefore, flaming does not provide control of root structures of perennials. However, some biennials and perennials, such as *Cirsium arvense* (L.) Scop. *Carduus crispus* L. and *Carduus acanthoides* L. are extremely sensitive to heat, and their above-ground tissue can be desiccated. A propane dose as low as 15 kg ha^{-1} turned leaves of *C. arvense* completely black 24 hours after flaming; however, the plants regrew from their below-ground meristems within two to three weeks. Therefore, flaming must be repeated several times during the season to control perennials and biennials (Knezevic et al. 2009a, 2009b, 2014a). Rifai et al. (1999) also reported less than 60% reduction in perennial weed numbers after one, two or three treatments of 35, 23 or 17 kg of propane ha^{-1} and only 75% after 320 and 216 kg ha^{-1} over four flaming treatments. Hakansson (2003), and Rask and Andreasen (2007) also reported that two to three treatments should be carried out after an initial regrowth, but before regrowing shoots become too large. Therefore, multiple flaming conducted within a year and repeated over several years can deplete root nutrient reserves and kill five- to six-year-old perennials (Knezevic et al. 2014a).

Ascard (1998) reported that plant size was also a critical factor in controlling annual weeds. It was much more important than plant density (plant number per area), with smaller weeds being more sensitive to flaming than larger ones. Larger plants could have leaves and stems with protective layers (the lignification level) and multiple growing points. Therefore, flaming controls most annual weeds best during the early growth stages (e.g. 5–10 cm tall) (Knezevic et al. 2014a).

Knezevic et al. (2009c, 2014a) determined that 60–80 kg/ha of propane controlled 90% of major broadleaf species – velvetleaf, ivyleaf morningglory (*Ipomoea hederacea* Jacq.), redroot pigweed (*Amaranthus retroflexus* L.), common waterhemp (*Amaranthus rudis* Sauer), lambsquarters, field bindweed (*Convolvulus arvensis* L.), kochia (*Kochia scoparia* (L.) Schrad.), and Venice mallow (*Hibiscus trionum* L.) – and 80% of several grass species – barnyardgrass (*Echinochloa crus-galli* (L.) Beauv.), green foxtail (*Setaria* glauca (L.) Beauv., and yellow foxtail (*Setaria pumila* (Poir.) Roemer and J.A. Schultes).

Crop tolerance (e.g. less injury) to heat varies with species and growth stages (Knezevic and Ulloa 2007; Domingues et al. 2008; Teixeira et al. 2008; Knezevic et al. 2009a, 2009b, 2009c; Knezevic, Stepanovic et al., 2013). For example, corn and sorghum flamed at earlier stages were more tolerant than soybean. Also, tolerance levels varied among corn types, their growth stages, and propane dose (Knezevic et al. 2009a).

Field corn flamed at the five-leaf stage (V5) was the most tolerant, while the two-leaf (V2) was the most susceptible with the highest crop injury and the largest yield loss (Ulloa et al. 2011). In most cases, the visual injury was transient because corn recovered within a few weeks. Most growth stages showed higher injury at early evaluation dates compared to later rating dates, suggesting that flamed plants were able to recover. A propane dose of 60 kg ha^{-1} was highly efficient in providing 80% to 90% control of many grasses and broadleaf weeds (Ulloa et al. 2010a, 2010b) and caused yield losses of 3%, 9%, and 13% for the V5, V7, and V2 growth stages, respectively. From a practical standpoint, the 3% yield reduction of field maize flamed broadcast at the V5 stage was below the arbitrary assigned acceptable yield reduction of about 5% (e.g. threshold level) and would be acceptable by producers.

In addition, Datta and Knezevic (2013) studied the effect of single and repeated flaming on yield components and yield of field corn. They concluded that flaming conducted once at V2, V4, or V6 growth stages resulted in 38%, 30%, and 20% crop injury, respectively, 7 DAT compared to significantly lower injury levels of 8%, 6%, and 6% for the corresponding growth stages by 28 DAT. Similar trends occurred when plants were flamed two times (e.g. V2 and V6 as well as V4 and V6), exhibiting an injury rating of about 6% 28 DAT. However, corn plants did not recover well over time when flamed three times (e.g. at V2, V4, and V6 growth stages), which resulted in over 50% injury 7 DAT and more than 30% injury by 28 DAT. Corn yields in their study ranged from 11.1 to 11.6 t ha^{-1} in the plots flamed once or two times, which were statistically similar to the yield obtained from the nonflamed control (11.7 t ha^{-1}). Flaming conducted three times (at V2, V4, and V6 growth stages) was the only treatment that resulted in significantly lower yield (9.9 t ha^{-1}) compared to the non-flamed control. Maize flamed three times produced about 8.5% lower yield than the non-flamed control. From a practical standpoint, an 8.5% yield reduction may not be acceptable. These results indicate that corn can tolerate a maximum of two flaming operations per season. Corn plants fully recovered when flamed twice under weed-free situations, suggesting that broadcast flaming two times in the season could be an acceptable alternative for weed control in corn.

Soybean flamed at the VC (unfolded cotyledons) stage was the most tolerant, whereas the VU (fully unrolled unifoliate leaves) stage was the most susceptible resulting in the highest crop injury and the largest yield loss (Ulloa et al. 2010c). Unlike maize, where the growing point is protected beneath the soil surface for several weeks, the growing point of soybean is between the cotyledons and moves above the soil surface at emergence. The growing point is partially protected by both cotyledons at the VC stage, resulting in the least effect of flaming. In contrast, the growing point was exposed at the VU stage resulting in severe damage. Consequently, many plants did not get the chance to regrow. Of all yield components, the number of plants m^{-2} was the most affected by broadcast flaming. A significant number of plants were lost when flaming was conducted at the VU and V2 stages. They also reported that a propane dose of 60 kg ha^{-1} caused yield losses of 6%, 67%, 30%, and 10% for the VC, VU, V2, and V5 growth stages, respectively.

From an economic standpoint, Ulloa et al. (2010c) also reported that the costs of a single flaming operation applied broadcast below the crop canopy could be \$30–\$40/ha, without taking into account the cost of the equipment and labor – current price of propane (\$0.5/kg × 60–80 kg)). Banded application (over the crop row) can cost \$12–\$20/ha due to lower propane use rates (30–40 kg/ha).

Flaming does not disturb soil structure as repeated mechanical weeding does. It has been reported that repeated cultivation promotes loss of organic matter and soil erosion induced by wind and heavy rains (Wszelaki et al. 2007; Stepanovic et al. 2016a, 2016b). Flaming can be carried out on wet or stony soils, does not disrupt the soil surface, and does not bring buried weed seeds to the soil surface (Ascard et al. 2007; Wszelaki et al. 2007).

Flame weeding is less expensive than hand weeding and organic herbicides (Nemming 1994), and there is no chance for weeds to develop resistance to high level of instant heat. Cost of hand weeding can range from \$100 to \$300 per acre (\$300 to \$800 per hectare).

The main disadvantage of weed flaming is fire risk in fields with heavy crop residue. Because heat is nonselective, it has the potential to injure healthy crops. The disadvantages of flame weeding when compared to conventional herbicides include higher cost of equipment compared to herbicide applicators, lack of selectivity for crop safety, low speed of application due to smaller coverage (e.g. most flamer can treat only 4–12 rows), and lack of residual weed control (Ascard 1998; Ascard et al. 2007). From a resource and environmental point of view, the high energy requirement, the ineffective use of fossil fuels, and the release of carbon emissions in relation to climate change could be seen as disadvantages. However, propane combustion is relatively clean compared to other fossil fuels, e.g. diesel (Ascard et al. 2007).

14.2 Making a Weed Control Decision

One of the most common dilemmas that farmers and practitioners face is how to make a decision on the method and timing of weed control. Before initiating weed control procedures, the following are some general guidelines to consider: field scouting and mapping weed patches and utilizing the concepts of critical period of weed control, weed threshold, and decision support computer models. Field scouting typically involves assessing the type and number of weeds to determine if and when weed control is necessary. Mapping and monitoring weed patches over time will also help assess the effectiveness of the control program.

Studies of crop-weed competition showed that yield loss is sensitive to small differences in the period between crop and weed emergence (Knezevic et al. 2002). It brings to light the importance of the concepts of critical period of weed control and economic thresholds.

14.2.1 Critical Period of Weed Control (CPWC)

The critical period of weed control (CPWC) is defined as a "span of time between that period after seeding or emergence, when weed competition does not reduce crop yield, and the time after which weed competition will no longer reduce crop yield" (Zimdahl 1988). Others defined the CPWC as the time interval when it is essential to maintain a weed-free environment to prevent crop yield loss. In recent years, university extension weed specialists have described the CPWC as a "window" in the crop growth cycle during which weeds must be controlled to prevent unacceptable yield losses (Knezevic et al. 2002).

In essence, the CPWC represents the time interval between two separately measured crop-weed competition components: the critical time of weed removal (CTWR) and the critical weed-free period (CWFP). During crop emergence, resources present in the environment may be sufficient to support both weed and crop growth. However, with continued competition between weeds and crops, the weeds are no longer tolerated due to negative effects on the crop; marking the beginning of the CPWC, which is also referred to as CTWR (Knezevic et al. 2002). The CTWR can be influenced by several factors including crop and weed characteristics (Tursun et al. 2016); environmental variables (Knezevic et al. 2007; Tursun et al. 2016); cropping practices such as crop planting density and row spacing (Adigun et al. 2014; Osipitan et al. 2016); soil nutrient (Knezevic et al. 2002; Evans et al. 2003; Odero and Wright 2013); and preemergence weed control program (Knezevic, Elezovic et al. 2013).

With the popularity of herbicide tolerant crops (HTC), dependence on total postemergence herbicide programs became prevalent in US cropping systems (Knezevic 2007). Such a shift in cropping practices highlights the importance for appropriately timed weed control measures. For example, since glyphosate tolerant soybeans have been widely adopted, the concept of CPWC became an important part of integrated weed management in answering some fundamental questions such as "if" and "when" to apply postemergence herbicides. A possible strategy for herbicide use, for example in glyphosate-tolerant soybean, would be to apply glyphosate tank-mixed with a short residual herbicide at the CTWR, which may provide adequate weed control during the entire critical period (Knezevic et al. 2020). Similar approach ca be also used in non-HTC crops by applying postemergence type herbicides with residual activity at the CTWR, which can provide weed control during entire critical period (Knezevic et al. 2020).

14.2.2 Weed Threshold

The concept of "threshold" was used first in the field of entomology and plant pathology (Cousens 1987). In the discipline of weed science, the weed threshold was also considered as a intregral component of an IWM system. It was defined as a point at which a particular weed density causes important crop losses. Knowledge of thresholds can help agriculturists make decisions on the timing of weed control operation (or herbicide applications), in deciding whether remedial weed control efforts are necessary or economically justified.

Economic threshold has been defined as "the weed density at which the cost of weed control equals the increased return on yield in the current year" (Headley 1972; Knezevic et al. 2017). Because they account for crop losses only in the current cropping season, economic thresholds are single-year measures of weed effects. In addition, economic thresholds are based on factors such as the price of the crop at harvest, herbicide, and application cost, anticipated crop yield and the yield loss – weed density relationships, which are a function of environmental factors (e.g. soil types and climate). Since major cause of yield reductions by weeds is through competition for growth limiting resources (light, water, and nutrients), the economic threshold is not therefore constant for particular weed-crop combinations and can differ within the same geographic region.

14.3 Computer-Based Models and Decision Support Systems

Computer-based models and decision support systems have also been developed to aid in making spray decisions. Computer modeling as a research tool is used much less in weed science than in other scientific fields. The reasons for this are not obvious but perhaps some are the relatively small number of researchers in that area and the straightforward use of herbicides marketed by large chemical corporations.

WeedSOFT is an example of a practical decision support software developed by the University of Nebraska (Neeser et al. 2004). This software can aid in selecting the most economical herbicide based on weed species, density, weed emergence time, herbicide efficacy, cost effectiveness, environmental safety, yield loss estimates, and corresponding weed management recommendations. WeedSOFT is used in the north central United States, including Illinois, Indiana, Kansas, Michigan, Missouri, Nebraska, Ohio, Pennsylvania, and Wisconsin.

In addition, models can have other important roles to play. In particular, models can help explain key mechanisms for weed control. For instance, modelers can develop scenarios that are not feasible to test through experiments, but through simulation optimal management strategies can be determined. Leibman and Nichols (2021), utilized long-term data from field experiments to show that diversifying crop sequences with additional crops species can improve weed suppression when either herbicides or cultivation serve as primary control tactics. They reported the results of modeling analyses that investigated how cropping system diversification would affect the population dynamics of giant ragweed (*Ambrosia trifida* L.), an annual dicotyledonous species that is problematic in the central United States for both conventional and organic farmers. They further reported that to prevent an increase in giant ragweed density, the minimum control efficacy needed from herbicides or cultivation used in corn (*Zea mays* L.) and soybean (*Glycine* max (L.) Merr.) would be 99.0% in a 2-year maize–soybean system, but 91.4% in a 5-year maize–soybean–rye (*Secale cereale* L.)–alfalfa (*Medicago sativa* L.) system. Thus, the diversified rotation would be better buffered against less-than-perfect weed control during maize and soybean phases. Further modeling analyses indicated that the weed suppression effect associated with greater rotation length was attributable not only to increased crop species richness but also to greater temporal variation in planting dates. A planting interval variation index (PIVI), calculated as the coefficient of variation in months between planting activities, was strongly associated with the weed suppressive ability of the rotations we modeled and may be a useful metric for designing other cropping systems. Overall, they concluded that diversified rotation systems based on both annual and perennial crops are likely to be valuable for managing problematic weed species (Leibman and Nichols 2021).

Redwitz and de Mol (2021), utilized simulation models that combine population dynamics and genetics to predict the impact of competing management options on weed density, allele frequency, and phenotypic resistance levels. The new statistical package (R package PROSPER) was utilized to provide functions for the forward simulation of weed population dynamics on a field scale, the selection of individuals according to their sensitivity to herbicides, and the recombination of alleles during reproduction. Objects are provided to enter and save model parameters in a clear structure, and to save output data for further processing in R. The basic functions are extensible with R code. PROSPER combines individual-based population dynamics with monogenic or polygenic diploid inheritance and flexible selection pressure. Stochasticity can be included at all model steps. Two examples of the population dynamics of two annual weed species with herbicide resistance were presented. All parameters and the models are available in PROSPER. In addition to simulation, PROSPER is intended for sharing and publishing population dynamic parameters and models, which is easily done utilizing statistical package R (Redwitz and de Mol (2021).

14.4 Documentation and Record Keeping

Documentation and record keeping is an essential part of an IWM program. Information on cropping practices and history of each field will help evaluate the weed control program over time. Information can be recorded on paper forms or directly on the computerized versions of those forms which can be developed as a database application. Data forms should have basic information including, site description, weed species composition, evaluation of herbicide performance, weed response to weed control methods, records of the amount, type and methods of herbicide applications, and other methods of weed control. Record-keeping pays off because knowing the weeds on the farm, taking notes, and watching for possible shifts in weed species may prevent costly surprises.

14.5 Ecologically Based Weed Management in Row Crops – Final Thoughts

Since there are many kinds of weeds with highly differing life cycles, they obviously cannot be managed by a single control method. However, if they are implemented in a systematic manner, significant advances in weed management can be achieved. Obviously, no one can use all of the above-described techniques at once. There are a number of ways to start developing weed management program. The easiest start will be to try one or two techniques and then add more practices as time goes on. After a few years, there will be a program of different techniques working together in an integrated approach. These options could act synergistically as part of ecologically based weed management, a sustainable system that can be used long term without harming the environment. Bottomline is that the use of a variety of weed control tools reduces the reliance on any specific tool, which means that those tools will still be effective in the years to come. The use of various weed control methods keeps weeds "off-balance" and prevents them from adapting to a particular control strategy. And remember, there is no such a thing as a "silver bullet" when it comes to weed control.

References

Adigun, J., Osipitan, A.O., Lagoke, S.T. et al. (2014). Growth and yield performance of cowpea (*Vigna unguiculata (L.) Walp*) as influenced by row-spacing and period of weed interference in South-West Nigeria. *Journal of Agricultural Science* 6: 188–198.

Ascard, J. (1998). Comparison of flaming and infrared radiation techniques for thermal weed control. *Weed Research* 38: 69–76.

Ascard, J., Hatcher, P.E., Melander, B., and Upadhyaya, M.K. (2007). Chapter 10, Thermal weed control. In: *Non-chemical Weed Management* (ed. M.K. Upadhyaya and R.E. Blackshaw), 155–175. London, UK: CAB International.

Carlson, M., Forcella, F., Worthman, S., and Clay, S. (2018). Using abrasive grit for weed management in field crops.

In: *Physical Methods for Stimulation of Plant and Mushroom Development.*

Cousens, R. (1987). Theory and reality of weed controlled thresholds. *Plant Protection Quarterly* 2: 13–20.

Datta, A. and Knezevic, S.Z. (2013). Flaming as an alternative weed control method for conventional and organic agronomic crop production systems: a review. *Advances in Agronomy* 118: 399–428.

Domingues, A.C., Ulloa, S.M., Datta, A., and Knezevic, S.Z. (2008). Weed response to broadcast flaming. *RURALS*. Web page: http://digitalcommons.unl.edu/rurals/vol3/iss1/2 (accessed 15 March 2012).

Evans, S.P., Knezevic, S.Z., Lindquist, J.L. et al. (2003). Nitrogen application influences the critical period for weed control in corn. *Weed Science* 51: 408–417.

Hakansson, S. (2003). *Weeds and Weed Management on Arable Land – An Ecological Approach.* Uppsala, Sweden: CABI Publishing.

Headley, J.C. (1972). Defining the economic threshold. In: *Pest Control Strategy for the Future*, 100–108. Washington, DC: National Academy of Science.

Knezevic, S. (2007). Herbicide tolerant crops: 10 years later. *Maydica* 52: 245–250.

Knezevic, S., Datta, A., Bruening, C., and Gogos, G. (2012). *Propane-fueled flame weeding in corn, soybean and sunflower training manual.* http://www.agpropane.com.

Knezevic, S., Stepanovic, S., and Datta, A. (2014a). Growth stage affects response of selected weed species to flaming propane. *Weed Technology* 28 (1): 233–242.

Knezevic, S., Stepanovic, S., Datta, A. et al. (2013). Soybean yield and yield components as influenced by the single and repeated flaming. *Crop Protection* 50: 1–5.

Knezevic, S.Z. (2009). Flaming: a new weed control tool in organic crops. *Crop Watch*, University of Nebraska-Lincoln Extension. Web page: http://cropwatch.unl.edu/archives/2009/crop17/organic_flaming.htm (accessed 12 March 2012).

Knezevic, S.Z., Costa, C.M., Ulloa, S.M., and Datta, A. (2009a). Response of maize (*Zea mays* L.) types to broadcast flaming. In: *Proceedings of the 8th European Weed Research Society Workshop on Physical and Cultural Weed Control* (ed. D.C. Cloutier), 92–97. Zaragoza, Spain.

Knezevic, S.Z., Dana, L., Scott, J.E., and Ulloa, S.M. (2007). Building a research flamer. In: *Proceedings of the North Central Weed Science Society Conference*(ed. R.G. Hartzler and A.N. Hartzler), 62:32. St. Louis, MO, USA.

Knezevic, S.Z., Datta, A., and Ulloa, S.M. (2009b). Growth stage impacts tolerance to broadcast flaming in agronomic crops. In: *Proceedings of the 8th European Weed Research Society Workshop on Physical and Cultural Weed Control* (ed. D.C. Cloutier), 86–91. Zaragoza, Spain.

Knezevic, S.Z., Datta, A., and Ulloa, S.M., (2009c). Tolerance of selected weed species to broadcast flaming at different growth stages. In:*Proceedings of the 8th European Weed Research Society Workshop on Physical and Cultural Weed Control* (ed. D.C. Cloutier), 98–103. Zaragoza, Spain.

Knezevic, S.Z., Elezovic, I., Datta, A. et al. (2013). Delay in the critical time for weed removal in imidaziolinone-resistant sunflower (Helianthus annuus) caused by application of pre-emergence herbicide. *International Journal of Pest Management* 59: 229–235.

Knezevic, S.Z., Evans, S.P., Blankenship, E.E. et al. (2002). Critical period for weed control: the concept and data analysis. *Weed Science* 50: 773–786.

Knezevic, S.Z., Jhala, A., and Datta, A. (2017). Integrated weed management. In: *Encyclopedia of Applied Plant Sciences*, vol. 3 (Editors in Chief B. Thomas, B.G. Murray, and D.J. Murphy), 459–462. Waltham, MA: Academic Press.

Knezevic, S.Z., Jhala, A., Klein, B. et al. (2020). *Guide for weed, disease and insect management in Nebraska.* Annual Extension Publication EC-130. 374 pages. Lincoln, Nebraska: University of Nebraska Lincoln.

Knezevic, S.Z., Streibig, J.C., and Ritz, C. (2007). Utilizing R software package for dose–response studies: the concept and data analysis. *Weed Technology* 21: 840–848.

Knezevic, S.Z. and Ulloa, S.M. (2007). Flaming: potential new tool for weed control in organically grown agronomic crops. *Journal of Agricultural Science* 52: 95–104.

Knezevic, Z.S., Evans, S., and Mainz, M. (2003). Row spacing influences critical time of weed removal in soybean. *Weed Technology* 17: 666–673.

Kristiansen, P.E., Sindel, B.M., and Jessop, R.S. (2001). The importance of diversity in organic weed management. The organic Challenge: Unity Through Diversity. *RIRDC Inaugural National Conference 2001*, Farling Harbour, Sydney, Australia, 188–198.

Liebman, M. and Nichols, V. (2021). *Cropping system redesign for improved weed management: a modelling approach illustrated with Giant ragweed.* Special issue of Agronomy Journal (ISSN 2073-4395). The Application of models for weed management in cropping systems, edited by Metcalfe H, Storkey J., and Milne A.

Liebman, M. and Ohno, T. (1998). Crop rotation and legume residue effects on weed emergence and growth: application for weed management. In: *Integrated Weed and Soil Management* (ed. J.L. Hartfield, D.D. Buhler, and B.A. Stewart), 181–221. Chelsea, MI: Ann Arbor Press.

Neeser, C., Dille, J.A., Krishnan, G. et al. (2004). WeedSOFT(R): a weed management decision support system. *Weed Science* 52: 115–122.

Nemming, A. (1994). Costs of flame cultivation. *Acta Horticulturae* 372: 205–212.

Odero, D.C. and Wright, A.L. (2013). Phosphorus application influences the critical period of weed control in lettuce. *Weed Science* 61: 410–414.

Osipitan, O.A., Adigun, J.A., and Kolawole, R.O. (2016). Row spacing determines critical period of weed control in crop: cowpea (*Vigna unguiculata*) as a case study. *Azarian Journal of Agriculture* 3: 90–96.

Osipitan, O.A., Dille, J.A., Assefa, Y., and Knezevic, S.Z. (2018). Cover crop for early-season weed suppression in crops: systematic review and meta-analysis. *Agronomy Journal* 110: 2211–2221.

Rask, A.M. and Andreasen, C. (2007). Influence of mechanical rhizome cutting, rhizome drying and burial at different developmental stages on the regrowth of *Calystegia sepium*. *Weed Research* 47: 84–93.

Redwitz, C. and de Mol, F. (2021). *The R package PROSPER: an environment for modelling weed population dynamics and the evolution of herbicide resistance*. Special issue of Agronomy Journal (ISSN 2073-4395). The Application of models for weed management in cropping systems, edited by Metcalfe H, Storkey J., and Milne A.

Rifai, M.N., Astatkie, T., Lacko-Bartosova, M., and Gadus, J. (2002). Effect of two different thermal units and three types of mulch on weeds in apple orchards. *Journal of Environmental Engineering and Science* 1: 331–338.

Rifai, M.N., Lacko-Bartosova, M., and Puskarova, V. (1996). Weed control for organic vegetable farming. *Rostl. Vyroba* 42: 463–466.

Rifai, N.M., Lacko-Bartosova, M., and Somer, R. (1999). Weed control by flaming and hot steam in apple orchards. *Plant Protection Science* 35: 147–152.

Ritchie, S.W., Hanway, J.J., and Benson, G.O. (1996). *How a corn plant develops*. Special Report 48. *Iowa State University Cooperative Extension Services*, Ames, IA, USA.

Shaw, W.C. (1982). Integrated weed management systems technology for pest management. *Weed Science* 30 (Suppl): 2–12.

Stepanovic, S., Datta, A., Neilson, B. et al. (2016a). The effectiveness of flame weeding and cultivation on weed control, yield and yield components of organic soybean as influenced by manure application. *Renewable Agriculture and Food Systems* 31 (4): 288–299.

Stepanovic, S., Datta, A., Neilson, B. et al. (2016b). Effectiveness of flame weeding and cultivation for weed control in organic maize. *Biological Agriculture & Horticulture* 32 (1): 47–62.

Stopes, C. and Millington, S. (1991). Weed control in organic farming systems. *Proceedings of the Brighton Crop Protection Conference-Weeds*, Brighton, UK, 185–192.

Swanton, C.J. and Weise, S.F. (1991). Integrated weed management: the rationale and approach. *Weed Technology* 5: 648–656.

Teixeira, H.Z., Ulloa, S.M., Datta, A., and Knezevic, S.Z. (2008). Corn (*Zea mays*) and soybean (*Glycine max*) tolerance to broadcast flaming. *RURALS*. Web page: http://digitalcommons.unl.edu/rurals/vol3/iss1/1 (accessed 16 March 2012).

Tursun, N., Datta, A., Budak, S. et al. (2016). Row spacing impacts the critical period for weed control in cotton (*Gossypium hirsutum*). *Phytoparasitica* 44: 139–149.

Ulloa, S.M., Datta, A., Bruening, C. et al. (2012). Weed control and crop tolerance to propane flaming as influenced by the time of day. *Crop Protection* 31: 1–7.

Ulloa, S.M., Datta, A., Bruening, C. et al. (2011). Maize response to broadcast flaming at different growth stages: effects on growth, yield and yield components. *European Journal of Agronomy* 34: 10–19.

Ulloa, S.M., Datta, A., and Knezevic, S.Z. (2010a). Growth stage influenced differential response of foxtail and pigweed species to broadcast flaming. *Weed Technology* 24: 319–325.

Ulloa, S.M., Datta, A., and Knezevic, S.Z. (2010b). Tolerance of selected weed species to broadcast flaming at different growth stages. *Crop Protection* 29: 1381–1388.

Ulloa, S.M., Datta, A., Malidza, G. et al. (2010c). Yield and yield components of soybean [*Glycine max* (L.) Merr.], are influenced by the timing of broadcast flaming. *Field Crops Research* 119: 348–354.

Wszelaki, A.L., Doohan, D.J., and Alexandrou, A. (2007). Weed control and crop quality in cabbage [*Brassica oleracea* (capitata group)] and tomato (*Lycopersicon lycopersicum*) using a propane flamer. *Crop Protection* 26: 134–144.

Zimdahl, R.L. (1988). The concept and application of the critical weed-free period. In: *Weed Management in Agroecosystems: Ecological Approaches* (ed. M.A. Altieri and M. Liebman), 145–155. Boca Raton, FL: CRC Press Inc.

15

Practical Vegetable and Specialty Crop Weed Management Systems

Katie Jennings[1] *and Steve Fennimore*[2]

[1] *Department of Horticultural Science, North Carolina State University, Raleigh, NC, USA*
[2] *Department of Plant Sciences, University of California, Davis, CA, USA*

15.0 Introduction

The ideal of an ecologically based weed management system is to promote a sustainable agroecosystem. Sustainable agroecosystems maintain the resource base on which they depend, are resilient and recover quickly from disturbances so that these systems are stable over long periods of time, without degrading the local environment or resources or are over-dependent on outside resources (Gliessman 2001). However, the goal of an individual grower is to make a living and remain economically viable in the long term. Here we explore the challenges and promises of a viable weed management system for specialty crops in the context of a sustainable agroecosystem.

Vegetable fields are low-stress environments designed to maximize crop yield, quality, and producer profits. The risk of crop damage from weed competition is a potential stress that must be managed to avoid loss (Dittmar et al. 2018). The challenge is how to protect the crop from weeds, maintain producer profits, and reduce negative environmental impacts? An ecologically based weed management system minimizes the effects of the weed management system on the local or regional environment such as reducing or eliminating groundwater contamination from herbicides, pesticide drift, fumigant emissions, or soil loss. An ecologically based weed management system should also be sustainable and stable in the long-term and less dependent on non-renewable inputs subject to interruption in availability (e.g. an herbicide label for a specific crop such as cycloate/spinach) (*Spinacia oleracea*), and make better utilization of renewable inputs (Lati et al. 2016). Ecologically based weed management systems would utilize more renewable inputs like locally generated energy. This chapter will address some unique weed management challenges faced by specialty crop producers as well as some of the unique opportunities that they have.

15.1 What Is Ecologically Based Weed Management in Specialty Crops?

An ecologically based method of weed management for specialty crops will do the following:

1) Allow the specialty crop grower to profitably produce high-quality marketable produce.
2) Minimize potential for crop loss to weeds and reduce weed seedbanks (Gallandt 2006; Riemens et al. 2007).
3) Reduce dependence on nonrenewable inputs like herbicides and plastic mulch (Barnes et al. 2009; Monteiro and Santos 2022).
4) Reduce dependence on hand-weeding labor.
5) Promote development of renewable weed control technologies such as automated weeders (Fennimore et al. 2016).

Many specialty crops are high-value crops with expensive weed control costs and low profit margins (Tourte et al. 2009; 2017a; 2017b; 2019). Cost-effective weed management systems that ensure timely removal of weeds but without being overly complex or time consuming to implement are needed. However, few herbicides are registered for use in specialty crops and those products may only provide partial weed control. New herbicides for specialty crops are rare. The lack of access to a choice of efficacious herbicides that control most weeds in specialty crops is a key distinction from agronomic crops that have many more herbicide options.

Ecologically Based Weed Management: Concepts, Challenges, and Limitations, First Edition. Edited by Nicholas E. Korres, Ilias S. Travlos, and Thomas K. Gitsopoulos.

15.2 Unique Challenges for Vegetables and Other Specialty Crops

Herbicide labels must list every registered crop by name. Vegetable crops consist of 1,097 species with thousands of varieties (Meldrum et al. 2018) which complicates the process of registration and labeling of herbicides for these crops (USEPA 2021). Vegetables and other specialty crops are primarily used for human food and are subject to the highest standards of appearance, taste, and safety such as absence of human pathogens and pesticide residues (Gil et al. 2015; Wanwimolruk et al. 2016). These crops are high value and require protection from weed competition to maintain produce quality, which often involves the expensive and labor-intensive process of hand weeding (Fennimore and Umeda 2003; Bolda et al. 2019). All together this creates very different challenges for specialty crops compared to agronomic crops, which have access to a wide array of effective and low-cost herbicides (Plastina 2021).

15.2.1 Weed Competition

Weeds damage crops through competition for scarce resources, as all crops are susceptible to weed competition during their respective critical periods of weed control (Fennimore and Umeda 2003; Chaudhari et al. 2016; McGowen et al. 2018; Zimdahl 2018; Bertucci et al. 2019; Smith et al. 2020). However, the threshold for tolerance to weeds is very different among agronomic and specialty crops. During stand establishment field corn (*Zea mays*) and lettuce (*Lactuca sativa*) are both susceptible to weed competition, which can result in reduced yields due to resource diversion to weed biomass. The objective of the weed management program is to reduce the impact of weed competition to acceptable levels. However, there are many differences in what constitutes "acceptable weed control" among various crops. Where there are weed control failures in agronomic crops, it still may be possible to salvage partial yield from the crop provided the value of the yield is greater than the harvest cost.

For example, if weeds are poorly controlled in a field corn planting, there is some salvageable yield even if it must be diverted to silage (Brackenrich and Roth 2020). A grain or silage crop used for animal feed is not subject to the aesthetic quality standards of specialty crops, which are chosen by consumers. There are few if any opportunities for salvage of specialty crops for animal feed. Lettuce is one such crop where the harvestable commodity is subject to high quality and appearance standards. Weeds that are not controlled in lettuce can distort the shape of the lettuce head causing the crop to be unmarketable resulting in 100% yield loss (Fennimore and Umeda 2003), or in other cases the cost of hand weeding in a lettuce crop may exceed the value of the crop. The threshold of specialty crop tolerance to weed competition is very narrow compared to agronomic crops – the specialty crop produce is either marketable or not, the value of the crop is either greater than the cost of hand weeding. Because weed control in specialty crops must be perfect, it is up to the producer to decide whether it is worth saving the crop from weeds or destroying the crop to reduce losses. Also, because of economic necessity, near-continuous cropping occurs in many fields, the value of the land and irrigation infrastructure means that it is less expensive to terminate a weed infested crop and start over with a new planting, rather than invest in an unprofitable crop.

15.2.2 Herbicides

Specialty crops consist of a large and diverse range of crop species and varieties, which makes labeling complex because herbicides are registered by crop or crop group. Root and tuber vegetables, for example, include 36 crops such as sweet potato (*Ipomoea batatas*) (Anonymous 2020; Table 15.1). The diversity of specialty crops means that it is more complex to register new herbicides on specialty crops than in a major agronomic crop such as field corn. For example, DCPA was used on less than 500 ha of radish (*Raphanus sativus*) in California, which is a small market to justify registration of a new herbicide, compared to 36

Table 15.1 Crop groups specific for specialty crops (Anonymous 2020).

Group	Description	Number of crops
1	Root and tuber vegetables	36
2	Leaves of root and tuber vegetables	18
3	Bulb vegetables	27
4	Leafy vegetables (non-brassicas)	27
5	Brassica leafy vegetables	17
6	Legume vegetables	14
7	Foliage of legume vegetables	14
8	Fruiting vegetables	22
9	Cucurbit vegetables	12
19	Herbs and spices	68
22	Stalk and stem and leaf petiole vegetables	20

million ha of field corn in the USA (McDougall 2016; Lohstroh and Koshlukova 2017). This is one of the reasons the IR-4 Project (IR-4) exists to facilitate registration of herbicides in some of the very limited acreage crops like vegetables, small fruit, and other specialty crops (IR-4 2021). However, herbicide development has slowed since the introduction of glyphosate-tolerant crops in 1996 (Duke 2012). With few new herbicides in development for large-acreage crops, fewer herbicides are available for development in specialty crops.

The risk/benefit equation for registration of herbicides on specialty crops is different than for agronomic crops. For example, the US market for herbicides in field corn in 2016 was $5.1 billion, while the herbicide market for processing tomatoes in California was $4.6 million (McDougall 2016; Fennimore and Cutulle 2019). Given the high value of specialty crops, the registrant exposure to liability due to crop injury from an herbicide may exceed the potential for herbicide sales in these crops. Registration of herbicides in specialty crops does not present the agricultural chemical industry with an attractive market for their products.

15.3 Compatibility of Specialty Crops with Ecologically Based Weed Management

Specialty crops are high-value, small-acreage crops and have some advantages compared to agronomic crops. Specialty crop growers by necessity spend more on weed control, including cultivation, herbicides, and hand weeding, than agronomic crop growers (Tables 15.2 and 15.3). The small specialty crop fields allow frequent scouting and management of emerging weed problems. The diversity of crops allows for a more complex crop rotation scheme than exists in the cornbelt. For example, in the Salinas Valley of California, two to four crops such as broccoli (*Brassica oleracea* var, *italica*), celery (*Apium graveolens*), lettuce (*Lactuca sativa*), spinach (*Spinacia oleracea*), and strawberry (*Fragaria* x *ananassa*) can be grown in any given field each calendar year (White et al. 2020). Short cycle crops like fresh-market spinach can be grown from seed to harvest in as little as 25 days, which is less time than most weeds need to set seed. This mix of crops means that the schedule and sequence of rotational crops can vary, which works to suppress weeds in contrast to crops grown in monoculture (Murphy et al. 2006; Blackshaw et al. 2015). Crop rotation reduces the opportunities for adaptation of weed species, creating a more diverse and ever-changing environment (Buhler et al. 1997).

The need to reduce the cost of hand weeding provides an incentive for development of new weed removal technology.

Table 15.2 Cost of hand weeding in selected crops in California and the percentage of cultural costs from hand weeding.

Crop/year	Hand-weed cost $/ha	% Cultural cost
Asparagus[1]	420	11.7
Blackberry[2]	1,494	3.0
Broccoli[3]	371	5.9
Lettuce[4]	739	7.5
Onion[5]	726	17.5
Pepper[6]	988	14.8
Spinach – organic[7]	1,087	10.1
Strawberry – organic[8]	9,129	20.5
Tomato – processing[9]	198	3.2

1) Asparagus (*Asparagus officinalis*) Aegeter et al. 2013;
2) Blackberry (*Rubus fruiticosus*) Bolda et al. 2018;
3) Broccoli (*Brassica oleracea var. italica*) Tourte et al. 2015;
4) Lettuce (*Lactuca sativa*) Tourte et al. 2019;
5) Onion (*Allium cepa*) Wilson et al. 2016;
6) Pepper (*Capsicum annuum*) Takele et al. 2013;
7) Spinach (*Spinacia oleracea*) Tourte et al. 2015;
8) Strawberry (*Fragaria ananassa*) Bolda et al. 2019;
9) Tomato (*Solanum lycopersicum*) Turini et al. 2018.

Table 15.3 Weed management costs in California lettuce and Iowa field corn.

Input	Lettuce[1]	Field corn[2]
	---------------- Costs $/ha ------------------	
Cultivation	40	0
Herbicide	247	98
Hand weeding	739	0
Total	1,026	98

1) Tourte et al. 2019;
2) Plastina 2021.

High hand-weeding costs and few herbicides creates an opening for new cost-effective weed technologies in specialty crops (Fennimore and Cutulle 2019; Tables 19.2; 19.3). There are a limited number of specialty herbicides, and almost none provide anything close to 100% weed control. About 1–2% of the specialty crops are produced organically and need organic-compliant weed-control techniques and technologies to reduce hand-weeding costs in those crops (Willer and Lernoud 2019). The lack of organic herbicides – currently limited to contact products – means there are business opportunities for weed control technologies in this market. What are the ecologically based weed removal technologies that are effective in specialty crops?

15.3.1 Physical Weed Control

Physical weed control methods include cultivation and hand weeding as well as thermal methods such as propane flaming and steam. Physical weed control is the primary weed-control method in many crops, especially organic crops. Mechanical control involves the uprooting and desiccation of the weeds (Fennimore 2014). Physical weed-control methods are among the more effective tools that hold promise as being compatible with sustainable cropping systems.

Physical pest control tools are classified by the USEPA as "pest control devices," most of which do not need to be registered in the same manner as pesticides (USEPA 2020). This means that the decision whether to use a pest control device rests with the grower, much as it would for the use of any other farm implement (i.e. direct adoption of the practice) is possible. The ability of a grower to directly adopt methods of physical pest control is a considerable advantage in speed of adoption compared to pesticides where registrants and regulators control the availability of the pesticide in a process that can take several years. Pest control devices can be adopted more readily by producers than pesticides and are more ecologically benign than herbicides.

15.3.1.1 Hand Weeding

Herbicides, cultivation, and cultural methods do not eliminate all weeds, especially those close to the crop (Kennedy et al. 2020). Therefore, hand weeding is a necessity in annual and perennial specialty crops (Table 19.2). Hand weeding removes weeds that escaped other methods of control or weeds that emerge after the initial weed management tactic is employed. Hand weeding employed to uproot weeds before viable seeds set reduces weed seed production, especially of those, which are aggressive competitors or prolific seed producers (Sosnoskie and Culpepper 2014). However, the cost of hand weeding and the availability of imported labor in the long term both suggest that this practice is unsustainable.

Weed-management costs in California lettuce averaged \$714 ha^{-1} (Tourte et al. 2019); with hand-weeding accounting for 48% of total weed management costs. Hand-weeding costs are increasing because of labor shortages and rising wages (Zahniser et al. 2018). Rural Mexico is the source for most agricultural labor in California, and the number of available workers is in decline. Stricter border enforcement, reduced birth rates, investments in rural education, a robust agricultural industry, and economic opportunities in the nonfarm sector all contribute to reduced labor availability (Charlton and Taylor 2016; Zahniser et al. 2018). New technology is needed to mitigate the effects of labor shortages and increasing costs

15.3.1.2 Mulches

Mulches are barriers to weed growth that consist of plastic or paper tarps, fiber mats, straw, or other organic material. Mulches serve multiple functions, such as warming the soil, protecting the crop foliage and fruit from the soil, enhancing moisture retention, and weed control. Clear tarps are most useful for warming the soil but can enhance weed growth in non-fumigated soils and are not recommended in organic systems. Synthetic mulches are available in many colors (Johnson and Fennimore 2005; Fennimore 2012). Black, and brown mulches block light effectively and are recommended for use in organic systems because of the excellent weed control they provide. The need to dispose of used plastic mulches in landfills render them a nonrenewable resource and disposal by burning creates air pollution. Double cropping with plastic mulches reduces the environmental impact of mulches by extending their use (Ngouajio and Ernst 2005). Recycling of used mulches is difficult due to the need to wash the soil off the plastic before processing. Use of more sustainable mulches is challenging due to the need for industrial production of standard mulches or biodegradable mulches. Biodegradable mulches are available but at a much higher cost than standard polyethylene (Moore and Wszelaki 2016). Organic mulches such as fabric, wood chips, chopped straw, and rice hulls are sometimes used to control weeds in home gardens and landscapes as well as small plantings of horticultural crops (Forcella et al. 2003). Organic mulch must be at least 5 cm deep and needs to block light from reaching the soil surface to control weeds, effectively. Organic mulches made from locally available materials may offer a more sustainable method of weed suppression, moisture retention, and soil conservation than use of plastic mulch. However, efficient mechanized methods to apply organic mulches are needed if these materials are to be used more widely in commercial agriculture (Ranjan et al. 2017).

For organic crops hand weeding is extremely important as there are no selective herbicides for use in these systems. Annual strawberry production systems conventional and organic all utilize plastic mulches (Strand 2008). Once mulch is installed, it is no longer possible to apply herbicide to soil under the plastic. At this point hand weeding, weeding with hoes or weed hooks to reach weeds under the plastic and near the strawberry plants is the only means of weed control available other than in the middle rows, where cultivation and directed sprays are still possible (Daugovish et al. 2008).

15.3.1.3 Cover Crops

Cover crops may be an effective and sustainable method of physically suppressing weeds in specialty crops (Mulvaney et al. 2011). These are rotational crops grown for the purpose of improving soil health, trapping nutrients such as nitrates to prevent groundwater pollution, preventing soil erosion by wind and water, increasing water infiltration, fixing nitrogen in the case of legumes, and suppressing pests, including weeds (Bugg and Smith 2011). Examples of cover crops are barley (*Hordeum vulgare*), bell bean (*Vicia faba*), buckwheat (*Fagopyrum esculentum*), cereal rye (*Secale cereal*), sorghum (*Sorghum bicolor*), sudangrass (*Sorghum bicolor* ssp. drummondii), clovers (*Trifolium* spp. and *Medicago* spp.), vetches (*Vicia* spp.), and many others. These crops can be planted at high densities and used in mixtures (Radosevich et al. 2007).

Cover crops that rapidly establish a canopy or can be grown at high densities are often called "smother crops" (Radosevich et al. 2007). They can be made to be more competitive with weeds by selecting mixtures of cover crops that are well adapted to the area and that rapidly establish a canopy (Brennan et al. 2011). Factors that enhance weed suppression in cover crops are listed in Table 15.4.

Organic specialty crop producers in California often use a cover crop mixture of 35% bell bean, 25% pea (*Pisum sativum*), 30% vetch (*Vicia* spp.), and 10% oat (*Avena sativa*) planted at 112 kg/ha, which results in a seed density of 151 plants m^2. This cover crop density results in poor weed suppression. Means of increasing the competitiveness of this mixture would be to: (i) add more oats to the mixture, (ii) increase the seeding rate to 220 kg/ha, and (iii) use a drill to ensure optimal plant spacing and planting depth (Brennan et al. 2011).

During cover crop growth, blind cultivation with a rotary hoe or tine weeder can be used to control up to 80% of the weeds in the cover crop with little damage to the cover crop (Brennan et al. 2011). Reduction of weed seed return in the cover crop can lead to less weed seed return to the weed seedbank (Brennan and Smith 2005). Cover crops provide a window of opportunity to reduce weed seedbanks further by blind cultivation utilizing a rotary hoe or tine weeder is unguided cultivation or "blind cultivation." Blind cultivation uproots shallowly rooted weeds while leaving more deeply rooted crops (Mallory and Gallandt 2021). This is a sustainable weed management practice that combines the predictability of cultivation with the ability of the cover crop to smother weeds. Cover crops would appear to be sustainable weed management inputs in that they improve soil quality.

15.3.1.4 Mechanical Cultivation

The small size of many specialty crop fields can be an advantage compared to large-scale grain farms. Mechanical cultivation of a 5-ha lettuce field can be performed in a few hours and is much more feasible than cultivation of a 250-ha corn field that may take many days. Mechanical cultivation is an area where considerable technological progress has been made recently and there is room for much more advancement (Fennimore and Cutulle 2019). Robotic intra-row cultivators guided by machine vision and digital systems are more effective in removing weeds from specialty crops than traditional inter-row cultivators (Melander et al. 2015; Fennimore et al. 2016). Intelligent cultivators can reduce hand weeding times by 35% to 50% (Kennedy et al. 2020). Machines like Robovator (F Poulsen Engineering, Hvalsoe, Denmark) are cultivators that detect the crop row and then weed in the intra-row space near the crop like what a laborer with a hoe would perform (Figure 15.1). Limitations of this system are high cost and wet soils. High costs of $109,000 to $155,000 per unit can create a

Figure 15.1 Robovator intelligent cultivator operating in a lettuce field. (Photo by Steve Fennimore).

Table 15.4 Effect of selected factors on weed suppression during cover crop production (Brennan et al. 2011).

Factor	Effect on weed suppression in the cover crop
Cover crop variety or mixture	Varieties and mixtures that rapidly develop a canopy are more weed suppressive.
Seeding rate	Up to a certain point, higher seeding rates are more weed suppressive than lower rates.
Planting date	Plant at the optimal date for rapid establishment of the cover crop.
Row spacing	Narrow spacing (15 cm) minimizes cover crop competition with itself and maximizes competition against weeds.
Irrigation	Irrigation can help establish cover crops and increase competition against weeds.
Planting method	Drilling in rows results in a more uniform cover crop planting than broadcast seeding.

challenge for small acreage growers. For example, the payback time for the Robovator used on 162 ha was 6.1 years, whereas when used on 486 ha the payback time is two years, which indicates the advantage to the use of these technologies on large farms (Tourte et al. in review). The return on investment from the intelligent cultivators is primarily due to labor savings and timely operations.

Timely cultivation can be challenging in areas with frequent rainfall as scheduling cultivation activities when field soils are dry enough for cultivation can be difficult. During prolonged rainy periods weeds will grow too large to remove by cultivation. Cultivation works well in arid irrigated locations like Arizona and California where field entry can be scheduled. But in the eastern half of N. America, entry of cultivators into the field can be limited during periods of prolonged rainy weather. However, machines such as the NAÏO Dino (NAÏO Technologies, Escalquens, France) autonomous cultivator is light weighing about 500 kg and can enter wet fields to cultivate much sooner than tractor mounted cultivators. Another possibility is use of non-tillage weeding tools such as flaming, mowing, electric, laser, abrasives, hot vegetable oil, or high-pressure water (Fennimore et al. 2016; Moore et al. 2021; Tensorfield 2021). These tools can be used to control emerged weeds without soil disturbance.

Small market size and high potential liability to herbicide manufacturers provides a disincentive for development of new herbicides for specialty crops compared to agronomic crops. However, specialty crops offer a greater incentive to companies interested in development of automated weeding technologies: (i) the crops are high-value and dependent on expensive hand weeding; (ii) the lack of adequate herbicides in these crops means there is less competition from inexpensive herbicides and greater unmet weed control needs. Because of the need for new weed control tools, specialty crops also provide a ready proving ground for development of innovative weed control technologies. Many of these new weeding technologies can be used indefinitely and have less impact on the environment.

Lettuce has benefited from intelligent equipment like Robovator and machine thinning (Lati et al. 2016; Mosqueda et al. 2017). There are about 121,457 ha of lettuce, most of which is in Arizona and California (USDA NASS 2020). Specialty crops are underserved by herbicide manufacturers and are dependent on hand-weeding laborers. This creates a niche where automated weeding and crop thinning technology has entered the market (Mosqueda et al. 2017). Blue River Technology (Blue River Technology, Sunnyvale, CA), started in lettuce where they demonstrated their technology for precision plant removal. This, in turn, led to their purchase by John Deere Company and adaptation of their technology for weed control in corn, cotton (*Gossypium hirsutum*), and soybean (http://www.naio-technologies.com/en/home/).

The Bourquin Organic Weed Puller/Roguing Machine (Organic Weed Puller Company) [Figure 15.2] can remove tall weeds over the canopy of crops with less total labor. The utility of the Weed Puller requires a height differential between the crop and the weed. Limited data is available on how the weed puller performs in vegetable crops. In sweet potatoes, the weed puller is effective in controlling Palmer amaranth 0.3 to 0.9 m tall but may not prevent sweet potato yield loss when used alone (Moore et al. 2021).

Electrification of farm vehicles and implements such as intelligent weeders allows use of locally generated energy sources for auto weeders (EPSRC UK RAS Network 2019). Locally generated renewable energy may reduce the need for nonrenewable herbicides – reducing the environmental impact. Renewable weed technology like battery-powered autonomous robotic weeders will be less dependent on external nonrenewable inputs than herbicides. Automated weeders such as the NAÏO from France (Figure 15.3) are battery powered electric vehicles (https://www.naio-technologies.com/en/agricultural-equipment/large-scale-vegetable-weeding-robot). Intelligent cultivators can be used repeatedly and thus are renewable weed control tools.

Automated weeders work best in environments where the position of each crop plant is predictable and regular – like in industrial settings. Precise placement of transplanted vegetables such as Plant Tape (Plant Tape, Salinas, CA) is a method of increasing the precision of planting, spacing (https://www.planttape.com) (Figure 15.4). It is in precision planted fields where the Robovator was first adopted in lettuce. The evolution of weeding technology from hand weeding, where imprecision in plant spacing is tolerated, to more precise planting like industrial settings,

Figure 15.2 The Bourquin Weed Puller. (Photo by Levi Moore).

(a) (b)

Figure 15.3 Two commercial autonomous cultivators used in California vegetable crops: (a) the NAÏO Dino weeder and (b) the Farmwise Titan weeder (FarmWise Labs, San Francisco, CA).

Figure 15.4 Plant Tape automated transplanter planting lettuce using uniform spacing that facilitate use of intelligent weeders. (Photo by Steve Fennimore).

results in better automated weeding performance (S. Fennimore personal observation).

15.3.1.5 Thermal Methods

Weeds and weed seeds are susceptible to heat. There are two broad categories of heat treatment to kill weeds: foliar and soil heating. Foliar heat utilizing flaming or foliar steam kill above ground weed foliage. Heat from solarization or steam injection into soil can kill weed seeds and perennial weed propagules such as nutsedge (*Cyperus* spp.) tubers – i.e. soil disinfestation (Samtani et al. 2011).

Flame Weeding. Flaming exposes plant tissues to heat from a propane burner and can be used for controlling weeds in both organic and conventional systems (Bond and Grundy 2001; Knezevic et al. 2017). Flaming controls weeds by heat disruption of plant tissues (Leroux et al. 2001). Weed flamers rapidly heat plant tissues by exposure to temperatures of up to 1,900 °C (Ascard 1998). Heat melts plant cell membranes, which results in loss of cell function (Lague et al. 2001) and eventually the weeds die or are severely injured. Unlike tillage, flaming does not disturb the soil surface and does not move weed seed to the soil surface where they can germinate. Flaming is useful in organic specialty crops to remove weeds from stale seedbeds prior to seeding or after seeding, but prior to crop emergence such as in onion, or prior to crop transplanting. In addition, flame weeding can be used after the crop is established by directing the flame away from the crop plant, by shielding the crop, or by directing between rows. Generally, the best weed control is obtained when the weeds are 2–5 cm tall and broadleaf weeds are more susceptible to flaming than grasses which have protected meristem. The potential for using biogas in weed flamers as a renewable source of fuel is an area that needs further evaluation. Flaming is limited to annual vegetable crops grown in bare ground production systems rather than on plasticulture raised beds. In addition, perennial fruit crops are not suited for flaming due to potential for injury to the trunks and at times there is high residue on the soil surface (Rohwer and Klodd 2019)

Electrical Weeding. Another example of thermal weed control is electrical weeding. The WeedZapper (Old School Manufacturing, LLC., 3127 S. Ingram Ave. Sedalia, MO 65301) uses electricity to selectively control weeds taller than the crop canopy as an alternative to hand weed removal (Figure 15.5). In a study including the WeedZapper in peanut and sweet potato Palmer amaranth control was at least 78% eight weeks after planting sweet potato or peanut. Palmer amaranth control from electrical treatments were like the hand removal treatments when applied before Palmer amaranth was 0.6 m in height (Moore et al. 2021). Electrical weed control should be incorporated into a weed management system for the best control. Prevention of damage to the crop with this treatment requires that weeds extend above the crop where the electrode contacts the weed but not the crop.

Figure 15.5 The WeedZapper treating Palmer amaranth in sweet potato in North Carolina. (Photo by Colton Blankenship).

Foliar Steam. Application of steam to target weeds causes cell membrane degradation of the weed foliage (Sirvydas et al. 2006). Kolberg and Wiles (2002) controlled common lambsquarters (*Chenopodium album*) and redroot pigweed (*Amaranthus retroflexus*) at the seedling stage using foliar steam similar to control by glyphosate at 560 g ai ha^{-1}. Seedling downy brome (*Bromus tectorum*) was not well controlled by steam presumably because the grass meristem was protected from the heat compared to the broadleaf weeds. Shrestha et al. (2013) found that foliar steam provided good initial weed control in an organic almond orchard, but that the weeds recovered after four weeks.

Leon and Ferreira (2008) evaluated the weed characteristics that correlated with susceptibility to foliar steam application. Leaf thickness was an important characteristic with thick leaves being less susceptible than thin leaves. Smaller weeds were more susceptible to steam than larger weeds, and in broadleaf weeds, those with wider leaves were more susceptible than those with narrow leaves.

Soil Steaming. Steam can be used to control weed seeds in the soil seedbank. Control of weed seedbanks, i.e., soil disinfestation, is accomplished by steam application to raise soil temperatures to the thermal death point for weed propagules as well as soil-borne pathogens – e.g., 70 °C for 20 minutes dwell time is lethal to weed seeds and soilborne pathogens like *Pythium ultimum* (Baker et al. 1957; Kim et al. 2021). Control of weeds and soil pathogens is achieved in the soil by depleting the soil pest population to levels that allow the crop to thrive. Energy requirements for soil steaming are very different from foliar steaming. Fennimore et al. (2014) determined that the energy requirements for soil steaming in the field were 163,533 kJ m^{-3}; in contrast Kolberg and Wiles (2002) reported foliar steam energy requirements of 890 kJ m^{-2}. While these comparisons reflect the difference in objectives, e.g., treating a three-dimensional soil volume for soil steaming vs. a two-dimensional surface area for foliar steam application, they do provide an idea of the large differences in energy consumption between the two methods of steaming. The targets of soil steaming are primarily soil pathogens, which must be controlled at a greater soil depth (e.g. 30 cm to protect root health during the production season). In the process of soil steaming to control soil pathogens, weed propagules are also controlled. However, because of the larger volumes of soil treated, the energy consumption of soil steaming is much greater than foliar steaming where the target weeds are at the soil surface (Fennimore et al. 2014; Kolberg and Wiles 2002). CO_2 production from fossil fuel steam generation is not trivial. However, as new methods of steam generation using fuel cells and or electricity become available, soil steam generators may be part of a renewable pest control technology and could contribute towards an ecological based weed management tool (Alabbadi 2012).

Table 15.5 Area and soil volume treated, fuel use, and cost per hectare for three steam application configurations.

Description	Cover	Volume treated	Diesel required[1]	Cost
	%	m^3	L/ha	$/ha
Broadcast	100	1,233	7,993	6,336
Band	20	123	799	634
Spot	8	49	320	253

1) Based on energy consumption to steam soil (Fennimore et al. 2014).

Precision Steam Placement. Our hypothesis is that it is not necessary to treat the entire surface area of a field with steam to realize the benefits of steam. Band steaming and spot steaming reduce the cost of steam compared to broadcast steam (Table 15.5). Band steam involves treatment of the seedline in a 10 cm band centered on the seedline. Spot steam involves treating 10 cm diameter circles, which are steamed and then planted after the soil cools. For band and spot steam alignment of the planter with the steam application zone is necessary so that the crop grows in the center of the steamed zone a potential ecologically compatible method of soil disinfestation providing steam can be produced using renewable energy.

Solarization. Heating weed seeds in the soil to a thermal death point denatures proteins in the seed embryo killing the seed. Weed seed in the seedbank can be killed by heat from solarization or steam. Clear plastic mulch is used with solarization to cover the field surface for four to six weeks during summer. This process heats the soil to the thermal death point killing weed seeds in the upper soil layer where heat time accumulates the necessary thresholds (Dahlquist et al. 2007). However, solarization is dependent on use of clear plastic film – a non-renewable resource.

Both solarization and steam treatment are soil disinfestation treatments for fallow fields. After treatment, the field starts out clean but weed seed from wind-blown species like common groundsel (*Senecio vulgaris*) can invade the field. The residual effect is due to maintenance of the clean condition of the field rather than any residual effect on weed germination such as is provided by soil active herbicides. Solarization and steam both provide control of many common soilborne pathogens in addition to weed seed.

15.3.2 Cultural Methods of Weed Control in Specialty Crops

Cultural methods of weed control are important to ecologically based weed management. These methods involve management of the crop to limit the effects of weed competition, prevent new weed infestations and reduce weed seed production. Examples of cultural controls include prevention, crop rotation, enhancing crop ability to out compete weeds, optimal crop management, companion crops, smother crops, optimal crop varieties and use of transplants rather than direct seeded crops.

15.3.2.1 Prevention and Sanitation

If a weed is not present in a field or geographic region, then preventing introduction of the pest is an efficient method of weed management. The scale of prevention effort can range from the field level to the country level (Norris et al. 2003). The priority of prevention is to limit the entry of new exotic weed species into the field. However, prevention must, whenever possible, limit the entry of any weed seeds into the seedbank whether it is already in the field or not. Every method of weed control in both conventional and organic systems can be used to prevent weeds from producing seeds. Tillage of a harvested field soon after harvest to

Figure 15.6 Use of irrigation to prepare a stale seedbed. Bed on the left is the control (no irrigation) and on the right is a bed that was irrigated 1 week before the weeds were removed with shallow tillage and planted afterward. Photos were taken three weeks after seeding (Photos by Steve Fennimore).

kill uncontrolled weeds is an example of preventing weed seed production. Given that specialty crop fields tend to be smaller and more intensively managed, prevention and sanitation have a good chance to succeed in preventing weed seed production.

Sanitation is a preventative weed management practice. Control of weeds in the peripheral areas surrounding a field such as drainage ditches, roads, and other noncrop areas is important to prevent invasion from areas surrounding the field. This is especially important for management of wind-blown weed seed from species like hairy fleabane (*Conyza bonariensis*) and horseweed (*Conyza canadensis*) (Daugovish et al. 2008). Sanitation is especially important following soil disinfestation with fumigants or solarization, which have no residual effects. Cleaning soil from equipment while moving from one field to another is another method of sanitation to prevent movement of weed seeds between fields.

15.3.2.2 Stale Seedbed

Annual weed seeds typically germinate and emerge from the top 2–5 cm of soil. Stale seedbed is a technique that can be used to deplete the upper soil layers of "germinable" weed seeds, i.e., nondormant seeds that germinate after watering. After forming raised beds, preplant irrigation of dry soil, followed by an interval of one week or as needed to dry the field, then followed by shallow tillage, flaming, or herbicide application to create a smooth planting bed. This practice kills many weeds that emerge during the interval between watering and the weed control strategy. The weed control objective of a stale seedbed is to promote weed germination before crop planting, when it is easier and more economical to control them, than during the crop production season when weed removal is more difficult and costlier due to the need to protect the crop. By depleting weed seeds in the upper soil layers, field soil acts as a mulch to inhibit emergence of germinable weed seeds from deeper soil layers. Therefore, it is important to not till the preirrigated beds too deeply, to avoid the risk of movement of germinable weed seed from deeper soil layers to near the soil surface where they can germinate. Studies conducted at Salinas, CA, found that stale seedbed system can provide up to 50% weed control (Figure 15.6) (Shem-Tov et al. 2006).

Fields can be reworked or sprayed with a nonselective herbicide prior to planting to ensure planting into a weed-free field. Some weeds, such as pigweed species (i.e. Palmer amaranth, redroot pigweed) and annual grasses, can reestablish after a cultivation, therefore requiring multiple cultivations or herbicide application followed by cultivation.

15.3.2.3 Subsurface Drip Irrigation

Drip tape buried 20 cm below the soil surface, "subsurface drip irrigation," can irrigate the crop but the soil surface remains dry, which reduces weed germination and growth (Grattan et al. 1988; Biasi et al. 2000). If properly managed, this technique can provide significant weed control during the dry months of the year, and it works well with transplanted crops such as tomato. This technique works best in arid areas where soil moisture can be controlled, and the soil surface can be kept dry so that weed seeds do not germinate.

15.3.2.4 Crop Rotation

Crop rotation is effective for weed management because varying patterns of crop production diversifies the weed management methods. Monoculture such as continuous wheat or rice tends to have the most difficult weed problems because the weeds become acclimatized to the field and tend to have similar growth habits such as the crop, e.g., black nightshade (*Solanum nigrum*) in tomato, and barnyardgrass (*Echinocloa crus-galli*) in rice (*Oryza sativa*). In contrast, crop rotation introduces diversification in the cropping cycle. The mechanism by which crop rotation reduces the size of the weed seedbanks is related to the use of crop sequences that employ varying patterns of resource competition, allelopathic interference, soil disturbance, and diverse weed management strategies. Proliferation of otherwise well-adapted weed species is reduced by these processes that produce a more diverse environment (Jordan et al. 1995; Murphy et al. 2006; Blackshaw et al. 2015; Zimdahl 2018).

Crop rotation also permits the use of different herbicides in the rotational crops. Crop rotation is a common practice in California vegetable production areas where cole crops, leafy vegetables, and strawberries are in frequent rotation, often several times per year (Daugovish et al. 2008; Le Strange et al. 2010; Koike et al. 2011; Smith et al. 2011). The short cycle of crops like spinach (e.g. 25 days) does not allow time for most weeds to set seed (Koike et al. 2011). Inclusion of strawberry into the crop rotation sequence includes a soil disinfestation treatment which reduces the weed seedbank. In other words, a diverse crop rotation is an effective way to reduce weed seed banks using sustainable methods.

15.3.2.5 Competition

Transplants are a way of starting field production of a specialty crop with much larger plants that will have a competitive advantage over later emerging weeds because the crop has a larger shoot or root mass than the weed (Bond and Grundy 2001). For example, one-month-old vegetable transplants will be much larger than emerging annual weeds and much less susceptible to weed competition. Use of transplants is commonly practiced in organic vegetable plantings where there are no herbicide options to remove weeds. The first-generation intelligent weeders also work better with transplanted vegetables. The larger size of the crop compared to emerging weeds allows the intra-row cultivator to detect and protect the crop while removing the much smaller weeds (Lati et al. 2016).

There is some potential to select for varieties tolerant to weed competition. For example, sweet potato variety "Carolina Bunch" has an erect habit that was more tolerant to weed competition than traditional spreading varieties like "Beauregard" (Harrison and Jackson 2011). Breeding for increased weed competition tolerance may be a viable strategy in some agronomic crops (Bond and Grundy 2001). Some vegetables such as beans, sweet corn, and cabbage have large leaves and can compete against weeds (Dittmar et al. 2018). These crops may have the potential for selection for greater tolerance to weed competition. Selection for early crop vigor, plant height, increased tillering, and leaf area have promise to increase the potential for wheat and rice to compete with weeds (McDonald and Gill 2009). However, it is not clear if breeding for weed competitive ability has more than a few limited applications in specialty crops.

15.4 Chemical Methods of Weed Control in Specialty Crops

Herbicides are nonrenewable and once consumed must be purchased again. Some herbicides can leach into groundwater, contaminate surface waters, and drift onto sensitive crops (Islam et al. 2018; Al-Khatib 2021). The inventory of herbicides labeled on vegetable crops is limited due to limited sales volumes and damage liabilities for the registrant (Fennimore and Tourte 2019). Many vegetable herbicides are old, such as pronamide, the major lettuce herbicide, which was registered in 1972. In the 49 years elapsed since pronamide was introduced, no other major lettuce herbicide has been registered (USEPA 1994; CDPR 2017). Except for bicyclopyrone, there are few new herbicides in development for specialty crops currently (Chen et al. 2018).

Pesticides have estimated external costs impacts on human health, regulatory, environmental, and other unforeseen costs and range from \$3.6 to \$7.8 billion; in some situations, costs can exceed \$12 billion in the US (Bourguet and Guillemaud 2016). In addition, recent lawsuits against the glyphosate registrant Bayer Crop Science may be settled for \$9.6 billion in claims (Zhang et al. 2019; Rosenblatt 2021). In contrast to the internal and external costs associated with herbicide development and regulation at the company level, research shows that pesticides can be a very good investment to the end user or producer. Zilberman et al. (1991) determined that "a \$1 increase in aggregate pesticide expenditure has been estimated to raise gross agricultural output from \$3 to \$6.50." The ratio of herbicide cost to benefit in lettuce was much less with \$1.36 savings in labor cost per \$1.00 in herbicide cost (Haar and Fennimore 2003). The conflict between the benefits of herbicides and the economic liability of maintaining these products in the specialty crop marketplace is challenging.

Despite some of the negative aspects of herbicides, they are among the most successful weed control tools available, and they are cost-effective, easy to apply and safe to humans and animals (Zimdahl 2018). Herbicides reduce hand-weeding costs and will continue to be used in specialty crops for the long-term. Application at the appropriate labelled rates limits the off-site movement of these products. Herbicides applied in bands on the crop rows limits the use of the product to only where it is needed (Haar and Fennimore 2003). The most promising area for introduction of new herbicides is in perennial crops like trees and vines where directed sprays and little worry of carryover make it easier to register new herbicides. However, the long-term trends in specialty crops are shifting towards organic crops, which are of increasing demand from consumers. Despite increased consumer demand for organic produce, there are few organic-compliant herbicides available. To date, all are contact herbicides with no soil activity (Hussein et al. 2009). Homeowners primarily use these products, and there is limited use in commercial agriculture due to high cost and weak activity. This may well change in the future as this is an active area of research. Potential organic-compliant herbicides might include acetic acid (vinegar), fatty acids, soaps, and natural oils such as clove oils (Boyd and Brennan 2006). Research reports indicate that acetic acid-based products can control weeds under some circumstances. However, the performance of acetic acid products is inconsistent (R. Smith and T. Miller, personal communication; Boyd et al. 2006). More research and development in organic-compliant herbicides is needed.

15.5 Conclusions and Suggestions for Future Research

The major challenges and limitations of sustainably managing weeds in specialty crops is due in part that these commodities are grown for food which means that they must meet the highest standards of appearance, quality and taste as well as be safe from human pathogens and pesticide residues. These high standards must be met with a limited number of herbicides and an expensive and declining hand-weeding labor pool. Given the high cost of developing new herbicides, and the considerable consolidation in the agricultural chemical industry, it seems improbable that a large number of new specialty crop herbicides will be introduced in the near term. Machine learning technology and intelligent weeders may offer some of the best options for these crops.

An ecologically based weed management system would be sustainable in the long term, repeatable from season to season and help maintain the health of the agroecosystem on the farm and the vicinity. It is a very difficult to eliminate all adverse effects on the environment yet provide cost-effective weed control in high-value crops. Renewable inputs like robotic weeders would seem to be a promising option because the cultivators can be powered by locally generated electrical sources and are not dependent on non-renewables such as herbicides, which can contaminate groundwater (Istok et al. 1993).

There are significant opportunities to reduce the impact of specialty crop weed management systems on the agroecosystem through increased use of renewable technologies such as intelligent weeders and locally generated renewable energy. It appears likely that farm labor availability will continue to decline while becoming more expensive and so there is little choice but to improve labor use efficiency. Management of weeds in specialty crops requires a great deal of skill. Demand for organic produce is increasing, and as some share of the industry shifts in that direction, management of weeds in organic fields will require even more skill and innovation than conventional production. Some suggestions to help meet these challenges are:

1) Develop intelligent cultivators that accurately distinguish between crops and weeds thus removing the weeds at field speeds of 1 m sec^{-1}.
2) Design intelligent weeders that use non-tillage methods to kill weeds: laser, precision flaming, hot oil, high pressure water, abrasives (sand blasting). This would allow intelligent weeders to enter fields earlier than when mechanical cultivation is possible (Fennimore et al. 2016).
3) Develop intelligent weeders for perennial tree and vine crops.
4) The intelligent weeders described above should be powered by rechargeable batteries powered by renewable sources.
5) Develop cost-effective biodegradable mulches that can last 12 months before decomposing.

References

Aegeter, B., Klonsky, K.M., and De Moura, R.L. (2013). Sample costs to establish and produce asparagus. University of California Cooperative Extension. https://coststudyfiles.ucdavis.edu/uploads/cs_public/91/0b/910b6b2a-1f01-4697-be7f-0ae88f5ee11b/asparagusvn2013.pdf (accessed 21 May 2021).

Alabbadi, S.A. (2012). Hydrogen oxygen steam generator integrating with renewable energy resource for electricity generation. *Energy Procedia* 29: 12–20.

Al-Khatib, K. (2021). Herbicide damage. University of California IPM. https://herbicidesymptoms.ipm.ucanr.edu/ (accessed 11 May 2021).

Anonymous (2020). 180.41 Crop group tables. Code of Federal Regulations. https://www.ecfr.gov/cgi-bin/text-idx?SID=46e0f15c31b8f5b1c0a348eac9f380e5&mc=true&node=se40.24.180_141&rgn=div8 (accessed 11 May 2021).

Ascard, J. (1998). Comparison of flaming and infrared radiation techniques for thermal weed control. *Weed Research* 38: 69–76.

Baker, K.F. (1957). The U.C. System for Producing Healthy Container-Grown Plants Manual 23 UC ANR Agric. Exp. Sta., Berkeley, CA.

Barnes, D.K.A., Galgani, F., Thompson, R.C., and Barlaz, M. (2009). Accumulation and fragmentation of plastic debris in global environments. *Philosophical Transactions of the Royal Society B* 364: 1985–1998.

Bertucci, M.B., Jennings, K.M., Monks, D.W. et al. (2019). Critical period for weed control in grafted and nongrafted watermelon grown in plasticulture. *Weed Science* 67: 221–228.

Biasi, W., Peratoner, C., Gasparinetti, P. et al. (2000). Underground drip lines for modern vineyard management. *Informatore-Agrario* 56: 85–94.

Blackshaw, R.E., Pearson, D.C., Larney, F.J. et al. (2015). Conservation management and crop rotation effects on weed populations in a 12-year irrigated study. *Weed Technology* 29: 835–843.

Bolda, M., Tourte, L., Murdock, J., and Sumner, D.A. (2018). Sample costs to produce and harvest fresh market blackberries. https://coststudyfiles.ucdavis.edu/uploads/cs_public/30/89/3089b129-9d64-4c00-9032-194edf41842e/bpblackberrycc2018-final-1.pdf (accessed 21 May 2021).

Bolda, M., Tourte, L., Murdock, J., and Sumner, D.A. (2019). Sample costs to produce and harvest organic strawberries. https://coststudyfiles.ucdavis.edu/uploads/cs_public/a9/5b/a95bfaa1-c387-4153-8021-05ad8bca327f/2019organicstrawberrycc-final-june_2019.pdf (accessed 21 May 2021).

Bond, W. and Grundy, A.C. (2001). Non-chemical weed management in organic farming systems. *Weed Research* 41: 383–405.

Bourguet, D. and Guillemaud, T. (2016). The hidden and external costs of pesticide use. In: *Sustainable Agriculture Reviews*, vol. 19 (ed. E. Lichfrouse), 35–120. SpringerLink. https://doi.org/10.1007/978-3-319-26777-7_2.

Boyd, N.B., Brennan, E.B.(2006). Burning nettle, common purslane, and rye response to a clove oil herbicide. *Weed Technology* 20: 646–650.

Boyd, N.B., Brennan, E.B., and Fennimore, S.A. (2006). Stale seedbed techniques for organic vegetable production. *Weed Technology* 20: 1052–1057.

Brackenrich, J. and Roth, G.W. (2020). Managing drought stressed corn. Penn State University Extension. https://extension.psu.edu/managing-drought-stressed-corn (accessed 24 May 2021).

Brennan, E.B., Daugovish, O., Fennimore, S.A., Smith, R.F. (2011). Weeds. In: *Cover Cropping for Vegetable Production* (ed. R.F. Smith, R.L. Bugg, M. Gaskell et al.), 1–2. University of California, Agriculture and Natural Resources. Publication 3517.

Brennan, E.B. and Smith, R.F. (2005). Winter cover crop growth and weed suppression on the Central Coast of California. *Weed Technology* 19: 1017–1024.

Bugg, R.L. and Smith, R.F. (2011). Introduction. In: *Cover Cropping for Vegetable Production* (ed. R.F. Smith, R.L. Bugg, M. Gaskell et al.), 1–2. University of California, Agriculture and Natural Resources. Publication 3517.

Buhler, D.D., Hartzler, R.G., and Forcella, F. (1997). Implications of weed seedbank dynamics to weed management. *Weed Science* 45: 329–336.

[CDPR] California Department of Pesticide Regulation (2017). Annual pesticide use report. Department of Pesticide Regulation, Sacramento, CA, USA. https://www.cdpr.ca.gov/docs/pur/pur17rep/statewide_site_2017.htm (accessed 11 May 2021).

Charlton, D. and Taylor, J.E. (2016). A declining farm workforce: analysis of panel data from rural Mexico. *American Journal Agricultural Economics* 98: 1158–1180.

Chaudhari, S., Jennings, K.M., Monks, D.W. et al. (2016). Critical period for weed control in grafted and nongrafted fresh market tomato. *Weed Science* 64: 523–530.

Chen, Y., Hu, C., and Doohan, D. (2018). Safety of bicyclopyrone on several vegetable crops and efficacy of weed control. *Weed Technology* 32: 498–505.

Christensen, C.M., Raynor, M.E., and McDonald, R. (2015). What is disruptive innovation? *Harvard Business Review* 93: 44–53.

Dahlquist, R.M., Prather, T.S., and Stapleton, J.J. (2007). Time and temperature requirements for weed seed thermal death. *Weed Science* 55: 619–625.

Daugovish, O., Fennimore, S.A., and Mochizuki, M.J. (2008). Integration of oxyfluorfen into strawberry (*Fragaria x ananassa*) weed management programs. *Weed Technology* 22: 685–690.

Dittmar, P., Boyd, N., and Kannisery, R. (2018). Weed management principles in commercial vegetable production. University of Florida. Publication HS717. http://edis.ifas.ufl.edu.

Duke, S.O. (2012). Why have no new herbicide modes of action appeared in recent years? *Pest Manag Science* 68: 505–512.

EPSRC UK-RAS Network (2019). UK-RAS19 Conference: "Embedded Intelligence: Enabling and Supporting RAS Technologies". Proceedings. https://doi.org/10.31256/Fu8Cq8D. ISSN: 2516-502X.

Fennimore, S., Martin, F.N., Miller, T.C. et al. (2014). Evaluation of a mobile steam applicator for soil disinfestation in California strawberry. *HortScience* 49: 1542–1549.

Fennimore, S.A. (2012). Weed control in organic strawberries. In: *Organic Strawberry Production Manual* (ed. S.T. Koike), 35–41. University of California ANR Publication 3531.

Fennimore, S.A. (2014). Cultural, mechanical and physical methods. In: *Principles of Weed Control*, 4e (ed. S.A. Fennimore and C.E. Bell), 51–66. California Weed Science Society Chapter 3.

Fennimore, S.A. and Cutulle, M. (2019). Robotic weeders can improve weed control options for specialty crops. *Pest Management Science* 75: 1767–1774.

Fennimore, S.A., Slaughter, D.C., Siemens, M.C. et al. (2016). Technology for automation of weed control in specialty crops. *Weed Technology* 30: 823–837.

Fennimore, S.A. and Tourte, L. (2019). Regulatory burdens on development of automated weeding machines and herbicides are different. *Outlooks on Pest Management* 30 (4): 147–152.

Fennimore, S.A. and Umeda, K. (2003). Time of glyphosate application in glyphosate-tolerant lettuce. *Weed Technology* 17: 738–746.

Forcella, F., Poppe, S.R., Hansen, N.C. et al. (2003). Biological mulches for managing weeds in transplanted strawberry (*Fragaria x ananassa*). *Weed Technology* 17: 782–787.

Gallandt, E. (2006). How can we target the seedbank? *Weed Science* 54: 588–596.

Gaskell, M., Daugovish, O., and Van Horn, M. (eds.) (2011). *Cover Cropping for Vegetable Production*. University of California, Agriculture and Natural Resources Publication 3517, 43–46.

Gil, M., Selma, M.V., Suslow, T. et al. (2015). Pre- and postharvest preventive measures and intervention strategies to control microbial food safety hazards of fresh leafy vegetables. *Critical Reviews in Food Science and Nutrition* 55 (4): 453–468.

Gliessman, S.R. (2001). *Agroecosystems Sustainability – Developing Practical Strategies*. 1. Boca Raton, FL: CRC Press.

Grattan, S.R., Schwankl, L.J., and Lanini, W.T. (1988). Weed control by subsurface drip irrigation. *California Agriculture* 42: 22–24.

Haar, M.J. and Fennimore, S.A. (2003). Evaluation of integrated practices for common purslane management in lettuce. *Weed Technology* 17: 229–233.

Harrison, H. and Jackson, M.D. (2011). Response of two sweet potato cultivars to weed interference. *Crop Protection* 30 (10): 1291–1296. https://doi.org/10.1016/j.cropro.2011.05.002.

Hussein, F.H., Ahmad, A., Omar, A.M. et al. (2009). Efficacy comparison of some new natural-product herbicides for Control at Two Growth Stages. *Weed Technol* 23: 431–437.

IR-4 (Interregional Project 4) (2021). IR-4 project mission. https://www.ir4project.org (accessed 11 May 2021).

Islam, F., Jian, W., Muhammad, A. et al. (2018). Potential impact of the herbicide 2,4-dichlorophenoxyacetic acid on human and ecosystems. *Environment International* 111: 332–335.

Istok, J.D., Smyth, J.D., and Flint, A.L. (1993). Multivariate geostatistical analysis of ground-water contamination: a case history. *Ground Water* 31 (1): 63–74.

Johnson, M.S. and Fennimore, S.A. (2005). Weed and crop response to colored plastic mulches in strawberry production. *HortScience* 40: 1371–1375.

Jordan, N., Mortensen, D.A., and Prenzlow, D.M. (1995). Simulation analysis of crop rotation effects on weed seedbanks. *American Journal of Botany* 82 (3): 390–398.

Kennedy, H., Fennimore, S., Slaughter, D. et al. (2020). Crop signal markers facilitate crop detection and weed removal from lettuce and tomato by an intelligent cultivator. *Weed Technology* 34: 342–350.

Kim, D.S., Kim, S., Fennimore, S.A. (2021). Evaluation of broadcast steam application with mustard seed meal in fruiting strawberry. *HortScience* 56: 500–505.

Knezevic, S., Fennimore, S.A., and Datta, A. (2017). Thermal weed control. In: *Encyclopedia of Applied Plant Sciences*, vol. 3 (ed. B. Thomas, B.G. Murray, and D.J. Murphy), 463–468. Amsterdam: Academic Press. https://doi.org/10.1016/B978-0-12-394807-6.00235-5.

Koike, S.T., Cahn, M., Cantwell, M. et al. (2011). *Spinach Production in California*. University of California Agriculture and Natural Resources Publication 7212.

Kolberg, R. and Wiles, L.J. (2002). Effect of steam application on cropland weeds. *Weed Technology* 16: 43–49.

Lague, C., Gill, J., and Peloquin, G. (2001). Thermal control in plant protection. In: *Physical Control Methods in Plant Protection* (ed. C. Vincent, B. Panneton, and F. Fleurat-Lessard). Berlin, Germany: Springer-Verlag.

Lati, R.N., Siemens, M.C., Rachuy, J.S., and Fennimore, S.A. (2016). Intra-row weed removal in broccoli and transplanted lettuce with an intelligent cultivator. *Weed Technology* 30: 655–663.

Le Strange, M., Cahn, M.D., Koike, S.T. et al. (2010). *Broccoli Production in California*. University of California Agriculture and Natural Resources Publication 7211.

Leon, R.G. and Ferreira, D.T. (2008). Interspecific differences in weed susceptibility to steam injury. *Weed Technology* 22: 719–723.

Leroux, G.D., Douheret, J., and Lanouette, M. (2001). Flame weeding in corn. In: *Physical Control Methods in Plant Protection* (ed. C. Vincent, B. Panneton, and F. Fleurat-Lessard), 47–60. Berlin, Germany: Springer-Verlag.

Lohstroh, P. and Koshlukova, S. (2017). *Evaluation of the potential human health effects from drinking groundwater containing Dacthal (DCPA) degradates*. Department of Pesticide Regulation – Human Health Assessment Branch. http://www.cdpr.ca.gov/docs/hha/memos/tpa%20in%20ground%20water%20reply%20final%2002232017%20complete%20executed.pdf (accessed 11 May 2021).

Mallory, E. and Gallandt, E. (2021). Blind cultivation for weed control in small grains. University of Maine. https://extension.umaine.edu/grains-oilseeds/topics/blind-cultivation-weed-control-small-grains (accessed 19 May 2021).

McDonald, G.K. and Gill, G.S. (2009). Improving crop competitiveness with weeds. In: *Crop Physiology* (ed. V. Sadras and D. Calderini), 449–488. Elsevier. Chapter 18.

McDougall, P. (2016). The cost of new agrochemical product discovery, development and registration in 1995, 2000, 2005–8 and 2010–2014. https://croplife.org/wp-content/uploads/2016/04/Cost-of-CP-report-FINAL.pdf (accessed 11 May 2021).

McGowen, S.J., Jennings, K.M., Chaudhari, S. et al. (2018). Critical period for Palmer amaranth (*Amaranthus palmeri*) control in pickling cucumber. *Weed Technology* 32: 586–591.

Melander, B., Lattanzi, B., and Pannacci, E. (2015). Intelligent versus non-intelligent mechanical intra-row weed control in transplanted onion and cabbage. *Crop Protection* 72: 1–8.

Meldrum, G., Padulosi, S., Lochetti, F. et al. (2018). Issues and prospects for the sustainable use and conservation of cultivated vegetable diversity for more nutrition-sensitive agriculture 8: 112. https://doi.org/10.3390/agriculture8070112.

Monteiro, A. and Santos, S. (2022). Sustainable approach to weed management: the role of precision weed management. *Agronomy* 12: 118. https://doi.org/10.3390/agronomy12010118.

Moore, J. and Wszelaki, A. (2016). Biodegradable Mulch. https://biodegradablemulch.tennessee.edu/ (accessed 11 May 2021).

Moore, L.D., Jennings, K.M., Monks, D. et al. (2021). Evaluating electrical and mechanical methods for Palmer Amaranth (*Amaranthus palmeri*) control. *Northeastern Plant, Pest, and Soil Conference.*

Mosqueda, E., Smith, R., Goorahoo, S., and Shrestha, A. (2017). Automated lettuce thinners reduce labor requirements and increase speed of thinning. *California Agriculture* 72: 114–119. https://doi.org/10.3733/ca.2017a0018.

Mujkic, S. and Andakudi Kesevan, K.N. (2020). Life cycle assessment of lettuce production systems: a case study performed at Hollbium. KTH Royal Institute of Technology School of Architecture and Built Environment Department of Sustainable Development, Environmental Science and Engineering SE-100 44, Stockholm, Sweden. https://www.diva-portal.org/smash/get/diva2:1463291/FULLTEXT01.pdf (accessed 11 May 2021).

Mulvaney, M.J., Price, A.J., and Wood, C.W. (2011). Cover crop residue and organic mulches provide weed control during limited-input no-till collard production. *Journal of*

Sustainable Agriculture 35 (3): 312–328. https://doi.org/10.1080/10440046.2011.554315.

Murphy, S.D., Clements, D.R., Belaussoff, S. et al. (2006). Promotion of weed species diversity and reduction of weed seedbanks with conservation tillage and crop rotation. *Weed Science* 54: 69–77.

Ngouajio, M. and Ernst, J. (2005). Changes in the physical, optical and thermal properties of polyethylene mulches during double cropping. *HortScience* 40: 94–97.

Norris, R.F., Caswell-Chen, E.P., and Kogan, M. (2003). *Concepts in Integrated Pest Management*. Prentice Hall, 1–13.

Plastina, A. (2021). Estimated costs of crop production in Iowa. Iowa State University. https://www.extension.iastate.edu/agdm/crops/html/a1-20.html (accessed 11 May 2021).

Radosevich, S.R., Holt, J.S., and Ghersa, C.M. (2007). *Ecology of Weeds and Invasive Plants: Relationship to Agriculture and Natural Resource Management*. John Wiley & Sons, 293–294.

Ranjan, P., Patatle, G.T., Prem, M., and Solanke, K.R. (2017). Organic mulching – a water saving technique to increase the production of fruits and vegetables. *Current Agriculture Research Journal* 5: 371–380.

Riemens, M.M., Groeneveld, R.M.W., Lotz, L.A.P., and Kropff, M.J. (2007). Effects of three management strategies on the seedbank, emergence and the need for hand weeding in an organic arable cropping system. *Weed Research* 47: 442–451.

Rohwer, C. and Klodd, A. (2019). Using a flame weeder in vegetable and fruit crops. https://blog-fruit-vegetable-ipm.extension.umn.edu/2019/01/using-flame-weeder-in-vegetable-and.html (accessed 17 May 2021).

Rosenblatt, J. (2021). Bayer won't ask U.S. Supreme Court to reverse roundup loss. Bloomberg. https://www.bloomberg.com/news/articles/2021-03-19/bayer-won-t-ask-u-s-supreme-court-to-reverse-first-roundup-loss (accessed 11 May 2021).

Samtani, J.B., Ajwa, H.A., Weber, J.B. et al. (2011). Evaluation of non-fumigant alternatives to methyl bromide for weed control and crop yield in California strawberries (*Fragaria ananassa* L.). *Crop Protection* 30: 45–51.

Shem-Tov, S., Fennimore, S.A., and Lanini, W.T. (2006). Weed management in lettuce (*Lactuca sativa*) with pre-plant irrigation. *Weed Technology* 20: 1058–1065.

Shrestha, A., Kurtural, S.K., Fidelibus, M.W. et al. (2013). Efficiency and cost-effectiveness of cultivators, steam, or an organic herbicide in weed control and vine productivity in organic vineyards in the San Joaquin Valley. *HortTechnology* 23: 99–108.

Siryvadas, A., Lazauskas, P., Stepanas, A. et al. (2006). Plant temperature variation in the thermal weed control process. *Journal of Plant Diseases and Protection* 20: 355–361.

Smith, R., Cahn, M., Daugovish, O. et al. (2011). *Leaf Lettuce Production in California*. University of California Agriculture and Natural Resources Publication 7216.

Smith, S.C., Jennings, K.M., Monks, D.W. et al. (2020). Critical timing of (*Amaranthus palmeri*) removal in sweet potato. *Weed Technology* 34: 547–551.

Sosnoskie, L. and Culpepper, A. (2014). Glyphosate-resistant Palmer amaranth (*Amaranthus palmeri*) increases herbicide use, tillage, and hand-weeding in Georgia cotton. *Weed Science* 62 (2): 393–402. https://doi.org/10.1614/WS-D-13-00077.1.

Strand, L. (2008). *Integrated Pest Management for Strawberries*. University of California Statewide Integrated Pest Management Project Publication 3351, 116.

Tensorfield (2021). https://tensorfield.ag/2021/02/28/robotic-chassis-field-testing (accessed 11 May 2021).

Takele, E., Daugovish, O., Vue, M. (2013). Costs and Profitability Analysis for Bell Pepper Production in the Oxnard Plain, Ventura County, 2012–13; Bell Pepper Production for Processing University of California Agricultural and Natural Resources Publication 7216.

Tourte, L., Smith, R.F., Klonsky, K.M., and De Moura, R.L. (2009). Sample costs to produce organic leaf lettuce, double cropped, Central Coast Region, Santa Cruz and Monterey counties. University of California Cooperative Extension. https://coststudies.ucdavis.edu/en/archived/commodity/lettuce (accessed 11 May 2021).

Tourte, L., Smith, R.F., Klonsky, K.M. et al. (2015). Sample costs to produce and harvest organic spinach https://coststudyfiles.ucdavis.edu/uploads/cs_public/79/02/79023ea8-80a8-4fba-b69e-5d60225dbf8b/2015_organicspinach-finaldraftjan29.pdf (accessed 21 May 2021).

Tourte, L., Smith, R.F., Murdock, J., and Sumner, D.A. (2017a). Sample costs to produce and harvest iceberg lettuce sprinkler irrigated – 40-inch beds. University of California Cooperative Extension. https://coststudies.ucdavis.edu/en/current/commodity/lettuce (accessed 11 May 2021).

Tourte, L., Smith, R.F., Murdock, J., and Sumner, D.A. (2017b). Sample costs to produce and harvest broccoli. University of California Cooperative Extension. https://coststudyfiles.ucdavis.edu/uploads/cs_public/e4/13/e413c195-5ddb-433b-9be8-a48c40b4063b/2017broccoli-final_5-25-2017.pdf (accessed 21 May 2021).

Tourte, L., Smith, R.F., Murdock, J., and Sumner, D.A. (2019). Sample costs to produce and harvest romaine hearts. University of California Cooperative Extension. https://coststudies.ucdavis.edu/en/current/commodity/lettuce (accessed 11 May 2021).

Turini, T., Stewart, D., Murdock, J., and Sumner, D.A. (2018). Sample costs to produce processing tomatoes. University of California Cooperative Extension. https://coststudyfiles.ucdavis.edu/uploads/cs_public/2e/7a/2e7a8cf0-b7fd-4207-b945-7b4b8a82625c/17processtomatofresnosdi-final_draft.pdf (accessed 21 May 2021).

UK RAS (United Kingdom Robotics and Autonomous Systems) (2018). Agricultural robotics: the future of robotic agriculture. https://www.ukras.org.uk/wp-content/

uploads/2021/01/UKRASWP_AgriculturalRobotics2018_online.pdf (accessed 11 May 2021).

USDA NASS (2020). Vegetable 2019 summary. Agricultural Statistics Board, NASS USDA, Washington, DC.

[USEPA] United States Environmental Protection Agency (1994). Registration Eligibility Document Pronamide, USA https://archive.epa.gov/pesticides/reregistration/web/pdf/0082fact.pdf (accessed 11 May 2021).

USEPA (2017). USEPA's National Organic Program Guidance. https://www.epa.gov/pesticide-registration/epas-national-organic-program-guidance (accessed 11 May 2021).

USEPA (2020). Pesticide devices. https://www.epa.gov/safepestcontrol/pesticide-devices-guide-consumers (accessed 11 May 2021).

USEPA (2021). Pesticide registration manual. https://www.epa.gov/pesticide-registration/pesticide-registration-manual (accessed 11 May 2021).

Wanwimolruk, A., Phopin, K., Boonpangrak, S., and Prachayasittikul, V. (2016). Food safety in Thailand 4: comparison of pesticide residues found in three commonly consumed vegetables purchased from local markets and supermarkets in Thailand. *PeerJ*. https://doi.org/10.7717/peerj.2432.

White, K.E., Brennan, E.B., Cavigelli, M.A., and Smith, R.F. (2020). Winter cover crops increase readily decomposable soil carbon, but compost drives total soil carbon during eight years of intensive, organic vegetable production in California. *PLoS ONE* 15 (2): e0228677. https://doi.org/10.1371/journal.pone.0228677.

Willer, H. and Lernoud, J. (2019). The World of Organic Agriculture. https://orgprints.org/id/eprint/34669/1/WILLER-LERNOUD-2018-final-PDF-low.pdf (accessed 25 May 2021).

Wilson, R., Sumner, D.A., Klonsky, K., and Stewart, D. (2016). Sample costs to produce onions. University of California Cooperative Extension. https://coststudyfiles.ucdavis.edu/uploads/cs_public/a7/1e/a71ed327-7d6c-4ae5-92a2-52854cb4195c/16_onionshydrostulelakefinaldraftmar22.pdf (accessed 21 May 2021).

Zahniser, S., Taylor, J.E., Hertz, T., and Charlton, D. (2018). Farm labor markets in the United States and Mexico pose challenges for U.S. Agriculture. United States Department of Agriculture – Economic Research Service. https://www.ers.usda.gov/publications/pub-details/?pubid=90831 (accessed 11 May 2021).

Zhang, L., Rana, I., Shaffer, R.M. et al. (2019). Exposure to glyphosate-based herbicides and risk for non-Hodgkin lymphoma: a meta-analysis and supporting evidence. Mutation Research-Reviews in Mutation Research. https://doi.org/10.1016/j.mrrev.2019.02.001 (accessed 11 May 2021).

Zilberman, D., Schmitz, A., Casterline, G. et al. (1991). The economics of pesticide use and regulation. *Science* 253: 518–522.

Zimdahl, R. (2018). *Fundamentals of Weed Science: Fifth Edition*. Elsevier.

16

The Need of Ecologically Based Weed Management Approaches in Orchard Crops

Victor Martins Maia[1,], Ignacio Aspiazú[1], Leandro Galon[2], Clevison Luiz Giacobbo[2], Germani Concenço[3], Alexandre Ferreira da Silva[3], Evander Alves Ferreira[4], and George Andrade Sodré[5]*

[1] *Department of Agricultural Sciences, State University of Montes Claros, Janaúba, MG, Brazil*
[2] *Universidade Federal da Fronteira Sul, Chapecó, SC, Brazil*
[3] *Brazilian Agricultural Research Corporation, Capão do Leão, Brazil*
[4] *Institute of Agricultural Sciences, Federal University of Minas Gerais, Montes Claros, MG, Brazil*
[5] *State University of Santa Cruz, Neighborhood Salobrinho, BA, Brazil*
[*] *Corresponding author*

16.0 Introduction

World fruit production, according to FAO data (2019), is characterized by the diversity of cultivated species and is formed mostly by temperate and subtropical fruits, produced and consumed mainly in the Northern Hemisphere. Worldwide, fruit production grows continuously every year, with more than 800 million tons of fruit.

Following global macro trends, it can be suggested that population and per capita income will continue to grow and the demand for fruit will increase. However, more and more consumers of fruit are looking for healthy products, with demands to reduce the negative impacts on human and environmental health. In this regard, the proper management of orchards, especially weeds, has undergone notable changes year after year, with this activity being extremely important for maintaining or increasing production. In the last decades, the search for better soil and weed managements began, since cultural interventions were, until then, aimed at maximizing production. Nowadays, there are new goals for improving the quality of the product and a greater integration between fruit production, environment and food quality (Marangoni et al. 1995; Souza and Maia 2012).

The quality of the fruit, through certification processes, has become a requirement of the importing markets since the end of the 1990s. These markets seek, in addition to the external aspect, the guarantee of internal quality, through specific programs and legislation that guarantee the permanent control and inspection of the entire production chain in the exporting country. These requirements have been intensified, with regulations, where the buyer or seller country or economic bloc determines norms and rules for fruit imports, generally related to minimizing the use of pesticides.

Another issue that has gained attention over the past few years is the increasing number of glyphosate-resistant biotypes of weeds. The increase in its use in conservationist production systems (no-tillage) and the greater flexibility for application of the product in genetically modified (transgenic) crops resulted in a greater risk of selection of resistant weed biotypes, as a result of the greater selection pressure imposed (Neve et al. 2003; Christoffoleti et al., 2008; Moreira et al. 2010). Several studies have shown that the intensive use of glyphosate, in addition to increasing the risk of selection of resistant plants, causes changes in weed flora due to the selection of species with greater tolerance to the product, such as morning glory (*Ipomoea* spp.), Benghal dayflower (*Commelina* spp.), and broadleaf buttonweed (*Borreria* spp.) (Monquero and Christoffoleti 2003; Culpepper 2006; Carvalho et al. 2008). Martinelli et al. (2022) demonstrated that the increase in glyphosate usage can occasionally increase weed control, but it can also decrease orchard development and its financial viability.

In this context, the management of weeds infesting tropical and subtropical fruit orchards using sustainable control methods becomes extremely important. Such management

Ecologically Based Weed Management: Concepts, Challenges, and Limitations, First Edition. Edited by Nicholas E. Korres, Ilias S. Travlos, and Thomas K. Gitsopoulos.

combines production with less aggression to the environment, generation of healthier fruits, and, thus, higher economic returns to the producers.

16.1 Ecologically Based Weed Management Approaches in Fruit Crops Species Grown in Tropical and Subtropical Environments

Considering the ecological aspects of tropical and subtropical fruit trees, the main tool for defining control tactics and management practices is the phytosociological survey of the weed community and the dynamics of this community over the time of cultivation.

Among the most used methods of sampling and analysis the most used are the one proposed by Braun-Blanquet (1979) and (Barbour et al. 1998). In these methods, in addition to identifying the weeds present in the area, the calculated indices indicate which species are most important and which of them compete most with the crop.

An approach to this data using fauna analysis and tools applied to animal ecology, with some adjustments, can also be used, even to follow the dynamics of the weed community over time. In this approach, the studies seek to identify the co-occurrence patterns of species that are useful to define the diversity of habitats or niche place of species (Fridley et al. 2007; Pannek et al. 2016; Costa et al. 2020).

Another important tool that can be used in the ecological management of weeds is precision agriculture. In this case, geostatistics and kriging can be used to define where in the orchard, with greater accuracy, are the places and areas most infested with weeds and which species occur in each of these areas, thus creating control tactics and strategies specific to each area or field (Lopes et al. 2020). Exclusion zones and buffer zones can also be created for those species that are more aggressive or more difficult to control, estimating the risk of infestation. This risk can even be estimated using machine learning techniques with Fuzzy Systems (Bressan et al. 2006).

The role of cover crop in suppressing weeds depends on the initial establishment of cover plants, population density, and residue produced (Carvalho and Vargas 2004). An alternative to natural mulching to prevent the growth of weeds, as already mentioned for other fruit trees, is the use of black plastic, widely used in strawberry fields (Medeiros and Santos 2003). Hoeing is another weed control method, but it requires a lot of labor and it leaves the soil bare, unprotected, and subject to erosion. In addition, the mechanical action of weeding can damage the root system

Table 16.1 Height and diameter of the trunk of apple plants at 24 and 36 months after planting, in response to the use of different soil covers. Vacaria-Rio Grande do Sul State, Brazil (2006).

	Plant Height		Plant Diameter	
Treatments	**24 months**	**36 months**	**24 months**	**36 months**
Control	1.37 e*	1.55 f	1.28 c	1.45 e
Hoeing	1.47 cde	2.05 abc	1.41 abc	2.58 bc
Black plastic	1.66 ab	2.23 a	1.81 a	3.23 a
Shading	1.55 cd	1.79 de	1.52 abc	2.22 cd
Pine sawdust	1.56 bc	1.94 bcd	1.52 abc	2.68 bc
Pine needles	1.74 a	2.07 ab	1.75 ab	3.00 ab
Andropogon bimaizeis straw	1.49 cde	1.87 bcde	1.44 abc	2.28 cd
Ryegrass	1.47 cde	1.77 def	1.41 abc	1.82 de
Black oats	1.45 cde	1.69 def	1.55 abc	1.90 de
Black oats+vetch	1.39 de	1.71 def	1.27 c	1.95 de
Black oats+ turnip	1.45 cde	1.68 ef	1.41 abc	1.73 de
Ryegrass + clover +natives	1.38 e	1.82 cde	1.44 abc	1.97 de
Moeing	1.45 cde	1.80 de	1.35 c	1.98 de
C.V. (%)	6.00	7.00	14.00	14.00

Averages followed by the same letter in the columns do not differ significantly, according to the Duncan test (p<0,05). *Source:* Adapted from Pelizza et al. (2009).

of apple trees. Among the management practices of the soil cover plants most used in orchards, mowing stands out (Matheis et al. 2006; Costa et al. 2020). However, the adoption of this practice needs to be well planned, especially the height of the cut, so that problems do not occur.

In a study by Pelizza et al. (2009), it was found that the most efficient treatments for suppressing weeds in the apple tree planting line, in an orchard during the installation phase, were hoeing, black plastic cover, pine needles, and barbas de indio (*Andropogon bimaizeis*). The authors also report that the presence of weeds in the planting line affects the growth of the apple tree, with reductions in plant height and diameter (Table 16.1).

Therefore, management in apple production areas must be done in order to maintain permanent soil coverage, which can be beneficial from an ecological point of view, but must avoid overgrowth of intermediary species, even mowing more frequently during the vegetative and fruiting period of the orchard if necessary (Pelizza et al. 2009).

Electricity, flaming, and microwaves, among others, can also be used as methods of weed control in orchards, or even the vegetation present between the lines of apple trees. This method has positive aspects, such as low environmental impact and contamination of the operator; its use is allowed in organic agriculture; and it can also be used in fields with weeds that are resistant to herbicides with different mechanisms of action.

16.2 Tropical Fruit Crop Species

16.2.1 Pineapple

In addition to considering aspects related to the phytosociological survey of the weed infesting community, as well as the spatial distribution of the species in the cultivation area, another point to consider is the interference periods. The main ones are the period before interference (PBI) and the total period of interference prevention (TPIP). The difference between the TPIP and the PBI is the critical period of interference prevention (CPIP) (Kuva et al. 2000). In order to determine the CPIP, it is important to define a level, usually in percentage (%), of tolerance in the reduction of productivity or in the quality of the fruits. This value depends on the cost of the final product, which in the case of fruit is usually high, and the cost of control. Usually a tolerance of less than 10% is used, but the ideal is to define it on a case-by-case basis.

In the case of irrigated pineapple grown in regions of dry tropics and considering the tolerance of 5% reduction in productivity, the critical period of interference prevention (CPIP) goes from 1.8; 1.3, and 2.1 months after planting to 8.8; 7.5, and 4.5 months after planting, for the cultivars IAC Fantástico, Pérola, and Smooth Cayenne, respectively (Santos 2019). The author also observed a reduction in fruit quality due to weed competition. The aerial view of this experiment is available in Figure 16.1A.

From these CPIP values, which in all cases were always prior to the artificial floral induction, tactics and control strategies can be defined and adjusted. However, another point to be considered is that, even after the CPIP has been defined, the presence of weeds in the area can hinder the execution of some cultural activities or practices, such as fertilization, artificial floral induction, protection of fruits against sunburn, and harvest.

Considering these aspects, some control practices eliminate weed interference on pineapple plants for practically the entire growing period. The main technologies that can be used are biodegradable plastic mulching (Figure 16.1B), the use of cover crops and the use of organic mulching of high C:N ratio, such as of sawmill residues of reforested wood, such as eucalyptus and pine.

Plastic mulching reduces the use of herbicides and labor cost, while promoting greater growth, reducing the production cycle, and the need for irrigation of the pineapple plants (Botrel et al. 1990; Carr 2012; Tachie-Menson et al. 2014). However, considering an ecological management aiming to reduce pollution caused by plastic waste, biodegradable plastic mulching is also an option. This product uses maize starch or polymers of vegetable origin such as sugarcane and carnauba wax as raw material (Oliveira et al. 2021); however, although already available on the market, the cost is still quite high. Some crops already adopt this technology, which is promising for the near future.

Alternatively or in joint use with the biodegradable plastic mulching, cover crops can be adopted for weed control (Eshetu et al. 2007; Matos et al. 2009). The management of cover crops consists of mowing at the blooming stage and chopping the cover crops, allowing their residues to remain on soil as mulch, and directly planting the cover crops as many times as necessary, depending on environmental conditions (Matos et al. 2009). Some species are already recommended for this type of management, such as Bermuda grass (*Cynodon dactylon* (L) Pers.) and *Pennisetum americanum* (Matos et al. 2009). Forage peanuts (*Arachis pintoi*) can be added to this list with the advantages of having low biological nitrogen fixation – therefore, with no need for desiccation. The use of cover crops also increases soil organic matter, as it contributes substantially to the improvement of the chemical and physical attributes of tropical soils, which are known to be poor in essential nutrients for the plants (Cordeiro et al. 2021).

Considering the organic mulching use, plant residues or crop remains can be used, depending on their availability (Eshetu et al. 2007; Alwis and Herath 2012), or dead cover with allelopathic plants (Ismail and Chong 2009).

Figure 16.1 Aerial view of trial on critical period of weed interference in pineapple (A), use of biodegradable plastic mulching and drip irrigation in pineapple (B), banana area with high weed infestation (C), banana area with excellent weed control and use of organic mulching (D), use of plastic mulching in cocoa (E), use of cover crop (*Arachis pintoi*) in cocoa (F) and use of organic mulching with residues from pruning in cocoa (G). (Sources: Image A: João Rafael Prudêncio dos Santos; Image B: Julien; Images C e D: Gilberto Cardoso dos Santos; Images E: George Andrade Sodré; Images F and G: Victor Martins Maia).

The *Dicranopteris linearis* (Burm. f.) Underw. is a tropical fern that inhibits germination and growth of several common weeds in Malaysia presenting allelopathic effects (Ismail and Chong 2009). Use of air-dried *Dicranopteris* at 3.0 kg m^{-2} reduced 99% of weed emerged and did not reduce the growth of pineapple plants and fruit weight (Chong et al. 2011). The application of organic mulch provides greater growth and productivity even compared to clean weeding all season (Eshetu et al. 2007). In addition, the use of sawmill residues from reforested wood with a high C:N ratio, such as eucalyptus or pine, applying 25–35 tons per hectare in total area after planting, leaves a 5–10 cm layer covering the soil. In addition to helping to maintain soil moisture, this cover helps to suppress weeds and probably has allelopathic effects (Sarmin et al. 2020). As an obstacle to the use of this technology, there is the need for the availability of these residues close to the cultivation area.

In addition, hand weeding or mowing are very common practices in low-tech crops and in developing countries with greater labor availability or in family farming. The use of a backpack mower increases the performance of this type of control (Maia et al. 2018). Those mechanical weed controls should be done frequently, every time weeds start growing and at least until the end of the critical period of weed interference (CPIP).

16.2.2 Bananas

Bananas are the most important cultivars in the world from the point of view of international market and planted area. Some common weed control practices mentioned for pineapple can be used for bananas as well. However, unlike the slow-growing pineapple, bananas grow quickly, completely shading the area and virtually eliminating competition with weeds in the first months or first year of cultivation (Santos et al. 2019). The exceptions are some species adapted to shading and with great dispersion capacity, such as the Benghal dayflower (*Commelina benghalensis*). Some species of this weed, like *C. diffusa*, have become a major problem in banana fields in several countries due to some control practices. Initially, it was even considered, erroneously, as a cover crop (Isaac et al. 2012).

Therefore, weed control in bananas needs more attention in the initial vegetative growth phase of the first year of cultivation (Figure 16.1C). The use of quality and vigorous seedlings also helps to reduce control time. However, the occurrence of competition in the early stages (i.e. in the first month after planting), it can reduce both the production and the quality of the fruits in the first and second production cycles (Santos 2019). However, there is a technical recommendation to leave the banana field free from competition for at least five months after planting. These differences in the recommendations can be justified by differences in weed communities, cultivars, types of seedlings planted, and cultural practices that accelerate the vegetative growth of the bananas.

Among the best practices to eliminate or reduce this initial competition are, in addition to manual or mechanized weeding with conventional or ecological brushcutters, the use of cover crops and/or biodegradable plastic mulching or organic mulching, including that produced by the banana plants themselves (Figure 16.1D). The adoption of localized irrigation such as drip, especially in dry, arid, or semi-arid regions in the tropics, in addition to reducing water consumption, reduces the wet area and, consequently, the germination of seeds from the weed seed bank present in the area (Sutton et al. 2006; Smith et al. 2007; Oliveira et al. 2020).

Intercropping the banana area with a fast, low-growing, shade-tolerant cover crop can be done to reduce weed interference. Some species are recommended such as *Mucuna pruriens* (velvet bean), tropical alfalfa, pigeonpea (*Cajanus cajan*), mung bean (*Vigna radiata*), cowpea (*V. unguiculata*), sunnhemp (*Crotalaria juncea*), creeping Indigo (*Indigofera endecaphylla*), *Phaseolus trinervius*, Carioca beans and sweet potato (Isaac et al. 2012; Tardy et al. 2017). All of these present rapid canopy coverage, which can suppress weed establishment (Isaac et al. 2012). *Arachis pintoi* is a tropical species that can be added to that list, with additional atmospheric nitrogen fixation. In addition, the practice of intercropping banana with common beans, okra, Crimson Sweet watermelon, and cowpea during the establishment and initial phases did not affect yield and quality of bananas, although cowpea delayed crop cycle (Souza et al. 2010).

16.2.3 Cocoa

Although commercial cocoa plantations are mostly conducted in shaded conditions, both under forest (agroforestry) and intercropped with other commercial species, there is an increasing tendency to plant cocoa monocultures under full sun conditions. Although in both situations competition with weeds is a problem for cocoa producers, in full sun crops this situation is aggravated. In the early years, right after planting, shading and organic residues that cover the soil after pruning are still not enough to control weeds. The treatments that ensured a weed-free environment to the cocoa over a long period were favorable to girth increments in the seedlings (Oppong et al. 2007).

To minimize this problem, some management practices already commonly adopted in other woody fruit crops (e.g. citrus, mango, peaches, and apples) can be adopted. Such practices may be the use of localized irrigation (drip); biodegradable plastic mulching (Figure 16.1E); intercropping or cover crops (Figure 16.1F); maintenance of organic residues (mulching) from pruning (Figure 16.1G); and using an ecological brushcutter; among other cultural practices that accelerate the shading of the area.

Of the species used as cover crops, those mentioned for both banana and citrus crops can be adopted for ecological management in cocoa plantations.

As previously mentioned, the phytosociological survey of the weed infestation community as well as the population dynamics of this community is also an important tool for choosing the different tactics and strategies for ecological weed control.

Although in shaded crops, competition with weeds is less of a problem than in crops under full sun conditions, some species that are adapted to shading such as those of the genus Commelina, as well as epiphytic species, can cause damage. These epiphytic species, which are plants that live in cacao trees, mainly in old plantations, use the plant only as a support and are not parasites. They hinder production by covering the floral cushions, making it difficult for the plant to emit new flowers on the trunk and branches. Some examples are mosses or slime, orchids, lichens, and bromeliads. In addition, there are also parasitic species that are plants that live by absorbing water and nutrients directly from the cacao tree, which can lead to death. Examples of these species are mistletoes (*Tripodanthus acutifolius, Struthanthus vulgaris, Struthanthus polyrhysus, Struthanthus uraguensis* and *Phoradendron linearifolium*) and dodder (*Cuscuta racemosa*). In both cases, the most effective form of control is mechanical control and pruning.

16.3 Subtropical Fruit Crop Species

16.3.1 Citrus

The ecological management of weeds in orchards begins even before the implantation of the crop. In addition to the phytosociological survey of weeds in the area, some previous management and implantation practices can be adopted to suppress the weed germination, which is aggravated by the rainy season (Carvalho et al. 2005).

If soil management practices such as liming or subsoiling have already been carried out, or even if there is no need to carry them out before planting citrus seedlings, some species that have an effect on improving physical, chemical characteristics, and biological soil can be sown. Examples are lab-lab-bean (*Dolichos lablab*), pigeonpea *Cajanus cajan*, sunnhemp (*Crotalaria juncea*), creeping indigo (*Indigofera hirsuta*), cowpea (*Vigna unguiculata*), common vetch (*Vicia sativa*), millet (*Pennisetum americanum*), ruzi and signal grass (*Urochloa* species), and alyceclover (*Alysicarpus vaginalis*) (Matheis and Victoria Filho 2005; Minatel et al. 2006; Linares et al. 2008; Abbas and Fares 2009; Kitis et al. 2011; Martinelli et al. 2017; Matheis et al. 2020). In choosing the ideal cover crops, the edaphoclimatic conditions of the location of the citrus orchard to be implanted must be taken into account. At the beginning of the flowering of these species, when they reach maximum vegetative growth, these plants are mowed in such a way that the plant residue covers the entire planting area. Following operations, citrus seedlings are planted by opening only the holes, leaving all organic mulching on the soil as a physical barrier and in some cases with allelopathic action. This type of planting can be called a no-tillage planting system. The no-tillage planting system, or even minimum cultivation, presents several benefits such as increasing SOC stocks and promoting greater root development and crop yield (Inagaki et al. 2021).

The use of species of the genus Urochloa, especially *Urochloa ruziziensis*, as cover crops and/or intercrop showed lower weed density (56% less) and greater fruit yield (up to 126% more) when combined with ecological mowing (Martinelli et al. 2017). The ecological mower cuts the inter-row area and launches the resulting biomass into the intra-row area (Martinelli et al. 2017).

As previously mentioned, biodegradable plastic mulching can be an additional option for controlling and/or reducing weed competition in the early years of the orchard. The use of this technology associated with others previously described and localized irrigation (drip) can promote a significant reduction in weed infestation in the intra-rows area. As the plants grow, shading will decrease competition, especially below the canopy, so weed-control practices must be used in intra-row areas.

Some species of economic interest can also be used for planting between the lines as a way of controlling weeds. Thus, in addition to reducing competition, the citrus grower can achieve some economic return in the early years. It is important that the intercropped plants are sown at least 0.5 m away from the citrus crown, alternating taller plants (e. g. maize or sorghum) and smaller plants (e.g. common beans) on the inter-row areas (Carvalho et al. 2005).

16.4 Ecologically Based Weed Management Approaches in Temperate Fruit Crops Species Growing in Tropical and Subtropical Environments

16.4.1 Peaches (*Prunus persica*)

Several factors can interfere in the productivity of the culture and the occurrence of weeds stands out as one of the most important, mainly for the competition for water, light, and nutrients. There are several weeds that infest peach orchards, but ryegrass (*Lolium multiflorum*) stands out as the main one due to the producers using this species as a cover plant in the orchard line (Figure 16.2A). This species has an annual development cycle, easy dispersion, and reproduction by natural reseeding. It produces a large

Figure 16.2 Ryegrass infestation before management (A) and after rolling (B) in peach orchard. Management of vegetation composed of vetch + ryegrass (C) and desiccated vegetation (D) in a fig orchard. Weed control in apple orchard using electricity (E and F). (Sources: A, B, C and D: Leandro Galon; E and F: Leandro Vargas).

amount of seeds and has a high ability to compete with crops (Kissmann and Groth 1997).

Ryegrass control in orchards is carried out using the herbicide glyphosate due to efficiency, practicality and lower cost compared to other methods. The practice has been in use for more than 20 years, and the number of applications in a field varies depending on the crop, the weeds present and the weather conditions. In some orchards, more than five applications are made during the production cycle, which is not economically and environmentally viable. In addition, in recent years, ryegrass control with glyphosate has been compromised due to the emergence of glyphosate-resistant biotypes (Roman et al. 2004); thus, there is a need to adopt more sustainable management practices (Westwood et al. 2018).

Studies conducted in peach orchards show viable forms of management in addition to the use of herbicides (Table 16.2). Practices such as lodging, mowing, and weeding, both in the planting row and between peach rows, have shown promising control and cost results in weed management, especially ryegrass.

Among these alternatives to the use of herbicides, rolling, and mowing show advantages by leaving the remains of vegetation or straw on the soil as protection against water loss and leaching. These remains also hamper the emergence of many positive photoblastic weeds, such as hairy beggarticks (*Bidens pilosa*), morning glory (*Ipomoea* spp.), Crabgrass (*Digitaria* spp.), common zarzabacoa (*Desmodium incanum*), and arrowleaf aids (*Sida rhombifolia*) (Figure 16.3).

The use of winter cover plants benefited productivity and had no effect on the diameter of the peach tree trunk in an

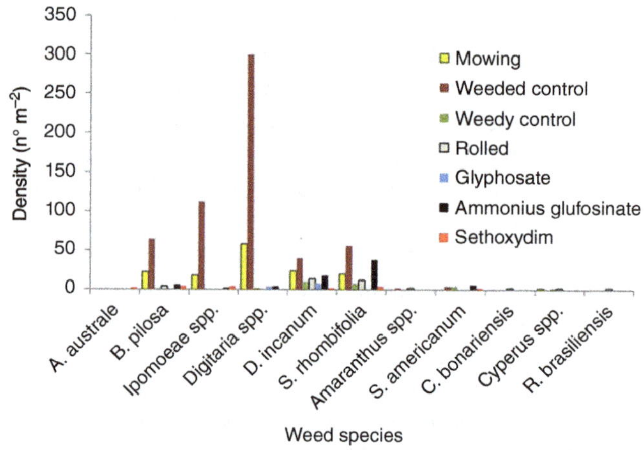

Figure 16.3 Species and number of weeds per square meter as a function of the application of management methods in a peach orchard. Maçambara – Rio Grande do Sul, Brazil.

orchard conducted in an integrated production system, evaluated until the fourth year of installation (Rufato et al. 2007).

16.4.2 Figs (*Ficus carica*)

The main fig-producing countries are Turkey and Egypt. Brazil stands out as the eighth largest producer, with an annual production of 25.8 tons and an average productivity of approximately 9,990 kg ha^{-1} (FAO 2019). Like all agricultural crops, fig trees can suffer from pest attacks. In this scenario, weeds are very important, because in addition to competition, they can host diseases and pests (Costa et al. 2020).

In a study conducted by Costa et al. (2020), the occurrence of 14 families and 23 weeds was observed, the most abundant families being Asteraceae and Poaceae, both with four

Table 16.2 Control (%) of ryegrass (*Lolium multiflorum*) in a peach orchard in Maçambara – Rio Grande do Sul, Brazil.

	Ryegrass Control (%)				
Treatments	**07 DAT[1]**	**14 DAT**	**21 DAT**	**28 DAT**	**66 DAT**
Rolling	95 a[2]	89 bc	83 c	85 c	100 a
Amonnium-glufosinate (400 g ha^{-1})	100 a	100 a	100 a	100 a	100 a
Glyphosate (1,440 g ha^{-1})	72 b	97 ab	100 a	100 a	100 a
Mowing	100 a	97 ab	92 b	93 b	100 a
Sethoxydim (230 g ha^{-1})	42 c	79 c	86 bc	93 b	86 b
Weeded control	100 a	100 a	100 a	100 a	97 a
Weed-infested control	0 d	0 d	0 d	0 d	0 c
General means	72.61	80.14	80.23	81.67	83.81
CV (%)	3.25	4.88	3.10	2.12	2.84

1 DAT: days after applying the treatments.
2 Means followed by the same letter in the line does not differ by Tukey's test at $p \leq 0.05$.

species each. Observing the phytosociological parameters, it was found that wild poinsettia (*Euphorbia heterophylla*) showed the greatest importance value index (IVI) and density (D), except for hoeing management (Figure 16.4; Table 16.3). Probably, the greater covering of the seeds with soil in the management with hoeing may have provided more germination since a larger IVI was found for arrowleaf.

Factors such as luminosity, thermal amplitude, and competitiveness may be related to the suppression of weeds imposed by soil cover (Zibilske and Makus 2009; Forte et al. 2018). De Rossi et al. (2007) observed that the daily amplitude of the temperature was lower in the treatments in which the black oats were kept on the soil surface of a pear orchard.

As there was no change in the total production of figs using different control methods (Figure 16.2 C and D), it is advisable to choose methods that reduce environmental risks and herbicide drift for fig plants or for areas with susceptible crops located close to the application (Figure 16.5) (Forte et al. 2018). These practices include soil cover, hoeing, rolling, or mowing. It is also recommended to employ weed control methods in fig orchards that contribute to reducing the impact of raindrops on the soil. In view of these factors, the

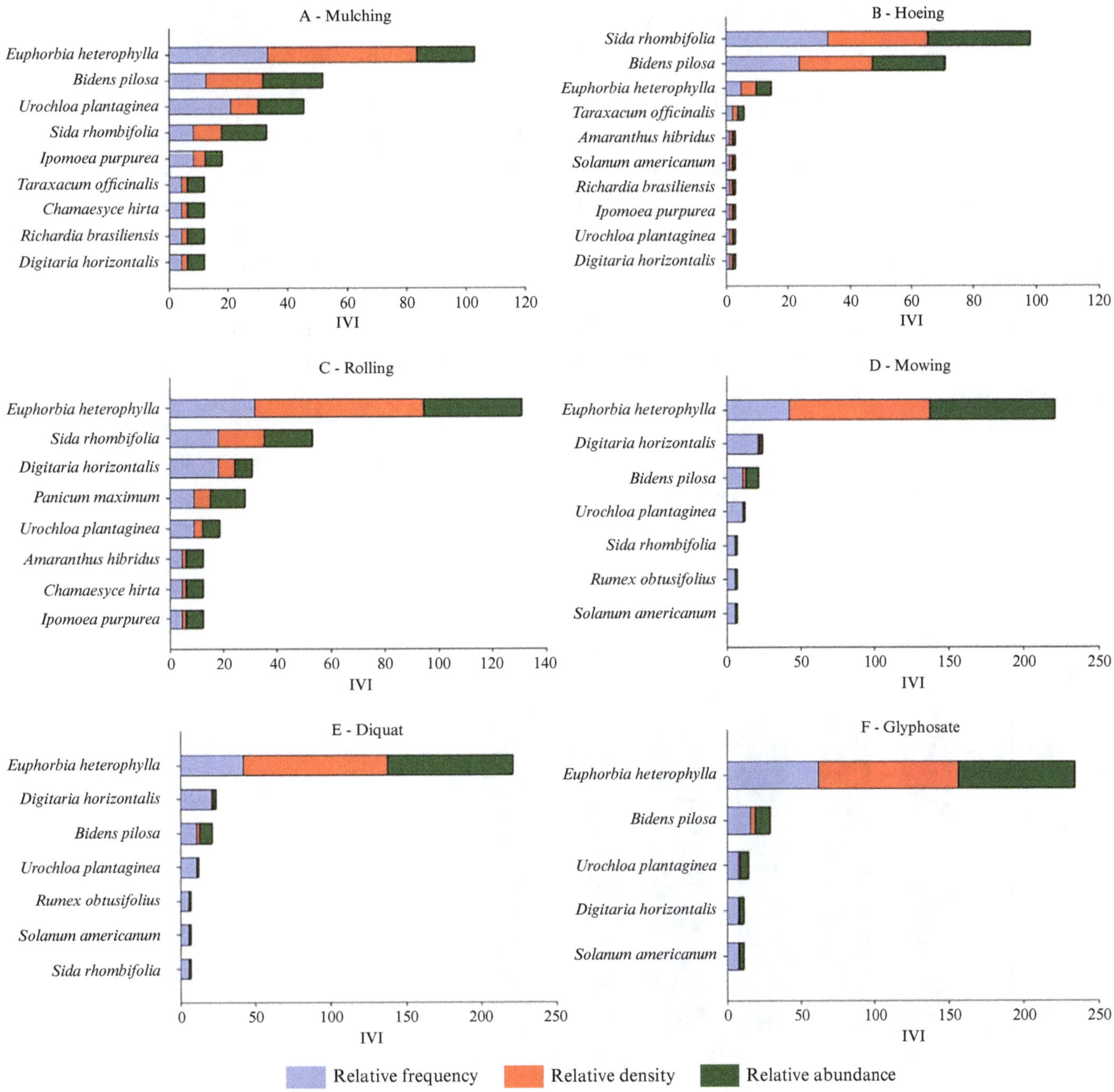

Figure 16.4 Importance value index (IVI) of weed on fig tree under the influence of different managements (A) cover maintenance, (B) hoeing, (C) rolling, (D) moeing, (E) diquat herbicide, and (F) glyphosate herbicide, in weed control. São Domingos – Santa Catarina, Brazil / SC. *Source:* Costa et al. (2020).

Table 16.3 Number of plants (NP) and dry matter of shoots (MS – g) per weed species according to the management method of the vegetal cover of the soil. São Domingos – Santa Catarina, Brazil.

	Cover Maintenance		Hoeing		Rolling		Mowing		Diquat		Glyphosate	
Species	**NP**	**MS**	**NP**	**MS**	**NP**	**MS**	**NP**	**MS**	**NP**	**MS**	**NP**	**MS**
Sida rhombifolia	0	0	15	3	0	0	5	0	0	0	1	0
Digitaria horizontalis	81	20	77	63	157	54	160	58	406	89	12	3
Urochloa plantaginea	11	5	27	3	179	61	35	6	25	9	23	40
Euphorbia heterophylla	14	5	9	30	51	80	27	34	24	58	68	208
Ipomoea purpurea	3	1	1	2	0	0	0	0	1	0	0	0
Bidens Pilosa	0	0	51	36	1	0*	28	61	0	0	3	1
Richardia brasiliensis	0	0	1	8	0	0	0	0	0	0	0	0
Chamaesyce hirta	1	0	1	3	4	0*	3	0	0	0	0	0
Amaranthus hibridus	0	0	5	4	2	13	7	6	0	0	0	0
Panicun maximum	11	59	9	44	0	0	155	207	10	140	0	0
Rumex obtusifolius	0	0	0	0	1	4	8	18	0	0	0	0
Sonchus oleraceus	0	0*	1	3	0	0	3	0	0	0	1	2
Raphanus sp.	3	1	18	28	1	0	17	94	55	81	77	155
Leonurus japonicus	0	0	1	1	0	0	0	0	0	0	0	0
Plantago spp.	0	0	0	0	0	0	5	7	0	0	0	0
Trifolium spp.	1	0*	0	0	0	0	0	0	0	0	0	0
Portulaca oleracea	0	0	3	0*	0	0	0	0	0	0	0	0
Dysphania ambrosioides	0	0	1	1	0	0	0	0	0	0	0	0
Total	125	89	220	228	396	213	453	491	521	376	186	408

*Did not reach dry mass of shoots of 1 g. Source: Costa et al. (2020).

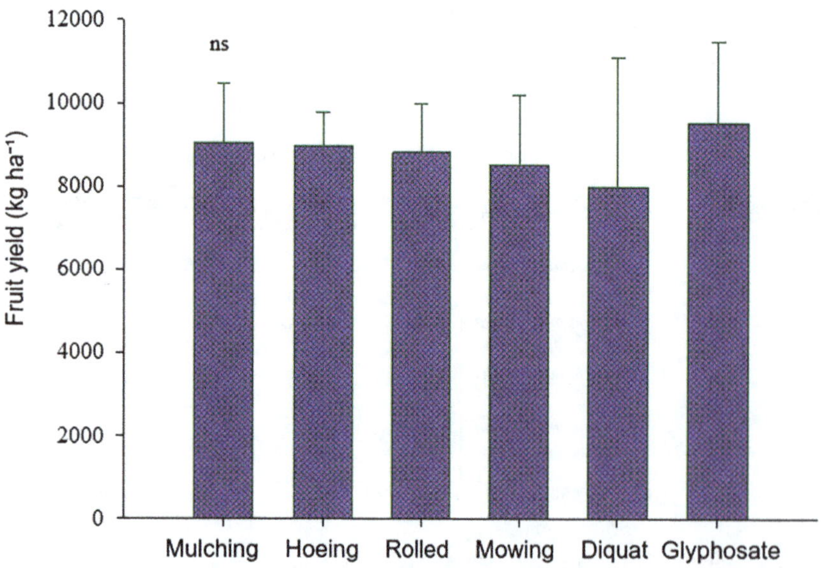

Figure 16.5 Total production of fig fruits (kg ha−1) in the 2012/13 harvest as a function of soil cover management in the winter period. ns = not significant at p≤0.05. *Source:* Costa et al. (2020).

maintenance of soil cover and rolling stand out as the most promising practices for weed management in fig orchards. In addition, as reported previously, electricity can be used as a method to control weeds in orchards, or even the vegetation present between the lines (Figure 16.2 E and F).

16.4.3 Grapevine (*Vitis vinifera*)

Weeds around the trunk not only compete with the grapevine but also develop an environment for organisms that feed on stems and roots, causing the death of the plant or serving as a gateway for pathogens harmful to their health add reference.

Weed control in vineyards must be carried out in the formation phase, since a vigorous initial development provides good formation of the arms, ensuring an adequate and flawless number of canes to explore the productive potential of the crop. In grapevine orchards with one or two annual cycles, the critical period for weed control is approximately 40 days after sprouting, since from that moment the leaf area of the crop ensures sufficient shade to suppress the development of weeds (Souza and Maia 2012).

Soil covers with plant residues are often used in grapevine orchards for their beneficial effects, improving soil characteristics, reducing erosion and improving wine quality (Oliveira Júnior and Constantin 2001). In the southern region of Brazil, the use of oats on grapevine orchards in the winter period is common; in tropical regions, other species can be used, such as legumes, like the pinto peanuts frequently used in the cerrado region of the state of Goiás, Brazil (Souza and Maia 2012). Other species can be used, such as black oats, white peas, or lupine, to maintain green cover during autumn and winter, followed by velvet beans, in spring and summer, without any harmful changes in commercial quality characteristics of fruits, such as the sizes of bunch and berries (Wutke et al. 2005).

Some studies have sought the response of temperate fruit trees to soil management, as reported by Dal Bó and Becker (1994), in relation to the use of winter cover plants in vineyards with 16 years of cultivation. In this case, the cover crops increased the grape productivity in relation to treatments with weeded soil, probably due to improvements in the soil. On the other hand, the cultivation of legumes in plum orchards in a mountainous region of India resulted in bigger and heavier fruits, but with a decrease in productivity when compared to the chemically desiccated treatment, in the sixth year of evaluation (Shylla and Chauhan 2004). More recently, it was found that the use of winter cover plants, except for forage turnip, increased the development of peach plants in an orchard in the region of Morro Redondo, in the state of Rio Grande do Sul, Brazil (Rufato et al. 2007).

The cultivation of grapevines, especially in tropical regions characterized by soils with low organic matter contents, requires organic fertilizer in greater quantity. The acquisition of organic fertilizer with complete fermentation contributes to the elimination of most propagules, which lose viability because of temperature and the action of microorganisms. However, even when purchasing good-quality organic fertilizer, the producer should take precautions regarding where it is stored on the property, which must be free of weeds. Those with vegetative dissemination forms, such as tropical spiderwort, nutsedge, Bermuda grass, panicum, and paspalum, among others, are the main concern.

16.5 Conclusions

Management of orchard-infesting weeds can be carried out in many ways, but the integration of methods provides more efficient control, less environmental contamination, and the possibility of controlling herbicide-resistant weed species. It should be noted that alternative methods to chemical control contribute to safer foods that are free of pesticide contamination. These products can cause health problems for their applicator, if not protected well, and can create more severe environmental impacts.

For soil cover practices to contribute effectively to more productive and less infested orchards, the critical period of competition with the weeds that infest fruit orchards must be considered; in general, this period occurs between the beginning of flowering and fruit harvesting.

The fruit grower, when opting for organic fertilization, must be aware of the quality of the material. It must be free of weeds, especially those that reproduce vegetatively.

When choosing weed management or control methods, the fruit grower must always consider the total cost (input, if any, plus operational costs of the application), combined with the effectiveness of the control and the economic return to the producer.

References

Abbas, F. and Fares, A. (2009). Best management practices in citrus production. *Tree and Forestry Science and Biotechnology* 3 (3): 1–11.

Alwis, A. and Herath, H. (2012). Impact of mulching on soil moisture, plant growth and yield of Mauritius Pineapple (Ananas comosus. L. Merr). *Journal of Food and Agriculture* 2 (1).

Barbour, M.G., Burk, J.H., Pitts, W.D. et al. (1998). *Terrestrial Plant Ecology*, 3e. New York: Benjamin Cummings.

Botrel, N., de Siqueira, D.L., d'Araújo Couto, F.A., and Ramos, V.H.V. (1990). Plantio de abacaxizeiro com cobertura de polietileno. *Pesquisa Agropecuária Brasileira* 25 (10): 1483–1488.

Braun-Blanquet, J. (1979). *Fitosociologia: bases para el estudio de las comunidades vegetales/Pflanzensoziologie* (trans. J. Lalucat Jo).

Bressan, G.M., Koenigkan, L.V., Oliveira, V.A. et al. (2006). Sistema de classificação fuzzy para o risco de infestação por plantas daninhas considerando a sua variabilidade espacial. *Planta daninha* 24 (2): 229–238. https://doi.org/10.1590/S0100-83582006000200004.

Carr, M. (2012). The water relations and irrigation requirements of pineapple (Ananas comosus var. comosus): a review. *Experimental Agriculture* 48. https://doi.org/10.1017/S0014479712000385.

Carvalho, J.E.B., Neves, C.S.V.J., Menegucci, J.L.P., and Silva, J.A.A. (2005). Práticas culturais. Citros. 1e. Matos Júnior, Dirceu; De Negri, José Dagoberto; Pio, Rose Mary; Pompeu Junior, Jorgino, Campinas, 449–482.

Carvalho, J.E.B. de and Vargas, L. (2004). Manejo e controle de plantas daninhas em frutíferas. In: *Manual de manejo e controle de plantas daninhas* (ed. L. Vargas and E.S. Roman), 481–517. Bento Gonçalves: Embrapa Uva e Vinho.

Carvalho, S.J.P., Dias, A.C.R., Damin, V. et al. (2008). Glifosato aplicado com diferentes concentrações de uréia ou sulfato de amônio para dessecação de plantas daninhas. *Pesquisa Agropecuária Brasileira* 43 (11): 1501–1508. https://doi.org/10.1590/S0100-204X2008001100008.

Chong, T.V., Nor Aris, A., and Amarudin Amran, A. (2011). Effect of dicranopteris linearis debris on weed emergence and pineapple growth in pineapple field. *Acta Horticulturae. International Society for Horticultural Science (ISHS)*, Leuven, Belgium, 377–380.

Christoffoleti, P.J. and López-Ovejero, R.F. (2008). Resistência das plantas daninhas a herbicidas: definições, bases e situação no Brasil e no mundo. In: *Aspectos de resistência de plantas daninhas a herbicidas*, 3e (Coord. P.J. Christoffoleti), 3–30. Piracicaba: HRAC-BR.

Cordeiro, C.F. dos S., Rodrigues, D.R., Rocha, C.H. et al. (2021). Glomalin and microbial activity affected by cover crops and nitrogen management in sandy soil with cotton cultivation. *Applied Soil Ecology* 167: 104026. https://doi.org/10.1016/j.apsoil.2021.104026.

Costa, T. da, Giacobbo, C.L., Galon, L. et al. (2020). Management of soil cover and its influence on phytosociology, physiology and fig production. *Comunicata Scientiae* 11: e3236–e3236. https://doi.org/10.14295/cs.v11i0.3236.

Culpepper, A.S. (2006). Glyphosate-induced weed shifts. *Weed Technology* 20 (2): 277–281.

Dal Bó, M.A. and Becker, M. (1994). Avaliação de sistemas de manejo do solo para a cultura da uva. *Pesquisa Agropecuária Brasileira* 29 (2): 263–266.

De Rossi, A., Rufato, L., Giacobbo, C.L. et al. (2007). Diferentes manejos da cobertura vegetal de aveia preta em pomar no sul do Brasil. *Bragantia* 66 (3): 457–463. https://doi.org/10.1590/S0006-87052007000300012.

Eshetu, T., Tefera, W., and Kebede, T. (2007). Effect of weed management on pineapple growth and yield. *Ethiopian Journal of Weed Management* 1: 29–40.

FAO, F. and A.O. of the U.N. (2019). FAOSTAT. Food and Agriculture Organization of the United Nations. http://faostat.fao.org/faostat/servlet (accessed 17 September 2020).

Forte, C.T., Beutler, A.N., Galon, L. et al. (2018). Soil physical properties and grain yield influenced by cover crops and crop rotation. *AJPS* 09 (04): 584–598. https://doi.org/10.4236/ajps.2018.94045.

Fridley, J.D., Vandermast, D.B., Kuppinger, D.M. et al. (2007). Co-occurrence based assessment of habitat generalists and specialists: a new approach for the measurement of niche width. *Journal of Ecology* 95 (4): 707–722. https://doi.org/10.1111/j.1365-2745.2007.01236.x.

Inagaki, T.M., de Moraes Sá, J.C., Tormena, C.A. et al. (2021). Mechanical and biological chiseling impacts on soil organic C stocks, root growth, and crop yield in a long-term no-till system. *Soil and Tillage Research* 211: 104993. https://doi.org/10.1016/j.still.2021.104993.

Isaac, W.-A., Brathwaite, R., and Ganpat, W. (2012). Weed management challenges in Fairtrade banana farm systems in the windward islands of the Caribbean. In: Herbicides – Environmental Impact Studies and Management Approaches (ed. R. Alvarez-Fernandez). InTech. https://doi.org/10.5772/1206.

Ismail, B. and Chong, T.-V. (2009). Allelopathic effects of Dicranopteris linearis debris on common weeds of Malaysia. *Allelopathy Journal* 23 (2).

Kissmann, K.G. and Groth, D. (1997). *Plantas infestantes e nocivas. Tomo I*, 2e. São Paulo: Basf.

Kitis, Y.E., Koloren, O., and Uygur, N. (2011). Evaluation of common vetch (Vicia sativa L.) as living mulch for ecological weed control in citrus orchards. *African Journal of Agricultural Research* 6: 1257–1264.

Kuva, M.A., Pitelli, R.A., Christoffoleti, P.J., and Alves, P.L.C.A. (2000). Períodos de interferência das plantas daninhas na cultura da cana-de-açúcar: I – Tiririca. *Planta Daninha* 18 (2): 241–251. https://doi.org/10.1590/S0100-83582000000200006.

Linares, J., Scholberg, J., Boote, K. et al. (2008). Use of the cover crop weed index to evaluate weed suppression by cover crops in organic citrus orchards. *HortScience* 43 (1): 27–34. https://doi.org/10.21273/HORTSCI.43.1.27.

Lopes, K.A. do L., da Silva, M.

de S., Costa, L. dos S. et al. (2020). Spatial distribution of weed seed banks in the agricultural field and anthropized Cerrado. *Journal of Agricultural Studies* 8 (2): 480–497.

Maia, V.M., Aspiazú, I., and Pegoraro, R.F. (2018). Sustainable weed control in pineapple. In: (Maia, V.M., Aspiazú, I., and Pegoraro, R.F.).*Weed Control: Sustainability, Hazards, and Risks in Cropping Systems Worldwide*, 1e, 470–484. New York: CRC Press.

Marangoni, B., Scudellari, D., and Tacliavini, M. (1995). Relazione tra nutrizione azotata e metabolismo delle piante. *Rivista di Frutticoltura e di Ortofloricoltura* LVII: 7–8.

Martinelli, R., Monquero, P.A., Fontanetti, A. et al. (2017). Ecological mowing: an option for sustainable weed management in young citrus orchards. *Weed Technology* 31 (2): 260–268. https://doi.org/10.1017/wet.2017.3.

Martinelli, R., Rufino, L.R., Jr., Alcántara-de la Cruz, R. et al. (2022). Glyphosate excessive use affects citrus growth and yield: the vicious (and unsustainable) circle in Brazilian orchards. *Agronomy* 12: 453. https://doi.org/10.3390/agronomy12020453.

Matheis, H.A.S.M., Azevedo, F.A., and Victória Filho, R. (2006). Adubação verde no manejo de plantas daninhas na cultura de citros. *Laranja* 27 (1): 101–110.

Matheis, H.A.S.M., Azevedo, F.A., and Victória Filho, R. (2020). Adubação verde no manejo de plantas daninhas na cultura de citros. *Citrus Research & Technology* 27 (1): 101–110.

Matheis, H.A.S.M. and Victoria Filho, R. (2005). Cover crops and natural vegetation mulch effect achieved by mechanical management with lateral rotary mower in weed population dynamics in citrus. *Journal of Environmental Science and Health, Part B* 40 (1): 185–190. https://doi.org/10.1081/PFC-200034321.

Matos, A.P., Sanches, N.F., Souza, L.F. da S. et al. (2009). Cover crops on weed management in integrated pineapple production plantings. *Acta Horticulturae. International Society for Horticultural Science (ISHS)*, Leuven, Belgium, 155–160.

Medeiros, A.R.M. and Santos, A.M. (2003). Práticas culturais. In: *Morango: produção*, 53–56. Pelotas: Embrapa Clima Temperado.

Minatel, A.L.G., Andrioli, I., Centurion, J.F., and Natale, W. (2006). Efeitos da subsolagem e da adubação verde nas propriedades físicas do solo em pomar de citros. *Engenharia Agrícola* 26 (1): 86–95. https://doi.org/10.1590/S0100-69162006000100010.

Monquero, P.A. and Christoffoleti, P.J. (2003). Dinâmica do banco de sementes em áreas com aplicação freqüente do herbicida glyphosate. *Planta Daninha* 21 (1): 63–69.

Moreira, M.S., Melo, M.S.C., Carvalho, S.J.P. et al. (2010). Herbicidas alternativos para controle de biótipos de *Conyza bonariensis* e *C. canadensis* resistentes ao glyphosate. *Planta Daninha* 28 (1): 167–175. https://doi.org/10.1590/S0100-83582010000100020.

Neve, P., Diggle, A.J., Smith, F.P., and Powles, S.B. (2003). Simulating evolution of glyphosate resistance in *Lolium rigidum*. II. Past, present and future of glyphosate use in Australian cropping. *Weed Research* 43 (2): 418–427.

Oliveira Júnior, R.S. and Constantin, J. (2001). *Plantas daninhas e seu manejo*. Guaíba: Agropecuária.

Oliveira, S.M., Carvalho, A., Aspiazú, I. et al. (2020). Phytosociological survey of weeds in coffee plants irrigated under different systems. *Journal of Agricultural Science* 12 (11): 179–189.

Oliveira, T.A. de, Mousinho, F.E.P., Barbosa, R. et al. (2021). Mulch films based on poly(butylene adipate-co-terephthalate)/carnauba wax/sugar cane residue: Effects on soil temperature and moisture. *Journal of Composite Materials*. https://doi.org/10.1177/00219983211011658.

Oppong, F.K., Osei-Bonsu, K., and Amoah, F.M. (2007). Appraisal of some methods of weed control during initial establishment of cocoa in a semi-deciduous forest zone of Ghana. *Ghana Journal of Agricultural Science* 40 (1): 67–72. https://doi.org/10.4314/gjas.v40i1.2155.

Pannek, A., Manthey, M., and Diekmann, M. (2016). Comparing resource-based and co-occurrence-based methods for estimating species niche breadth. *Journal of Vegetation Science* 27 (3): 596–605.

Pelizza, T.R., Mafra, Á.L., Amarante, C.V.T. do et al. (2009). Coberturas do solo e crescimento da macieira na implantação de um pomar em sistema orgânico de produção. *Revista Brasileira de Fruticultura* 31 (3): 739–748. https://doi.org/10.1590/S0100-29452009000300017.

Roman, E.S., Vargas, L., Rizzardi, M.A., and Mattei, R.W. (2004). Resistência de azevém (Lolium multiflorum) ao herbicida glyphosate. *Planta daninha* 22 (2): 301–306. https://doi.org/10.1590/S0100-83582004000200018.

Rufato, L., Rufato, A.D.R., Kretzschmar, A.A. et al. (2007). Coberturas vegetais no desenvolvimento vegetativo de plantas de pessegueiro. *Revista Brasileira de Fruticultura* 29 (1): 107–109. https://doi.org/10.1590/S0100-29452007000100023.

Santos, G.C., Maia, V.M., Aspiazú, I. et al. (2019). Weed interference on 'Prata-Anã' banana production. *Planta Daninha* 37: 1–9. https://doi.org/10.1590/S0100-83582019370100150.

Santos, J.R.P. dos (2019). Fitossociologia e interferência de plantas daninhas na cultura do abacaxizeiro. Master's degree. Janaúba: Plant Production in the Semiarid Region), State University of Montes Claros.

Sarmin, I.J., Rahman, M.S., Amin, M.H., and Ahmed, K. (2020). Effects of bark and stem exudates of eucalyptus on three crop plants. *Journal of Science and Technology* 17: 24.

Shylla, B. and Chauhan, J.S. (2004). Influence of orchard floor management practices on cropping and quality of santa rosa plum grown under mid hill conditions. *Acta Horticulturae. International Society for Horticultural Science (ISHS)*, Leuven, Belgium, 213–216.

Smith, R.W., Lanini, T., Mitchell, J. et al. (2007). Weed management for organic crops. http://anrcatalog.ucdavis.edu/pdf/7250.pdf.

Souza, R.T. and Maia, J.D.G. (2012). Manejo de plantas daninhas em videira. In: *O cultivo da videira Niágara no Brasil* (ed. Maia, J.D.G. and Camargo, U.A.), 176–186. Brasília: Embrapa.

Souza, I. de, Pereira, M.C.T., Ribeiro, R.C.F. et al. (2010). Irrigated banana resistant to black Sigatoka, with annual intercrops. *Revista Brasileira de Fruticultura* 32 (1): 172–180. https://doi.org/10.1590/S0100-29452010005000007.

Sutton, K.F., Lanini, W.T., Mitchell, J.P. et al. (2006). Control, yield, and quality of processing tomato production under different irrigation, tillage, and herbicide systems. *Weed Technology* 20: 831–838.

Tachie-Menson, J.W., Sarkodie-Addo, J., and Carlson, A.G. (2014). Assessment of the impact of some common weed management methods on the growth and yield of pineapples in Ghana. *Journal of Science and Technology (Ghana)* 34 (3): 1–10. https://doi.org/10.4314/just.v34i3.1.

Tardy, F., Damour, G., Dorel, M., and Moreau, D. (2017). Trait-based characterisation of soil exploitation strategies of banana, weeds and cover plant species. *PLOS ONE* 12 (3): e0173066. https://doi.org/10.1371/journal.pone.0173066.

Westwood, J.H., Charudattan, R., Duke, S.O. et al. (2018). Weed management in 2050: perspectives on the future of weed science. *Weed Science* 66 (3): 275–285. https://doi.org/10.1017/wsc.2017.78.

Wutke, E.B., Terra, M.M., Pires, E.J.P. et al. (2005). Influência da cobertura vegetal do solo na qualidade dos frutos de videira 'Niagara Rosada'. *Revista Brasileira de Fruticultura* 27 (3): 6.

Zibilske, L.M. and Makus, D.J. (2009). Black oat cover crop management effects on soil temperature and biological properties on a Mollisol in Texas, USA. *Geoderma* 149 (3): 379–385. https://doi.org/10.1016/j.geoderma.2009.01.001.

17

Application of Ecologically Based Weed Management in Pastures

Jonathan W. McLachlan and Brian M. Sindel*

School of Environmental and Rural Science, University of New England, Armidale, NSW, Australia
** Corresponding author*

17.0 Introduction

Grasslands cover up to about 40% of the world's total land area (Suttie et al. 2005; Gibson 2009). These complex plant communities are generally comprised of a range of species including grasses, legumes, shrubs, and/or other forbs (Figure 17.1) (Allen et al. 2011), and are used primarily for grazing livestock such as cattle, sheep, and goats (Sanderson et al. 2004). The nomenclature associated with grasslands varies around the world and is largely influenced by the level of management needed to achieve some level of productive potential (Gibson 2009). For example, rangelands are generally comprised of indigenous vegetation that is sparingly grazed with little other management. In contrast, intensively managed pastures are likely to include improved species and agrochemical inputs, and are specifically devoted to the production of forage for harvest by either grazing or cutting. Furthermore, grazing is generally carefully managed to maintain satisfactory levels of plant persistence and animal productivity (Sindel 2006; Allen et al. 2011). For the purposes of this chapter, all plant communities that are grazed and have some level of management will be referred to as pastures.

Weeds can be defined as unwanted plants that are growing out of place (Harlan and deWet 1965; Ghanizadeh and Harrington 2019). In arable and horticultural cropping systems, weeds are generally classified as anything other than the established crop. In contrast, the complex nature of pastures can make the classification of weeds much more difficult (Smith and Martin 1995). In particular, plants other than desirable pasture species can be of relatively high nutritive value (Harrington et al. 2006) and can provide palatable forage at some stage of growth (Smith and Martin 1995), although forage value is often time and/or situation dependent (Marten et al. 1987). For example, *Lolium rigidum* (annual or rigid ryegrass) is considered a weed in arable cropping systems due to its competitive growth, which is exacerbated due to problems associated with herbicide resistance (Powles and Matthews 1992; Owen et al. 2014). Nevertheless, this grass is highly productive and palatable, and provides good forage value for grazing livestock (Edgar 1992). Similarly, other species such as *Hordeum leporinum* (barley grass) and *Echium plantagineum* (Paterson's curse) provide moderate forage value in certain circumstances or for certain classes of livestock (Popay and Field 1996; Piltz et al. 2019). However, care must be taken when grazing these species; *H. leporinum* produces heavily awned seed heads that can irritate the mouths and eyes of grazing livestock (Smith 1968; Piltz et al. 2019) while *E. plantagineum* contains pyrrolizidine alkaloids that can cause chronic liver damage if large quantities are consumed over prolonged periods (Edgar 1992; Huwer et al. 2005).

The difficulties associated with classifying weeds in pastures may be why they have been tolerated to a greater extent in grazing systems than in arable cropping systems (Frame and Laidlaw 2011). Nevertheless, the negative impacts of pasture weeds are well established. For example, weeds in pastures are estimated to reduce the productivity of Australian cattle and sheep enterprises by 5% on average (McLeod 2018), but the percentage losses will be much higher in many situations. This lost productivity can be attributed to undesirable species that reduce pasture utilization due to animal avoidance, compete for limited resources such as nutrients, water, and light (Froud-Williams 2002), or are injurious/poisonous to livestock or damage livestock products (Ghanizadeh and Harrington 2019). Depending on the type of weed, they can either be

Ecologically Based Weed Management: Concepts, Challenges, and Limitations, First Edition. Edited by Nicholas E. Korres, Ilias S. Travlos, and Thomas K. Gitsopoulos.

Figure 17.1 Pastures comprise a range of species that vary in their palatability and value to grazing livestock such as sheep.

managed or removed from the pasture. However, targeted weed control in pastures is generally only considered when a particular plant reaches an abundance that reduces livestock production, is detrimental to animal health, or competes too effectively with desirable species for limited resources (Dowling et al. 2000). It is for these reasons that the impact of weeds in grazing systems must consider both pasture and livestock productivity.

Our aim in this chapter is to outline some of the problems caused by weeds in pastures, and to provide an overview of how pasture weeds can be effectively managed. As the focus is on ecologically based weed management principles, we will highlight the importance of understanding pasture ecology and weed life cycles. These factors determine the most appropriate control strategies for managing weeds while maximising pasture and livestock production. In most circumstances, these control strategies will involve a combination of "weed removing" and "pasture improving" techniques (Sindel 2006). It must be recognized that pasture weed management can be difficult, due to the permanent nature of many pastures in which weed ingress can occur at any point of time during the year, and because herbicides and other weed control strategies (e.g. grazing) can cause collateral damage to desirable pasture species (Sindel 2006). These problems highlight the need for an integrated weed management approach, whereby a variety of strategies are used to increase the likelihood of successful, long-term weed control (Zimdahl 2007; Liebman et al. 2016). Throughout the chapter, there will be a focus on the temperate pastures of southern Australia, of which we are most familiar. This region is characterized by cool, wet winters where the majority of pasture growth occurs during winter and spring (Puckridge and French 1983). Commonly grown desirable species include *Phalaris aquatica* (phalaris), *Lolium perenne* (perennial ryegrass), *Dactylis glomerata* (cocksfoot), *Festuca arundinacea* (tall fescue), *Trifolium repens* (white clover), *Trifolium subterraneum* (subterranean clover), and *Medicago* spp. (medics). There are also some important native species belonging to the *Eragrostis*, *Bothriochloa*, and *Austrodanthonia* genera (Dowling et al. 2000). Where possible, other examples will also be considered.

17.1 Ecology of Pasture Systems

Pasture ecology describes the interactions that occur between pasture components (i.e. plants and animals) and their surrounding environment within the grazing system (Michalk and Dowling 1996). An important factor that influences pasture ecology is the perenniality of the species within a pasture sward. For example, swards that are comprised of annual species often have periods of limited groundcover, which means that weed ingress can occur, whereas swards that are comprised of perennial species generally maintain higher levels of groundcover throughout the year, reducing weed ingress (Sindel 2006). Pastures can be further defined as either native (dominated by indigenous or naturally occurring species), naturalized (dominated by introduced species that have become established and persist long term), or introduced (dominated by introduced species that are productive but require more management than either native or naturalized species) (Whalley and Bellotti 1997; Allen et al. 2011). Among these pasture types, there are inherent differences in the ability of desirable species to compete with weeds for limited resources.

17.1.1 Composition of Pastures

Pastures comprise a variety of species that fill a range of ecological niches (Tozer et al. 2017). The growth and persistence of each species within a pasture sward is largely determined by climate, as environmental factors such as rainfall, temperature, and solar radiation directly influence plant growth (Johnson 1997). In particular, these factors determine which species are most suited to a particular region, and whether the growth of annual or perennial species can be supported. For example, annual legumes such as *T. subterraneum* and *Medicago* spp. have been widely adopted in the Mediterranean environments of southern Australia because the hot, dry summers cannot support the growth of perennial legumes (Puckridge and French 1983; Dear and Virgona 1996). However, plant growth and persistence are also influenced by soil type. For example, the deep-rooted perennial legume *Medicago sativa* (lucerne or alfalfa) is suited to the climate of southern Australia, yet it requires deep, free-draining soils of neutral to alkaline pH (Humphries and Auricht 2001). These examples highlight the need for suitable pasture species that are adapted to the prevailing environment and soil conditions (Blair 1997).

The "mixed" nature of pasture swards is important in grazing systems, as both grasses and legumes contribute to the nutritional requirements of grazing livestock (Roca-Fernandez et al. 2016). In addition, legumes increase soil fertility by fixing atmospheric nitrogen (Davies 2001), provided that a legume content of ~30–40% is maintained (Kemp et al. 1999; Graham 2017). However, interspecific competition influences the botanical composition of pasture swards that contain a variety of species (McKenzie et al. 1999; Sheppard 2000). In particular, grasses may outcompete legumes for limited resources such as nutrients and water (Cech et al. 2010). Thus, pastures need to be carefully managed so that the botanical composition remains relatively stable.

Although the diversity of species can make pasture management more difficult, multiple desirable species can extend the growing season of a pasture sward (Humphries 2012), which maximizes its competitive ability. By filling a large range of ecological niches, a mixed pasture sward should have few "open" niches for weeds to occupy (Zimdahl 2007).

17.1.2 Pasture Growth

The growth of a pasture sward that is dominated by grasses can be divided into three distinct phases (Davies 2001; Graham 2017), as shown in Figure 17.2:

- Phase 1 – Slow growth until a critical leaf area is achieved.
- Phase 2 – Rapid growth due to optimal leaf area and efficient acquisition of water.
- Phase 3 – Slow growth as assimilates are directed toward reproduction and as leaves senesce.

Under most circumstances, pasture swards should be maintained in Phase 2 as long as possible. In this growth phase, pasture quality and quantity are optimal for livestock production, while plant persistence is adequate and groundcover is maintained to reduce the ingress of weeds into the pasture sward. Nevertheless, there are times when grazing during the first and third phases is desirable – for example, when competing species or dead leaf material must be removed from the pasture sward to facilitate the growth of desirable species (Phase 1) or to promote a favorable environment for the germination of desirable species following a period of "lax grazing," which allows for seed set and replenishment of seed reserves in the soil (Phase 3) (Graham 2017).

The growth of a pasture sward is determined by plant acquisition of, and competition for, above- and below-ground resources (Ryser and Eek 2000). Plants acquire resources such as carbon dioxide and light energy from above-ground, which means that they must develop sufficient leaf area to remain competitive (Donald 1963). In particular, it is paramount that pasture swards are managed to maximize the acquisition of light, through the maintenance of a dense, photosynthetically active canopy (i.e. Phase 2 of the pasture growth curve shown in Figure 17.2; Graham 2017). However, differences in plant height and leaf morphology mean that some plants such as legumes are likely to be outcompeted over time within the pasture sward (Jennings and Foster 2020).

Plants acquire resources such as nutrients and water from below-ground, which means that they must develop sufficient root length and distribution to remain

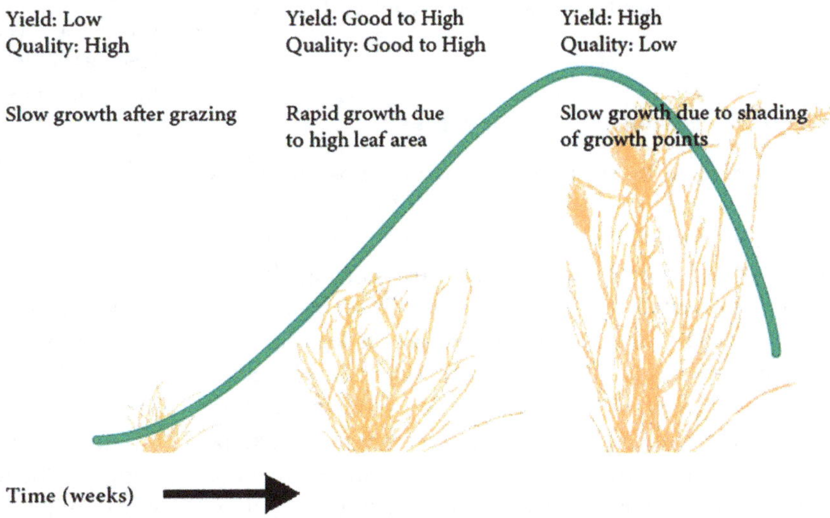

Figure 17.2 Growth curve of a pasture sward dominated by grasses which includes three distinct phases of growth, simplified from the PROGRAZE Manual, 9th ed. (Graham 2017). © Meat & Livestock Australia and State of New South Wales through Department of Regional NSW.

competitive (Donald 1963). Among desirable pasture species, it is well established that the fibrous root system of grasses (including long roots and long root hairs) have greater foraging potential than the taproot system of legumes (including short roots and short root hairs) (Evans 1977). These differences mean that grasses generally outcompete legumes for nutrients and, to a lesser extent, water.

To account for differences in leaf and root morphology, pasture swards need to be managed so that grasses do not outcompete legumes, by controlling defoliation (Graham 2017) and by providing adequate nutrition to maintain the legume component (Ozanne et al. 1976).

17.1.3 Influence of Defoliation on Pasture Composition

Interspecific competition, which influences pasture composition, can be modified and managed by defoliation. This predominately occurs via herbivory by grazing livestock (McKenzie et al. 1999). Grazing should be managed to ensure that pasture productivity is maximized and botanical composition is maintained, because there are inherent differences among pasture species in terms of persistence/grazing resistance and competitive ability (Blair 1997). Furthermore, long periods of heavy grazing should be avoided as these are likely to reduce plant carbohydrate reserves and may lead to reduced plant persistence (Earl 2014), and there are often differences in palatability, which means that some species (generally the legume component) will be preferentially grazed (McCaskill et al. 2019). Poor grazing management, particularly overgrazing, can lead to gaps within the pasture canopy that fill in with competitive weeds (Friend and Kemp 2000). It is therefore important to consider the grazing requirements of pasture species, the minimum residual biomass that remains after a grazing event, and the rest period required for pasture recovery.

17.1.4 Weed Infestations in Pastures

Weeds are likely to become established wherever there is a gap in the pasture canopy (Figure 17.3) (Scott 2000). Pasture production and persistence generally decline with time due to factors such as drought, overgrazing, disease and pest pressure, and a lack of fertilizer (Tozer et al. 2011). This can lead to the loss of species such as legumes and perennial grasses, resulting in gaps that can be exploited by competitive weeds (Dowling et al. 2000). As the proportion of bare ground increases, a pasture becomes more susceptible to weed ingress. This is why annual pastures are more prone to invasion by weeds than pastures based on perennial species (Sindel 2006). Furthermore, annual weeds can

Figure 17.3 A gap in a *Cynodon dactylon* (couch) pasture allows a *Carduus* sp. (slender thistle) plant to establish.

be self-sustaining once they are established, in that, when they die, they leave a gap that can then be exploited by the same species because of previously high levels of seed production (Dowling et al. 2000; Sindel 2006).

Weed ingress can indicate a problem with pasture production and management, particularly if the desirable perennial species are not as competitive as expected. This scenario may be indicative of underlying problems such as inappropriate grazing strategies, nutrient deficiencies, or disease and pest pressure (Tozer et al. 2011). When limited resources – light, nutrients, and water – cannot meet the needs of both desirable and undesirable species, then resource capture (i.e. competition) by weeds will reduce the growth of the desirable species (Froud-Williams 2002).

Within pasture systems, graziers will often only initiate weed control after weed populations have exceeded perceived thresholds that are deemed to impact the growth of desirable species. This may be because thresholds are poorly defined, or because individual farmer perception plays a large part in determining whether a species is desirable or undesirable (based on palatability and contribution to the feed base). However, it is clear that pasture ecology and weed life cycles must be understood so that weed control practices can be implemented before weeds become problematic, particularly given their potential to multiply rapidly.

17.1.5 Weed Seedbanks in Pastures

Weed seedbanks are the build-up of weed seeds in the soil, which can germinate in subsequent seasons. Weed species that produce prolific amounts of seed, such as *Sporobolus indicus* (giant Parramatta grass) and *Nassella neesiana* (Chilean needlegrass), contribute to the production of large seedbanks that can remain problematic for many years (Whalley et al. 1999). It is important for weeds to be managed to reduce seed set, whereby the soil seedbank is progressively reduced over time (Bastiaans et al. 2008). It is also important to identify the weeds that produce prolific amounts of seed or have long-lived seed, and how they can be managed. It has been reported that species composition in the seedbank can be very different from the above-ground composition of the pasture sward, with the amount of weed seeds generally being much higher than the seeds of desirable species (Tozer et al. 2010). This means that there is a higher chance of weed seeds germinating and seedlings establishing, when compared with desirable species, in response to a disturbance in groundcover (Sindel 2006).

17.2 Impacts of Weeds in Pastures

The primary focus of pasture management is the maintenance of a productive pasture sward. Hence, the impact of weeds can be considered in terms of deleterious effects on pasture quality, quantity, and palatability. However, the end goal of pasture management is often livestock production. Therefore, the impact of weeds can also be considered in terms of deleterious effects on animal health (injury and poisoning) and production (reduced quality and/or quantity of animal products) (Sindel 2006).

17.2.1 Weed Impacts on Livestock Production

Weeds that are harmful to animals can severely reduce livestock productivity. For example, poisonous plants can be toxic due to the presence of chemicals such as alkaloids (Edgar 1992). In general, the toxicity of different weed species depends on a variety of factors including palatability, the availability of alternative feed, the age/stage of plant growth, and variation in animal tolerance (Sindel 2006). An example of a potentially toxic weed is *Senecio madagascariensis* (fireweed), which contains a pyrrolizidine alkaloid that causes liver damage but is usually unpalatable and not normally grazed. However, *S. madagascariensis* can be grazed if insufficient alternative feed is available. There are also differences in potential animal toxicity, as sheep and goats are much less susceptible to poisoning by *S. madagascariensis* than cattle (Sindel and Michael 1988). Harmful weeds can also include plants that cause mechanical injury. For example, the heavily awned seeds of *H. leporinum* may irritate the mouths and eyes of livestock (Smith 1968) while *Chondrilla juncea* (skeleton weed) presents a choking hazard to cattle (Jacobs and Goodwin 2009). Other weeds negatively impact the quality of animal products. For example, *Xanthium occidentale* (Noogoora burr) and *Xanthium spinosum* (Bathurst burr) make shearing of sheep and goats difficult and downgrade the value of wool (Sindel 2006), while meat and milk products can be tainted by pungent weeds such as *Anthemis cotula* (stinking chamomile) and *Arctotheca calendula* (capeweed) (Ghanizadeh and Harrington 2019). In addition to these direct effects, the aforementioned weeds can lead to avoidance by grazing livestock, which reduces pasture utilization and animal productivity.

17.2.2 Weed Impacts on Pasture Production

Weeds that are unpalatable and avoided by livestock compete with desirable species for limited resources (Sindel 2006). For example, *Rubus fruticosus* (blackberry) has large spines on its stems and leaves that interfere with the grazing of livestock. Even after *R. fruticosus* plants are killed by herbicide, the canes must be cleared from pastures, or they will continue to prevent grazing (Figure 17.4) (Popay and Field 1996). Similarly, the perennial grass *Nassella trichotomoa* (serrated tussock) is largely unpalatable due to extremely low crude protein and high fiber content, meaning that it generally remains ungrazed (Humphries and Florentine 2021). These types of weeds occupy ground that could be used for better pasture production. Some of these weeds can be grazed to some extent during parts of their life cycle, such as the early vegetative stage of both grasses and broadleaf weeds. However, these weeds inevitably capture resources that are available for desirable species (Bastiaans et al. 2008). There are also transitional weeds that can be beneficial at certain stages of growth and problematic at other stages. These weeds provide valuable forage during vegetative stages of growth but are undesirable when mature (e.g. *H. leporinum*). Some weeds can also act as hosts to various plant diseases and pests, which can exacerbate pasture degradation and lead to further weed infestations.

Figure 17.4 Even once killed by herbicide, this woody weed *Rubus fruticosus* (blackberry) prevents grazing of the underlying pasture.

17.3 Weed Management Principles for Pastures

Ecologically based weed management should center on cultural weed control practices. The objective of cultural control is to reduce the establishment of seedlings from the weed seedbank, alter the competitive relations between desirable/undesirable species, and gradually reduce the weed seedbank (Bastiaans et al. 2008). However, other strategies, grouped as biological, chemical, and physical strategies, may be required if severe weed infestations already exist within a pasture. The following sections focus on cultural control methods for long-term pasture management, although the other methods of control are considered in terms of weed removal.

17.3.1 Pasture Monitoring and Weed Prevention

Pastures should be regularly monitored for weed ingress, as monitoring allows the extent of weed infestations to be quantified. It also facilitates timely control. Land managers should be aware of what weed species are likely to be present within a pasture and how to identify them, particularly during the early stages of growth when weed management strategies are likely to be most successful. It is also important for accurate records to be maintained to track infestations and removal.

Wherever possible, weed ingress should be prevented (DiTomaso 2000) by practicing good farm hygiene, both between different farms and different fields within a farm (Gill et al. 2018). For example, the movement of fodder (e.g. hay) can introduce new weed species on a farm, particularly during droughts when fodder is moved long distances between different regions. Under these circumstances, it is important to use "sacrifice" fields for feeding out fodder, so that any new weed infestations can be detected, contained, and controlled quickly (Gill et al. 2018). Weed seeds can also be transported between fields by livestock. For example, the seeds of *S. indicus* adhere to livestock, particularly when the seeds are wet. This means that livestock should not be moved early in the morning when there is dew on the vegetation or during/shortly after rainfall events (Sindel 2006).

17.3.2 Selection of Suitable Species

Pasture species that suit the prevailing environmental conditions (including climatic and soil conditions, but also expected management) should be grown (Blair 1997) as they are more likely to compete vigorously with undesirable weed species for limited resources. Wherever possible, pastures should be based on perennial species (Campbell 1997). In general, perennial pasture species grow more slowly in their establishment phase than annual species, but compete more effectively with undesirable weed species because they maintain ground cover for a longer period of time than annual pasture species. A range of desirable species is also needed as they fill a variety of ecological niches (Tozer et al. 2017). However, complementary species should be selected so that the botanical composition of a pasture can be maintained over time. For example, *L. perenne* and *T. repens* are suitable companion species because they have similar grazing requirements within a pasture sward (Davies 2001).

17.3.3 Pastures Competing with Weeds

Selected pasture species should not only be adapted to the prevailing conditions but also be competitive. These species should be established when competition from weeds is likely to be low, and when the sown species are likely to achieve fast canopy closure (Scott 1997). Once established, the competitive ability of a pasture is maximized by maintaining a vigorously growing, dense sward. Established pastures should be managed carefully by spelling at critical times, for example prior to peak weed establishment periods to increase competition and during pasture seeding to increase the soil seedbank of the desirable pasture species (Tozer et al. 2010). Fertilizer application is also an important management strategy, with fertilizer best applied when the desirable pasture species are actively growing. In addition, pasture growth can be improved by controlling pasture diseases and pests, which can weaken pastures and facilitate weed ingress. Management should be directed toward maintaining a competitive pasture sward that has minimal gaps in the canopy as any form of disturbance that results in bare ground provides an opportunity for weeds to invade the pasture sward.

17.3.4 Grazing Management

Grazing management is an important component of any grazing system and should generally aim to maintain the desirable pasture species in a vegetative stage of growth. Pastures should be managed so that a minimum groundcover of ~1000 kg dry matter/ha is maintained throughout the year (Graham 2017). However, this minimum groundcover should be increased to ~1500 kg dry matter/ha when weed recruitment is expected (Burton et al. 2006). This amount of groundcover provides a dense pasture canopy that reduces light for weed germination, emergence, and establishment. However, defoliation by grazing modifies the pasture canopy. Furthermore, preferential grazing of the desirable species within a pasture sward can increase the chance of weed ingress (Ghanizadeh and Harrington

2019). Livestock selectivity can be controlled, to some extent, by manipulating the timing of grazing, the amount of grazing pressure, the method of grazing, and the type of livestock (Friend and Kemp 2000). The primary aim should be to encourage the growth of desirable plants.

Increasing grazing pressure can also be used to encourage livestock to consume weeds that are relatively unpalatable using mob stocking (very high stocking rates for short periods of time) (Campbell 1997). Grazing of potentially poisonous species may even be appropriate if the toxicity of a particular plant is relatively low (e.g. during specific phenological growth stages of *Hypericum perforatum* – St. John's wort; Dowling et al. 2000) or when a particular class of livestock is more tolerant of the toxins (e.g. goats and sheep grazing *S. madagascariensis*; Sindel and Michael 1988). Using different animal species to impose different grazing pressures should also be considered. For example, cattle generally prefer palatable grasses whereas sheep prefer forbs, while goats will graze many unpalatable weed species (Popay and Field 1996). In addition, different grazing strategies can also be implemented. For example, rotational grazing (where grazing livestock are moved between fields to allow regular recovery periods) can minimize the effect of animal traffic within a field, thereby reducing the likeliness of gaps in the canopy, while reducing selectivity of grazers (Andrews 1997). However, grazing alone will rarely eradicate a weed infestation.

17.3.5 Manipulating the Soil Seedbank

The weed seedbank should be depleted by increasing the losses and/or by reducing the inputs of seeds (Bastiaans et al. 2008). For example, the seedbank can be reduced by strategic tillage to promote germination of weeds that can then be destroyed, either by follow-up cultivation or herbicide application (Campbell 1997). This method is particularly important when preparing for pasture establishment. Alternatively, a pasture should be managed to ensure that problematic weeds do not set seed. This can be achieved by maintaining sufficient grazing pressure of palatable species or by strategically cutting unpalatable species. For example, Andrews et al. (1996) determined that timely defoliation of *S. indicus* by hand clipping resulted in the seedbank being reduced by 34% over 7 months, whereas the seedbank continued to increase when there was no control of grass growth.

17.3.6 Removing Problematic Weeds

Weed removal often relies on a combination of biological, chemical, and/or physical control strategies.

Biological techniques include the use of biological control agents (often insects or fungal pathogens) that are known to attack specific weeds (Briese 2000). This strategy requires careful management of the weed species so that the biological control agent survives to provide continual pressure. It also requires an understanding of the life cycle of biological control agents. However, biological control is rarely effective in controlling weeds alone, nor is the aim to eradicate the target weed (DiTomaso 2000). Rather, the purpose of biological control is to reduce the presence of specific weeds to more manageable levels. Most biological control agents have been used to target broadleaf weeds of pastures and rangelands (i.e. perennial systems, where the host is present for the majority of the year). The most successful biological control agent is possibly *Cactoblastis cactorum* (cactoblastis moth) that was released to control *Opuntia stricta* (common prickly pear), which once infested over 24 M ha of grazing land in Australia but is no longer a significant weed (Sheppard 2013).

Chemical strategies provide the most immediate control of weed infestations, but they should not be considered as the only solution for weed management as control is often only temporary. Chemical weed control includes broadscale application (for large weed infestations, using either nonselective or selective herbicides) or spot-spraying (when specific weeds are scattered and in relatively low abundance) (Campbell 1997). When considering chemical control options, it is important that the undesirable species are selectively targeted without weakening the pasture sward. In general, herbicides should be applied when weeds are growing vigorously and are more susceptible, particularly in their early stages of growth. However, the use of herbicides is difficult in mixed pasture swards, where both grasses and legumes are present. When clover is present, it is possible to use "clover-safe" herbicides such as 2,4-DB or MCPB, rather than 2,4-D and MCPA (Ghanizadeh and Harrington 2019). It is also possible to use nonselective herbicides if the desirable pasture species are dormant or if there is a height difference between desirable and undesirable species (e.g. grazing the desirable species before applying herbicide to weeds that are taller than the pasture using a weed wiper) (Harrington and Ghanizadeh 2017). When herbicides are applied and weeds are successfully removed from the pasture, it is important that the space is then occupied by desirable pasture species, otherwise further weed infestations are likely to occur.

Alternative chemical techniques are also available, and include spray topping (where low rates of herbicide are applied to annual grass species prior to, or during, early flowering, to reduce viable seed set and hence reduce the weed seedbank) (Medd et al. 1992) and spray grazing (where sublethal doses of herbicides are used to modify the metabolism and palatability of broadleaf weeds, to encourage grazing by livestock) (Dellow 1997). In both cases, the timing of herbicide application and subsequent grazing management is critical.

Physical techniques include cutting weeds, but the effectiveness of this strategy is variable and is largely dependent on the weed species. Defoliation by cutting changes the balance between pasture species and weeds, and can improve livestock access to a field. However, timing is critical to ensure weed seed production is not encouraged and seed dispersal is not increased. Regular cutting can also selectively favor weeds with a prostrate growth habit. The aim of cutting should be to deplete root carbohydrate reserves. Other physical techniques include fodder conservation (e.g. cutting for hay or silage, to maintain control of pasture biomass), cultivation (the potential negative impact on soil structure and stimulation of buried weed seeds to germinate must be considered), hand weeding/chipping (to immediately remove newly identified weed infestations or sparsely scattered plants), and burning (important for woody weeds, particularly in extensive grazing systems, and for stimulating fresh growth for grazing). Burning can also make new growth palatable for grazing, but care must be taken so that fire tolerant species such as *Imperata cylindrica* (blady grass) do not become a problem (Graham 2017).

17.4 Application of Weed Management Principles

The following sections outline how the aforementioned weed management principles can be applied within a pasture system.

17.4.1 Weed Control When Establishing a Pasture

Weed control is the single most important factor that determines the success of pasture establishment (Brouwer and Ison 2014). This is because pasture seedlings are generally slow to establish and are easily outcompeted by other plants, in particular annual "pioneer" weeds that are adapted to the prevailing conditions and are often stimulated by the disturbance associated with pasture establishment processes. For this reason, preestablishment weed control should commence at least one year, and preferably two years, before planting. This increases the chance of successful weed control and pasture establishment, particularly if a heavy weed burden is present initially. The following steps are a suggested guide for weed control prior to, and during, pasture establishment.

Step 1 – Pasture Assessment

Pastures should be assessed for current weed burden, with weed species identified and the extent of infestations quantified. Species identification can be achieved by using resources such as weed identification books or internet guides. Species quantification can be achieved by undertaking a transect across a field, counting the number of times a particular weed species is encountered (e.g. at the tip of a shoe when walking 100 paces). The main types of weeds will determine the most suitable management options.

Step 2 – Weed Removal

Weed species should be removed in preparation for planting. This could include a fast-growing forage crop that competes with weeds and allows for the use of selective herbicide. For example, *Avena sativa* (forage oats) can be grown to outcompete undesirable species while allowing the use of selective herbicides for the control of broadleaf weeds.

Step 3 – Seedbed Preparation

Successful pasture establishment requires a weed-free seedbed, which can be achieved by either cultivation or a double-knock spray (the first spray kills the existing pasture, while the second spray kills any emerging weed/volunteer seedlings) (Thom et al. 2011). If a pasture is to be oversown, heavy grazing and/or cutting prior to planting can sufficiently open the existing pasture canopy to allow the new pasture species to germinate and establish without too much competitive pressure from the existing pasture sward.

Step 4 – Species/Variety Selection

Adapted species, which are suited to the prevailing environmental conditions (including climate and soil), should be selected for pasture establishment. The selection of specific varieties should also be considered, as advances in breeding mean that new varieties are more competitive under certain conditions. Regardless of the species/variety, high-quality weed-free seed should be used, with seeds planted at an appropriate depth and an adequate sowing rate to ensure rapid germination and canopy closure.

Step 5 – Soil Fertility

Constraints to pasture production such as nutrient deficiencies should be identified and ameliorated where possible. For example, starter fertilizer should be considered as this will improve the growth of the desirable species, provided weed control has been successful.

Step 6 – Planting Time

Pastures should be planted at the correct time, to ensure that the pasture species germinate and establish quickly. This leads to rapid canopy closure that maximizes the competitive ability of the pasture sward. Competition from weeds at the time of planting and emergence should also be considered, based on an understanding of weed life cycles.

Step 7 – Monitoring and Grazing

An establishing pasture sward should be monitored for weeds, particularly during the first six weeks following planting. Pastures should also be monitored for disease and pest pressure, as these factors can negatively impact pasture establishment and lead to weed ingress. Grazing must be delayed until the pasture is well established and can tolerate grazing. After establishment, grazing should be used to maintain the botanical composition of the established pasture, by encouraging sown grasses to tiller and to place pressure on any germinating weeds.

17.4.2 Weed Control in Established Pastures

Weed control in established pastures is largely determined by the predominate species (i.e. annual or perennial). Nevertheless, the following steps are a suggested guide to control weeds in an established pasture.

Step 1 – Pasture Monitoring
Pastures should be regularly monitored for weed ingress. This requires an understanding of potential weed species, including their identification and ecology/life cycle.

Step 2 – Farm Hygiene
Farm hygiene should be maintained to prevent the ingress of weeds onto a farm or into a field. Weeds around field boundaries should be managed by regular cutting of fence lines to prevent weed establishment and spread.

Step 3 – Pasture Improvement
Pasture competitiveness should be maximized by alleviating any constraints to production. This could include minimizing disease and pest pressure, as well as ameliorating any nutritional constraints to production.

Step 4 – Control Grazing
Grazing should be carefully controlled to maintain pasture composition and productivity. In many circumstances, grazing management will be the main form of weed control in pasture systems. Overgrazing should be avoided, to reduce gaps in the pasture sward for infestation by weeds, while grazing pressure can be manipulated to reduce weed seed set, depending on the palatability of the weed species present, while allowing seed set of desirable pasture species.

Step 5 – Weed Removal
Weed infestations should be controlled by removing the undesirable plants as needed. This will require the use of biological, chemical, and/or physical methods of weed control.

17.4.3 Target Weed Groups and Suggested Control

Pasture weeds and their associated control can be grouped into four general categories, as outlined by Earl (2014):

- Annual grasses, which include species such as *L. rigidum* and *H. leporinum*. These species have good forage value from autumn until early spring, but they commonly invade bare space before other more desirable pasture species commence growing. Due to their palatability and forage value, they can be grazed heavily to control growth and to limit seed production. Because they produce prolific amounts of seed, it is important that the soil seedbank is reduced where possible. This may involve planting perennial species if climate and soil type allow.
- Perennial grasses, which include species such as *N. trichotoma* and *Eragrostis curvula* (African lovegrass). These species generally have lower forage value than annual grasses and are avoided by grazing livestock, so they persist throughout the year. Where possible, these types of species should be kept short and growing actively to promote palatability. If grazing options are restricted, seed production could be reduced by cutting. Herbicide application can also be considered, for example, by applying a broad-spectrum herbicide such as glyphosate to the ungrazed plants using a weed wiper.
- Broadleaf weeds, which include species such as *Cirsium vulgare* (spear thistle) and *Marrubium vulgare* (horehound). These species rapidly invade gaps in the pasture canopy, produce prolific amounts of seed, and generally have low palatability and poor digestibility. The best form of weed control is through the maintenance of a highly productive pasture sward, although selective herbicides can be used in grass dominant pastures.
- Perennial woody weeds, which include species such as *R. fruticosus* and *Ulex europaeus* (gorse). Woody weeds generally have low palatability and poor digestibility once mature. However, young seedlings can be palatable meaning that strategic grazing at an early stage is an appropriate control measure. Control may also be achieved by maintaining a competitive pasture, or through selective herbicide applications. In more extensive systems, control can be achieved by using strategic burning or grazing with goats.

17.4.4 Integrated Weed Management

The implementation of individual weed management strategies is rarely successful in isolation, particularly because many pastures are based on perennial species that persist for several years. For this reason, the management of pasture weeds must have a long-term focus and be based on a variety of strategies. Integrated weed management (IWM) is therefore suggested for use within pasture systems (Campbell 1997; Dowling et al. 2000). Integrated weed management can be defined as an "holistic approach to weed management that integrates different methods of weed control" (Harker and O'Donovan 2017). In many circumstances, the most important objective of IWM is to establish and/or promote a competitive pasture sward (i.e. cultural control). This is often achieved by growing perennial species, which maximize groundcover throughout the year. Careful grazing management is then required, both to promote a productive sward and to limit weed ingress. However, an integrated weed management strategy is also likely to involve biological, chemical, and/or physical techniques to remove existing mature weeds, before maintaining a competitive pasture sward. Farm hygiene should also be a major focus as this can be easily implemented in conjunction with other strategies.

The difficulty associated with integrated weed management is in determining the potential benefit when different management techniques are combined and therefore deciding on what combinations of management techniques are likely to be most successful. Indeed, there are limited documented examples of successful integrated weed management strategies that have been adopted in pastures. This may be because integrating weed management strategies is difficult to achieve, or because it is not seen to be worthwhile. In particular, the long-term benefits associated with integrated weed management can quickly diminish due to climatic extremes such as drought (which causes major disturbance to groundcover) and floods (which facilitate the movement of weeds onto a farm). Nevertheless, many of the aforementioned control strategies are used on a regular basis as part of standard pasture management. For example, producers carefully manage grazing, maintain soil fertility to ensure competitive pasture growth, and remove weed infestations as required using a range of control strategies. This highlights how weed management may not be the main aim of pasture production, yet it is an outcome of successful pasture management.

17.4.5 Example of an Integrated Weed Management Strategy

Echium plantagineum (Paterson's curse) is a winter annual weed (germinating in autumn and flowering in spring) that is widespread throughout the temperate pasture zone of southern Australia. The weed competes with pastures and can reduce livestock carrying capacity by up to 30%. It also produces high concentrations of alkaloids that can lead to chronic livestock poisoning and, in some cases, even death (Burton et al. 2006).

Echium plantagineum can be successfully controlled by integrating a variety of control strategies, as depicted in Figure 17.5. The implementation of these control strategies is determined by pasture ecology and weed life cycle, and focuses on either reducing the recruitment and seed set of *E. plantagineum* or boosting the production of desirable species. Cultural control strategies are most important, particularly through the improvement of pasture production by ameliorating constraints (e.g. fertilizer application) and careful grazing management. This allows a competitive sward to be maintained, particularly during times of seedling recruitment (autumn) and seed set (summer). Weed removal includes the use of biological (several biological control agents have been released), chemical (tactical herbicide application during early stages of growth), and physical (cutting to reduce seed set) techniques (Burton et al. 2006).

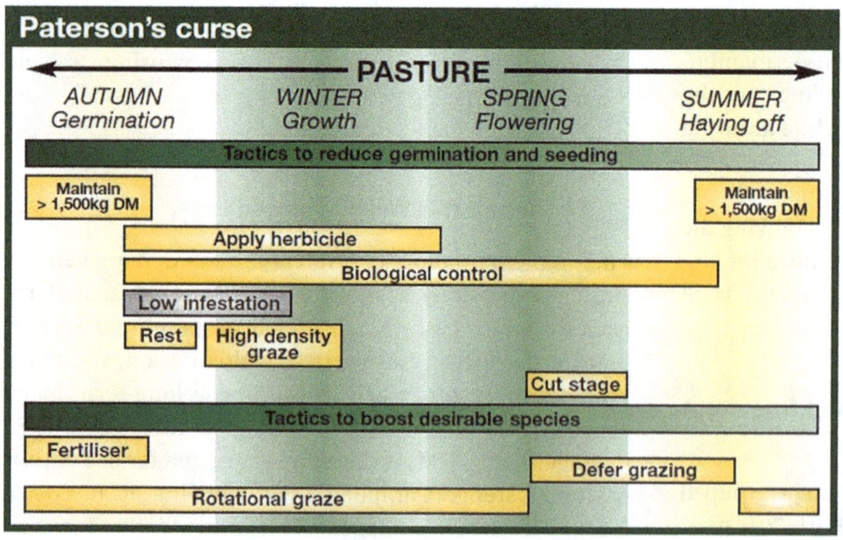

Figure 17.5 Suggested integration and timing of weed management strategies for *Echium plantagineum* (Paterson's curse) (Burton et al. 2006). © Meat & Livestock Australia.

17.5 Future Perspectives

It is clear that a variety of weed management strategies is required for the long-term control of pasture weeds. However, the adoption of integrated weed management has been relatively poor across many agricultural and non-agricultural ecosystems (Liebman et al. 2016). Within pasture systems, this may be because there is limited information about the success of integrating different strategies. For example, there is a lack of information about thresholds for many weeds, meaning that producers do not know at what population weeds should be controlled. Nevertheless, continual improvements in pasture management are likely to have contributed to the control of weeds in many grazing systems. Undoubtedly, there has been a tendency to focus on weed management by herbicide application because the effects of this control strategy are seen immediately, whereas other strategies, particularly those based on cultural control methods, may take longer to yield results (Moss 2019). It is likely that herbicides will remain an important component for weed management in pastures into the future, but they should not replace the maintenance of a competitive pasture sward that reduces weed seedling recruitment and careful grazing management that reduces weed seed-set.

Change into the future may also pose threats to the adoption of integrated weed management in pasture systems. In particular, the predicted change in climate may influence the growth and distribution of pastures (Cullen et al. 2009; Pembleton et al. 2020) and weeds (Peterson et al. 2008). For example, research has suggested that there is potential for tropical pasture grasses such as *Digitaria eriantha* (digit grass) and *Chloris gayana* (Rhodes grass) to be grown in more mild parts of southern Australia (Boschma et al. 2010). Similarly, modeling has suggested that the distribution of the highly invasive weed *Lantana camara* (lantana) will expand into more southern areas of Victoria, South Australia, and Tasmania, with a concomitant decline in northern, tropical parts of Australia, in response to predicted future climatic trends (Taylor et al. 2012). These findings indicate that pasture and weed ecology/life cycles will change over time, and our current understanding must adapt to ensure adequate weed control is achieved into the future.

17.6 Conclusions

Weeds are ubiquitous in grazing systems due to regular disturbances that cause gaps in the pasture canopy. Under these circumstances, weeds can be a symptom of disease and pest pressure, poor grazing management resulting in overgrazing of the pasture sward, or drought conditions that reduce pasture productivity. However, weeds can also be a cause of pasture degradation because they can promote gaps in the canopy, particularly if the weed is a highly competitive annual species that cannot be controlled by grazing. Ecologically based weed management strategies should focus on maintaining a competitive pasture sward that remains vegetative for as long as possible. This will involve a variety of pasture improvement techniques that will ensure the pasture sward is highly productive, to reduce the likeliness of weed infestations. For example, growing the most suitable species, ameliorating nutrient deficiencies, reducing disease and pest pressure, and controlling grazing. At the same time, weed removal techniques may also be required to reduce infestations to acceptable levels that can then be managed by pasture improvement strategies. These observations highlight the need for an integrated weed management strategy that incorporates a variety of techniques to maximize the chance of long-term weed control. However, the most successful strategy for individual situations will be determined by the ecology and life cycles of the pasture and associated weed species, along with other environmental and local factors.

References

Allen, V.G., Batello, C., Berretta, E.J. et al. (2011). An international terminology for grazing lands and grazing animals. *Grass and Forage Science* 66: 2–28.

Andrews, A. (1997). Feed year planning and grazing management. In: *Pasture Production and Management* (ed. J.V. Lovett and J.M. Scott). Marrickville, Australia: Inkata Press.

Andrews, T.S., Whalley, R.D.B., and Jones, C.E. (1996). Seed production and seedling emergence of Giant Parramatta grass on the North Coast of New South Wales. *Australian Journal of Experimental Agriculture* 36: 299–308.

Bastiaans, I., Paolini, R., and Baumann, D.T. (2008). Focus on ecological weed management: what is hindering adoption? *Weed Research* 48: 481–491.

Blair, G. (1997). Matching pastures to the Australian environment. In: *Pasture Production and Management* (ed. J.V. Lovett and J.M. Scott). Marrickville, Australia: Inkata Press.

Boschma, S.P., Lodge, G.M., and McCormick, L.H. (2010). Recent tropical perennial grass research and their potential role in maintaining production in a variable and changing climate. *Proceedings of the 25th Annual Conference of the Grassland Society of NSW*, 85–92.

Briese, D.T. (2000). Classical biological control. In: *Australian Weed Management Systems* (ed. B.M. Sindel). Melbourne, Australia: R.G. and F.J. Richardson.

Brouwer, D. and Ison, M. (2014). Pastures in a farming system. Paterson, Australia: NSW Department of Primary Industries.

Burton, J., Dowling, P., Smyth, M., and Huwer, R. (2006). Managing Paterson's curse to boost pasture production. Sydney, Australia: Meat and Livestock Australia.

Campbell, M. (1997). Pasture weeds. In: *Pasture Production and Management* (ed. J.V. Lovett and J.M. Scott). Marrickville, Australia: Inkata Press.

Cech, P.G., Edwards, P.J., and Venterink, H.O. (2010). Why is abundance of herbaceous legumes low in African savanna? A test with two model species. *Biotropica* 42: 580–589.

Cullen, B.R., Johnson, I.R., Eckard, R.J. et al. (2009). Climate change effects on pasture systems in south-eastern Australia. *Crop and Pasture Science* 60: 933–942.

Davies, A. (2001). Competition between grasses and legumes in established pastures. In: *Competition and Succession in Pastures* (ed. P.G. Tow and A. Lazenby). Wallingford, United Kingdom: CABI Publishing.

Dear, B.S. and Virgona, J.M. (1996). Legumes in low-input perennial pastures of southern Australia: historical role and future development. *New Zealand Journal of Agricultural Research* 39: 579–589.

Dellow, J.J. (1997) The 'spray-graze' technique. A method of pasture manipulation. In: *Proceedings of the 9th Biennial Noxious Weeds Conference* (ed. M. Michelmore), 78–80. Dubbo, Australia. http://www.nswweedsoc.org.au/conferences/past_proceedings.

DiTomaso, J.M. (2000). Invasive weeds in rangelands: species, impacts, and management. *Weed Science* 48: 255–265.

Donald, C.M. (1963). Competition among crop and pasture plants. In: *Advances in Agronomy*, vol. 15 (ed. A.G. Norman), 1–118. Academic Press.

Dowling, P.M., Michalk, D.L., and Sindel, B.M. (2000). Weed management in pasture systems. In: *Australian Weed Management Systems* (ed. B.M. Sindel). Melbourne, Australia: R.G. and F.J. Richardson.

Earl, J. (2014). Grazing and pasture management and utilisation in Australia. In: *Beef Cattle Production and Trade* (ed. D. Cottle and L. Kahn). Collingwood, Australia: CSIRO Publishing.

Edgar, J.A. (1992). Impact of weeds on animal productivity. *Proceedings of the First International Weed Control Congress*, Melbourne, Australia, 140–144. https://caws.org.nz.

Evans, P.S. (1977). Comparative root morphology of some pasture grasses and clovers. *New Zealand Journal of Agricultural Research* 20: 331–335.

Frame, J. and Laidlaw, A.S. (2011). *Improved Grassland Management*. Ramsbury, England: Crowood Press.

Friend, D.A. and Kemp, D.R. (2000). Grazing management methods. In: *Australian Weed Management Systems* (ed. B.M. Sindel). Melbourne, Australia: R.G and F.J. Richardson.

Froud-Williams, R.J. (2002). Weed competition. In: *Weed Management Handbook* (ed. R.E.L. Naylor), 16–38. Cornwall, England: Blackwell Science Ltd.

Ghanizadeh, H. and Harrington, K.C. (2019). Weed management in New Zealand pastures. *Agronomy* 9: 448–465.

Gibson, D.J. (2009). *Grasses and Grassland Ecology*. New York, United States of America: Oxford University Press.

Gill, N., Graham, S., Cross, R., and Taylor, E. (2018). Weed hygiene practices in rural industries and public land management: variable knowledge, patchy implementation, inconsistent coordination. *Journal of Environmental Management* 223: 140–149.

Graham, P. (2017). PROGRAZE – profitable, sustainable grazing. Australia: NSW Department of Primary Industries.

Harker, K.N. and O'Donovan, J.T. (2017). Recent weed control, weed management, and integrated weed management. *Weed Technology* 27: 1–11.

Harlan, J.R. and deWet, J.M.J. (1965). Some thoughts about weeds. *Economic Botany* 19: 16–24.

Harrington, K.C. and Ghanizadeh, H. (2017). Herbicide application using wiper applicators – a review. *Crop Protection* 102: 56–62.

Harrington, K.C., Thatcher, A., and Kemp, P.D. (2006). Mineral composition and nutritive value of some common pasture weeds. *New Zealand Plant Protection* 59: 261–265.

Humphries, A.W. (2012). Future applications of lucerne for efficient livestock production in southern Australia. *Crop and Pasture Science* 63: 909–917.

Humphries, A.W. and Auricht, G.C. (2001). Breeding lucerne for Australia's southern dryland cropping environments. *Australian Journal of Agricultural Research* 52: 153–169.

Humphries, T. and Florentine, S.K. (2021). A comparative review of six invasive *Nassella* species in Australia with implications for their management. *Plants* 10: 1–17.

Huwer, R.K., Briese, D.T., Dowling, P.M. et al. (2005). Can an integrated management approach provide a basis for long-term prevention of weed dominance in Australian pasture systems? *Weed Research* 45: 175–192.

Jacobs, J. and Goodwin, K. (2009). Ecology and management of rush skeletonweed (*Chondrilla juncea* L.). United States Department of Agriculture.

Jennings, J. and Foster, J. (2020). Legume structure and morphology. In: *Forages: The Science of Grassland Agriculture*, vol. II (ed. K.J. Moore, M. Collins, C.J. Nelson, and D.D. Redfearn). Chichester, United Kingdom: John Wiley & Sons Ltd.

Johnson, I. (1997). Climate and pasture production. In: *Pasture Production and Management* (ed. J.V. Lovett and J.M. Scott). Marrickville, Australia: Inkata Press.

Kemp, P.D., Condron, L.M., and Matthew, C. (1999). Pastures and soil fertility. In: *New Zealand Pasture and Crop Science* (ed. J. White and J. Hodgson). Melbourne, Australia: Oxford University Press.

Liebman, M., Baraibar, B., Buckley, Y. et al. (2016). Ecologically sustainable weed management: how do we get from proof-of-concept to adoption? *Ecological Applications* 26: 1352–1369.

Marten, G.C., Sheaffer, C.C., and Wyse, D.L. (1987). Forage nutritive value and palatability of perennial weeds. *Agronomy Journal* 79: 980–986.

McCaskill, M.R., Mitchell, M.L., Zollinger, R. et al. (2019). Dry matter and nutritive value responses of native, naturalised and sown pasture species to soil Olsen P. *Crop and Pasture Science* 70: 1097–1109.

McKenzie, B.A., Valentine, I., Matthew, C., and Harrington, K.C. (1999). Plant interactions in pastures and crops. In: *New Zealand Pasture and Crop Science* (ed. J. White and J. Hodgson). Melbourne, Australia: Oxford University Press.

McLeod, R. (2018). Annual costs of weeds in Australia. Canberra, Australia: Center for Invasive Species Solutions.

Medd, R.W., McMillan, M.G., and Cook, A.S. (1992). Spray-topping of wild oats (*Avena* spp.) in wheat with selective herbicides. *Plant Protection Quarterly* 7: 62–65.

Michalk, D.L. and Dowling, P.M. (1996). Cultural weed management of vulpia. *Plant Protection Quarterly* 11: 207–210.

Moss, S. (2019). Integrated weed management (IWM): why are farmers reluctant to adopt non-chemical alternatives to herbicides? *Pest Management Science* 75: 1205–1211.

Owen, M.J., Martinez, N.J., Powles, S.B., and Iannetta, P. (2014). Multiple herbicide-resistant *Lolium rigidum* (annual ryegrass) now dominates across the Western Australian grain belt. *Weed Research* 54: 314–324.

Ozanne, P.G., Howes, K.M.W., and Petch, A. (1976). The comparative phosphate requirements of four annual pastures and two crops. *Australian Journal of Agricultural Research* 27: 479–488.

Pembleton, K.G., Cullen, B.R., Rawnsley, R.P., and Ramilan, T. (2020). Climate change effects on pasture-based dairy systems in south-eastern Australia. *Crop and Pasture Science* 72: 666–677.

Peterson, A.T., Stewart, A., Mohamed, K.I., and Araujo, M.B. (2008). Shifting global invasive potential of European plants with climate change. *PLoS One* 3: 1–7.

Piltz, J.W., Flinn, S.J., and Weston, L.A. (2019). Comparative effects of grazing, herbicide or forage conservation on barley grass content in *Trifolium subterraneum* L. clover-based pasture. *Crop and Pasture Science* 70: 800–806.

Popay, I. and Field, R. (1996). Grazing animals as weed control agents. *Weed Technology* 10: 217–231.

Powles, S.B. and Matthews, J.M. (1992). Multiple herbicide resistance in annual ryegrass (*Lolium rigidum*): a driving force for the adoption of integrated weed management. In: *Resistance' 91: Achievements and Development in Combating Pesticide Resistance* (ed. I. Denholm, A. Devonshire, and D. Hollomon), 75–87. Dordrecht, Netherlands: Springer.

Puckridge, D.W. and French, R.J. (1983). The annual legume pasture in cereal-ley farming systems of southern Australia: a review. *Agriculture, Ecosystems & Environment* 9: 229–267.

Roca-Fernandez, A.I., Peyraud, J.L., Delaby, L., and Delagarde, R. (2016). Pasture intake and milk production of dairy cows rotationally grazing on multi-species swards. *Animal* 10: 1448–1156.

Ryser, P. and Eek, L. (2000). Consequences of phenotypic plasticity vs. interspecific differences in leaf and root traits for acquisition of aboveground and belowground resources. *American Journal of Botany* 87: 402–411.

Sanderson, M.A., Skinner, R.H., Barker, D.J. et al. (2004). Plant species diversity and management of temperate forage and grazing land ecosystems. *Crop Science* 44: 1132–1144.

Scott, J. (1997). Pasture establishment. In: *Pasture Production and Management* (ed. J.V. Lovett and J.M. Scott). Marrickville, Australia: Inkata Press.

Scott, J.K. (2000). Weed invasion, distribution and succession. In: *Australian Weed Management Systems* (ed. B.M. Sindel). Melbourne, Australia: R.G and F. J. Richardson.

Sheppard, A. (2013) Using integrated weed management to minimize production and environmental impacts in grasslands: an Australian perspective. *Proceedings of the 22nd International Grassland Congress*, Sydney, Australia.

Sheppard, A.W. (2000). Weed ecology and population dynamics. In: *Australian Weed Management Systems* (ed. B.M. Sindel). Melbourne, Australia: R.G and F.J. Richardson.

Sindel, B.M. (2006). Management of weeds in pasture systems. In: *Handbook of Sustainable Weed Management* (ed. H.P. Singh, D.R. Batish, and R.K. Kohli). Binghamton, United States of America: The Haworth Press.

Sindel, B.M. and Michael, P.W. (1988). Survey of the impact and control of fireweed (*Senecio madagascariensis* Poir.) in New South Wales. *Plant Protection Quarterly* 3: 22–28.

Smith, A.E. and Martin, L.D. (1995). Weed management systems for pastures and hay crops. In: *Handbook of Weed Management Systems* (ed. A.E. Smith). New York, United States of America: Marcel Dekker.

Smith, D.F. (1968). The growth of barley grass (*Hordeum leporinum*) in annual pasture. 1. Germination and establishment in comparison with other annual pasture species. *Australian Journal of Experimental Agriculture and Animal Husbandry* 8: 478–483.

Suttie, J.M., Reynolds, S.G., and Batelloa, C. (2005). *Grasslands of the World*. Rome, Italy: Food and Agriculture Organization of the United Nations (FAO).

Taylor, S., Kumar, L., and Reid, N. (2012). Impacts of climate change and land-use on the potential distribution of an invasive weed: a case study of *Lantana camara* in Australia. *Weed Research* 52: 391–401.

Thom, E.R., Fraser, T.J., and Hume, D.E. (2011). Sowing methods for successful pasture establishment – a review.

New Zealand Grassland Association Pasture Persistence Symposium, Hamilton, New Zealand. Grassland Research and Practice Series No. 15, 31–38. https://www.nzgajournal.org.nz.

Tozer, K.N., Barker, G.M., Cameron, C.A., and James, T.K. (2010). Relationship between seedbank and above-ground botanical composition during spring. *New Zealand Plant Protection* 63: 90–95.

Tozer, K.N., Bourdot, G.W., and Edwards, G.R. (2011). What factors lead to poor pasture persistence and weed ingress? *New Zealand Grassland Association Pasture Persistence Symposium*, Hamilton, New Zealand. Grassland Research and Practice Series No. 15, 129–138. https://www.nzgajournal.org.nz.

Tozer, K.N., Minnee, E.M.K., Greenfield, R.M., and Cameron, C.A. (2017). Effects of pasture base and species mix complexity on persistence and weed ingress in summer-dry dairy pastures. *Crop and Pasture Science* 68: 561–573.

Whalley, W. and Bellotti, B. (1997). Natural pastures and rangelands. In: *Pasture Production and Management* (ed. J.V. Lovett and J.M. Scott). Marrickville, Australia: Inkata Press.

Whalley, W., Gardener, M., and Earl, J. (1999). Pasture management of reproductively efficient grassy weeds. *Twelfth Australian Weeds Conference*, Hobart, Australia. https://caws.org.nz.

Zimdahl, R.L. (2007). *Fundamentals of Weed Science*. Burlington, United States of America: Academic Press.

Index

Note: page numbers in italics refer to figures; those in bold to tables.

Ecologically Based Weed Management: Concepts, Challenges, and Limitations, First Edition. Edited by Nicholas E. Korres, Ilias S. Travlos, and Thomas K. Gitsopoulos.

www.ingramcontent.com/pod-product-compliance
Lightning Source LLC
LaVergne TN
LVHW082115200926
848484LV00008B/1957
9781119709664